ENZYMOLOGY AND MOLECULAR BIOLOGY OF CARBONYL METABOLISM 4

ADVANCES IN EXPERIMENTAL MEDICINE AND BIOLOGY

Recent Volumes in this Series

Volume 320
THE BIOLOGY AND PREVENTION OF AERODIGESTIVE TRACT CANCERS
Edited by Guy R. Newell and Waun Ki Hong

Volume 321
PANCREATIC ISLET CELL REGENERATION AND GROWTH
Edited by Aaron I. Vinik

Volume 322
EXERCISE, CALORIES, FAT, AND CANCER
Edited by Maryce M. Jacobs

Volume 323
MECHANISMS OF LYMPHOCYTE ACTIVATION AND IMMUNE REGULATION IV: Cellular Communications
Edited by Sudhir Gupta and Thomas A. Waldmann

Volume 324
PROSTATE CANCER AND BONE METASTASIS
Edited by James P. Karr and Hidetoshi Yamanaka

Volume 325
RECOVERY FROM BRAIN DAMAGE: Reflections and Directions
Edited by F. D. Rose and D. A. Johnson

Volume 326
THE USE OF RESEALED ERYTHROCYTES AS CARRIERS AND BIOREACTORS
Edited by Mauro Magnani and John R. DeLoach

Volume 327
GENETICALLY ENGINEERED VACCINES
Edited by Joseph E. Ciardi, Jerry R. McGhee, and Jerry M. Keith

Volume 328
ENZYMOLOGY AND MOLECULAR BIOLOGY OF CARBONYL METABOLISM 4
Edited by Henry Weiner, David W. Crabb, and T. Geoffrey Flynn

Volume 329
DENDRITIC CELLS IN FUNDAMENTAL AND CLINICAL IMMUNOLOGY
Edited by Eduard W. A. Kamperdijk, Paul Nieuwenhuis, and Elisabeth C. M. Hoefsmit

ENZYMOLOGY AND MOLECULAR BIOLOGY OF CARBONYL METABOLISM 4

Edited by

Henry Weiner

Purdue University
West Lafayette, Indiana

David W. Crabb

Indiana University School of Medicine and
Veterans Administration Medical Center
Indianapolis, Indiana

and

T. Geoffrey Flynn

Queens University
Kingston, Ontario, Canada

SPRINGER SCIENCE+BUSINESS MEDIA, LLC

Library of Congress Cataloging in Publication Data

Enzymology and molecular biology of carbonyl metabolism 4 / edited by Henry Weiner, David W. Crabb, and T. Geoffrey Flynn.
p. cm. — (Advances in experimental medicine and biology; v. 328)
"Proceedings of the Sixth International Workshop on Enzymology and Molecular Biology of Carbonyl Metabolism, held June 28–July 1, 1992, in Dublin, Ireland" — T.p. verso.
Includes bibliographical references and index.
ISBN 978-0-306-44357-2 ISBN 978-1-4615-2904-0 (eBook)
DOI 10.1007/978-1-4615-2904-0
1. Aldehyde dehydrogenase — Congresses. 2. Aldose reductase — Congresses. 3. Carbonyl reductase — Congresses. 4. Alcohol dehydrogenase — Congresses. 5. Carbonyl compounds — Metabolism — Congresses. I. Weiner, Henry. II. International Workshop on Enzymology and Molecular Biology of Carbonyl Metabolism (6th: 1992: Dublin, Ireland) III. Series.
[DNLM: 1. Alcohol Dehydrogenase — physiology — congresses. 2. Alcohol Oxidoreductases — physiology — congresses. 3. Aldehyde Dehydrogenase — physiology — congresses. 4. Aldehyde Reductase — physiology — congresses. W1 AD559 no.328 1993 / QU 140 E61 1992] QP603.A35E573 1993
599'.019'258 — dc20
DNLM/DLC 92-48366
for Library of Congress CIP

Proceedings of the Sixth International Workshop on Enzymology and Molecular Biology of Carbonyl Metabolism, held June 28–July 1, 1992, in Dublin, Ireland

ISBN 978-0-306-44357-2

Originally published by Plenum Press New York in 1993

The collected papers on aldo-keto reductases presented in this volume are dedicated to the memory of Dr. Clyde Doughty who died at the age of 67 on June 12, 1992.

PREFACE

The Sixth International Workshop on the Enzymology and Molecular Biology of Carbonyl Metabolism was held outside of Dublin, Ireland at the end of June, 1992. Prof. Keith Tipton, Chairman of the Biochemistry Department at Trinity College, kindly agreed to host the meeting. On behalf of all of us who attended I wish to extend our sincere thanks to the whole Tipton family for making us feel so welcome in Ireland.

It has been a decade since the first workshop was held in Bern, Switzerland. The scope of the meetings reflected somewhat the changes that have occurred in biochemistry during the past decade. At the first meeting primarily enzymes and their properties were discussed. At this last meeting many of the talks centered on gene regulation as well as more traditional aspects of enzymology and metabolism. During the past decade site directed mutagenesis to probe for the active site of an enzyme has become part of traditional enzymology; this was virtually unheard of at our first meeting. Many of the presenters now used this tool to study some aspect of structure and function of one of the three carbonyl metabolizing enzymes.

I wish to thank all those who attended the meeting and for making it an enjoyable experience for me. Special thanks must be extended to my two co-editors for their help and to Raven Twitchell whose administrative skills made this the easiest meeting for me to co-organize. We never planned to hold a second, let alone a sixth workshop. Now plans are underway for the seventh to be held in Palmerston North, New Zealand in June, 1994. I invite scientists interested in attending this workshop to contact me.

HENRY WEINER
W. Lafayette, Indiana

CONTENTS

ALDEHYDE DEHYDROGENASE

ALDOSE REDUCTASE

CARBONYL REDUCTASE

ALCOHOL DEHYDROGENASE

ALDEHYDE DEHYDROGENASES -- THE 1992 PERSPECTIVE

Ronald Lindahl

Department of Biochemistry and Molecular Biology
The University of South Dakota School of Medicine
Vermillion, South Dakota 57069

The past decade has been an exciting one for researchers studying the family of enzymes collectively known as the aldehyde dehydrogenases. Ten years ago, if aldehyde dehydrogenases were even known to scientists outside the small group who actively studied the enzyme, it was for their role in acetaldehyde oxidation. Although multiple forms of mammalian aldehyde dehydrogenase were known, there was little understanding of the physiological roles played by the various enzymes. Today, the field has expanded to such a degree that a role in ethanol metabolism is only one of several important physiological functions for this multigene family. It is apparent that aldehyde dehydrogenases are primarily responsible for the metabolism of a wide variety of endogenous and exogenous aldehydes.

Berne, Switzerland was the site of the inaugural International Workshop on the Enzymology of Carbonyl Metabolism in 1982. At that meeting seventeen papers were presented that dealt with aldehyde dehydrogenases. All the presentations related to the enzymology of the enzymes, ranging from structure to physiological roles. Interest in enzymology also dominated the twelve presentations at the Second Workshop held in Kingston, Ontario, Canada in 1984.

The 1986 meeting in Espo, Finland signaled the beginning of a significant change in direction for the Workshop. At that meeting two of the fifteen presentations dealt with the characterization of cDNAs for two different mammalian aldehyde dehydrogenases. The name of the workshop changed to Enzymology and Molecular Biology of Carbonyl Metabolism, reflecting the growing interest in the molecular aspects of the aldehyde dehydrogenases.

The next two meetings, in 1988 in Gifu, Japan and 1990 in West Lafayette, Indiana witnessed the increasing application of the tools of molecular biology to the aldehyde dehydrogenases. From the approximately dozen presentations at each of these meetings emerged a standardized nomenclature for the mammalian aldehyde dehydrogenase and the concept of the aldehyde dehydrogenases as a gene family composed of several related members.

This years meeting in Dublin, Ireland reflected the continued use of a variety of approaches, from traditional to molecular biological, to study various aspects of aldehyde dehydrogenase biology and chemistry. The meeting also affirmed a renewed interest in the physiological roles for the various aldehyde dehydrogenases. Progress in all areas since 1990 has been dramatic. The thirty 1992 presentations reporting on some aspect of the aldehyde dehydrogenases was almost twice that of any previous workshop. Of these, twenty papers described work on some aspect of the Class 1 and/or 3 aldehyde dehydrogenases. Two major themes were apparent, mechanisms of regulation of aldehyde dehydrogenase gene expression and physiological roles for the various enzyme forms. These dominated the discussions and it is clear that much has been learned in the past two years. However, many questions remain.

With respect to physiological roles for the aldehyde dehydrogenases the major focus was on the involvement of the Class 1 enzyme in retinoic acid metabolism. Five papers, those of Yoshida et al., Pereira et al., Sladek and Lee, McCaffery and Drager, and Duester presented convincing evidence from a variety of systems that Class 1 aldehyde dehydrogenase is important in the oxidation of retinaldehyde to retinoic acid under physiological conditions. The aldehyde dehydrogenase-catalyzed synthesis of retinoic acid may be important in modulating cell differentiation, especially during embryonic development. Moreover, either changes in Class 1 aldehyde dehydrogenase gene expression or competition between retinaldehyde and other aldehyde substrates, especially acetaldehyde or biogenic amines, may be important factors in the development of human diseases as diverse as neoplasia and androgen-insensitivity and fetal alcohol syndromes.

Several other roles for the aldehyde dehydrogenases continue to emerge. Ambroziak and Pietruszko presented evidence for a fourth class of mammalian aldehyde dehydrogenases structurally distinct from the Classes 1, 2 and 3, which has kinetic properties consistent with a physiological role in the oxidation of aldehydes generated from biogenic amine metabolism. Messiha presented data demonstrating competition between ethanol and various biogenic amine aldehydes for liver aldehyde dehydrogenases. Interest also continues in the role for the various aldehyde dehydrogenases in the metabolism of aldehydes derived from lipid peroxidation. From the work of Canuto and colleagues and Hartley and Petersen it is apparent that cell lines differing in aldehyde dehydrogenase activity are differentially sensitive to the cytotoxic effects of 4-hydroxynonenal and that substrate competition between various aldehydes can have a significant effect on lipid aldehyde detoxification.

The role of aldehyde dehydrogenase in antitumor drug metabolism continues to be of interest. Yoshida et al. presented data indicating that certain human tumor cell lines can develop acquired resistance to cyclophosphamide toxicity by increased transcription of the Class 1 aldehyde dehydrogenase gene. Sreerama and Sladek presented evidence for a novel Class 3 aldehyde dehydrogenase in a cyclophosphamide-resistant human mammary tumor cell line. This Class 3 enzyme appears to be smaller than the prototypical Class 3 form, but exhibits many of the physical and functional properties of the typical Class 3 enzyme.

Another area of considerable interest continues to be the relationship of the aldehyde dehydrogenases to other aldehyde-metabolizing enzymes and other proteins, such as the crystallins. Crabb et al. expanded the aldehyde dehydrogenase gene family to include coenzyme A-dependent methylmalonate semialdehyde dehydrogenase. Interestingly, this enzyme is more closely related to a number of aldehyde dehydrogenases than to other semialdehyde dehydrogenases such as glutamic semialdehyde dehydrogenase. Also of interest is that with inclusion of the methylmalonate semialdehyde enzyme in the aldehyde dehydrogenase family, conservation of putative critical residues is maintained with the exception of glutamate 268, which is an asparagine in the semialdehyde dehydrogenase. Agarwal and colleagues provided primary sequence data confirming that the aldehyde dehydrogenase 4 gene product is glutamic semialdehyde dehydrogenase.

Three papers, those of Algar and Holmes, Lee et al., and Cooper and colleagues presented molecular evidence for the identity of the major cornea protein and/or certain crystallins and Class 3 and Class 1 aldehyde dehydrogenases. Algar and Holmes also discussed evidence for a second major aldehyde dehydrogenase transcript in the cornea. Nucleotide sequencing indicated this transcript to be identical to an uncharacterized aldehyde dehydrogenase, ALDH-X, identified as an aldehyde dehydrogenase transcript from a human testis cDNA library. Lee et al. presented additional data indicating that the zeta-crystallin/Class 1 aldehyde dehydrogenase may be produced by alternative transcript splicing and that the Class 1 aldehyde dehydrogenase gene may possess an antioxidant response element. These observations strongly suggest that aldehyde dehydrogenases play both an enzymatic role and structural role in the cornea and lens. Their functional role is in the oxidation of lipid aldehydes and their structural role is as a crystallin.

Interest in the regulation of aldehyde dehydrogenase gene expression has also increased dramatically since 1990. In 1992, seven presentations directly addressed aspects of aldehyde dehydrogenase gene expression. Several other presentations, for example those related to the eye aldehyde dehydrogenases and some discussing retinaldehyde metabolism, also included considerable information about aldehyde dehydrogenase gene regulation. The majority of the molecular work has focused on the regulation of Class 3 aldehyde dehydrogenase gene expression because this enzyme exhibits both tissue-specific constitutive and tissue-specific inducible expression. Vasiliou and colleagues and Asman et al. presented evidence that the Class 3 aldehyde dehydrogenase gene is a member of the Ah receptor-mediated gene battery. Evidence was presented for transcriptional up- and down-regulation by both Ah receptor-mediated and Ah receptor-independent mechanisms. The structure of the 5' untranslated region of the rat, but apparently not the human (Hsu et al.), Class 3 aldehyde dehydrogenase genes is complex with an intron interrupting the 5' untranslated region of the rat Class 3 gene. The 5' regulatory region of the rat Class 3 gene possesses a variety of putative DNA-binding protein motifs, included a xenobiotic-response element predicted from the dioxin-inducibility of the gene.

Meier-Tackmann et al. and Marselos et al. presented additional biochemical data on for the aromatic hydrocarbon-induced expression of Class 3 aldehyde dehydrogenase. Although controversial, Yin et al. suggest that the multiple banding patterns observed in isoelectric focusing of human gastric Class 3 aldehyde dehydrogenase is due to a two gene with polymorphism genetic model.

As noted above, Lee et al. presented data for alternative splicing and the presence of an antioxidant-response element in the Class 1 aldehyde dehydrogenase gene. Both Yoshida et al. and Pereira et al. presented evidence for a role for androgens in regulation of Class 1 aldehyde dehydrogenase gene expression, including the presence of glucocorticoid/androgen response elements in the 5' region of this gene. The role of Class 1 aldehyde dehydrogenase in retinoic acid metabolism also suggests that perhaps a retinoic acid response element might ultimately be identified upstream of the Class 1 and perhaps the Class 3 aldehyde dehydrogenase genes.

Several papers discussed various aspects of the enzymology of the aldehyde dehydrogenases. Zheng and Weiner employed a variety of techniques to increase expression of the rat Class 2 aldehyde dehydrogenase cDNA in E. *coli.* Although expression could be increased by modifying the cDNA and the resulting polypeptides, the majority of the expressed protein was found in inclusion bodies. Hurley et al. reported the crystallization of bovine Class 2 aldehyde dehydrogenase.

Pietruszko et al. and Blackwell et al. both described the active sites of the aldehyde dehydrogenase and esterase activities. Pietruszko suggested the active sites for the two reactions are largely, but not entirely separate, perhaps sharing some critical residues. Blackwell et al. argue that the active sites for the dehydrogenase and esterase activities are clearly separate. Kitson and colleagues and Woenckhaus et al. have used disulfiram and a photoaffinity NAD+ analog to probe enzyme function.

The 1992 meeting also provided an excellent opportunity to update the mammalian aldehyde dehydrogenase nomenclature. The original 1988 classification scheme was based on the primary structure relationships of the various aldehyde dehydrogenases and adopted the human aldehyde dehydrogenase gene designations. Consistent with this, it is clear that a fourth distinct class of aldehyde dehydrogenases exist as represented by the human liver E3 protein, which is tentatively assigned as human aldehyde dehydrogenase gene 6.

It is also apparent that the aldehyde dehydrogenase designated ALDH-4 is in fact glutamic semialdehyde dehydrogenase possessing its own enzyme commission number 1.5.1.12. Thus, glutamic semialdehyde dehydrogenase should no longer be considered a member of the broad substrate aldehyde dehydrogenases, but may perhaps be considered a structurally-related member of the superfamily as is methylmalonate semialdehyde dehydrogenase.

The accompanying table presents the updated mammalian aldehyde dehydrogenase nomenclature. Three issues related to nomenclature will require attention in the near future. These are the resolution of the human and mouse aldehyde dehydrogenase gene designation differences. The second is establishing the relationship of aldehyde dehydrogenase-X to the known members of the family.

Lastly, as interest in aldehyde dehydrogenases brings new investigators to the field, the continued, consistent use of a single, widely accepted terminology will greatly simplify communication among those interested in this enzyme family.

Given the progress that has occurred since the 1988 Workshop when the impact of the application of molecular biology approaches to the aldehyde dehydrogenases first indicated that this enzyme system was indeed a rather large gene family, one can only look forward with eager anticipation to the work to be presented in 1994 in New Zealand. One thing is certain, the progress in the next two years is certain to exceed that of the past two in several very interesting and important areas.

METABOLIC ROLE OF ALDEHYDE DEHYDROGENASE

Wojciech Ambroziak[1] and Regina Pietruszko[2]

[1]Institute of General Food Chemistry
Technical University of Lodz
Lodz, Poland

[2]Center of Alcohol Studies
Rutgers University
Piscataway, New Jersey 08855-0969, U.S.A.

INTRODUCTION

Aldehyde dehydrogenase (EC 1.2.1.3), an enzyme with a broad substrate specificity and low Km values for short chain aliphatic aldehydes utilizes NAD as coenzyme, is universally distributed in mammalian livers and also at lower concentrations in other organs. The enzyme is a homotetramer of MW of ca. 220,000 (see review by Pietruszko, 1989). Because of its broad substrate specificity, it is frequently considered to be an enzyme of detoxication which functions in the organisms in oxidation of toxic aldehydes ingested in foodstuffs (see review by Jakoby and Ziegler, 1990). Metabolism of ethanol derived acetaldehyde, which is known to be catalyzed by this enzyme (Parrilla et al., 1974) is an example of detoxication role of aldehyde dehydrogenase. In the human liver the enzyme occurs as three known isozymes, E1, E2 and E3; other isozymes not yet identified may exist.

Our more recent experiments (Kurys et al., 1989; Ambroziak and Pietruszko, 1991; Ambroziak et al., 1991) strongly suggest that in addition to detoxication the enzyme has other functions, some of which may be crucial in metabolism. Some of the work presented in this Symposium by other investigators indicates that E1 isozyme may be specifically involved in retinal oxidation. Retinoic acid is an important hormone in development and differentiation and E1 isozyme has been demonstrated to play a role in polarity of development of the mammalian eye (McCaffery et al., 1991). Although the E2 isozyme has for some time been considered to play an important role in metabolism of aldehydes derived from monoamines (Tank and Weiner, 1979) its apparent absence from some Oriental individuals (Agarwal et al., 1981) suggests that this role can be duplicated by another enzyme or isozyme.

The list of aldehyde dehydrogenase substrates has been extended considerably since purification and characterization of E3 isozyme (Kurys et al., 1989; Ambroziak and Pietruszko, 1991). The naturally occurring substrates of all three isozymes, in addition to

Enzymology and Molecular Biology of Carbonyl Metabolism 4
Edited by H. Weiner, Plenum Press, New York, 1993

aldehydes derived from monoamines (MacKerell et al., 1986) and retinaldehyde now include aldehyde metabolites of diamines and polyamines (Kurys et al., 1989; Ambroziak and Pietruszko, 1991). Monoamines are physiologically important as neurotransmitters (Cooper et al., 1974) while diamines and polyamines are essential for cell growth, cell division, and are involved in development, differentiation and in cancer (Tabor and Tabor, 1984). Involvement of aldehyde dehydrogenase in dehydrogenation of τ-aminobutyraldehyde connects metabolism of the diamine, putrescine with that of τ-aminobutyric acid - a well known inhibitory neurotransmitter. Putrescine is a precursor of the polyamines, spermidine and spermine. An increase of polyamine levels has been described in rapidly growing tissues, such as embryonic tissues, regenerating liver, target organs after hormone administration, human and animal tumors, and in chemical carcinogenesis (Tabor and Tabor, 1984; Cohen, 1971; Boggust and O'Connel, 1984).

The aminoaldehydes which are products of polyamine oxidation, are toxic to a variety of cell types and induce metabolic effects opposite to those of polyamines themselves (Mondovi et al., 1988). Oxidized polyamines have been shown to inactivate various bacteria (Bachrach and Persky, 1964), and viruses Bachrach et al., 1963 and 1965), to inhibit growth and cell division (Gaugas and Dewey, 1978 and 1981), to kill intracellular plasmodial parasites (Morgan et al., 1981; Morgan et al., 1986; Ferrante et al., 1983), and to change the ultrastructure of eukaryotic cells (Bachrach et al., 1987). Moreover, an immuno-suppressive role has been proposed for these aldehydes (Lahib and Tomassi, 1981). Thus, oxidation of polyamine derived aldehydes to corresponding acids is an important physiological function of aldehyde dehydrogenase.

For many years, investigators of diamine and polyamine metabolism knew about the existence in mammalian liver of a disulfiram insensitive aldehyde dehydrogenase which metabolized aldehydes derived from diamines and polyamines to corresponding acids (Tabor and Tabor, 1984). The identity of the enzyme catalyzing this reaction was unknown until E3 isozyme was purified from human liver by Kurys et al., (1989) and characterized (Kurys et al., 1989; Ambroziak and Pietruszko, 1991). The E3 isozyme was subsequently shown to be localized in the cytoplasmic fraction (Ambroziak et al., 1991) of mammalian liver.

MATERIALS AND METHODS

The procedures employed for determination of kinetic constants with different substrates, and preparation or synthesis of substrates are described in detail by Ambroziak and Pietruszko (1987 and 1991). Commercially available substrates were also employed. Using the coupled reaction system employing diamines and polyamines as substrates and diamine and polyamine oxidases, it was possible to characterize all isozymes with respect to substrate specificity using aldehydes derived from: putrescine, cadaverine, agmatine, spermidine and spermine.

RESULTS AND DISCUSSION

Substrate Specificity of E3 Isozyme: Comparison with E1 and E2

In Table 1. Michaelis constants and maximal velocities of a variety of substrates of E3 isozyme are presented. The kcat/Km ratios are also included to facilitate identification of the best substrates. Properties of E3 isozyme demonstrate functional similarity to E1 and

Table 1. Substrate Specificity of E3 Isozyme.

Substrate	Km (μM)	kcat (IU)	kcat/Km
<u>Aliphatic Aldehydes:</u>			
Formaldehyde	410	0.19	0.001
Acetaldehyde	57	0.29	0.005
Propionaldehyde	9.5	0.30	0.03
Butyraldehyde	2.8	0.32	0.11
Pentanaldehyde	1.4	0.37	0.26
Hexanaldehyde	0.6	0.39	0.65
<u>Aldehyde Metabolites of Monoamines:</u>			
Imidazoleacetaldehyde	59	0.92	0.16
3,4-Dihydroxyphenyl-acetaldehyde	2.6	0.19	0.07
5-Hydroxyindoleacetaldehyde		active (low Km)	
<u>Aldehyde Metabolites of Diamines:</u>			
τ-Aminobutyraldehyde	4.6	1.53	0.33
τ-Guanidinobutyraldehyde	8.3	0.8	0.1
5-Aminovaleraldehyde	18.5	1.1	0.06
<u>Aldehyde Metabolites of Polyamines:</u>			
Aldehyde from Spermidine	7.1	1.9	0.27
Monoaldehyde from Spermine	16.4	1.7	0.1
Dialdehyde from Spermine	2.2	-	-
Acrolein	4.9	0.38	0.08

IU = μmol/min/mg; all values were determined in 50mM sodium phosphate buffer pH 7.0 - 7.4 (Ambroziak and Pietruszko, 1991).

E2 isozymes of human aldehyde dehydrogenase. Like E1 and E2 isozymes, the E3 isozyme has broad substrate specificity: the Km values for short chain aliphatic aldehydes are low (Table 1). In fact the E3 isozyme substrate specificity profile with aliphatic aldehydes (1-6 carbon atom chain length) is indistinguishable from that of E1 isozyme (Ambroziak and Pietruszko, 1991).

Table 2. Comparison of Some Properties of E1, E2 and E3 Isozymes.

Property	Isozyme of ALDH		
	E1	E2	E3
Similarities:			
Native Molecular Weight[1]	240,000	220,000	240,000
Subunit Molecular Weight[2]	54,000	54,000	54,000
Broad Substrate Specificity	+	+	+
Aminoaldehydes as Substrates	+	+	+
NAD as Coenzyme	+	+	+
Imidazoleacetaldehyde Km (μM)	40	38	59
3,4-Dihydroxyphenylacetaldehyde Km (μM)	0.4	1	2.6
Differences:			
Bound Coenzyme	-	-	+
Disulfiram Inhibition	I	P	P
Effect of Mg^{++}	Inhibition	Activation	No Effect
Acetaldehyde Km (μM)	50	1	57
τ-Aminobutyraldehyde Km (μM)	760	512	5
Aldehyde from Spermidine Km (μM)	High Km	High Km	7

[1] = Molecular weights determined by gradient gel electrophoresis; [2] = subunit molecular weights determined by electrophoresis on SDS gels; I = instantaneously inhibited; P = partially inhibited.

Its Km and kcat value for acetaldehyde (Tables 1 and 2) are identical to that of E1 isozyme (Km = 50 μM; kcat = 0.25) when determined under the same experimental conditions. The E3 isozyme shows high affinity for 3,4-dihydroxyphenylacetaldehyde (aldehyde metabolite of dopamine) and is active at low concentrations with 5-hydroxyindoleacetaldehyde (a metabolite of serotonin). With regard to these two substrates the E3 isozyme is similar to both E1 and E2 isozymes. The kinetic constants for imidazoleacetaldehyde (a metabolite of histamine) are almost identical for all three isozymes (Ambroziak and Pietruszko, 1991). Thus, with respect to metabolites of monoamines, the properties of E1, E2 and E3 isozymes are indistinguishable. Moreover, all three isozymes can utilize aminoaldehydes (which are diamine metabolites) as substrates.

Differences between isozymes can be only seen in the values of Michaelis constants for acetaldehyde and for aminoaldehydes (Table 2). The Michaelis constant for acetaldehyde of the E2 isozyme is almost two orders of magnitude lower than that for E1 and E3 isozymes. Michaelis constants for aminoaldehydes are low micromolar values for E3 isozyme (Table 1); they are much larger (about two orders of magnitude, Table 2) for E1 and E2 isozymes. Among the substrates listed in Table 1, τ-aminobutyraldehyde (a metabolite of putrescine) and aminoaldehyde from spermidine are the best natural substrates for E3 isozyme as defined by the highest kcat/Km ratios. These values are comparable with those of pentanaldehyde and hexanaldehyde which are also good substrates.

2,3-Enoic aldehyde (acrolein) which is also a polyamine metabolite, is also a substrate for E3 isozyme which is utilized with catalytic efficiency comparable to that of butyraldehyde (see Table 1). Moreover, acrolein does not appear to covalently react with E3 isozyme in the manner it reacts with E1 and E2 isozymes (Ferencz-Biro and Pietruszko, 1984); the E3 isozyme remains unchanged following catalysis of acrolein dehydrogenation; both E1 and E2 isozymes are damaged by acrolein.

Isoelectric Focusing of Liver Homogenates and Staining with Naturally Occurring Aldehydes

All three isozymes have similar pI values and are visualized on isoelectric focusing gels in the region of pH 5-5.4. Isoelectric focusing of liver homogenates was attempted (Figure 1) to see if E1, E2 and E3 isozymes could be detected with natural substrates and also if other aldehyde dehydrogenases could be visualized with those substrates. The following substrates were used: τ-aminobutyraldehyde (100 μM); 3,4-dihydroxyphenylacetaldehyde (100 μM); indole-3-acetaldehyde (100 μM). With these substrates the same bands were visualized as with low concentrations of propionaldehyde (100 μM). All bands were localized within pH 5 to pH 5.4 region of the isoelectric focusing gel suggesting that aldehyde dehydrogenase (EC 1.2.1.3) is the only enzyme in the human liver homogenate with activity with the above substrates at low concentrations.

The results of isoelectric focusing do not totally exclude the presence of other aldehyde dehydrogenases specific for those substrates. The reason for this is that a majority of liver proteins occur in the area of pI 5-8 and there are a large number of superimposable proteins and enzymes of opposing catalytic activity. However, our gel electrophoresis experiments suggest the possibility that aldehyde dehydrogenase (EC 1.2.1.3) is the only enzyme in the human liver capable of metabolizing putrescine and dopamine metabolites.

Subcellular Localization

The E2 isozyme is mitochondrial while both E1 and E3 (Pietruszko, 1989; Ambroziak et al., 1991) are cytoplasmic. The locus of diamine and polyamine catabolism

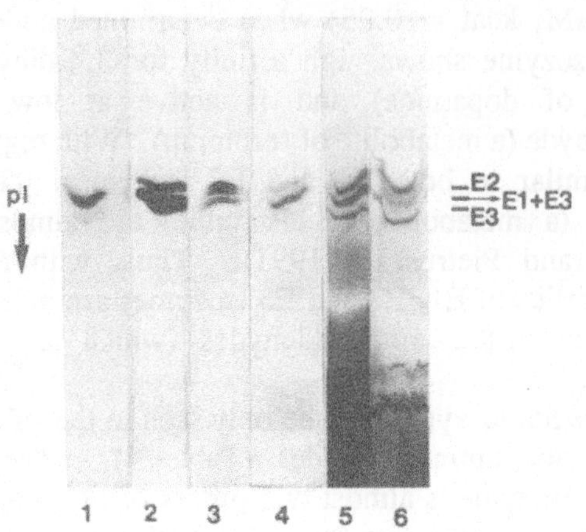

Figure 1. Isoelectric focusing gel of aldehyde dehydrogenases in the human liver homogenate and localization with natural substrates. The arrow shows direction of increase of pI; positions of E1,E2 and E3 isozymes during isoelectric focusing are shown diagrammatically. Lane 1:homogeneous E3 control developed with τ-aminobutyraldehyde, lane 2:purified E1 and E2 mixture developed with propionaldehyde. Lanes 3-6 = liver homogenate, lane 3:developed with propionaldehyde (100μM), lane 4:developed with τ-aminobutyraldehyde (100μM), lane 5:developed with 3,4-dihydroxy-phenylacetaldehyde (100μM), lane 6:developed with indole-3-acetaldehyde (100μM).

in mammalian cells is not yet established but it appears likely to be cytoplasmic. Both diamine and polyamine oxidases occur in the cytoplasmic fraction of mammalian cells. Subcellular distribution of El, E2 and E3 isozymes may be connected with that of specialized amine oxidases (monoamine oxidase in the mitochondria, diamine oxidase in the cytoplasm and polyamine oxidase in the cytoplasm and in blood plasma) and ready accessibility of biogenic amines and polyamines as substrates for naturally occurring coupled enzyme systems.

Postulated Metabolic Function

Michaelis constants of all three isozymes with aldehyde metabolites of dopamine and serotonin are low micromolar values. Thus, the isozymes have properties which make them suitable for metabolism of these aldehydes. The low Km values for aldehydes derived from dopamine and serotonin suggest that low micromolar or even nanomolar concentrations of 3,4-dihydroxyphenylacetaldehyde and 5-hydroxyindoleacetaldehyde could be metabolized by all three isozymes. Metabolism of monoamine metabolites has been reported to occur in the mitochondria (Tank and Weiner, 1979), thereby suggesting that enzymes like the mitochondrial E2 isozyme would be involved in that function. However, apparent absence of this isozyme in Orientals (Agarwal et al., 1981) makes such an assignment rather difficult. It appears more likely that the metabolic role in monoamine metabolism is interchangeable with all three isozymes. The actual *in vivo* concentrations of aldehydes derived from dopamine and serotonin are unknown.

It was of interest to note during this investigation that properties of all three isozymes were indistinguishable with regard to imidazoleacetaldehyde (a metabolite of histamine). The almost identical Michaelis constants with imidazoleacetaldehyde are a unique feature of all three isozymes which seems to define their identity as imidazoleacetaldehyde dehydrogenases. The Km value for imidazoleacetaldehyde is considerably larger than for aldehydes derived from serotonin and dopamine (Table 2). The level of circulating histamine is known to be much larger than that of dopamine and serotonin; whether the level of imidazoleacetaldehyde is also larger is unknown.

In living organisms, putrescine arises from ornithine by decarboxylation and is converted to spermidine and spermine which function in development and differentiation (Figure 2). Conversion of putrescine to τ-aminobutyraldehyde is catalyzed by diamine oxidase ((Fogel, 1986; Seiler and Eichentopf, 1975; Shaff, and Beaven, 1976), also known as histaminase. Diamine oxidase is a cytoplasmic enzyme present in considerable amounts in mammalian liver and intestine. Its amount in placenta increases about 1000 fold during pregnancy (Gahl et al., 1982; Swanberg, 1950) but the reason for this increase is unknown.

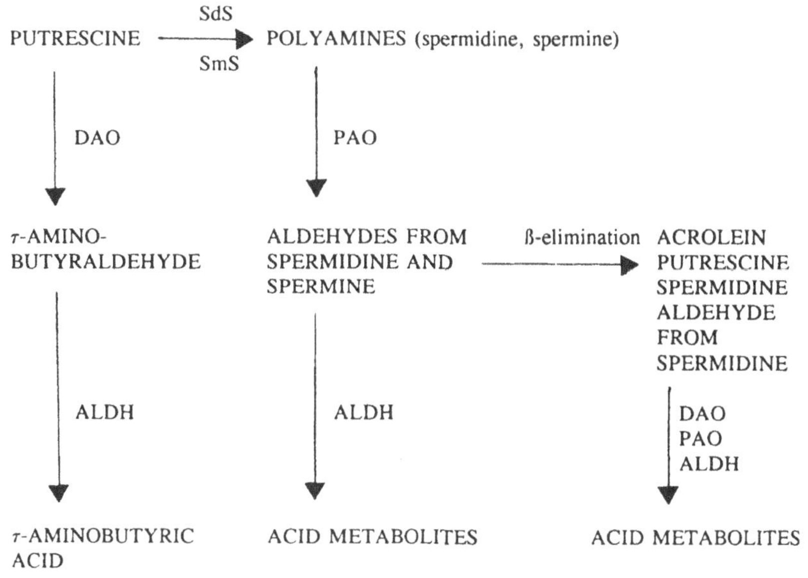

Figure 2. Metabolism of putrescine and suggested physiological role of E3 isozyme. PAO - polyamine oxidase; DAO - diamine oxidase; ALDH - E3 isozyme; SmS - synthetic pathway of spermidine; SdS - synthetic pathway of spermine; ß-elimination is a chemical reaction.

Putrescine and polyamines promote cell growth and proliferation, while corresponding aminoaldehydes have potent antiproliferation effects (see above). The fact that all metabolites of the putrescine cycle are excellent substrates for E3 isozyme, together with gel staining results, indicates that a physiological role of E3 isozyme may be in the metabolism of aldehydes produced by amine oxidases from putrescine, diamines and polyamines (Figure 2). The fact that the highest monoamine oxidase and diamine oxidase activities are localized, like aldehyde dehydrogenase, in the liver and tissues with the highest blood supply (and therefore free in and out flow) also suggests physiological importance of an amine oxidase:aldehyde dehydrogenase coupled system in metabolism of neurotransmitters and hormones.

The products of aminoaldehyde dehydrogenation by E3 isozyme are acid metabolites (Figure 2) whose physiological roles, except in one case, are unknown. The known role is that of γ-aminobutyric acid (Figure 2) which is a well characterized negative neurotransmitter. Thus, via E3 isozyme putrescine is converted to γ-aminobutyric acid. It is speculated that this pathway could be of major importance in peripheral neurotransmission, suggesting a new and exciting metabolic role for aldehyde dehydrogenase.

Thus, aldehyde dehydrogenase (EC 1.2.1.3) appears to have a specialized role in metabolism of active biogenic metabolites with important physiological function. Among these metabolites are aldehydes derived from monoamines, diamines and polyamines as well as retinaldehyde, all functioning in development and differentiation. Recent studies in insects where aldehyde dehydrogenase functions in inactivation of pheromones which are important in sex recognition and survival (Prestwich, 1987), also support hormone metabolizing role for aldehyde dehydrogenase. Although absence of mitochondrial E2 isozyme has been reported (Agarwal et al., 1981) in some Orientals and rare cases of absence of the cytoplasmic E1 isozyme have been reported (Eckey et al, 1986), there has been no single report of the absence of both isozymes, suggesting that aldehyde dehydrogenase may be essential for survival.

The broad substrate specificity of E1, E2 and E3 isozymes suggests that all three isozymes are suited for the metabolic function of detoxication. All three isozymes can easily metabolize aldehydes ingested in foodstuffs, or alcohols that are ingested and converted to aldehydes via alcohol dehydrogenase. The enzyme most likely performs a dual role: that of metabolism of hormones and neurotransmitters, and that of detoxication. Due to the fact that rates of metabolism of aldehyde metabolites of neurotransmitters and hormones would most likely be influenced by composition of foodstuffs derived from surrounding environment, the enzyme may perform a larger role of adaptation to surrounding environment.

Metabolic Function and Ethanol Metabolism

In living organisms enzymes normally function at substrate concentrations far below Km values where the system is sensitive to changes in substrate concentration. Metabolism of ethanol derived acetaldehyde via the mitochondrial isozyme (Parilla et al., 1974) presumably proceeds at enzyme saturation, since the Km value of E2 isozyme for acetaldehyde is only ca. 1 μM while the concentration of acetaldehyde in liver during alcohol metabolism is reported to be ca. 100 μM (Erickson et al., 1984). Thus, during ethanol metabolism total E2 isozyme has to be present in the liver in the form of productive ternary complex: enzyme-acetaldehyde-NAD (Segel, 1973). At 100 μM acetaldehyde, both E1 and E3 isozymes (Km = ca. 50 μM) must also largely exist as enzyme:acetaldehyde:NAD complexes (70% site occupation calculated from Michaelis-Menten equation). Thus, only a third of the E1 and E3 isozymes would be available for metabolism of natural substrates.

Interaction of dopamine, serotonin and norepinephrine metabolism with that of ethanol at the level of aldehyde dehydrogenase is well documented (Davis et al., 1967a,b; Davis et al., 1970; Feldstein et al., 1967). The information about interaction of putrescine (Diehl et al., 1992) metabolism and that of ethanol is only very recent. In Figure 3 a scheme is presented to show how interaction of metabolism of ingested substrates with that of naturally occurring metabolites can occur at the level of aldehyde dehydrogenase.

Properties of E3 isozyme suggest that metabolism of alcohol and other substrates ingested in foodstuffs which are detoxicated by aldehyde dehydrogenase may also interact with that of putrescine and polyamines (Figure 3). The quantities of naturally occurring aldehydes ingested at a given time would be relatively small and would occupy aldehyde

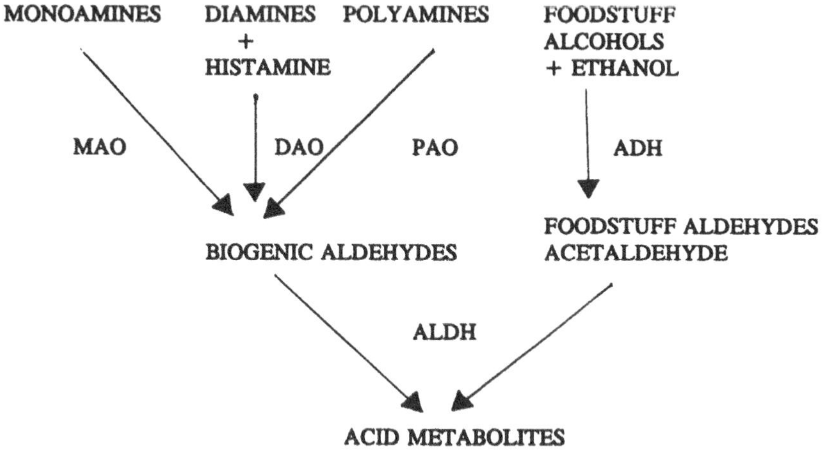

Figure 3. Interaction of biogenic amine metabolism with foodstuff alcohols and aldehydes and with ethanol. MAO - monoamine oxidase; DAO - diamine oxidase; PAO - polyamine oxidase; ADH - alcohol dehydrogenase; ALDH - aldehyde dehydrogenase.

dehydrogenase isozymes for only a short time. During ethanol metabolism, however, the E1 and E3 isozymes would be ca. 70% occupied by acetaldehyde almost continuously, or for long continuous time periods. This is expected in the case of E3 isozyme to result in aminoaldehyde (which are biologically active in a variety of ways, see above) accumulation. The inhibitory effect of ethanol on metabolism of putrescine to τ-aminobutyric acid has been recently demonstrated in animal experiments (Diehl et al., 1992). Interaction of ethanol metabolism with that of putrescine and polyamines could be of particular importance in fast-growing tissues such as embryos.

ACKNOWLEDGEMENTS

Financial support of Fullbright Fellowship to W. Ambroziak, Research Scientist Award (K05 AA00046) to R. Pietruszko and NIAAA Grant R02 AA00186 is acknowledged.

REFERENCES

Agarwal, D.P., Harada, S., and Goedde, H.W., 1981, Racial differences in biological sensitivity to ethanol: the role of alcohol dehydrogenase and aldehyde dehydrogenase isozymes, Alcohol. Clin. Expl. Res. 5:12.

Ambroziak, W., and Pietruszko, R., 1987, Human aldehyde dehydrogenase: metabolism of putrescine and histamine, Alcohol. Clin. Expl. Res. 11:528.

Ambroziak, W. and Pietruszko, R., 1991, Human aldehyde dehydrogenase: activity with aldehyde metabolites of monoamines, diamines and polyamines, J. Biol. Chem. 266:13011.

Ambroziak, W., Kurys, G., and Pietruszko, R., 1991, Aldehyde dehydrogenase (EC 1.2.1.3): comparison of subcellular localization of the third isozyme that dehydrogenates τ-aminobutyraldehyde in rat, guinea pig and human liver, Comp. Biochem. Physiol. 100B:321.

Bachrach, U., Ash, I., and Rahamim, E., 1987, Effect of microinjected amine and diamine oxidases on the ultrastructure of eukaryotic cultured cells, Tissue Cell 19:39.

Bachrach, U., and Persky, S., 1964, Antibacterial action of oxidized spermine, J. Gen. Microbiol. 37:195.

Bachrach, U., Rabina, S., Loebenstein, G., and Eilon, G., 1965, Antiviral action of oxidized spermine - inactivation of plant viruses, Nature 208:1095.
Bachrach, U., Tabor, C.W., and Tabor, H., 1963, Inactivation of bacteriophages by oxidized spermine, Biochim. Biophys. Res. Commun. 78:768.
Boggust, N.A., and O'Connell, C., 1984, Polyamines, mono and diacetyl putrescine, oxidation and proliferation in tumor cells, IRCS Med. Sci. 12:267.
Cohen, S.S., 1971, Polyamines in Eukaryotic Organisms, Prentice Hall, Inc.
Cooper, J.R., Bloom, F.E., and Roth, R.H., 1974, The Biochemical Basis of Neuropharmacology, Oxford University Press, New York, London, Toronto.
Davis, V.E., Brown, H., Huff, J.A., and Cashaw, J., 1967a, The alteration of serotonin metabolism to 5-hydroxytryptophol by ethanol ingestion in man, J. Lab. Clin. Med. 69:132.
Davis, V.E., Brown, H., Huff, J.A., and Cashaw, J., 1967b, Ethanol induced alterations in norepinephrine metabolism in man, J. Lab. Clin. Med. 69:787.
Davis, V.A., Walsh, M.J., and Yamanaka, Y., 1970, Augmentation of alcohol formation from dopamine by alcohol and acetaldehyde *in vivo*, J. Pharmacol. Expl. Ther. 174:401.
Diehl, A.M., Yang, S.O., Brown, N., Smith, J., Ralford, D., Gordon, R., and Casero, R., 1992, Ethanol-associated alterations in the kinetics of putrescine uptake and metabolism by the regenerating liver, Alcohol. Clin. Expl. Res. 16:5.
Eckey, R., Agarwal, D.P., Saha, N., and Goedde, H.W., 1986, Detection and partial characterization of a variant form of cytosolic aldehyde dehydrogenase isozyme, Hum. Genet. 72:95.
Eriksson, C.J.P., Atkinson, N., Petersen, D.R., and Deitrich, R.A., 1984, Blood and liver acetaldehyde concentrations during ethanol oxidation in C57 and DBA mice, Biochem. Pharmacol. 33:2213.
Feldstein, A., Hoagland, H., Freeman, H., and Williamson, O., 1967, The effect of ethanol ingestion on serotonin-C14 metabolism in man, Life Sci. 6:53.
Ferencz-Biro, K., and Pietruszko, R., 1984, Inhibition of human aldehyde dehydrogenase isozymes by propiolaldehyde, Alcohol. Clin. Expl. Res. 8:302.
Ferrante, A., Rzepczyk, M.C., and Allison, A.C., 1983, Polyamine oxidase mediates intra-erythrocytic death of Plasmodium falciparum. Trans. Roy. Soc. Trop. Med. Hyg. 77:789.
Fogel, W.A., 1986, in GABAergic Mechanisms in the Mammalian Periphery, pp 35-56, eds. S.L. Erdo and N.G. Bowery, Raven Press, New York.
Gahl, W.A., Vale, A.M., and Pitot, H.C., 1982, Spermidine oxidase in human pregnancy serum: probable identity with diamine oxidase, Biochem. J. 201:161.
Gaugas, J.M., and Dewey, D.L., 1978, Evidence for serum binding of oxidized spermine and its potent G-1 phase inhibition of cell proliferation, Br. J. Cancer 39:548.
Gaugas, J.M., and Dewey, D.L., 1981, Hog kidney diamine oxidase conversion of biogenic diamines to inhibitors of cell proliferation, J. Pathol. 134:243.
Jakoby, W.B., and Ziegler, D.M., 1990, The enzymes of detoxication, J. Biol. Chem. 265:20715.
Kurys, G., Ambroziak, W., and Pietruszko, R., 1989, Human aldehyde dehydrogenase: purification and characterization of a third isozyme with low Km for τ-aminobutyraldehyde, J. Biol. Chem., 264:4715.
Lahib, R.S., and Tomassi, T.B., 1981, Enzymatic oxidation of polyamines, relationship to immunosuppressive properties, Eur. J. Immunol. 11:266.
MacKerell, A.D., Blatter, E.E., and Pietruszko, R., 1986, Human aldehyde dehydrogenase: kinetic identification of the isozyme for which biogenic aldehydes and acetaldehyde compete, Alcohol. Clin. Expl. Res. 10:266.
McCaffery, P., Tempst, P.P., Lara, G., and Drager, U.C., 1991, Aldehyde dehydrogenase is a positional marker in the retina, Development 112:693.
Mondovi, B., Riccio, P., and Agostinelli, E., 1988, Experimental Medicine and Biology, Progress in Polyamine Research, 250:147, eds. V. Zappia and A.E. Pegg, Plenum Press, New York, London.
Morgan, D.M.L., Christensen, J.R., and Allison, A.C., 1981, Polyamine oxidase and the killing of intracellular parasites, Biochem. Soc. Trans. 9:563.
Morgan, D.M.L., Bachrach, R., Assaraf, Y.G., Harari, E., and Golenser, J., 1986, The effect of purified aminoaldehydes produced by polyamine oxidation on the development in vitro of Plasmodium falciparum in normal and glucose-phosphate-dehydrogenase deficient erythrocytes, Biochem. J. 236:97.
Parrilla,R., Okhawa, K., Lindros, K.O., Zimmerman, U.-I.P., Kobayashi, K and Williamson, J.R., 1974, Functional compartmentation of acetaldehyde oxidation in rat liver, J. Biol. Chem. 249:4926.
Pietruszko, R., 1989, Aldehyde dehydrogenase (EC 1.2.1.3), in Biochemistry and Physiology of Substance Abuse, R.R. Watson, ed., CRC Press Inc., Boca Raton, Florida.

Prestwich, G.D., 1987, Chemistry of pheromone and hormone metabolism in insects, Science 237:999.
Segel, I.H., 1973, Enzyme Kinetics, John Wiley and Sons, New York, Chichester, Brisbane, Toronto.
Seiler, N., and Eichentopf, B., 1975, 4-Aminobutyrate in mammalian putrescine metabolism, Biochem. J. 152:201.
Shaff, R.E., and Beaven, M.A., 1976, Turnover and synthesis of diamine oxidase (DAO) in rat tissues, studies with heparin and cycloheximide, Biochem. Pharmacol. 25:1057.
Swanberg, 1950, Histaminase in Pregnancy, Acta Physiol. Scand. Supplement 79.
Tabor, C.W., and Tabor, H., 1984, Polyamines, Annu. Rev. Biochem. 53:749.
Tank, A.W., and Weiner, H., 1979, Ethanol-induced alteration of dopamine metabolism in rat liver, Biochem. Pharmacol. 28:3139.

EFFECTS OF ALDEHYDE PRODUCTS OF LIPID PEROXIDATION ON THE ACTIVITY OF ALDEHYDE METABOLIZING ENZYMES IN HEPATOMAS

Rosa A. Canuto, *Margherita Ferro, Giuliana Muzio, *Anna M. Bassi, Gabriella Leonarduzzi, Marina Maggiora, *Daniela Adamo, Giuseppe Poli and **Ronald Lindahl

Dept. Experimental Oncology and Medicine, University of Turin Corso Raffaello 30 - 10125 Turin (Italy). *Institute of General Pathology, University of Genoa. **Dept. Biochemistry and Molecular Biology, University of South Dakota, Vermillion, S.D. 57069

INTRODUCTION

In general, lipid peroxidation in hepatomas and hepatoma cell lines is lower than in normal tissue (Dianzani et al., 1986; Poli et al., 1986). It is known that lipid peroxidation, which often occurs in response to oxidative stress, produces a great diversity of aldehydes, when lipid hydroperoxides break down in biological systems. Some of these aldehydes are highly reactive, and may be considered as second toxic messengers, which disseminate and augment initial free radical events. The aldehydes most intensively studied are 4-hydroxynonenal (4-HNE), 4-hydroxyhexenal, and malondialdehyde (Esterbauer et al., 1988). 4-HNE has been demonstrated to inhibit various biochemical processes in liver and tumour cells (Canuto et al., 1985).

In hepatomas an increased level of some enzymes metabolizing aldehydes produced during the lipid peroxidation process is also found (Huang and Lindahl, 1990; Canuto et al., 1990). The enzymes involved in this increase are especially aldehyde dehydrogenase (ALDH) and, to a lesser extent, aldehyde reductase. The increase seems to be directly correlated with the degree of deviation of the tumors. During chemically-induced hepatocarcinogenesis, these enzymes are increased in nodules, and the increases are accentuated in the resulting hepatomas (Lindahl and Evces, 1987; Canuto et al., 1989).

The question of whether the decrease in lipid peroxidation is related to the increase in aldehyde metabolizing enzymes is clearly of interest; likewise, it would be interesting to establish whether either or both are linked to tumor growth, and whether agents capable of inducing lipid peroxidation are able to induce aldehyde metabolizing enzymes.

Our research has been aimed at answering these questions by studying Yoshida ascites hepatoma AH-130, which has a low level of lipid peroxidation and a low level of aldehyde metabolizing enzymes, and also by studying various hepatoma cell lines, with differing enzyme activities and low levels of lipid peroxidation. 4-HNE was used as substrate for enzyme activity determination, because of its importance among the products of lipid peroxidation.

MATERIALS AND METHODS

Enrichment Of Hepatoma Cells With Arachidonic Acid. The cells used were those of Yoshida ascites hepatoma AH-130 . The enrichment with arachidonic acid, the measure of lipid peroxidation and of the mortality of these cells were done in accordance with a previous paper (Canuto et al., 1991).

Culture Conditions For Rat Hepatoma Cell Lines. The cells (7777, JM2, H4IIEC3, HTC) were grown routinely under an atmosphere of 5% CO_2 and 95% air, as monolayers at 37 C in 25 cm^2 tissue culture flasks containing 5.0 ml of DMEM/F 12 medium, supplemented with 0.5% newborn calf serum, 1% antibiotic/antimycotic solution, 1% glutamine, 1% vitamin solution, 1% non-essential aminoacid solution and 1% ITS (insulin-transferin-selenite).

Lipid Peroxidation. Lipid peroxidation was measured as malondialdehyde production by the spectrophotometric thiobarbituric acid assay (Slater and Sawyer, 1971).

Cytotoxicity On Hepatoma Cells Lines. Cytotoxicity of 4-HNE on cells was determined by two tests: 1) release of lactate dehydrogenase: after growth in the presence of different concentrations of 4-HNE for several hours, cell monolayers were lysed in 0.1% Triton X-100, and lactate dehydrogenase was measured according to Wroblevski (1955); 2) colony forming ability was determined as in a previous paper (Ferro et al., 1988). After 24 hours of cell growth in the presence of complete medium, cells were maintained in medium with different concentrations of 4-HNE, or only with solvent for 1 or 6 hours. After these times, the cells were grown in complete medium for 7 more days, to allow the survivors to form colonies. H4IIEC3 were unsuitable for this test, since they did not form good colonies.

Induction Of Enzymes. Clofibrate prepared in DMSO was added to flasks for 4 days at 125 nmoles/10^6 cells/day or 500 nmoles/10^6 cells/day. Control cultures received solvent alone. At 4 days after addition of clofibrate, cells were harvested by trypsin treatment and used for enzyme assays.

Enzyme Assays. Cells were homogenized and suspended as described in Canuto (1990). Diluted homogenates were centrifuged at 134000xg for 45 min to obtain cytosolic fractions. ALDH, aldehyde reductase, alcohol dehydrogenase and glutathione transferase activities were determined as described in Canuto (1990). Proteins were determined by a biuret procedure (Gornal et al., 1949).

RESULTS AND DISCUSSION

It is possible to increase lipid peroxidation in hepatoma cells by enriching them with arachidonic acid. In fact, one cause of decreased susceptibility to lipid peroxidation is the lack of peroxidable substrate. If we add arachidonic acid to Yoshida ascites hepatoma AH-130 cells, this makes them susceptible to lipid peroxidation.

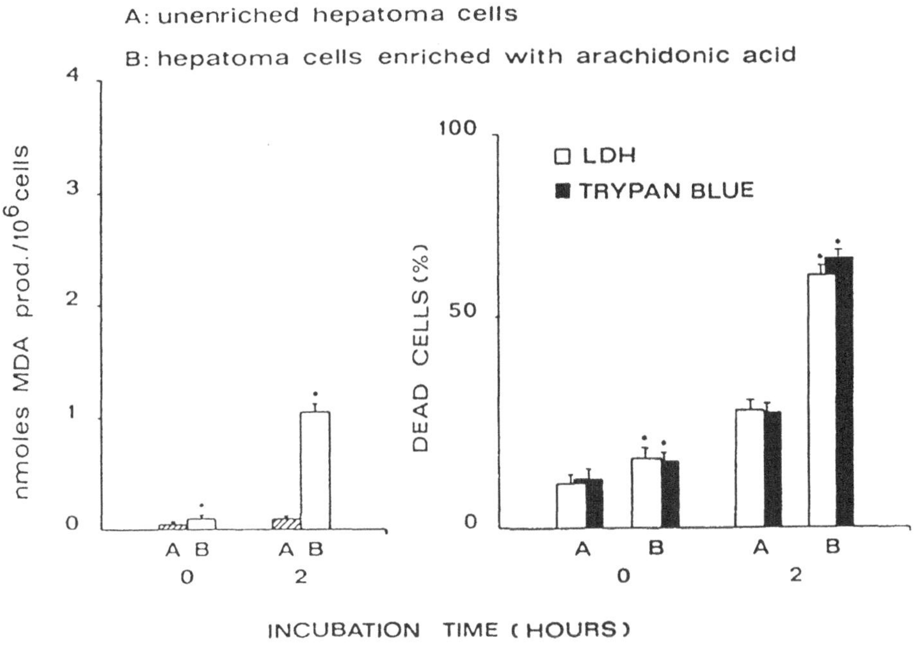

Figure 1. On the left is lipid peroxidation expressed in terms of malondialdeyde production. On the right is the mortality of the cells measured by trypan blue exclusion, expressed as percentage of cells stained, and by lactate dehydrogenase (LDH) release, expressed as a percentage of total activity present in the cells before 2 hours of incubation. t-test: * $p < 0.001$ (hepatoma AH-130 cells enriched with arachidonic acid vs unenriched hepatoma AH-130 cells).

An important consequence of increased lipid peroxidation in arachidonic acid-treated hepatoma cells is a decrease of cell viability, due to the toxic effect of lipid aldehyde on these cells.

Figure 1 reports the cell survival and lipid peroxidation values at 0 and 2 hours of incubation, with the lipid peroxidation inducer ascorbate/iron sulphate, in control hepatoma AH-130 cells and in hepatoma AH-130 cells enriched with arachidonic acid. Arachidonic acid-enriched cells have an increased lipid peroxidation and the highest mortality. An interesting consideration is that in these cells aldehyde metabolizing enzymes remain low (data not shown).

Various rat hepatoma cell lines, with differing activities of ALDH, are available. It is of interest to know whether cell lines differing in ALDH activity have correspondingly differing levels of lipid peroxidation. 7777 cells have a much lower activity of ALDH than JM2 cells when assayed with 4-HNE as substrate (Table 1). As expected from earlier work, the 4-HNE ALDH activity is cytosolic and appears to be class 3 ALDH.

7777 and JM2 cells also have differing levels of lipid peroxidation (Table 2). Three prooxidants, cumene hydroperoxide, ferric nitrilotriacetate (NTA), and clofibrate, were able to induce a production of malondialdehyde in 7777 cells, but at very low level. The lipid peroxidation level is even lower or non existent in JM2, with all prooxidants used. Therefore the levels of lipid peroxidation are inversely proportional to ALDH activity.

The sensitivity of 7777 and JM2 hepatoma cell lines to 4-HNE was then examined. Two other lines of hepatoma cells were also examined, H4IIEC3, which is similar to 7777 in ALDH activity, and HTC, which is similar to JM2. Two tests were done: release of lactate dehydrogenase (all cell lines) and colony-forming ability (7777, HTC and JM2), as shown in Table 3 and 4. Overall, 7777 and H4IIEC3, cells with low ALDH activities, are more sensitive to 4-HNE than JM2 or HTC cells, which have much higher ALDH activity.

Table 1. Aldehyde dehydrogenase activity in 7777 and JM2 rat hepatoma cells with 0.1 mM 4-hydroxynonenal as substrate.

Cells	Coenzyme	Homogenate	Cytosol
7777	NAD	4.45 ± 0.54	9.21 ± 1.05
	NADP	2.15 ± 0.32	3.37 ± 0.45
JM2	NAD	6.52 ± 0.63*	30.16 ± 3.64*
	NADP	4.88 ± 0.25*	17.25 ± 2.70*

Data are expressed as nmoles of NAD(P) reduced/min/mg of protein. t-test: * $p < 0.05$ (JM2 cells vs 7777 cells).

Table 2. Malondialdehyde production in 7777 and JM2 rat hepatoma cells.

PROOXIDANT	T_0	T_{2h}	T_{6h}
		7777 cells	
none	0.042	0.068	0.133
CuOOH (0.5 mM)	0.073	0.150	0.253
Ascor./$FeSO_4$ (0.5 mM/0.1 mM)	0.069	0.068	0.133
ADP/$FeCl_3$ (2.5 mM/0.1 mM)	0.069	0.085	0.093
Fe^{3+}/NTA (0.1 mM)	0.071	0.131	0.227
$FeSO_4$/Istidine (0.1 mM/1 mM)	0.096	0.107	0.127
Clofibrate (0.5 mM)	0.053	0.080	0.247
		JM2 cells	
none	0.060	0.073	0.067
CuOOH (0.5 mM)	0.073	0.096	0.083
ADP/$FeCl_3$ (2.5 mM/0.1 mM)	0.076	0.105	0.109
Ascor./$FeSO_4$ (0.5 mM/0.1 mM)	0.087	0.133	0.107
Clofibrate (0.5 mM)	0.047	0.047	0.057

The cells reaching confluence were detached with trypsin, resuspended in a serum-free medium, and incubated with prooxidants for times up to 6 hours. Data, means of 2 experiments, are expressed as nmoles of malondialdehyde for 10^6 cells.

There were significant differences in cell viability, measured as lactate dehydrogenase content of cells, in cell lines exposed to 4-HNE. After six hours of incubation with 100 uM 4-HNE, 7777 cells had only 29% of lactate dehydrogenase activity of unexposed 7777 cells, and H4IIEC3 had 11%. JM2 had 81% of its control cell lactate dehydrogenase, and HTC cells had 93%. After longer incubation, the release of lactate dehydrogenase continued to increase much more in 7777 and H4IIEC3 than in JM2 or HTC. The loss of colony-forming ability was also greater in 7777 than in the other two cell lines, and was time-dependent. For example, after six hours of incubation with 10 uM 4-HNE, 7777 were no longer able to form colonies; HTC still formed 11% of colonies and JM2 formed 34% of colonies.

Therefore the toxicity of 4-HNE, externally added or induced by prooxidants, is higher in hepatoma cells with lower ALDH content than in those with higher content.

Table 3. Effect of 4-hydroxynonenal on lactate dehydrogenase activity in hepatoma cell lines.

HNE µM	Time hours	HTC	JM2	7777	H4IIEC3
10	6	98 ± 4	95 ± 7	97 ± 3	94 ± 8
	12	98 ± 4	94 ± 8	98 ± 2	94 ± 6
	24	96 ± 6	97 ± 7	98 ± 3	83 ± 7
25	6	96 ± 5	87 ± 12	98 ± 2	85 ± 7
	12	95 ± 9b	91 ± 11	91 ± 6	77 ± 9
	24	97 ± 4b	88 ± 16	85 ± 9	70 ± 9
50	6	95 ± 10b	81 ± 8b	96 ± 4	49 ± 9
	12	86 ± 9ab	84 ± 2ab	28 ± 2	19 ± 3
	24	77 ± 10ab	85 ± 16ab	25 ± 5	20 ± 7
100	6	93 ± 10ab	81 ± 3ab	29 ± 6	11 ± 2
	12	58 ± 8ab	80 ± 11ab	23 ± 6	3 ± 3
	24	34 ± 9ab	70 ± 12ab	20 ± 3	3 ± 1

Data, means ± S.D. of 4 experiments, are expressed as percentages of lactate dehydrogenase in untreated cells.
a = $p < 0.001$ (HTC or JM2 cells vs 7777 cells)
b = $p < 0.001$ (HTC or JM2 cells vs H4IIEC3 cells).

Table 4. Effect of 4-hydroxynonenal on colony formation efficiency of hepatoma cell lines.

HNE µM	Time hours	HTC	JM2	7777
1	1	99 ± 1*	85 ± 6	75 ± 10
	6	90 ± 7*	84 ± 9*	66 ± 7
5	1	95 ± 5*	83 ± 8	76 ± 13
	6	33 ± 4	45 ± 9	32 ± 12
10	1	94 ± 7*	64 ± 7	60 ± 7
	6	11 ± 4*	34 ± 10*	2 ± 1
50	1	12 ± 4*	25 ± 10*	0
	6	0	0	0

The cells were exposed to 4-HNE at doses ranging from 1 to 50µM. Data, means ± S.D. of 4 experiments, are expressed as percentages of colony formation in untreated cells. t-test: * $p < 0.001$ (HTC or JM2 cells vs 7777 cells).

To see whether this difference is actually due to differences in ALDH activity, attempts were made to increase ALDH activity in 7777 cells, with the aim of determining the effect of 4-HNE on these cells. For this purpose, we used clofibrate (known to induce peroxisomal proliferation and cytochrome P-452 production in the liver) and determined not only ALDH, but also the activity of other enzymes involved in the metabolism of lipid peroxidation products, in order to obtain an overview of aldehyde catabolism induction (Table 5).

Table 5. Induction by clofibrate of 4-hydroxynonenal metabolizing enzymes in cytosol of hepatoma cells.

Enzymes	7777	7777+clof.1	7777+clof.2
ALDH-NAD	4.74±0.99	4.78±0.66	3.07±1.08
ALDH-NADP	1.47±0.28a	2.78±0.22b	3.32±0.32c
ADH-NADH	1.17±0.34a	2.71±0.47b	4.83±0.43c
ALRED	4.07±0.39a	7.16±0.86b	10.17±1.23c
GST	149.46	126.41	220.79

Clofibrate was added to the cells each day for 4 days, at 125 (clof.1) or 500 nmoles (clof.2) per one million of cells per day. Data are means ± S.D. of 5 experiments for aldehyde dehydrogenase (ALDH), alcohol dehydrogenase (ADH) and aldehyde reductase (ALRED), except for glutathione transferase (GST) (2 experiments). They are expressed as nmoles of coenzyme reduced or oxidized, or 4-HNE consumed per min per mg of protein. Substrate was 0.1 mM 4-HNE. Means of the three groups with different letters are statistically different ($p < 0.05$) from one another as determined by variance analysis followed by the Newman-Keuls test.

For ALDH with 4-HNE as substrate, only the activity of Class 3 ALDH-NADP is stimulated by clofibrate, and its increase is progressive with clofibrate concentration. Aldehyde reductase and the reduction activity of alcohol dehydrogenase were also both increased to about two- or three-fold in a dose-dependent manner by clofibrate. Glutathione transferase activity is also increased about two-fold by clofibrate.

We therefore conclude that the enzymes that metabolize aldehyde products of lipid peroxidation can be induced by clofibrate, but the induction appears to involve all enzymes involved in lipid aldehyde metabolism.

What is now needed is to try to induce these enzymes more, by using other prooxidants such as ascorbate/iron sulphate, and to determine whether cells with higher levels of one or more of these enzymes are correspondingly more resistant to 4-HNE toxicity.

It will be interesting to know whether enzyme induction occurs at the level of transcription and, if so, what regulates the response.

Acknowledgments

We are grateful to Prof. H. Esterbauer for providing 4-HNE. This paper was supported by grants from the Italian Ministry for University, Scientific and Technological Research, from the National Research Council Bilateral and Targeted Project "Clinical Applications in Tumor Research", from the Italian Association for Cancer Research, and by NIH Grant CA 21103.

REFERENCES

Canuto, R.A., Biocca, M.E., Muzio, G., Garcea, R., and Dianzani, M.U., 1985, The effect of various aldehydes on the respiration of rat liver and hepatomas AH-130 cells, Cell Biochem. Funct., 3:3.

Canuto, R.A., Muzio, G., Biocca, M.E., and Dianzani, M.U., 1989, Oxidative metabolism of 4-hydroxy-2,3-nonenal during diethyl-nitrosamine-induced carcinogenesis in rat liver, Cancer Lett., 46:7.

Canuto, R.A., Muzio, G., Bassi, A.M., Biocca, M.E., Poli, G., Esterbauer, H., and Ferro, M., 1990, Metabolism of 4-hydroxynonenal in hepatoma cell lines, Adv. Exp. Med. Biol., 284:75.

Canuto, R.A., Muzio, G., Biocca, M.E., and Dianzani, M.U., 1991, Lipid peroxidation in rat AH-130 hepatoma cells enriched in vitro with arachidonic acid, Cancer Res., 51:4603.

Dianzani, M.U., Poli, G., Canuto, R.A., Rossi, M.A., Biocca M.E., Biasi, F., Cecchini, G., Muzio, G., Ferro, M., and Esterbauer, H., 1986, New data on kinetics of lipid peroxidation in experimental hepatomas and preneoplastic nodules, Toxicol. Pathol., 14:404.

Esterbauer, H., Zollner, H., and Schaur, R.J., 1988, Hydroxyalkenals: cytotoxic products of lipid peroxidation, ISI Atlas of Science, 1:311.

Ferro, M., Bassi, A.M., Penco, S., Piana, S., Ravere, G., Nanni, G., 1992, Use of cultured hepatoma cell lines in the assessment of aldehyde metabolism, ATLA, 22:77.

Gornal, A.G., Bardawill, C.J., and David, M., 1949, Determination of serum proteins by means of biuret reaction, J. Biol. Chem., 177:751.

Huang, M., and Lindahl, R., 1990, Aldehyde dehydrogenase heterogeneity in rat hepatic cells, Arc. Biochem. Biophys., 277:296.

Lindahl, R., and Evces, S., 1987, Changes in aldehyde

dehydrogenase activity during diethylnitrosamine-initiated rat hepatocarcinogenesis, Carcinogenesis, 8:785.
Poli, G., Cecchini, G., Biasi, F., Chiarpotto, E., Canuto, R.A., Biocca, M.E., Muzio, G., Esterbauer, H., and Dianzani, M.U., 1986, Resistance to oxidative stress by hyperplastic rat liver tissue monitored in terms of unpolar and medium polar carbonyls, Biochim. Biophys. Acta, 883:207.
Sharma, R., Lake, B.G., Foster, J., and Gibson, G.G., 1988, Microsomal cytochrome P-452 induction and peroxisome proliferation by hypolipidaemic agents in rat liver. A mechanistic inter-relationship, Biochem. Pharmacol., 37:1193.
Slater, T.F., and Sawyer, B.C., 1971, The stimulatory effects of carbon tetrachloride and other halogenoalkanes on peroxidative reactions in rat liver fractions in vitro. I. General features of the systems used, Biochem. J., 123:805.
Wroblevski, F., and LaDue J.S., 1955, Lactic dehydrogenase activity in blood, Proc. Soc. Exper. Biol., 90:210.

METABOLIC INTERACTIONS OF 4-HYDROXYNONENAL, ACETALDEHYDE AND GLUTATHIONE IN ISOLATED LIVER MITOCHONDRIA

Dylan P. Hartley and Dennis R. Petersen

School of Pharmacy and Alcohol Research Center
University of Colorado Health Sciences Center
Denver, Colorado 80262

INTRODUCTION

It is well documented that the metabolism of ethanol occurs through a series of pyridine nucleotide-linked enzymatic reactions resulting in the production of acetaldehyde and acetate. The factors regulating hepatic acetaldehyde oxidation are of critical importance, since ethanol-derived acetaldehyde has been implicated in hepatic lipid peroxidation. Isolated hepatocytes incubated in high concentrations of acetaldehyde exhibit significant increases in lipid peroxidation as measured by malondialdehyde formation (Stege, 1982); and the infusion of acetaldehyde to perfused rat liver also stimulates lipid peroxidation as measured by an increased formation of ethane (Muller and Sies, 1983). Recently an alternative pathway of ethanol metabolism has been identified which may also contribute to initiation of hepatic lipid peroxidation (Reinke, et al., 1991; and Knecht et al. 1990). Occurring independent of the pyridine nucleotide-linked dehydrogenase enzymes, this enzymatic pathway involves a P-450 mediated, one electron oxidation of ethanol resulting in the production of an ethanol-derived free radical which could serve as an initiator of lipid peroxidation.

One of the major products of lipid peroxidation is *trans*-4-hydroxynonenal. The metabolism of this α,ß-unsaturated aldehyde has recently been studied in detail and has been described as a substrate for the pyridine nucleotide-dependent alcohol and aldehyde dehydrodrogenase enzymes. In hepatocyte suspensions and subcellular fractions enriched with NADH, 4-hydroxynonenal metabolism is mediated by alcohol dehydrogenase (ADH; E.C.1.1.1.1.), generating the corresponding alcohol metabolite, 4-hydroxy-2-nonen (Esterbauer et al., 1985). In addition 4-hydroxynonenal was shown to be a substrate for oxidative metabolism by both hepatic cytosolic and mitochondrial aldehyde dehydrogenase (ALDH; E.C.1.2.1.3.) isozymes (Mitchell and Petersen, 1987). Interestingly, it was demonstrated that 4-hydroxynonenal is a superior substrate for the mitochondrial high-affinity form of aldehyde dehydrogenase (Mitchell and Petersen, 1987). More recently, it has been shown that in highly purified preparations of the high-affinity form of Class 2 ALDH, 4-hydroxynonenal is a potent competitive inhibitor of

Enzymology and Molecular Biology of Carbonyl Metabolism 4
Edited by H. Weiner, Plenum Press, New York, 1993

acetaldehyde oxidation (Mitchell and Petersen, 1991). This observation suggests that as a cosubstrate for ALDH, 4-hydroxynonenal may effectively decrease the total amount of enzyme available for acetaldehyde oxidation, thereby potentiating the toxic effects of ethanol-derived acetaldehyde.

Four-hydroxynonenal, a strong electrophile, has the propensity to react with cellular nucleophiles, such as glutathione. Glutathione reacts chemically, via a Michael addition, with 4-hydroxynonenal in a pH dependent manner producing a glutathione-4-hydroxynonenal conjugate molecule (Esterbauer *et al.*, 1975). In addition to the spontaneous chemical conjugation of glutathione with 4-hydroxynonenal, glutathione S-transferase enzymes mediate a more efficient conjugation of glutathione with 4-hydroxynonenal in biological systems (Alin *et al.*, 1985). As a result of this apparent reactivity, numerous investigators have reported the complete depletion of glutathione by 4-hydroxynonenal in a variety of experimental cellular systems (Cadenas *et al.*, 1983; and Ishikawa *et al.*, 1986). The ability of 4-hydroxynonenal to deplete glutathione, and the potential for 4-hydroxynonenal to alter acetaldehyde metabolism, could further potentiate events initiating lipid peroxidation. Therefore, it was the purpose of these studies to investigate and quantitate the interactions of acetaldehyde, 4-hydroxynonenal, and glutathione in isolated respiring mouse liver mitochondria.

METHODS

Chemicals and Materials

All chemicals were prepared in deionized and distilled water. *trans*-4-hydroxy-2-nonenal was liberated from the diacetyl form, as outlined elsewhere (Mitchell and Petersen, 1991). Acetaldehyde was purchased from Aldrich Chemical Co. (Milwaukee, WI) and redistilled prior to use. All other chemicals were from Sigma Chemical Co. (St. Louis, MO) and were reagent grade or better.

Animals and Treatments

All experiments used male C57BL/6Ibg mice (60-90 days old) obtained from the Institute for Behavioral Genetics, Boulder, CO. Mice were maintained (5/cage) on a 12 hour day/night cycle with free access to water and food (Wayne Sterilizable Lab Blox, Golden K, Denver, CO). For those experiments requiring inhibition of ALDH cyanamide was dissolved in phosphate buffered saline, pH 7.4 and administered intraperitoneally at a dose of 80 mg/kg to mice one hour prior to sacrifice (Hellstrom and Tottmar, 1982).

Mitochondrial Isolation and Metabolism of Acetaldehyde and 4-Hydroxynonenal

The procedure for isolation of hepatic mitochondria using differential centrifugation was according to that described by Little and Petersen (1983). Incubations were performed at 37°C in 25 ml Erlenmeyer flasks containing 4.0 ml of mitochondrial incubate which was shaken continuously. Incubations contained mitochondrial protein (1.0 mg/ml), acetaldehyde (10 µM or 100 µM) and/or 4-HNE (10 µM or 25 µM), suspended in a buffer consisting of 20 mM mannitol, 70 mM sucrose, 2.5 mM KH_2PO_4, 2.0 mM HEPES, 0.5 mM Na_2EDTA, and KOH to pH to 7.4. At various time points 1.0 ml of the incubation was withdrawn and combined with 0.1 ml of 70% perchloric acid which was followed by rapid centrifugation for removal of precipitated protein. Acetaldehyde concentration was determined by adding 0.1 ml acidified supernatant to 0.9 ml water in a glass test tube. Samples were capped with air tight rubber septums and heated for 15 min at 60°C in a water bath. Sample gas (250 μl) was withdrawn and injected with a gas-tight syringe into a Hewlett-Packard 5710A gas chromatograph (injection port 150°C, detector 250°C, oven 85°C). The g.c. separation was performed on a six foot column packed with 60/80 Carbopak B (Supelco, Inc., Bellefonte, PA). Helium was used as carrier gas with a flow rate of 40 ml/min. Acetaldehyde was detected at a retention time of 0.54 min. Peak identification and quantification was based on standard solutions of acetaldehyde ranging in concentrations from 1 μM to 100 μM.

Analysis of 4-HNE and the 4-hydroxy-2-nonenoic acid metabolite of 4-HNE was by reversed-phase HPLC with a Shimadzu SPD-6A UV- spectrophotometric detector (set to 224 nm and 0.08 full scale absorbance units) linked to a Shimadzu CR601 chromatopac (chart speed 2.0 mm/min) and Shimadzu LC-600 dual pump system (flow rate 1.0 ml/min). Supernatant (20 μl) from acidified samples was injected onto a 250 mm x 4.0 mm Lichrosorb RP-18 (5 μm) column (Hibar, Darmstadt, Germany) and eluted with a mobile

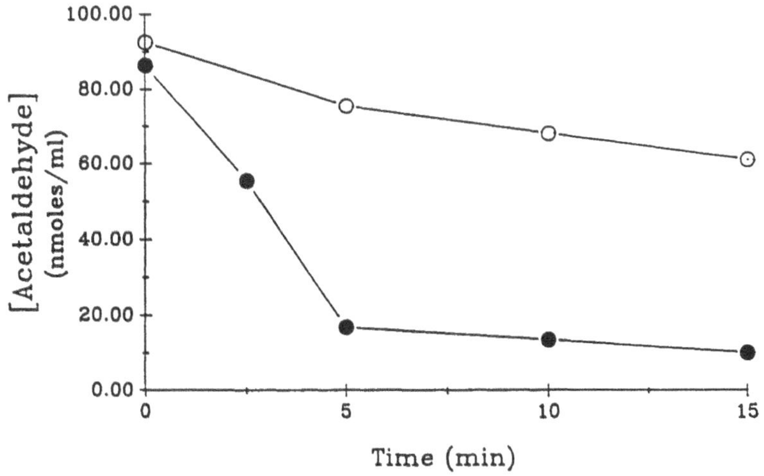

Figure 1. Time course for mitochondrial oxidation of 100 μM acetaldehyde in the absence (filled circles) and in the presence (open circles) of 10 μM 4-hydroxynonenal. Acetaldehyde concentrations for each time point represent the mean ± the standard error of the mean for five experiments.

phase consisting of 30% acetonitrile and 70% phosphate buffer, pH 2.7. Peak identification and quantitation of 4-HNE was based on reference chromatograms of standard solutions (4 μM to 50 μM). Peak identification and quantitation of 4-hydroxy-2-nonenoic acid was determined through the production of standards verified by GC-MS analysis.

Analysis and Quantitation of Mitochondrial Glutathione

Mitochondrial glutathione (oxidized and reduced forms) was determined by reversed-phase HPLC analysis as outlined previously (Reed *et al.*, 1980). Mitochondrial samples retained for GSH analysis were prepared from acid precipitates of incubations containing mitochondrial protein (10 mg/ml), acetaldehyde (100 μM), 4-HNE (25 μM), or both acetaldehyde and 4-HNE. Mitochondrial protein concentrations in all incubations were determined by the Biuret method (Gornall *et al.*, 1949).

RESULTS

Figure 1 presents the rate of acetaldehyde oxidation by isolated respiring mouse liver mitochondria in the absence and presence of 4-hydroxynonenal. Isolated mitochondrial preparations oxidized 100 μM acetaldehyde to near completion within 15 minutes of initiation of the reaction and the initial rate of acetaldehyde metabolism, calculated from the first five minutes of the time course, was 13.2 nmoles/minute/mg of mitochondrial protein.

The data presented in Figure 1 also demonstrate that 4-hydroxynonenal significantly decreased mitochondrial acetaldehyde oxidation. The addition of 10 μM 4-hydroxynonenal to the mitochondrial incubations inhibited the rate of acetaldehyde oxidation by 76 percent of that observed in mitochondrial incubations containing only acetaldehyde. This degree of inhibition is remarkable given that acetaldehyde concentrations (100 μM) were 10-fold greater greater than the 4-hydroxynonenal concentration. These results support the notion that in isolated mitochondria, 4-hydroxynonenal inhibits acetaldehyde oxidation through interactions with the high-affinity form of Class 2 ALDH.

As a result of the data in Figure 1 demonstrating that 4-hydroxynonenal is an inhibitor of mitochondrial acetaldehyde oxidation, studies on the mitochondrial oxidation of 4-hydroxynonenal in the absence and presence of acetaldehyde were performed. The metabolism of 4-hydroxynonenal was monitored in mitochondrial incubations using a reversed-phase HPLC-UV detection system for quantitation of the oxidative metabolite, 4-

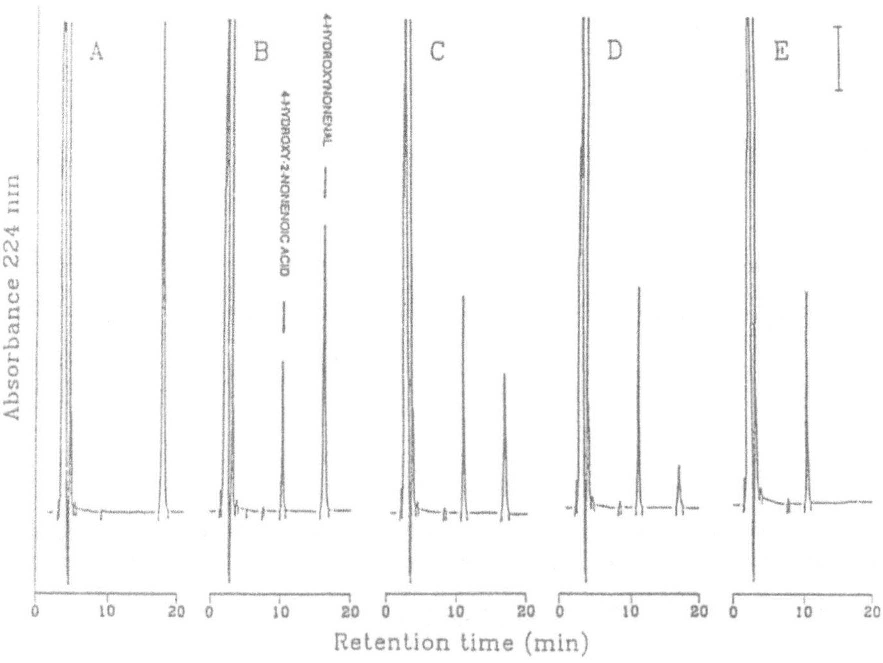

Figure 2. Reversed-phase HPLC analysis of 4-hydroxynonenal oxidation by hepatic mitochondria. Chromatograms represent a 30 minute incubation of 25 μM 4-hydroxynonenal with 1.0 mg/ml mitochondrial protein. Incubation periods are indicated by chromatograms A) 0 min; B) 5 min; C) 10 min; D) 15 min; and E) 30 min. The retention times of 4-hydroxynonenal and 4-hydroxy-2-nonenoic acid are 16 and 10 min, respectively. Scale is 0.01 absorbance units.

hydroxynonenoic acid. A representative analysis of 4-hydroxynonenal oxidation is depicted in the HPLC chromatograms in Figure 2 A-E which presents a series of chromatographs resulting from incubation of 25 μM 4-hydroxynonenal with respiring mitochondria for 30 minutes. These chromatograms illustrate the oxidative metabolism of 4-hydroxynonenal by the concomitant, linear formation of the oxidative metabolite, 4-hydroxy-2-nonenoic acid, during the 30 minute incubation period.

It is apparent in Figure 3 that mitochondrial oxidation of 10 μM 4-hydroxynonenal is rapid with complete depletion of the substrate occurring within 2.5 minutes after initiation of the reaction. The initial rate of 10 μM 4-hydroxynonenal metabolism, determined from the first 2.5 minutes of the reaction, was 3.2 nmoles/min/mg mitochondrial protein. Of the total concentration of 4-hydroxynonenal added to the incubation (10 nmoles/ml), an estimated 30 percent (3.20 nmoles) was detected as the acid metabolite (Figure 3., inset). Consequently, other pathways mediating the metabolism of this aldehyde or its oxidative product warrant further investigation.

It is also evident from Figure 3 that mitochondria incubated with 10 μM 4-hydroxynonenal and 100 μM acetaldehyde display an approximate 60% decrease in the rate of 4-hydroxynonenal oxidation as compared to the rates of oxidation by mitochondria incubated only with 4-hydroxynonenal. In addition, the formation of 4-hydroxy-2-nonenoic acid was also reduced and accounted for approximately fifteen percent (1.5 nmoles/ml) of the initial 4-hydroxynonenal concentration added to the incubation. Collectively, these observations provide additional evidence that both acetaldehyde and 4-hydroxynonenal are

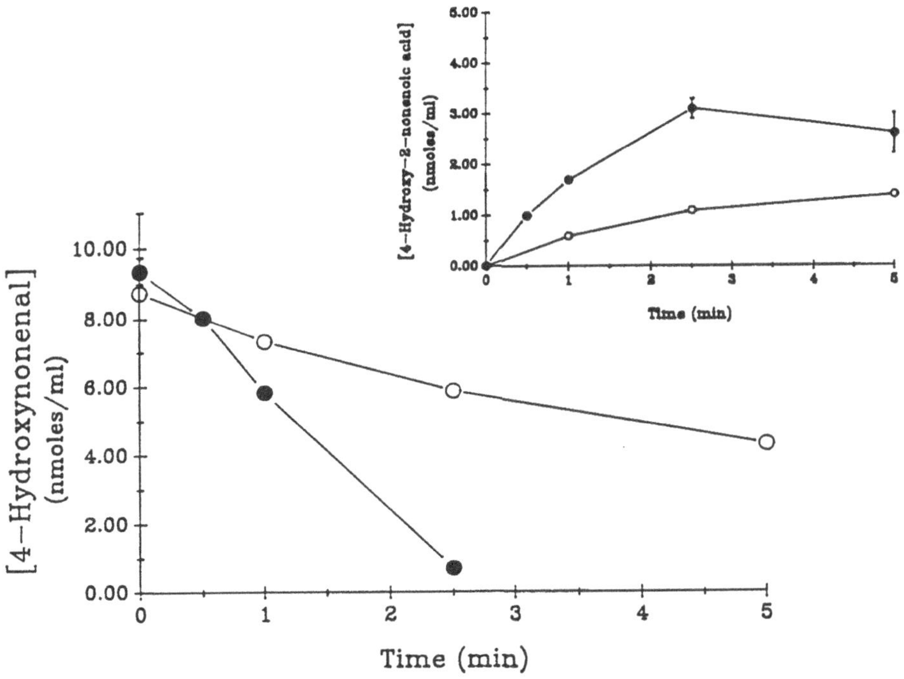

Figure 3. Time course for the mitochondrial metabolism of 10 μM 4-hydroxynonenal in the absence (filled circles) and in the presence (open circles) of 100 μM acetaldehyde. Inset plot represents the time course for the formation of the oxidative metabolite of 4-hydroxynonenal, 4-hydroxy-2-nonenoic acid, in the absence (filled circles) and in the presence (open circles) of 100 μM acetaldehyde. Time points represent the mean ± the standard error of the mean for five experiments.

Table 1. Cyanamide inhibition of hepatic mitochondrial ALDH-mediated oxidation of acetaldehyde and 4-hydroxynonenal.

	Aldehyde oxidation rate (nmoles/minute/mg mitochondrial protein)		
Substrate	Control	+Cyanamide	% Inhibition[a]
Acetaldehyde (100 μM)	13.2±0.3[b]	1.5±1.2	89
4-Hydroxynonenal (25 μM)	3.2±0.4	1.1±0.2	67

[a]Inhibition of substrate oxidation by cyanamide treatment is determined by comparisons of treated to untreated control oxidation rates.
[b]Values are for three to five experiments ± the standard error of the mean.

co-metabolized by the same high-affinity Class 2 ALDH and that each aldehyde inhibits the oxidation of the other.

To further investigate the possibility that acetaldehyde and 4-hydroxynonenal are metabolized by the same mitochondrial ALDH, cyanamide, a specific inhibitor of ALDH, was used to inhibit the oxidation of acetaldehyde and 4-hydroxynonenal (Table 1).

As expected, cyanamide pretreatment of mice resulted in an approximate 90% inhibition of acetaldehyde oxidation in mitochondrial incubations. In addition, 4-hydroxynonenal oxidation was also inhibited approximately 70% by cyanamide (Table 1). The observation that cyanamide pretreatment did not inhibit the capacity of hepatic mitochondria to oxidize 4-hydroxynonenal to the extent of that observed for acetaldehyde oxidation suggest that some 4-hydroxynonenal oxidation might occur by an alternative ALDH-mediated pathway in mitochondria which is not sensitive to inhibition by cyanamide.

Due to the previously reported effects of 4-hydroxyalkenals on cellular glutathione, experiments were performed to determine the glutathione status of mitochondria incubated with 4-hydroxynonenal, acetaldehyde or 4-hydroxynonenal and acetaldehyde. After a ten minute incubation, neither aldehyde nor a combination of both aldehydes caused a significant depletion of mitochondrial glutathione (Table 2). Therefore, these results indicate that in intact mitochondria, the enzymatic or non enzymatic interactions of 4-hydroxynonenal with glutathione are minimal.

Table 2. Glutathione status of mitochondria incubated in the presence of acetaldehyde and 4-hydroxynonenal.

Substrate	Glutathione Concentration (nmoles/mg mitochondrial protein)[a]
Control	4.9±1.4
Acetaldehyde (100 μM)	4.6±0.8
4-Hydroxynonenal (25 μM)	5.3±1.3
Acetaldehyde (100 μM) + 4-Hydroxynonenal (25 μM)	4.8±1.3

[a]Glutathione values were determined after a 10 minute incubation and represent the mean± the standard error for three experiments.

DISCUSSION

In a previous report (Mitchell and Petersen, 1991) we presented data demonstrating that 4-hydroxynonenal is a potent (K_i = 0.5μM) competitive or mixed-type inhibitor of acetaldehyde oxidation by highly purified preparations of the high-affinity ALDH isoform present in rat liver mitochondria. The data presented here (Figures 1 & 3) extend our previous observations by confirming that 4-hydroxynonenal also inhibits acetaldehyde oxidation by isolated, respiring mouse liver mitochondria. The fact that 10 μM 4-hydroxynonenal inhibited mitochondrial oxidation of 100 μM acetaldehyde by nearly 80% suggests that this aldehydic product of lipid peroxidation is also a relatively potent inhibitor of Class 2 ALDH in its native intramitochondrial environment. These observations are consistent with data presented in Table 1 demonstrating that pretreatment of mice with the ALDH-specific inhibitor cyanamide, significantly inhibits the oxidation of 4-hydroxynonenal or acetaldehyde 70 and 90% respectively.

Results of the present study also provide new insight concerning the extent to which the co-oxidation of acetaldehyde affects the oxidation of 4-hydroxynonenal by the Class 2 high-affinity ALDH. The mitochondrial ALDH-mediated oxidation of 4-hydroxynonenal was decreased approximately 60% by addition of 100 μM acetaldehyde. This observation provides additional evidence that these two aldehydes are oxidized by the same Class 2 ALDH. Collectively, the data presented in Figures 1 and 3 suggest that hepatocellular lipid peroxidation initiated by ethanol or its metabolism would result in accumulation of 4-hydroxynonenal as well as ethanol-derived acetaldehyde. Consistent with this proposition is the observation that malondialdehyde, also a major product of lipid peroxidation, accumulates in the livers of rats chronically ingesting alcohol (Kawase *et al.*, 1989). Interestingly, it has been demonstrated that malondialdehyde is a relatively potent irreversible inhibitor of the rat liver Class 2 high-affinity isoform of ALDH (Hjelle *et al.*, 1982). While the accumulation of 4-hydroxynonenal in livers of animals or humans consuming ethanol has not yet been reported, it is documented that human alcoholics display higher blood acetaldehyde concentrations than nonalcoholic controls following identical doses of ethanol (Korsten *et al.*, 1975) suggesting that alcoholic subjects display compromised acetaldehyde oxidation. Clearly, the relationship between *in vivo* ethanol-induced hepatocellular lipid peroxidation and acetaldehyde metabolism requires further investigation.

The chemical reactivity of α,β-unsaturated aldehydes such as 4-hydroxynonenal with glutathione is well documented (Esterbauer *et al.* 1975). For instance, the calculated rate constant for formation of the Michael addition product of glutathione and 4-hydroxynonenal is 1.09 M^{-1} sec^{-1}. In addition, the glutathione *S*-transferases have been reported to catalyze formation of the glutathione-hydroxynonenal conjugate 300-600 times faster than this reported spontaneous rate (Danielson and Mannervik, 1985). Given these reports, the data presented in Table 2 are rather surprising in that the glutathione content of isolated mouse liver mitochondria incubated with 4-hydroxynonenal and acetaldehyde for up to 10 minutes was not reduced significantly as compared to control mitochondria incubated for the same period of time in the absence of both aldehydes. The apparent absence of reports describing mitochondrial-specific forms of glutathione S-transferases suggest that a transferase-catalyzed formation of the 4-hydroxynonenal -GSH conjugate is unlikely and consistent with the results reported here. In addition, the inability of 25 μM 4-hydroxynonenal to deplete mitochondrial GSH suggests the rate of spontaneous formation of the 4-hydroxynonenal-GSH conjugate must be relatively slow compared to the rate at which 4-hydroxynonenal is oxidized to the respective acid by the high-affinity form of Class 2 ALDH. The enzymatic efficiency of this Class 2 ALDH is remarkable when one considers that mitochondrial GSH concentration did not change significantly in the presence of 100 μM acetaldehyde and 25 μ

M 4-hydroxynonenal suggesting that this enzyme may play an important protective role against depletion of intramitochondrial GSH by electrophilic α,β-unsaturated aldehydes such as 4-hydroxynonenal.

The potential of ethanol to initiate hepatocellular lipid peroxidation raises some interesting questions concerning the role of aldehydic products of lipid peroxidation, such as 4-hydroxynonenal, in ethanol induced liver damage. The chemical reactivity and cellular toxicities of 4-hydroxynonenal are well documented and are discussed in two recent comprehensive reviews (Schaur *et al.*, 1991; Esterbauer *et al.*, 1991). Briefly, these reviews describe the potential of this α,β-unsaturated aldehyde to rapidly deplete cellular glutathione levels, function as a potent inhibitor of a number of enzymes and form adducts with various cellular constituents. In addition, our results suggest that 4-hydroxynonenal and ethanol-derived acetaldehyde are competitive substrates for the high-affinity form of Class 2 ALDH and would accumulate in hepatocytes as a result of ethanol-induced lipid peroxidation and acetaldehyde generated from the oxidation of ethanol. Interestingly, it was proposed sometime ago that decreased hepatic acetaldehyde oxidation in human alcoholics was partially attributable to impaired mitochondrial function which in turn impacts the rate at which acetaldehyde is removed (Lieber, 1977). The accumulation of acetaldehyde would lead to further mitochondrial damage establishing a "viscous cycle" further promoting alcohol-induced liver injury. A comprehensive review of the more recent literature (Lieber, 1991) implicates the involvement of acetaldehyde in ethanol induced liver injury through formation of acetaldehyde protein adducts, glutathione depletion, initiation of lipid peroxidation and stimulation of hepatic collagen production and fibrosis. It is reasonable to assume that the same cellular systems would also be adversely affected by the chemically reactive 4-hydroxynonenal and other α,β-unsaturated products of lipid peroxidation. This proposition is presented in Figure 4.

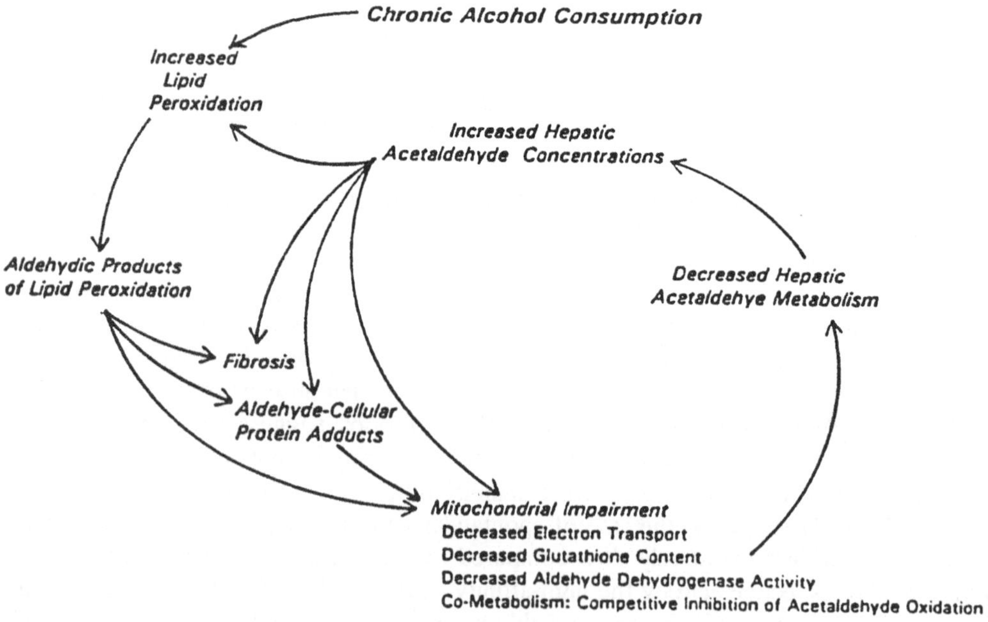

Figure 4. Hypothetical scheme illustrating the possible roles of aldehydic products of lipid peroxidation in alcohol-induced liver injury.

This scheme depicts lipid peroxidative events associated with chronic alcohol consumption which are possibly initiated from production of ethanol-derived free radicals or acetaldehyde. Separately, or in tandem, acetaldehyde and 4-hydroxynonenal could initiate fibrosis through stimulation of hepatic collagen synthesis. Both aldehydes could form aldehyde-cellular protein adducts interfering with important cellular process (mitochondrial impairment) and/or resulting in antibody production initiating immune mediated cellular injury. Mitochondrial impairment could be manifested in decreased electron transport, decreased glutathione content due to total cellular glutathione depletion, and decreased oxidation of 4-hydroxynonenal and acetaldehyde by the high-affinity form of Class 2 ALDH. Accumulation of acetaldehyde and 4-hydroxynonenal, as well as other aldehydic products of lipid peroxidation, could then amplify the initial processes of alcohol induced liver injury. Currently, the involvement of 4-hydroxynonenal and other chemically reactive products of lipid peroxidation in these events at the cellular or organ level is speculative. However, the results presented here demonstrating the metabolic interactions of 4-hydroxynonenal and acetaldehyde in isolated mitochondria can now be systematically evaluated at the level of isolated hepatocytes and eventually in animal models developed to study alcohol-induced liver injury.

ACKNOWLEDGMENTS

This work was supported by Grants AA 03527 and AA00106 to DRP

REFERENCES

Alin, P., Danielson, U.H., Mannervik, B., 1985, 4-Hydroxyalk-2-enals are substrates for glutathione transferase, *FEBS lett.*, **179**:267.

Cadenas, E., Muller, A., Brigelius, R., Esterbauer, H., Sies, H., 1983, Effects of 4-hydroxynonenal on isolated hepatocytes: studies on chemiluminescence response, alkane production, and glutathione status, *Biochem. J.*, **214**:479.

Esterbauer, H., Zollner, H., Scholz, N., 1975, Reaction of glutathione with conjugated carbonyls, *Z. Naturforsch.*, **30**:466.

Esterbauer, H., Zollner, H., Lang J., 1985, Metabolism of lipid peroxidation product 4-hydroxynonenal by isolated hepatocytes and by liver subcellular fractions, *Biochem. J.*, **228**:363.

Esterbauer, H., Schaur, R.J., Zollner, H., 1991 Chemistry and biochemistry of 4-hydroxynonenal, malonaldehyde, and related aldehydes, *Free Rad. Biol. Med.*, **11**:81.

Gornall, A.G., Bardawill, L.J., David, M.M., 1949, Determination of serum proteins by means of a biuret reaction, *J. Biol. Chem.*, **247**:260.

Hellstrom, E., Tottmar, O., 1982, Effects of acetaldehyde dehydrogenase inhibitors on enzymes involved in the metabolism of biogenic aldehydes in rat liver and brain, *Biochem. Pharmacol.*, **31**:3899.

Hjelle, J., Grubbs, J. and Petersen, D.R., 1982, Inhibition of mitochondrial aldehyde dehydrogenase by malondialdehyde, Toxicol. Letters., **41**:436

Ishikawa, T., Esterbauer, H., Sies, H., 1986, Role of cardiac glutathione transferase and of glutathione S-conjugate export system in biotransformation of 4-hydroxynonenal in the heart, *J. Biol. Chem.*, **261**:1576.

Kawase, T., Kato, S., Lieber, C.S., 1989, Lipid peroxidation and antioxidant defense systems in rat liver after chronic ethanol feeding, *Hepatology*, **10**:815.

Knecht, K.T., Bradford, B.U., Mason, R. P., Thurman, R.G., 1990, In vivo formation of a free radical metabolite of ethanol, *Molec. Pharmacol.*, **38**:26.

Korsten, M.A., Matsuzaki, S., Fieinman, L., Lieber, C.S., 1975, High blood acetaldehyde levels after ethanol administration. Differences between alcoholic and non-alcoholic subjects. *N. Engl. J. Med.*, **292**, 386.

Lieber, C.S., 1991, Hepatic, metabolic and toxic effects of ethanol: 1991 update, *Alcoholism: Clin. Exp. Res.*, **15**:573.

Little, R.G., Petersen, D.R., 1983, Subcellular distribution and kinetic parameters of HS mouse liver aldehyde dehydrogenase, *Comp. Biochem. Physiol.*, **74**:271.

Mitchell, D.Y., Petersen, D.R., 1987, The oxidation of α,ß-unsaturated aldehydic products of lipid peroxidation by rat liver aldehyde dehydrogenases, *Toxicol. Appl. Pharmacol.*, 87:403.

Mitchell, D.Y., Petersen, D.R., 1991, Inhibition of rat hepatic mitochondrial aldehyde dehydrogenase-mediated acetaldehyde oxidation by trans-4-hydroxy-2-nonenal, *Hepatology*, 13:728.

Muller, A., Sies, H., 1983, Ethane release during metabolism of aldehydes and monoamines in perfused rat liver, *Br. J. Biochem.*, 134:599.

Reinke, L.A., Kotake, Y., McCay, P.B., Janzen, E.G., 1991, Spin-trapping studies of hepatic free radicals formed following the acute administration of ethanol to rats: in vivo detection of 1-hydroxyethyl radicals with PBN, *Free Rad. Med. Biol.*, 11:31.

Reed, D. J., Babson, J R., Beatty, P.W., Brodie, A.E., Ellis, W.W., Potter, D.W., 1980, High-performance liquid chromatography analysis of nanomole levels of glutathione, glutathione disulfide, and related thiols and disulfides, *Anal. Biochem.*, 106:55.

Schaur, R.J., Zollner, H., Esterbauer, H., 1991, Biological effects of aldehydes with particular attention to 4-hydroxynonenal and malonaldehyde, in: "Membrane Lipid Oxidation, Vol. III.," C. Vigo-Pelfrey, ed., CRC Press, Inc., Boca Raton.

Stege, T.E., 1982, Acetaldehyde-induced lipid peroxidation in isolated hepatocytes, *Res. Commun. Chem. Pathol. Pharmacol.*, 36:287.

BIOLOGICAL ROLE OF HUMAN CYTOSOLIC ALDEHYDE DEHYGROGENASE 1: HORMONAL RESPONSE, RETINAL OXIDATION AND IMPLICATION IN TESTICULAR FEMINIZATION

A. Yoshida, L.C. Hsu and Y. Yanagawa

Department of Biochemical Genetics
Beckman Research Institute of the City of Hope
Duarte, CA 91010

INTRODUCTION

A number of aldehyde dehydrogenase isozymes are distinguished based on the separation by physicochemical methods,, tissue and subcellular distributions, and enzymatic properties. Although ALDH isozymes are usually assayed with short chain aliphatic aldehydes as substrate, they exhibit relatively broad substrate specificities and can oxidize various biogenic and xenobiotic aliphatic and aromatic aldehydes including dopaldehyde and aminaldehydes. However, their physiological substrates are not identified and the primary biological roles of individual ALDH isozymes are not yet clear.

This paper describes the retinal oxidation activity of the human ALDH1, the existence of glucocorticoid response elements in the ALDH1 gene and the biological role of ALDH1, including the pathogenesis of the X-linked, androgen receptor-negative testicular feminization.

EXPERIMENTS AND RESULTS

1) Retinal oxidation activity: It has been known for some time that a crude rodent ALDH preparation can oxidize retinal (vitamin A aldehyde) to retinoic acid (vitamin A acid) (Elder and Topper, 1962). Available data suggest that mitochondrial isozymes have no (or very low) activity for retinal oxidation while cytosolic isozyme(s) exhibit the activity in rodents (Connor and Smit, 1987; Lee et al., 1991). Detailed kinetic studies using purified, homogeneous rodent's enzymes were not carried out and the reported Km values for retinal are widely divergent, i.e., 19 μM at pH 8 in one report (Connor and Smit, 1987), 0.7 μM at pH 8.2 in another report for mouse enzyme(s) (Lee et al., 1989), and 66 μM at pH 8. 5 for rat enzyme (s)

(Leo et al., 1989). The Km f or the "apparent homogeneous" rat enzyme was 300 μM at pH 7.7 (Moffa et al., 1970). Since the retinal concentration in tissues is estimated to be about 0.1 μM or less (McCormick and Napali, 1982), it is questionable whether or not the rodent's cytosolic ALDH can play a major role in the retinoic acid generation. In the study of human ALDH isozymes, Sladek, et al., (1990) reported that the cytosolic ALDH1, but not other ALDH isozymes, oxidized retinal. The Km value for retinal was estimated to be <1 μM. By contrast, Ambroziak and Pietruszko (1991) reported that the purified homogeneous human cytosolic ALDH1, mitochondrial ALDH2 and γ-aminobutyraldehyde dehydrogenase all oxidized retinal. Km values of their preparations were not reported.

We purified the cytosolic ALDH1 and the mitochondrial ALDH2 to homogeneity from an autopsy human liver by the method reported previously (Ikawa et al., 1983). Homogeneity of the preparations was confirmed by the starch gel electrophoresis and polyacrylamide gel electrophoresis.

The cytosolic ALDH3 was prepared from the E. coli cells transformed with the expression construct containing the full-length ALDH3 cDNA (Hsu et al., 1991). Since the host E. coli, extract had no detectable aldehyde dehydrogenase activity, the ALDH3 partially purified through DEAE cellulose column was used for kinetic study.

The enzyme activity was spectrophotometrically assayed. Activity for retinal oxidation was calculated from the superimposed increment of absorbency at 340 nm using the known molar absorbency of the reactants and products; i.e. NAD=0, NADH=6620, retinal=22800, and retinoic acid=39200 (Elder and Topper, 1962). one unit of enzyme activity was defined as the amount of enzyme which converts one μ mole of substrate per minute at 25° C, and the specific activity was expressed as units/mg protein.

Determination of initial velocity in the presence of less than 1 μM retinal in the reaction mixture was not reliable and the conventional initial rate method could not be used for the estimation of the extremely low Km value. The single reaction progress curve method (Yun and Suelter, 1977) was applied for estimation of the Km for retinal.

Among the three human ALDH isozymes examined, only the cytosolic ALDH1 oxidized retinal. The mitochondrial ALDH2 oxidized acetaldehyde and propionaldehyde equally well, and the cytosolic ALDH3 oxidized propionaldehyde, but not acetaldehyde effectively, as previously observed (Table 1).

Table 1. Kinetic Constants of Aldehyde Dehydrogenase Isozymes

Isozyme	Specific Activity (Vmax/mg)		kM (μM)			Catalytic Efficiency (Vmax/KM)		Effect of Dihy-* drotestosterone	
	Acetal.	Ret.	NAD	Acet.	Ret.	Acet.	Ret.	Acet.	Ret.
ALDH1	0.22	0.22	5**	36	0.06	0.006	3.7	1	1.3
ALDH2	0.3	0	16**	3.5**	-	0.086	-	-	-

Enzyme activity was assayed in Tris-HCl buffer at pH 7.5.
ALDH3 had activity for oxidation of propionaldehyde and benzaldehyde, but no activity for retinal.
* Ratio of enzyme activity in the presence and absence of 5 μM dihydrotestosterone.
** Previous data determined under the same conditions (Ikawa et al., 1983).

The ALDH1, ALDH2 and ALDH3 all oxidized dopaldehyde and the Km values at pH 7.5 were: 4.4 μM, 0.52 μM and 1.7 μM respectively.

The Km of ALDH1 for retinal was estimated to be about 0.06 μM (ranging from 0.04 to 0.07 μM in five assays) by analysis of single reaction curves at pH 7.5. The Km for acetaldehyde was about 36 μM at pH 7.5, as previously reported.

The specific activities of ALDH1 for retinal oxidation and that for acetaldehyde oxidation were almost equal, i.e. 0.22 u/mg at pH 7.5. Since the Km for retinal is much lower than that for acetaldehyde, the catalytic efficiency (V max/Km) of ALDH1 is about 600-fold higher than that for acetaldehyde oxidation. Disulfiram, 10 μM, completely inhibited acetaldehyde and retinal oxidation activities of ALDH1.

2) Existence of glucocorticoid response elements: The gene and cDNA for the human ALDH1 were previously cloned and characterized (Hsu et al., 1985; Hsu et al., 1989). The *ALDH1* gene, which is assigned to human chromosome 9, spans about 53 kb, and contains 13 exons which encode 501 amino acid residues (Hsu et al., 1989).

Nucleotide sequence analysis of the extended 5'-region of the gene revealed the existence of two hormone response elements upstream from the transcription initiation site (Fig. 1). The distal and proximal elements are in the forward and the reverse orientation, respectively. There is a 9 out of 12 match with the consensus glucocorticoid response element sequence (5'-GGTACANNNTGTTCT-31) in both elements.

3) Hormonal Response of ALDH1 Expression and Implication in Testicular Feminization: The androgen receptor (AR)-negative testicular feminization (McKusick MIM No. 323700) is a well-documented X-linked disorder. The affected males have female external genitalia and breast development, absent uterus and female adnexa, abdominal testes and a normal male (XY) karyotype. Recent studies demonstrated that ALDH1 is present in the normal genital skin cells, but absent in the cells of a majority of AR-negative TF patients (Pereira et al., 1991). Hormone receptors, including AR, are generally considered as transcriptional regulators, and thus it is conceivable that the genetically defective ARs fail to activate the *ALDH1* gene which plays an essential role in testicular development. However, the

```
-900  GTACTTAAAA TGGGAGTAAC ACTGTTTAAA AAATTGTCCT TCAGCTAAAT ATTAATTTAA GAACTTGAAT TGTTTGGAAG CCCTGCTTAA ATTTAGTCTT
-800  TGTACACTGC CTATGTTGAT AAATACAGAC AACTTCCAAA ACCAGGATTA CTTTCATTTT AAATGAGTAT TAATAGATGA TATTGCCATA TTTCTAATTG
-700  TGGTGATTGT GTGTGACAGT GTGTTCCGAA TTCCCTAAAA GTCCTGCTGG CTTTTCTGTT CACATATAGA AAATAAAGAT AATTTAGGGC TTCTGAGATC
-600  ACAGTAGGTC TACTTACCCA GCACTGAAAA TACACAAGAC TGATACGATA TTTTAAAACT AACTTAGGGT AGGGTGTAGA TAAAGGGCCT TTCTTCCCCA
-500  AACAGCACCT TGATTTTCTG GGAGATGGAC TGATTTCCTG AAAGCCTTGT CCTGAAGACA CCTGGCCAGG GTTCTCTCCT CACCAGCTTC TACTGAGAAC
-400  AAGTTCCCTT TTAGACTCTT TTCAATCCTC AAATTCTCTG ATTCCAAGTC TGTCAGAGAA CAGAAAGTTA CATAGTAGCA TTAAAAAGCA TGAGAAGTCA
-300  AAAAAATAAT AACTGGCCTT AGTGGCCAGA GCAGCTGCTG CATACACTTA TCACAGGTTT CGGCTTTGTA AATTAATTCA TCTGCAAATA GTGCACTGTC
-200  TCCAGGTACA AATTCGATGC TGGAGCACTG GTTTCTTAAG GATTTAAGTT TAAAGTCAAA GGCTTCCTGC CCTAGGTGTT ACAAATAAGT AGTGTCGTTT
-100  CCTTTTTTTG CTCTGAGTTT GTTCATCCAA TCGTATCCGA GTATGCAAAT AAACTTTAGC CCGTGCAGAT AAAAAGGAA CAAATAAAGC CAAGTGCTCT
+1    ATCAGAACCA AATTGCTGAG CCAGTCACCT GTGTTCCAGG AGCCGAATCA GAAATG,TCA,TCC,TCA,GGC,ACG----
                                                                Met Ser Ser Ser Gly Thr
```

Figure 1. Nucleotide sequence of the 5'-flanking and 5'-untranslated region of the human ALDH1 gene. The negative numbers on the left indicate the nucleotide position upstream of the transcription initiation site. Sequences homologous to consensus GRE sequence (5'-GGTACANNNTGTTCT-31) are double-underlined. The distal and proximal GREs of the gene are in the forward and the reverse orientation, respectively. The translational initiation codon ATG is underlined.

genital skin cells of some AR-negative TF patients contained ALDH1 (Nickel et al., 1988), and therefore, the correlation between the lack of ALDH1 expression and the pathogenesis of TF is not clear.

In order to solve this enigma, we examined expressions of the ALDH1 gene as well as several other genes in the normal and AR-negative TF patients' genital cells and other cells.

Western blot, Northern blot and PCR-mediated hybridization analysis using specific antibodies and nucleotide probes showed that the ALDH1 gene was strongly expressed in normal genital fibroblast cells, but shut off in the cells obtained from AR-negative TF patients and in normal non-genital skin fibroblasts (Fig. 2 and 3) . The expression of marker genes examined (i.e. ALDH2, ALDH3, alcohol dehydrogenase Class I, glucose-6

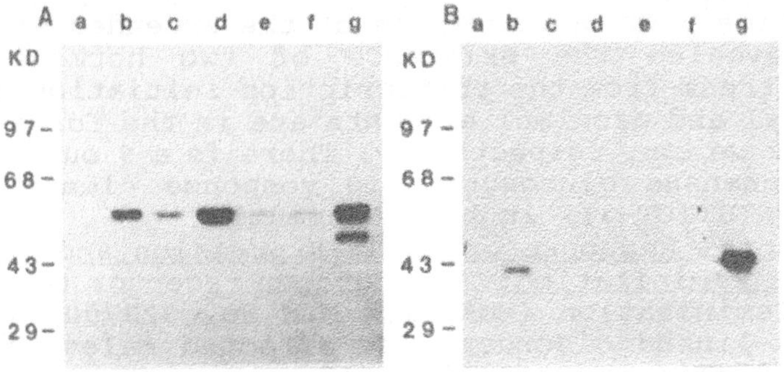

Figure 2. Western blot hybridization. Cellular extracts (about 50 μg protein per channel) were subjected to SDS-PAGE (10% acrylamide) electrophoresis, blotted into nitrocellulose filters, and hybridized with specific antibodies.

A: Hybridized with rabbit anti-human ALDH1 IgG, and peroxidase-conjugated goat antiserum against rabbit IgG.

B: Hybridized with rabbit anti-human ADH (class I) IgG, and peroxidase-conjugated goat antiserum against rabbit IgG.

Lanes a: AR-negative TF patient genital skin fibroblast (GM02299) ; b: AR-negative TF patient genital skin fibroblast (GM02715) ; c, d, e and f: normal genital skin fibroblasts; g: liver.

The filters were also hybridized with anti-human ALDH2, anti-human ALDH3, and anti-human G6PD (figures are not shown).

phosphate dehydrogenase, and actin) was not different in the normal and patient's genital fibroblast cells. As previously reported (Nickel et al., 1988), genital fibroblast cells from a patient (GM02715) produced ALDH1. However, the expression of other genes, such as Class I ADH gene(s), which are under the tissue specific developmental control (Ikuta et al., 1986), are also drastically different from that of other genital and non-genital fibroblast cells (Fig. 2 and 3). These results indicate that the mode of tissue-and/or development-dependent regulatory mechanism has been altered in this particular cell line (GM02715), and the cell line does not represent the original patient's genital cells.

DISCUSSION

The expression of the human ALDH1 gene is under tissue-specific and developmental dependent regulation mediated by hormone receptor(s) and presumably other regulatory elements(s) (Fig. 1, 2 and 3).

The human cytosolic ALDH1 has an extremely high affinity (Km=0.06 μM) for retinal (Table 1). Therefore, the enzyme is capable of producing retinoic acid from retinal efficiently even in tissues where the retinal concentration is below 0.1 μM.

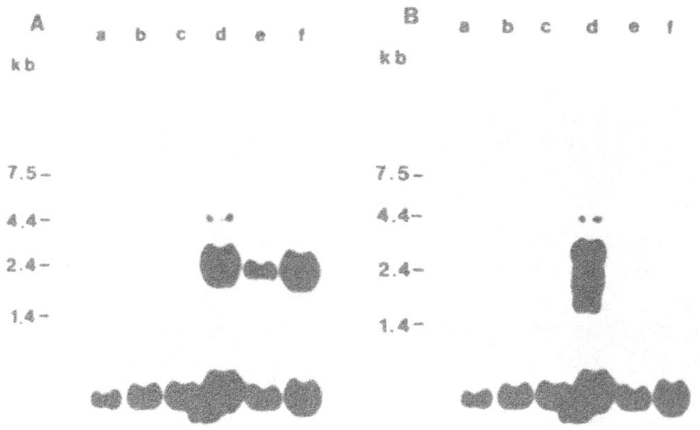

Figure 3. Northern blot hybridization. Cellular RNAs were subjected to agarose gel (1%) electrophoresis, blotted into a nitrocellulose filter, and successively hybridized with human ALDH1 cDNA probe and human ADH1 probe.

A. Hybridized with ALDH1 cDNA; B: hybridized with ADH (class I) cDNA probe. In order to compare the amount of DNA samples placed in each channel, the filter was also hybridized with human actin cDNA probe (shown at the bottom of the figure).

Lanes a and b: Non-genital skin fibroblast; c: ARnegative TF patient genital skin fibroblast (GM02299); d: ARnegative TF patient genital skin fibroblast (GM02715); e and f: normal genital skin fibroblast.

Retinoic acid binds to several receptors (RAR-α, RAR-β, RAR-γ and RAR-X), and the complexes thus produced act as extremely potent transcriptional regulators for the genes which are involved in embryonal development and cell differentiation (Ragsdale and Brockes, 1991).

Based on these available informations, we propose the following model for the pathogenesis of testicular feminization. Defective mutant AR gene on the X-chromosome produces functionally defective AR protein. Thus, the ALDH1 gene on chromosome 9 cannot be expressed in the patient's genital cells, resulting in diminished formation of retinoic acid which is essential to activate the male-determining gene(s) such as the testis-determining gene (TDF). The testicular feminization is a consequence of the lack of positive trans-activation of the autosomal ALDH1 gene by the

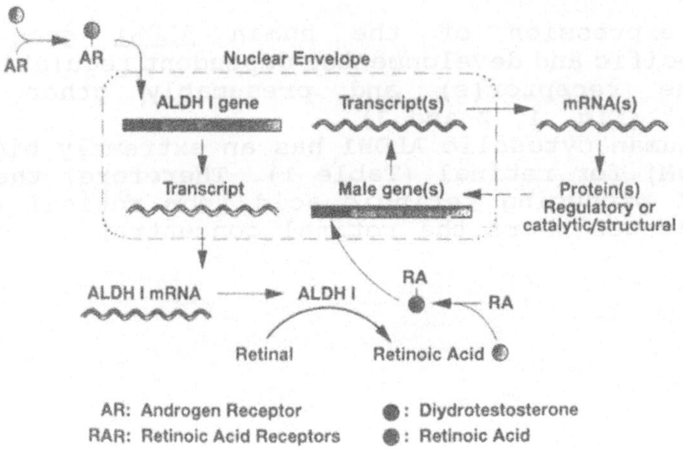

Figure 4. Proposed model for the retinod-mediated testicular differentiation.

positive trans-activation of the autosomal ALDH1 gene by the -linked AR gene. Although many genes must be involved in sex determination, the activation of ALDH1 by the AR-dihydrotestosterone complex and the subsequent activation of the male gene(s) by the retinoic acid receptor(s) provoke the testicular differentiation, i.e., the series of gene expression is AR gene (X- chromosome) → ALDH1 gene (chromosome 9) → male specific gene(s) (Y-chromosome) (Fig. 4).

Further studies, such as CAT assay, DNAse foot printing and mobility shift assay, are required for elucidation of the mode of interaction of the hormone response elements in the ALDH1 gene and AR. The specific retinoic acid receptor (i.e., RAR-α, β, γ, or X, etc.) and the male-specific gene(s) responding to the RAR remain to be identified and characterized.

Acknowledgment. This work was supported by the U.S. Public Health Service Grant HL-29515 and AA05763.

REFERENCES

Ambroziak W, Pietruszko R:. Human aldehyde dehydrogenase activity with aldehyde metabolites of monoamines, diamines and polyamines. J. Biol. Chem., 1991; 266:13011-13018.

Connor MJ, Smit MH: Terminal-group oxidation of retinol by mouse epidermis: Inhibition in vitro and in vivo. Biochem. J., 1987;244:489-492.

DeLuca LM: Retinoids and their receptors in differentiation, embryogenesis and neoplasia. FASEB Jour., 1991; 5:2924-2933.

Duester G, Shean ML, McBride MS, Stewart MJ: Retinoic acid response element in the human alcohol dehydrogenase gene ADH3: Implications for regulation of retinoic acid synthesis. Mol. Cell. Biol., 1991;11:1638-1646.

Elder TD, Topper YJ: The oxidation of retinine (vitamin A1 aldehyde) to vitamin A acid by mammalian steroid-sensitive aldehyde dehydrogenase. Bioch. Biophys. Acta, 1962;64:430-437.
Evans RM: The steroid and thyroid hormone receptor superfamily. Science, 1988;240:889-895.
Hsu LC, Chang W-C: Cloning and characterization of a new functional human aldehyde dehydrogenase gene. J. Biol. Chem., 1991;266:12257-12265.
Hsu LC, Chang W-C, Shibuya A, Yoshida A: Human stomach aldehyde dehydrogenase cDNA and genomic cloning, primary structure and expression in E.coli. J. Biol. Chem. 1991;267:3030-3037.
Hsu LC, Chang W-C, Yoshida A: Genomic structure of the human cytosolic aldehyde dehydrogenase gene. Genomics 1989;5:857865.
Ikawa M, Impraim CC, Wang G, Yoshida A: Isolation and characterization of aldehyde dehydrogenase isozymes from usual and atypical human livers. J. Biol. Chem. 1983;258:6282-6287.
Ikuta T, Yoshida A: mRNA for the three human alcohol dehydrogenase subunits: size heterogeneity and developmental changes. Bioch. Biophys. Res. Commun. 1986;140:1020-1027.
Lee M-O, Manthey CL, Sladek NE: Identification of mouse liver aldehyde dehydrogenases that catalyze the oxidation of retinaldehyde to retinoic acid. Bioch. Pharmacol. 1991; 42:1279-1285.
Leo MA, Kim C-I, Lowe N, Lieber CS: Increased hepatic retinal dehydrogenase activity after phenobarbital and ethanol administration. Biochem. Pharmacol. 1989;38:97-103.
Lowry OH, Rosebrough NJ, Farr AL, Randall RJ: Protein measurement with the Folin phenol reagent. J Biol Chem 1951;193:265-275.
Maxwell ES, Topper YJ: Steroid-sensitive aldehyde dehydrogenase from rabbit liver. J Biol Chem 1961;236:10321037.
McCormick AM, Napali JL: Identification of 5,6-epoxy-retinoic acid as an endogeneous retinol metabolite. J Biol Chem 1982;257:1730-1735.
Moffa DJ, Lotspeich FJ, Krause RF: Preparation and properties of retinal-oxidizing enzyme from rat intestinal mucosa. J Biol Chem 1970;245:439447.
Napali JL: Retinal metabolism in Lic-PKI cells: Characterization of retinoic acid synthesis by an established mammalian cell line. J Biol Chem 1986;261:13592-13597.
Nickel B, Schwartz A, Rosemann E, Kaufman M, Pinsky L, Wrogemann K: A study of androgen-resistant subjects indicates that the 6.7 pI/56 KDa protein in genital skin fibroblasts is related to the androgen receptor. Clin Inv Med 1988;11:23-33.
Pereira F, Rosemann E, Nylen E, Kaufman M, Pinsky L, Wrogemann K: The 56 KDa androgen, binding protein is an aldehyde dehydrogenase. Biochem Biophys Res Commun 1991; 175:831-838.
Ragsdale CW and Brockes JP: Retinoids and their targets in vertebrate development. Current opinion in Cell Biol, 1991; 3:928-934.

Yamamoto KR: Steroid receptor regulated transcription of specific genes and gene networks. Annu Rev Genet 1985; 19:209-252.
Yoshida A, Hsu LC, Yasunami M: Genetics of human alcohol-metabolizing enzymes. Prog Nucleic Acid Res and Mol Biol 1991;40:255-287.
Yun SL, Suelter CH: A simple method for calculating Km and V from single enzyme reaction progress curve. Bioch Biophys Acta 1977;480:1-13.

HUMAN CYTOSOLIC ALDEHYDE DEHYDROGENASE IN ANDROGEN INSENSITIVITY SYNDROME

Fred A. Pereira, Eduardo Rosenmann, Edward G. Nylen and Klaus Wrogemann

Department of Biochemistry and Molecular Biology
University of Manitoba
Winnipeg, Manitoba, Canada, R3E 0W3

INTRODUCTION

The androgen insensitivity syndromes (AIS) form a clinical spectrum of abnormalities that range from phenotypic women with primary amenorrhea (complete testicular feminization) to undervirilized men, all of whom are genotypic males (French et al., 1990). The majority of them are caused by defects of the androgen receptor gene which is located on the long arm of the X-chromosome at Xq11-12 (Brown et al., 1989). Patients with AR defects form the X-linked subgroup of the androgen insensitivity syndromes. The type of defects in the AR are very heterogenous, in agreement with the wide clinical spectrum of androgen insensitivity (McPhaul et al., 1991). It is becoming clear that heterogenous defects at the gene level underlie the diverse receptor abnormalities (Griffin, 1992).

The numerous genes whose expressions must be altered to result in this very different phenotype as a result of the primary cause, the AR mutation, are unknown to date. We have found that aldehyde dehydrogenases 1 (ALDH1) is underexpressed in genital skin fibroblasts of most patients with complete androgen insensitivity syndrome. ALDH1 appears to be able to oxidize retinal to the morphogen retinoic acid. We postulate that failure to produce sufficient concentrations of this morphogen is part of the pathogenetic mechanism and that the enzyme plays a critical role in the development of the different sexual phenotypes. An attraction of this hypothesis is that it can be tested in an animal model of complete AIS, the *Tfm* mouse (Charest et al., 1991; He et al., 1991).

RESULTS AND DISCUSSION

In a comparison of proteins between genital (GSF) and non-genital skin fibroblasts (non-GSF) by two-dimensional gel electrophoresis we have identified in genital skin fibroblasts (GSF) a 56 kDa protein (Thompson et al., 1983) that is not expressed in non-GSF. Under denaturing conditions this protein has an apparent molecular weight of 56 kDa and isoeclectric points of 6.5 and 6.7 (see Fig. 1A). Using protein staining methods this protein is not detected in GSF of most patients with complete AIS (Nickel et al., 1988), although more sensitive detection methods indicate that it is often underexpressed rather than missing. Our 56 kDa protein has turned out to be an additional marker in this syndrome, as the protein is underexpressed in 14 out of 16 unrelated patients with complete X-linked AIS (Nickel et al., 1988), studied to date, Table 1.

Table 1. Androgen receptor (AR) and the 56 kDa protein status in genital skin fibroblasts from normal and patients with complete androgen insensitivity syndrome.

Cell Strain	AR-Binding (fmol/mg protein)	56 kDa Protein Status	Number of Analyses
598	5	-	4
2379	5	-	8
6779	1	-	8
8481	1	-	7
31082	2	-	5
A5879	2	-	18
CVL	0	-	8
JRL	0	-	13
LEL	1	-	4
TML	1	-	6
NHL	1	4+	11
TRL	1	3+	10
30284	17	-	4
KIL	11	-	4
DB	1	1+	15
8812	0	1+	10
Normal Controls			
MCH6	24	4+	>100
MCH49	24	3+	>25

This protein has androgen binding activity, as detected by covalent labeling with the androgen analogs methyltrienolone (Wrogemann et al., 1988; Pereira et al., 1990), mibolerone (Belsham et al., 1989) and with dihydrotestosterone-bromoacetate (Kovacs & Turney, 1988; Pereira et al., 1991), Fig. 1.

Because of this activity others have mistakenly identified this protein as the androgen receptor itself or a portion of it (Kovacs & Turney, 1988). Under native conditions the labelled protein migrates as a complex of greater than 200 kDa. We prepared a polyclonal antiserum from excised protein spots of two-dimensional gels and isolated cDNA clones from an expression library of GSF. These clones identified a 2.1 kb message, on Northern blots, which paralleled the expression of the 56 kDa protein in fibroblast cell strains from a variety of patients and controls. Sequencing these clones revealed that they are 100% identical with those published for aldehyde dehydrogenase 1 (ALDH1) (Hsu et al. 1989; Pereira et al., 1991) and unpublished data. Thus, the 56 kDa protein turns out to be ALDH1; it has androgen binding activity, it is expressed in GSF, but not in non-GSF; and it is underexpressed in most patients with complete AIS.

Preliminary experiments and the literature indicate that this enzyme can oxidize retinaldehyde to retinoic acid (Lee et al. 1991), and references therein. Already in 1962 a "steroid-sensitive" aldehyde dehydrogenase was described which could oxidize retinene to retinoic acid (Futterman, 1962; Elder & Topper, 1962). It is likely that the enzyme described 30 years ago is the same ALDH1 with androgen binding activity identified by us. As detailed below, we postulate that this enzyme plays a role in the pathogenesis of AIS.

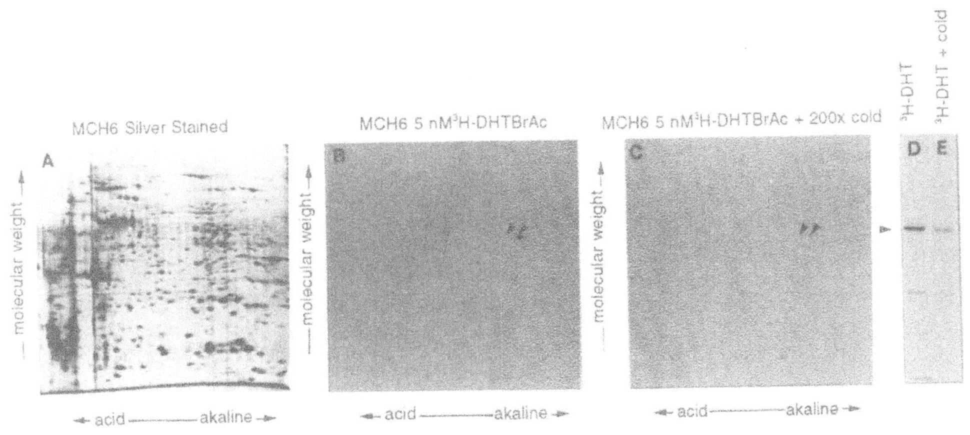

Figure 1. Specific androgen binding activity of the 56 kDa in normal GSF. A, two-dimensional protein map of cytosol, silver stained; B, fluorogram of a map from cells labelled covalently with 5 nM [^{3}H]-dihydrotestosterone-bromoacetate; C, with inclusion of 200-fold cold ligand; D and E, as B and C, but separated on one-dimensional SDS gels. (From Biochem. Biophys. Res. Commun. 175:831-838 (1991), with permission).

It is obvious from the AIS phenotype that numerous genes must have an altered expression to result in the female phenotype in contrast to the wild-type male. Overall, we think that altered transcription of the many unknown genes is brought about by a cascade of transcription factors altered in their level of expression and/or activities. Some of these factors may also regulate each others' expression level as exemplified by the first transcription factor which regulates this hierarchical cascade, the androgen receptor. In complete AIS the mutant receptor results in an altered expression of specific genes involved in sexual differentiation or other transcription factors cumulatively resulting in a different sexual phenotype, namely female. We know virtually nothing about which genes might be involved. However, one possible gene is ALDH1 which we have found to be underexpressed in GSF of AIS patients. This is presented in the central column of the scheme below in Fig. 2. As this enzyme could produce retinoic acid, a known morphogen, the altered production of retinoic acid could be responsible for the altered transcription of many other genes through mediation of retinoic acid receptors.

The androgen binding activity of the enzyme is of possible significance as well. If the previous data are valid for our enzyme (Elder & Topper, 1962) one can expect inhibition of retinoic acid formation by androgens. As androgens are generally elevated in AIS patients, it would have the effect to further curtail the production of the morphogen retinoic acid and thus compound the effect of underexpression of the enzyme.

The above scheme thus leads to the following hypothesis:

Aldehyde dehydrogenase is required for the development of the male (female) phenotype by producing the morphogen retinoic acid in precursor tissues at the appropriate time during development.

From Table 1 it is apparent that two AIS patients do express the protein normally, yet they do exhibit the full clinical picture of complete AIS. If the expression of ALDH1 in GSF faithfully reflects the situtation in the anlagen for the external genitalia during the critical stages of development, then the hypothesis presented above is untenable. However, at this time we do not understand the mechanism by which mutations in the androgen receptor gene lead to the underexpression of ALDH1. The most obvious mechanism, one of conventional

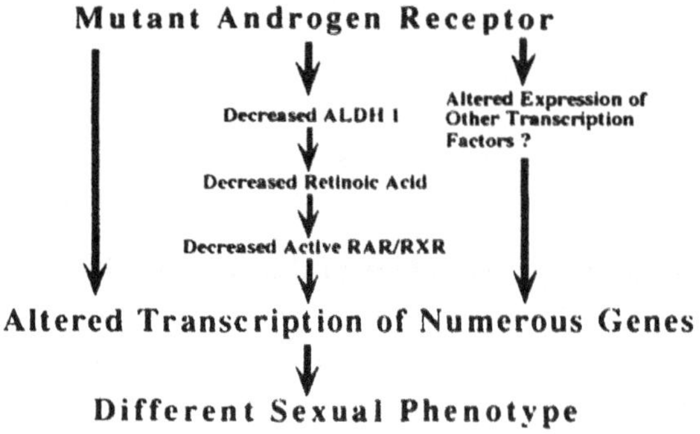

Figure 2. Speculations on the molecular pathogenesis of complete AIS

androgen induction, does not appear to operate in GSF, as we could never influence the expression of ALDH1 by culture conditions with androgens, antiandrogens or serum-free media. Furthermore, the normal expression of ALDH1 in the two exceptional patients (Table 1) provides strong genetic evidence against a simple mechanism of androgen dependence for the expression of ALDH1. The 5'-promoter region is currently being analyzed for sequence-specific DNA-binding proteins by gel mobility shift assays and DNA footprinting. Differences in band shifts between cell strains that express ALDH1 and those that do not have been observed in preliminary experiments. Thus, we speculate that the control of expression is influenced by other transcription factors, some of which may be controlled by the androgen receptor, and possibly other epigenetic mechanisms not yet understood.

The attraction of the hypothesis that ALDH1 plays a pivotal role in the development of the altered phenotype in complete androgen insensitivity is that it can be tested. It would be difficult to test in man, because it would be relatively meaningless to try this in human genital skin fibroblasts, where we fortuitously found the underexpression of ALDH1 (Nickel et al.,

1988; Pereira et al., 1991). Rather, it is important to test this in the precursor tissue or anlagen for the internal and external genitalia at just the right time during embryogenesis. We therefore shall test this in an animal model of complete AIS, the *Tfm* mouse. Ultimately, we hope to achieve this using *in situ* hybridization techniques. Of special interest will be the anlagen for external genitalia identified as the genital tubercle (Dollé et al., 1990), the sexually undifferentiated precursor of the male and female external genitalia. In support of our scheme above is the recent finding that the various retinoic acid receptor transcripts, especially for the α-isoform can be found in these precursor structures at 12.5 to 14.5 days post-coitum (Dollé et al., 1990). It should be possible to test whether ALDH1 transcripts can be found in these structures at the same time and whether clear differences will be noticeable in *Tfm* vs. wild-type and male vs. female animals. Furthermore, if the results are positive, they would also lend support to the *in vitro* observations that ALDH1 may be one ALDH isoform that can produce retinoic acid *in vivo*.

ACKNOWLEDGMENTS

This work was supported by the Manitoba Health Research Council, Children's Hospital of Winnipeg Research Foundation and the Medical Research Council of Canada.

REFERENCES

Belsham, D.D., Rosenmann, E., Pereira, F.A., Williams, S.G., Turney, M.K., Kovacs, W.J., Faber, L.E. & Wrogemann, K. (1989). The 56 kDa protein of human skin fibroblasts is identical to that radiolabelled by [^{3}H]-dihydrotestosterone 17 β-bromoacetate. J.Steroid Biochem. 33:389-394.

Brown, C.J., Goss, S.G., Lubahn, D.B., Joseph, D.R., Wilson, E.M., French, F.S. & Willard, H.F. (1989). Androgen receptor locus on the human X Chromosome: Regional localization to Xq11-12 and description of a DNA polymorphism. Am.J.Hum.Genet. 44:264-269.

Charest, N.J., Zhou, Z.-X, Lubahn, D.B., Olsen, K.L., Wilson, E.M. & French, F.S. (1991). A frameshift mutation destabilizes androgen receptor messenger RNA in the tfm mouse. Mol.Endocrinol. 5:573-581.

Dollé, P., Ruberte, E., Leroy, P., Morris-Kay, G. & Chambon, P. (1990). Retinoic acid receptors and cellular retinoid binding proteins. I. A systematic study of their differential pattern of transcription during mouse organogenesis. Development 110:1133-1151.

Elder, T.D. & Topper, Y.J. (1962). The oxidation of retinene (vitamin A1 aldehyde) to vitamin A acid by mammalian steroid sensitive aldehyde dehydrogenase. Biochim.Biophys.Acta 64:430-437.

French, F.S., Lubahn, D.B., Brown, T.R., Simental, J.A., Quigley, C., Yarbrough, W.G., Tan, J.-A., Sar, M., Joseph, D.R., Evans, B.A.J., Hughes, I.E., Migeon, C.J. & Wilson, E.M. (1990). The molecular basis of androgen insensitivity. Recent.Prog.Horm.Res. 46:1-42.

Futterman, S. (1962). Enzymatic oxidation of vitamin A aldehyde to vitamin A acid. J.Biol.Chem. 237:677-680.

Griffin, J.E. (1992). Androgen resistance — the clinical and molecular spectrum. N.Engl.J.Med. 326:611-617.

He, W.W., Kumar, M.V. & Tindall, D.J. (1991). A frame-shift mutation in the androgen receptor gene causes complete androgen insensitivity in the testicular feminized mouse. Nucleic.Acids.Res. 19:2373-2378.

Hsu, L.C., Chang, W.-C. & Yoshida, A. (1989). Genomic structure of the human cytosolic aldehyde dehydrogenase gene. Genomics 5:857-865.

Kovacs, W.J. & Turney, M.K. (1988). High efficiency covalent radiolabeling of the human androgen receptor. Studies in cultured fibroblasts using dihydrotestosterone 17β-bromoacetate. J.Clin.Invest. 81:342-348.

Lee, M.O., Manthey, C.L. & Sladek, N.E. (1991). Identification of mouse liver aldehyde dehydrogenases that catalyze the oxidation of retinaldehyde to retinoic acid. Biochem.Pharmacol. 42:1279-1285.

McPhaul, M.J., Marcelli, M., Tilley, W.D., Griffin, J.E. & Wilson, J.D. (1991). Androgen resistance caused by mutations in the androgen receptor gene. FASEB J. 5:2910-2915.

Nickel, B., Schwartz, A., Rosenmann, E., Kaufman, M., Pinsky, L. & Wrogemann, K. (1988). A study of androgen-resistant subjects indicates that the 6.7/56 kD protein in genital skin fibroblasts is related to the androgen receptor. Clin.Invest.Med. 11:22-33.

Pereira, F., Belsham, D., Duerksen, K., Rosenmann, E., Kaufman, M., Pinsky, L. & Wrogemann, K. (1990). The 56 kDa androgen-binding protein in human genital skin fibroblasts: its relation to the human androgen receptor. Mol.Cell.Endocrinol. 68:195-204.

Pereira, F., Rosenmann, E., Nylen, E., Kaufman, M., Pinsky, L. & Wrogemann, K. (1991). The 56 kDa androgen binding protein is an aldehyde dehydrogenase. Biochem.Biophys.Res.Commun. 175:831-838.

Thompson, R.G., Nickel, B., Finlayson, B., Meuser, R., Hamerton, J.L. & Wrogemann, K. (1983). 56K fibroblast protein not specific for Duchenne muscular dystrophy but for skin biopsy site. Nature 304:740-741.

Wrogemann, K., Pereira, F., Belsham, D., Kaufman, M., Pinsky, L. & Rosenmann, E. (1988). An abundant 56 kD protein with low affinity androgen binding: another member of the steroid/thyroid receptor family? Biochem.Biophys.Res.Commun. 155:907-913.

THE USE OF IMMORTALIZED MOUSE L1210/OAP CELLS ESTABLISHED IN CULTURE TO STUDY THE MAJOR CLASS 1 ALDEHYDE DEHYDROGENASE-CATALYZED OXIDATION OF ALDEHYDES IN INTACT CELLS

Norman E. Sladek and Mi-Ock Lee

Department of Pharmacology, University of Minnesota
Minneapolis, Minnesota 55455

INTRODUCTION

NAD(P)-linked aldehyde dehydrogenases are a group of enzymes that catalyze the oxidation of a wide variety of aldehydes to their corresponding acids (Sladek *et al.*, 1989; Goedde and Agarwal, 1990). Based on their primary structure, and/or subcellular distribution and kinetic, physical and immunological properties, these enzymes have been divided into three classes, *viz.*, 1, 2, and 3 (Lindahl and Hempel, 1991). The major class 2 enzyme has been studied quite extensively both *in vivo* and in cell-free experiments, in large part because of its importance in ethanol metabolism. More recently, class 3 aldehyde dehydrogenases have, because of their association with tumorigenesis, also been the subject of extensive investigation both *in vivo* and in cell-free experiments. In contrast, class 1 aldehyde dehydrogenases have not been as extensively studied, especially *ex* or *in vivo*. This is probably because their importance in catalyzing the oxidation of several biologically and/or pharmacologically important aldehydes, *e.g.*, aldophosphamide, has only recently been recognized (Sladek *et al.*, 1989).

Retinaldehyde appears to be another important substrate for class 1 aldehyde dehydrogenases (Lee *et al.*, 1991; Dockham *et al.*, 1992). Oxidation of retinol, and metabolic cleavage of β-carotene, gives rise to retinaldehyde which, in turn, can be further oxidized to retinoic acid (Olson, 1967; Frolik, 1984; Allen and Bloxham, 1989; Olson, 1989; Lakshman *et al.*, 1989; Kim *et al.*, 1992). Enzyme-catalyzed oxidation of retinaldehyde to retinoic acid can be viewed as a bioactivation since retinoic acid is the most potent form of the naturally occurring retinoids in regulating cell growth and differentiation (Breitman *et al.*, 1980; Lotan, 1980; Williams and Napoli, 1985). As judged by cell-free experiments, oxidation of retinaldehyde, at least that in the liver, appears to be catalyzed largely by the major class 1 aldehyde dehydrogenase, *viz.*, AHD-2 in mice and ALDH-1 in humans, present in this organ (Lee *et al.*, 1991; Dockham *et al.*, 1992).

The experimental conditions utilized in cell-free experiments usually differ substantially from those ordinarily considered to be physiological. For example, in our laboratory, the pH

Enzymology and Molecular Biology of Carbonyl Metabolism 4
Edited by H. Weiner, Plenum Press, New York, 1993

of the incubation media is usually 8.2, and high concentrations of EDTA (1 mM), glutathione (5 mM), sodium pyrophosphate (32 mM) and NAD (4 mM) are routinely included along with an inhibitor of alcohol dehydrogenase, *viz.*, pyrazole (0.1 mM).

Cultured mouse L1210/0 lymphocytic leukemia cells lack AHD-2 but large amounts of it are expressed in a subline, *viz.*, L1210/OAP, that is resistant to various antineoplastic agents. Thus, these cell lines offer an excellent opportunity to evaluate AHD-2-catalyzed oxidation of aldehydes to their corresponding acids in intact cells. Hence, the oxidation of retinaldehyde to retinoic acid by these cells was examined.

It was noted in the course of these studies, that intact L1210 cells contained at least one enzyme that catalyzed the reduction of retinaldehyde to retinol. Enzyme-catalyzed reduction of retinaldehyde to retinol has been shown to occur in several mammalian tissues, but the responsible enzyme has not always been conclusively identified. Thus, the reaction has been shown to be catalyzed by a relatively substrate non-specific cytosolic "aldehyde reductase" present in rat intestine (Fidge and Goodman, 1968), by an NAD(P)H-dependent enzyme present in the microsomal fractions of rat liver, rat small intestine and human intestinal Caco-2 cells (Leo *et al.*, 1987; Kakkad and Ong, 1988; Quick and Ong, 1990), by an "alcohol dehydrogenase" present in the cytosol of rat ocular tissue (Julià *et al.*, 1986), by a Class I alcohol dehydrogenase present in the cytosol of rat liver (Zachman and Olson, 1961), and by a Class IV alcohol dehydrogenase present in human tissues (Dr. Xavier Parès, personal communication). Initial experiments attempting to identify the operative enzyme(s) in intact L1210 cells are also reported herein.

MATERIALS AND METHODS

Cultured mouse L1210/0 and L1210/OAP lymphocytic leukemia cells, were originally obtained from Southern Research Institute, Birmingham, AL. Culture conditions and growth of these cells were as described previously (Sladek and Landkamer, 1985). Cells in exponential growth were harvested by centrifugation at 100 x g for 10 min, and were washed with a phosphate-buffered saline solution, pH 7.4 (Sladek and Landkamer, 1985), before further use.

Aldose reductase and aldehyde reductase isolated from human placenta were provided by Dr. H. Parekh, University of Minnesota, Minneapolis, MN. The method of Vander Jagt *et al.* (1990) had been used, with some modification, to purify these enzymes to homogeneity.

Purified to homogeneity but unidentified human liver 36 kDa aldo-keto reductase, and semipurified but unidentified human liver ~65 kDa aldo-keto reductase, were also provided by Dr. Parekh. It has been established that these enzymes are not aldose reductase, aldehyde reductase or any of the alcohol dehydrogenases; they have yet to be characterized in terms of substrate specificity.

Rabbit antibody to bovine lens aldose reductase was kindly provided by Dr. J. M. Petrash, Washington University School of Medicine, St. Louis, MO.

Preparation of L1210 Subcellular Fractions

L1210/0 and L1210/OAP soluble (105,000 x g supernatant), and solubilized particulate, fractions were prepared essentially as described before (Kohn *et al.*, 1987) except that 0.3% Lubrol, rather than 0.3% deoxycholate, was used to solubilize the particulate fraction. Briefly, the cell pellet was resuspended in homogenization medium (1.15% KCl and 1.0 mM EDTA in aqueous solution, pH 7.4) and sonicated to lyse the cells. The homogenate was then centrifuged at 105,000 x g and 4°C for 60 min. The resultant supernatant (soluble) fraction was saved for analysis. The pellet was washed once with homogenization medium and was then resuspended in homogenization medium containing 0.3% Lubrol. This preparation was centrifuged as above and the resultant supernatant (solubilized particulate) fraction was saved for assay.

Solubilized microsomal fractions of L1210/0 and L1210/OAP cells were prepared essentially as described before (Manthey *et al.*, 1990) except that 0.3% Lubrol, rather than 0.3% deoxycholate, was used to solubilize the microsomal fraction. Briefly, the cell pellet was resuspended in an aqueous solution of KCl (1.15%) and was centrifuged at 9000 x g and 4°C for 20 min. The supernatant fraction was removed and centrifuged at 105,000 x g and 4°C for 60 min. The pellet resulting from the second centrifugation was washed once by resuspending it in KCl solution and centrifuging at 105,000 x g and 4°C for 60 min. It was then resuspended in KCl solution supplemented with 0.3% Lubrol, and was again centrifuged at 105,000 x g and 4°C for 60 min. The resultant supernatant (solubilized microsomal) fraction was saved for assay.

In vitro Enzyme Assays

Enzyme-catalyzed oxidation of retinaldehyde to retinoic acid was quantified as described previously (Lee *et al.*, 1991). The reaction mixture (1 ml, pH 8.2) contained 4.0 mM NAD, 32 mM sodium pyrophosphate, 0.1 mM pyrazole, 5.0 mM glutathione, 1.0 mM EDTA, 25 mM retinaldehyde and a subcellular fraction. The reaction was initiated by the addition of retinaldehyde. Incubation was at 37°C for 5 min. The reaction was stopped by placing the reaction mixture into an ice-water bath. Retinoids were extracted and quantified by HPLC/spectrophotometric analysis as described previously (Lee *et al.*, 1991), except that the internal standard, *viz.*, tetraphenylethylene, was omitted from the extraction solvent. Retinoic acid, retinol and retinaldehyde concentrations were each estimated by comparing experimentally-obtained peak areas with those produced by known concentrations of the authentic retinoid.

Reduction of retinaldehyde to retinol was quantified by a modification of the method of Vander Jagt *et al.* (1990). The reaction mixture (1 ml, pH 7.0) consisted of 0.16 mM NADPH, 100 mM sodium phosphate, 25 μM retinaldehyde and a subcellular fraction or (semi)purified enzyme. The reaction was initiated by the addition of retinaldehyde. Incubation was at 37°C for 5 min ((semi)purified enzyme) or 20 min (subcellular fraction). The reaction was stopped by placing the reaction mixture into an ice-water bath. Retinoids were extracted and quantified by HPLC/spectrophotometric analysis as described above. All the procedures were done in dim light. The smallest detectable rate was approximately 0.03 nmol/min.

Enzyme-catalyzed reduction of glyceraldehyde (10 mM) and glucuronate (10 mM) was quantified spectrophotometrically at 37°C by monitoring the disappearance of NADPH at 340 nm.

Intact Cell Retinaldehyde Metabolism Assay

The incubation mixture (1 ml) consisted of L1210 cells suspended in the phosphate-buffered saline solution supplemented with 10% charcoal/dextran-treated fetal bovine serum, and 10 μM retinaldehyde. Incubation was at 37°C for 5 to 180 min. Following incubation, cells were lysed in an ice-bath by submitting them to sonication. Retinoids were extracted from the lysate and quantified by HPLC/spectrophotometric analysis as described above.

Complete and rapid separation of cells from the incubation medium was achieved by a modification of the method of Wohlhueter and Plagemann (1989), that is, by layering the suspended cells above a silicone oil/mineral oil mixture (4:1) and centrifuging at 9000 x g and 4°C for 50 sec. The cell-free aqueous layer (incubation media) remaining on top of the oil layer after centrifugation was saved for analysis. The cell pellet was resuspended in the phosphate-buffered saline solution supplemented with 10% charcoal/dextran-treated fetal bovine serum. Suspended cells were then lysed as described above. Extraction and quantification of retinoids in the incubation medium and cell lysate were also as described above. All the procedures were done in dim light.

Immunoblot Analysis

Soluble (105,000 x g supernatant) fractions obtained from L1210 cells were submitted to sodium dodecyl sulfate-polyacrylamide gel electrophoresis (SDS-PAGE) and were then electrophoretically transferred to an Immobilon-PVDF transfer membrane as described by others (Laemmli, 1970; Sambrook *et al.*, 1989). Immunodetection of aldose reductase on the blotted membrane was as previously described (Petrash *et al.*, 1991) except that 5% nonfat dried milk, rather than 3% gelatin, was used as the blocking solution.

RESULTS AND DISCUSSION

Enzyme-catalyzed retinaldehyde oxidation was first quantified in subcellular fractions of L1210/0 and L1210/OAP cells, Table 1. NAD-independent enzyme-catalyzed oxidation of retinaldehyde to retinoic acid was not observed with any of the fractions evaluated indicating that none of them contained an oxidase capable catalyzing the reaction. NAD-dependent enzyme-catalyzed oxidation of retinaldehyde to retinoic acid was observed but only when soluble fraction obtained from L1210/OAP cells was used as the enzyme source. This was as expected given that AHD-2 is present only in the cytosol of L1210/OAP cells. Assuming that 1 x 10^9 L1210/OAP cells weigh about 1 g, the amount of activity, *viz.*, 422 nmol/min, present in the soluble fraction obtained from 1 g of these cells approximately equals the amount of activity, *viz.*, 196 and 450 nmol/min, present in soluble fractions obtained from 1 g of murine or human liver, respectively (Lee *et al.*, 1991; Dockham *et al.*, 1992).

Table 1. Retinaldehyde oxidizing activity in the soluble (105,000 x g supernatant), and solubilized particulate, fractions of L1210/0 and L1210/OAP cells.

Cell Line	Fraction[1]	Retinoic Acid, pmol/min/10^8 cells	
		+ NAD	- NAD
L1210/0	Soluble	0	0
	Solubilized Particulate	0	0
L1210/OAP	Soluble	42,200	0
	Solubilized Particulate	0	0

[1]Soluble, and solubilized particulate, fractions were prepared from L1210 cells, and were assayed for retinaldehyde oxidizing activity as described in Materials and Methods. Reaction mixtures contained a soluble, or solubilized particulate, fraction prepared from 5 x 10^6 L1210/0 cells, a soluble fraction obtained from 1 x 10^6 L1210/OAP cells, or a solubilized particulate fraction prepared from 5 x 10^6 L1210/OAP cells. The retinaldehyde concentration was 25 μM.

Figure 1 shows some of the HPLC chromatograms obtained when the metabolism of retinaldehyde in intact L1210/0 and L1210/OAP cells was investigated. About 40 and 10% of the retinaldehyde was converted to retinoic acid and retinol, respectively, when retinaldehyde (10 µM) was incubated with intact L1210/OAP cells (2 x 10^6 cells/ml) for 1 hr. About 16% of the retinaldehyde was converted to retinol when incubation was with L1210/0 cells for 1 hr; no retinoic acid was generated in this case. That reduction of retinaldehyde to retinol was less in L1210/OAP cells is as expected given that the competition for retinaldehyde by the oxidative pathway is exclusive to these cells.

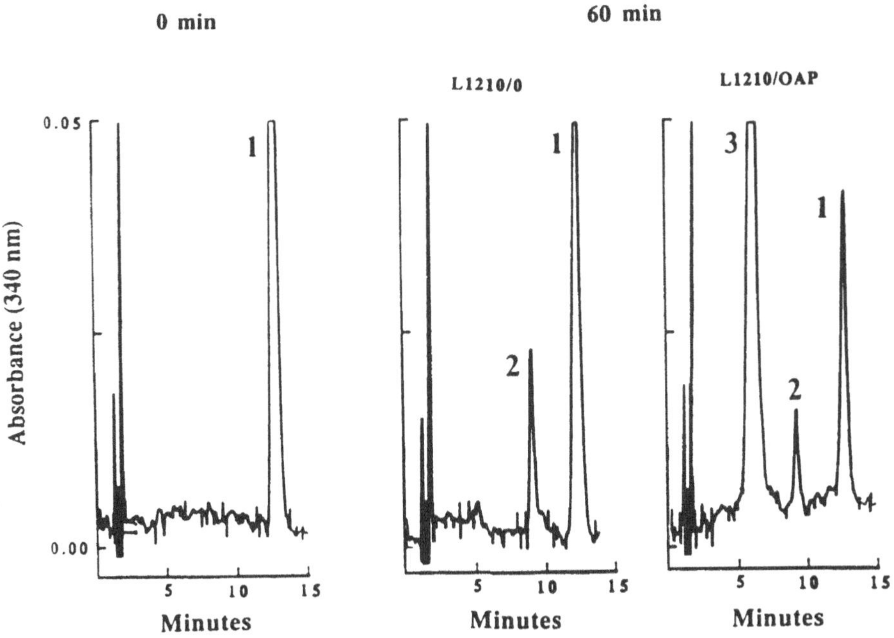

Figure 1. Metabolism of retinaldehyde in intact L1210 cells: HPLC/Spectrophotometric chromatograms. The incubation mixture contained 2 x 10^6 L1210/0, or L1210/OAP, cells and 10 µM retinaldehyde, and was incubated at 37°C for the indicated time period. Cells were then lysed, the resultant mixture of incubation medium and lysed cells was extracted with an organic solvent, and HPLC/spectrophotometric analysis was performed as described in Materials and Methods to quantify retinoids present in the extracts. Peaks 1, 2, and 3 are retinaldehyde, retinol, and retinoic acid, respectively. The detector setting was 0.05 AUFS.

Figure 2 shows the oxidation and reduction of retinaldehyde in intact L1210/0 and L1210/OAP cells as a function of incubation time. As in the previous experiment, reduction of retinaldehyde occurred in both cell lines whereas oxidation of retinaldehyde occurred only in L1210/OAP cells; retinoic acid formation reached a plateau at about 2 hr. The estimated initial velocity was about 16 nmol/min/2 x 10^6 cells. This compares to a value of about 49 nmol/min/2 x 10^6 cells obtained in the cell-free experiments. Probably accounting for part of this discrepancy is that incubation was at different pH's, *viz.*, 8.2 and, most likely, about 7.4, for the cell-free and intact cell studies, respectively. Supporting this notion is the observation that, in cell-free experiments, the rate at which a soluble fraction obtained from L1210/OAP cells catalyzed the oxidation of retinaldehyde at pH 7.4 was about 65% of the rate at which it did so at pH 8.2.

The distribution dynamics of newly generated retinoic acid and retinol between the cells and the incubation media were also studied. As presented in Figure 3 and Table 2, measurable amounts of the newly synthesized retinol and retinoic acid were released into the incubation medium. However, at approximate equilibrium, intracellular retinol and retinoic acid concentrations were, respectively, approximately 100- and 25-fold higher than those in incubation medium. These observations are consistent with those of Edwards *et al.* (1992); they reported that the intracellular retinoic acid concentration was higher than that in the incubation medium when cultured rabbit Müller cells were incubated with retinol. Intracellular accumulation of retinoic acid in Müller cells might be due to the presence of retinoic acid binding proteins (CRABP) in these cells (Gaur *et al.*, 1990; Milam *et al.*, 1990). L1210 cells, however, apparently lack such proteins (Lotan, 1980).

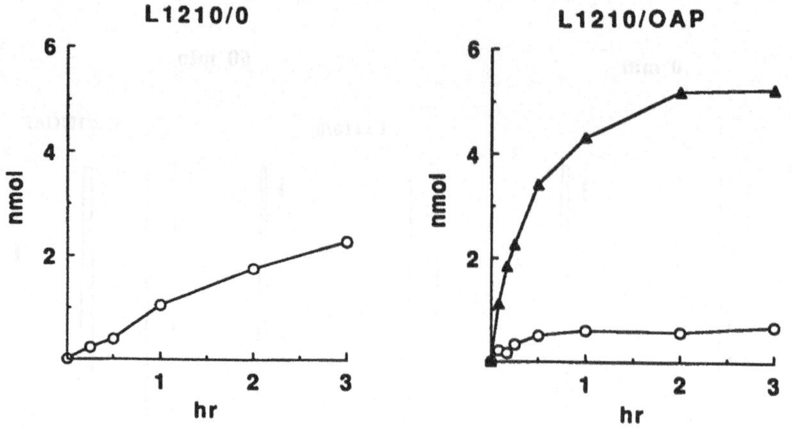

Figure 2. Metabolism of retinaldehyde in intact L1210 cells: Time course. The incubation mixture contained 2 x 10^6 L1210/0, or L1210/OAP, cells and 10 μM retinaldehyde, and was incubated at 37°C for the indicated time period. Cells were then lysed, the resultant mixture of incubation medium and lysed cells was extracted with an organic solvent, and HPLC/spectrophotometric analysis was performed as described in Materials and Methods to quantify retinoids present in the extracts. Results are nmol of retinol (○) and retinoic acid (▲) present in the extract (cells plus incubation medium) at the indicated time points. Not shown in the figure for reasons of clarity is that retinoic acid was not generated when L1210/0 cells were incubated with retinaldehyde.

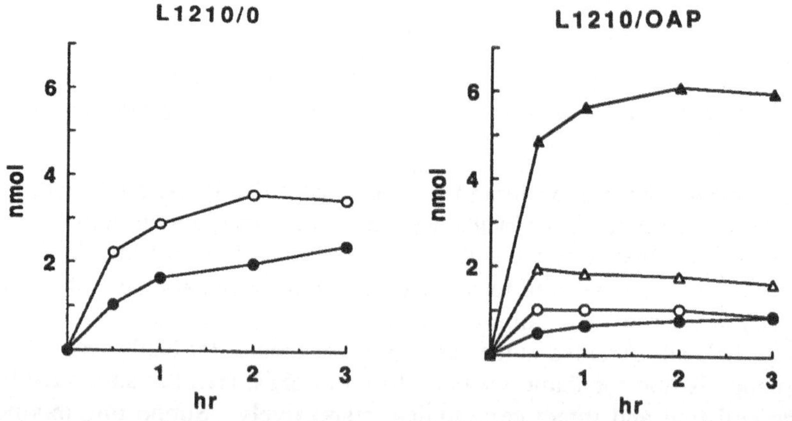

Figure 3. Metabolism of retinaldehyde in intact L1210 cells: Time course and the distribution of the metabolites in cells and the incubation medium. The incubation mixture contained 1 x 10^7 L1210/0, or L1210/OAP, cells and 10 μM retinaldehyde, and was incubated at 37°C for the indicated time period. Cells were then separated from the incubation medium by passage through oil as described in Materials and Methods. Retinoids present in the incubation media, and those present in the cells, were separately extracted and quantified by HPLC/spectrophotometric analysis as described in Materials and Methods. Results are nmol of retinol (●) and retinoic acid (▲) released into the incubation medium, and retinol (○) and retinoic acid (△) present in cells. Not shown in the figure for reasons of clarity is that retinoic acid was not generated when L1210/0 cells were incubated with retinaldehyde.

Table 2. Metabolism of retinaldehyde in intact L1210 cells: Concentrations of parent compound and metabolites in the cell and the incubation medium at 3 hr.

Retinoids	pmol/μl[1]			
	L1210/0		L1210/OAP	
	Intracellular	Extracellular	Intracellular	Extracellular
Retinol	343	2	88	1
Retinaldehyde	166	2	15	1
Retinoic acid	0	0	163	6

[1]The incubation mixture contained 1 x 10^7 L1210/0, or L1210/OAP, cells and 10 μM retinaldehyde, and was incubated at 37°C for 3 hr. Cells were then separated from the incubation medium by passage through oil as described in Materials and Methods. Retinoids present in the incubation media, and those present in the cells, were separately extracted and quantified by HPLC/spectrophotometric analysis as described in Materials and Methods. The volume of 1 x 10^7 L1210 cells was assumed to be 10 μl and the volume of the incubation media was 1 ml.

Several drugs, *e.g.*, disulfiram, cyanamide, chloral hydrate, and chloramphenicol are known to inhibit some aldehyde dehydrogenases including, at adequate concentrations, the major cytosolic (class 1) aldehyde dehydrogenase (Deitrich and Hellerman, 1963; Vallari and Pietruszko, 1981; Brien and Loomis, 1985). Others give rise to metabolites that are aldehydes and thus substrates for some of the aldehyde dehydrogenases. Included are cyclophosphamide which gives rise to aldophosphamide, and procarbazine which gives rise to N-isopropyl-p-formylbenzamide; each of these metabolites is a good substrate for the major cytosolic (class 1) aldehyde dehydrogenase (Manthey *et al.*, 1990; Maki and Sladek, 1991; Dockham *et al.*, 1992). Not unexpectedly, then, considerable inhibition of retinaldehyde oxidation to retinoic acid was observed in intact L1210/OAP cells when the aforementioned inhibitors and alternative substrates were included in the incubation medium, Table 3. The potential biological and/or pharmacological significance of such drug interactions, *viz.*, inhibition of retinaldehyde oxidation by various drugs, and, possibly, *vice versa* if the drug is an aldehyde, is obvious and needs to be evaluated.

Reduction of retinaldehyde to retinol catalyzed by subcellular fractions obtained from the two L1210 cell lines was also studied, Table 4. In contrast to the catalysis of retinaldehyde oxidation which occurred exclusively in the soluble fraction of L1210/OAP cells, that of retinaldehyde reduction occurred at approximately equal rates in the soluble, and solubilized particulate, fractions of both L1210/0 and L1210/OAP cells. Further investigation revealed that all of the enzyme-catalyzed retinaldehyde reduction that was observed with the solubilized particulate fractions of L1210/0 and L1210/OAP cells could be accounted for by an enzyme present in the microsomal fraction. Microsomal enzymes that catalyze the reduction of retinaldehyde have been found in rat and human tissues as well (Leo *et al.*, 1987; Kakkad and Ong, 1988; Quick and Ong, 1990).

The identity of the operative reductase(s) present in the soluble fraction was pursued in additional experiments.

Table 3. Inhibitors of enzyme-catalyzed retinaldehyde oxidation to retinoic acid in intact L1210/OAP cells.

Inhibitor (μM)	% Inhibition[1]
Disulfiram (10)	90
Cyanamide (30)	45
Chloral hydrate (100)	78
Chloramphenicol (2,000)	90
N-Isopropyl-p-formylbenzamide (200)	100
Mafosfamide[2] (100)	41

[1]The incubation mixture contained 2 x 10^6 L1210/OAP cells and 10 μM retinaldehyde, and was incubated at 37°C for 1 hr in the absence or presence of the indicated concentration of inhibitor as described in Materials and Methods. L1210/OAP cells were incubated with disulfiram and chloral hydrate for 30 min prior to the addition of retinaldehyde. The other agents were added at the same time that retinaldehyde was. Chloramphenicol and N-isopropyl-p-formylbenzamide were dissolved in 95% ethanol and were added in a volume of 40 μl and 20 μl, respectively. Disulfiram was dissolved in methanol and added in a volume of 10 μl. The other agents were dissolved in the phosphate-buffered saline solution. These amounts of vehicle were essentially without effect on retinaldehyde metabolism. The amount of retinoic acid generated in the absence of any inhibitor ranged from 3.5 to 4.9 nmol.

[2]Mafosfamide gives rise to aldophosphamide without benefit of enzyme involvement (Sladek, 1988).

Table 4. Retinaldehyde reducing activity in the soluble (105,000 x g supernatant), and solubilized particulate, fractions of L1210/0 and L1210/OAP cells.

Cell Line	Fraction[1]	Retinol, pmol/min/10^8 cells	
		+ NADPH	- NADPH
L1210/0	Soluble	75	0
	Solubilized Particulate	106	0
L1210/OAP	Soluble	135	0
	Solubilized Particulate	185	0

[1]Soluble, and solubilized particulate, fractions were prepared from L1210 cells, and were assayed for retinaldehyde reducing activity as described in Materials and Methods. Reaction mixtures contained a soluble, or solubilized particulate, fraction prepared from 5 x 10^7 cells. The retinaldehyde concentration was 25 μM.

Four (semi)purified aldo-keto reductases obtained from human tissues, *viz.*, aldose reductase, aldehyde reductase, unidentified 36 kDa aldo-keto reductase, and unidentified ~65 kDa aldo-keto reductase, were examined for their ability to catalyze the reduction of retinaldehyde to retinol, Table 5. Of the four, only aldose reductase was capable of doing so. The estimated apparent Km value for aldose reductase-catalyzed reduction of retinaldehyde was < 1 μM (data not presented). Aldose reductase is not likely to be the human counterpart of the rat intestinal, substrate non-specific, aldehyde reductase of Fidge and Goodman (1968) even though both are found in the cytosol and catalyze the reduction of retinaldehyde to retinol, because the two enzymes differ with respect to cofactor specificity (aldose reductase preferred NADPH as cofactor whereas the substrate non-specific enzyme did not show a preference for NADPH over NADH), molecular weight (35 *vs* 60 - 80 kDa, respectively), and apparent Km (retinaldehyde) values (<1 *vs* 20 μM, respectively).

Table 5. Reduction of retinaldehyde: Human enzymes.

Operative	Not Operative	Don't Know
Aldose reductase[1]	Aldehyde reductase[2]	Class I alcohol dehydrogenase[3]
Microsomal retinaldehyde reductase[4]	Unidentified 36 kDa aldo-keto reductase[5]	Class II alcohol dehydrogenase
Class IV alcohol dehydrogenase[6]	Unidentified ~65 kDa aldo-keto reductase[7]	Class III alcohol dehydrogenase[8]
		Class V alcohol dehydrogenase

[1]Initial rates for aldose reductase-catalyzed reduction of retinaldehyde (25 μM), glucuronate (10 mM), and glyceraldehyde (10 mM) were 1.9, 281, and 366 nmol/min/mg protein, respectively.

[2]Measurable catalysis of retinaldehyde (25 μM) reduction was not observed with an amount of enzyme sufficient to catalyze the reduction of glucuronate (10 mM) and glyceraldehyde (10 mM) at initial rates of 203 and 87 nmol/min, respectively.

[3]It has been reported that human class I alcohol dehydrogenase catalyzes the oxidation of retinol to retinaldehyde (Mezey and Holt, 1971; Duester, 1991). Since, all alcohol dehydrogenase-catalyzed reactions appear to be reversible (Vallee and Bazzone, 1983), reduction of retinaldehyde catalyzed by this enzyme is predicted. Supporting this notion is the report that a class I alcohol dehydrogenase present in rat liver catalyzes the reduction of retinaldehyde to retinol (Zachman and Olson, 1961; Duester, 1991).

[4]Demonstrated by Quick and Ong (1990).

[5]Measurable catalysis of retinaldehyde (25 μM) reduction was not observed with an amount of enzyme sufficient to catalyze the reduction of glucuronate (10 mM) at an initial rate of 76 nmol/min.

[6]Dr. Xavier Parés, personal communication

[7]Measurable catalysis of retinaldehyde (25 μM) reduction was not observed with an amount of enzyme sufficient to catalyze the reduction of glyceraldehyde (10 mM) at an initial rate of 51 nmol/min.

[8]It has been reported that human class III alcohol dehydrogenase does not catalyze the oxidation of retinol to retinaldehyde (Beisswenger *et al.*, 1985). Thus, in view of the comments made in footnote 3, the inability of this enzyme to catalyze the reverse reaction is predicted.

Soluble fractions from L1210/0 and L1210/OAP cells were fractionated by SDS-PAGE and visualized by immunoblotting with an antibody raised to bovine lens aldose reductase to determine whether aldose reductase was present in these cells. The results of these experiments indicate that it is present in both L1210/0 and L1210/OAP cells, Figure 4. L1210/0 and L1210/OAP soluble fractions were unable to catalyze the oxidation of ethanol to acetaldehyde, and retinol was not converted to retinaldehyde (or retinoic acid) by either intact, L1210/0, or L1210/OAP, cells (data not presented), indicating that these cells do not contain an alcohol dehydrogenase that catalyzes the reduction of retinaldehyde to retinol. It is highly likely, therefore, that reduction of retinaldehyde in the soluble fractions of L1210/0 and L1210/OAP cells is catalyzed, in part, if not totally, by aldose reductase.

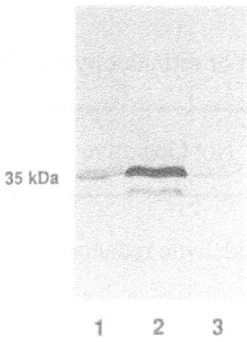

Figure 4. Immunoblot visualization of aldose reductase present in L1210/0 and L1210/OAP cells. Immunoblotting was as described in Materials and Methods. Rabbit antibody to bovine lens aldose reductase was used to visualize the enzyme. Lane 1, soluble (105,000 x g supernatant) fraction obtained from 5.6×10^6 L1210/0 cells; lane 2, aldose reductase (4 μg) purified from human placenta; lane 3, soluble (105,000 x g supernatant) fraction obtained from 5.6×10^6 L1210/OAP cells.

As judged by the failure to induce the expression of surface immunoglobulin, *i.e.*, conversion of immature B-lymphocyte to mature B-lymphocyte, the retinoids are incapable of inducing the differentiation of L1210 cells (Lee and Sladek, unpublished observations). Therefore, any role that AHD-2 might have in regulating retinoid-mediated cellular differentiation could not be studied in these cells. Any role that AHD-2 might have in regulating retinoid-mediated inhibition of tumor cell growth could also not be studied because the growth of both L1210/0 and L1210/OAP cells was unaffected by continuous exposure to physiological concentration (≤ 0.5 μM) of retinaldehyde or retinoic acid.

Several conclusions can be made from these studies. First, the major cytosolic aldehyde dehydrogenase, AHD-2, is meaningfully active in intact cells, at least in catalyzing the oxidation of retinaldehyde to retinoic acid, and the activity and behavior of this enzyme in cell-free systems reflects that in intact cells. Second, aldose reductase and a microsomal retinaldehyde reductase apparently catalyze the reduction of retinaldehyde to retinol in L1210 cells. Third, for reasons that are yet to be identified, retinoic acid and retinol are largely retained in L1210 cells following their synthesis therein; the physiological significance of this retention also remains to be determined. Fourth, an immortalized cultured cell line, *viz.*, L1210/OAP, may be useful for the study, in intact cells, of aldehyde oxidation catalyzed by the major, at least in the liver, murine class 1 aldehyde dehydrogenase, *viz.*, AHD-2. Finally L1210, especially L1210/0, cells may also be useful for the study of aldo-keto reductase activities in intact cells.

ACKNOWLEDGMENTS

Financial support was provided by Bristol-Meyers Squibb Company Grant 100-R220.

REFERENCES

Allen, J. G., and Bloxham, D. P., 1989, The pharmacology and pharmacokinetics of the retinoids, *Pharmac. Ther.*, **40**:1.

Beisswenger, T. B., Holmquist, B., and Vallee, B. L., 1985, χ-ADH is the sole alcohol dehydrogenase isozyme of mammalian brains: Implications and inferences, *Proc. Natl. Acad. Sci., USA*, **82**:8369.

Brien, J. F., and Loomis, C. W., 1985, Aldehyde dehydrogenase inhibitors as alcohol-sensitizing drugs: A pharmacological perspective, *TIPS*, **6**:477.

Breitman, T. R., Selonick, S. E., and Collins, S. J., 1980, Induction of differentiation of the human promyelocytic leukemia cell line (HL-60) by retinoic acid, *Proc. Natl. Acad. Sci., USA*, **77**:2936.

Deitrich, R. D., and Hellerman, L., 1963, Diphosphopyridine nucleotide-linked aldehyde dehydrogenase: II. Inhibitors, *J. Biol. Chem.*, **238**:1683.

Dockham, P. A., Lee, M.-O., and Sladek, N. E., 1992, Identification of human liver aldehyde dehydrogenases that catalyze the oxidation of aldophosphamide and retinaldehyde, *Biochem. Pharm.*, **43**:2453.

Duester, G., 1991, A hypothetical mechanism for fetal alcohol syndrome involving ethanol inhibition of retinoic acid synthesis at the alcohol dehydrogenase step, *Alcoholism: Clin. Exp. Res.*, **15**:568.

Edwards, R. B., Adler, A. J., Dev, S., and Claycomb, R. C., 1992, Synthesis of retinoic acid from retinol by cultured rabbit Müller cells, *Exp. Eye Res.*, **54**:481.

Fidge, N. H., and Goodman, D. S., 1968, The enzymatic reduction of retinal to retinol, *J. Biol. Chem.*, **243**:4372.

Frolik, C. A., 1984, Metabolism of retinoids, in "The Retinoids," Vol. 2, M. B. Sporn, A. B. Roberts, and D. S. Goodman, eds., Academic Press, Orlando, p. 177.

Gaur, V. P., De Leeuw, A. H., Milam, A. H., and Sarri, J. C., 1990, Localization of cellular retinoic acid-binding protein to amacrine cells of rat retina, *Exp. Eye Res.*, **50**:505.

Goedde, H. W., and Agarwal, D. P., 1990, Pharmacogenetics of aldehyde dehydrogenase (ALDH), *Pharmac. Ther.*, **45**:345.

Julià, P., Farrés, J., and Rapés, X., 1986, Ocular alcohol dehydrogenase in the rat: Regional distribution and kinetics of the ADH-1 isozyme with retinol and retinal, *Exp. Eye Res.*, **42**:305.

Kakkad, B. P., and Ong, D. E., 1988, Reduction of retinaldehyde bound to cellular retinol-binding protein (type II) by microsomes from rat small intestine, *J. Biol. Chem.*, **263**:12916.

Kim, C.-I., Leo, M. A., and Lieber, C. S., 1992, Retinol forms retinoic acid via retinal, *Arch. Biochem. Biophys.*, **294**:388.

Kohn, F. R., Landkamer, G. J., Manthey, C. L., Ramsay, N. K. C., and Sladek, N. E., 1987, Effect of aldehyde dehydrogenase inhibitors on the *ex vivo* sensitivity of human multipotent and committed hematopoietic progenitor cells and malignant blood cells to oxazaphosphorines, *Cancer Res.*, **47**: 3180.

Lakshman, M. R., Mychkovsky, I., and Attlesey, M., 1989, Enzymatic conversion of all-trans-β-carotene to retinal by a cytosolic enzyme from rabbit and rat intestinal mucosa, *Proc. Natl. Acad. Sci., USA* , **86**:9124.

Laemmli, U. K., 1970, Cleavage of structural proteins during the assembly of the head of bacteriophage T4, Nature, **226**:680.

Lee, M.-O., Manthey, C. L., and Sladek, N. E., 1991, Identification of mouse liver aldehyde dehydrogenases that catalyze the oxidation of retinaldehyde to retinoic acid, *Biochem. Pharm.*, **42**:1279.

Leo, M. A., Kim, C.-I., and Lieber, C. S., 1987, NAD^+-dependent retinol dehydrogenase in liver microsomes, *Arch. Biochem. Biophys.*, **259**:241.

Lindahl, R., and Hempel, J., 1991, Aldehyde dehydrogenase: What can be learned from a baker's dozen sequences?, *Adv. Exp. Med. Biol.*, **284**:1.

Lotan, R., 1980, Effects of vitamin A acid and its analogs (retinoids) on normal and neoplastic cells, *Biochim. Biophys. Acta*, **605**:33.

Maki, P. A., and Sladek, N. E., 1991, Potentiation of the cytotoxic action of mafosfamide by N-isopropyl-p-formylbenzamide, *Cancer Res.*, **51**:4170.

Manthey, C. L., 1988, Resolution and characterization of the aldehyde dehydrogenases important in cyclophosphamide metabolism, Ph.D. Dissertation, University of Minnesota, Minneapolis.

Manthey, C. L., Landkamer, G. J., and Sladek, N. E., 1990, Identification of the mouse aldehyde dehydrogenases important in aldophosphamide detoxification, *Cancer Res.*, **50**:4991.

Mezey, E., and Holt, P., 1971, The inhibitory effect of ethanol on retinol oxidation by human liver and cattle retina, *Exp. Mol. Path.*, **15**:148.

Milam, A. H., De Leeuw, A. M., Gaur, V. P., and Sarri, J. C., 1990, Immunolocalization of cellular retinoic acid-binding protein to Müller cells and/or a subpopulation of GABA-positive amacrine cells, *J. Comp. Neurol.*, **296**:123.

Olson, J. A., 1967, The metabolism of vitamin A, *Pharmacol. Rev.*, **19**:559.

Olson, J. A., 1989, Provitamin A function of carotenoids: The conversion of β-carotene into vitamin A, *J. Nutr.*, **119**:105.

Petrash, J. M., Delucas, L. J., Bowling E., and Egen, N., 1991, Resolving isoforms of aldose reductase by preparative isoelectric focusing in the rotofor, *Electrophoresis*, **12**:84.

Quick, T. C., and Ong, D. E., 1990, Vitamin A metabolism in the human intestinal Caco-2 cell line, *Biochemistry*, **29**:11116.

Sambrook, J., Fritsh, E. F., and Maniatis, T., 1989, Transfer of proteins from SDS-polyacrylamide gels to solid supports: Immunological detection of immobilized proteins (western blotting), in "Molecular Cloning: A Laboratory Manual," Vol. 3, C. Nolan, ed., Cold Spring Harbor Laboratory Press, Plainview, p. 18.60.

Sladek, N. E., and Landkamer, G. J., 1985, Restoration of sensitivity to oxazaphosphorines by inhibitors of aldehyde dehydrogenase activity in cultured oxazaphosphorine-resistant L1210 and cross-linking agent-resistant P388 cell lines, *Cancer Res.*, **45**:1549.

Sladek, N. E., 1988, Metabolism of oxazaphosphorines, *Pharmac. Ther.*, **37**:301.

Sladek, N. E., Manthey, C. L., Maki, P. A., Zhang, Z., and Landkamer, G. J., 1989, Xenobiotic oxidation catalyzed by aldehyde dehydrogenases, *Drug Metab. Rev.*, **20**:697.

Vallari, R. C., and Pietruszko, R., 1981, Kinetic mechanism of the human cytoplasmic aldehyde dehydrogenase E1, *Arch. Biochem. Biophys.*, **212**:9.

Vallee, B. L., and Bazzone, T. J., 1983, Isozymes of human liver alcohol dehydrogenase, in "Isozymes: Current Topics in Biological and Medical Research," Vol. 8, M. C. Rattazzi, J. G. Scandalios, and G. S. Whitt, eds., Alan R. Liss, New York, p. 219.

Vander Jagt, D. L., Hunsaker, L. A., Robinson, B., Stangebye, L. A., and Deck, L. M., 1990, Aldehyde and aldose reductases from human placenta: Heterogeneous expression of multiple enzyme forms, *J. Biol. Chem.*, **265**:10912.

Williams, J. B., and Napoli, J. L., 1985, Metabolism of retinoic acid and retinol during differentiation of F9 embryonal carcinoma cells, *Proc. Natl. Acad. Sci., USA*, **82**: 4658.

Wohlhueter, R. M., and Plagemann, P. G., 1989, Measurement of transport versus metabolism in cultured cells, in "Methods in Enzymology," Vol. 173, S. L. Berger and A. R. Kimmel, eds., Academic Press, Orlando, p. 714.

Zachman, R. D., and Olson, J. A., 1961, A comparison of retinen reductase and alcohol dehydrogenase of rat liver, *J. Biol. Chem.*, **236**:2309.

ENHANCED TRANSCRIPTION OF THE CYTOSOLIC ALDH GENE IN CYCLOPHOSPHAMIDE RESISTANT HUMAN CARCINOMA CELLS

Akira Yoshida[1], Vibha Davé[1], Hong Han[2], Kevin J. Scanlon[2]

[1]Department of Biochemical Genetics
[2]Biochemical Pharmacology, Medical Oncology
City of Hope National Medical Center
Duarte, CA 91010

INTRODUCTION

Cyclophosphamide, an important cancer chemotherapeutic agent, has been shown to be effective against some types of cancer (Hilton, 1984a; Manthey and Sladek, 1989). However, both natural and acquired resistance to cyclophosphamide limits its utility in the clinic. Several investigators have observed a dramatic increase (about 200-fold) of cytosolic aldehyde dehydrogenase (ALDH, EC 1.2.1.3) activity in a cyclophosphamide-resistant murine leukemia cell line. It has been proposed that ALDH which can oxidize aldophosphamide, an intermediate metabolite produced from cyclophosphamide, prevents the generation of tissue toxic phosphamide mustard within cells (Manthey and Sladek, 1989; Russo et al., 1989). The drug resistance of the murine cell line with high ALDH activity can be explained by this mechanism.

The cytosolic ALDH isozyme is also strongly induced by phenobarbital, and a specific enhancement of transcription of the cytosolic <u>ALDH</u> gene was found in the liver of the responsive rat strain (Dunn et al., 1989). Recently, a similar transcriptional activation of the cytosolic ALDH gene was observed in the cyclophosphamide-resistant murine leukemic cell line (Radin et al., 1991).

The cytosolic ALDH activity is widely differed among tissues and cells. In this study we tried to elucidate the molecular basis for both natural and acquired cyclophosphamide resistance: i) whether or not a carcinoma cell line with a high ALDH activity is more drug resistant than that with a low enzyme activity; ii) whether or not a cell line with a high ALDH activity and that with a low activity both can develop

Enzymology and Molecular Biology of Carbonyl Metabolism 4
Edited by H. Weiner, Plenum Press, New York, 1993

drug-resistance in culture; and iii) whether or not the acquired drug-resistance is associated with specific transcriptional enhancement of expression of the cytosolic ALDH gene.

MATERIALS AND METHODS

Human Carcinoma Cells and Induction of Drug Resistance

The human ovarian carcinoma cell line, A2780, was obtained from Dr. R. Ozols, Fox Chase Cancer Center (Eva et al., 1982). The human colon carcinoma cell line, HCT8, was obtained from Dr. J.R. Bertino, Memorial Sloan Kettering (Tompkins et al., 1974). These cells were made resistant to cyclophosphamide by weekly continuous exposures to increasing concentrations (up to 600 μM) of cyclophosphamide for twelve weeks, and maintained as previously described (Lu et al., 1988). For growth curves and cytotoxicity determinations, 200 cells were inoculated in 60 mm tissue culture dishes in 2 ml of growth medium. Twenty-four hours later, the cells were treated with the drug. After each treatment, the cells were incubated at 37°C for nine days. The media was aspirated from the dishes, the colonies were washed with phosphate buffered saline (PBS, Gibco), fixed in methanol and then stained with Giemsa dye (Sigma). The colonies were counted with at least 50 cells per colony, and 50% lethal doses (EC_{50}) were estimated.

Enzyme and Protein Assay

The cultured cells (in a 25 cm_2 flask) were harvested by trypsin treatment and washed twice with phosphate buffered saline (137 mM NaCl, 2.7 mM KCl, 4.3 mM Na_2HPO_4, 1.4 mM KH_2PO_4, pH 7.3). The washed-packed cells were mixed with an approximately equal volume of 0.01 M phosphate buffer, pH 7.0, 1 mM EDTA, 1 mM 2-mercaptoethanol, 50 μM NAD and 50 μM NADP, and disintegrated by freezing-thawing several times. The centrifuge extracts (10,000 r.p.m. for 10 min. in a microcentrifuge) were used for enzyme assay, protein assay and SDS-PAGE electrophoresis.

Activities of three enzymes, i.e. ALDH, glucose-6phosphate dehydrogenase (G6PD, EC 1.1.1.49) and phosphoglycerate kinase (PGK, EC 2.7.2.3) were measured as previously described (Ikawa et al., 1983; Yoshida, 1966; Yoshida and Watanabe, 1972). Protein was assayed by the method of Lowry et al. (1951) using bovine serum albumin as standard.

mRNA Assay

Cellular RNA was prepared from the cultured cells by extraction with guanidine thiocyanate, and density gradient centrifugation in a cesium chloride solution, followed by ethanol precipitation (Chirgwin et al., 1979). The RNA preparation was further purified by treatment with chloroform/butanol, and re-precipitation with ethanol. The purified RNA preparation was dissolved in deionized water to give concentration about 500 μg/ml. For estimation of relative quantities of mRNA components specific for $ALDH_1$, G6PD and PGK, slot blot hybridization was carried out (Sambrook et al., 1989). Briefly, the RNA preparations of serial double fold dilutions (about 0.1–4 μg/slot) were blotted onto a nitrocellulose filter using a slot blotter (Schleicher & Schuell). The filter was hybridized successively with the $ALDH_1$ specific probe, G6PD specific probe and PGK specific probe in the following conditions: hybridization in 50% formamide, 3x SSC, 1x Denhardt's solution, 200 μg/ml salmon(Tani et al. , 1985). These probes were ^{32}P labeled by the random prime method (Sambrook et al., 1989).

Since the tissues might contain mRNAs for other ALDH isozymes which are homologous and could cross-hybridize with the $ALDH_1$ cDNA probe in the slot blot and the Northern blot hybridization (Yoshida et al., 1991), the quantities of $ALDH_1$ mRNA were also estimated by a more specific and reliable method, as follows. An approximately one μg of RNA was amplified by the polymerase chain reaction (PCR) (Saiki et al., 1988) using the following three sets of primers: $ALDH_1$ GGATCAACAGAGGTTGGCAAG (5'-primer) and GTAGAATACCCATGGTGTGC (3'-primer), which were expected to produce 153 bp product hybridizable with the $ALDH_1$ probe (Hsu et al., 1985). G6PD: CCATTTGGATCATCCGGGACG (5'-primer) and TCGCAAAAGTGGCGCTTGGTGGA (3'-primer), which were expected to produce 270 bp fragment hybridizable with the G6PD probe (Takizawa et al., 1986). PGK: TCCTTAGAGCCAGTTGCTGT (5'-primer) and TTCTGTGTGGCAGATTGACTCC (3'-primer), which were expected to produce 323 bp fragment hybridizable with the PGK probe (Michelson et al., 1983).

One fifth of the amplification product was electrophoresed in an agarose gel (2%) and blotted onto a nitrocellulose filter. The filter was hybridized successively with the specific $ALDH_1$, G6PD and PGK cDNA probes as described above.

Southern Blot Hybridization

The standard SDS-proteinase K treatment was used to extract DNA from the cells harvested by trypsin treatment (Sambrook et al., 1989). DNA preparation was purified by phenol/chloroform/isoamylalcohol treatment, and ethanol precipitation.

An approximately 10 μg of DNA was completely digested by MspI (100 units). The digests were separated by agarose gel electrophoresis (0.8%) and subjected to Southern blot hybridization using the $ALDH_1$, G6PD and PGK specific cDNA probes successively.

RESULTS

Induction of Drug Resistance

Both ovarian and clone cells were more sensitive to 4-hydroperoxy cyclophosphamide than to cyclophosphamide. After exposure to cyclophosphamide, the cells became resistant to 4-hydroperoxy cyclophosphamide as well as to cyclophosphamide (Table 1).

Activities of Three Enzymes in Original and Drug-resistant Cancer Cell Lines

A several-fold increase of ALDH and G6PD activities was observed in the drug-resistant ovarian A2780 cell line (Table 1). Such a remarkable increase was not observed in the naturally resistant or the acquired resistant colon HCT8 cells.

Table 1. Resistance Induced by Cyclophosphamide (CPA)

Cell Line		Drug Sensitivity, EC 50, μM	
		CPA	4-Hyproperoxy CPA
Human Ovarian (A2780)	Before	145	18
	After	379	89
Human Colon (HCT8)	Before	409	8.5
	After	589	70

Table 2. Enzyme Activities in Original and Drug-resistant Cancer Cell Lines

Cell Lines		Enzyme Activity				
		unit/mg			ratio	
		ALDH	G6PD	PGK	ALDH/PGK	G6PD/PGK
A2780[a]	Original	0.0015	0.050	2.57	0.0006	0.0195
(ovarian)	Resistant	0.0069	0.62	2.40	0.0029	0.255
HCT8[a]	Original	0.0030	0.425	2.51	0.0014	0.169
(colon)	Resistant	0.0071	0.54	3.50	0.0020	0.154
A2780[b]	Original	0.0007	0.095	2.56	0.0003	0.037
(ovarian)	Resistant	0.006	1.05	2.60	0.0023	0.404
HCT8[b]	Original	0.0033	0.69	2.78	0.0012	0.25
(colon)	Resistant	0.005	0.95	2.72	0.0018	0.35

a: Cells harvested on September 24, 1990.

b: Cells harvested on October 3, 1990.

The enzyme activity is expressed μmol of substrate used per min at 25°C. ALDH activity was assayed using propionaldehyde as substrate at pH 8.6 in the presence of pyrazole. G6PD activity was assayed at pH 8.6, and PGK activity was assayed at pH 7.5.

However, in comparison to the original ovarian cell line, the original colon cell line had higher ALDH and G6PD enzyme activities. PGK activity was nearly constant in the original and drug-resistant ovarian and colon carcinoma cell lines (Table 2).

Human and other mammalian tissues contain two major ALDH with low Km values (micromolar range) for acetaldehyde and propionaldehyde, i.e. the cytosolic $ALDH_1$, (designated as

ALDH-2 in case of mouse) and the mitochondrial $ALDH_2$ (designated as ALDH-1 in case of mouse). The $ALDH_1$ is strongly inhibited by a low concentration of disulfiram, while the $ALDH_2$ is insensitive to the reagent (Greenfield and Pietruszko, 1977). By measuring the enzyme activity in the presence and absence of disulfiram (30 μM) in the reaction mixture, it was roughly estimated that 50-70% of total ALDH activity in the original ovarian cell line, and about 90% of that of the drug-resistant cell line were from the cytosolic $ALDH_1$; i.e. most, if not all, increase of ALDH activity can be attributed to the elevation of $ALDH_1$ activity.

mRNAs for the Three Enzymes

Patterns of slot blot hybridization indicated that contents of mRNA for $ALDH_1$ and that for G6PD were substantially increased in the drug-resistant ovarian cell line. By contrast, such increase was not observed in the drug-resistant colon cell line (Fig. 1) . The content of PGK mRNA was not changed in both cell lines before and after developing drug-resistance.

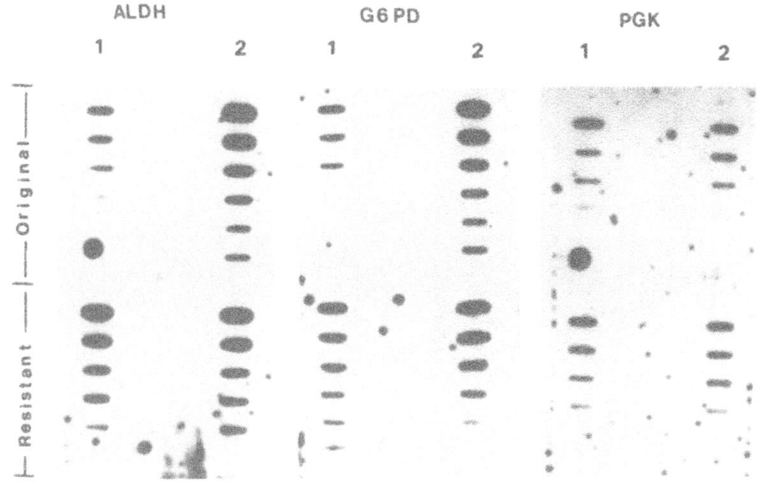

Figure 1. Slot blot hybridization of cellular RNA with specific cDNA probes. RNAs of serial double-fold dilutions (about 0.1-4 μg/slot) were blotted onto a nitrocellulose filter. The filter was hybridized, de-probed and re-hybridized successively with the $ALDH_1$ CDNA probe, G6PD cDNA probe and PGK cDNA probe. Lane 1: RNA prepared from the ovarian cell line; and lane 2: RNA prepared from the colon cell line.

More specific analysis of mRNA components using PCR products confirmed the increase of ALDH mRNA and G6PD mRNA in the drug-resistant ovarian cell line. In comparison to 323 bp product hybridizable with the PGK cDNA probe, which served as an internal reference, relative quantity of 153 bp product hybridizable with the $ALDH_1$ probe and that of 270 bp product hybridizable with the G6PD probe were much higher in the drug-resistant ovarian cell line than in the original ovarian cell

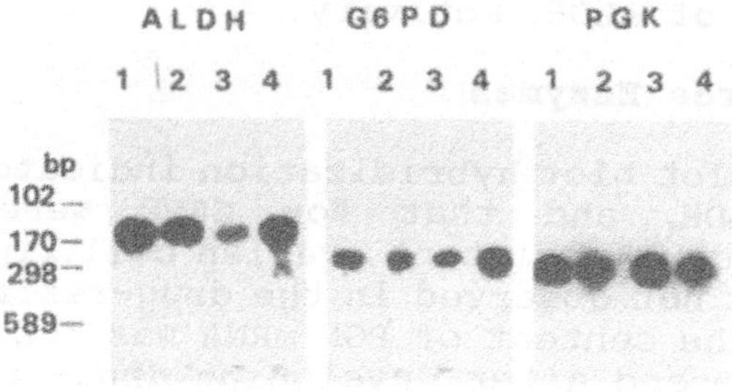

Figure 2. Northern blot hybridization of PCR amplification products with specific CDNA probes. The RNA preparations were amplified using the three sets of primers (see materials and Methods). The products were separated by agarose gel electrophoresis and transferred onto a filter. The filter was hybridized, de-probed, and re-hybridized successively with the ALDH$_1$ cDNA probe, G6PD cDNA probe and PGK-cDNA probe. Lane 1: RNA from the original colon cell line; lane 2: RNA from the drug-resistant colon cell line; lane 3: RNA from the original ovarian cell line; and lane 4: RNA from the drug-resistant ovarian cell line.

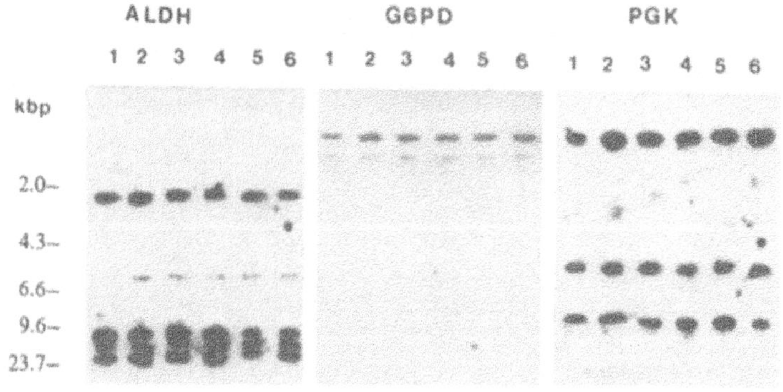

Figure 3. Southern blot hybridization of genomic DNAs extracted from original and drug-resistant cells. DNA samples digested by MspI were size fractionated by agarose gel electrophoresis, transferred onto a filter, and hybridized, deprobed and re-hybridized successively with the ALDH cDNA probe, G6PD cDNA probe and PGK cDNA probe. Lanes 1 and 2: DNA from healthy individuals; lane 3: DNA from the original colon cell line; lane 4: DNA from the drug-resistant colon cell line; lane 5: DNA from the original ovarian cell line; and lane 6: DNA from the drug-resistant ovarian cell line.

line (Fig. 2). Such increases were not observed in the drug-resistant colon cell line.

Southern Blot Hybridization

Southern blot hybridization of the chromosomal DNAs did not show any indication of multiplication of the <u>ALDH</u>1 gene and the <u>G6PD</u> gene in the drug-resistant ovarian cell line (Fig. 3).

Protein Components

Examination of extractable cellular proteins by SDS-PAGE electrophoresis showed differences of several protein components, which did not correspond to $ALDH_1$ between the original and the drug-resistant ovarian cell line. By contrast, the protein patterns did not appear to be changed in the drug-resistant colon cancer cell line (figure not shown).

DISCUSSION

The mode of response to cyclophosphamide is different in the human ovarian A2780, the human colon HCT8, and the murine leukemic L1210 cell lines (Table 1). Substantial increase of ALDH1 and G6PD activities, and proportional increase of $ALDH_1$ mRNA and G6PD mRNA contents were observed in the drug-resistant ovarian cell line (Table 2, Figs. 1 and 2). Since the <u>$ALDH_1$</u> gene and <u>G6PD</u> gene did not amplify in the drug-resistant cells, as evidenced by the Southern blot hybridization (Fig. 3), the change occurred at the level of transcription. None of these changes were observed in the drug-resistant colon cell line. A very large increase (about 200-fold) of cytosolic ALDH activity, which was observed in the drug-resistant murine leukemic cell line L1210 (Hilton, 1984a), did not occur in either human ovarian or colon carcinoma cell lines.

In the case of ovarian cell line, which had low ALDH and G6PD activities, the elevation of both ALDH and G6PD occurred in developing acquired drug-resistance. The elevation of ALDH activity may be beneficial for the cell growth in the presence of the drug as proposed in the case of murine leukemic cell line (Manthey and Sladek, 1989; Russo et al., 1989). An enhanced pentose phosphate synthesis and an elevation of G6PD activity are commonly observed in various types of carcinoma tissues (Weber et al., 1975; Bannasch et al., 1983). G6PD also plays a primary role for NADPH generation in the pentose phosphate shunt pathway. Thus the elevation of G6PD can increase the generation of reduced glutathione which has been implicated in protection of the cells against alkylating agents including cyclophosphamide (Green et al., 1984; Crook et al., 1986). The enhanced transcription of the cytosolic ALDH and <u>G6PD</u> genes could account for the development of drug-resistance in the ovarian cell line.

Compared to the ovarian cells, the human colon carcinoma cells were more resistant to cyclophosphamide, and the acquired drug-resistance was only moderate (about 1.4-fold) in HCT8 cell line (Table 1). However, the original HCT8 cells were sensitive ($EC5_0$ = 8.5, μM) to 4-hydroperoxycyclophosphamide, which can be spontaneously converted to 4-hydroxycyclophosphamide, and the cells became highly resistant (EC_{50} = 70 μM) to the drug after exposing the cells to cyclophosphamide.

The colon carcinoma cell line had several-fold higher ALDH and G6PD activities than the ovarian cells, and notable elevation of G6PD and ALDH was not observed in the drug-resistant cells (Table 1). The acquired high resistance to 4-hydroperoxycyclophosphamide but not to cyclophosphamide might be due to the development of resistance against the 4-hydroperoxycyclophosphamide transport in the cell membranes.

The activity of ALDH was extremely low in the murine leukemic cell line in comparison to other tissues, and the cell line was very sensitive to the drug (EC_{50} <25μM) (Hilton, 1984b). Over hundred-fold elevation of the ALDH activity would be required for developing the drug-resistance (EC_{50}) = 150μM) in this particular cell line L1210.

ACKNOWLEDGMENTS

The authors thank Shionogi Co. & Ltd., Osaka, Japan, for providing us with 4-hydroperoxycyclophosphamide.

This work was supported by Public Health Service grants HL-29515, AAO-57763 and CA50618, and American Cancer Society grant CH483.

REFERENCES

Bannasch, P., Hacker, H. J., Klimek, F., and Mayer, P., 1983, Hepatocellular glycogenesis and related pattern of enzymatic changes during hepatocarcinogenesis, Adv. Enzyme Req., **22**:97.

Chirgwin, J. J., Przbyla, A. E., MacDonald, R. J., and Rutter, W. J., 1979, Isolation of biologically active ribonucleic acid from sources enriched in ribonuclease, Biochem., **18**:5294.

Crook, T. R., Souhami, R. L., Whyman, G. D., and McLean, A. E.M., 1986, Glutathione depletion as a determinant of sensitivity of human leukemic cells to cyclophosphamide, Cancer Res., **46**:5035.

Dunn, T.J., Koleske, A.J., Lindahl, R., and Pitot, H.C. 1989, Phenobarbital-inducible aldehyde dehydrogenase in the rat: cDNA sequence and regulation of the mRNA by phenobarbital in responsive rats. J. Biol. Chem. **264**:13057.

Eva, A., Robbins, K. C., Andersen, P. R., Srinivasan, A., Tronick, S. R., Reddy, E. P., Ellmore, N. W., Galen, A. T., Lantenberger, J. A., Papas, T. S., Westin, E. H., Wong-Stall, F. , Gallo, R. C. , and Anderson, S. A., 1982, Cellular genes analogous to retroviral onc genes as transcribed in human tumor cells, Nature, **295**:116.

Green, J. A., Vistica, D. T., Young, R. C., Hamilton, T. C., Rogan, A. M., and Ozols, R. F., 1984, Melphalan resistance in human ovarian cancer: Characterization of drug-resistant cell lines and potentiation of melphalan cytotoxicity by glutathione depletion, Cancer Res., **44**:5427.

Greenfield, N. J., and Pietruszko, T., 1977, Two aldehyde dehydrogenases from human liver: Isolation via affinity chromatography and characterization of the isozymes, Bioch. Bioshys. Acta, **483**:35.

Hilton, J., 1984a, Role of aldehyde dehydrogenase in cyclophosphamide-resistant L1210 leukemia, Cancer Res. **44**:5156.

Hilton, J., 1984b, Deoxyribonucleic acid crosslinking by 4-hydro-peroxycyclo-phosphamide in cyclophosphamide-sensitive and resistant L1210 cells, Bioch. Pharm., **33**:1867.
Hsu, L. C., Tani, K., Fujiyoshi, T., Kurachi, K., and Yoshida, A., 1985, Cloning of cDNAs for human aldehyde dehydrogenases 1 and 2, Prog. Natl. Acad. Sci. USA, **82**:2703.
Ikawa, M., Impraim, C. C., Wang, G., and Yoshida, A., 1983, Isolation and characterization of aldehyde dehydrogenase isozymes from usual and atypical human livers, J. Biol. Chem., **258**:6282.
Lowry, O. H., Rosebrough, N. J., Farr, A. L., Randall, R. J., 1951, Protein measurement with the Folin phenol reagent, J. Biol. Chem., **193**:265.
Lu, Y., Han, J., and Scanlon, K. J., 1988, Biochemical and molecular properties of cisplatin-resistant A2780 cells grown in folinic acid, J. Biol. Chem., **263**:4891.
Manthey, C. L., and Sladek, N. E., 1989, Aldehyde dehydrogenasecatalyzed bioinactivation of cyclophosphamide, in "Enzymology and Molecular Biology of Carbonyl Metabolism," H. Weiner, ed. , Alan R. Liss, New York, p. 49.
Michelson, A. M., Markham, A. F., and Orkin, S. H., 1983, Isolation and DNA sequence of a full-length CDNA clone for human X chromosome-encoded phosphoglycerate kinase, Proc. Natl. Acad. Sci. USA, **80**:472.
Radin, A. I., Zhoa X-L., Woo, T. H., Colvin, M., and Hilton, J., 1991, Structure and expression of the cytosolic aldehyde dehydrogenase gene in cyclophosphamide-resistant murinBe leukemia L1210 cells, Biochem. Pharm., **42**:1933.
Russo, J. E. , Hilton, J. , and Colvin, O. M. , 1989, The role of aldehyde dehydrogenase isozymes in cellular resistance to the alkylating agent cyclophosphamide, in "Enzymology and Molecular Biology of Carbonyl Metabolism," H. Weiner, ed., Alan R. Liss, New York, p. 65.
Saiki, R. K., Gelfand, D. H., Stoffel, S., Scharf, S. J., Higuchi, R., Horn, G. T., Mullis, K. B., and Erlich, H. A., 1988, Primer-directed enzymatic amplification of DNA with a thermostable DNA polymerase, Science, **239**:487.
Sambrook, J., Fritsch, E. F., and Maniatis, T., 1989, "Molecular Cloning: A Laboratory Manual," 2nd ed., Cold Spring Harbor Lab Press, New York.
Takizawa, T., Huang, I-Y., Ikuta, T., and Yoshida, A., 1986, Human glucose-6-phosphate dehydrogenase: Primary structure and CDNA cloning, Proc. Natl. Acad. Sci. USA, **83**:4157.
Tani, K., Takizawa, T., and Yoshida, A., 1985, Normal mRNA content in a phosphoglycerate kinase variant with severe enzyme deficiency, Am. J. Hum. Genet., **37**:931.
Tompkins, W. A. F., Watrach, A. M., Schmale, J. D., Schultz, R. M., and Harris, J. A., 1974, Cultural and genetic properties of newly established cell strains derived from adenocarcinomas of the human colon and rectum, J. Natl. Cancer Inst., **52**:1101.
Weber, G. , Prajda, N. , and Williams, J. C., 1975, Biochemical strategy of the cancer cell: malignant transformation-linked enzymatic imbalance, Adv. Enzyme Reg., **13**:3.
Yoshida, A., 1966, Glucose-6-phosphate dehydrogenase of human erythrocytes, J. Biol. Chem., **241**:4966.

Yoshida, A., and Watanabe, S., 1972, Human phosphoglycerate kinase I: Crystallization and characterization of normal enzyme, J. Biol. Chem., 247:440.
Yoshida, A., Hsu, L.C., and Yasunami, M., 1991, Genetics of human alcohol-metabolizing enzymes, Prog. Nucleic Acid Res. Mol. Biol., 40:255.

ATTEMPTS TO INCREASE THE EXPRESSION OF RAT LIVER MITOCHONDRIAL ALDEHYDE DEHYDROGENASE IN *E. coli* BY ALTERING THE mRNA

Chao-Feng Zheng and Henry Weiner

Department of Biochemistry
Purdue University,
West Lafayette, IN 47907

INTRODUCTION

We have recently cloned and expressed the cDNA coding for rat and human liver mitochondrial class 2 aldehyde dehydrogenase (Farres *et al.*, 1989; Jeng and Weiner, 1991, Zheng *et al.*, submitted). The recombinantly expressed enzymes were found to have properties virtually identical to those of the native enzyme. Having an expression system has allowed us to perform site directed mutagenesis to study the role of amino acid residues such as glutamate 487, histidine 235, cysteine 302, and serine 74, to mention just a few. Though we could isolate the enzyme using conventional procedures (Jeng and Weiner, 1991) or a hydroxyacetophenone-Sepharose affinity matrix (Ghenbot and Weiner, submitted) the amount of enzyme produced was very low. Using the plasmid pT7-7 others have reported that as much as 20% of the protein expressed was from their cDNA (see Marston, 1986 for a review) while we could obtain just 1-2 mg of ALDH per liter of cultured cells using the same system. In order for us to study chemical and physical properties of the recombinantly expressed enzymes it will be necessary for us to produce much more enzyme.

The reason for the low level of expression of aldehyde dehydrogenase was not completely apparent. It was suspected that the reason for this low expression may be a result of an inhibitory effect on translation caused by the secondary structure at the 5'-end of the mRNA. Evidence accumulated from several translation systems demonstrated that secondary structure around the ribosome binding site and the initiation ATG codon has a dramatic effect on the translation efficiency in both prokaryotic and eukaryotic systems (Hartz *et al.*, 1991). We previously reported that there appeared to be a stem-loop at the 5'-end of the mRNA coding for mitochondrial ALDH and showed that the rate of translation of the message could be related to the strength of this stem loop (Guan and Weiner, 1989).

We attempted to increase the level of expression of ALDH by altering the mRNA coding for the enzyme. All alterations were based upon the assumption that a decrease in the stability of a stem-loop structure at the 5'-end would cause an increased rate of translation to occur. Two major approaches were undertaken. In one, every degenerate codon containing a G/C was replaced by an A/T while maintaining the same amino acid composition. The second approach was to delete a number of amino acids from the N-terminal end of the protein or to make chimeric proteins by fusing the cDNA coding for ALDH with the one for T-7 gene 10 protein (BH). In this report we showed that it was possible to dramatically increase the expression of ALDH though most of the expressed ALDH was found to be insoluble or in inclusion bodies.

MATERIALS AND METHODS

Reagents

All the restriction and modifying enzymes were purchased from New England Biolab, Boeringer Mannheim, Promega or BRL. Sequenase version II (U. S. Biochemical) was used for all the DNA sequencing work. AmpliTaq DNA polymerase from Perkin Elmer-Cetus was used for all PCR work. Prestained molecular weight standard for SDS-PAGE was from Bio-Rad. All other reagents were commercially available analytical reagents.

Construction Of Expression Plasmids

All the constructs used in this study, except for BH, were made by polymerase chain reaction using a C-terminal primer containing a BamHI site and a N-terminal primer containing a NdeI site which contains a methionine codon which serves as the translation initiation codon. These amplified DNA fragments were cloned into the NdeI and BamHI site of plasmid pT7-7. BH was constructed by first introducing a BamHI site in front of the serine codon (position 2) of mature ALDH and then being inserted into the BamHI site of pT7-7. This resulted in an ALDH containing 8 additional amino acids at its N-terminal end. The precursor of ALDH which contains the 19 amino acid signal peptide was also included in this study as a control. Point mutations were performed to change the base at postion 4 from the A in the initiation AUG codon.

In vivo Labeling Of Newly Synthesized Proteins

In order to measure the rate of synthesis of recombinantly expressed ALDH, the pulse-chase method was employed as described (Jeng and Weiner, 1990). Briefly, 0.2 ml of an overnight culture of *E. coli* bearing the plasmid containing different ALDH cDNAs was inoculated into 10 ml of M9 media. When the OD_{600} of the culture reached 0.5, 0.4mM IPTG was added. After 1 hr induction, rifampicin, a bacterial protein synthesis inhibitor, was then added to a final concentration of 0.2 mg/ml. Thirty minutes later, the cells were pelleted and resuspended in 1/10 volume of M9 media and 1 μl of [^{35}S]-methionine (10 mCi/ml) was added. The labeling was terminated after 5 minutes by isolating the cells by centrifugation and washing once with fresh media.

mRNA Secondary Structure Prediction

Secondary structure of mRNA was predicted by a program created by Dr. P. T. Gilham (Department of Biological Sciences, Purdue University), using a window of 144 base pairs, beginning at the transcription start nucleotide of T7 promoter. Therefore, the initiation AUG codon begins at position 81. The most stable stem-loop structure involving the ribosome binding site and/or the initiation AUG codon was chosen to compare among different constructs.

Gel Electrophoresis

SDS-PAGE and Western blotting were performed as described (Jeng and Weiner, 1990). The autoradiography was scanned using an Ultrascan XL Laser densitometer (LKB).

RESULTS AND DISCUSSION

Expression of ALDH in E. coli

All of the constructs employed expressed proteins with the predicted size, as measured by SDS-PAGE, and cross reacted with anti-ALDH antibody as shown by Western blotting as shown in Figure 1. The amounts of expressed ALDH protein from different constructs showed the following pattern: precursor4G and Δ56 > SSD, BH, Δ20 and Δ20G > precursor > mature > mature4G (Figure 1). The ALDH amounts expressed by

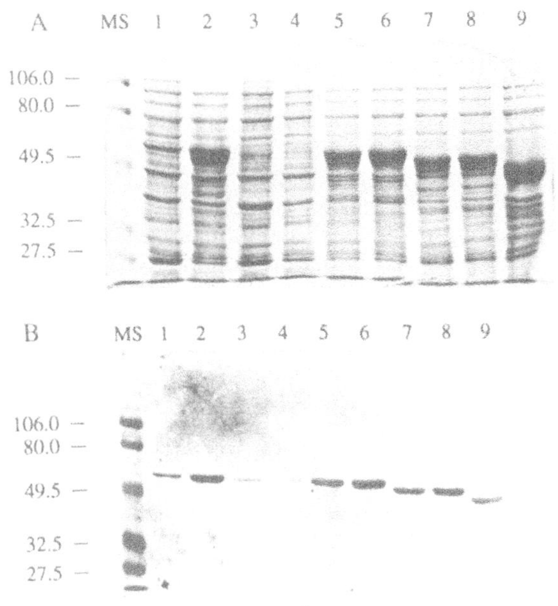

Figure 1. Expression of different constructs of ALDH cDNA in *E. coli.* cDNA encoding rat class 2 ALDH were expressed in *E. coli* cells as described in Materials and Methods. After 3 hr induction by IPTG, 100 μl cell culture was pelleted, dissolved in 1xSDS treatment buffer and loaded on a SDS-PAGE which was either stained by Commassie Brilliant Blue R-250 (Panel A) or transfered onto nitrocellulose filter paper and blotted using anti-ALDH antibody (Panel B). MS represents molecular weight standards. Numbers from 1 to 9 on the top of the gel stand for precursor, precursor4G, mature, mature4G, BH, SSD, Δ20, Δ20G and Δ56 respectively, where precursor is the mature ALDH but containing the natural leader sequence of 19 amino acids, mature is the form of the enzyme that is isolated from liver, precursor4G, mature4G and Δ20G are constructs where the base at position 4 was changed to a G, BH is a fusion protein between 8 amino acids of the T-7 gene 10 protein and mature ALDH, Δ20 and Δ56 are constructs for the mature ALDH that have the first 20 or 56 amino acids deleted, SSD is the construct for a mature enzyme but has codons changed so that the secondary structure of the mRNA could be disrupted.

Table 1. In vivo translation rate.

Constructs	Translation Rate
precursor4G	250
Δ56	250
BH	160
precursor	100
SSD	80
Δ20G	60
Δ20	55
mature	15
mature4G	1.5

The amount of ALDH protein present in the cell at different times within 5 minutes were determined by scanning the band on the autoradiograph. The rate of synthesis was calculated from the linear portion of the time curve. Unit of the values given in this table are normalized arbitrary units/minutes precursor4G and Δ56 were over 160 times greater than that by mature4G. Δ56 reacted much poorer with the antibody used for the Western blotting, possibly due to the deletion of 56 amino acids, and thus the amount of protein is underestimated by Western blotting (Figure 1, compare Panel A and Panel B).

Initial Rate Of Protein Synthesis In *E. coli*

The initial rates of synthesis were determined by labeling the newly synthesized protein with ^{35}[S]-methionine for less than 5 minutes, sampling at different times and then disrupting the cells and measuring the amount of ALDH present. The initial rate of synthesis of the various constructs varied by 160-fold. The results are presented in Table 1. Kozak (1986), among others, has shown that translation was faster if the residue at position +4 from the A in the initiation codon AUG was a purine, especially a guanidine. The nature of the base at position +4 was found not to be responsible for all the rate differences presented in Table 1. It could be noted that both the fastest and the slowest entries each had a G. Within some individual pairs of constructs a dramatic effect was observed, such as when mature is compared to mature4G. Here the rate was decreased while with precursor and precursor4G it was increased. A G could contribute to stabilizing secondary structure by hydrogen bonding to a C.

The construct used to express the mature ALDH (mature) had the lowest rate of synthesis except for mature4G. Constructs which had additional amino acids at the N-terminus, such as precursor and BH, as well as those which had fewer amino acids than mature, including Δ20 and Δ56, were expressed at rates ranging from 4- to 17-fold faster than that for mature. Since the *E. coli* cell, plasmid, promoter and ribosome binding site were identical for all these constructs it could be concluded that the 5'-end 30 or so base pairs of the mRNA must have had some inhibitory effect on the rate of expression of ALDH.

To more precisely test for the possible inhibitory effect of the stem-loop structures in the mRNA we disrupted the structure by replacing as many G/C with A/T within the 5'-end 30 base pairs. The rate of translation of this new construct (SSD), which still coded for the same amino acid sequence, was increased 5-fold compared with that of mature.

Table 2. Codon usage of the N-terminal 10 amino acids

Construct	Position 1	2	3	4	5	6	7	8	9	10
precursor	ATG	CTG	CGC	GCC	GCA	CTC	AGC	ACC	GCC	CGC
	26.2	56.7	21.4	22.5	20.9	9.3	11.3	23.8	22.5	21.4
precursor4G	ATG	GTG	CGC	GCC	GCA	CTC	AGC	ACC	GCC	CGC
	26.2	24.5	21.4	22.5	20.9	9.3	11.3	23.8	22.5	21.4
mature	ATG	TCC	GCC	GCC	GCC	ACC	AGC	GCG	GTG	CCA
	26.2	10.1	22.5	22.5	22.5	23.8	11.3	33.7	24.5	7.4
mature4G	ATG	GCC	GCC	GCC	GCC	ACC	AGC	GCG	GTG	CCA
	26.2	22.5	22.5	22.5	22.5	23.8	11.3	33.7	24.5	7.4
SSD	ATG	TCT	GCA	GCA	GCA	ACA	AGT	GCA	GTA	CCA
	26.2	11.3	20.9	20.9	20.9	5.5	6.3	20.9	13.2	7.4
BH	ATG	GCT	AGA	ATT	CGC	GCC	CGG	GGA	TCC	GCC
	26.2	19.3	1.4	26.0	21.4	22.5	3.6	5.5	10.1	22.5
Δ20	ATG	ATC	TTC	ATT	AAC	AAT	GAG	TGG	CAT	GAT
	26.2	30.2	19.1	26.0	25.1	14.8	19.0	11.1	19.0	30.8
Δ20G	ATG	GTC	TTC	ATT	AAC	AAT	GAG	TGG	CAT	GAT
	26.2	30.2	19.1	26.0	25.1	14.8	19.0	11.1	19.0	30.8
Δ56	ATG	GAG	GAC	GTA	GAC	AAG	GCA	GTG	AAG	GCC
	26.2	19.0	23.4	13.2	23.4	12.2	20.9	24.5	12.2	22.5

The codon usage (per thousand) listed in this table are from Aota *et al.*, 1988.

One possible explanation for the low rate of translation of ALDH compared to other mammalian enzymes could be related to the codon usage. It is known that some codons are rarely used in *E. coli* (Aota *et al.*, 1988). In Table 2 are listed the actual codons for the first 10 amino acids in each of the constructs. Below the codon is listed the frequency (per thousand) *E. coli* uses that particular codon for the amino acid in question. Though it was impossible to present a quantitive value for this information, it could be noted that SSD uses more rare codons than did mature, yet it was translated 6 times more rapidly. Thus codon usage is most likely not an explanation for the differences in rates of translation shown in Table 1.

Partition Of The Expressed ALDHs

More than 95% of the expressed mature ALDH was in a soluble tetrameric form. In contrast, BH, Δ20, Δ56 and SSD, which all were expressed to a greater extent than was mature ALDH, were almost exclusively found in an insoluble form or in inclusion bodies. Thus, although we could increase the rate of expression of protein, we were not able to isolate active ALDH. No systematic attempt was made to find conditions to solubilize the aggregates except for growing the cells at a lower temperature (25°C). It was unexpected to have found that whenever the rate of expression increased, the protein was found in the insoluble portion. It was recently reported by Hsu *et al.* (1992) that 85% of the recombinantly expressed human class 3 ALDH was insoluble.

The insolubility was not an innate property of ALDH. It is possible to concentrate purified ALDH protein up to 5-10 mg/ml. Thus the formation of insoluble ALDH was not due to its insolubility but must have been related to the rate of folding compared to its rate of aggregation in *E. coli*. The formation of inclusion body is not considered to be a random process. It has been reported to be due to one or more off-pathway steps during protein folding and/or association process which involves specific folding intermediate(s) (Mitraki and King, 1989). Therefore, it could be rationalized that there is a very slow step in the folding and/or subunit assembly pathway for ALDH. If the synthesis rate was too high, the folding/assembly process may not be able to keep pace with it and hence result in formation of aggregates instead of correctly folded ALDH.

Expression Level, Translation and Secondary Structure

The plasmid, promoter, ribosome binding site, initiation codon and the spacing among these possible transcription and translation regulatory elements for all the constructs used in this study were the same. Therefore the initial protein synthesis rate should be able to reflect the translational rate of the individual mRNA. It has been shown with many genes, such as T4 lysozyme, lamB (Hall *et al.*, 1982), Tn10 (Davis *et al.*, 1985), erythromycin-resistance (Horinouchi and Weisblum, 1980) and lacZ (Munson *et al.*, 1984), to mention a few, that secondary structure around the ribosome binding site and the initiation codon in mRNA has a profound effect on the rate of translation. Secondary structure predictions (Table 3) showed that there is a strong stem-loop structure at the 5'-end of the mature part of ALDH with a free energy of -13.8 kcal/mol and mature4G has one of -18.6 kcal/mol. Consistent with this calculation was the observation that mature4G had a 10-fold lower rate of expression compared to mature. If we disrupt this secondary structure by changing all possible G/C into A/T within 30bp area downstream the initiation AUG (SSD), the synthesis or the translation rate increased by 5 times (Table 1) while the calculated free energy decreased to -8.7 kcal. These results demonstrated that this secondary structure following the initiation codon in the mature part of ALDH mRNA resulted in the poor expression of mature ALDH in *E. coli*. Consistent with this conclusion, all the other higher expression constructs have less stable secondary structures in this area compared to mature, with Δ56 having the least stable structure and the highest expression. These conclusions supported what we previously suggested to explain the different rates of translation for rat and beef mRNA (Guan and Weiner, 1989). It has been reported that the expression of human IFN-gamma (Tessier *et al.*, 1984) and immunoglobulin heavy chain (Wood *et al.*, 1984) in *E. coli* could be increased over 50-and 90-fold, respectively, by abolishing the secondary structure at the 5'-end of their mRNAs. We did not find a linear relationship between the rates of expression and the stability of the stem-loop structure. Futhermore, we could not offer any explanation for the very high rate of expression of precursor4G which appears to have a relatively stable stem-loop structure.

Table 3. Predicted secondary structures involving the ribosome binding sequences and the initiation AUG codon.

Constructs	Predicted stability(-ΔG)	Loop size	Stem size	Start position
precursor4G	10.9	13	8	83
Δ56	<1.0			
BH	6.3	10	3	83
precursor	8.7	5	4	82
SSD	8.7	11	5	85
Δ20G	7.2	49	5	87#
Δ20	6.7	9	6	69,84*
mature	13.8	5	9	86
mature4G	18.6	5	13	82

Refer to Materials and Methods for details. The stability values are -ΔG/mol at 37°C. *This stem-loop is in front of the initiation codon and ends at position 84. # This structure has too large a loop and possibly would not actually form or will be very unstable.

CONCLUSIONS

It can be concluded from these results that the poor expression of mature ALDH in *E. coli* was due to the low translation rate caused by the presence of a stable stem-loop structure following the initiation AUG. Disruption of this structure will increase the expression level while stablizing it will decrease the expression level. On the other hand, codon usage, base composition and the nucleotide right behind initiation AUG codon have little effect on the translation efficiency in *E. coli* cell. Unfortunately, higher expression of ALDH resulted in the formation of protein aggregates instead of soluble tetrameric ALDH.

ACKNOWLEDGEMENTS

This work was supported in part by a grant from the National Institute of Alcohol Abuse and Alcoholism #AA05812. H.W. is a recipient of a Senior Scientist Award #AA00028 from the same Institute. mRNA secondary structure prediction was performed on the computer system of the AIDS Center for Computational Biochemistry, Purdue University, which is supported by grant A127713 from NIH. We thank Prof. P. T. Gilham for his help in determining the structures of the mRNAs.

REFERENCES

Aota, S., Gojobori, T., Ishibashi, F., Maruyama, T., and Ikemura, T., 1988, Codon usage tabulated from the GenBank genetic sequence data, *Nucl. Acids Res.* 16:r315-r402.

Davis, M. A., Simons, R. W., and Kleckner N., 1985, Tn10 protects itself at two levels from fortuitous activation by external promoters, *Cell.* 43:379-387.

Farres, J., Guan, K. G., and Weiner, H., 1989, Primary structure of rat and bovine liver mitochondrial aldehyde dehydrogenases deduced from cDNA sequences, *Eur. J. Biochem.* 180: 67-74.

Ghenbot, G., and Weiner, H., Purification of liver aldehyde dehydrogenase by p-hydroxyacetophenol-Sepharose affinity chromatography and the co-elution of chloramphenicol acetyl transferase, (Submitted).

Guan, K. G., and Weiner, H., 1989, Influence of the 5'-end region of aldehyde dehydrogenase mRNA on translational efficiency. Potential secondary structure inhibition of translation *in vitro*, *J. Biol. Chem.* 264:17764-17769.

Hall, M. N., Gabay, J., Debarbouille, M., and Schwartz, M., 1982, A role for messenger-RNA secondary structure in the control of translation initiation, *Nature* (London). 295:616-618.

Hartz, D., McPheeter, D. S., and Gold, L., 1991, Influence of mRNA determinants on translation in *Escherichia coli*, *J. Mol. Biol.* 218:83-97.

Horinouchi, S., and Weisblum, B., 1980, Post translational modification of mRNA information: mechanism that regulates erythromycin-induced resistance, *Proc. Natl. Acad. Sci., U.S.A.* 77:7079-7083.

Hsu, L. C., Chang, W. C., Shibuya, A., and Yoshida, A., 1992, Human stomach aldehyde dehydrogenase cDNA and genomic cloning, primary structure, and expression in *Escherichia coli*, *J. Biol. Chem.*, 287:3030-3037.

Jeng, J., and Weiner, H., 1991, Purification and characterization of catalytically active precursor of rat liver mitochondrial aldehyde dehydrogenase expressed in *Escherichia coli*, *Arch. Biochem. Biophys.* 289:214-222.

Kozak, M., 1986, Point mutations define a sequence flanking the AUG initiator codon that modulates translation by eukaryotic ribosomes, *Cell.* 44:283-292.

Marston, F. A. O., 1986, The purification of eukaryotic polypeptides synthesized in *Escherichia coli*, *Biochem. J.* 240:1-2.

Mitraki, A., and King, J., 1989, Protein folding intermediates and inclusion body formation, *Bio/Technology*. 7:690-697.

Munson, L. M., Stormo, G. D., Niece, R. L., and Reznikoff, W. S., 1984, LacZ translation initiation mutations, *J. Mol. Biol.* 177:663-684.

Tessier, L. H., Sondermeyer, P., Faure, T., Dreyer, D., Benavente, A., Villeral, D., Courtney, M., and Lecocq, J. P., 1984, The influence of mRNA primary and secondary structure on human IFN-gene expression in *E. coli*, *Nucl. Acid Res.* 12:7663-7675.

Wood, C. R., Boss, M. A., Patel, T. P., and Emtage, J. S., 1984, The influence of messenger RNA secondary structure on expression of an immunoglobulin heavy chain in *Escherichia coli*, *Nucl. Acid Res.* 12:3791-3937.

Zheng, C. F., Wang, T. T. Y., and Weiner, H., Cloning and expression of the full length cDNAs encoding human liver class 1 and class 2 Aldehyde dehydrogenase, submitted.

PRELIMINARY CHARACTERIZATION OF THE RAT CLASS 3 ALDEHYDE DEHYDROGENASE GENE

David C. Asman[1], Koichi Takimoto[2], Henry C. Pitot[2] and Ronald Lindahl[1]

[1]The University of South Dakota School of Medicine
Department of Biochemistry and Molecular Biology
414 E. Clark Vermillion, SD 57069-2390
[2]McArdle Laboratory for Cancer Research, Department of Oncology and Pathology, University of Wisconsin Medical School, 1400 University Ave. Madison WI 53706

INTRODUCTION

Class 3 aldehyde dehydrogenase (ALDH) is unique among the three ALDH classes because its expression is either constitutive or inducible depending on the tissue. Class 3 ALDH is induced in liver and urinary bladder during tumorigenesis (Lindahl, 1992). Expression of class 3 ALDH in tumor cells is due to initiator-induced, stable, genetic changes which occur within a subpopulation of initiated cells (Jones et al., 1984). This phenomenon occurs early in tumorigenesis when the normally repressed class 3 ALDH gene becomes activated. Class 3 ALDH is constitutively expressed in the cornea (Evces and Lindahl, 1989), stomach (Yin et al., 1989), and lung (Dunn, et al., 1988). At least for the stomach and lung, the activity is also inducible, as activity increases after exposure to inducing agents.

Class 3 ALDH can also be induced in normal rat liver by treatment with 2,3,7,8 - tetrachlordibenzo-p-dioxin (TCDD) or other aromatic hydrocarbons such as 3 methylcholanthrene (3MC) (Deitrich et al., 1977; Torronen et al., 1981; Marselos et al., 1987). Presently, the exact mechanism for the modulation of the class 3 ALDH gene by TCDD and PAHs is unknown. However, it has been postulated that xenobiotic induction of class 3 ALDH may occur via an Ah-receptor mediated pathway, through binding of the Ah ligand to the cytoplasmic Ah receptor, followed by translocation of the ligand-receptor complex from the cytosol to the nucleus (Nebert, et al., 1990; Lindahl, 1992). However, the inducibility of the ALDH3 gene by TCDD does not correlate well with the tissue distribution of the Ah receptor (Carlstedt-Duke, 1979), and the time course of TCDD induction of the class 3 ALDH gene in liver is distinctly different from that of other TCDD-inducible genes (Dunn, et al., 1988). For example, for TCDD-inducible genes such as *Nmo-1* and *CYP1A1*, transcription rates are maximal within the first 12 hours following TCDD exposure. In contrast, transcription of the ALDH3 gene does not begin until 24 to 48 hours following TCDD exposure and is not maximal until several days later. Therefore, control of class 3 ALDH gene regulation must include both Ah-receptor mediated and Ah-receptor independent mechanisms.

A full-length cDNA encoding a catalytically functional class 3 ALDH derived from rat hepatoma cell line HTC, has been cloned into pUC8 and completely characterized (pTALDH; Jones et al., 1988). The class 3 ALDH cDNA contains 173 nucleotides of 5' non-coding sequence, 1359 bp of coding region and 248 bp of 3'

Enzymology and Molecular Biology of Carbonyl Metabolism 4
Edited by H. Weiner, Plenum Press, New York, 1993

non-coding sequence including the polyadenylation signal. Interestingly, in the absence of any manipulation of this cDNA, *E. coli* transformed with pTALDH produce class 3 ALDH protein which is enzymatically active.

Rat class 3 ALDH has a cloned human homologue in the form of the human stomach ALDH-3 gene (Hsu et al., 1992). The degree of identity between the coding regions of rat ALDH-3 and human stomach ALDH-3 has been shown to be 81% at both the nucleotide and amino acid levels. This is similar to homologies seen between different class 1 or class 2 ALDH genes and their products from various species. The nucleotide sequence 5' to the putative human class 3 ALDH coding region has been examined for possible regulatory elements, and two putative transcriptional control elements were identified; a TATA box between positions -79 and -76 and a CCAAT box located between positions -117 and -114. No other regulatory elements were observed in this 5' regulatory region (Hsu et al., 1992).

Recently, a partial class 3 ALDH genomic clone (NL2) was isolated from normal rat liver and characterized. NL2 contains all of the class 3 ALDH coding exons. However, this clone lacks all of the associated 5' regulatory sequences.

At this time, we report the cloning and partial characterization of the 5' regulatory region of the rat class 3 ALDH gene. Southern analysis and sequence data indicate that striking structural differences exist between the human and rat ALDH-3 genes.

EXPERIMENTAL APPROACH

A lambda Dash rat genomic library was screened for the pTALDH untranslated region (UTR) and its associated 5' flanking domain. A three tiered screening approach utilizing three different DNA probes was utilized (Figure 1). The probes were as follows: Probe one was a SacI-BamHI fragment of approximately 1500 bp from the most 5' extreme region of the previously isolated partial ALDH3 genomic clone NL2. This probe contained the ALDH-3 exon 2 which includes the translational start site and seven bp of the 173 bp upstream sequence. In addition, probe one contained approximately 800 bp of DNA subsequently identified as intron one. Probe two was an 800 bp EcoRI-BglII fragment from the 5' end of the cDNA pTALDH. This probe was used to screen the clones selected in the first screening and to identify them as ALDH clones. Probe three was an 140 bp EcoRI-PvuII fragment from the untranslated region of pTALDH. This DNA was not contained within the genomic clone NL2. Probe three was used in all tertiary screens to select for only those clones which contained the missing regulatory region. Of eighteen clones selected by probe two, three showed positive hybridization with probe three.

The three ALDH3 gene clones selected were initially digested with BamHI and EcoRI, and analyzed by Southern blotting. One clone, ALDH-UTR-1, contained the largest insert (approximately 16kb) and demonstrated the strongest hybridization. This clone was further characterized via restriction digestion and southern analysis.

RESULTS AND DISCUSSION

Using appropriate oligonucleotides as probes, restriction fragments containing the DNA corresponding to the pTALDH 5' UTR and the translational start site were identified. Oligonucleotide 1 corresponded to positions -38 through -60 in the pTALDH cDNA UTR, and oligonucleotide 2 corresponded to the pTLADH translational start site position +1 through +18. Southern hybridization on HindIII-digested ALDH-UTR-1 indicated that the translational start site (oligonucleotide 2) was in a single 4kb HindIII fragment whereas the UTR (oligonucleotide 1) resided within a 5kb HindIII fragment (Figure 2). Both HindIII fragments were gel-isolated and subcloned into pUC118 for confirmation by DNA sequencing.

Sequence analysis of these two fragments confirmed that the translational start site plus 5 bp of the UTR, and the remaining UTR sequence are encoded by

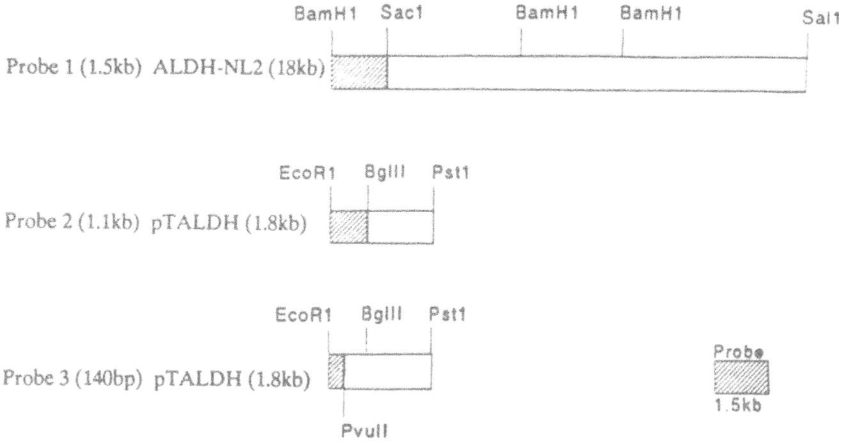

Figure 1. Cloning strategy for the isolation of the 5' flanking region of the rat class 3 ALDH gene.
Probe 1 - 1.5kb BamH1-Sac1 fragment from the extreme 5' end of genomic ALDH clone NL2.
Probe 2 - 1.1kb EcoR1-BglII fragment from cDNA pTALDH.
Probe 3 - 140bp EcoR1-PvuII fragment from pTALDH.

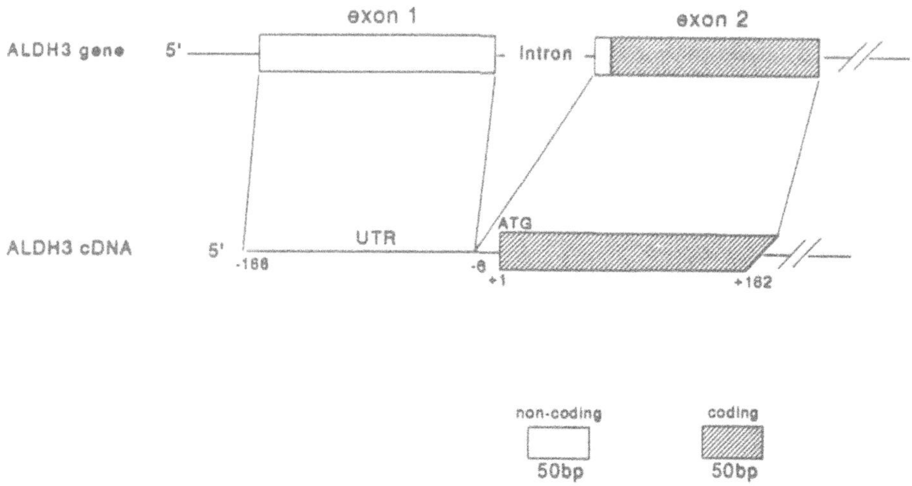

Figure 2. Exons 1 and 2 of the rat class 3 ALDH gene. Exon one consists of the first 160 bp of pTALDH UTR and is entirely non-coding. Exon 2 consists of 5 bp of UTR adjacent to coding sequence for the first 54 amino acids of class 3 ALDH.

two distinct exons separated by an intron of over 2 KB (Figure 2). An additional 500 bp of class 3 ALDH sequence upstream of the UTR has been sequenced. A DNA analysis program was used for open reading frame (ORF) analysis, and no ORFs of any significant length were located in the sequence 5' to exon one, or in the intron residing between exon one and exon two.

Exon one is 160 bp in length, and contains nucleotides corresponding to -166 through -6 from the 173bp pTALDH UTR. We know from primer extension experiments from isolated rat liver RNA that the majority of class 3 ALDH transcripts start at approximately -80 (Asman et al., unpublished). Perhaps the cDNA, pTALDH having been derived from a hepatoma cell line, represents a rare extended transcript.

Exon two is 166 bp long and consists of pTALDH UTR nucleotides 169 through 173 along with coding sequence to 335, and includes the translation start site at position 174 (Figure 2). Therefore, exon one is a non-peptide coding exon and exon two encodes the first 54 codons of class 3 ALDH mRNA. In addition, the intron/exon junctions for exon one and two conform to the accepted consensus sequences specific for intron donor and acceptor splice signals (Breathnach and Chambon, 1981). Structurally, the region 5' to exon one contains several putative regulatory elements (Figure 3). A CCATT motif (GGCCCATCT) is present upstream of exon one at position -91. In addition, at position -283 through -269, resides a putative xenobiotic receptor element (XRE) which shares a very close homology to those XREs described by Fujisawa-Sehara et al., (1987) residing upstream of the mouse *CYP1A1* gene. The core sequence of the putative XRE consists of a short alternate purine/ pyrimidine stretch 5'-CACGC(A)-3' shows exact homology to that of the *CYP1A1* gene.

```
agg ctg tga tct gta cct cag tgg acc ctt gcc cct gtg cat gtg

gca gtg cac gaa tgg gat gtg att ttt ctg cct ttt ctg att gcc

gcg ccg cCT GCG TGA ctg cag ctT GcC cga tgt aga cct tac ata
         XRE
aca ggc agg act cac aca aca gca gct cgt aaa aag cag gaa atg

tgt gag ctt aca tgc caa gct gaa tta ctc agt ggc ccc cac acc

cag ggc aaa cac agg ctt ctt ttg agg gct ggc taa gag ccc aca

caa act ttc aca GGC CCA TCT tgc att tca gaa cag aac aaa aca
                CCATT
gac agc cca agc cgc aat ctt caa aac ctg tgt gaa gca gag gtc

cgg att aTG CTG CAC CCA TAT TTG AAT ATT GTT TTA TCC TCA TCA
        Exon one
```

Figure 3: 5' organization of the rat ALDH3 gene showing positional relationships of putative regulatory elements to exon one.

The organization of the 5' region of the rat class 3 ALDH gene is substantially different from the human class 3 ALDH gene reported by Hsu et al., (1992). Although the rat and human ALDH-3 share a high degree of similarity in their coding regions, these two homotopic isozymes appear to be very different at the level of gene organization.

In the human Class 3 ALDH gene the UTR is reported to be contiguous with the translational start codon in a single first exon. This is clearly not the case in rat, as an intron of approximately 2-3 kb resides between the bulk of the UTR (exon one) and the translational start site (exon two). Additionally, the human gene has clearly defined CAATT (-117 through -114) and TATA (-79 through -76) regulatory sequences, whereas these sequences are either absent in the rat or much less well defined. Finally, based upon the published sequence, the 5' flanking region of the human gene lacks the XRE regulatory sequence found in the 5' flanking region of the rat ALDH-3 gene.

For the rat ALDH-3 gene, the presence of an XRE confers Ah receptor-mediated inducibility, and its presence is consistent with proposed models for class 3 ALDH gene regulation (Nebert et al., 1990; Lindahl, 1992). In addition, the single XRE observed is compatible with the organization of two related TCDD-inducible *genes Nmo-1* and *Gst-1* (Nebert et al., 1990).

Based on the above information concerning the 5' regulatory region of the rat class 3 ALDH gene, and in the absence of additional sequence information for rat class 3 ALDH genes expressed constitutively, it is possible to speculate that two different mechanisms of transcriptional control exist for inducible vs constitutive expression of the ALDH-3 gene. Perhaps in inducible tissues, ie. liver, RNA polymerases are utilizing the controlling elements (XREs, sp1, CAAT) present upstream of exon 1 for Ah-mediated transcriptional control of the ALDH-3 gene. In constitutively expressing tissues, ie. stomach and cornea, perhaps a second set of controlling elements (CAAT and TATA) present 5' to exon two are used to initiate transcription. This is consistent with both the data for the human class 3 ALDH organization and the fact that the controlling elements in this region 5' to exon two are similar to those found 5' to housekeeping genes.

ACKNOWLEDGEMENT: Supported by NIH Grant CA21103 to R.L.

REFERENCES

Breathnach, R., and Chambon, P., 1981, Organization and expression of eucaryotic split genes coding for proteins, Annu. Rev. Biochem. 50:349.

Carlstedt-Duke, J.M.B., 1979, Tissue distribution of the receptor for 2,3,7,8-tetrachlorodibenzo-p-dioxin in the rat, Cancer Res. 38:3172.

Deitrich, R.A., Bludeau, P., Stock, T., and Roper, M., 1977, Induction of different supernatant aldehyde dehydrogenase by phenobarbital and tetrachlorodibenzo-p-dioxin, J. Biol. Chem. 252:6169.

Dunn, T.J., Lindahl, R., and Pitot, H.C., 1988, Differential gene expression in response to 2,3,7,8-tetrachlorodibenzo-p-dioxin (TCDD). Noncoordinate regulation of a TCDD-inducible aldehyde dehydrogenase and cytochrome P-450c in the rat, J. Biol. Chem. 263:10878.

Evces, S. and Lindahl, R., 1989, Characterization of rat cornea aldehyde dehydrogenase, Arch. Biochem. Biophys. 274:518.

Fujisawa-Sehara, A., Sogawa, K., Yamane, M., and Fujii-Kuriyama, Y., 1987, Characterization of xenobiotic responsive elements upstream from the drug-metabolizing cytrochrome P450c gene: a similarity to glucocorticoid regulatory elements, Nucleic Acids Res. 15:4179.

Hsu, L.C., Chang, W-C., Shibuya, A., Yoshida, A., 1992, Human stomach aldehyde dehydrogenase cDNA and genomic cloning, primary structure, and expression in *Escherichia coli*, J. Biol. Chem. 267:3030.

Jones, D.E., Brennan, M.D., Hempel, J., and Lindahl, R., 1988, Cloning and complete nucleotide sequence of a full-length cDNA encoding a catalytically functional tumor-associated aldehyde dehydrogenase, Proc. Natl. Acad. Sci. U.S.A. 85:1782.

Jones, D.E., Evces., and Lindahl, R., 1984, Expression of tumor-associated aldehyde dehydrogenase during rat hepatocarcinogenesis using the resistant hepatocyte model, Carcinogenesis 5:1679.

Lindahl, R., 1992, Aldehyde dehydrogenases and their role in carcinogenesis, Crit. Rev. Biochem. Molec. Biol. 27:in press.

Marselos, M., Strom, S.C., and Michalopoulus, G., 1987, Effect of phenobarbital and 3-methylcholanthrene on aldehyde dehydrogenase activity in cultures of HepG2 cells and normal human hepatocytes, Chem. Biol. Interactions 62:75.

Nebert, D.W., Petersen, D.D., and Fornace, A.J., 1990, Cellular responses to oxidative stress: the [*Ah*] gene battery as a paradigm, Environ. Health Perspect 88:13.

Torronen, R., Nousianinen, U., and Hanninen,O., 1981, Induction of aldehyde dehydrogenase by polycyclic aromatic hydrocarbons in rats, Chem. Biol. Interactions 36:33.

Yin, S.-J., Liao, C.-S, Wu, C.-W., Cheng, T.-C., and Yin, S.-J., 1989, Kinetic evidence for human liver and stomach aldehyde dehydrogenase-3 representing a unique class of isozymes, Biochem. Genet. 27:321.

HUMAN HIGH-K_m ALDEHYDE DEHYDROGENASE (ALDH3): MOLECULAR, KINETIC AND STRUCTURAL FEATURES

Shih-Jiun Yin,[1] Sung-Ling Wang,[1] Chin-Shya Liao,[1] and Hans Jörnvall[2]

[1]Department of Biochemistry
National Defense Medical Center
Taipei, Taiwan, Republic of China

[2]Department of Chemistry I
Karolinska Institutet
Stockholm, Sweden

INTRODUCTION

Human aldehyde dehydrogenases (ALDH), catalyzing irreversible oxidation of various aliphatic and aromatic aldehydes to the corresponding carboxylic acids, constitute a complex enzyme family (Pietruszko, 1983; Hempel and Jörnvall, 1989; Hempel and Lindahl, 1989; Yoshida et al., 1991). There are at least five ALDHs that have been purified and characterized from human liver or stomach: ALDH1 and ALDH2 (E1 and E2; Greenfield and Pietruszko, 1977), ALDH3 (Yin et al., 1989; 1991), ALDH4 (E4, glutamic γ-semialdehyde dehydrogenase; Forte-McRobbie and Pietruszko, 1986), and γ-aminobutyraldehyde dehydogenase (E3, Kurys et al., 1989). On the basis of Michaelis constants for acetaldehyde, human ALDHs can be divided into high-K_m (mM range) and low-K_m (μM range) groups. The low K_m forms comprise ALDH1 (Km = 30 μM), ALDH2 (3 μM), and γ-aminobutyraldehyde dehydrogenase (50 μM) (Greenfield and Pietruszko, 1977; Kurys et al., 1989). Saliva ALDHs (106 μM; Harada et al., 1989) and brain-specific ALDH2 (<1 μM; Ryzlak and Pietruszko, 1987) also appear to belong to this low K_m group. Human ALDH3 (83 mM) and ALDH4 (5 mM) (Yin et al., 1989; Forte-McRobbie and Pietruszko, 1986), and rat microsomal ALDH (2.4 mM) (Lindahl and Evces, 1984a) are the high K_m forms. The sequence homologies at the amino acid and/or nucleotide level between the low K_m ALDH1 and ALDH2 (Hempel et al., 1984a, 1985; Hsu et al., 1985) or between the high K_m human ALDH3 and rat microsomal ALDH (Yin et al., 1991; Hsu et al., 1992; Miyauchi et al., 1991) are 65-70%. However, the degree of positional identity between the high-K_m and low-K_m group enzymes is only 24-30%.

Enzymology and Molecular Biology of Carbonyl Metabolism 4
Edited by H. Weiner, Plenum Press, New York, 1993

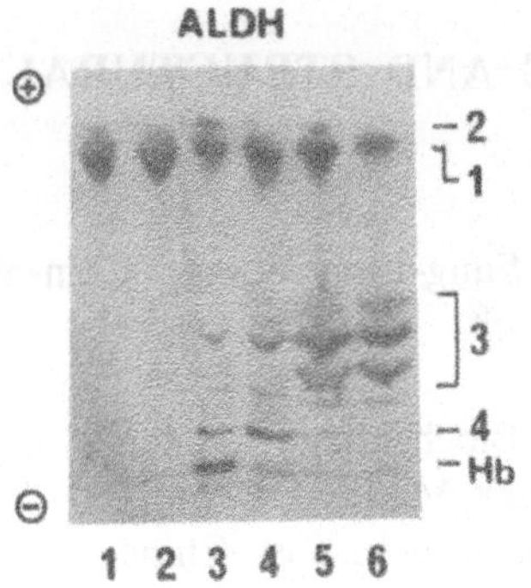

Figure 1. Distribution of ALDHs in liver, stomach, and duodenum of Chinese subjects. The agarose isoelectric focusing gels were stained for enzyme activity. Samples 1 and 2 are endoscopic biopsies from the first (bulb) and second portion of the duodenum from the same individual. Samples 3, 4 and samples 5, 6 are autopsy livers and surgical stomach biopsies, respectively. Hb, hemoglobin.

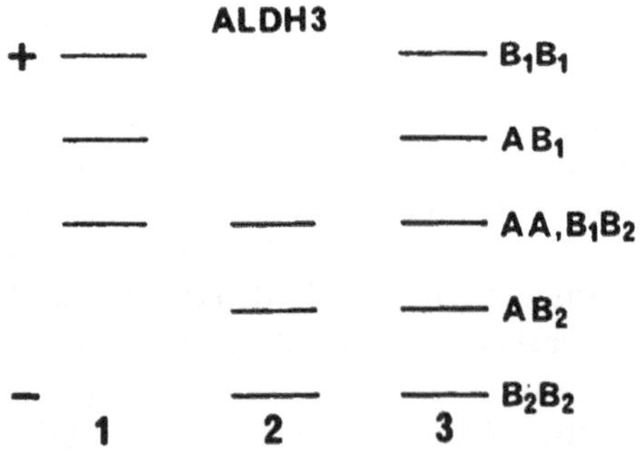

Figure 2. Scheme of the banding patterns and the assigned subunit compositions of ALDH3 isozymes after agarose isoelectric focusing.

Recently, a new human ALDH gene, designated $ALDH_x$, has been cloned and exhibits 65-72% resemblance to ALDH1/2 at the deduced amino acid sequence (Hsu and Chang, 1991), suggesting it might be a low K_m form.

Approximately 50% of Orientals lack ALDH2 activity, which can be attributed to a single nucleotide exchange resulting in a glutamic acid to lysine substitution at the 14th position from the carboxyl terminus (Yoshida et al., 1984; Hempel et al., 1984b). This deficiency has been implicated in the alcohol-flush reaction and alcoholic liver disease and has been considered as a negative biological risk factor for alcoholism (Mizoi et al., 1979; Enomoto et al., 1991; Thomasson et al., 1991). Since the K_m value for acetaldehyde for human ALDH3 is very high, it does not contribute to ethanol metabolism in vivo. This high K_m enzyme has been proposed to be involved in the detoxification of the aldehyde products of lipid peroxidation (Algar and Holmes, 1989; Evces and Lindahl, 1989; Lindahl and Petersen, 1991). In this paper, we report recent advances about the tissue distribution, genetic, kinetic and structural studies of human ALDH3.

TISSUE DISTRIBUTION

The distribution of ALDH1 and ALDH2 appears to be ubiquitous in human tissues, with the highest activity in liver, except that ALDH2 is not present in erythrocytes (Harada et al., 1980; Duley et al., 1985; Santisteban et al., 1985). ALDH4 occurs in a limited number of tissues and is most abundant in liver and kidney (Harada et al., 1980; Duley et al., 1985; Santisteban et al., 1985). γ-Aminobutyraldehyde dehydrogenase has been identified and purified from liver (Kurys et al., 1989) and its tissue distribution remains to be elucidated.

Human ALDH3 is found in liver, stomach, esophagus, gingiva, lung, cornea, and lens with high activity in stomach, esophagus and cornea (Harada et al., 1980; Yin et al., 1988, 1992a,b; Holmes, 1988). The expression of ALDH3 is low in liver and lens and is often comparatively low also in lung, as detected in 17 of 39 surgical lung specimens studied (Yin et al., 1992b). It is notable that ALDH3 is abundantly present in the upper digestive tract, i.e., gingiva, esophagus and stomach, but absent in the lower digestive tract, i.e. duodenum, jejunum and colon (Fig.1). ALDH3 also exhibits regional distribution - it is visible in the esophageal mucosal layer but not in the muscular layer (Yin et al., 1992a). A considerably increased expression of ALDH3 was found in human hepatocellular carcinoma (Agarwal et al., 1989). Similar observations have been reported for rat liver and bladder carcinomas (Lindahl and Evces, 1987; Campbell et al., 1989).

MOLECULAR MODEL

Genetic polymorphism at the stomach ALDH3 locus in a Chinese population was first reported by Teng (1981) and later confirmed by others for Caucasian populations (Hopkinson et al., 1985; Duley et al., 1985). A proposed single-gene model was essentially based on the triple-banded observations using starch gel electrophoresis. Among 71 Chinese autopsy stomach samples examined (Teng, 1981), most showed only one ALDH3 activity band, designated phenotype $ALDH_3$ 1-1; the remaining four samples with multiple bands were heterozygous $ALDH_3$ 2-1. This gastric variant three-banded pattern of ALDH3 in autopsy stomach was also seen at a frequency of about 1 in 25 in European populations (Hopkinson et al., 1985). However, Meier-

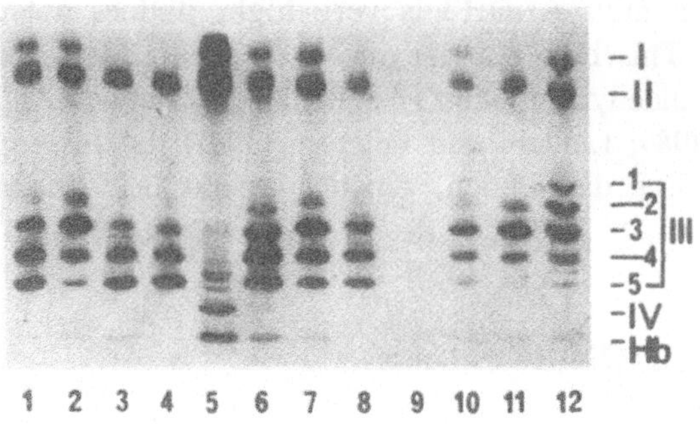

Figure 3. Agarose isoelectric focusing of ALDHs from human gastroendoscopic biopsies. Sample 5 is an autopsy liver. Sample 6 is a surgical gastric mucosa amd remaining samples are endoscopoc biopsies. Sample 9 is invisible due to low activity. I ,II,III,IV represent ALDH 2,1,3,4, respectively. Hb, hemoglobin. The gel figure is from Yin et al. (1988) and used with permission.

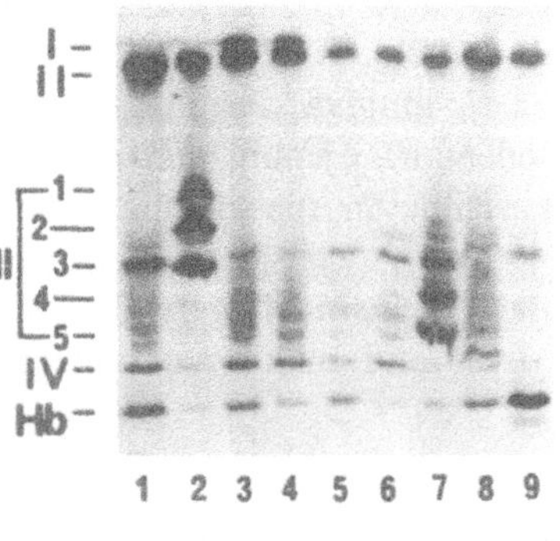

Figure 4. Comparison of agarose isoelectric focusing of ALDHs from surgical samples of gastric mucosa (lanes 2 and 7) and postmortem (lanes 1, 3-6) and surgical (lanes 8 and 9) samples of liver. I ,II,III,IV represent ALDH2,1,3,4, respectively. Hb, hemoglobin. Gel figure from Yin et al. (1988), with permission.

Tackmann et al. (1984) argued that the multiplicity of stomach ALDH3 forms was due to post-translational modification rather than to genetic variation.

Later, Yin et al. (1988) used gastroendoscopic biopsies and an agarose isoelectric focusing procedure which resolved ALDH3 into five activity bands. Based on the banding patterns and a dimeric molecule, they proposed a two-gene model with polymorphism at one of the genes. $ALDH_{3a}$ and $ALDH_{3b}$ would then code for A and B polypeptide chains, respectively, and the $ALDH_{3b}$ locus would exhibit polymorphism, with B_1 subunits encoded by the allele $ALDH^1_{3b}$ and B_2 by the allele $ALDH^2_{3b}$. This scheme for the gastric ALDH3 activity band patterns and the assigned subunit compositions are shown in Fig. 2. This genetic model was inferred from the isoelectric focusing data of 55 gastroendoscopic biopsies and the correlations with the enzyme pattern for liver (Figs. 3 and 4). Two gastric biopsy samples exhibited three-banded pattern 1 (Fig. 2, lane 1), 42 three-banded pattern 2 (Fig. 2, lane 2), and 11 five -banded pattern 3 (Fig. 2, lane 3). In contrast to the single activity band found in autopsy stomach (Teng, 1981; Hopkinson et al., 1985), all the surgical and endoscopic gastric biopsies exhibited three to five bands. The liver specimens showed only a single band of ALDH3 (Fig. 4), which was designated a homodimer AA. Furthermore, surgical esophagus and gingiva displayed predominantly a single band with differnt electrophoretic mobility, which was designated homodimer BB (Fig. 5). Thus, AA and BB forms appeared to be expressed in liver and esophagus/gingiva, respectively. However, both AA and BB and the heterodimer AB was found in stomach. Although minor bands anodic to the ALDH3 isozymes were found in some stomach specimens after storage at -70°C for 1 year or at 30°C for 1 hr, the major band patterns remained unchanged (Yin et al., 1988). These results appear not to support the arguments of Meier-Tackmann et al. (1984) that the multiple ALDH3 forms were due to secondary isozyme formation such as proteolytic degradation or deamidation.

Several lines of biochemical evidence are consistent with the proposed subunit composition for the banding patterns of stomach ALDH3 (Fig. 2). Activity bands 1-5 of ALDH3 are separated with equal pI interval on isoelectric focusing gels (Figs. 3 and 4). The proportions of the staining density of activity bands appear to be compatible with a binomial distribution of three dimers containg two different subunits. The banding patterns of ALDH3 remain unchanged in diseases such as gastritis and gastric ulcer or in the different areas of the stomach (Fig. 6). It should be pointed out that these different banding patterns can either result from genetic variations or from post-translational or epigenetic mechanisms, such as alternative splicing. To clarify this issue, further isolation of all gastric ALDH3 forms, identification of the subunit composition of these forms by dissociation-recombination experiments, determination of their primary structures as well as the gene structures of various ALDH3 subunits, and pedigree study of the inheritance of ALDH3 isozymes or genes, are necessary. Recently, a partial amino acid sequence and the cDNA sequence of stomach ALDH3 has been determined (Yin et al., 1991; Hsu et al., 1992), indicating that ALDH3 is a unique class of ALDHs. It is noteworthy that equivalents of the proposed $ALDH_{3a}$ and $ALDH_{3b}$ forms could not be distinguished in the genomic DNA by Southern analysis using seven different retriction enzymes and a common ALDH3 cDNA probe (Hsu et al., 1992). Therefore, if two genes occur, as has been proposed (Yin et al., 1988), the degree of nucleotide sequence identity between them must be very high. Since subunits A and B can form a heterodimer (Fig. 2), and by analogy to human class I α-, β-, and γ-ADH subunits which can also form heterodimers (Yoshida et al., 1991), we estimate that the sequence homology between coding and possibly noncoding regions of the two separate structural genes $ALDH_{3a}$ and $ALDH_{3b}$ should be greater than 90% and the difference between the

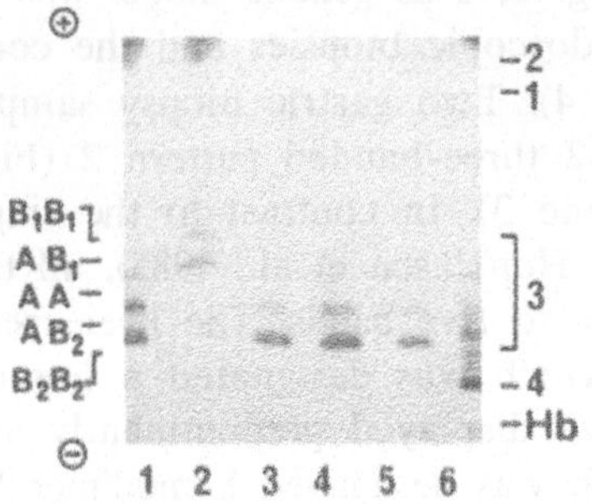

Figure 5. Agarose isoelectric focusing of ALDH3 isozymes from human esophageal mucosa and gingiva. Lanes 3 and 4 are surgical samples of esophageal mucosa and lane 5, surgical sample of gingiva. For comparison, lanes 1, 2 represent surgical samples of gastric mucosa and sample 6, autopsy liver. Hb, hemoglobin.

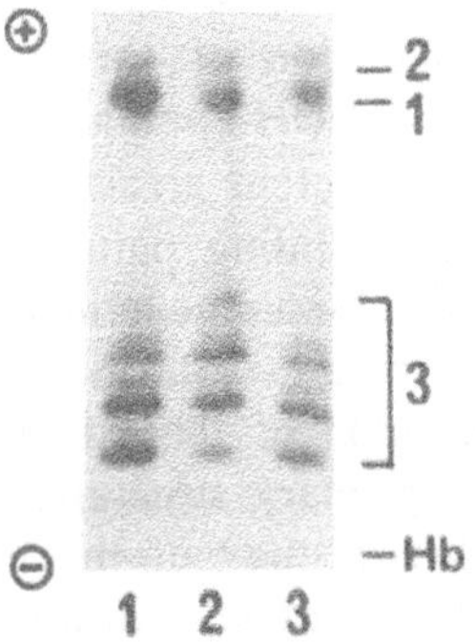

Figure 6. Distribution of ALDH3 isozymes in different areas of human stomach. Samples 1, 2, 3 are gastroendoscopic biopsies from the fundus, corpus, and antrum, respectively, of the same individual. Hb, hemoglobin.

allelic genes $ALDH^1_{3b}$ and $ALDH^2_{3b}$ would be only one or two nucleotides. The ALDH3 gene has been assigned to chromosome 17, differing from that of ALDH1 (chromosome 9), ALDH2 (chromosome 12) and ALDHX (chromosome 9) (Hopkinson et al., 1985; Hsu et al., 1986; Hsu and Chang, 1991).

MOLECULAR AND KINETIC PROPERTIES

Subunit molecular weights of both the liver ALDH3 AA form and the stomach isozyme mixture of AB_1, AA and AB_2 are 55,000 daltons, while the native molecular weight of the liver enzyme is 100,000 ,suggesting a dimeric molecule with identical subunits (Wang et al., 1990; Wang and Yin, 1992). The pI values of the isozymes B_1B_1, AB_1, AA, AB_2, and B_2B_2 are 5.9, 6.1, 6.3, 6.5, and 6.7 respectively, intermediate between that of ALDH1 (5.2)/ALDH2 (5.0) and ALDH4 (6.9).

Kinetic properties of stomach and liver ALDH3 are similar but unique when compared with those of other ALDHs (Yin et al., 1989; Wang et al., 1990). Table 1 shows the assay conditions which can distinguish human ALDH1, ALDH2, ALDH3, and ALDH4. ALDH3 is characteristic in its extremely high K_m for acetaldhyde.

Table 1. Assay Conditions to Distinguish Human ALDH Isozymes*

Assay condition	ALDH			
	1	2	3	4
Propionaldehyde				
0. 1 mM	+	+	-	-
15 mM	+	+	+	+
2-Furaldehyde				
0. 1 mM	+	+	-	-
15 mM	+	+	+	-
Heptaldehyde, 0. 1 mM	+	+	+	-
Succinic semialdehyde, 0. 5 mM	+	+	-	+
Glutamic γ-semialdehyde, 0. 5 mM	-	-	-	+
$NADP^+$, 1 mM	-	-	+	+
Disulfiram, 20 μM	+	-	-	-

*Assayed in 0.1 M sodium pyrophosphate at pH 8.5 and 25℃, for details see Yin et al. (1989). + represents active substrates, coenzymes, or inhibitor (i.e., disulfiram), and - represents inactive or far less active substrates, coenzymes, or inhibitor.

Similar high-K_m ALDH3 forms have been found in rat stomach, hepatoma and cornea (Koivusalo et al., 1989; Lindahl and Evces, 1984b; Evces and Lindahl, 1989). Comparisons of kinetic properties of human and rat ALDH3 are shown in Table 2. Substrate specificity studies of purified human liver ALDH3 AA form suggest that the substrate binding site constitutes a hydrophobic pocket. This is based on the kinetic constants of liver ALDH3, showing that the K_m values decrease successively for long straight-chain aliphatic aldehydes (C2 up to C10) and that the V_{max}/K_m values increase significantly (Wang and Yin, 1992). The ratio of the V_{max}/K_m value for decanal to that for acetaldehyde is 18,000. The V_{max} values remain relatively unchanged regardless of chain lengths, suggesting that the rate-limiting step in enzyme catalysis may be at the dissociation of NADH, which is a similar finding that for human alcohol dehydrogenases (Yin et al., 1984). Furthermore, long chain aldehydes such as nonanal and decanal (K_m = 2.8 μM; K_i = 300 μM) exhihit substrate inhibition, suggesting that the binding of the substrate to the enzyme in the enzyme-aldehyde -NADH ternary complex is tight.

Table 2. Kinetic Properties of High-K_m Aldehyde Dehydrogenases (ALDH3) from Human and Rats*

	K_m (mM)				
	Acetaldehyde	Propionaldehyde	Benzaldehyde	NAD^+	$NADP^+$
Human					
Liver	83	28	—	0. 013	0. 86
Stomach	88	24	0. 12	0. 015	0. 94
Rat					
Hepatoma	—	2. 2	0. 8	0. 074	0. 42
Stomach	29	4. 5	0. 030	0. 015	1. 7
Cornea	—	5. 7	0. 3	0. 003	0. 26

*ALDH3 isolated from human liver (Yin et al., 1989) and stomach (Wang et al., 1990), and from rat hepatoma (Lindahl and Evces, 1984b), stomach (Koivusalo et al., 1989) and cornea (Evces and Lindahl, 1989), was assayed in sodium phosphate or pyrophosphate buffer at pH 8.0-8.5. —, not determined.

STRUCTURAL FEATURES

The primary structure of human ALDH3 determined by amino acid (Yin et al., 1991) and genomic nucleotide (Hsu et al.,1992) sequence analysis shows a 81% homology with that of rat liver ALDH induced by 2,3,7,8-tetrachlorodibenzo-p-dioxin (TCDD) and that from HTC rat hepatoma cells (Jones et al., 1988; Hempel et al., 1989). The high-K_m dimeric ALDH3 (452 amino acid residues) exhibits, however, a maximum positional identity of 25-30% with the low-K_m tetrameric ALDH1/2 (500 amino acid residues; Hempel et al., 1984a,1985) by placing 11 gaps into the two structures. Recently, partial cDNA sequencing of a major soluble bovine corneal protein, 54K (BCP54), has revealed that it is a homologue of rat hepatoma ALDH (~80% identity), suggesting a gene sharing of this enzyme protein (Cooper et al., 1991). Furthermore, the η-crystallins of the lens in elephant shrews and the Ω -crystallin of octopus have also been structurally related to the low-K_m ALDH1/2 (Wistow and Kim, 1991; Tomarev et al., 1991). These findings demonstrate that there are multiple functions for ALDHs in vertebrates and invertebrates.

Alignment of the ALDH3 and ALDH1/2 sequences indicates that the conserved residues Glu-209/268, Cys-243/302, Gly-187/245, Gly-192/250 are structurally and/or functionally important. Previous studies have shown that Cys-302 and Glu-268 reside in the active site and Gly-245 and Gly-250 in the coenzyme binding domain (Hempel et al., 1985,1989, and references cited therein). It is interesting to note that the flanking sequences for these four conserved residues are also conserved in ALDH1, ALDH2 and ALDH3 (Hempel et al., 1984a,1985; Hsu et al.,1992). Furthermore, the putative active site residues of Glu-209 and Cys-243 are located in the longest conserved stretch of 71 residues between the human and rat ALDH3. These structural data are compatible with the kinetic similarities of this homotopic pair of human and rat enzymes (Table 2).

The degree of sequence differences between pairs of human/rat alcohol -metabolizing enzymes, e.g. alcohol dehydrogenase (ADH) and aldehyde dehydrogenase, as well as sorbitol dehydrogenase (SDH) and glucose-6-phosphate dehydrogenase (G-6-P DH), falls into two distinct ranges (Table 3) (Yin et al., 1991, and references cited therein; Hsu et al., 1992). ALDH3, ALDH1, class I ADH, and SDH are variable enzymes with 17-19% positional differences. ALDH2, class III ADH, and G-6-P DH belong to a group of constant enzymes with only 4-6% differences. These data imply that the sequence of the constant enzymes is influenced by stricter requirements on important properties than in the case of the variable enzymes. Indeed, in addition to the oxdiation of ethanol, the mammalian ADH/ALDH pair has been implicated in the metabolism of biogenic amines, steroids, retinol, and peroxidic aldehydes (Weiner, 1989; Ambroziak and Pietruszko, 1991). Since class III ADH and ALDH1 are located in the cytosol and appear to be ubiquitously distributed, it may be speculated that one of the physiological roles of this enzyme pair is the formation of retinoic acid, a tissue differentiation factor. The finding of a hydrophobic substrate binding pocket for ADH and ALDH is compatible with this notion.

Table 3. Differences Between Primary Structures of Human and Rat Pairs of Several Dehydrogenases

Rat/human enzyme pair	Positions with differences	
	'Constant' enzymes	'Variable' enzymes
ALDH3		85 of 452 (19%)
ALDH2	18 of 500 (4%)	
ALDH1		86 of 500 (17%)
Class I ADH		66 of 374 (18%)
Class III ADH	21 of 373 (6%)	
Sorbitol DH		63 of 356 (18%)
G-6-P DH	33 of 513 (6%)	

CONCLUSIONS

Human high-K_m aldehyde dehydrogenase (ALDH3) isozymes exhibit a limited tissue distribution with high activity in esophagus, stomach and cornea. The multiplicity of ALDH3 forms can be explained by a two-chain model for the enzyme with additional multiplicity for one of the chains. We do not know the explanation for the occurrence of the two types of chain, whether through genetic, epigenetic, or post -translational mechanisms. ALDH3 isolated from stomach and liver, using NAD^+ as preferred coenzyme, displays broad substrate specificity, suggesting a hydrophobic binding pocket for the substrate. The primary structure of ALDH3 shows approximately 30% homology with those of ALDH1 and ALDH2, but 80% identity with that of rat tumor-derived ALDH. Thus, within the complex human aldehyde dehydrogenase family, ALDH3 isozymes are unique in high K_m values for acetaldehyde, dimeric subunit structures and distinctive amino acid sequences.

ACKNOWLEDGMENTS

The work performed in the authors' laboratories was supported by grants from the National Science Council of the Republic of China (78-0412-B016-27 and 79-0412-B 016-43), the Swedish Medical Research Council (Project O3X-3532), the Knut and Alice Wallenberg Foundation, and by a visiting scientist fellowship to S.-J.Y. from Karolinska Institutet.

REFERENCES

Agarwal, D. P., Eckey, R., Rudnay, A.-C., Volkens, T., and Goedde, H. W., 1989, "High -K_m" aldehyde dehydrogenase isozymes in human tissues: Constitutive and tumor -associated forms, in "Enzymology and Molecular Biology of Carbonyl Metabolism 2: Aldehyde dehydrogenase, alcohol dehydrogenase, and aldo-keto reductase," H. Weiner and T. G. Flynn, eds., Alan R. Liss, New York, p. 119.

Algar, E. M., and Holmes, R. S., 1989, Purification and properties of mouse stomach aldehyde dehydrogenase. Evidence for a role in the oxidation of peroxidic and aromatic aldehydes, Biochim. Biophys. Acta, 995:168.

Ambroziak, W., and Pietruszko, R., 1991, Human aldehyde dehydrogenase. Activity with aldehyde metabolites of monoamines, diamines, and polyamines, J. Biol. Chem., 266:13011.

Campbell, P., Irving, C. C., and Lindahl, R., 1989, Changes in aldehyde dehydrogenase during rat urinary bladder carcinogenesis, Carcinogenesis, 10:2081.

Cooper, D. L., Baptist, E. W., Enghild, J. J., Isola, N. R., and Klintworth, G. K., 1991, Bovine corneal protein 54K (BCP54) is a homologue of the tumor-associated (class 3) rat aldehyde dehydrogenase (RATALD), Gene, 98:201.

Duley, J. A., Harris, O., and Holmes, R. S., 1985, Analysis of human alcohol- and aldehyde-metabolizing isozymes by electrophoresis and isoelectric focusing, Alcohol. Clin. Exp. Res., 9:263.

Enomoto, N., Takase, S., Takada, N., and Takada, A., 1991, Alcoholic liver disease in heterozygotes of mutant and normal aldehyde dehydrogenase-2 genes, Hepatology, 13:1071.

Evces, S., and Lindahl, R., 1989, Characterization of rat cornea aldehyde dehydrogenase, Arch. Biochem. Biophys., 274:518.

Forte-McRobbie, C. M., and Pietruszko, R., 1986, Purification and characterization of human liver "high K_m" aldehyde dehydrogenase and its identification as glutamic γ -semialdehyde dehydrogenase, J. Biol. Chem., 261:2154.

Greenfield N. J., and Pietruszko, R., 1977, Two aldehyde dehydrogenases from human liver. Isolation via affinity chromatography and characterization of the isozymes, Biochim. Biophys. Acta, 483:35.

Harada, S., Agarwal, D. P., and Goedde, H. W., 1980, Electrophoretic and biochemical studies of human aldehyde dehydrogenase isozymes in various tissues, Life Sci., 26:1773.

Harada, S., Muramatsu, T., Agarwal, D. P., and Goedde, H. W., 1989, Polymorphism of aldehyde dehydrogenase in human saliva, in "Enzymology and Molecular Biology of Carbonyl Metabolism 2," H. Weiner and T. G. Flynn, eds., Alan R. Liss, New York, p. 133.

Hempel, J., von Bahr-Lindstrom, H., and Jornvall, H., 1984a, Aldehyde dehydrogenase from human liver. Primary structure of the cytoplasmic isoenzyme, Eur. J. Biochem., 141:21.

Hempel, J., Kaiser, R., and Jornvall, H., 1984b, Human liver mitochondrial aldehyde dehydrogenase. A C-terminal segment positions and defines the structure corresponding to the one reported to differ in the Oriental enzyme variant, FEBS Lett., 173:367.

Hempel, J., Kaiser, R., and Jornvall, H., 1985, Mitochondrial aldehyde dehydrogenase from human liver. Primary structure, differences in relation to the cytosolic enzyme, and functional correlations, Eur. J. Biochem., 153:13.

Hempel, J., and Jornvall, H., 1989, Aldehyde dehydrogenases - Structure, in "Human Metabolism of Alcohol," Vol II, K. E. Crow and R. D. Batt, eds., CRC, Boca Raton, p. 77.

Hempel, J., Lindahl, R., 1989, Class III aldehyde dehydrogenase from rat liver: Superfamily relationship to classes I and II and functional interpretions, in "Enzymology and Molecular Biology of Carbonyl Metabolism 2," H. Weiner and T. G. Flynn, eds., Alan R. Liss, New York, p. 3.

Hempel, J., Harper, K., and Lindahl, R., 1989, Inducible (class 3) aldehyde dehydrogenase from rat hepatocellular carcinoma and 2,3,7,8-tetrachlorodibenzo-p-dioxin-treated liver: Distant relationship to the class 1 ans 2 enzymes from mammalian liver cytosol /mitochondria, Biochemistry, 28:1160.

Holmes, R. S., 1988, Alcohol dehydrogenases and aldehyde dehydrogenases of anterior eye tissues from humans and other mammals, in "Biomedical and Social Aspects of Alcohol and Alcoholism," K. Kuriyama, A. Takada and H. Ishii, eds., Elsevier, Amsterdam, p. 51.

Hopkinson, D. A., Santisteban, I., Povey, S., and Smith, M., 1985, Biochemical genetic analysis of human and rodent aldehyde dehydrogenase, Alcohol, 2:73.

Hsu, L. C., Tani, K., Fujiyoshi, T., Kurachi, K., and Yoshida, A., 1985, Cloning of cDNAs for human aldehyde dehydrogenases 1 and 2, Proc. Natl. Acad. Sci. USA, 82:3771.

Hsu, L. C., Yoshida, A., and Mohandas, T., 1986, Chromosomal assignment of the genes for human aldehyde dehydrogenase-1 and aldehyde dehydrogenase-2, Am. J. Hum. Genet., 38:641.

Hsu, L. C., and Chang, W.-C., 1991, Cloning and characterization of a new functional human aldehyde dehydrogenase gene, J. Biol. Chem., 266:12257.

Hsu, L. C., Chang, W.-C., Shibuya, A., and Yoshida, A., 1992, Human stomach aldehyde dehydrogenase cDNA and genomic cloning, primary structure, and expression in Escherichia coli, J. Biol. Chem., 267:3030.

Jones, Jr., D. E., Brennan, M. D., Hempel, J., and Lindahl, R., 1988, Cloning and complete nucleotide sequence of a full-length cDNA encoding a catalytically functional tumor -associated aldehyde dehydrogenase, Proc. Natl. Acad. Sci. USA, 85:1782.

Koivusalo, M., Aarnio, M., Baumann, M., and Rautoma, P., 1989, NAD(P)-linked aromatic aldehydes preferring cytoplasmic aldehyde dehydrogenases in the rat. Constitutive and inducible forms in liver, lung, stomach and intestinal mucosa, in "Enzymology and Molecular Biology of Carbonyl Metabolism 2," H. Weiner and T. G. Flynn, eds., Alan R. Liss, New York, p. 19.

Kurys, G., Ambroziak, W., and Pietruszko, R., 1989, Human aldehyde dehydrogenase. Purification and characterization of a third isozyme with low Km for γ-aminobutyrald ehyde, J. Biol. Chem., 264:4715.

Lindahl, R., and Evces, S., 1984a, Rat liver aldehyde dehydrogenase. I. Isolation and characterization of four High K_m normal liver isozymes, J. Biol. Chem., 259:11986.

Lindahl, R., and Evces, S., 1984b, Rat liver aldehyde dehydrogenase. II. Isolation and characterization of four inducible isozymes, J. Biol. Chem., 259:11991.

Lindahl, R., and Evces, S., 1987, Changes in aldehyde dehydrogenase activity during diethylnitrosamine-initiated rat hepatocarcinogenesis, Carcinogenesis, 8:785.

Lindahl, R., and Petersen, D. R., 1991, Lipid aldehyde oxidation as a physiological role for class 3 aldehyde dehydrogenases, Biochem. Pharmacol., 41:1583.

Meier-Tackmann, D., Agarwal, D. P., Saha, N., and Goedde, H. W., 1984, Aldehyde dehydrogenase isozymes in stomach autopsy specimens from Germans and Chinese, Enzyme, 32:170.

Miyauchi, K., Masaki, R., Taketani, S., Yamamoto, A., Akayama, M., and Tashiro, Y., 1991, Molecular cloning Seqencing, and expression of cDNA for rat liver microsomal aldehyde dehydrogenase, J. Biol. Chem., 266:19536.

Mizoi, Y., Ijiri, I., Tatsuno, Y., Kijima, T., Fujiwara, S., and Adachi, J., 1979, Relationship between facial flushing and blood acetaldehyde levels after alcohol intake, Pharmacol. Biochem. Behav., 10:303.

Pietruszko, R., 1983, Aldehyde dehydrogenase isozymes, Isozymes Curr. Top. Biol. Med. Res., 8:195.

Ryzlak, M. T., and Pietruszko, R., 1987, Purification and characterization of aldehyde dehydrogenase from human brain, Arch. Biochem. Biophys., 255:409.

Santisteban, I., Povey, S., West, L. F., Parrington, J. M., and Hopkinson, D. A., 1985, Chromosome assignment, biochemical and immunological studies on a human aldehyde dehydrogenase, ALDH3, Ann. Hum. Genet., 49:87.

Teng, Y.-S., 1981, Stomach aldehyde dehydrogenase: Report of a new locus, Hum. Hered., 31:74.

Thomasson, H. R., Edenberg, H. J., Crabb, D. W., Mai, X.-L., Jerome, R. E., Li, T.-K., Wang, S.-P., Lin, Y.-T., Lu, R.-B., and Yin, S.-J., 1991, Alcohol and aldehyde dehydrogenase genotypes and alcoholism in Chinese men, Am. J. Hum. Genet., 48:677.

Tomarev, S. I., Zinovieva, R. D., and Piatigorsky, J., 1991, Crystallins of the octopus lens. Recruitment from detoxification enzymes, J. Biol. Chem., 266:24226.

Wang, S.-L., Wu, C.-W., Cheng, T.-C., and Yin, S.-J., 1990, Isolation of high-K_m aldehyde dehydrogenase isoenzymes from human gastric mucosa, Biochem. Int., 22:199.

Wang, S.-L., and Yin, S.-J., 1992, in preparation.

Weiner, H., 1989, Role of alcohol and aldehyde dehydrogenase in vivo: Speculations on their natural substrates, in "Human Metabolism of Alcohol," Vol II, K. E. Crow and R. D. Batt, eds., CRC, Boca Raton, p. 147.

Wistow, G., and Kim, H., 1991, Lens protein expression in mammals: Taxon-specificity and the recruitment of crystallins, J. Mol. Evol., 32:262.

Yin, S.-J., Bosron, W. F., Magnes, L. J., and Li, T.-K., 1984, Human liver alcohol dehydrogenase: Purification and kinetic characterization of the $\beta_2\beta_2$, $\beta_2\beta_1$, $\alpha\beta_2$ and $\beta_2\gamma_1$ "Oriental" isoenzymes, Biochemistry, 23:5847.

Yin, S.-J., Chang, T.-C., Chang C.-P., Chen, Y.-J., Chao, Y.-C., Tang, H.-S., Chang, T.-M., and Wu, C.-W., 1988, Human stomach alcohol and aldehyde dehydrogenase (ALDH): A genetic model proposed for ALDH III isoenzymes, Biochem. Genet., 26:343.

Yin, S.-J., Liao, C.-S., Wang, S.-L., Chen, Y.-J., and Wu, C.-W., 1989, Kinetic evidence for human liver and stomach aldehyde dehydrogenase-3 representing a unique class of isozymes, Biochem. Genet., 27:321.

Yin, S.-J., Vagelopoulos, N., Wang, S.-L., and Jornvall, H., 1991, Structural features of stomach aldehyde dehydrogenase distinguish dimeric aldehyde dehydrogenase as a 'variable' enzyme. 'Variable' and 'constant' enzymes within the alcohol and aldehyde dehydrogenase families, FEBS Lett., 283:85.

Yin, S.-J., Chou, F.-J., Chao, S.-F., Tsai, S.-F., Liao, C.-S., Wang, S.-L., Wu, C.-W., and Lee, S.-C., 1992a, Alcohol and aldehyde dehydrogenases in human esophagus. Comparison with the stomach enzyme activities, submitted.

Yin, S.-J., Liao, C.-S., Chen, C.-M., Fan, F.-T., and Lee, S.-C., 1992b, Genetic polymorphism and activities of human lung alcohol and aldehyde dehydrogenases: Implications for ethanol metabolism and cytotoxicity, Biochem. Genet., 30:203.

Yoshida, A., Huang, I. Y., and Ikawa, M., 1984, Molecular abnormality of an inactive aldehyde dehydrogenase variant commonly found in Orientals, Proc. Natl. Acad. Sci. USA, 81:258.

Yoshida, A., Hsu, L. C., and Yasunami, M., 1991, Genetics of human alcohol-metabolizing enzymes, Progr. Nucleic Acid Res. Mol. Biol., 40:255.

OVEREXPRESSION OR POLYCYCLIC AROMATIC HYDROCARBON-MEDIATED INDUCTION OF AN APPARENTLY NOVEL CLASS 3 ALDEHYDE DEHYDROGENASE IN HUMAN BREAST ADENOCARCINOMA CELLS AND ITS RELATIONSHIP TO OXAZAPHOSPHORINE-SPECIFIC ACQUIRED RESISTANCE

Lakshmaiah Sreerama and Norman E. Sladek

Department of Pharmacology, University of Minnesota
Minneapolis, MN 55455

INTRODUCTION

Oxazaphosphorines such as cyclophosphamide are widely used in the treatment of certain neoplasms (Sladek, 1988). Because they are, *per se*, without cytotoxic activity, their metabolism, Figure 1, has been the subject of intensive investigation. In the course of these investigations, it was established that certain aldehyde dehydrogenases catalyze the irreversible detoxification of the oxazaphosphorines when they catalyze the oxidation of aldophosphamide to carboxyphosphamide. Class 1 aldehyde dehydrogenases, *e.g.*, mouse AHD-2 and human ALDH-1, are particularly important in this regard (Manthey *et al.*, 1990; Dockham *et al.*, 1992). Other "aldehyde" dehydrogenases, *e.g.*, human ALDH-2 and succinic semialdehyde dehydrogenase, also catalyze the reaction albeit less well as judged by Km values (Dockham *et al.*, 1992). Still others, *e.g.*, human ALDH-4, ALDH-5 and betaine aldehyde dehydrogenase, do not catalyze the reaction at all (Dockham *et al.*, 1992). The pharmacological upshot is that a relative oxazaphosphorine-insensitivity is conferred on those cells in which constitutive or induced expression of the relevant enzyme(s) occurs.

Clinically, intrinsic and acquired resistance to the oxazaphosphorines on the part of neoplastic cells is frequently encountered (Sladek, 1992). In some cases, resistance is only to the oxazaphosphorines, *i.e.*, the resistance is oxazaphosphorine-specific. In other cases, it extends to other antineoplastic agents. Oxazaphosphorine-specific resistance could, in theory, be due to any of several possibilities, the most probable of which is an overexpression of a relevant aldehyde dehydrogenase (Sladek, 1992). Supporting this notion is that over-expression of a class 1 aldehyde dehydrogenase, *viz.*, AHD-2, accounts for the oxazaphosphorine-specific acquired resistance exhibited by the mouse L1210 and P388/CLA lymphocytic leukemia cell lines (Hilton, 1984; Sladek and Landkamer, 1985; Sladek *et al.*, 1985; Sladek, 1988, 1992; Radin *et al.*, 1991).

Enzymology and Molecular Biology of Carbonyl Metabolism 4
Edited by H. Weiner, Plenum Press, New York, 1993

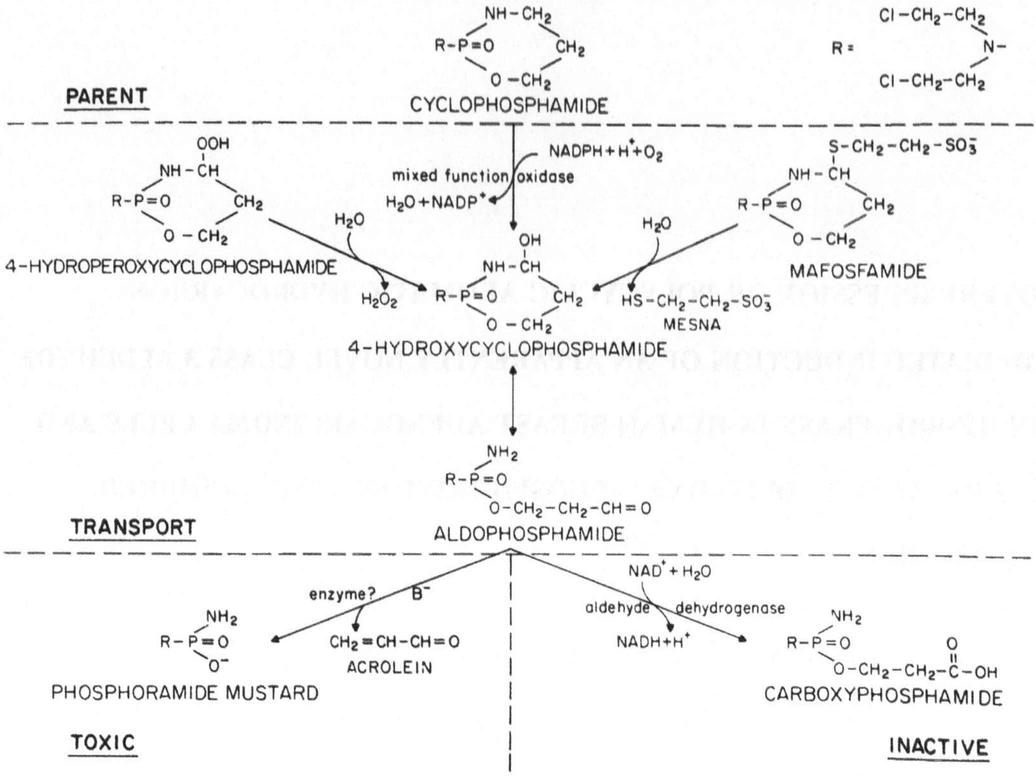

Figure 1. Salient features of oxazaphosphorine metabolism. Oxazaphosphorines such as cyclophosphamide, mafosfamide and 4-hydroperoxycyclophosphamide are prodrugs, *i.e.*, *per se*, they are without cytotoxic activity. Each gives rise to 4-hydroxycyclophosphamide which exists in equilibrium with its ring-opened tautomer, aldophosphamide. Hydroxylation of cyclophosphamide to 4-hydroxycyclophosphamide proceeds only with the catalytic participation of mixed-function oxidases. Hydrolysis of mafosfamide or 4-hydroperoxycyclophosphamide to 4-hydroxycyclophosphamide occurs rapidly without any enzyme involvement. 4-Hydroxycyclophosphamide and aldophosphamide are, themselves, also without cytotoxic activity. However, aldophosphamide gives rise to acrolein and phosphoramide mustard, each of which is cytotoxic; the latter effects the bulk of the therapeutic action effected by the oxazaphosphorines (Sladek, 1988). Alternatively, aldophosphamide can be further oxidized to carboxyphosphamide by certain aldehyde dehydrogenases (Sladek, 1988; Manthey *et al.*, 1990; Dockham *et al.*, 1992). Carboxyphosphamide is without cytotoxic activity nor does it give rise to a cytotoxic metabolite. Therefore, aldehyde dehydrogenase-catalyzed oxidation of aldophosphamide to carboxyphosphamide is properly viewed as an enzyme-catalyzed detoxification of the oxazaphosphorines. The reader should consult Sladek (1988) for a comprehensive review of oxazaphosphorine metabolism.

A subline of cultured human MCF-7/0 breast adenocarcinoma cells, *viz.*, MCF-7/OAP, also exhibits oxazaphosphorine-specific acquired resistance (Frei *et al.*, 1988; Sreerama and Sladek, 1991). The resistance is of a substantial magnitude. LC_{90} values (concentration of drug required to render 90% of the cells incapable of further indefinite proliferation) for mafosfamide of 50 µM (MCF-7/0) and >2 mM (MCF-7/OAP) were obtained in our laboratory (Sreerama and Sladek, 1991). The expectation was that overexpression of one or more of the human aldehyde dehydrogenases known to catalyze the oxidation of aldophosphamide to carboxyphosphamide, *viz.*, ALDH-1, ALDH-2 and/or succinic semialdehyde dehydrogenase, would account for the oxazaphosphorine-specific acquired resistance exhibited by these cells. Evidence supporting or refuting this notion was sought at the outset of the investigations reported herein.

RESULTS AND DISCUSSION

NAD(P)-dependent enzyme-catalyzed oxidation of several aldehydes, *viz.*, benzaldehyde, octanal and 4-pyridinecarboxaldehyde, was indeed hugely elevated in MCF-7/OAP cells, Table 1. However, NADP-dependent enzyme-catalyzed oxidation of aldophosphamide and acetaldehyde was not observed, and NAD-dependent enzyme-catalyzed oxidation of these substrates was only minimally elevated and, even then, did not proceed very rapidly. Thus, the notion that aldehyde dehydrogenase-catalyzed oxidation (detoxification) of aldophosphamide to carboxyphosphamide accounts for the oxazaphosphorine-specific acquired resistance exhibited by the MCF-7/OAP cells is highly problematical. However, inclusion of benzaldehyde, octanal or 4-pyridinecarboxaldehyde in the drug-exposure medium largely restored the sensitivity of MCF-7/OAP cells to mafosfamide whereas the inclusion of acetaldehyde did not (Sladek *et al.*, manuscript in preparation). These latter observations strongly indicate that the enzyme that catalyzes the oxidation of benzaldehyde, octanal and 4-pyridinecarboxaldehyde does, in some way, account for the oxazaphosphorine-specific resistance exhibited by MCF-7/OAP cells, *vide infra*.

Table 1. Aldehyde dehydrogenase activity in the soluble (105,000 x g supernatant) fractions of MCF-7/0 and MCF-7/OAP cells[a]

Substrate (concentration)	Cofactor (4 mM)	mIU/10^7 cells	
		MCF-7/0	MCF-7/OAP
Aldophosphamide (160 μM)	NAD	0	2.8
	NADP	0	0
Acetaldehyde (4 mM)	NAD	1.4	6.6
	NADP	0	0
Benzaldehyde (4 mM)	NAD	1.7	110
	NADP	1.9	254
Octanal (100 μM)	NAD	1.1	75
	NADP	1.3	152
4-Pyridinecarboxaldehyde (500 μM)	NAD	12.6	112
	NADP	1.8	213

[a]Monolayer cultures of MCF-7/0 and MCF-7/OAP cells, originally obtained from Dr. Beverly Teicher, Dana-Farber Cancer Institute, Boston, MA, were propagated at 37°C in Dulbecco's modified Eagle medium supplemented with L-glutamine (2 mM), sodium bicarbonate (3.7 g/L), gentamicin (50 mg/L) and fetal bovine serum (10%); the atmosphere of 5% CO_2 in air was fully humidified. Cells in asynchronous exponential growth were submitted to trypsinization and then harvested by low-speed centrifugation (500 x g for 10 min). Soluble (105,000 x g supernatant) fractions were prepared as described by Kohn *et al.* (1987). Aldehyde dehydrogenase activity was quantified at pH 8.1 and 37°C as described by Manthey *et al.* (1990). Each value is the mean of duplicate determinations on each of one to three samples. Preliminary experiments established that particulate (105,000 x g pellet) fractions lacked aldehyde dehydrogenase activity.

Several properties of the enzyme activity observed in MCF-7/OAP cells suggested that it might be due to the class 3 aldehyde dehydrogenase, *viz.*, ALDH-3, known to be constitutively present in several human tissues, most notably stomach mucosa (Goedde and Agarwal, 1990). Included are the facts that 1) it was localized to the cytosol, 2) aromatic, and long-chain aliphatic, aldehydes appeared to be the preferred substrates, 3) at high concentrations, *viz.*, 4 mM, NADP was the "better" cofactor when metabolism of benzaldehyde, octanal or 4-pyridinecarboxaldehyde was quantified, and 4) pyridine nucleotide-dependent enzyme-catalyzed oxidation of benzaldehyde by MCF-7/OAP cytosol was not inhibited by disulfiram (50 µM) or chloral hydrate (100 µM). Additionally, neither MCF-7/OAP cytosol, nor a whole homogenate of MCF-7/OAP cells, was able to catalyze the oxidation of glutamic-γ-semialdehyde, succinic semialdehyde, betaine aldehyde or glyceraldehyde 3-phosphate thus eliminating the "aldehyde" dehydrogenases that catalyze the oxidation of these substrates as possibilities.

In an attempt to further ascertain whether the MCF-7/OAP enzyme was ALDH-3, a soluble (105,000 x g supernatant) fraction of human stomach mucosa obtained from a middle aged Caucasian male who died of severe coronary heart disease, and whole homogenates of MCF-7/0 and MCF-7/OAP cells, were submitted to isoelectric focusing after which gels were stained for NAD-dependent enzymes that catalyzed the oxidation of benzaldehyde, Figure 2. Clear from the results of this experiment was that whereas the MCF-7 enzyme isoelectric focused (4 bands; pI values ranging from 6.0 to 6.45) in a manner very similar to the stomach mucosa enzyme (5 bands; pI values ranging from 5.75 to 6.35), it did not do so in an identical manner. At this point, we began to entertain the notion that the MCF-7 enzyme was indeed a class 3 aldehyde dehydrogenase but one that was different from the ALDH-3 found in human stomach mucosa and other tissues. Evidence for this hypothesis was pursued in the next series of experiments.

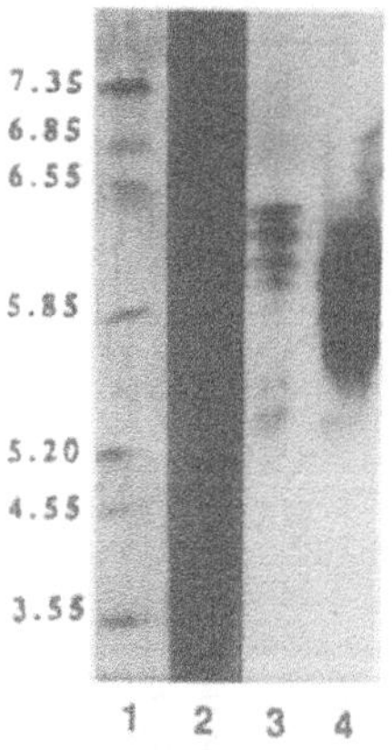

Figure 2. Isoelectric focusing of aldehyde dehydrogenases present in the soluble (105,000 x g supernatant) fraction of human stomach mucosa (lane 4) and in Lubrol®-treated whole homogenates of MCF-7/0 and MCF-7/OAP cells (lanes 2 and 3, respectively). Stomach mucosa was homogenized in 25 mM sodium phosphate buffer, pH 7.5, supplemented with 1 mM EDTA. The homogenate was centrifuged at 105,000 x g and 4°C for 1 h and the resultant 105,000 x g supernatant fraction was harvested and saved for assay. MCF-7 cells in asynchronous exponential growth were harvested as described in the footnote to Table 1, washed twice, and resuspended in an aqueous solution, pH 7.4, containing 1.15% (w/v) KCl and 1 mM EDTA. The suspended cells were then placed into an ice-bath and lysed by submitting them to sonication; the resultant preparations were adjusted to 0.3% Lubrol®, vortexed, and centrifuged at 105,000 x g and 4°C for 1 h to get Lubrol®-treated whole homogenates. An aliquot of the human stomach mucosa soluble fraction sufficient to generate 5.0 nmol NADH/min when benzaldehyde was the substrate, Lubrol®-treated whole homogenates obtained from 1×10^7 MCF-7/0, or 1.5×10^5 MCF-7/OAP, cells, and pI marker proteins (lane 1) were loaded onto the gel and subjected to isoelectric focusing essentially as described by Manthey *et al.* (1990) except that commercially available Ampholine PAGplates® (pH 3.5 - 9.5) were used. The lane containing pI standards was stained for proteins with Coomassie Brilliant Blue R-250. The rest of the gel was stained for aldehyde dehydrogenase activity using a nitroblue tetrazolium-coupled enzyme activity stain; benzaldehyde (4 mM) was the substrate and NAD (4 mM) was the cofactor. The overall dark appearance of lane 2 is because MCF-7/0 cells contain comparatively very little of the enzyme necessitating that this part of the gel be left in the staining solution for a relatively much longer time period in order to visualize it.

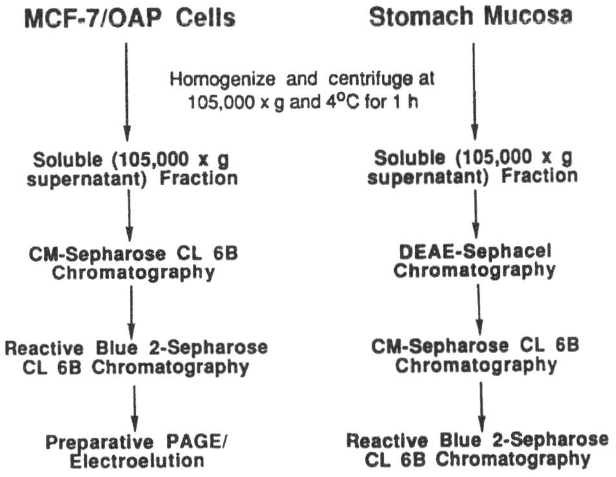

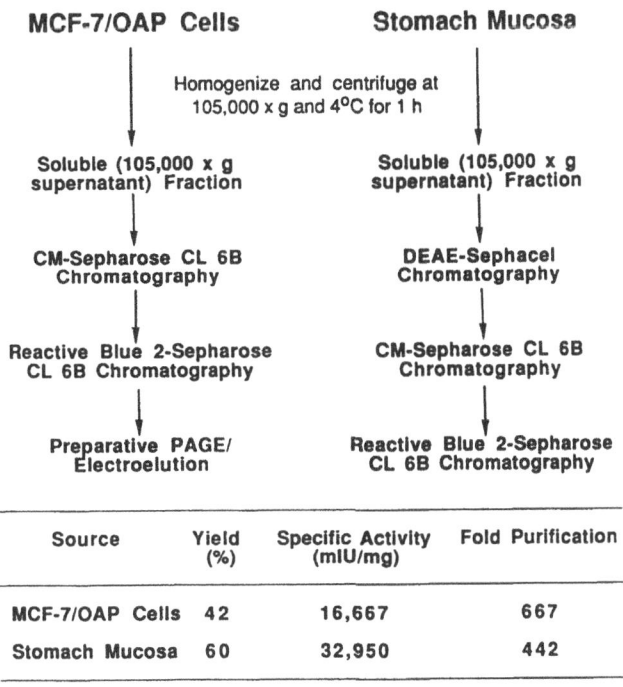

Source	Yield (%)	Specific Activity (mIU/mg)	Fold Purification
MCF-7/OAP Cells	42	16,667	667
Stomach Mucosa	60	32,950	442

Figure 3. Schematic presentation of the protocols used to purify class 3 aldehyde dehydrogenases from human MCF-7/OAP cells and stomach mucosa. Preparative PAGE was essentially according to the method of Davis (1964).

Table 2. Comparison of the properties of class 3 aldehyde dehydrogenases isolated from human stomach mucosa and MCF-7/OAP cells[a]

Parameter	Stomach Mucosa	MCF-7/OAP Cells
Subcellular location	Cytosol	Cytosol
Substrate preference	Benzaldehyde	Benzaldehyde
Cofactor preference	NAD/NADP	NAD/NADP
Catalysis of aldophosphamide oxidation	No	Yes
Molecular weight	108 kDa	125 kDa
Subunit molecular weight	54.5 kDa	40 kDa
Functional form	Dimer	?
IEF banding pattern	Multiple bands	Multiple bands
pI	5.75 - 6.35	6.0 - 6.45
Thermal stability	Labile	Labile
Optimal pH stability	7.5 - 9.0	6.5 - 7.5
Inhibited by disulfiram (50 μM)	≤ 30%	≤ 2%
Inhibited by chloral hydrate (1 mM)	≤ 7%	≤ 5%

[a]Each of the enzymes was purified to homogeneity utilizing the protocols presented in Figure 3. Homogeneity was pronounced because, in each case, enzyme activity and protein bands exactly corresponded after isoelectric focusing and after linear-gradient polyacrylamide gel electrophoresis (PAGE), and because, again in each case, SDS-polyacrylamide gels stained for proteins showed only a single band.

Human stomach mucosa ALDH-3 and the enzyme present in MCF-7/OAP cytosol were purified to homogeneity utilizing the protocols presented in Figure 3, and the physical and catalytic properties of the purified enzymes were compared, Table 2. The two enzymes differed from each other in several important ways, *viz.*, native molecular weight, subunit molecular weight, pI values, optimal pH stability, and in their ability to catalyze the oxidation of aldophosphamide. With regard to the latter, stomach mucosa ALDH-3 did not catalyze the oxidation of aldophosphamide regardless of whether NAD or NADP was used as the cofactor. In contrast, the MCF-7/OAP enzyme did catalyze this reaction (Km = 640 μM) but only when NAD was used as the cofactor. Mouse stomach AHD-4, putatively identified as a class 3 aldehyde dehydrogenase and as the mouse homolog of human ALDH-3, also catalyzes the reaction (Manthey *et al.*, 1990). Further evidence supporting the notion that stomach mucosa ALDH-3 and MCF-7/OAP enzyme are different entities was the observation that antibodies to stomach mucosa ALDH-3 recognized both the native and denatured forms of this enzyme but recognized only the native state of the MCF-7/OAP enzyme. Isoelectric focusing of purified stomach mucosa ALDH-3 and the MCF-7/OAP enzyme along with several other purified human "aldehyde" dehydrogenases showed that, in addition to not being identical to stomach mucosa ALDH-3, the MCF-7/OAP enzyme was not ALDH-1, ALDH-2, glutamic-γ-semialdehyde dehydrogenase (ALDH-4), succinic semialdehyde dehydrogenase, or betaine aldehyde dehydrogenase, Figure 4.

Collectively, these observations led us to conclude that whereas the MCF-7 enzyme is indeed a class 3 aldehyde dehydrogenase, it is somewhat different from the previously characterized stomach mucosa ALDH-3 and thus a novel ALDH-3. It is referred to hereafter in this paper as Type-2 ALDH-3 to distinguish it from the stomach mucosa enzyme referred to hereafter in this paper as Type-1 ALDH-3. Small amounts of Type-2 ALDH-3 are also present in normal breast tissue obtained from either pre- or post-menopausal donors, Figure 5.

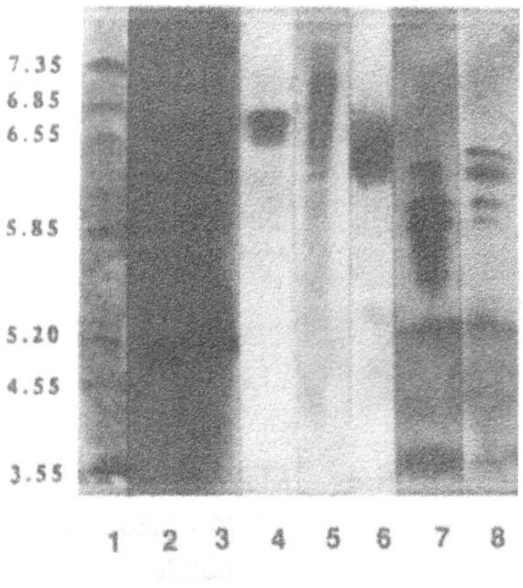

Figure 4. Isoelectric focusing of purified human aldehyde dehydrogenases. Authentic aldehyde dehydrogenases were subjected to isoelectric focusing and were visualized as described in the legend to Figure 2. The amount of each purified enzyme loaded onto the gel was sufficient to generate 1.5 - 5.0 nmol NADH/min as determined by spectrophotometric assay using the substrate that was ultimately used to visualize the enzyme on the gel. Lane 1 (pI standards) was stained for proteins. Lanes 2 - 8 were stained for aldehyde dehydrogenase activity. Enzymes and substrates were: Lane 2, ALDH-2 and acetaldehyde (4 mM); lane 3, ALDH-1 and acetaldehyde (4 mM); lane 4, glutamic-γ-semialdehyde dehydrogenase (ALDH-4) and glutamic-γ-semialdehyde (500 μM); lane 5, succinic semialdehyde dehydrogenase and succinic semialdehyde (100 μM); lane 6, betaine aldehyde dehydrogenase and betaine aldehyde (100 μM); lane 7, stomach mucosa ALDH-3 and benzaldehyde (4 mM); and lane 8, MCF-7/OAP ALDH-3 and benzaldehyde (4 mM). The cofactor was NAD (4 mM) in all cases. ALDH-1, ALDH-2, ALDH-4, succinic semialdehyde dehydrogenase and betaine aldehyde dehydrogenase were prepared (Dockham *et al.*, 1992) and provided by Dr. P. A. Dockham of our laboratory.

Given that aldophosphamide is a relatively poor substrate for Type-2 ALDH-3, the notion that Type-2 ALDH-3-catalyzed oxidation of aldophosphamide is the underlying mechanism accounting for the oxazaphosphorine-specific acquired resistance exhibited by the MCF-7/OAP cell line, *vide supra*, remains highly problematical. It may be that Type-2 ALDH-3 accounts for this resistance in some other way, *e.g.*, by binding aldophosphamide with great affinity thus effectively removing it from the cell (Sladek, 1992).

ALDH-3 is one of at least six "drug-metabolizing enzymes" that are coded for by genes that are members of the so-called polycyclic aromatic hydrocarbon-responsive gene battery (Lindahl, 1992). The others are cytochrome P450s IA1 and IA2, glutathione S-transferase, NAD(P)H:menadione oxidoreductase (DT-diaphorase) and UDP-glucuronyl transferase. Activation of the gene battery is believed to involve a series of receptor-mediated events initiated by the binding of an appropriate agonist (ligand) to a cytosolic protein termed the Ah receptor (Lindahl, 1992). Known agonists for the Ah receptor include the polycyclic aromatic hydrocarbons, 3-methylcholanthrene, benzpyrene, 9,10-dimethyl-1,2-benzanthracene, and especially, TCDD (Lindahl, 1992). One or more of these agents have been shown to induce ALDH-3 in rat and human, hepatocytes and hepatoma cells (Marselos *et al.*, 1986, 1987; Huang and Lindahl, 1990). Moreover, polycyclic aromatic hydrocarbons are known to induce cytochrome P450 IA1 in MCF-7 cells (Vickers *et al.*, 1989). Thus, it seemed likely that these agents would also induce the expression of ALDH-3 in these cells. This notion was tested in the next series of experiments.

3-Methylcholanthrene, benzpyrene and 9,10-dimethyl-1,2-benzanthracene each markedly induced aldehyde dehydrogenase activity in MCF-7/0 cells when they were grown in the presence of 3 μM of these agents, Figure 6. Given the identity of the substrate and cofactor used to demonstrate induction, and the fact that polycyclic aromatic hydrocarbons are known to induce ALDH-3 but not other aldehyde dehydrogenases (Lindahl, 1992), the expectation is that the increase in NADP-dependent aldehyde dehydrogenase-catalyzed oxidation of benzaldehyde to benzoic acid can be fully accounted for by an increased expression of ALDH-3; preliminary isoelectric focusing experiments have confirmed this expectation.

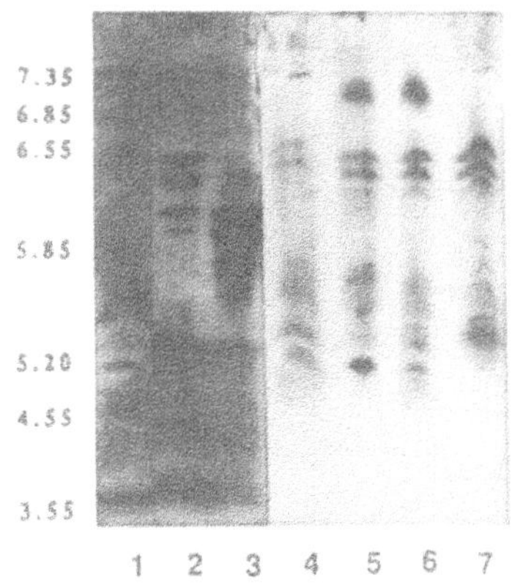

Figure 5. Isoelectric focusing of aldehyde dehydrogenases present in a Lubrol®-treated whole homogenate of MCF-7/OAP cells (lane 2), and in soluble (105,000 x g supernatant) fractions prepared from stomach mucosa (lane 3) and normal breast tissue samples donated by a 61- (lane 4), a 59- (lane 5), a 51- (lane 6) and a 19- (lane 7) year-old patient. All soluble (105,000 x g supernatant) fractions were prepared as described for stomach mucosa in the legend to Figure 2. Placed on the gel were a whole homogenate prepared from 2.5 x 10^5 MCF-7/OAP cells, a soluble fraction prepared from stomach mucosa sufficient to generate 5.0 nmol NADH/min as determined by spectrophotometric assay using benzaldehyde as the substrate, and soluble fractions prepared from 1 g of breast tissue obtained from each of the four patients. Electrophoresis and visualization of proteins/enzymes was as described in the legend to Figure 2. Lane 1 (pI standards) was stained for protein. Lanes 2 - 7 were stained for aldehyde dehydrogenase activity; benzaldehyde (4 mM) was the substrate and NAD (4 mM) was the cofactor.

Whether it is a Type-1 or Type-2 ALDH-3 remains to be established; the latter is the obvious expectation. Glutathione S-transferase activity was also induced by the polycyclic aromatic hydrocarbons, Figure 6, but the magnitude of the induction was much less, *viz.*, 3- to 5-fold *versus* 150- to 350 fold. NAD(P)H:menadione oxidoreductase activity was also induced (about 3-fold with 3-methylcholanthrene) but NADPH-dependent enzyme-catalyzed reduction of glucuronate (10 mM) or glyceraldehyde (10 mM), measured at 37°C as described by Ris and Wartburg (1973) except that the reaction mixture contained 0.1 mM pyrazole, was not (data not shown).

As judged by experiments with 3-methylcholanthrene, induction of ALDH-3 and glutathione S-transferase is concentration-dependent, Figure 7. That there is a lack of parallelism between the induction of ALDH-3 and that of glutathione S-transferase as a function of inducing agent, Figure 6, or concentration thereof, Figure 7, has been noted. Also noted was that at concentrations of 3 μM, each of the polycyclic aromatic hydrocarbons not only induced ALDH-3 and glutathione S-transferase activities, but also inhibited the proliferation of MCF-7/0 cells, Figure 6. Attempts to achieve induction without inhibiting cell proliferation, and *vice versa*, were unsuccessful, Figure 7. ALDH-3 and glutathione S-transferase were measurably induced, and cellular proliferation was measurably inhibited, by concentrations of 3-methylcholanthrene as low as 0.01 μM; neither was observed when the 3-methylcholanthrene concentration was lowered to 0.001 μM (data not shown). Inhibition

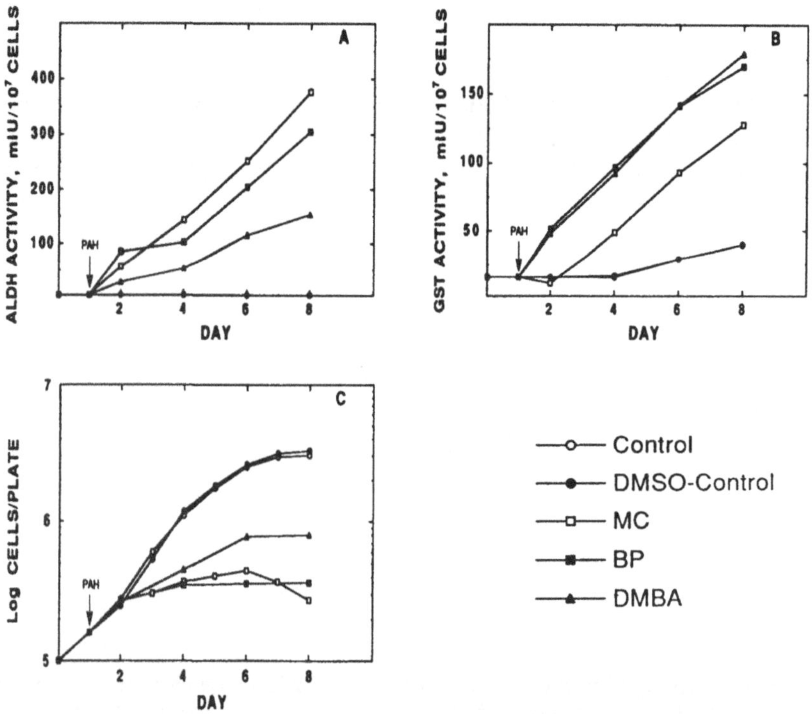

Figure 6. Induction of aldehyde dehydrogenase (ALDH) and glutathione S-transferase (GST) activities in MCF-7/0 cells by polycyclic aromatic hydrocarbons (PAHs). Exponentially growing MCF-7/0 cells were continuously exposed to 3 μM 3-methylcholanthrene (MC), benzpyrene (BP), or 9,10-dimethyl-1,2-benzanthracene (DMBA) for 7 days. Dimethyl sulfoxide (DMSO) was used as the vehicle for these agents. Enzyme activity was quantified in Lubrol®-treated whole homogenates prepared as described in the legend to Figure 2. ALDH activity was quantified as described in the footnote to Table 1; benzaldehyde (4 mM) and NADP (4 mM) were used as substrate and cofactor , respectively, Panel A. The method of Habig *et al.* (1974) was used to quantify GST activity, Panel B. Growth of MCF-7/0 cells in the presence of the test agents is shown in Panel C.

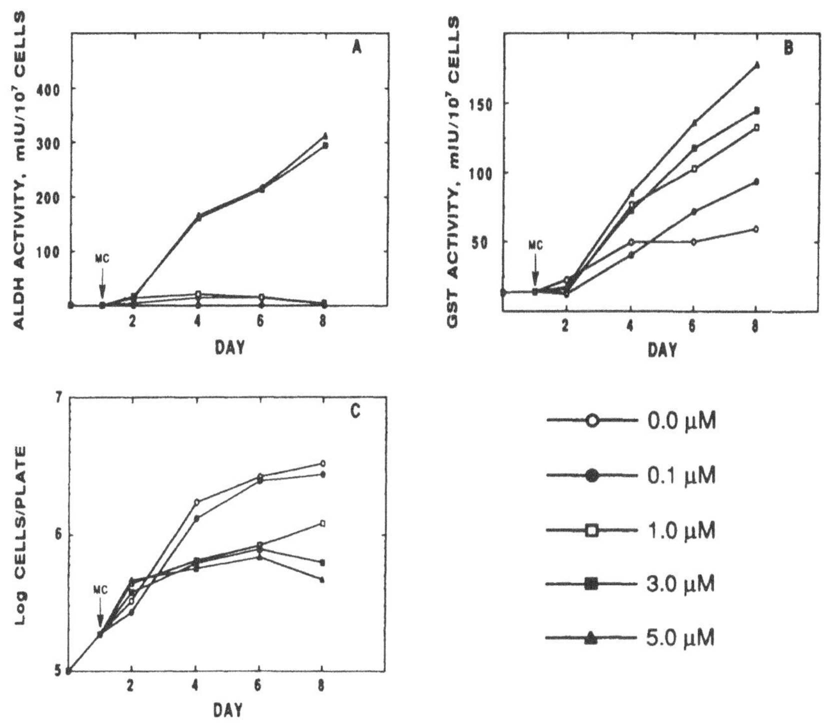

Figure 7. Induction of aldehyde dehydrogenase (ALDH) and glutathione S-transferase (GST) activities in MCF-7/0 cells by 3-methylcholanthrene (MC): Concentration dependence. Experimental details are as described in the legend to Figure 6 except that only one agent, *viz.*, 3-methylcholanthrene, was evaluated, and several concentrations of this agent were tested.

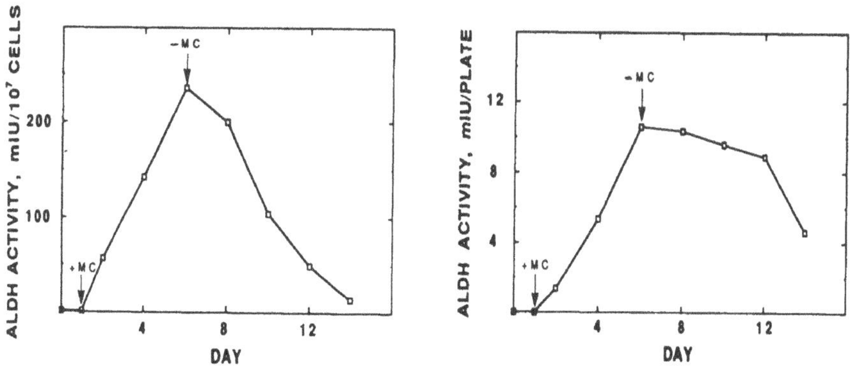

Figure 8. Effect of adding, and then removing, 3-methylcholanthrene (MC) on aldehyde dehydrogenase (ALDH) activity in MCF-7/0 cells. Exponentially growing MCF-7/0 cells were grown in the presence of 3 µM MC for 5 days. At the end of this time, cells were harvested, washed, resuspended in MC-free growth medium and cultured for an additional 8 days. ALDH activity was quantified at the times indicated as described in the legend to Figure 6.

of cell proliferation was virtually immediate and complete at 10 μM 3-methylcholanthrene; in fact, the number of cells present 24 h after 3-methylcholanthrene was first introduced was substantially less than the number present when it was first added (data not shown). However, conclusive evidence demonstrating that enzyme induction and inhibition of cell proliferation by 3-methylcholanthrene are inexorably, or perhaps even causally, related remains to be obtained.

ALDH-3- and glutathione S-transferase-catalyzed metabolism was not induced when MCF-7/0 cells were grown in the presence of 3 μM 2-acetylaminofluorene, diethylnitrosamine, ethionine, cyclophosphamide, mafosfamide or phenobarbital for 7 days, nor was it induced when they were grown in the presence of 300 μM 2-acetylaminofluorene, 300 μM cyclophosphamide or 5 μM mafosfamide for 16 weeks (data not shown).

ALDH-3 activity returned to basal levels when the inducing agent was removed from the growth medium, Figure 8. Shown is that, if the fact that the number of cells per plate increases daily when the inducer is removed (data not presented) is taken into account, the return of enzyme activity to basal levels begins in earnest about 3 days after the inducing agent is removed from the growth medium. This is in contrast to our experience with MCF-7/OAP cells where ALDH-3 levels have remained elevated after the many cell divisions (population doubling time is about 30 h) that occur during a time period of at least 2 months (data not presented). Thus, ALDH-3 levels can be constitutively low, *e.g.*, in normal breast tissue, constitutively high, *e.g.*, in MCF-7/OAP cells, or transiently high, *e.g.*, in polycyclic aromatic hydrocarbon-treated MCF-7/0 cells. The molecular biology underlying the production of ALDH-3, and regulation thereof, is under investigation.

Associated with the induction of ALDH-3 activity in the MCF-7/0 cell line by 3-methylcholanthrene was a loss of sensitivity to mafosfamide by these cells, Figure 9. LC_{90} values for mafosfamide were 60 μM (vehicle-treated MCF-7/0 cells) and >2 mM (3-methylcholanthrene-treated MCF-7/0 cells). The acquired resistance exhibited by the latter was oxazaphosphorine-specific as judged by the fact that, as compared to vehicle-treated MCF-7/0 cells, 3-methylcholanthrene-treated MCF-7/0 cells were only slightly less sensitive to phosphoramide mustard; LC_{90} values for phosphoramide mustard were 860 μM (vehicle-treated MCF-7/0 cells) and 1.43 mM (3-methylcholanthrene-treated MCF-7/0 cells).

Whether increased expression of ALDH-3 accounts for the oxazaphosphorine-specific acquired resistance induced by 3-methylcholanthrene is uncertain, but, once again, this hypothesis appears to be the most attractive.

Reported herein are two different models in which a markedly elevated level of ALDH-3 activity is associated with an oxazaphosphorine-specific resistance of substantial magnitude. In the one case, MCF-7/0 cells were grown in the presence of increasing concentrations of 4-hydroperoxycyclophosphamide over a period of many months until a resistant subline, MCF-7/OAP, emerged (Frei *et al.*, 1988). Resistance and elevated ALDH-3 levels were relatively stable; each was retained over a period of at least two months when MCF-7/OAP cells were grown in the absence of the mutagenic selecting agent. In the other case, MCF-7/0 cells were grown in the presence of a low concentration of 3-methylcholanthrene for only 5 days. Elevated ALDH-3 levels returned to basal levels in a few days when these cells were grown in the absence of the inducing agent; not known at this time is whether resistance to oxazaphosphorines is also lost at this time.

The possibility that oxazaphosphorine-specific acquired resistance may be, in some cases, short-term (transient) rather than long-term or permanent is of obvious relevance to the design of optimal cancer chemotherapeutic strategies.

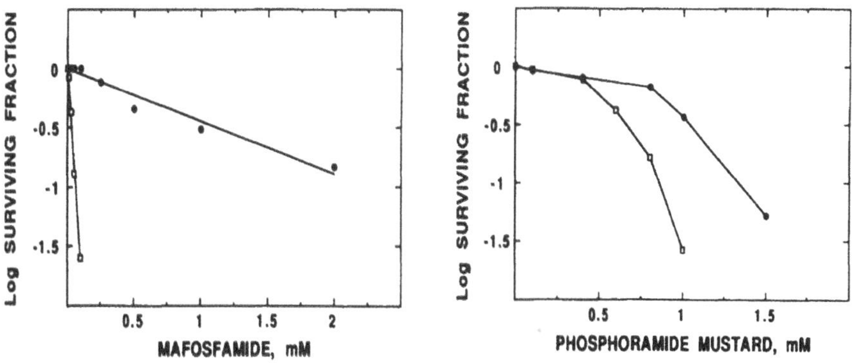

Figure 9. Oxazaphosphorine-specific acquired resistance induced by 3-methylcholanthrene in human MCF-7/0 breast adenocarcinoma cells. MCF-7/0 (□) and MCF-7/0 cells grown in the presence of 3-methylcholanthrene (3 μM) for 5 days (●) were submitted to trypsinization and then harvested by low-speed centrifugation. After washing once, they were resuspended in drug-exposure medium (1 x 10^5 cells/ml), pH 7.4, and exposed to mafosfamide or phosphoramide mustard for 30 min at 37°C. At the end of the 30 min incubation period, the cells were placed in an ice-bath and allowed to chill for 5 min. They were then harvested by low-speed centrifugation, washed with drug-free growth-medium (see footnote to Table 1 for composition), and resuspended in growth medium at concentrations that allowed the transfer, in triplicate, of 1000 and 100 cells in the case of MCF-7/0 cells (plating efficiency ~50%), and 2000 and 200 cells in the case of MCF-7/0 cells treated with 3-methylcholanthrene (plating efficiency ~20%), each in a volume of 1 ml, to 60 x 15 mm petri dishes containing 4 ml of growth medium. The cells were allowed to grow at 37°C in a fully humidified 5% CO_2 in air atmosphere for 15 days in the case of MCF-7/0 cells, and 20 - 25 days in the case of MCF-7/0 cells treated with 3-methylcholanthrene, after which time the medium was removed, cells were stained with methylene blue dye, and colonies (≥ 50 cells) were counted. NADP-linked aldehyde dehydrogenase activities in Lubrol®-treated whole homogenates of the MCF-7/0, and 3-methylcholanthrene-treated MCF-7/0, cells used in the above experiment were 1 and 310 nmol benzaldehyde oxidized/min/10^7 cells, respectively.

Reiterating, Type-2 ALDH-3 may very well be an important determinant of cellular sensitivity to the oxazaphosphorines but this remains to be proven. It follows that Type-1 ALDH-3 may also function in this manner although this possibility is even more remote since Type-1 ALDH-3 does not catalyze the oxidation of aldophosphamide, and an inverse relationship between Type-1 ALDH-3 levels and cellular sensitivity to oxazaphosphorines has yet to be reported. Never-the-less, the idea is attractive, especially since the Type-1 enzyme may have a much wider tissue distribution than does the Type-2 enzyme, Table 3. Equally relevant is that ALDH-3 is not found in all normal tissue nor is it found in all

Table 3. Tissue distribution of class 3 aldehyde dehydrogenases.

Tissue	Type 1[a]	Type 2[b]	?[c]	Reference[d]
Stomach mucosa	++++			1-9
Cornea	++++			10-11
Small intestine mucosa	++			6, 12-13
Lung	++/+/-			1, 5-6, 8-9, 12, 14
Testis	+			6, 8
Spleen	±			1, 6, 8
Skin	±			7-8
Kidney	±			1, 6, 8- 9, 12
Liver	±			1, 3, 5-9, 12, 15
PAH-treated cultured hepatocytes			++++	16
PAH-treated cultured hepatoma cells (Hep G2)			++++	16
Primary hepatocellular carcinoma	++++			17
Breast		+		9, 18, 19
Cultured breast adenocarcinoma cells (MCF-7/0)		+		9, 18, 19
PAH-treated cultured breast adenocarcinoma cells (MCF-7/0)			++++	19
Cultured OSAR breast adenocarcinoma cells (MCF-7/OAP)		++++		9, 18, 19
Primary colon adenocarcinoma			+++	20
Cultured colon adenocarcinoma cell lines HCT 116 & 116a	+			19
Cultured colon adenocarcinoma cell line C	++++			19
Cultured ovarian cystadenocarcinoma cells (JAM)	+++			7

[a]Type 1 ALDH-3: pI 5.75 - 6.35; Mr 108 kDa; subunit Mr 54.5 kDa.
[b]Type 2 ALDH-3: pI 6.0 - 6.45; Mr 125 kDa; subunit Mr 40 kDa.
[c]Information available is insufficient to permit assignment to a specific type.
[d]1 - Harada *et al.*, 1978; 2 - Teng, 1981; 3 - Ricciardi *et al.*, 1983; 4 - Meier-Tackmann *et al.*, 1984; 5 - Duley *et al.*, 1985; 6 - Santisteban *et al.*, 1985; 7 - Parsons *et al.*, 1990; 8 - Goedde and Agarwal, 1990; 9 - Sreerama and Sladek, 1992; 10 - Holmes, 1988; 11 - Gondhowiardjo *et al.*, 1991; 12 - Hopkinson *et al.*, 1985; 13 - Agarwal and Goedde, 1989; 14 - Yin *et al.*, 1992; 15 - Forte-McRobbie and Pietruszko, 1985, 16 - Marselos *et al.*, 1987; 17 - Agarwal *et al.*, 1989; 18 - Sreerama and Sladek, 1991; 19 - Present investigation; 20 - Marselos and Michalopoulos, 1987.

malignant tissue, Table 4. Moreover, the polycyclic aromatic hydrocarbons do not (transiently) induce ALDH-3 in all tissues, whether malignant or nonmalignant. It may be that these agents induce the enzyme in a given cell only if it is already constitutively expressed, however poorly, in that cell. Finally, whether cells that ordinarily do not express ALDH-3 (even at low levels) can be caused to do so on a permanent basis by exposing them to a mutagen such as one of the oxazaphosphorines, also remains to be determined.

An endogenous substrate for ALDH-3 has yet to be identified and its biological role is unknown. Perhaps its *raison d'etre* is to detoxify xenobiotics. Supporting this notion is that tissues which come in direct contact with xenobiotics present in food, water and air, *e.g.*, stomach mucosa, intestinal mucosa and lungs, contain large amounts of this enzyme. Apparently inconsistent with this notion is the high level of ALDH-3 in the cornea although a similar role, *viz.*, protection against UV-light induced cellular damage, has been proposed for ALDH-3 in this tissue (Holmes, 1988). Alternatively, ALDH-3 levels, as well as the levels of other products of the Ah receptor/agonist-activated gene battery, may be high in tissues that constitute "ports of entry" because they come in direct contact with relevant inducing agents that are present in the environment. The elevated ALDH-3 levels may then serve to protect such cells from any further damage by these agents.

Table 4. Human tissues lacking a class 3 aldehyde dehydrogenase.

Tissue	Reference[a]
Muscle	1-4
Placenta	3, 5-6
Heart	2, 4, 7-8
Brain	1-4
Prostate	2
Erythrocytes	1-4
Hair roots	2, 4, 7
Cultured fibroblasts	2, 8
Adrenals	9
Ovary	10
Cultured ovarian tumor cell lines GG, CI-80-13S, OVCAR-3 & OVCAR-432	10, 11
PAH-treated cultured ovarian tumor cell lines OVCAR-3 & OVCAR-432	11
Cultured myeloma cell lines 8226/S & 8226/LR-5	11
PAH-treated cultured myeloma cell line 8226/S	11
Cultured melanoma cell lines MM96L, MM253cl & MM418	10
Cultured HeLa cells (carcinoma of the cervix)	10
Rhadomyosarcoma xenografts HxRh10, HxRh12, HxRh18, HxRh28 & HxRh30	11
Primary B-Cell and T-Cell ALL	11
Primary AML, CLL, CML & NHL	11

[a]1 - Harada *et al.*, 1978; 2 - Santisteban *et al.*, 1985; 3 - Goedde and Agarwal, 1990, 4 - Harada *et al.*, 1980; 5 - Sreerama and Sladek, 1992; 6 - Meier-Tackmann *et al.*, 1985; 7 - Goedde *et al.*, 1980; 8 - Goedde *et al.*, 1979; 9 - Duley *et al.*, 1985; 10 - Parsons *et al.*, 1990; 11 - Sladek, Sreerama and Dockham, unpublished observations.

ACKNOWLEDGEMENTS

Financial support was provided by Bristol-Meyers Squibb Company Grant 100-R220. The authors gratefully acknowledge the technical assistance provided by Mr. Matthew Hedge.

REFERENCES

Agarwal, D. P., and Goedde, H. W., 1989, Human aldehyde dehydrogenases: Their role in alcoholism, *Alcohol*, 6:517.

Agarwal, D. P., Eckey, R., Rudnay, A. C., Volkens, T., and Goedde, H. W., 1989, "High Km" aldehyde dehydrogenase isozymes in human tissues: Constitutive and tumor-associated forms, *Prog. Clin. Biol. Res.*, 290:119.

Davis, B. J., 1964, Disc electrophoresis-II method and application to human serum proteins, *Ann. N. Y. Acad. Sci.*, 121:404.

Dockham, P. A., Lee, M. O., and Sladek, N. E., 1992, Identification of human liver aldehyde dehydrogenases that catalyze the oxidation of aldophosphamide and retinaldehyde, *Biochem. Pharmacol.*, 43:2453

Duley, J. A., Harris, O., and Holmes, R. S., 1985, Analysis of human alcohol- and aldehyde-metabolizing isozymes by electrophoresis and isoelectric focusing, *Alcoholism: Clin. Exp. Res.*, 9:263.

Forte-McRobbie, C. M., and Pietruszko, R., 1985, Aldehyde dehydrogenase content and composition of human liver, *Alcohol*, 2:375.

Frei, E. III., Teicher, B. A., Holden, S. A., Cathcart, K. N. S., and Wang, Y., 1988, Preclinical studies and correlation of the effect of alkylating dose, *Cancer Res.*, 48:6417.

Goedde, H. W., Agarwal, D. P., and Harada, S., 1979, Alcohol metabolizing enzymes: Studies of isozymes in human biopsies and cultured fibroblasts, *Clin. Genet.*, 16:29.

Goedde, H. W., Agarwal, D. P., and Harada, S., 1980, Genetic studies on alcohol-metabolizing enzymes: Detection of isozymes in human hair roots, *Enzyme*, 25:281.

Goedde, H. W., and Agarwal, D. P., 1990, Pharmacogenetics of aldehyde dehydrogenase (ALDH), *Pharmac. Ther.*, 45:345.

Gondhowiardjo, T. D., van Haeringen, N. J., Hoekzema, R., Pels, L., and Kijlstra, A., 1991, Detection of aldehyde dehydrogenase activity in human corneal extracts, *Curr. Eye Res.*, 11:1001.

Habig, W. H., Pabst, M. J., and Jakoby, W. B., 1974, Glutathione S-transferases, *J. Biol. Chem.*, 249:7130.

Harada, S., Agarwal, D. P., and Goedde, H. W. , 1978, Isozyme variations in acetaldehyde dehydrogenase (EC 1.2.1.3) in human tissues, *Hum Genet.*, 44:181.

Harada, S., Agarwal, D. P., and Goedde, H. W., 1980, Electrophoretic and biochemical studies of human aldehyde dehydrogenase isozymes in various tissues, *Life Sci.*, 26:1773.

Hilton, J., 1984, Role of aldehyde dehydrogenase in cyclophosphamide-resistant L1210-leukemia, *Cancer Res.*, 44: 5156.

Holmes, R. S., 1988, Alcohol dehydrogenases and aldehyde dehydrogenases of anterior eye tissues from humans and other mammals, in "Biomedical and Social Aspects of Alcohol and Alcoholism," K. Kuriyama, A. Takada, and H. Ishii, eds., Elsevier Science Publishers, Amsterdam, The Netherlands, p. 51.

Hopkinson, D. A., Santisteban, I., Povey, S., and Smith, M., 1985, Biochemical genetic analysis of human and rodent aldehyde dehydrogenase (ALDH), *Alcohol*, 2:73.

Huang, M., and Lindahl, R., 1990, Effects of hepatocarcinogenic initiators on aldehyde dehydrogenase gene expression in cultured rat hepatic cells, *Carcinogenesis*, 11:1059.

Kohn, F. R., Landkamer, G. J., Manthey, C. L., Ramsay, N. K. C., and Sladek, N. E., 1987, Effect of aldehyde dehydrogenase inhibitors on the *ex vivo* sensitivity of human multipotent and committed hematopoietic progenitor cells and malignant blood cells to oxazaphosphorines, *Cancer Res.*, 47:3180.

Lindahl, R., 1992; Aldehyde dehydrogenases and their role in carcinogenesis, *Crit. Rev. Biochem. Mol. Biol.*, in press.

Manthey, C. L., Landkamer, G. J., and Sladek, N. E., 1990, Identification of the mouse aldehyde dehydrogenases important in aldophosphamide detoxification, *Cancer Res.*, 50:4991.

Marselos, M., and Michalopoulos, G., 1987, Changes in the patterns of aldehyde dehydrogenase activity in primary and metastatic adenocarcinomas of human colon, *Cancer Letters*, 34:27.

Marselos, M., Strom, S. C., and Michalopoulos, G., 1986, Enhancement of aldehyde dehydrogenase activity in human and rat hepatocyte cultures by 3-methylcholanthrene, *Cell Biol. Toxicol.*, 2:257

Marselos, M., Strom, S. C., and Michalopoulos, G., 1987, Effect of phenobarbital and 3-methylcholanthrene on aldehyde dehydrogenase activity in cultures of HepG2 cells and normal human hepatocytes, *Chem. Biol. Interactions*, 62:75.

Meier-Tackmann, D., Agarwal, D. P., Saha, N., and Goedde, H. W., 1984, Aldehyde dehydrogenase isozymes in stomach autopsy specimens from Germans and Chinese, *Enzyme*, 32:170.

Meier-Tackmann, D., Korenke, G. C., Agarwal, D. P., and Goedde, H. W., 1985, Human placental aldehyde dehydrogenase, *Enzyme*, 33:153.
Parsons, P. G., Lean, J., Kable, E. P. W., Favier, D., Khoo, S. K., Hurst, T., Holmes, R. S., and Bellet, A. J. D., 1990, Relationships between resistance to cross-linking agents and glutathione metabolism, aldehyde dehydrogenase isozymes and adenovirus replication in human tumor cell lines, *Biochem Pharmacol.*, 40:2641.
Radin, A. I., Zhoa, X. L., Woo, T. H., Colvin, O. M., and Hilton, J., 1991, Structure and expression of the cytosolic aldehyde dehydrogenase gene in cyclophosphamide-resistant murine leukemia L1210 cells, *Biochem. Pharmacol.*, 42:1933.
Ricciardi, B. R., Saunders, J. B., Williams, R., and Hopkinson, D. A., 1983, Hepatic ADH and ALDH isozymes in different racial groups and in chronic alcoholism, *Pharmcol. Biochem. Behav.*, 18:61.
Ris, M. M., and von Wartburg, J. P., 1973, Heterogeneity of NADPH-dependent aldehyde reductase from human and rat brain, *Eur. J. Biochem.*, 37:69.
Santisteban, I., Povey, S., West, L. F., Parrington, J. M., and Hopkinson, D. A., 1985, Chromosome assignment, biochemical and immunological studies on a human aldehyde dehydrogenase, ALDH3, *Ann. Hum. Genet.*, 49:87.
Sladek, N. E., 1988, Metabolism of oxazaphosphorines, *Pharmac. Ther.*, 37: 301.
Sladek, N. E., 1992, Oxazaphosphorine-specific acquired resistance, in "Mechanisms of Drug Resistance in Oncology," B. A. Teicher, ed., Marcel Dekker, New York and Basel, in press.
Sladek, N. E., and Landkamer, G. J., 1985, Restoration of sensitivity to oxazaphosphorines by inhibitors of aldehyde dehydrogenase activity in cultured oxazaphosphorine-resistant L1210 and cross-linking agent resistant P388 cell lines, *Cancer Res.*, 45:1549.
Sladek, N. E., Low, J. E., and Landkamer, G. J., 1985, Collateral sensitivity to cross-linking agents exhibited by cultured L1210 cells resistant to oxazaphosphorines, *Cancer Res.*, 45:625.
Sreerama, L., and Sladek, N. E., 1991, Apparently resistance-irrelevant elevated aldehyde dehydrogenase-3 activity in a breast adenocarcinoma cell line exhibiting oxazaphosphorine-specific acquired resistance, *Proc. Am. Assoc. Cancer. Res.*, 32:352.
Sreerama, L., and Sladek, N. E., 1992, Identification of a novel class 3 aldehyde dehydrogenase over-expressed in a human breast adenocarcinoma cell line (MCF-7/OAP) exhibiting oxazaphosphorine-specific acquired resistance, *Proc. Am. Assoc. Cancer Res.*, 33:447.
Teng, Y.-S., 1981, Stomach aldehyde dehydrogenase: Report of a new locus, *Hum. Hered.*, 31:74.
Vickers, P. J., Dufresne, M. J., and Cowan, K. H., 1989, Relation between cytochrome P450IA1 expression and estrogen receptor content of human breast ceancer cells, *Mol. Endocrinol.*, 3:157.
Yin, S.-J., Liao, C.-S., Chen, C.-M., Fan, F.-T., and Lee, S.-C., 1992, Genetic polymorphism and activities of human lung alcohol and aldehyde dehydrogenases: Implications for ethanol metabolism and cytotoxicity, *Biochem. Gent.*, 30:203.

TUMOR-ASSOCIATED ALDEHYDE DEHYDROGENASE (ALDH3): EXPRESSION IN DIFFRENT HUMAN TUMOR CELL LINES WITH AND WITHOUT TREATMENT WITH 3-METHYLCHOLANTHRENE

Doris Meier-Tackmann, Rolf Eckey,
Christoph Wolff*, Ulrich v. Eitzen,
Dharam P. Agarwal, and H. Werner Goedde

Institute of Human Genetics,
University of Hamburg, Hamburg, F.R.G.

INTRODUCTION

Many past studies from various laboratories have shown that several isozymes of aldehyde dehydrogenase (ALDH, EC 1.2.1.3) can be induced in animals by various xenobiotics and carcinogens such as 2,3,7,8-tetrachlorodibenzo-p-dioxin (TCDD), 3-methylcholanthrene and phenobarbital (Feinstein et al., 1976; Deitrich et al., 1977). In cultures of normal human hepatocytes and human hepatoma cell line Hep G2, the total ALDH activity has been shown to be increased after application of phenobarbital or 3-methylcholanthrene (Marselos et al., 1987).

In a previous study we have reported the occurence of a unique aldehyde dehydrogenase (ALDH3) in a human primary liver tumor that is generally expressed at a very low level in normal mammalian liver, but at high level in lung, cornea, lens, and gastric mucosa (Agarwal et al., 1989). This tumor-associated ALDH was found to be immunologically identical to the human gastric class 3 ALDH (ALDH3) but differed substantially from the human class 1 and class 2 ALDHs. Moreover, the kinetic data indicated a significant parallel to the rat liver tumor-associated ALDH (Feinstein, 1975). In constrast to the major liver ALDH isozymes (ALDH1 and ALDH2), which are NAD-dependent and predominantly oxidize aliphatic aldehydes, ALDH3 prefers aromatic aldehydes as substrate and can use both NAD^+ and $NADP^+$ as coenzyme.

Resistance of certain tumor cells to cyclophosphamide is attributed to the activity of ALDH, the enzyme capable of oxidizing aldophosphamide, the cytotoxic metabolite of cyclophosphamide (Lin and Lindahl, 1987; Manthey and Sladek, 1989; Sladek et al., 1989; Dockham et al., 1992).

*A substantial part of this work will be included in the MD dissertation to be submitted by C. Wolff to the Medical Faculty, University of Hamburg.

Enzymology and Molecular Biology of Carbonyl Metabolism 4
Edited by H. Weiner, Plenum Press, New York, 1993

In the present study we have examined the expression and immunological properties of ALDH isozymes in human tumor cell lines. Specifically, the expression and function of tumor-associated ALDH3 in 4 different tumor cell lines and in normal cells of the corresponding tissues were studied.

MATERIALS AND METHODS

Autopsy samples of human tissues were obtained from the Department of Legal Medicine, University of Hamburg. The samples were washed in water and kept frozen at -20°C until used. Extracts were prepared by homogenizing the tissues in a 30 mM sodium phosphate buffer, pH 6.0, (1 ml buffer/g tissue).

The human hepatoma cell line Hep G2 was derived from a liver tumor, the HT-1080 tumor cell line was derived from a fibrosarcoma and the skin fibroblast cell line (#236) was established by Dr. I. Willers (Institute of Human Genetics, University of Hamburg) from a healthy donor. A549 cells, derived from a lung adenocarcinoma, and UMSCC2 cells (recently isolated from a human pharynx carcinoma by Dr. T. Carey, University of Michigan) were a gift from E. Schuuring, The Netherland Cancer Institute, Amsterdam.

The tumor cell lines were grown in minimal essential medium (MEM, Gibco) supplemented with 10% fetal calf serum (FCS), 2.2 g/l $NaHCO_3$, and 1.25 ml/l gentamycin. The cultures were kept at 37°C, 5% CO_2 in air condition, and 90% humidity. The fibroblast cell line was grown in the same medium composition supplemented with 15% FCS and was kept at 37°C in a humidified atmosphere of 5% CO_2 and 5% O_2. 3-methylcholanthrene dissolved in DMSO was added to the culture medium at final concentrations of 2.5 or 5 μM or 1 mM. After 3 or 4 days the cells were washed in phoshate-buffered saline solution (PBS: 0.01 M phosphate buffer, 0.9% NaCl) and harvested by scraping the cells in the same solution. Washed and packed cells were stored at -20°C. Cell extracts were prepared by thawing and homogenization in a micro-Potter followed by centrifugation at 27,000 x g for 10 minutes.

ALDH activity in the supernatants was determined spectrophotometrically at 340 nm and 25°C in a 0.1 M pyrophosphate buffer, pH 9.5, using either 5 mM acetaldehyde and 2 mM NAD^+ for ALDH1, ALDH2 and ALDH4 or 6 mM 3-nitrobenzaldehyde (3-NBA) and 2 mM $NADP^+$ for ALDH3. The ratio of the activity with 3-NBA + $NADP^+$ to the activity with acetaldehyde + NAD^+ was determined. A ratio below 1.0 indicated a very low or missing ALDH3 activity.

Isoelectric focusing was performed on 0.75% agarose gels, containing 12% D-sorbitol, 1% polyethylene glycol 6000, and 7% carrier ampholytes, pH range 5-8 or 3.5-9.5. After focusing at constant voltage of 400 V with a maximum of 20 W for 2 hours the gels were stained for ALDH activity with an agarose overlay gel using propionaldehyde/NAD^+ or 3-NBA/$NADP^+$.

The preparation of monospecific rabbit antibodies to human stomach ALDH3 and Ouchterlony double diffusion tests were carried out as previously described (Eckey et al., 1990).

The protein concentration was estimated according to Macart and Gerbaut (1982) with bovine serum albumin as the standard.

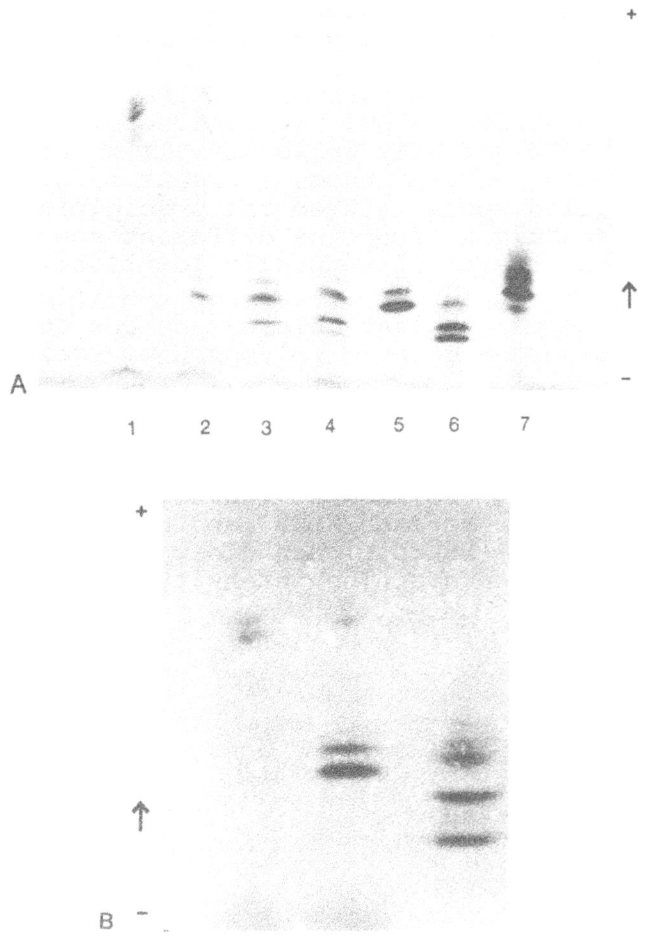

Figure 1. ALDH isoenzyme pattern of different human tissue and tumor cell line extracts after isoelectric focusing in a pH range of 3.5-9.5 (A) or 3.5-8.0 (B) and ALDH-specific staining with 3-NBA/NADP$^+$. Samples in A: 1, liver; 2, stomach (autopsy); 3, stomach (biopsy); 4, pharynx; 5, UMSCC2 cells; 6, A549 cells; 7, purified stomach ALDH3.
Samples in B: 1, Hep G2 cells; 2, Hep G2 cells after exposure to 3-methylcholanthrene; 3, A549 cells.

RESULTS AND CONCLUSIONS

On isoelectric focusing with subsequent ALDH-specific staining, the extracts of UMSCC2 and A549 cells showed 2 or 4 activity bands with isoelectric points ranging between 6.0 and 6.5, similar to the ALDH3 bands of stomach and pharynx mucosa (Fig. 1A). Most intense staining of these bands was obtained with 3-NBA as substrate and $NADP^+$ as coenzyme. By contrast, in diluted homogenates of normal liver (1:10) and in extracts of Hep G2 cells, no corresponding activity bands were detected. After incubation with 3-methylcholanthrene, however, Hep G2 cells showed a banding pattern similar to that of stomach mucosa (Fig. 1B).

In Ouchterlony double diffusion tests, monospecific antibodies to human stomach ALDH3 formed precipitin lines with extracts of UMSCC2 and A549 cells as well as with purified stomach ALDH3 and pharynx mucosa extracts (Fig. 2). The absence of so-called spurs between the precipitin lines indicates that the antigens from the different sources are closely related or even structurally identical.

The observed resolution of stomach-specific ALDH into several bands on pH gradient gels may be due to posttranslational alterations or genetic polymorphism (Meier-Tackmann et al., 1984).

A close structural relationship between the enzyme proteins composed of several activity bands was apparent in immunoprecipitation studies. When extracts of stomach mucosa, UMSCC2 cells and A549 cells were incubated with monospecific rabbit antibodies to stomach ALDH3 and subsequently centrifuged, all the ALDH3-bands either disappeared or showed a substantially reduced staining intensity (Fig. 3).

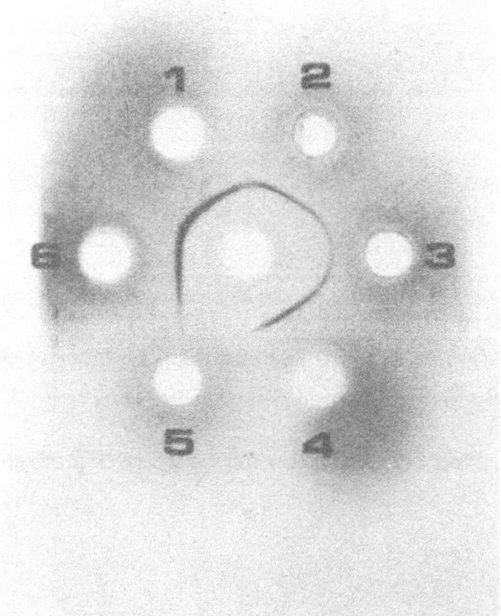

Figure 2. Ouchterlony double-diffusion test with antibodies to human stomach ALDH (center well) and different tissue and cell extracts. Wells 1 and 4, purified stomach ALDH3; well 2, UMSCC2 cells; well 3, pharynx (1:2 diluted); well 5, liver (1:2 diluted); well 6, A549 cells.

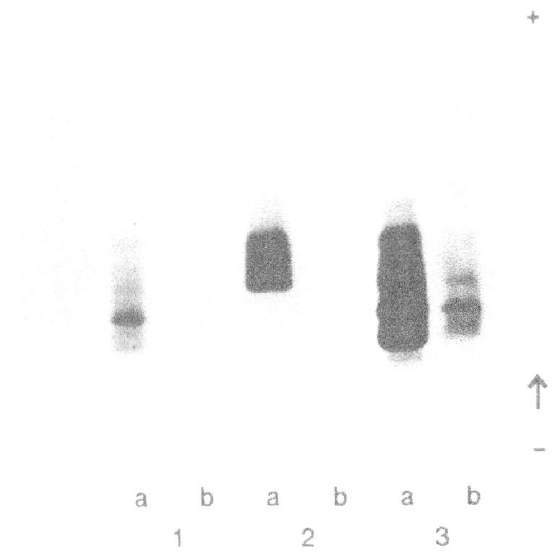

Figure 3. ALDH isoenzyme patterns of stomach mucosa (1), UMSCC2 cells (2), and A549 cells (3) before (a) and after (b) immunoprecipitation with antibodies to human ALDH3.

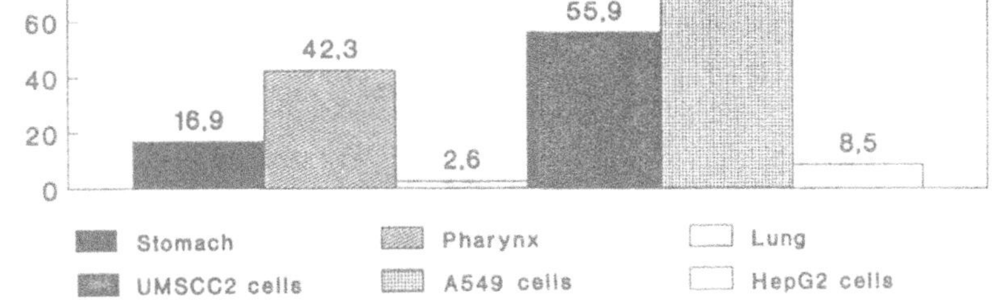

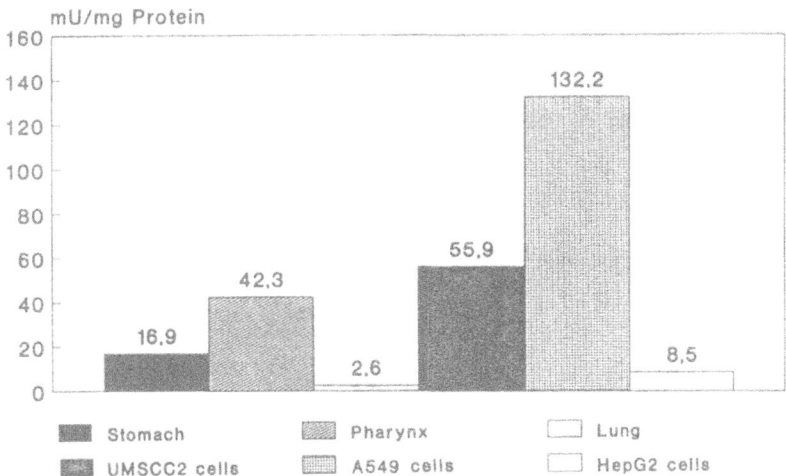

Figure 4. Specific ALDH activity in different tissues and tumor cells determined with 3-nitrobenzaldehyde as substrate and $NADP^+$ as coenzyme.

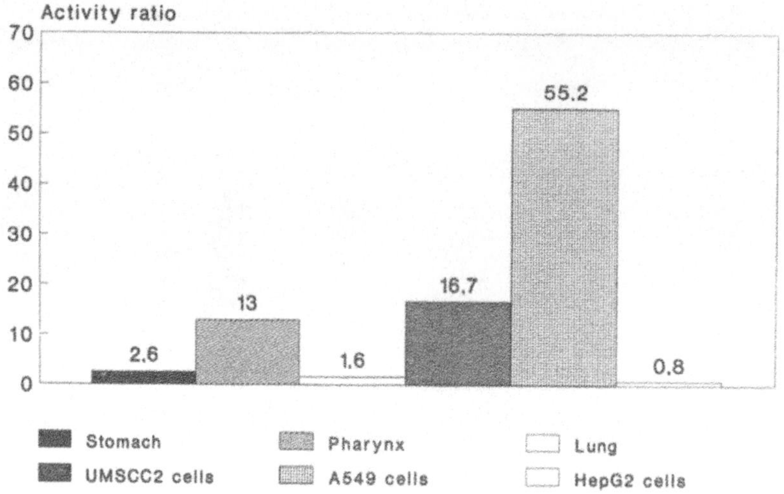

Figure 5. ALDH activity ratio in different tissues and tumor cells calculated from the activities assayed with 3-nitrobenzaldehyde + $NADP^+$ and acetaldehyde + NAD^+.

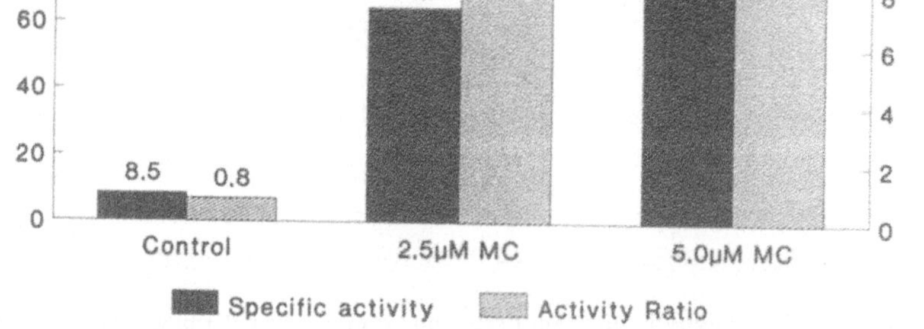

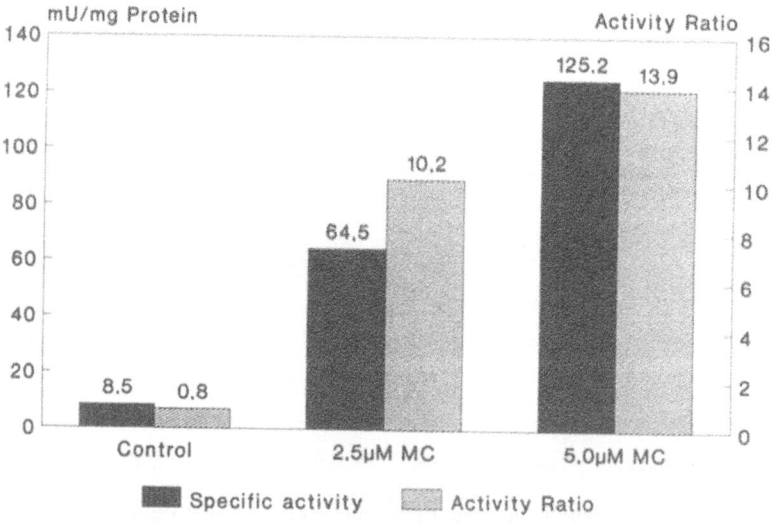

Figure 6. Enhancement of the specific ALDH activity determined with 3-nitrobenzaldehyde + $NADP^+$ and the activity ratios of Hep G2 cells after exposure to 2.5 and 5.0 µM 3-methylcholanthrene.

Specific ALDH activities in extracts of stomach, pharynx and lung tissues as well as in extracts of UMSCC2, A549, and Hep G2 cells were determined spectrophotometrically using 3-NBA + $NADP^+$ (Fig. 4). In the normal tissues, pharynx mucosa showed the highest specific activity towards 3-NBA + $NADP^+$, followed by stomach mucosa and lung tissue. The specific activities in extracts of A549 and UMSCC2 cells were several times higher as compared to extracts of the corresponding normal tissues (Fig. 4).

Expression of ALDH3 in different tissues and cells was reflected by the ratio of the activities determined with 3-NBA + $NADP^+$ and with acetaldehyde + NAD^+ (Fig. 5). In tissues and cells with a substantial ALDH3 expression, activity ratio values of up to 55 were observed (Fig. 5).

In Hep G2 cells, the values for specific enzyme activity and the activity ratio were found to be relatively low due to the absence of ALDH3. After incubation with 2.5 or 5.0 μM 3-methylcholanthrene, however, both the values were enhanced several times (Fig. 6). By contrast, no significant enhancement of the enzyme activity was observed, when UMSCC2 or A549 cells were grown in the presence of 3-methylcholanthrene (results not shown here).

In fibroblast and HT-1080 tumor cells no ALDH activity was detected before and after incubation for 3 days with neither 5 μM nor with 1 mM 3-methylcholanthrene.

These results unequivocally indicate that ALDH3 is inducible in different cells during chemical carcinogenesis - suggesting a role for ALDH3 in the oxidation of aldophosphamide leading to the acquired cyclophosphamide resistance in carcinoma cells.

ACKNOWLEDGMENTS

The authors are indebted to Prof. Dr. K. Püschel, Institute of Legal Medicine, University of Hamburg, for kindly providing autopsy specimens. They also thank Mrs. E. Losenhausen for skillful technical assistance.

REFERENCES

Agarwal, D.P., Eckey, R., Rudnay, A.-C., Volkens, T., and Goedde, H.W., 1989, "High-Km" aldehyde dehydrogenase isozymes in human tissues: constitutive and tumor-associated forms, *in*: Progress in Clin. Biol. Res., vol. 290: "Enzymology and Molecular Biology of Carbonyl Metabolism 2," H. Weiner, T.G. Flynn, eds., Alan R. Liss, Inc., New York, pp 119-31.

Deitrich, R.A., Bludeau, P., Stock, T., and Roper, M., 1977, Induction of different rat liver supernatant aldehyde dehydrogenase by phenobarbital and tetrachlorodibenzo-p-dioxin, *J. Biol. Chem.* 252:6169-76.

Dockham, P.A., Lee, M.-O., and Sladek, N.E., 1992, Identification of human liver aldehyde dehydrogenases that catalyze the oxidation of aldophosphamide and retinaldehyde, *Biochem. Pharmacol.* 43:2453-69.

Eckey, R., Timmann, R., Hempel, J., Agarwal, D.P., and Goedde, H.W., 1990, Biochemical, immunological and molecular characterization of a "high Km" aldehyde dehydrogenase, *in*: Advances in Experimental Medicine and Biology, vol.284: "Enzymology and Molecular Biology of Carbonyl Metabolism 3," H. Weiner, B. Wermuth, D.W. Crabb, eds., Plenum Press, New York, London, pp 43-52.

Feinstein, R.N., 1975, Aldehyde dehydrogenase isozymes in certain hepatomas, *in*:"The Isozymes," vol. III, C.L. Markert, ed., Academic Press, New York, pp 969-86.

Feinstein, R.N., Fry, R.J.M., Cameron, E.C., Paraino, C., and Morris, H.P., 1976, New aldehyde dehydrogenase in chemically induced liver tumors in the rat, *Proc. Soc. Exptl. Biol. Med.* 152:463-8.

Lin, K.-H. and Lindahl, R., 1987, Role of aldehyde dehydrogenase activity in cyclophosphamide metabolism in rat hepatoma cell lines, *Biochem. Pharmacol.* 36:3305-07.

Macart, M. and Gerbaut, L., 1982, An improvement of the Coomassie blue dye binding method allowing an equal sensitivity to various proteins: application to cerebrospinal fluid, *Clin. Chim. Acta* 122:93-101.

Manthey, C.L. and Sladek, N.E., 1989, Aldehyde dehydrogenase-catalyzed bioinactivation of cyclophosphamide, *in*: Progress in Clin. Biol. Res., vol. 290: "Enzymology and Molecular Biology of Carbonyl Metabolism 2," H. Weiner, T.G. Flynn, eds., Alan R. Liss, Inc., New York, pp 49-63.

Marselos, M., Strom, S.C., and Michalopoulos, G., 1987, Effect of phenobarbital and 3-methylcholanthrene on aldehyde dehydrogenase activity in cultures of Hep G2 cells and normal human hepatocytes. *Chem. Biol. Int.* 62:75-88.

Meier-Tackmann, D., Agarwal, D.P., Saha, N., and Goedde, H.W., 1984, Aldehyde dehydrogenase isozymes in stomach autopsy specimens from Germans and Chinese. *Enzyme* 32:170-7.

Sladek, N.E., Manthey, C.L., Maki, P.A., Zhang, Z., and Landkammer, G.J., 1989, Xenobiotic oxidation catalyzed by aldehyde dehydrogenases, *Drug Metab. Rev.* 20:697-720.

SEXUAL DIFFERENTIATION IN THE INDUCTION OF THE CLASS 3 ALDEHYDE DEHYDROGENASE

Maria Karageorgou[1], Constantine Papadimitriou[2], and Marios Marselos[1]

[1]Department of Pharmacology
Medical School, University of Ioannina
[2]Department of Pathology
Medical School, University of Thessaloniki
Greece

INTRODUCTION

Various xenobiotics, also known for their ability to induce a number of different cytochrome P-450 isozymes, can increase the activity of aldehyde dehydrogenase (ALDH, EC 1.2.1.3) in the soluble fraction of rat liver (Deitrich, 1971; Marselos and Haninnen, 1974). It is well known that this induction involves at least two different isozymes which, depending on the type of the inducer, selectively react (Deitrich et al., 1977; Marselos et al., 1979).

Phenobarbital (PB) and other inducers of various P-450 isozymes, such as CYP2B1 and CYP2B2, increase the activity of PB-ALDH. The increase in activity which can only be detected in specific genetically defined strains of rats, is best measured with phenylacetaldehyde and NAD. 2,3,7,8-tetrachlorodibenzo-p-dioxin (TCDD) and other aryl hydrocarbon (Ah) receptor ligands, on the other hand, induce the activity of another isozyme, ALDH3, in all of the rat strains tested (Deitrich et al., 1978; Marselos et al., 1979). This induction process requires a functional Ah receptor, as well as translocation of the ligand-receptor complex to the nucleus (Vasiliou et al., 1992). The ALDH3 enzyme activity is barely detectable in normal liver, although it can be dramatically increased (up to 1000 -fold) by Ah receptor ligands. The activity of this isozyme is best measured with benzaldehyde and NADP (Feinstein and Cameron, 1972; Torronen et al., 1981).

During the past years, our work has focused on various factors which may regulate the expression of inducible ALDH isozymes. Besides TCDD, potent inducers of ALDH3 are planar compounds which possess aromatic rings in their structure, such as methylcholanthrene (MC) (Vasiliou et al., 1988; Marselos and Vasiliou, 1991). After a single dose of MC (20-50 mg/kg), the induction of ALDH3 reaches a peak following a lag period of about five days. While this effect of MC is blocked by inhibitors of microsomal oxidations, it is enhanced when the animals have been pretreated with inducers of microsomal mixed-function oxidases (Marselos et al., 1987; Vasiliou and Marselos, 1989a; b). This suggests that the eventual inducer of ALDH3 may be one or more of the metabolites of MC.

Sex differences in drug metabolism have been well documented in various experimental animals, and is in general agreement with the different levels of activity of the enzymes concerned. Depending upon the substrate, metabolism may be more efficient in either the male or female. Similarly, the induction of the microsomal enzymes is often more pronounced in one sex compared to the other (Burke et al., 1978). Sex differences in the microsomal, mixed-function oxidase isystem appears at puberty, a phenomenon defined as "imprinting". Presumably, imprinting involves various neuroendocrinological mechanisms, where the balance of gonadal sex hormones appear to play an important role (Skett, 1987 and references therein).

Antiestrogenic effects of TCDD have been described in several experimental paradigms. TCDD inhibits estrogen-induced uterine imbibition in weanling CD1 mice (Gallo et al., 1986). C57BL/6 female mice treated with TCDD had greatly reduced uterine weights, histopathological changes in the uterus and irregular estrus (Umbreit et al., 1987). When administered subchronically, TCDD and MC also reduce the number of estrogen receptors (ER) in the liver (Romkes et al., 1987). To date, it is unknown whether the administration of a nonsteroidal xenobiotic, like MC, can modify the phenomenon of imprinting.

In this chapter, we show for the first time that the induction of ALDH3 is differentially expressed in adult male and female rats. In addition, we present data which demonstrate that treatment with MC during puberty abolishes sex differences in the biological response to MC later in adult life.

MATERIALS AND METHODS

Chemicals

All chemicals used were reagent grade. Pyrazole was obtained from Fluka (Germany), and propionaldehyde and benzaldehyde from Ferak (Germany). All other chemicals were obtained from Sigma Chemical Co. (U.S.A.).

Animals

Wistar rats (weighing 200-250 g), originating from the Animal Center of the University of Kuopio, Finland, of the substrain Mol/Io/rr were used. Animals were housed in groups of 5 in plastic cages (Makrolon) with wood bedding (Populus sp.), under constant temperature (20 C) and a light-dark cycle of 12h. Pelleted chow (EL.VI.Z., Greece) and tap water were available *ad libitum.*

In order to establish possible differences in the degree of induction of ALDH3 due to sex and age, animals of both sexes were sacrificed 5 days following a single, i.p. injection of MC. Separate groups of rats were treated at the ages of 20, 30, 40, 50, 60 and 90 days (N = 6/group). Controls were treated with vehicle (olive oil, Minerva, Greece).

Adult female animals (2 groups of 6 animals) were treated acutely with MC (50 mg/kg, i.p., for 2 days) and were sacrificed 5 days later, for the determination of estrogen receptor content in the liver and the uterus. Olive oil was given to the controls.

In another series of experiments, 12 rats of each sex were treated with MC (20 mg/kg, i.p., every second day) during puberty (between 20 and 30 days of age in this species). An equal number of animals were treated with olive oil to serve as controls. At 90 days of age, half of the animals (6 males and females pretreated with MC at puberty, as well as 6 males and females pretreated with olive oil) were given a single dose of MC (50 mg/kg). The other half of the animals were given an injection of olive oil. All animals were sacrificed 5 days later, in order to measure ALDH3 activity.

Tissue Preparation

Immediately following decapitation, the livers were disected and homogenized with a teflon pestle in 3 vol (w/v) of an ice-cold 0.25 M sucrose solution. The homogenate was first spun at 10,000 x g for 30 min at 0-4 C. The supernatant was then centrifuged at 100,000 x g for 1h at 0-4 C and the soluble fraction was collected and used for the enzyme activity and estrogen receptor assays.

Enzyme Activity

The class 3 ALDH is best measured using benzaldehyde and NADP (B/NADP activity) (Torronen et al., 1981). In the present series of experiments ALDH3 activity was assessed by measuring B/NADP activity in cytosol, as previously described (Marselos et al., 1986;

1987). Protein determination was carried out using the biuret method with bovine serum albumin as the standard (Gornall et al., 1949).

Estrogen Receptors

The concentration of estrogen receptors was determined in hepatic and uterine cytosol. An Enzyme Immunoassay Antibody Monoclonal System (Abbott GmbH Giagnostika, Wiesbaden-Denkelheim, Germany) was used according to the detailed instructions supplied by the manufacturer.

Statistical Analysis

Results were analyzed by Analysis of Variance (ANOVA) or Student's t-test, using a PC software package (STATGRAPHICS, Statistical Graphics Corporation, U.S.A.).

RESULTS

Basal ALDH3 activity is very low in both sexes and at all ages. The activity induced by MC, however, had a different profile and increased with age. Although until 20 days of age activity was similar in both sexes, after puberty, females consistently demonstrated a greater amount of ALDH3 activity inducibility compared to males (Table 1).

TABLE 1. Effect of age and sex on the induction of ALDH3

Age (days)	**ALDH3 Activity** (nmoles NAD(P)H/min/mg protein)	
	Males	Females
10	20.5 ± 5.1	18.4 ± 3.8
20	49.3 ± 3.1	50.4 ± 2.5
30	126.4 ± 8.6	274.4 ± 15.4 **
40	143.3 ± 15.6	317.9 ± 26.7 **
50	243.2 ± 22.6	377.0 ± 18.2 **
60	269.0 ± 23.7	404.1 ± 32.4 **
90	334.8 ± 27.5	439.8 ± 23.6 **

Animals received a single i.p. injection of MC (50 mg/kg) and were sacrificed after 5 days. Each value represents the mean ± S.D. of 5 animals. The two sexes differed significantly after 30 days of age. (**, P<0.001)

Administration of MC to male rats during puberty was found to influence the expression of the basal ALDH3 activity even later in life. At 90 days of age, which is about 3 months following the initial dose, basal ALDH3 activity in male rats was still greater than controls (8-fold). In females, ALDH3 activity at 90 days of age had returned to control levels, indicating that the inductive effect had dissipated (Figure 1).

Acute treatment with MC in adult rats increased ALDH3 activity to a greater extent in females (190-fold), compared to males (75-fold). In contrast, ALDH3 activity in males and females was similar when pretreated at puberty with MC. This abolition of differences between the two sexes was related to the increased inducibility of ALDH3 activity in males (Figure 1).

Acute administration of MC in adult female animals (without any pretreatment at puberty), produced a significant decrease (47%) in the number of cytosolic estrogen receptors in the liver (10.2±1.4 vs. 18.7±3.8 fmol/mg protein in treated and control animals, respectively; p<0.001). No difference was observed, however, in uterine estrogen receptor content.

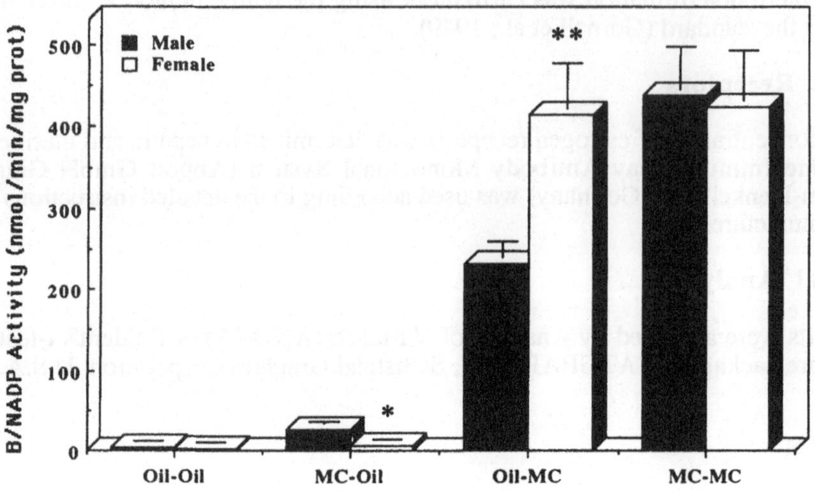

Figure 1. Effect of neonatal MC treatment on the induction of B/NADP activity. Animals were treated at 25 and then 90 days of age with either vehicle (oil) or MC (50 mg/kg) as indicated under each pair of columns. (*, $p<0.05$; **, $p<0.001$)

DISCUSSION

In addition to the rat, induction of the ALDH3 isozyme by MC or TCDD has been demonstrated in cultures of human and mouse hepatoma cell lines, as well as in primary cultures of human or rat hepatocytes (Marselos and Michalopoulos, G., 1986; Marselos et. al., 1986; 1987; Vasiliou et al., 1992). Furthermore, relatively high levels of the ALDH3 activity could be detected in some chemically induced hepatomas (Feinstein and Cameron, 1972; Lindahl 1977; Lindahl et al., 1982). Although the exact role of this isozyme still remains unclear, it may participate in the oxidation of some aldehydes generated during the peroxidation of lipids (Marselos and Lindahl, 1988).

The present series of experiments demonstrate that the induction of ALDH3 by MC in female rats is greater than that found in males. This sex difference becomes evident after 25 days of age. Depriving male rats of testosterone by castration during the prepubertal period, prevents the development of this difference, hence producing an identical response to MC treatment in males and females (unpublished findings). Replacement of testosterone to castrated males re-establishes this difference (unpublished findings), thus suggesting a testosterone-dependent inhibition of ALDH3-induction by MC.

It is possible that the differential response of ALDH3 between the two sexes reflects a different degree of metabolic activation to MC necessary for the induction of ALDH3 (Marselos et al., 1987; Vasiliou and Marselos, 1989a, b). The inducibility of ethoxyresorufin O-deethylase activity which reflects the activity of the isozyme of cytochrome P-450 (CYP1A1), as well as MC metabolism, is initially low in young Wistar rats, but increases with age (Lum et al., 1985). Furthermore, the inducibility of CYP1A1 is greater in females and also appears to be regulated by exposure to physiological concentrations of androgens via a pituitary-mediated mechanism (Burke et al., 1978).

Sex specific P450s have been purified from rats. Best studies are the male-specific, P-450h (CYP2C11), and the female-specific, P-450i (CYP2C12) (Ryan et al., 1982; 1984). CYP2C11 is expressed only in adult male rats. This expression is apparently dependent on the exposure to androgens during the neonatal period and/or adulthood. Castration of neonates abolishes the expression of CYP2C11, while castration of adults results in only a partial decrease. This partial effect in adults is eliminated after the rats are replaced with physiological concentrations of testosterone, implicating a direct role for testosterone in CYP2C11 expression in adults. Similarly, testosterone injections to castrated male rats during the first 3 days of life results in a significant increase in CYP2C11 expression later in adulthood, consistent with the hypothesis that this enzyme is subjected to neonatal imprinting (Waxman et al., 1985; Dannan et al., 1986).

MC, on the other hand, has been shown not only to switch on CYP1A1, but also to decrease CYP2C11 mRNA levels by 50-60% in the liver following daily administration of MC for three days (Yeowell et al., 1987). In this regard, it is interesting to note that MC and TCDD have also been reported to produce a significant decrease in serum testosterone concentrations (Kleeman et al., 1990). In contrast, MC does not appear to affect either CYP2C12 expression or serum estradiol levels (Yeowell et al., 1987). Our results clearly suggest that gonadectomy results in the feminization of the ALDH3-pattern of induction in adult males, and that administration of MC during the prepubertal period produces similar results as gonadectomy.

Although other mechanism(s) cannot be excluded, it appears that CYP2C11 negatively regulates the induction of ALDH3. TCDD has been shown to down-regulate hepatic estrogen receptors (ER) in female Long Evan (Romkes et al., 1987) and Sprague-Dawley rats (Hruska and Olson, 1988), as well as in congenic mice (Lin et al., 1991). These observations are compatible with our results which show that MC treatment results in a 47% decrease in the number of ER in the liver of the Wistar rat. Recent studies demonstrate that the sensitivity to ER down-regulation by TCDD is related to the Ah locus (Lin et al., 1991). The Ah receptor is an intracellular protein which is present in many tissues, including liver and uterus, and which binds to TCDD (Romkes et al., 1987). The Ah receptor is functionally similar to the receptors for steroid hormones, thyroid hormones, vitamin D_3 and retinoic acid, which act as ligand-activated transcriptional regulators and are defined as being part of the erb-A superfamily of receptors (Evans, 1988). Although the members of this receptor superfamily are functionally similar, TCDD does not appear to bind to other steroid hormone receptors, nor do steroid hormones appear to bind to the Ah receptor (Neal et al., 1979; Romkes and Safe, 1988). Interestingly, several members of this family of proteins are known to play a role in the regulation of other receptors in the group (Clarke et al., 1990). Although MC is also capable of specifically binding to the Ah receptor, it is not clear whether hepatic ER down-regulation is mediated by an Ah receptor mechanism.

In conclusion, the difference in the response to polycyclic aromatic hydrocarbons between male and female animals may be related to sexually determined differences in the generation of active metabolites. In this regard, a better understanding of the complex nature of steroid and xenobiotic metabolism may also result in a better understanding of carcinogen activation and inactivation.

ACKNOWLEDGMENTS

The authors wish to thank Drs. V. Vasiliou and E. Johnson for their valuable criticism of the manuscript. Supported in part by a grant from the Research Committees of the Greek Ministries of Health and Education, and by a grant from the Commission of the European Communities (STRIDE/Hellas).

REFERENCES

Burke, M., Orrenious, S., and Gustafsson, J., 1978, Pituitary involvement in the sexual differentiation and 3-methylcholanthrene induction of rat liver microsmal monooxygenases, Biochem. Pharmacol., 27:1125.

Clarke, C.L., Roman, S.D., Graham, J., Koga, M., and Sutherland, R.L., 1990, Progesterone receptor regulation by retinoid acid in the human breast cancer cell line T-47D, J. Biol. Chem., 265:12694.

Dannan, G.A., Guengerich, F.P., and Waxman, D.J., 1986, Hormonal regulation of rat liver microsomal enzymes: role of gonadal steroids in programming, maintenance, and suppression of D4-steroid 5a-reductase, flavin-containing monoxygenase, and sex specific cytochromes P450, J. Biol. Chem., 261:10728.

Deitrich, R.A., 1971, Genetic aspects of increase in rat liver aldehyde dehydrogenase induced by phenobarbital, Science, 173:334.

Deitrich, R.A., Bledeau, P., Stock, T., and Roper, M., 1977, Induction of different rat liver supernatant aldehyde dehydrogenase by phenobarbital and tetrachlorodibenzo-p-dioxin, J. Biol. Chem., 252:6169.

Deitrich, R.A., Bledeau, P., Roper M., and Schmuk, J., 1978, Induction of aldehyde dehydrogenase. Biochem. Pharmacol., 27:2343.

Evans, R.M., 1988, The steroid and thyroid hormone receptor superfamily, Science, 240:889.

Feinstein, R.N. and Cameron, E.C., 1972, Aldehyde dehydrogenase activity in rat hepatomas, Biochem. Biophys. Res. Commun., 48:1140.

Gallo, M.A., Hesse, E.J., MacDonald, G.J., and Umbreit, T.H., 1986, Interactive effects of estradiol and 2,3,7,8-tetrachlorodibenzo-p-dioxin and hepatic cytochrome P-450 and mouse uterus, Toxicol Lett., 32:123.

Gornall, A.G., Bardawill, C.J., and David, M.M., 1949, Determination of serum proteins by means of the biuret reaction, J. Biol. Chem., 177:751.

Harris, M., Zacharewski, T., and Safe, S., 1990, Effects of 2,3,7,8-tetrachlorodibenso-p-dioxin and related compounds on the occupied nuclear estrogen receptor in MCF-7 human breast cancer cells. Cancer Res., 50:3579.

Hruska, R.E., and Olson, J.R., 1988, Species differences in estrogen receptors and in the response to 2,3,7,8-tetrachlorodibenzo-p-dioxin exposure. Tocicol. Let., 48:289.

Kleeman, J., Moore, R., and Peterson, R., 1990, Inhibition of testicular steroidogenesis in 2,3,7,8-tetrachlorodibenzo-p-dioxin treated rats: Evidence that the key lesion occurs prior to or during pregnenolone formation, Tox. Appl. Pharmacol., 106:112.

Lin, F.H., Stohs, S.I., Birnbaum, L.S., Clark, G., Lucier, G.W., and Goldstein, J.A., 1991, The effects of 2,3,7,8-tetrachlorodibenzo-p-dioxin (TCDD) on hepatic estrogen and glucocorticoid receptors in congenic strains of Ah responsive and Ah nonresponsive C57BL/6J mice, Toxicol. Appl. Pharmacol., 108:129.

Lindahl, R., 1977, Aldehyde dehydrogenase in 2-acetamido-fluorene-induced rat hepatomas. Ontogeny and evidence that the new isoenzymes are not due to normal gene derepression, Biochem. J., 164:119.

Lindahl, R., Evces, S., and Sheng, W.-L., 1982, Expression of the tumor aldehyde dehydrogenase phenotype during 2-acetylamino-fluorene-induced rat hepatocarcinogenesis. Cancer Res., 42:577.

Lum, P., Walker, S., and Ioannides, C., 1985, Foetal and neonatal development of cytochrome P-450 and cytochrome P-448 catalysed mixed function oxidases in the rat: induction by 3-methylcholanthrene, Toxicol. 35:307.

Marselos, M., and Hanninen, O., 1974, Enhancement of D-Glucuronolactone dehdrogenase and acetaldehyde dehydrogenase activities by inducers of drug metabolism, Biochem. Pharmacol., 23:1457.

Marselos, M., Torronen, R., Koivula, T., and Koivusalo, M., 1979, Comparison of phenobarbital anc carcinogen-induced aldehyde dehydrogenases in the rat, Biochim. Biophys. Acta., 583:110.

Marselos, M., Strom, S.C., and Michalopoulos, G., 1986, Enhancement of aldehyde dehydrogenase activity in human and rat hepatocyte cultures by 3-methylcholanthrene. Cell Biol. Toxicol., 2:257.

Marselos, M., Michalopoulos, G., 1986, Phentobarbital enhances the aldehyde dehydrogenase activity in rat hepatocytes in vitro and in vivo, Acta Pharmacol. et Toxicol., 59:403.

Marselos, M., Strom, S.C., and Michalopoulos, G., 1987, Effect of phenobarbital and 3-methylcholanthrene on aldehyde dehdrogenase activity in cultures of HepG2 cells and normal human hepatocytes, Chem.-Biol. Interact., 62:75.

Marselos, M., and Lindahl, R., 1988, Substrate preference of a cytosolic aldehyde dehydrogenase inducible in the rat by treatment with methylcholanthrene. Toxicol. Appl. Pharmacol., 95:339.

Marselos, M., and Vasiliou, V., 1991, Effect of various chemicals on the aldehyde dehydrogenase activity of the rat liver cytosol, Chem.-Biol. Interact., 79:79.

Neal, R.A., Beatty, P.W., and Gasiewicz, T.A., 1979, Studies on the mechanisms of toxicity of 2,3,7,8-tetrachlorodibenzo-p-dioxin, Ann. NY Acad. Sci., 320:204.

Romkes, M., Piskorska-Pliszczynska, J., and Safe, S., 1987, Effects of 2,3,7,8-tetrachlorodibenzo-p-dioxin on hepatic and uterine estrogen receptor levels in rats. Toxicol. Appl. Pharmacol., 87:306.

Ritter, E., and Eriksson, L.C., 1987, Aldehyde dehydrogenase activities in hepatocyte nodules and hepatocellular carcinomas from Wistar rats. Carcinogenesis, 6:1683.

Ryan, D.E., Thomas, P.E., Reik, L.M., and Levin, W., 1982, Purification characterization, and regulation of five rat hepatic microsomal cytochrome P-450 isozymes, Xenobiotics, 11:727.

Ryan, D.E., Shinji, I., Wood, A.W., Thomas, P.E., Lieber, C.S., and Levin, W., 1984, Characterization of three highly purified cytochrome P-450 from the hepatic microsomes of adult male rats, J. Biol. Chem., 259:1239.

Skett, P., 1987, Hormonal regulation and sex differences of xenobiotic metabolism, in "Progress in Drug Metabolism, vol 10" J.W. Bridges, L.F. Chasseaud and S.G. Gibson, eds., Taylor & Francis, New York, p. 85.

Torronen, R., Nousiainen, U., and Haninnen, O., 1981, Induction of aldehyde dehydrogenase activity by polycyclic aromatic hydrocarbons, Chem. -Biol. Interact., 36:33.

Vasiliou, V., Athanasiou, K., and Marselos, M., 1988, The use of ALDH induction as a carcinogenic risk marker with a typical in vitro mutagenicity system. in: "Biologically Based Methods for Cancer Risk Assessment," C.C. Travis, ed., Plenum Press, New York, p. 231.

Vasiliou, V., and Marselos, M., 1989a, Changes in the inducibility of a hepatic aldehyde dehydrogenase by various effectors, Arch. Toxicol., 63:221.

Vasiliou, V., and Marselos, M., 1989b, Tissue distribution of inducible aldehyde dehydrogenase activity in the rat after treatment with phenobarbital or methylcholanthrene. Pharmacol. Toxicol., 64:39.

Vasiliou, V., Puga, A., and Nebert, D.W., 1992, Negative regulation of the muirine cytosolic aldehyde dehydrogenase-3 (ALDH-3c) gene by functional CYP1A1 and CYP1A2 proteins, in press.

Waxman, D.J., Dannan, G.A., and Guengerich, F.P., 1985, Regulation of the hepatic cytochrome P-450: age-dependent expression, hormonal imprinting and xenobiotic inducibility of sex specific isozymes, Biochem., 24:4409.

Yeowell, H., Waxman, D., Wadhera, A., and Goldstein, J., 1987, Suppression of the constitutive, male-specific rat hepatic cytochrome P-450 2c and its mRNA by 2,3,7,8-tetrachlorodibenzo-p-dioxin and 3-methylcholanthrene, Mol. Pharmacol., 32:340.

Umbreit, T.H., Hesse, E.J., and Gallo, M.A., 1987, Reproductive toxicity in female mice of dioxin-contaminant soils form a 2,4,5-trichlorophenoxyacetic acid manufacturing site, Arch. Enciron. Contam. Toxicol., 16:461.

MOUSE CLASS 3 ALDEHYDE DEHYDROGENASES: POSITIVE AND NEGATIVE REGULATION OF GENE EXPRESSION

Vasilis Vasiliou, Alvaro Puga and Daniel W. Nebert

Department of Environmental Health, University of Cincinnati Medical Center, Cincinnati, OH 45267-0056

INTRODUCTION

Class 3 aldehyde dehydrogenases include the 2,3,7,8-tetrachlorodibenzo-*p*-dioxin (TCDD; dioxin)-inducible cytosolic form (ALDH3c) and the constitutive microsomal form (ALDH3m). Using the rat ***ALDH3c*** cDNA as a probe, we have cloned and sequenced the murine microsomal aldehyde dehydrogenase-3 (***Aldh-3m***) cDNA; the gene is located on mouse chromosome 11, and alignment with the rat ALDH3m amino acid sequence (Miyauchi et al., 1991) shows 95% identity [V. Vasiliou, C.A. Kozak, R. Lindahl and D.W. Nebert, manuscript in preparation]. We have also cloned and sequenced the murine ***Aldh-3c*** cDNA [V.V. and D.W.N., manuscript in preparation]. Using a variety of genetically different mice and cell culture lines, we have recently compared the positive and negative regulatory mechanisms of ***Aldh-3c*** and ***Aldh-3m*** gene expression (Vasiliou et al., 1992). Although both genes are TCDD-inducible, we have determined that the murine ***Aldh-3c***, but not ***Aldh-3m***, can be classified as a ***bona fide*** member of the murine aromatic hydrocarbon-responsive [***Ah***] battery. How we arrived at this classification is the principal subject of this Chapter.

THE [*Ah*] GENE BATTERY

Drug-metabolizing enzymes appear to be involved in maintaining the steady-state levels of ligands used in ligand-modulated transcription of genes effecting growth, differentiation, homeostasis, and neuroendocrine functions. An evolutionary argument has been proposed, suggesting that particular subsets of drug-metabolizing enzymes are coordinately regulated in response to abnormal concentrations of various foreign chemicals that mimic endogenous ligands effecting growth, differentiation and homeostasis (Nebert, 1991). Distinctly different subsets of these enzymes are induced, for example, by TCDD, steroids, ethanol, peroxisome proliferators, and phenobarbital (Nebert and Gonzalez, 1987).

The best studied example of coordinately regulated genes encoding drug-metabolizing enzymes is the murine [*Ah*] battery (Nebert and Gonzalez, 1987; Nebert et al., 1990). This battery comprises at least six genes that are coordinately induced by TCDD and polycyclic aromatic hydrocarbons such as benzo[a]pyrene. In addition to two Phase I cytochrome P450 genes, ***Cyp1a-1*** and ***Cyp1a-2***, there are four Phase II genes: the "tumor-specific" aldehyde dehydrogenase (***Aldh-3c***); NAD(P)H:menadione oxidoreductase (***Nmo-1***); a UDP glucuronosyltransferase having 4-methyl-umbelliferone as substrate (***Ugt1*06***); and a glutathione transferase having 2,4-dinitro-1-chlorobenzene as substrate (GST Ya, ***Gst1a-1***).

Each of these genes is positively regulated by the aromatic hydrocarbon-responsive (Ah) receptor, involving induction by foreign chemicals such as TCDD and benzo[a]pyrene. This Ah receptor-dependent induction process uses one or more aromatic hydrocarbon-responsive elements (AhREs), which have so far been identified in the 5' regulatory domain of all six mammalian [*Ah*] genes, with the exception of ***Ugt1*06*** (Nebert and Gonzalez, 1987; Bayney et al., 1989; Rushmore et al., 1990; Takimoto and Pitot, 1992). In addition, at least one electrophilic response element (EpRE) has been found upstream of rat (Rushmore et al., 1990) and mouse (Friling et al., 1992) ***GST1A1*** and rat (Rushmore et al., 1991) and human (Jaiswal, 1991) ***NMO1***. Finally, a negative response element (NRE) appears to be present upstream of the murine and human ***CYP1A1*** gene (Gonzalez and Nebert, 1985; Hines et al., 1988). For reasons outlined in this Chapter, we predict that we will find AhREs, an EpRE and an NRE in regulatory regions of the mouse ***Aldh-3c*** gene, but perhaps only the AhREs and EpRE upstream of the mouse ***Aldh-3m*** gene.

Characterization of the Aromatic Hydrocarbon-Responsive Elements (AhREs)

The action of the Ah receptor is known to involve nuclear translocation and binding to specific DNA sequence motifs in the proximity of the ***CYP1A1*** promoter (***reviewed in*** Nebert and Gonzalez, 1987; Nebert and Jones, 1989; Whitlock, 1990; Landers and Bunce, 1991). Three of these motifs, termed aromatic hydrocarbon (or xenobiotic or dioxin)-responsive elements (AhREs, XREs, DREs), are located between positions -1100 and -896 upstream of the murine ***Cyp1a-1*** transcription initiation site and define a positive, *cis*-acting Ah receptor-dependent aromatic hydrocarbon-responsive domain (AhRD), or enhancer. The core sequence is known to be **5'-TA/TGCGTG-3'**. Two or more AhREs in tandem appear to strengthen the enhancer activity (Fujisawa-Sehara et al., 1987). Gel mobility shift and methylation interference assays have demonstrated the cooperative interaction of a functional Ah receptor with more than one AhRE (Saatcioglu et al., 1990).

Identification of the Negative Response Element (NRE)

Another element in the upstream ***Cyp1a-1*** regulatory sequences involves a negative autoregulatory loop. In the absence of a functional CYP1A1 enzyme in P_1^- cells, there is a derepression of constitutive ***Cyp1a-1*** transcription (Gonzalez and Nebert, 1985; RayChaudhuri et al., 1990). The negative response element was localized between -823 and -389 upstream of the murine ***Cyp1a-1*** gene (Gonzalez and Nebert, 1985) and between -821 and -735 upstream of the human ***CYP1A1*** gene (Hines et al., 1988). The human NRE core sequence has recently been proposed to be an inverted repeat with five spacer nucleotides: **5'-GTGCTCTGCCAATC AAAGCAC-3'** (Boucher et al., 1992). The exact role (if any) of the AhREs and Ah receptor in this negative regulatory mechanism involving the NRE is not yet understood.

Characterization of the Electrophile-Responsive Element (EpRE)

To date, the rat and mouse homologous ***GST1A1*** genes (Rushmore et al., 1990; Friling et al., 1992) and the rat and human homologous ***NMO1*** genes (Rushmore et al., 1991; Jaiswal, 1991) have been found to have both an AhRE and an EpRE, the latter being also termed an "antioxidant response element, ARE." The EpRE core sequence has variously been reported as **5'-puGTGACNNNGC-3'** (Rushmore et al., 1991) or **5'-GTGCTC/ATG-3'** (Takimoto and Pitot, 1992) in the rat and **5'-TGACAT/AT/AGC-3'** in the mouse (Friling et al., 1992). For this regulatory sequence, therefore, the final consensus is not yet firmly established, and the relationship of the EpRE (if any) to the AhRE and Ah receptor is not yet known. Interestingly, the position of the AhRE relative to the EpRE appears not to be critical; for example, the AhRE is 5'ward of the EpRE in the rat ***NMO1*** gene but 3'ward in the human ***NMO1*** gene (Rushmore et al., 1991; Jaiswal, 1991).

Inbred Mice and Cell Lines Used in Our Studies

This laboratory has dissected the [***Ah***] gene battery carefully, through the use of inbred mouse lines and cell culture lines having important genetic differences in [***Ah***] gene battery expression. For example, the C57BL/6 (B6) mouse and DBA/2 (D2) mouse have high- and low-

Table 1. The hepatoma Hepa-1 *wt* and mutant cell lines.

wt, wild-type Hepa-1c1c7
c37, CYP1A1 metabolism-deficient (P_1^-)
c4, chromatin binding-defective (*cb*$^-$), also called nuclear translocation-impaired (*nt*$^-$)
c2, Ah receptorless (*r*$^-$) containing <10% of normal functional receptor levels
CX4, ***c37*** stable transfectant in which the AhRD in its native orientation is ligated (*cis* configuration) to SV40 promoter driving murine ***Cyp1a-1*** cDNA
CE1, ***c37*** stable transfectant in which AhRD in its opposite orientation is ligated (*cis* configuration) to SV40 promoter driving murine ***Cyp1a-1*** cDNA
CP47, ***c37*** stable transfectant in which AhRD is on separate plasmid (*trans* configuration) to that containing SV40 promoter and murine ***Cyp1a-1*** cDNA
CA3.4, ***c37*** stable transfectant in which AhRD in its native orientation is ligated (*cis* configuration) to SV40 promoter driving human ***CYP1A2*** cDNA

affinity Ah receptors, respectively (Nebert and Gonzalez, 1987; Nebert, 1989). We have also developed D2.B6-***Ah***^{b} and B6.D2-***Ah***^{d} congenic lines; the first having the high-affinity Ah receptor in a >98% D2 genomic background, and the latter having the low-affinity Ah receptor in a >98% B6 genomic background (Bigelow et al., 1989). We have studied the wild-type (*wt*) mouse hepatoma cell culture line Hepa-1c1c7 and three of its mutant lines (Table 1): the CYP1A1 metabolism-deficient derivative *c37* (P_1^-); the receptorless *c2* (*r*$^-$), which actually has normal Ah receptor at 5-10% wild-type levels; and the chromatin binding-defective *c4* (*cb*$^-$). Recently we have also developed the following four ***c37*** stable transfectants (Table 1): **CX4**, in which the AhRD in its native orientation, ligated in the *cis* configuration to the SV40 promoter, drives the murine ***Cyp1a-1*** cDNA; **CE1**, in which the AhRD in its opposite orientation, ligated in the *cis* configuration to the SV40 promoter, drives the murine ***Cyp1a-1*** cDNA; **CP47**, in which the AhRD is on a separate plasmid (*trans* configuration) to that containing the SV40 promoter and murine ***Cyp1a-1*** cDNA; and **CA3.4**, in which the AhRD in its native orientation, ligated in the *cis* configuration to the SV40 promoter, drives the human ***CYP1A2*** cDNA (RayChaudhuri et al., 1990). In addition, we have used the ***14CoS/14CoS*** mutant mouse (homozygous for a 1.2-centiMorgan deletion on chromosome 7) and the ***ch/ch*** (wild-type) mouse (Gluecksohn-Waelsch, 1979; Nebert et al., 1990), as well as SV40-transformed cell lines derived from newborn liver of these mice.

Either Functional CYP1A1 or CYP1A2 Protein Appears to be a Negative Regulator of the Endogenous *Cyp1a-1* Gene and [*Ah*] Phase II Genes

Expression of the murine ***Cyp1a-1*** gene is transcriptionally derepressed in the ***c37*** cell line that contains a ***Cyp1a-1*** gene encoding a nonfunctional CYP1A1 protein (Kimura et al. 1987). This derepression also includes the ***Nmo-1*** gene (Robertson et al., 1987; Nebert and Gonzalez, 1987; RayChaudhuri et al., 1990). The ***Cyp1a-1*** gene product appears to be necessary in a mechanism of feedback regulation, not only of the ***Cyp1a-1*** gene (Puga et al., 1990) but also of the NMO1 mRNA levels (RayChaudhuri et al., 1990). By introducing expression plasmids carrying the appropriate cDNA into the mutant ***c37*** P_1^- cells, this laboratory has shown that expression of a functional, exogenous murine CYP1A1 protein, or a human CYP1A2 protein, is sufficient to restore the repression of the endogenous ***Cyp1a-1*** gene, as well as to restore the repression of the NMO1 mRNA levels, and that this effect for ***Cyp1a-1*** takes place primarily at the level of transcription (RayChaudhuri et al., 1990).

This laboratory has found that the Ah receptor-mediated coordinate induction (positive regulation) involves all six [*Ah*] battery genes. On the other hand, a gene on chromosome 7 is necessary for the up-regulation of the ***Nmo-1*** and ***Aldh-3c*** genes and this appears to occur via a (negative) mechanism quite independent of ***Cyp1a-1*** and ***Cyp1a-2*** induction (Nebert and Gonzalez, 1987; Nebert et al., 1990; Vasiliou et al., 1992). Whether this chromosome 7-mediated regulatory mechanism is related to the CYP1A1/CYP1A2 protein-dependent mechanism of repression is not yet known.

ALDH3c and ALDH3m enzyme activities and mRNA Levels in Murine Cell Culture Lines but Not in the Intact Mouse

Expression of basal ALDH3c activity, and its inducibility by polycyclic aromatic compounds, is known to exist in rat liver but not mouse liver (Vasiliou et al., 1989). ALDH3c enzyme activity was surprisingly measurable and inducible by TCDD in the murine Hepa-1 ***wt***, ***ch/ch***, and ***14CoS/14CoS*** cell lines, as measured by increases in the B/NADP to P/NAD ratio (Table 2). Whereas the ***wt*** and ***ch/ch*** ratios went from control values of about 1.0 to TCDD-induced values of 3.4, the ***14CoS/14CoS*** ratio went from a control value of 2.7 to a TCDD-induced value of 4.0 (Table 2). Hence, ALDH3c basal activity is markedly elevated in the ***14CoS/14CoS*** cells having the homozygous chromosome 7 deletion, suggesting that loss of both copies of a gene on chromosome 7 results in removal of negative control and leads to derepression of the ALDH3c activity. A slight increase was also found in the ALDH3m enzyme activity in ***wt*** cells (Table 2).

These enzyme activities paralleled the ALDH3c and ALDH3m mRNA levels (Table 3). In the Hepa-1 ***wt*** cell line, the ALDH3c mRNA basal levels are negligible and the mRNA is highly induced by TCDD; the ALDH3m mRNA basal levels are elevated but also significantly induced by TCDD (Table 3). In addition to hybridizing to the ALDH3m 3.1-kb mRNA, the murine ***Aldh-3m*** cDNA probe cross-hybridizes weakly with the ALDH3c 1.9-kb mRNA (Vasiliou et al., 1992). In addition to hybridizing to the murine ALDH3c 1.9-kb mRNA, the rat ***ALDH3c***

Table 2. Effect of TCDD on ALDH3c and ALDH3m enzyme activities in Hepa-1 ***wt***, ***ch/ch*** and ***14CoS/14CoS*** cells in culture.

	B/NADP	P/NAD	B/NADP to P/NAD Ratio
Cytosol:			
wt, control	1.3 ± 0.4	1.3 ± 0.2	1.0
wt, TCDD	16 ± 2.1	4.8 ± 0.6	3.4
ch/ch, control	1.9 ± 0.7	1.7 ± 0.3	1.1
ch/ch, TCDD	8.6 ± 0.1	2.5 ± 0.2	3.4
14CoS/14CoS, control	8.8 ± 0.3	3.3 ± 0.5	2.7
14CoS/14CoS, TCDD	32 ± 1.3	8.1 ± 0.4	4.0
Microsomes:			
wt, control	0.4 ± 0.1	1.5 ± 0.2	0.27
wt, TCDD	3.1 ± 0.6	3.1 ± 0.7	1.0

Cells were treated with TCDD (20 nM) for 24 h, and cytosolic and microsomal fractions were prepared. B/NADP, $NADP^+$-dependent benzaldehyde oxidation. P/NAD, NAD^+-dependent propionaldehyde oxidation. Enzyme assays were carried out by monitoring NAD(P)H production at 340 nm and 37 °C. To measure the NAD^+-dependent oxidation of propionaldehyde (P/NAD activity), the assay mixture contained sodium pyrophosphate buffer (75 mM, pH 8.0), 1 mM pyrazole (to inhibit alcohol dehydrogenase), 1 mM NAD^+, and 5 mM propionaldehyde. To measure the $NADP^+$-dependent oxidation of benzaldehyde (B/NADP activity), assay conditions were the same except benzaldehyde (5 mM originally in 20% methanol) was substituted for propionaldehyde, and the coenzyme $NADP^+$ (2.5 mM) was used instead of NAD^+. In either enzyme assay, the reaction was started by adding the substrate subsequent to a 5-min preincubation, and a blank was run without the substrate (Vasiliou and Marselos, 1989a). Protein was measured by the bicinchoninic acid method, according to details supplied by Pierce Chemical Company. Units denote nmoles of NAD(P)H formed per min; specific activities are expressed in units/mg protein. Specific activities are expressed as units/mg protein in mean ± standard deviation, from duplicates in three separate experiments. A ratio of B/NADP to P/NAD of >1.00 is generally believed to reflect dioxin-inducible ALDH3c activity (Vasiliou and Marselos, 1989b).

Table 3. Effect of TCDD and *tert*-butylhydroquinone (tBHQ) on ALDH3c and ALDH3m mRNA levels in the Hepa-1 *wt* and the three mutant cell lines.

Cell line, mRNA	Control	TCDD	tBHQ
wt, ALDH3c	±	++++	++
c37, ALDH3c	+++	++++	++++
c2, ALDH3c	±	++	++
c4, ALDH3c	o	o	++
wt, ALDH3m	+	++	++
c37, ALDH3m	+	+++	++
c2, ALDH3m	+	++	++
c4, ALDH3m	+	+	++

TCDD treatment (20 nM) of the cell lines was 12 h. RNA was extracted by the acid guanidinium thiocyanate method (Chomczynski and Sacchi, 1987). Total RNA (10 μg) was separated in formaldehyde-agarose gels and transferred to Nytran. Prehybridizations and hybridizations were carried out at 42 °C in a solution containing 50% deionized formamide, 6X SSC (SSC = 0.9 M NaCl and 0.09 M sodium citrate, pH 7.0), 2.5X Denhardt's solution, 0.5% sodium dodecylsulfate (SDS), and denatured salmon sperm DNA (0.1 mg/ml). Radioactively labeled probes were prepared by random priming (Feinberg and Vogelstein, 1983), with the use of [α-^{32}P]dCTP (3000 Ci/mmole, New England Nuclear/DuPont) as the labeled precursor, and were added to the hybridization solutions at 5×10^6 to 10×10^6 cpm/ml. After hybridization for 16-20 h, the filters were washed twice in 2X SSC and 0.1% SDS for 10 min at room temperature and then twice in 0.1X SSC and 0.1% SDS for 30 min at 50 °C. The filters were then exposed for 48 h to Kodak XAR-5 film at -70 °C with intensifying screens. Probes included the murine *Aldh-3c* and *Aldh-3m* full-length cDNAs, and mRNA quantification was made by comparing the mRNA in each lane to the amount of β-actin-1 mRNA.

cDNA probe cross-hybridizes weakly with the murine ALDH3m 3.1-kb mRNA (Vasiliou et al., 1992). The ALDH3c, but not ALDH3m, mRNA is strikingly elevated in the CYP1A1-deficient *c37* line, and both ALDH3c and ALDH3m mRNAs are increased by TCDD treatment of *c37* cells (Table 3). These results indicate that, although both genes are TCDD-inducible, only the *Aldh-3c* gene is dramatically derepressed in the P_1^- mutant line--similar to what had previously been reported for the NMO1 mRNA (RayChaudhuri et al., 1990)

In the untreated *14CoS/14CoS* cells, basal ALDH3c mRNA levels are markedly elevated, relative to that in *ch/ch* cells (Table 4). The ALDH3m enzyme activity and mRNA levels were unchanged in both *14CoS/14CoS* and *ch/ch* animals and cell lines (data not shown). These

Table 4. Effect of TCDD on ALDH3c mRNA levels in *ch/ch* and *14CoS/14CoS* liver from the intact newborn and in the corresponding SV40-derived cell lines.

Newborn liver	Control	TCDD
ch/ch	o	o
ch/14CoS	o	o
14CoS/14CoS	o	o
(Mother)	o	o
Cell cultures		
ch/ch	±	+
14CoS/14CoS	++	++++

TCDD (10 μg/kg) was given to the pregnant mother 1 day before birth. TCDD treatment (20 nM) of the cell lines was 12 h. RNA isolation and hybridization conditions are described in Table 3.

Table 5. Effect of TCDD on ALDH3c enzyme activity in ***ch/ch*** and ***14CoS/14CoS*** liver from the intact newborn.

	B/NADP	P/NAD	B/NADP to P/NAD Ratio
ch/ch, control	0.6 ± 0.5	4.2 ± 0.4	0.14
ch/ch, TCDD	0.4 ± 0.3	3.6 ± 1.1	0.11
14CoS/14CoS, control	< 0.3	3.1 ± 1.2	< 0.10
14CoS/14CoS, TCDD	0.6 ± 0.2	3.1 ± 0.4	0.19

TCDD (10 μg/kg) was given to the pregnant mother 1 day before birth. Specific activities and abbreviations are the same as those described in Table 2. Specific activities are expressed as units/mg protein in mean ± standard deviation, from duplicates in three separate experiments. A ratio of B/NADP to P/NAD of < 0.20 is regarded as nondetectable ALDH3c activity (Marselos and Vasiliou, 1991).

results suggest that the deletion homozygote might have lost both copies of a regulatory gene encoding a factor that is a negative effector for the ***Aldh-3c*** gene.

Although we found large amounts of ALDH3c in mouse cells in culture, neither elevated basal levels nor TCDD-inducible ALDH3c mRNA or enzyme activity could be detected in ***14CoS/14CoS*** or ***ch/ch*** newborn liver of the intact mouse (Tables 4 & 5). These data thus suggest that, in mouse liver cell culture or in hepatoma cell culture, either an inhibitor of ***Aldh-3c*** gene expression is extinguished or an activator of ***Aldh-3c*** gene expression is activated, as compared with hepatic ALDH3c activity in the intact mouse.

ALDH3c and ALDH3m mRNA Levels in *c37* Stable Transfectants

As described earlier, the CX4 and CE1 cell lines are derivatives of ***c37*** cells in which the murine ***Cyp1a-1*** cDNA expressing a functional murine CYP1A1 protein is stably expressed (Table 1). The ALDH3c mRNA basal levels--which are dramatically elevated in ***c37*** cells (Tables 3 & 6)--are decreased to negligible levels by the exogenously expressed CYP1A1 enzyme in the CX4 and CE1 cell lines, and the ALDH3c mRNA is clearly elevated by TCDD treatment (Table 6). In contrast, the ALDH3m mRNA basal levels are present but not derepressed in the ***c37*** cells; in addition, the ALDH3m mRNA appears virtually unchanged in the CX4 and CE1 lines, and a slight increase by TCDD is observed in the CX4 line (Table 6).

The ALDH3c mRNA basal levels, dramatically elevated in ***c37*** cells, are also decreased to negligible levels by the exogenously expressed human CYP1A2 enzyme in the CA3.4 cells, and slight TCDD inducibility can also be seen in this line (Table 6). In contrast, the ALDH3m mRNA basal levels are not decreased in the CA3.4 line, while slight TCDD inducibility is seen.

Table 6. ALDH3c and ALDH3m mRNA levels by Northern blot analysis of control and TCDD-treated ***c37***, CX4, CE1, CP47 and CA3.4 cell lines.

mRNA, treatment	c37	CX4	CE1	CP47	CA3.4
ALDH3c, control	+++	±	±	+++	±
ALDH3c, TCDD	++++	++	++	++++	+
ALDH3m, control	+	±	±	+	++
ALDH3m, TCDD	+	+	±	+	+++

TCDD treatment (20 nM) of the cell lines was 12 h. RNA isolation and hybridization conditions are described in Table 3.

Lastly, the CP47 cell line contains the AhRD enhancer in *trans* to the SV40 promoter and murine ***Cyp1a-1*** cDNA and has no functional CYP1A1 enzyme (RayChaudhuri et al., 1990). The expression of ALDH3c and ALDH3m mRNA levels was indistinguishable from that seen in the ***c37*** line from which CP47 had been derived (Table 6).

This laboratory has previously shown that either low or high expression of the murine CYP1A1 or the human CYP1A2 functional enzyme in these *c37* stable transfectants is capable of repressing the endogenous NMO1 mRNA levels (RayChaudhuri et al., 1990); Table 6 thus demonstrates that ALDH3c, but not ALDH3m, follows the same pattern as NMO1.

Effect of *tert*-Butylhydroquinone on Expression of the Aldehyde Dehydrogenase-3 Genes

The effect of electrophilic metabolites, such as menadione or *tert*-butylhydroquinone (tBHQ), on the [***Ah***] battery genes has been studied in several laboratories. We and others have found that menadione or tBHQ increases GST1A1 or NMO1 mRNAs and enzyme activities (Friling et al., 1990; Rushmore et al., 1991) but not CYP1A1 or CYP1A2 mRNAs or enzyme activities (V. Vasiliou, A.J. Fornace, Jr. and D.W. Nebert, unpublished data). We found that tBHQ increased both ALDH3c and ALDH3m mRNA levels (Table 3). We therefore would predict to find an EpRE in the regulatory regions of both the ***Aldh-3c*** and ***Aldh-3m*** genes. Because of TCDD inducibility for both genes, we also would expect to find an AhRE in the regulatory regions of both the ***Aldh-3c*** and ***Aldh-3m*** genes.

CONCLUSIONS

The [***Ah***] battery comprises at least two Phase I genes (***Cyp1a-1*** and ***Cyp1a-2***) and four Phase II genes (***Aldh-3c***, ***Nmo-1***, ***Ugt1*06***, and ***Gst1a-1***). When a cell "senses" abnormally elevated concentrations of a ligand necessary for a critical life function, or concentrations of a foreign compound that mimics this important ligand (Nebert, 1991), the [***Ah***] battery--as a particular subset of drug-metabolizing enzymes--responds to this danger to the cell.

We have identified at least three distinct mechanisms by which the [***Ah***] battery can respond. (i) The Ah receptor, acting via AhREs, is a positive regulator of all six [***Ah***] battery genes. (ii) There appears to be a chromosome 7-mediated repression--acting via the EpRE--of the [***Ah***] Phase II genes but not ***Cyp1a-1*** or ***Cyp1a-2***. (iii) There appears to be a CYP1A1 /CYP1A2-dependent repression, acting via the negative response element (NRE), of ***Cyp1a-1*** expression and of ***Nmo-1*** and ***Aldh-3c*** expression (Nebert et al., 1991). Using the CX4, CE1, CP47, and CA3.4 cell lines, we have found the same to be true for ***Ugt1*06*** and ***Gst1a-1*** [V. Vasiliou and D.W. Nebert, in preparation].

These three distinct mechanisms are consistent with the present-day concept of modularity in promoters and enhancers (Dynan, 1989). The promoters and enhancers that control transcription comprise multiple genetic elements, or modules, which are generally upstream of the gene being controlled but can be downstream or inside the gene. The cellular transcriptional machinery is able to gather and integrate the regulatory information conveyed by each module, and this information can be developmental-, sex- or tissue-specific, as well as specific for environmental signals. These integrated signals allow different sets (or batteries) of genes to evolve distinct, often complex, patterns of transcriptional regulation.

The definition of a gene battery in eukaryotes is a group of (nonlinked) genes having an intricate interrelationship with regard to up- and down-regulation in response to specific regulatory proteins whose binding may be combinatorial in nature (Britten and Davidson, 1969; McKnight and Tjian, 1986). We therefore propose that the necessary criterion for classification of a gene in the [***Ah***] battery is (iii) above, ***i.e.*** CYP1A1- and/or CYP1A2-dependent repression, acting via the negative response element (NRE). Hence, in addition to [***Ah***] battery genes, other genes may be TCDD-inducible and contain AhREs in their regulatory regions (***e.g.*** c-***Fos***, c-***Jun***, ***Aldh-3m***, etc.). Other genes may be tBHQ-, menadione- or H_2O_2-inducible and contain an EpRE in their regulatory regions [***e.g.*** the eukaryotic ***Aldh-3m*** shown here; ***sodA***, ***nfo*** and other genes in *E.coli* (Demple and Amábile-Cuevas, 1991)]. TCDD inducibility *per se*, or induction by potent electophiles, are therefore not sufficient to classify a gene as a member of the [*Ah*] battery. How, or if, the Ah receptor is involved in this intriguing CYP1A1/CYP1A2 metabolism-dependent process, will require further study.

ACKNOWLEDGMENTS

We thank our colleagues for valuable discussions and a critical reading of this manuscript. Vasilis Vasiliou is a Fogarty Scholar at the University of Cincinnati. This work was supported in part by NIH Grants R01 AG09235 and P30 ES06096.

REFERENCES

Bayney, R.M., Morton, M.R., Favreau, L.V., and Pickett, C.B., 1989, Rat liver NAD(P)H:quinone reductase. Regulation of quinone reductase gene expression by planar aromatic compounds and determination of the exon structure of the quinone reductase structural gene, *J. Biol. Chem.* **264**:21793-21797.

Bigelow, S.W., Collins, A.C., and Nebert, D.W., 1989, Selective mouse breeding for short ethanol sleep time has led to high levels of the hepatic aromatic hydrocarbon (Ah) receptor, *Biochem. Pharmacol.* **38**:3565-3572.

Boucher, P.D., Ruch, R.J., and Hines, R.N., 1992, Characterization of the human *CYP1A1* negative regulatory element, *The Toxicologist* **12**:241.

Britten, R.J., and Davidson, E.H., 1969, Gene regulation for higher cells: A theory, *Science* **165**:349-357.

Chomczynski, P., and Sacchi, N., 1987, Single-step method of RNA isolation by acid guanidinium thiocyanate-phenol-chloroform extraction, *Anal. Biochem.* **162**:156-159.

Demple, B., and Amábile-Cuevas, C.F., 1991, Redoc redux: The control of oxidative stress responses, *Cell* **67**:837-839.

Dynan, W.S., 1989, Modularity in promoters and enhancers, *Cell* **58**:1-4.

Feinberg, A.P., and Vogelstein, B., 1983, A technique for radiolabeling DNA restriction endonuclease fragments to high specific activity, *Anal. Biochem.* **132**:6-13.

Friling, R.S., Bensimon, A., Tichauer, Y., and Daniel, V., 1990, Xenobiotic-inducible expression of murine glutathione S-transferase Ya subunit gene is controlled by an electrophile-responsive element, *Proc. Natl. Acad. Sci. U.S.A.* **87**:6258-6262.

Friling, R.S., Bergelson, S., and Daniel, V., 1992, Two adjacent AP-1-like binding sites form the electrophile-responsive element of the murine glutathione S-transferase Ya subunit gene, *Proc. Natl. Acad. Sci. U.S.A.* **89**:668-672.

Fujisawa-Sehara, A., Sogawa, K., Yamane, M., and Fujii-Kuriyama, Y., 1987, Characterization of xenobiotic responsive elements upstream from the drug-metabolizing cytochrome P-450c gene: A similarity to glucocorticoid regulatory elements, *Nucleic Acids Res.* **15**:4179-4791.

Gluecksohn-Waelsch, S., 1979, Genetic control of morphogenetic and biochemical differentation: Lethal albino deletions in the mouse, *Cell* **16**:225-237.

Gonzalez, F.J., and Nebert, D.W., 1985, Autoregulation plus upstream positive and negative control regions associated with transcriptional activation of the mouse P_1450 gene, *Nucleic Acids Res.* **13**:7269-7288.

Hines, R.N., Mathis, J.M., and Jacob, C.S., 1988, Identification of multiple regulatory elements on the human cytochrome P450IA1 gene, *Carcinogenesis* **9**:1599-1605.

Jaiswal, A.K., 1991, Human NAD(P)H:quinone oxidoreductase (NQO_1) gene structure and induction by dioxin, *Biochemistry* **30**:10647-10653.

Kimura, S., Smith, H.H., Hankinson, O., and Nebert, D.W., 1987, Analysis of two benzo[a]pyrene-resistant mutants of the mouse hepatoma Hepa-1 P_1450 gene via cDNA expression in yeast, *EMBO J.* **6**:1929-1933.

Landers, J.P., and Bunce, N.J., 1991, The Ah receptor and the mechanism of dioxin toxicity, *Biochem. J.* **276**:273-287.

Marselos, M., and Vasiliou, V., 1991, Effect of various chemicals on the aldehyde dehydrogenase activity of the rat liver cytosol, *Chem.-Biol. Interactions* **79**:79-89.

McKnight, S., and Tjian, R., 1986, Transcriptional selectivity of viral genes in mammalian cells, *Cell* 46:795-805.

Miyauchi, K., Masaki, R., Taketani, S., Yamamoto, A., Akayama, M., and Tashiro, Y., 1991, Molecular cloning, sequencing, and expression of cDNA for rat liver microsomal aldehyde dehydrogenase, *J. Biol. Chem.* 266:19536-19542.

Nebert, D.W., 1989, The Ah locus: Genetic differences in toxicity, cancer, mutation and birth defects, *CRC Crit. Rev. Toxicol.* R.O. McClellan, ed., Vol. 20, pp. 153-174, CRC Press, Inc., Boca Raton, Florida.

Nebert, D.W., 1991, Proposed role of drug-metabolizing enzymes: Regulation of steady state levels of the ligands that effect growth, homeostasis, differentiation, and neuroendocrine functions, *Mol. Endocrinol.* 5:1203-1214.

Nebert, D.W., and Gonzalez, F.J., 1987, P450 genes: Structure, evolution and regulation, *Annu. Rev. Biochem.* 56:945-993.

Nebert, D.W., and Jones, J.E., 1989, Regulation of the mammalian cytochrome P_1450 (*CYP1A1*) gene, *Int. J. Biochem.* 21:243-252.

Nebert, D.W., Petersen, D.D., and Fornace, A.J. Jr., 1990, Cellular responses to oxidative stress: The [*Ah*] gene battery as a paradigm, *Environ. Health Perspect.* 88:13-25.

Nebert, D.W., Petersen, D.D., and Puga, A., 1991, Human *AH* polymorphism and cancer: Inducibility of *CYP1A1* and other genes by combustion products and dioxin, *Pharmacogenetics* 1:68-78.

Puga, A., RayChaudhuri, B., Salata, K., Zhang, Y.-H., and Nebert, D.W., 1990, Stable expression of mouse *Cyp1a-1* and human *CYP1A2* cDNAs transfected into mouse hepatoma cells lacking detectable P450 enzyme activity, *DNA Cell Biol.* 9:425-436.

RayChaudhuri, B., Nebert, D.W., and Puga, A., 1990, The *Cyp1a-1* gene negatively regulates its own transcription and that of other members of the aromatic hydrocarbon-responsive [*Ah*] gene battery, *Mol. Endocrinol.* 4:1773-1781.

Robertson, J.A., Hankinson, O., and Nebert, D.W., 1987, Autoregulation plus positive and negative elements controlling transcription of genes in the [*Ah*] battery, *Chem. Scripta* 27A:83-87.

Rushmore, T.H., King, R.G., Paulson, K.E., and Pickett, C.B., 1990, Regulation of glutathione S-transferase Ya subunit gene expression: Identification of a unique xenobiotic-responsive element controlling inducible expression by planar aromatic compounds, *Proc. Natl. Acad. Sci. U.S.A.* 87:3826-3830.

Rushmore, T.H., Morton, M.R., and Pickett, C.B., 1991, The antioxidant responsive element. Activation by oxidative stress and identification of the DNA consensus sequence required for functional activity, *J. Biol. Chem.* 266:11632-11639.

Saatcioglu, F., Perry, D.J., Pasco, S., and Fagan, J.B., 1990, Multiple DNA-binding factors interact with overlapping specificities at the aryl hydrocarbon response element of the cytochrome P450IA1 gene, *Mol. Cell. Biol.* 10:6408-6416.

Takimoto, K., and Pitot, H.C., 1992, Identification of 2,3,7,8-tetrachlorodibenzo-*p*-dioxin-responsive elements in the aldehyde dehydrogenase gene, *The Toxicologist* 12:195.

Vasiliou, V., Puga, A., and Nebert, D.W., 1992, Negative regulation of the murine cytosolic aldehyde dehydrogenase-3 (*Aldh-3c*) gene by functional CYP1A1 and CYP1A2 proteins, *Biochem. Biophys. Res. Commun.* in press.

Vasiliou, V., and Marselos, M., 1989a, Changes in the inducibility of a hepatic aldehyde dehydrogenase by various effectors, *Arch. Toxicol.* 63:221-225.

Vasiliou, V., and Marselos, M., 1989b, Tissue distribution of inducible aldehyde dehydrogenase activity in the rat after treatment with phenobarbital or methylcholanthrene, *Pharmacol. Toxicol.* 64:39-42.

Vasiliou, V., Törrönen, R., Malamas, M., and Marselos, M., 1989, Inducibility of liver cytosolic aldehyde dehydrogenase activity in various animal species, *Comp. Biochem. Physiol.* 94C:671-675.

Whitlock, J.P. Jr., 1990, Genetic and molecular aspects of 2,3,7,8-tetrachlorodibenzo-*p*-dioxin action, *Annu. Rev. Pharmacol. Toxicol.* 30:251-277.

HUMAN STOMACH ALDEHYDE DEHYDROGENASE, $ALDH_3$

Lily C. Hsu and Akira Yoshida

Department of Biochemical Genetics
Beckman Research Institute of the City of Hope
Duarte, CA 91010

INTRODUCTION

Aldehyde dehydrogenase (ALDH; aldehyde:NAD^+ oxidoreductase, EC 1.2.1.3) activities are observed in most tissues with the highest activity in the liver (Deitrich, 1966). In humans, five liver ALDH isozymes have been purified and characterized and one isozyme, $ALDH_x$, has been identified by reverse genetics (Hsu and Chang, 1991). Cytosolic $ALDH_1$ mitochondrial $ALDH_2$, and γ-aminobutyraldehyde dehydrogenase, exhibit low Km values (μM range), while cytosolic $ALDH_3$ and mitochondrial $ALDH_4$ have high Km values (mM range) toward acetaldehyde. $ALDH_3$ is active for oxidation of heptaldehyde and benzaldehyde, and $ALDH_4$ is most active for glutamic γ-semialdehyde as substrate (Forte-McRobbie and Pietruszko, 1986).

The amino acid and cDNA sequences of human $ALDH_1$ and $ALDH_2$ have been determined (Hempel et al., 1984; Hempel et al., 1985; Hsu et al., 1985). Their genes were also cloned, characterized, and mapped to chromosomes 9q21 and 12q24, respectively (Hsu et al., 1986; Hsu et al., 1988; Hsu et al., 1989; Raghunathan et al., 1988). Both isozymes are widely distributed among various tissues and generally constitutively expressed. They are homotetramers consisting of subunits of 54,000 D.

Human $ALDH_3$ isozymes have been purified from stomach and liver tissues (Santisteban et al., 1985, Agarwal et al., 1989; Yin et al., 1989; Wang et al., 1990). A partial amino acid sequence of human stomach $ALDH_3$ was recently reported (Yin et al., 1991). Unlike $ALDH_1$ and $ALDH_2$, $ALDH_3$ probably has dimeric structure with subunit of 55,000 D (Teng, 1981; Santisteban et al., 1985; Yoshida et al., 1991). $ALDH_3$ is expressed at a high level in stomach, esophagus and lung, but generally at a low level in the liver and kidney. The activity was not detected in many other tissues tested (Harada et al., 1980; Santisteban et al., 1985; Duley et al., 1985; Yin et al., 1988).

Several studies have demonstrated that human stomach $ALDH_3$ exhibited multiple bands on isoelectric focussing gel with various banding patterns in different individuals (Teng, 1981; Duley et al., 1985; Hopkinson et al., 1985; Santisteban et al.,

1985; Yin et al., 1988; Yin et al., 1989). One group attributed the multiplicity to post-translational modification rather than to genetic variation (Meier-Tackmann et al., 1984). Other investigators proposed a single-gene model (Teng, 1981; Duley et al., 1985; Hopkinson et al., 1985). Recently, based on the isoelectrofocussing patterns of human stomach and liver samples, a two-gene model was proposed by Yin et al.(1988). In this model human $ALDH_3$ isozyme is encoded by two separate gene loci, $\underline{ALDH_{3a}}$ and $\underline{ALDH_{3b}}$ for A and B subunits, and two common alleles, $\underline{ALDH_{3b}^1}$ and $\underline{ALDH_{3b}^2}$ exist at the $\underline{ALDH_{3b}}$ locus. Both single-gene and two-gene models lack supporting pedigree analysis data, possibly due to the difficulty in obtaining stomach specimens.

ALDH isozymes with similar enzymatic and molecular properties to that of the human stomach $ALDH_3$ were obtained from stomach of mouse (Algar and Holmes, 1989), rat (Koivusalo et al., 1989), opossum (Holmes et al., 1991) and baboon (Holmes and Vandeberg, 1986). The animal stomach $ALDH_3$ was found also in urinary bladder and ocular/corneal tissues (Lindahl, 1986;Holmes et al, 1988; Evces and Lindahl, 1989; Holmes et al., 1991). Variant $ALDH_3$ bands were found in mouse and opossum tissues, and the mode of inheritance was compatible to expression of two alleles existing at a single autosomal locus (Mather and Holmes, 1984; Holmes et al., 1988; Holmes et al., 1991). Epigenetic modifications could cause further complexity in zymogram patterns (Rout and Holmes, 1984).

Rat $ALDH_3$ exists in hepatocellular carcinomas (HCC) and can be induced in the normal liver by treatment with various carcinogenic and xenobiotic reagents (Lindahl and Feinstein, 1976; Lindahl and Evces, 1984; Marselos and Lindahl, 1988; Koivusalo et al., 1989). ALDH cDNA which can encode 435 amino acid residues was cloned from an HTC rat hepatoma cell line (Jones et al., 1988). Judging from their kinetic, electrophoretic, and immunological properties, human and rat hepatoma ALDHs are probably identical to their stomach $ALDH_3$, respectively (Agarwal et al., 1989).

In this study, we used reverse genetic approach to clone and identify human stomach $\underline{ALDH_3}$ gene. The defined genomic structure can provide some insight into the possibility of the gene duplication, tissue dependent expression, and the mechanism of inducibility of the $ALDH_3$ in HCC.

EXPERIMENTAL PROCEDURES

Isolation and characterization of $ALDH_3$ cDNA clones from human stomach cDNA library: The screening probe (PCR#7) was obtained by the reverse transcription-polymerase chain reaction (RT-PCR) procedure which was described in detail previously (Hsu et al., 1992). Briefly, human stomach poly$(A)^+$ RNA was used as a template for the first strand cDNA synthesis and followed by PCR with a pair of degenerate oligonucleotide primers, primer 1 (5'-GAGGTACCCTGGAGCTKGGRGGWAARAGCCC-3') and primer 2 (5'-TCGAATTCACWGGYCCRAARATCTCCTC-3'). These sequences correspond, respectively, to the sense and antisense sequences of the first and second selected conserved regions among nucleotide sequences of the following ALDH: Cytosolic $ALDH_1$, mitochondrial $ALDH_2$ and mitochondrial $ALDH_x$ from human (Hsu et al., 1985; Hsu and Chang, 1991), $ALDH_2$ from bovine and rat

(Farrés et al., 1989), $ALDH_1$ (Ahd-2) from mouse (Rongnoparut and Weaver, 1991), phenobarbital-induced ALDH from rat (Dunn et al., 1989), tumor associated ALDH from rat (Jones et al., 1988). The positions of all the primers described in this publication are shown in Fig. 1. The PCR product was purified on a gel and subcloned for nucleotide sequencing and for preparation of the PCR probe. The radioactive-labeled probe was used in a standard plaque hybridization procedure (Sambrook et al., 1989) to screen a human stomach cDNA library constructed in λgt11 vector (Clontech Lab.)

Three independent positive phage clones (λ5, λ35 and λ38) were isolated, and their inserts were mapped, subcloned in pBluescript KS(+) vector and sequenced, respectively.

Construction of full-length cDNA of human stomach $ALDH_3$: The above-three partial cDNA clones which had an overall length of 1083 bp, do not contain a full-length cDNA sequence of human stomach $ALDH_3$. In order to construct a full-length cDNA, the 5'- and 3'- ends of the $ALDH_3$ transcript were amplified with PCR technique according to the method described by Frohman et al. (1988).

To prepare 5'-end of $ALDH_3$ cDNA fragment, first strand cDNA of human stomach poly(A)$^+$RNA was synthesized with primer 1-20 (5'-GGGTCACAGAGGATGTAGTC-3') tailed with poly(A) at the 3'-end by terminal deoxynucleotidyl transferase and followed by second strand cDNA synthesis with primer amp-dT-36 (5'-CTAGGATCCTCGAGTCGAC$(T)_{17}$-3'). The resulting double-stranded cDNA was amplified by PCR with the primer amp-19 (5'-CTAGGATCCTCGAGTCGAC-3'; same sequence as amp-dT-36 except no poly(dT) tail, and the primer 5-21(5'-GTAGAGCTCGTCCTGCTGAGT-3'). To prepare 3'-end of $ALDH_3$ cDNA fragment, the stomach first strand cDNA was synthesized with oligo$(dT)_{12\text{-}18}$ primer and amplified with two PCR primers; primer 2-21 (5'-GGACTATGGAAGAATCATTAG-3') and primer amp-19. The 5'-end and 3'-end PCR products were gel-purified, subcloned and identified by $ALDH_3$-specific hybridization probes. DNA sequence determination of one positive 5'-end clone (PCR-N, which had the longest 5'-end sequence) and one positive 3'-end clone (PCR-C) showed that the two clones overlapped with λ38cDNA and λ5cDNA at their 5' and 3'-end, respectively.

Full-length $ALDH_3$ cDNA was constructed by the stepwise ligation of XhoI/SstI fragment, SstI/SmaI fragment, SmaI/BamHI fragment and BamHI fragment of the insert of PCR-N clone, λ38 clone, λ5 clone and PCR-C clone, respectively. The structure of the full-length cDNA was confirmed by mapping with various restriction endonucleases and by DNA sequencing.

Isolation and characterization of human stomach $ALDH_3$ genomic clones: A human genomic library constructed in λDASH was screened by PCR#7 probe. The restriction maps of two positive genomic clones (λ4 and λ29) were determined and the restricted fragments containing exonic sequence were identified by Southern blot and sequence analysis.

Purification of $ALDH_3$ and preparation of rabbit antibody: $ALDH_3$ was purified to homogeneity from human HCC cells by stepwise fractionation through CM-Sephadex column chromatography, Bio-gel filtration, affinity chromatography with 5'-AMP Sepharose 4B, and chromato-focussing. Details of the procedure will be reported elsewhere. An antibody was produced in rabbits by immunizing the animals with the purified $ALDH_3$ together with Freund's adjuvant as previously described (Ikawa et al., 1983).

```
         1                                                                                                                        30
Human    Met Ser Lys Ile Ser Glu Ala Val Lys Arg Ala Pro Ala Ala Phe Ser Ser Gly Arg Thr Arg Pro Leu Gln Phe Arg Ile Gln Gln Leu
         ATG AGC AAG ATC AGC GAG GCC GTG AAG CGC GCC CCC GCC GCC TTC AGC TCG GGC AGG ACC CGT CCG CTG CAG TTC CGG ATC CAG CAG CTG
Rat      --- --- -GT --- --T --C A-- --- --A --G --- AGG -AG --- --T -A- --C --- -A- --T --A T-- --- --- --- --A --- --- --- T--
         --- --- Ser --- --- Asp Thr --- Lys --- --- Arg Glu --- --- Asn --- --- Lys --- --- Ser --- --- --- --- --- --- --- ---

         31                                                                                                                       60
Human    Glu Ala Leu Gln Arg Leu Ile Gln Glu Gln Glu Gln Glu Leu Val Gly Ala Leu Ala Ala Asp Leu His Lys Asn Glu Trp Asn Ala Tyr
         GAG GCG CTG CAG CGC CTG ATC CAG GAG CAG GAG CAG GAG CTG GTG GGC GCG CTG GCC GCA GAC CTG CAC AAG AAT GAA TGG AAC GCC TAC
Rat      --- --- --- --- --- A-- --- A-T --- A-C CT- A-A AGC A-C TCT --G --- --- --- T-T --- --- GG- --- --C --- --- -C- T-- --A
         --- --- --- --- --- Met --- Asn --- Asn Leu Lys Ser Ile Ser --- --- --- --- Ser --- --- Gly --- --- --- --- Thr Ser ---

         61                                                                                                                       90
Human    Tyr Glu Glu Val Val Tyr Val Leu Glu Glu Ile Glu Tyr Met Ile Gln Lys Leu Pro Glu Trp Ala Ala Asp Glu Pro Val Glu Lys Thr
         TAT GAG GAG GTG GTG TAC GTC CTA GAG GAG ATC GAG TAC ATG ATC CAG AAG CTC CCT GAG TGG GCC GCG GAT GAG CCC GTG GAG AAG ACG
Rat      --- --- --A --- -CT C-- --A --G --- --- C-T --T AC- -CA --T A-- G-- --- --- --T --- --T -A- --- --- --T --- -CC --- --T
         --- --- --- --- Ala His Val --- --- --- Leu Asp Thr Thr --- Lys Glu --- --- Asp --- --- Glu --- --- --- --- Ala --- ---

         91          (5-21)                                                                                                      120
Human    Pro Gln Thr Gln Gln Asp Glu Leu Tyr Ile His Ser Glu Pro Leu Gly Val Val Leu Val Ile Gly Thr Trp Asn Tyr Pro Phe Asn Leu
         CCC CAG ACT CAG CAG GAC GAG CTC TAC ATC CAC TCG GAG CCA CTG GGC GTG GTC CTC GTC ATT GGC ACC TGG AAC TAC CCC TTC AAC CTC
Rat      -G- --- --C --- --- --T --C --- --- --- --- --- --- --C --- --T --- --- --T --- --A --T G-T --- --- --- --- --- --- ---
         Arg --- --- --- --- --- Asp --- --- --- --- --- --- --- --- --- --- --- --- --- --- --- Ala --- --- --- --- --- --- ---

         121                                                                                                                     150
Human    Thr Ile Gln Pro Met Val Gly Ala Ile Ala Ala Gly Asn Ser Val Val Leu Lys Pro Ser Glu Leu Ser Glu Asn Met Ala Ser Leu Leu
         ACC ATC CAG CCC ATG GTG GGC GCC ATC GCT GCA GGG AAC TCA GTG GTC CTC AAG CCC TCG GAG CTG AGT GAG AAC ATG GCG AGC CTG CTG
Rat      --- --- --- --- --- --- --T --- G-T --- --- --A --- G-- --- A-- --- --- --- --A --A G-- --C -G- C-- --- --A GA- --- --A
         --- --- --- --- --- --- --- --- Val --- --- --- --- Ala --- Ile --- --- --- --- --- Val --- Gly His --- --- Asp --- ---

         151                                                                                                                     180
Human    Ala Thr Ile Ile Pro Gln Tyr Leu Asp Lys Asp Leu Tyr Pro Val Ile Asn Gly Gly Val Pro Glu Thr Thr Glu Leu Leu Lys Glu Arg
         GCT ACC ATC ATC CCC CAG TAC CTG GAC AAG GAT CTG TAC CCA GTA ATC AAT GGG GGT GTC CCT GAG ACC ACG GAG CTG CTC AAG GAG AGG
Rat      --G --A C-- --- --T --- --T A-- --- C-- A-- --- --- -T- --G G-- --A --- --- --- --- --A --- --- --- --- --- --A --- ---
         --- --- Leu --- --- --- --- Met --- Gln Asn --- --- Leu --- Val Lys --- --- --- --- --- --- --- --- --- --- --- --- ---

         181                                                                                                                     210
Human    Phe Asp His Ile Leu Tyr Thr Gly Ser Thr Gly Val Gly Lys Ile Ile Met Thr Ala Ala Ala Lys His Leu Thr Pro Val Thr Leu Glu
         TTC GAC CAT ATC CTG TAC ACG GGC AGC ACG GGG GTG GGG AAG ATC ATC ATG ACG GCT GCT GCC AAG CAC CTG ACC CCT GTC ACG CTG GAG
Rat      --T --- --C --- A-- --- --T --G --- --A -CC --A --- --- --T G-T --- G-C --- --C --- --- --T --- --T --- --- --C --- ---
         --- --- --- --- Met --- --- --- --- --- Ala --- --- --- --- Val --- Ala --- --- --- --- --- --- --- --- --- --- --- ---

         211 (Primer 1)                                                                                                          240
Human    Leu Gly Gly Lys Ser Pro Cys Tyr Val Asp Lys Asn Cys Asp Leu Asp Val Ala Cys Arg Arg Ile Ala Trp Gly Lys Phe Met Asn Ser
         CTG GGA GGG AAG AGT CCC TGC TAC GTG GAC AAG AAC TGT GAC CTG GAC GTG GCC TGC CGA CGC ATC GCC TGG GGG AAA TTC ATG AAC AGT
Rat      --T --- --- --A --C --T --T --- --- --- --- G-- --- --T --A --T --- --T --- A-G --T --- --- --- --- --- --T --- --- ---
         --- --- --- --- --- --- --- --- --- --- --- --- --- --- --- --- --- --- --- --- --- --- --- --- --- --- --- --- --- ---

         241                             (1-20)                                                                                  270
Human    Gly Gln Thr Cys Val Ala Pro Asp Tyr Ile Leu Cys Asp Pro Ser Ile Gln Asn Gln Ile Val Glu Lys Leu Lys Lys Ser Leu Lys Glu
         GGC CAG ACC TGC GTG GCC CCA GAC TAC ATC CTC TGT GAC CCC TCG ATC CAG AAC CAA ATT GTG GAG AAG CTC AAG AAG TCA CTG AAA GAG
Rat      --- --- --- --T --- --- --- --- --- --- --- --- --- --- -GC --T --- --- --- --C --- --- --- --- --- --- --- --C --- --T
         --- --- --- --- --- --- --- --- --- --- --- --- --- --- --- --- --- --- --- --- --- --- --- --- --- --- --- --- --- Asp

         271                                     (2-21)                                                                          300
Human    Phe Tyr Gly Glu Asp Ala Lys Lys Ser Arg Asp Tyr Gly Arg Ile Ile Ser Ala Arg His Phe Gln Arg Val Met Gly Leu Ile Glu Gly
         TTC TAC GGG GAA GAT GCT AAG AAA TCC CGG GAC TAT GGA AGA ATC ATT AGT GCC CGG CAC TTC CAG AGG GTG ATG GGC CTG ATT GAG GGC
Rat      --- --T --- --- --- --- --- C-G --- --T --T --- --G --G --- --C -A- -A- --T --- --- --- C-- --C -AA --- --- --- --C AA-
         --- --- --- --- --- --- --- Gln --- --- --- --- --- --- --- --- Asn Asp --- --- --- --- --- --- Lys --- --- --- Asp Asn

         301                                                                                                                     330
Human    Gln Lys Val Ala Tyr Gly Gly Thr Gly Asp Ala Ala Thr Arg Tyr Ile Ala Pro Thr Ile Leu Thr Asp Val Asp Pro Gln Ser Pro Val
         CAG AAG GTG GCT TAT GGG GGC ACC GGG GAT GCC GCC ACT CGC TAC ATA GCC CCC ACC ATC CTC ACG GAC GTG GAC CCC CAG TCC CCG GTG
Rat      --- --A --A --C C-- --A --- --T T-- --C CAG T-- T-A --A --- --- --T --A --- --- --G GT- --T --- --- --- --- --- --A ---
         --- --- --- --- His --- --- --- Trp --- Gln Ser Ser --- --- --- --- --- --- --- --- Val --- --- --- --- --- --- --- ---

         331      (Primer 2)                                                                                                     360
Human    Met Gln Glu Glu Ile Phe Gly Pro Val Leu Pro Ile Val Cys Val Arg Ser Leu Glu Glu Ala Ile Gln Phe Ile Asn Gln Arg Glu Lys
         ATG CAA GAG GAG ATC TTC GGG CCT GTG CTG CCC ATC GTG TGC GTG CGC AGC CTG GAG GAG GCC ATC CAG TTC ATC AAC CAG CGT GAG AAG
Rat      --- --G --- --- --- --T --- --G --- A-- --- --T --- --T --T --A --- T-- --- --- --- --T --A --T --- --- --- --- --- ---
         --- --- --- --- --- --- --- --- --- Met --- --- --- --- --- --- --- --- --- --- --- --- --- --- --- --- --- --- --- ---

         361                                                                                                                     390
Human    Pro Leu Ala Leu Tyr Met Phe Ser Ser Asn Asp Lys Val Ile Lys Lys Met Ile Ala Glu Thr Ser Ser Gly Gly Val Ala Ala Asn Asp
         CCC CTG GCC CTC TAC ATG TTC TCC AGC AAC GAC AAG GTG ATT AAG AAG ATG ATT GCA GAG ACA TCC AGT GGT GGG GTG GCG GCC AAC GAT
Rat      --- --- --A --- --T G-- --- --- -A- --T --G --- --- --C --- --A --- --C --- --- --- --- --C --- --A --- A-A --- --T --C
         --- --- --- --- --- Val --- --- Asn --- Glu --- --- --- --- --- --- --- --- --- --- --- --- --- --- --- Thr --- --- ---

         391                                                                                                                     420
Human    Val Ile Val His Ile Thr Leu His Ser Leu Pro Phe Gly Gly Val Gly Asn Ser Gly Met Gly Ser Tyr His Gly Lys Lys Ser Phe Glu
         GTC ATC GTC CAC ATC ACC TTG CAC TCT CTG CCC TTC GGG GGC GTG GGG AAC AGC GGC ATG GGA TCC TAC CAT GGC AAG AAG AGC TTC GAG
Rat      --- --T --T --- --- --- G-- -C- A-- T-- --- --T --T --T --- --- --- --- --- --- --G G-- --- --- --- --- --A --- --- ---
         --- --- --- --- --- --- Val Pro Thr --- --- --- --- --- --- --- --- --- --- --- --- Ala --- --- --- --- --- --- --- ---

         421                                                                                                                     450
Human    Thr Phe Ser His Arg Arg Ser Cys Leu Val Arg Pro Leu Met Asn Asp Glu Gly Leu Lys Val Arg Tyr Pro Pro Ser Pro Ala Lys Met
         ACT TTC TCT CAC CGC CGC TCT TGC CTG GTG AGG CCT CTG ATG AAT GAT GAA GGC CTG AAG GTC AGA TAC CCC CCG AGC CCG GCC AAG ATG
Rat      --C --- --C --- --- --- --- --- --- --- -A- T-- --- T-- --- --A --- -CT -AC --- -C- --G --T --- --A --- --A --- --- ---
         --- --- --- --- --- --- --- --- --- --- Lys Ser --- Leu --- Glu --- Ala His --- Ala --- --- --- --- --- --- --- --- ---

         451
Human    Thr Gln His ***
         ACC CAG CAC TGA
Rat      C-- -G- --- ---
         Pro Arg --- ---
```

Figure 1. Comparison of amino acid and nucleotide sequences of human stomach $ALDH_3$ with those of rat tumor ALDH (Jones et al. 1988). The corresponding amino acid sequence is shown above the nucleotide sequence for the human and beneath the nucleotide sequence for the rat cDNA. Dashes mean identity. The rat tumor ALDH cDNA also encodes 453 amino acid residues. Hence, no gap is introduced for the alignment. Arrows indicate the exon-intron junction sites of the gene. The positions of the primers used in this study are underlined.

Expression of human stomach $ALDH_3$ in a bacterial expression system: DNA coding for the open reading frame of human $ALDH_3$ was cloned into the EcoRI site of bacterial expression plasmid pkk223-3 (Pharmacia), after removing 5'- and 3'-UT region of full-length $ALDH_3$ by PCR amplification. Bacterial clones (89-36 and 89-33) carrying the recombinant construct with insert in the correct and opposite orientation, respectively, were induced with 1 mM IPTG in their midlog growth phase and analyzed on a SDS-polyacrylamide gel. The supernatant of the lysate was partially purified and the ALDH activity was assayed spectrophotometrically at 340 nm.

RESULTS

Isolation and characterization of human stomach $ALDH_3$ cDNA: The constructed full-length cDNA has an overall length of 1,624 bp, containing an open reading frame encoding for 453 amino acid residues. The complete nucleotide sequence, the deduced amino acid sequence of the human stomach $ALDH_3$ and the sequence of rat tumor $ALDH_3$ are shown in Fig. 1. The sequence around ATG start codon is highly similar to the Kozak consensus sequence ((GCC)GCCA/GCCATGG) for higher eukaryotes (Kozak, 1987). The 3'-UT region is 234 bp long and is followed by a poly(A) tail. The consensus polyadenylation signal (AATAAA) (Birnstiel et al., 1985) is located 27 bp upstream of the poly(A)$^+$ tract.

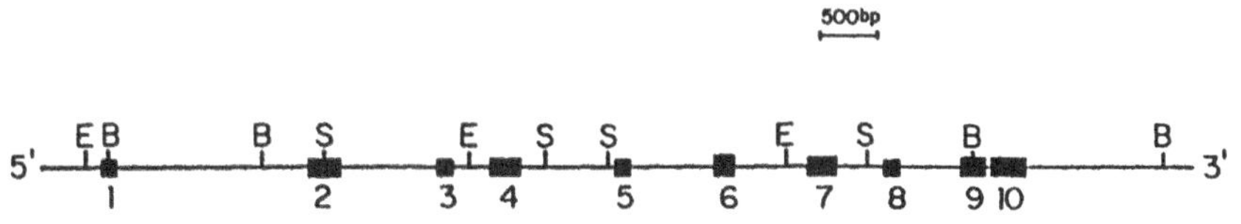

Figure 2. Restriction map of the human stomach $ALDH_3$ gene. Boxes indicate the extent of exons and lines connecting the exons denote the length of introns. Single letters symbolize restriction sites: E, EcoRI; S, SstI; and B, BamHI.

Isolation and characterization of human stomach $ALDH_3$ gene: From approximately $1x10^6$ bacteriophage plaques screened with probe #7, five positive independent genomic clones were obtained. One of the bacteriophage clones (λ4),which included the entire $ALDH_3$ gene, was subjected to extensive characterization. The restriction map of $ALDH_3$ gene is shown in Fig. 2. This restriction map was consistent with that obtained from Southern blot analysis of total genomic DNA digested by seven restriction enzymes, EcoRI, SstI, BamHI, HindIII, XbaI, PvuII and BglII (Fig. 3). Hence, the genomic DNA clone has retained the sequence organization of the cellular DNA. The human $ALDH_3$ gene spans approximately 8 Kb, and consists of 10 exons ranging in size between 86 and 249 bp with an average of 148+51 bp, and 9 introns ranging in size between 0.09 to 1.55 Kb. The intron/exon junctions conform to the consensus sequences for intronic donor and acceptor splice signals (Breathnach and Chambon, 1981). About half of the introns (5 of 9) interrupt the coding sequence between the codons. The nucleotide sequences of the exons were in complete agreement with that of the cDNA.

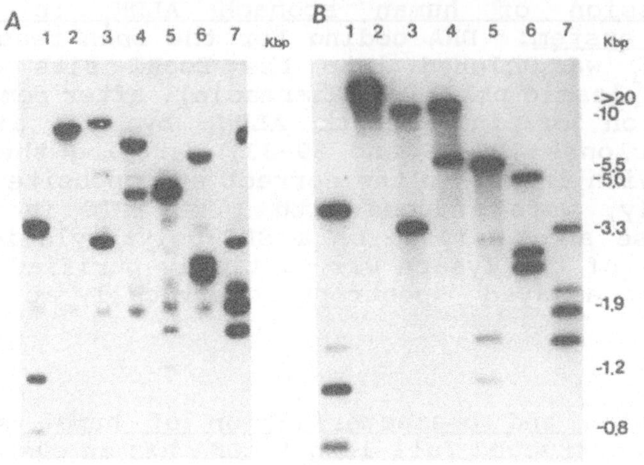

Figure 3. Southern blot analysis of (A) human genomic DNA and (B) human stomach $\underline{ALDH_3}$ genomic clone (λ4). Genomic and cloned DNA were treated with (1) PvuII, (2) HindIII, (3) XbaI, (4) BglII, (5) BamHI, (6) EcoRI, and (7) SstI, respectively. The blots were hybridized with full-length human $ALDH_3$ cDNA. The sizes of restricted fragments were shown at the right of each panel. Although genomic clone λ4 contains the entire $\underline{ALDH_3}$ gene, EcoRI (5.5 Kb) and XbaI (5.0 Kb) fragments originated from the 3'-region of the clone are truncated and shorter than the corresponding fragments (9.5 Kb and >20 Kb) originating from the total genomic DNA. Another genomic clone, λ29, had an extended 3'-sequence and produced longer fragments (9.5 Kb and >20 Kb) corresponding to those originating from the total genomic DNA (figure is not shown).

The 5'-upstream sequence to the ATG initiation codon was examined to identify possible regulatory sequence elements. Two canonical transcriptional control elements were found in this region. A TATA box and a CCAAT box are located between -79 and -76 and between -117 and -114, respectively. The two boxes were in the consensus distance from the 5'-end of the longest $ALDH_3$ cDNA clone obtained. Therefore, the putative transcription initiation site was assigned at -49 position from the ATG initiation codon.

<u>Analysis of mRNA</u>: Northern blot analysis detected the existence of $ALDH_3$ mRNA in two independent stomach samples but not in six normal liver samples (data not shown). The size of

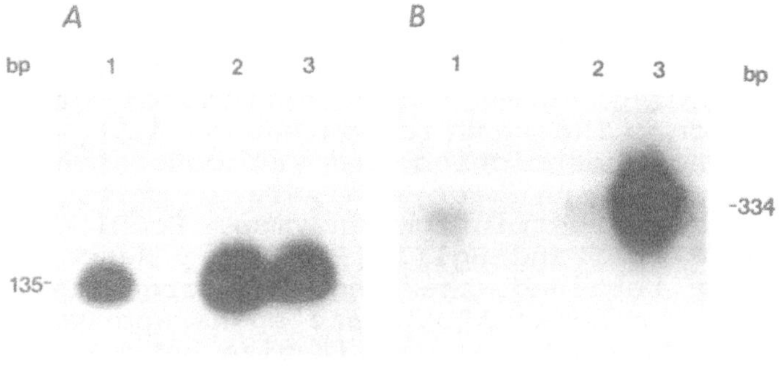

Figure 4. Hybridization of PCR amplification products with specific cDNA probes. Poly(A)$^+$ RNA preparations were amplified using two sets of primers, one specific for $ALDH_3$ and another specific for phosphoglycerate kinase (PGK). The products were separated by agarose gel electrophoresis and transferred onto a nitrocellulose filter. The filter was hybridized, de-probed, and re-hybridized successively with: (A) $ALDH_3$ cDNA probe; and (B) PGK cDNA probe. Lane 1, RNA from HCC cells; lane 2, RNA from stomach; lane 3, RNA from normal liver. In order to detect $ALDH_3$ mRNA, a larger amount of PCR products of the normal liver RNA was used as shown by the strong hybridization band (lane 3) for PGK which served as an internal reference.

the hybridization band, 1.7 Kb, is compatible with the cDNA and genomic data. However, when liver RNA samples were amplified using a pair of specific $ALDH_3$ primers and subjected to separation by agarose gel electrophoresis followed by hybridization with the $ALDH_3$ cDNA probe, a hybrid band with the expected size, 135 bp, was observed in normal liver RNA as well as stomach RNA (Fig. 4). The relative abundance of mRNA estimated by PCR using PGK mRNA as an internal reference, is much higher in HCC cells than in the normal livers (Fig. 4).

Expression of human $ALDH_3$ in E.coli system: The expression plasmid (89-36) containing complete coding region of $ALDH_3$ in the expression orientation with respect to the promoter produced a novel protein of 55,000 D, which can be immunostained with rabbit anti-human HCC ALDH antiserum. While the plasmid (89-33) with the same insert in the opposite orientation did not produce any novel protein, more than 85% of the induced protein was present as a precipitate in inclusion bodies.

The kinetic properties of the expressed human $ALDH_3$ were characterized using various substrates and cofactors. The Km values for propionaldehyde, hexanal, benzaldehyde, NAD^+ and $NADP^+$ match well with the published Km values for $ALDH_3$ purified from human liver or stomach (Yin et al., 1989; Wang et al., 1990). The expressed human $ALDH_3$ exhibited an optimal pH range for oxidation of propionaldehyde at alkaline range above pH 8.5 (data not shown).

DISCUSSION

We present the full sequence of the cDNA and the genomic organization of human stomach $ALDH_3$ gene in this study. The human stomach $ALDH_3$ expressed in bacteria exhibited similar kinetic properties to that of the purified human stomach and liver $ALDH_3$ (Yin et al., 1989; Wang et al., 1990), and reacted with rabbit antiserum against human $ALDH_3$ prepared from HCC cells.

Rat hepatoma ALDH is the homologue of human stomach $ALDH_3$ (for review, see Yoshida et al., 1991). The degree of identity between the coding regions of human stomach $ALDH_3$ and that of rat tumor ALDH is 81.7% at the nucleotide level and 80.6% at the amino acid level (Fig. 1). A high degree of similarity is observed in homotopic pairs of isozymes, i.e. about 80% to 90% identity between any two $ALDH_1$, and between $ALDH_2$ isozymes from various mammalia species, while only <70% identity is found between $ALDH_1$ and $ALDH_2$ isolated from the same species.

The structure of $ALDH_3$ (453 residues) is substantially different from the 500-residue $ALDH_1$ and $ALDH_2$ structure. The rat tumor ALDH shows a maximum positional identity of 29.8% with human $ALDH_1$ by placing a total of 11 gaps into the two structures (Hempel et al., 1989). The alignments showed the conservation of Glu-209/268, Cys-243/302, Gly-187/245 and Gly-192/250 residues, which have been previously implicated in the active sites of ALDH isozymes by the studies of selective chemical modifications of $ALDH_1$ and/or $ALDH_2$ isozymes. In this study, the positional alignments between human $ALDH_3$ and rat tumor ALDH supports the importance of these previously identified specific residues since all are clearly conserved (Glu-210, Cys-244, Gly-188 and Gly-193 in Fig. 1).

The flanking sequence of Glu-210 residue, Val-Thr-Leu-Glu-Leu-Gly-Gly-Lys-Ser-Pro, is conserved among different ALDH sequences described so far including human stomach $ALDH_3$. The two conserved Gly residues (Gly-188 and Gly-193) of human stomach $ALDH_3$ are also included in a consensus sequence, Thr-Gly-Ser-Thr-X-Val/Ile-Gly, followed by a basic amino acid and then an alkyl-branched amino acid residue as other known mammalian ALDH. Cys-244 of human stomach $ALDH_3$ is also flanked by a conserved sequence, Gly-Gln-X-Cys-X-Ala. Furthermore, the putative active site residue, Glu-210 and Cys-244, are located in the longest conserved stretch (71 residues) as shown in Fig. 1. The conservation of flanking sequences of specific residues mentioned above can account for the functional similarities of structurally distinct $ALDH_1$/$ALDH_2$ and $ALDH_3$ isozymes.

A homologous stretch (5'-CCAGCTGCT/AGGA/TGA/CA/TG/CG/TCTGT-3') was found at the 5'-UT sequences of human $ALDH_3$ and rat tumor ALDH mRNA. It has been suggested that 5'-UT sequences play a role in translation efficiency of mRNA, possibly via the formation of secondary structures (Guan and Weiner, 1989; Ito et al., 1990). The significance of $ALDH_3$ 5'-UT homologous stretch was not clear, although the 5'-end of the stretch contains the sequence for a potential hairpin structure. The 5'-UT regions of human, mouse, and phenobarbital-induced rat $ALDH_1$ mRNAs show 87% positional identity (Dunn et al., 1989; Hsu et al., 1989; Rongnoparut and Weaver, 1991), while human and rat mitochondrial $ALDH_2$ mRNAs have only 16% positional identity at the 5'-UT region, and do not have a homologous stretch. These different rates of divergence at the 5'-UT regions of the three ALDH isozymes might correlate with their biological requirements. The 5'-UT sequences of $ALDH_1$, $ALDH_2$ and $ALDH_3$ mRNA from other mammalian species may provide some insight into this possibility.

Partial amino acid sequence (total 215 residues) of human stomach ALDH3 was reported recently by Yin et al. (1991). The reported sequence is compatible with the present deduced sequence except that discrepancies were found at positions 170 and 436, i.e. Val and Asp in the deduced sequence and Leu and Glu in the protein sequence, respectively. By using the PCR method, it can be distinguished whether or not these differences are due to the polymorphism of $\underline{ALDH_3}$ locus.

Southern blot analysis of total genomic DNA argues against the two gene model for the $ALDH_3$ isozymes proposed by Yin et al. (1989). If $ALDH_3$ isozymes are homo- and hetero-dimers of two types of subunits encoded by two separate genes, i.e. $\underline{ALDH_{3a}}$ and $\underline{ALDH_{3b}}$, as proposed, it is anticipated that the two genes are homologous in their coding sequences and hybridizable with a common $ALDH_3$ cDNA probe. Since the flanking UT-regions and intron sequences of the duplicated genes are less similar than their coding regions, Southern blot hybridization patterns originating from $\underline{ALDH_{3a}}$ and $\underline{ALDH_{3b}}$ genes cannot be always identical. However, in our Southern blot analysis using seven different restriction enzymes, the hybridization bands of total human genomic DNA always coincided with the genetic map of the genomic clone (Figs. 2 and 3). Thus, the two gene model is not likely to be the case, unless the duplication unit is larger than the size detected by the seven restriction enzymes used and, in addition, identical restriction sites had been preserved in the duplicated genes. It should be pointed out that, even in the case of the recently duplicated human globin α locus, the $\underline{\alpha_1}$ and $\underline{\alpha_2}$ genes with identical coding sequences can be detected by restriction analysis (Orkin, 1978).

Human $ALDH_3$ isozyme either cannot be detected or is

expressed at low levels in the normal liver, but at high levels in the stomach and HCC cells (Harada et al., 1980; Teng, 1981; Duley et al., 1985; Hopkinson et al., 1985; Santisteban et al., 1985; Yin et al., 1988; Agarwal et al., 1989; Yin et al., 1989; Wang et al., 1990). Our Northern blot analysis and Southern blot hybridization of PCR products indicated that the content of $ALDH_3$ mRNA in the normal liver is much lower than that in the stomach and HCC cells (Fig. 4). However, such higher $ALDH_3$ expression was found only in 30-40% of the HCC patient samples and it was not related to the severity of the cancerous stages (Shibuya, unpublished observation). It is evident that the transcription of $\underline{ALDH_3}$ gene is suppressed in normal livers, but it is either highly induced or de-repressed in HCC cells of certain patients. In contrast, the expression of $\underline{ALDH_3}$ in stomach and lung tissue is constitutive. Tissue-specific transcription factor(s) must be involved in the expression of the $\underline{ALDH_3}$ gene.

REFERENCES

Agarwal, D.P., Eckey, R., Rudnay, A.C., Volkens, T., and Goedde, H.W., 1989, "High Km" aldehyde dehydrogenase isozymes in human tissues: constitutive and tumor-associated forms, Progr. Clin. Biol. Res., 290:119.

Algar, E.M., and Holmes, R.S., 1989, Purification and properties of mouse stomach aldehyde dehydrogenase. Evidence for a role in the oxidation of peroxidic and aromatic aldehydes, Biochi. Biophys. Acta, 995:168.

Birnstiel, M.L., Busslinger, M., and Strub, K., 1985, Transcription termination and 3' processing: the end is in site, Cell, 41:349.

Breathnach, R., and Chambon, P., 1981, Organization and expression of eukaryotic split genes coding for proteins, Annu. Rev. Biochem., 50:349.

Dietrich, R.A., 1966, Tissue and subcellular distribution of mammalian aldehyde-oxidizing capacity, Biochem. Pharmacol. 15:1911.

Duley, J.A., Harris, O., and Holmes, R.,S., 1985, Analysis of human alcohol-and aldehyde-metabolizing isozymes by electrophoresis and isoelectric focusing, Alcoholism: Clin. Exp. Res., 9:263.

Dunn, T.J., Koleske, A.J., Lindahl, R., and Pitot, H.C., 1989, Phenobarbital-inducible aldehyde dehydrogenase in the rat, cDNA sequence and regulation of the mRNA by phenobarbital in responsive rats, J. Biol. Chem., 264:13057.

Evces, S., and Lindahl, R., 1989, Characterization of rat cornea aldehyde dehydrogenase, Arch. Biochem. Biophys., 274:518.

Farrés, J., Guan, K.-L., and Weiner, H., 1989, Primary structures of rat and bovine liver mitochondrial aldehyde dehydrogenases deduced from cDNA sequences, Eur. J. Biochem., 180:67.

Forte-McRobbie, C.M., and Pietruszko, R., 1986, Purification and characterization of human liver "high Km" aldehyde dehydrogenase and its identification as glutamic γ-semi aldehyde dehydrogenase, J. Biol. Chem., 261:2154.

Frohman, M.A., Dush, M.K., and Martin, G.R., 1988, Rapid production of full-length cDNAs from rare transcripts: amplification using a single gene-specific oligonucleotide

primer, Proc. Natl. Acad. Sci. USA, 85:8998.
Guan, K., and Weiner, H., 1989, Influence of the 5'-end region of aldehyde dehydrogenase mRNA on translational efficiency, J. Biol. Chem., 264:17764.
Harada, S., Agarwal, D.P., and Goedde, H.W., 1980, Electrophoretic and biochemical studies of human aldehyde dehydrogenase isozymes in various tissues, Life Sci., 26:1773.
Hempel, J., von Bahr-Lindström, H., and Jörnvall, H., 1984, Aldehyde dehydrogenase from human liver, primary structure of the cytoplasmic isoenzyme, Eur. J. Biochem., 141:21.
Hempel, J., Kaiser, R., and Jörnvall, H., 1985, Mitochondrial aldehyde dehydrogenase from human liver: primary structure, differences in relation to the cytosolic enzyme, and functional correlations, Eur. J. Biochem., 153:13.
Hempel, J., Harper, K., and Lindahl, R., 1989, Inducible (class 3) aldehyde dehydrogenase from rat hepatocellular carcinoma and 2,3,7,8-tetrachlorodibenzo-p-dioxin-treated liver: distant relationship to the class 1 and 2 enzymes from mammalian liver cytosol/mitochondria, Biochemistry, 28:1160.
Holmes, R.S., and Vandeberg, J.L., 1986, Aldehyde dehydrogenases, aldehyde oxidase and xanthine oxidase from baboon tissues: phenotypic variability and subcellular distribution in liver and brain, Alcohol, 3:205.
Holmes, R.S., Popp, R.A., and Vandeberg, J.L., 1988, Genetics of ocular NAD^+-dependent alcohol dehydrogenase and aldehyde dehydrogenase in the mouse: evidence for genetic identity with stomach isozymes and localization of Ahd-4 on chromosome 11 near trembler, Biochem. Genet., 26:191.
Holmes, R.S., Oorschot, R.A.H., and Vandeberg, J.L., 1991, Aldehyde dehydrogenase (ALDH) isozymes in the gray short-tailed opossum (Monodelphis domestica): tissue and subcellular distribution and biochemical genetics of $ALDH_3$, Biochem. Genet., 29:163.
Hopkinson, D.A., Santisteban, I., Povey, S., and Smith, M., 1985, Biochemical genetic analysis of human and rodentaldehyde dehydrogenase (ALDH), Alcohol, 2:73.
Hsu, L.C., Tani, K., Fujiyoshi, T., Kurachi, K., and Yoshida, A., 1985, Cloning of cDNAs for human aldehyde dehydrogenase 1 and 2, Proc. Natl. Acad. Sci. USA, 82:3771.
Hsu, L.C., Yoshida, A., and Mohandas, T., 1986, Chromosomal assignment of the genes for human aldehyde dehydrogenase-1 and aldehyde dehydrogenase-2, Am. J. Hum. Genet., 38:641.
Hsu, L.C., Bendel, R.E., and Yoshida, A., 1988, Genomic structure of the human mitochondrial aldehyde dehydrogenase gene, Genomics, 2:57.
Hsu, L.C., Chang, W.-C., and Yoshida, A., 1989, Genomic structure of the human cytosolic aldehyde dehydrogenase gene, Genomics, 5:857.
Hsu, L.C., and Chang, W.-C., 1991, Cloning and characterization of a new functional human aldehyde dehydrogenase gene, J. Biol. Chem., 266:12257.
Hsu, L.C., Chang, W.-C., Shibuya, A., and Yoshida, A., 1992, Human stomach aldehyde dehydrogenase cDNA and genomic

cloning, primary structure, and expression in *Escherichia coli*, *J. Biol. Chem.*, 267:3030.
Ikawa, M., Impraim, C.C., Wang,G., and Yoshida, A., 1983, Isolation and characterization of aldehyde dehydrogenase isozymes from usual and atypical human livers, *J. Biol. Chem.*, 258:6282.
Ito, K., Kashiwagi, K., Watanabe, S., Kameji, T., Hayashi, S., and Igarashi, K., 1990, Influence of the 5'-untranslated region of ornithine decarboxylase mRNA and spermidine on ornithine decarboxylase synthesis, *J. Biol. Chem.*, 265:13036.
Jones, D.E., Jr., Brennan, M.D., Hempel, J., and Lindahl, R., 1988, Cloning and complete nucleotide sequence of a full-length cDNA encoding a catalytically functional tumor-associated aldehyde dehydrogenase, *Proc. Natl. Acad. Sci. USA*, 85:1782.
Koivusalo, M., Aarnio, M., Baumann, M., and Rautoma, P., 1989, NAD (P)-linked aromatic aldehydes preferring cytoplasmic aldehyde dehydrogenases in the rat. Constitutive and inducible forms in liver, lung, stomach and intestinal mucosa, *Progr. Clin. Biol. Res.*, 290:19.
Kozak, M., 1987, An analysis of 5'-noncoding sequences from 699 vertebrate messenger RNAs, *Nucleic Acids Res.*, 15:8125.
Lindahl, R., and Feinstein, R.N., 1976, Purification and immunochemical characterization of aldehyde dehydrogenase from 2-acetyl aminofluorene-induced rat hepatomas, *Biochim. Biophys. Acta*, 452:345.
Lindahl, R., and Evces, S., 1984, Rat liver aldehyde dehydrogenase II. Isolation and characterization of four inducible isozymes, *J. Biol. Chem.*, 259:11991.
Lindahl, R., 1986, Identification of hepatocarcinogenesis-associated aldehyde dehydrogenase in normal rat urinary bladder, *Cancer Res.*, 46:2502.
Marselos, M., and Lindahl, R., 1988, Substrate preference of a cytosolic aldehyde dehydrogenase inducible in rat liver by treatment with 3-methylcholanthrene, *Toxicol. Appl. Pharmacol.*, 95:339.
Mather, P.B., and Holmes, R.S., 1984, Biochemical genetics of aldehyde dehydrogenase isozymes in the mouse: evidence for stomach- and testis-specific isozymes, *Biochem. Genet.*, 22:981.
Meier-Tackmann, D., Agarwal, D.P., Saha, N., and Goedde, H.W., 1984, Aldehyde dehydrogenase isozymes in stomach autopsy specimens from Germans and Chinese, *Enzyme*, 32:170.
Orkin, S.H., 1978, The duplicated human α globin genes lie close together in cellular DNA, *Proc. Natl. Acad. Sci. USA*, 75:5950.
Raghunathan, L., Hsu, L.C., Klisak, I., Sparkes, R.S., Yoshida, A., and Mohandas, T., 1988, Regional localization of the human genes for aldehyde dehydrogenase-1 and aldehyde dehydrogenase-2, *Genomics*, 2:267.
Rongnoparut, P., and Weaver, S., 1991, Liver cytosolic aldehyde dehydrogenase of the mouse: isolation and characterization of the cDNA, *Gene*, 101:261.
Rout, U.K., and Holmes, R.,S., 1985, Isoelectric focusing studies of aldehyde dehydrogenases from mouse tissues: variant phenotypes of liver, stomach and testis

isozymes, Comp. Biochem. Physiol., 81B:647.
Sambrook, J., Fritsch, E.F., and Maniatis, T., 1989, in "Molecular cloning, a laboratory manual," Cold Spring Harbor Laboratory, Cold Spring Harbor, NY.
Santisteban, I., Povey, S., West, L.F., Parrington, J.M., and Hopkinson, D.A., 1985, Chromosome assignment, biochemical and immunological studies on a human aldehyde dehydrogenase, $ALDH_3$, Ann., Hum. Genet., 49:87
Teng, Y-S., 1981, Stomach aldehyde dehydrogenase: report of a new locus, Hum. Hered., 31:74.
Wang, S.-L., Wu, C.-W., Cheng, T.-C., and Yin, S.-J., 1990, Isolation of high-Km aldehyde dehydrogenase isoenzymes from human gastric mucosa, Biochem. Int., 22:199.
Yin, S.-J., Cheng, T.-C., Chang, C.-P., Chen, Y.-J., Chaio, Y.-C., Tang, H.-S., Chang, T.-M., and Wu, C.-W., Human stomach alcohol and aldehyde dehydrogenases (ALDH): A genetic model proposed for ALDH III isozymes, Biochem. Genet., 26:343.
Yin, S.-J., Liao, S.-L., Wang, Y.-J., Chen, Y.-J., and Wu, C.-W., 1989, Kinetic evidence for human liver and stomach aldehyde dehydrogenase-3 representing a unique class of isozymes, Biochem. Genet., 27:321.
Yin, S.-J., Vagelopoulos, N., Wang, S.-L., and Jörnvall, H., 1991, Structural features of stomach aldehyde dehydrogenase distinguish dimeric aldehyde dehydrogenase as a 'variable' enzyme, FEBS, 283:85.
Yoshida, A., Hsu, L.C., and Yasunami, M., 1991, Genetics of human alcohol-metabolizing enzymes, Prog. in Nucleic Acid Res. and Mol. Biol., 40:255.

BOVINE CORNEAL ALDEHYDE DEHYDROGENASES: EVIDENCE FOR MULTIPLE GENE PRODUCTS (ALDH3 AND ALDHX)

Elizabeth M. Algar,[1,2] Brenda Cheung,[1] Jodie Hayes,[1]
Roger S. Holmes,[1,3] and Ifor R. Beacham[1]

[1]Faculty of Science and Technology, Griffith University, Brisbane, Qld 4111, Australia
[2]Present address: Department of Pathology, University of Queensland Medical School, Brisbane, Qld 4006, Australia
[3]Address for correspondence

INTRODUCTION

The mammalian cornea is a rich source of the enzyme, aldehyde dehydrogenase (ALDH; EC1.2.1.3), exhibiting a much higher level of specific activity (per mg of protein or per g of tissue) than liver (Holmes and VandeBerg, 1986; Holmes, 1988). Moreover, bovine (Abedinia *et al*, 1990), baboon (Algar *et al*, 1990) and opossum (Holmes *et al*, 1990) corneal ALDHs represent a major portion of the total soluble protein in each case, and a dual role has been proposed, involving both structural and catalytic properties of this enzyme. It has been well established (see Boettner and Walters, 1962; Zigman, 1983) that the mammalian cornea absorbs damaging UVR, particularly in the wavelength range 290-320nm, and corneal protein has been implicated in this role (Cogan and Kinsey, 1946; see Ringvold, 1980). As a major soluble protein, corneal ALDH may therefore assist in this UV-B absorbing role, as well as perform a catalytic function in the detoxification of UVR-induced peroxidic aldehydes.

Independent earlier studies (Alexander *et al*, 1981; Silverman *et al*, 1981) had reported the isolation and characterization of the major soluble protein of bovine cornea, designated BCP54 (Bovine Corneal Protein, apparent MW 54,000). BCP54 comprises approximately 30 percent of bovine corneal soluble protein, and about 50 percent of bovine corneal epithelial soluble protein. Recent amino acid sequence (Cooper *et al*, 1990) and molecular genetic studies (Cooper *et al*, 1991; Cooper and Baptise, 1991) demonstrated the identity of BCP54 with bovine corneal ALDH, and extensive sequence homology with rat class 3 ALDH (Jones *et al*, 1988). The partial sequence for this enzyme was 78 percent identical at the nucleotide level, and 82 percent identical in amino acid sequence, with rat class 3 ALDH.

Agarose-isoelectric focusing (IEF) studies of human and bovine corneal ALDHs, however, have revealed additional forms of this enzyme (Holmes, 1988; King and Holmes, 1992). Human corneal ALDH, for example, showed several forms of ALDH3 activity (corresponding in pI values to the major stomach ALDH), and a separate form, which had a similar pI value to the major human liver cytosolic enzyme, ALDH1. Bovine corneal ALDH also exhibited a complex sub-banding pattern on agarose-IEF, corresponding to the major corneal ALDH described by Abedinia *et al* (1990), as well as additional forms as yet uncharacterized.

In this paper, we report the presence of at least two ALDH classes among bovine corneal transcripts. A partial amino acid sequence for one of these ALDHs is homologous (85 percent positional identity) to human ALDHx, recently reported (Yoshida et al, 1991; Hsu and Chang, 1991).

MATERIALS AND METHODS

DNA Amplification

Total RNA was isolated from a pool of ten bovine corneas using the method of Chomczynski and Sacchi, (1987). Poly(A^+) RNA was then isolated by affinity separation using biotinylated oligo(dT) (Promega), and reverse transcribed using oligo(dT) as primer. The resulting cDNA was then subjected to PCR amplification using primers CA1 plus CA2. The sequences and positions of these primers are as follows (amino acid residue numbers are derived from the human sequences, as indicated in Figs.1 and 2):

```
                       266                   272
                       ThrLeuGluLeuGlyGlyLys
CA1        5' GCATCGATACCCTCGAICTCGGCGGCAA 3'
                 ClaI     I  I     I  I  I

                       405                   399
                       MetValProGlyPheIleGlu
CA2        5' GACTCGAGCATCACCGGCCCIAAIATCTC 3'
                 XhoI      I  I  I        T
```

Amplification was performed using Taq polymerase and an annealing temperature of 55°C. Amplified DNA was isolated following agarose electrophoresis, and cloned into pSK (Promega) using either restriction sites incorporated into the primers or by 'T-vector' cloning (Marchuk et al, 1991).

Nucleotide sequencing

Dideoxy sequencing of amplified, and cloned, DNA was performed using supercoiled templates. The sequence was determined entirely on both strands, and from three independent PCR amplifications in order to ensure the absence of errors arising during amplification.

RESULTS AND DISCUSSION

In order to probe for the presence of ALDH transcripts in bovine cornea, oligonucleotides were synthesised using regions of amino acid sequence which were conserved among ALDH classes 1, 2, and 3. Poly(A^+) RNA isolated from bovine cornea was reverse transcribed, and the resulting cDNA used for DNA amplification using primers CA1 and CA2. It was expected that the sequence corresponding to the abundant bovine corneal ALDH, now known to be ALDH 3, would be amplified, but this was not the case. The amplified sequence shows 85% sequence identity to human ALDHx and clearly belongs to this class (see Figs. 1, 2 and Table 1). A pair of primers, based on the bovine ALDH class 3 amino acid sequence (Algar and Holmes, unpublished data), readily amplifies a bovine corneal class 3 sequence (data not shown).

These results demonstrate that ALDH class x is synthesised in bovine cornea. Since the human testis cDNA-derived ALDHx amino acid sequence reveals a 17 residue leader peptide, (with homology to the class 2 leader peptide), it has been proposed to be a mitochondrial enzyme (Yoshida et al, 1991). The relationship of this subcellular localisation to the postulated

```
G AGC CTG AGC ATC GTG TTG GCT GAT GCC GAC ATG GAC CAT GCC GTG GAG CAG CGC
  Ser Leu Ser Ile Val Leu Ala Asp Ala Asp Met Asp His Ala Val Glu Gln Arg  290

  CAA GAG GCA CTG TTC TTC AGC ATG GGC CAG TGC TGC TGT CCG GGT TCC TGG ACC
  Gln Glu Ala Leu Phe Phe Ser Met Gly Gln Cys Cys Cys Pro Gly Ser Trp Thr  308

  TTC ATT GAA GAA TCC ATC TAT GAT GAG TTT CTG GAG AGA ACG GTG GAG AAA GCT
  Phe Ile Glu Glu Ser Ile Tyr Asp Glu Phe Leu Glu Arg Thr Val Glu Lys Ala  326

  AAG CAG AGG AGA GTC GGG AAC CCA TTT GAT CTG GAC ACC CAA CAG GGG CCC CAG
  Lys Gln Arg Arg Val Gly Asn Pro Phe Asp Leu Asp Thr Gln Gln Gly Pro Gln  344

  GTG GAC AGG GAA CGG TTC GAA CGA ATC CTG GGC TAT ATC CAG CTT GGC CAG AAG
  Val Asp Arg Glu Arg Phe Glu Arg Ile Leu Gly Tyr Ile Gln Leu Gly Gln Lys  362

  GAG GGG GCA AAA CTT CTC TGC GGT GGG GAG CAT TTC AGA CAA CAA TGT TTC TTC
  Glu Gly Ala Lys Leu Leu Cys Gly Gly Glu His Phe Arg Gln Gln Cys Phe Phe  380

  ATC AAG CCC ACC GTC TTT GGT GGT GTG CAA GAT GAC ATG AGG ATC GCT AGG GAG
  Ile Lys Pro Thr Val Phe Gly Gly Val Gln Asp Asp Met Arg Ile Ala Arg Glu  398
```

Figure 1: Nucleotide sequence, and derived amino acid sequence, of an amplified cDNA product from bovine corneal mRNA. Numbers correspond to amino acid residues in the human ALDH 1, 2 and x sequences (Yoshida et.al., 1991).

Table 1. Amino acid positional identity amongst human ALDHs (1,2,3 and x), bovine corneal ALDH (BCP54) and bovine corneal ALDHx.

	ALDH 2	ALDH 3	BCP54	ALDH x	Bov cornea ALDHx
ALDH1	60%	25%	25%	59%	53%
ALDH2		24%	22%	66%	63%
ALDH3			90%	25%	23%
BCP54				25%	22%
ALDHx					85%

```
HumALDH1           IVPGYGPTAGAAISSHMDIDKVAFTGSTEVGKLIKEAAGKSNLKRVTLELGGKSPCIVLA
HumALDH2           IVPGFGPTAGAAIASHEDVDKVAFTGSTEIGRVIQVAAGSSNLKRVTLELGGKSPNIIMS
HumALDH3           VINGGVPETTELLKER--FDHILYTGSTGVGKIIMTAAAK-HLTPVTLELGGKSPCYVDK
BCP54              -----------------------------------------------------NPHYVDK
HumALDHx           IITGYGPTAGAAIAQHMDVDKVAFTGSTEVGHLIQKAAGDSNLKRVTLELGGKSPSIVLA
BovcorneaALDHX     -----------------------------------------------------SLSIVLA

HumALDH1           D-ADLDNAVEFAHHGVFYHQGQCCIAASRIFVEESIYDEFVRRSVERAKKYILGNPLTPG
HumALDH2           D-ADMDWAVEQAHFALFFNQGQCCCAGSRTFVQEDIYDEFVVRSVARAKSRVVGNPFDSK
HumALDH3           NC-DLDVACRRIAWGKFMNSGQTCVAPDYILCDPSIQNQIVEKLKKSLKEF-YGEDAKKS
BCP54              DR-DLDIACRRIAWGKFMNSGQTCVAPDYILCDPSIQSQVVEKLKKSLKEF-YGEDAKKS
HumALDHx           D-ADMEHAVEQCHEALFFNMGQCCCAGSRTFVEESIYNEFLERTVEKAKQRKVGNPFELD
BovcorneaALDHX     D-ADMDHAVEQRQEALFFSMGQCCCPGSWTFIEESIYDEFLERTVEKAKQRRVGNPFDLD
                      *   *        *   ** *           *            *    *

HumALDH1           VTQGPQIDKEQYDKILDLIESGKKEGAKLECGGGPWGNKGYFVQPTVFSNVTDEMRIAKE
HumALDH2           TEQGPQVDETQFKKILGYINTGKQEGAKLLCGGGIAADRGYFIQPTVFGDVQDGMTIAKE
HumALDH3           RDYGRIISARHFQRVMGLI-----EGQKVAYGGTG-DAATRYIAPTILTDVDPQSPVMQE
BCP54              RDYGRIINSRHFQRVMGLL-----EGQKVAYGGTG-DATTRYIAPTILTDVDPESPVMQE
HumALDHx           TQQGPQVDKEQFERVLGYIQLGQKEGAKLLCGGERFGERGFFIKPTVFGGVQDDMRIAKE
BovcorneaALDHX     TQQGPQVDRERFERILGYIQLGQKEGAKLLCGGEHFRQQCFFIKPTVFGGVQDDMRIARE
                      *                    *  *   **          **    *        *
```

Figure 2: Alignment of ALDH amino acid sequences using the program ClustalV. With the exception of the bovine corneal ALDH, sequences were obtained from GenBank, and the following references: Human ALDH 1 and 2, Hsu et.al.,1985; Human ALDH 3, Hsu et.al., 1992; BCP54, Cooper et.al.,1991; Human ALDHx, Hsu and Chang,1991. (*) designates conserved residue for all sequences.

role of the abundant class 3 enzyme, in the detoxification of peroxidic aldehydes in the cornea, remains to be determined. Likewise, the specificity and kinetic properties need to be established; however, a specificity for short chain aliphatic aldehydes is indicated (Yoshida et al 1991).

Recent kinetic studies on the purified ALDH Class 3 enzyme from bovine, baboon and human cornea (Abedinia et al, 1990; Algar et al, 1990; King and Holmes, 1992) have shown that medium-chain aliphatic and aromatic aldehydes are the preferred substrates for this enzyme, and peroxidic aldehydes such as 4-hydroxynonenal and D-2-hexenal are likely in vivo substrates. This enzyme is inactive, however, with another lipid peroxidation by-product, malondialdehyde, and the presence of ALDH1 activity, reported in human cornea (Holmes, 1988), may assist in this detoxifying role.

The deduced partial amino acid sequence for bovine corneal ALDHx corresponds, in terms of alignment with the human ALDH2 and ALDHx sequences (Hsu and Chang, 1991), to the 274-398 amino acid residues of ALDH (Fig. 2). This region incorporates the active site for ALDH isozymes at cys 302, which together with gly 299 and gln 300, are conserved. Moreover, the -cys-cys-cys- sequence (301-303), which is common to human ALDH2 and ALDHx sequences (Hsu and Chang, 1991), as well as rat and bovine ALDH2 homologues (Farres et al, 1989), is also conserved in the bovine ALDHx. Hsu and Chang (1991) have assigned possible subunit interaction regions for human ALDHx at positions 249-255, 335-340 and 390-394, for which the latter two sites are found in the partial bovine ALDHx sequence reported (Fig. 2). With the exception of a single conservative substitution [glu (human)/asp (bovine)] at position 336, the sequences are identical in both species, for these regions.

Table 1 summarizes the extent of amino acid positional identity of human ALDHs 1,2,3 and x, with bovine corneal ALDH3 (BCP54) [based on the partial sequence of Cooper et al (1991)] and with bovine corneal ALDHx. These enzymes have 18 identical amino acids for the 274-398 ALDH sequence, including several glycine residues. Human and bovine ALDHx sequences are 85 percent identical in amino acid sequence, which compares with 86 percent for the corresponding ALDH3 sequences (Table 1), and 94 percent for the human and bovine ALDH2 sequences (Farres et al, 1989). In contrast, inter-class comparisons for these enzymes revealed much lower levels of positional identity, with Classes 1,2 and x showing 52-58 percent identity, whereas Class 3 ALDH is more divergent in sequence (20-25 percent identical) from the other classes.

In summary, we have isolated and sequenced a bovine corneal cDNA, generated from poly(A^+) RNA following reverse transcription and amplification using primers conserved among ALDH classes, 1,2 and 3. The deduced amino acid sequence corresponded to ALDH2 residues 274-398, and was highly homologous (85 percent) with human ALDHx. We have concluded that bovine cornea contains multiple ALDH transcripts from at least two distinct classes, ALDHx and ALDH3, for which the latter represents the major soluble protein with a possible dual role in protecting the cornea from UV-B induced damage. The kinetic and biochemical properties of bovine corneal ALDHx, however, await further investigation.

ACKNOWLEDGEMENTS

This research was supported in part by grants from the Australian Research Council and the National Health and Medical Research Council of Australia.

REFERENCES

Abedinia, M.A., Pain, T., Algar, E.M., and Holmes, R.S., 1990, Bovine corneal aldehyde dehydrogenase: the major soluble corneal protein with a possible dual protective role for the eye, Exp. Eye Res., 51: 419-426.

Alexander, R.J., Silverman, B., and Henley, W.L., 1981, Isolation and characterization of BCP54, the major soluble protein of bovine cornea, Exp. Eye Res. 32: 205-216.

Algar, E.M., Abedinia, M., VandeBerg, J.L., and Holmes, R.S., 1990, Purification and properties of baboon corneal aldehyde dehydrogenase: proposed UVR protective role, in Weiner, H., Wermuth, B., and Crabb, D.W., eds., Enzymology and Molecular Biology of Carbonyl Metabolism 3, New York: Plenum Press, pp. 53-60.
Boettner, E.A., and Wolters, J.R., 1962, Transmittance of the ocular media, Invest. Ophthalmol., 1: 775-783.
Chomczynski, P., and Sacchi, N, 1987, Single-step method of RNA isolation by acid guanidinium thiocyanate-phenol-chloroform extraction, Anal. Biochem. 162: 156-159.
Cogan, D.G., and Kinsey, V.E., 1946, Action spectrum of keratitis produced by ultraviolet radiation, Arch. Ophthalmol., 55: 670-677.
Cooper, D.L., Baptist, E.W., Enghild, J., Lee, H., Isola, N., and Klintwoth, G.K., 1990, Partial amino acid sequence determination of bovine corneal protein 54K (BCP54), Current Eye Res., 9: 781-786.
Cooper, D.L., Baptist, E.W., Enghild, J.J., Isola, N.R., and Klintworth, G.K., 1991, Bovine corneal protein 54K (BCP54) is a homologue of the tumour-associated (class 3) rat aldehyde dehydrogenase (RATALD), Gene 98, 201-207.
Farres, J., Guan, K-L., and Weiner, H., 1989, Primary structures of rat and bovine liver mitochondrial aldehyde dehydrogenases deduced from cDNA sequences, Eur. J. Biochem. 180: 67-74.
Holmes, R.S., and VandeBerg, J.L., 1986, Ocular NAD dependent alcohol dehydrogenase and aldehyde dehydrogenase in the baboon, Exp. Eye. Res., 43: 383-396.
Holmes, R.S., van Oorschot, R.A.H., and VandeBerg, J.L., 1990, Genetics of alcohol dehydrogenase and aldehyde dehydrogenase from Monodelphis domestica cornea: further evidence for identity of corneal aldehyde dehydrogenase with a major soluble protein, Genetical Res., 56: 259-265.
Holmes, R.S., 1988, Alcohol dehydrogenases and aldehyde dehydrogenases of anterior eye tissues from humans and other mammals, in Kuriyama, K., Takada, A., and Ishii, H., eds. Biomedical and Social Aspects of Alcohol and Alcoholism, Amsterdam: Elsevier Science Publishers, pp. 51-57.
Hsu, L.C., and Chang, W-C., 1991, Cloning and characterisation of a new functional aldehyde dehydrogenase gene,J. Biol., Chem., 266: 12257-12265.
Hsu, L.C., Tani, K., Fujiyoshi, T., Kurachi, K., and Yoshida, A., 1985, Cloning of cDNAs for human aldehyde dehydrogenases 1 and 2, Proc. Natl. Acad. Sci., 82: 3771-3775.
Hsu, L.S., Chang, W-C., Shibuya, A., and Yoshida, A., 1992, Human stomach aldehyde dehydrogenase cDNA and genomic cloning, primary structure, and expression in *Escherichia coli*. J. Biol. Chem., 267: 3030-3037.
King, G., and Holmes, R.S., 1992, Purification and properties of human corneal aldehyde dehydrogenase, Proceedings of the 1992 ISBRA Congress.
Marchuk, D., Drumm, M., Saulino, A., and Collins, F.S., 1991, Construction of T-vectors, a rapid and general system for direct cloning of unmodified PCR products. Nucleic Acids Res., 19: 1154.
Ringvold, A., 1980, Cornea and ultraviolet radiation, Acta. Ophthalmol. 58: 63-68.
Silverman, B., Alexander, R.J., and Henley, W.L., 1981, Tissue and species specificity of BCP54, the major soluble protein of bovine cornea, Exp. Eye Res. 33: 19-29.
Yoshida, A., Hsu,L.C., and Yasunami, M., 1991, Prog. Nucleic Acid Res. Mol. Biol., 40: 255-287.
Zigman, S., 1983, The role of sunlight in human cataract formation, Survey Ophthalmol. 27: 317-326.

CARBONYL-METABOLIZING ENZYMES AND THEIR RELATIVES RECRUITED AS STRUCTURAL PROTEINS IN THE EYE LENS

Douglas C. Lee, Pedro Gonzalez, P. Vasantha Rao,
J. Samuel Zigler, Jr. and Graeme J. Wistow

National Eye Institute,
NIH, Bethesda, MD 20892 USA

ABSTRACT

The refractive properties of the eye lens are determined by abundant soluble structural proteins known as crystallins. While some crystallins are common to most vertebrates, others are abundant only in groups of related species. These taxon-specific crystallins all turn out to be enzymes, apparently recruited by modification of gene expression without prior gene duplication. They include η-crystallin, accounting for up to 25% of protein in elephant shrew lenses and apparently identical to cytoplasmic aldehyde dehydrogenase; ρ-crystallin from frog lenses, a member of the same superfamily as aldose and aldehyde reductases; and ζ-crystallin, found in guinea pig and camel lenses, which is structurally related to alcohol dehydrogenase (ADH). Unlike ADH, ζ-crystallin requires NADPH rather than NAD+/NADH as cofactor. Molecular modelling of ζ-crystallin shows that amino-acid changes around the co-factor binding site are responsible for this change in affinity. Purified guinea pig lens ζ-crystallin has a substrate preference for orthoquinones which are reduced by a single electron transfer mechanism. cDNA sequencing of ζ-crystallin suggests that the expression in lens as a crystallin depends on a different gene promoter from that used predominantly in liver. The putative guinea pig ζ-crystallin lens promoter has now been assayed for function in transfection studies. Elements with positive and negative effects on transcription, at least one of which has tissue preferred function, have been defined. When introduced into transgenic mice this promoter exhibits tissue-specific expression in the lens. This is the first identification of a lens-specific, alternative promoter in an enzyme crystallin gene.

INTRODUCTION

Crystallins are highly abundant soluble proteins of the eye lens (Bloemendal, 1981; Berman, 1991). They constitute most of the refractive structure of the lens and are essential for normal function. Surprisingly, many crystallins are not specialized structural proteins. Instead they have been directly recruited from stress proteins or enzymes, serving dual roles generally without gene duplication (Carper et al., 1987; Wistow et al., 1987; Wistow

Enzymology and Molecular Biology of Carbonyl Metabolism 4
Edited by H. Weiner, Plenum Press, New York, 1993

and Piatigorsky, 1987, 1988; Mulders et al., 1988; de Jong et al., 1989; Piatigorsky and Wistow, 1991; Wistow and Kim, 1991). Enzyme crystallins are particularly interesting, having been recruited relatively recently in specific lineages. These taxon-specific crystallins may have arisen by increased expression in lens of enzymes already serving a catalytic function in the tissue. The increase in expression may serve to modify the refractive properties of the lens in species adapting to new visual environments, such as moving from water to land or from nocturnal to diurnal habits. However, the enormous overexpression of functional enzymes in the lens may also have important consequences for metabolism, affecting the abundance of cofactor and substrate molecules in the tissue. Three vertebrate enzyme crystallins, η, ρ and ζ-crystallins, are either identical or at least closely related to enzymes of carbonyl metabolism. All three are capable of binding pyridine nucleotide cofactors.

η-CRYSTALLIN/ALDH1

The major component of the lenses of elephant shrews, primitive insectivorous mammals, is η-crystallin, a protein of 54kDa subunit size (Wistow and Kim, 1991). η-Crystallin accounts for as much as a quarter of total lens protein in some species. This protein was isolated from the lens of *Elephantulus rufescens* and subjected to tryptic digestion and microsequencing. All peptides obtained closely matched the sequence of human cytoplasmic aldehyde dehydrogenase (ALDH). Furthermore in western blots η-crystallin reacted strongly with antisera specific for the cytoplasmic form of ALDH but not with similar antisera for mitochondrial ALDH. Much lower levels of cytoplasmic ALDH, more typical of an enzyme than a structural protein were observed in the lenses of most other mammals. Extracts of lenses from elephant shrews were assayed for ALDH activity but none was observed. The animals were natural casualties from zoos and at least several years old, suggesting that enzyme activity may have been lost through aging or post-mortem modification. A partial cDNA sequence has now been obtained for η-crystallin/ALDH1 from *Macroscelides proboscideus* lens (J. Hodin and G. Wistow unpublished).

Other proteins related to ALDH are also found overexpressed in other transparent tissues. A protein previously named Ω-crystallin is moderately abundant in the lens of the octopus (Chiou, 1988). Sequence analysis has now shown that this protein is just over 50% identical to both ALDH1 and ALDH2 of mammals (Tomarev et al., 1991). A more distantly related member of the ALDH superfamily, designated L-crystallin (Montgomery and McFall-Ngai, 1992), is also found as the major constituent of the "lens" of the squid light organ, a remarkable luminescent organ with superficial organizational similarities to an eye. Finally, the major soluble protein of the mammalian corneal endothelium, a thin layer of cells which maintain the transparency of the corneal stroma, is identical to tumor-inducible ALDH3 (Abedinia et al., 1990; Cooper et al., 1991).

In spite of these coincidences, it seems unlikely that there is anything inherently transparent about aldehyde dehydrogenases. Their common presence in such diverse transparent tissues may instead reflect common functional requirements. For example transparency requires close control of the balance between protein and water content, as shown by osmotic cataracts in mammals (Harding and Crabbe, 1984). Carbonyl metabolizing enzymes could perhaps play a role in manipulating osmolytes in such tissues. Indeed, it has been suggested that osmoregulation may be a common factor linking many of the proteins recruited as crystallins (Wistow 1990; Wistow and Kim, 1991).

ρ-CRYSTALLIN

10% or more of the total lens protein in frogs of the genus *Rana* is a novel protein designated ρ-crystallin (Zigler and Sidbury, 1976; Tomarev et al., 1984) (previously called frog ε-crystallin). When rat lens aldose reductase (AR), an enzyme implicated in diabetic complications resulting from osmotic stress, was cloned and sequenced it became apparent that AR and ρ-crystallin were closely related (Carper et al., 1987). Further comparisons extended this relationship to include aldehyde reductase, prostaglandin-F synthase, *Corynebacterium* 2,5-diketo-D-gluconate reductase and yeast GCY protein (Carper et al., 1989) and later to human liver chlordecone reductase (Winters et al., 1990) and mouse vas deferens androgen-dependent protein (Pailhoux et al., 1990). It appears that these proteins constitute a widespread superfamily of well-conserved NADPH-dependent aldo-keto reductases. Recently the tertiary structure of pig lens aldose reductase has been determined by x-ray crystallography (Rondeau et al., 1992), showing that members of this superfamily adopt the familiar αβ barrel first observed in triose phosphate isomerase. No enzyme activity has yet been defined for ρ-crystallin, although in addition to conserved co-factor binding the lens protein conserves all the amino-acid residues common to eukaryotic and bacterial members of the superfamily (Carper et al., 1989).

ζ-CRYSTALLIN

Distribution

ζ-Crystallin was first observed in the lenses of guinea pigs, hystricomorph rodents (Huang et al., 1987), where it accounts for about 10% of total protein. In other cases, the high expression of taxon-specific crystallins is restricted to closely related species, so the distribution of ζ-crystallin might also be indicative of close phylogenetic relationships. This has become more interesting as the phylogenetic relatedness of guinea pigs has been questioned (Graur et al., 1991). Three other South American hystricomorphs have now been examined. Π-crystallin is detectable in the lens of the rock cavy (*Kerodon rupestris*) and the degu (*Octodon degus*) (figure 1). However, under the same conditions, ζ-crystallin is undetectable in the lens of the coypu (*Myocastor coypus*), an hystricomorph, or in the lenses of murine rodents or most other mammals (figure 1). This is consistent with a close relationship between guinea pigs, rock cavies and degus but, perhaps, more distant relationships between these animals and coypus. However, ζ-crystallin, or a very similar protein, occurs again as an abundant crystallin in a very surprising place, in the lenses of camels, animals not obviously related to hystricomorph rodents (Garland et al., 1991) (figure 1). The most likely explanation for this is that the recruitment of crystallins is a partially neutral process (Mulders et al., 1988; Wistow et al., 1990) and that the same choice was made in different lineages. Both guinea pigs and camels evolved in South America, so it is conceivable that similar environmental pressures, perhaps resulting from common dietary toxins, may have selected for the parallel recruitments of an enzyme to protect the lens. Alternatively, ζ-crystallin may be an ancestral, neutral mammalian feature which has been retained by chance in disparate modern genera.

Identification and Enzyme Activity

Tryptic peptide analysis followed by cDNA cloning produced the complete sequence of guinea pig ζ-crystallin (Rodokanaki et al., 1989). Comparison of peptides, and later the full predicted sequence, with the sequence databases revealed a distant similarity with enzymes of the alcohol dehydrogenase superfamily, including the enoyl reductase moiety of the fatty-acid synthase complex. More recently another related protein, the membrane associated VAT-1 of cholinergic synaptic vesicles in marine rays, has been identified (Linial et al., 1989).

Since other taxon-specific crystallins are identical to active enzymes, and given the similarity to ADH, the enzymatic properties of ζ-crystallin were investigated. ADH, and indeed most of the taxon-specific enzyme crystallins, require nicotine adenine dinucleotide cofactors. ζ-crystallin was the only major guinea pig lens protein to be retained on a blue-sepharose column under high salt conditions. Cofactor specificity was tested by elution with NAD, NADH, NADP and NADPH. Surprisingly, in view of the similarity with ADH, ζ-crystallin was preferentially eluted by NADPH, suggesting an enzymatic role as a reductase (Rao and Zigler, 1991).

Isolated lens ζ-crystallin was tested with a variety of carbonyl compounds (Table 1). While it had no activity as an alcohol dehydrogenase, ζ-crystallin was found to be an active quinone reductase with a preference for orthoquinones (Rao et al., 1992). There is no similarity between the sequence of ζ-crystallin and previously known quinone reductases. Its reaction mechanism was investigated further and was found to proceed by a single electron transfer mechanism, in contrast to the two-electron mechanism of other cytosolic quinone reductases (Rao et al., 1992).

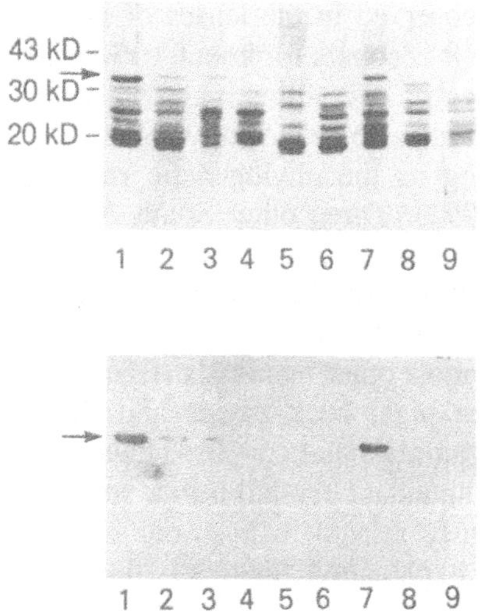

Figure 1. a) SDS polyacrylamide gel electrophoresis and b) ζ-crystallin western blot (see ref Garland et al., 1991 for methods) of lens protein extracts. Species: 1: Guinea pig; 2:rock cavy (*Kerodon rupestris*); 3:degu (*Octodon degus*); 4:coypu (*Myocastor coypu*); 5:mouse; 6:rat; 7:camel (*Camelus dromedarius*);8:domestic cow; 9:human.

Molecular model-building

Alignment of the sequences of ζ-crystallin and ADH shows that these proteins are well conserved in the N-terminal half of the molecule (Borras et al., 1989), with the exception of a large deletion corresponding to a loop in ADH which contains a zinc-binding site. The second half of the ζ-crystallin sequence is much less similar to ADH and consequently hard to model in a rational way. Using the molecular graphics package Quanta (Polygen, Waltham MA), the N-terminal half of the ζ-crystallin sequence was built into one of the experimentally determined x-ray structures for horse ADH (Eklund et al., 1984), Protein DataBank (PDB) entry 5ADH, to investigate the general features of the structure. The large deletion was engineered fairly easily by deleting residues 94 to 140 of the ADH structure and joining the peptide backbone between residues 93 and 141, residues which are close in the template structure. Between positions 25 and 270 of the ADH structure residues were replaced by their aligned equivalents in Π-crystallin. A relative insertion of two residues in a surface loop was accomplished by addition of a threonine following position 57 of ADH, and leucine preceding the conserved proline at position 62. The loop was extended and energy minimized. Side chains were manipulated to avoid collisions or to maintain obvious interactions. Other insertions/ deletions were accomplished in a similar way. All changes from ADH to Π-crystallin sequence made structural sense confirming that primary structure similarity does reflect similarity in tertiary structure. For example, increased side chain size, as in (ADH)Ser193->(Π)Glu was accommodated by a reciprocal change of Phe264->Val, while in the change of Ala183->Leu collision is avoided because of Cys211->Ala.

Table 1. Catalytic properties of ζ-crystallin/quinone reductase.

Selected Substrate	Km (uM)	Vmax (U/mg)	Relative velocity %
1,2-naphthoquinone	-	-	800
9,10-phenanthrenequinone	13	16.6	400
1,4-benzoquinone	143	5.7	122
5-OH-1,4 naphthoquinone	27	4.5	100
2,6-dichlorophenol-indophenol	37	1.0	30
NADPH	5	-	-

Selected Inhibitors	Type	Ki(uM)
Dicumarol	competitive	13
Nitrofuran	competitive	14

Two different versions of possible cofactor-binding were modelled into the structure, based on ADP-ribose from structure 5ADH and NAD from 6ADH. There are some differences in atomic positions between the two, but in general terms both versions fit similarly into the same groove. Interestingly, the ribose-binding site of ADH showed changes corresponding to the alteration in cofactor affinity. Specifically, Asp223 of ADH is close to the 2'-hydroxyl of the ribose ring of NAD, preventing NADP(H) binding by both steric hindrance and charge repulsion. In ζ-crystallin the aspartate is replaced by alanine. The peptide backbone around the binding pocket is also lengthened by insertion of serine before conserved Gly201. The net effect is to remove the block to phosphate attached to the ribose ring. No unambiguous counter-ion to the phosphate moiety is obvious in this part of the structure although His198 which replaces the phenylalanine of ADH is a candidate for this function.

Gene Structure and The Recruitment of ζ-Crystallin

Although ζ-crystallin was first noticed as an abundant structural component of the eye lens (Huang et al., 1987) it is also detectable in other tissues, such as liver and kidney, at much lower levels more appropriate for an enzyme (Huang et al., 1990). Like other taxon-specific crystallins it seems that ζ-crystallin is indeed an enzyme normally expressed in a variety of tissues which, in certain evolutionary lineages, has acquired greatly elevated gene expression specifically in the lens. An important insight into the mechanism of recruitment of this enzyme as a crystallin was gained when a cDNA for ζ-crystallin was cloned from guinea pig liver (Hernandez-Calzadilla, et al, in preparation). The sequence of this clone was identical to that previously obtained from the lens except for the most 5' region. When the gene for guinea pig ζ-crystallin was cloned alternative first exons, separately encoding the liver and lens 5' regions, were identified (Gonzalez et al, submitted). Relative to the common second exon containing the translation start site, the lens first exon was found about 2kbp upstream, with the liver first exon another 3kb farther upstream. This suggests that lens expression of ζ-crystallin depends on a promoter contained in what would be the first intron of the gene as transcribed from the liver promoter (figure 2).

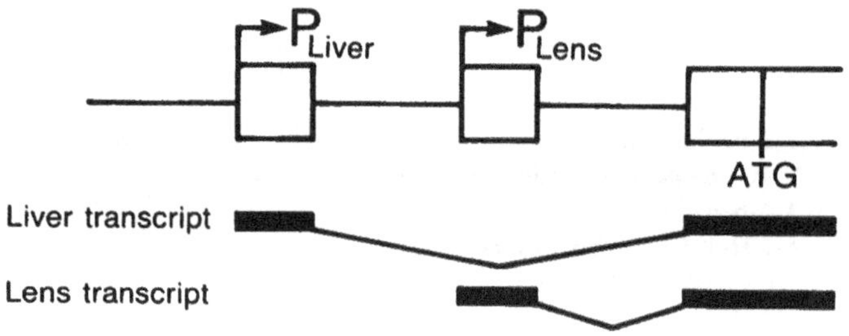

Figure 2. The Structure of the Promoter Region of the Guinea Pig ζ-crystallin gene. Alternative promoters give rise to tissue-specific versions of exon 1.

The putative liver and lens promoters have been cloned and sequenced. Primer extension identifies the single start site for transcription from the lens promoter about 30 bp downstream from a TAATAA sequence (Gonzalez et al, submitted). Close by this TATA-box are consensus binding sites for the transcription factors AP4 and Sp1. Further upstream are potential anti-oxidant response elements common to other quinone reductase genes (Favreau and Pickett, 1991) and a T-rich stretch similar to sequences found in many crystallin gene promoters. In contrast, the putative liver promoter contains no identifiable TATA or CCAAT boxes and resembles the kind of promoter typical of housekeeping enzyme genes.

The function of the putative lens promoter has now been verified and analyzed in some detail (Lee et al, in preparation). A fragment corresponding to sequences -1575 to +69 of the 5' end of the lens transcription unit was cloned into an expression vector upstream of a reporter gene encoding bacterial chloramphenicol acetyltransferase (CAT). This recombinant construction was then transfected into cultured N1003A cells, derived from rabbit lens epithelia (Reddan et al., 1986), and its promoter function estimated by CAT assay. In the forward orientation the ζ-crystallin lens promoter was highly active in lens-derived cells but had no activity in the reverse orientation (figure 3). Primer extension analysis of mRNA from transfected cells gave an identical start site to that seen in guinea pig lens. When tested in mouse fibroblasts the same promoter was inactive, suggesting the presence of cell-specific *cis* elements with a preference for lens.

Subsequent dissections (Lee et al, in preparation) have shown that the ζ-crystallin promoter contains several distinct *cis* elements, probably binding sites for different transcription factors. Somewhat surprisingly, constructions with up to 185bp upstream of the transcription start site containing several binding sites common to other crystallin promoters including an Sp1 site close to the TATA box, are not active. Sequences between -1575 and -380 contain positive elements, causing an increase in transcriptional activity. However there may be a negative element, or possibly only part of a positive element which acts negatively in isolation, between -527 and -380. Strong positive elements or enhancers lie between -751 and -527, while between -1575 and -751 lie other elements which decrease activity by about 50% from the maximum.

The best test of tissue-specificity is to express the promoter in intact tissues. Therefore the -751bp-CAT construction was introduced into transgenic mice (Lee et al, in preparation). Two independent founder mice were examined for CAT activity in various tissues. In both mice there was strong CAT activity in the lens, but none in liver, heart,

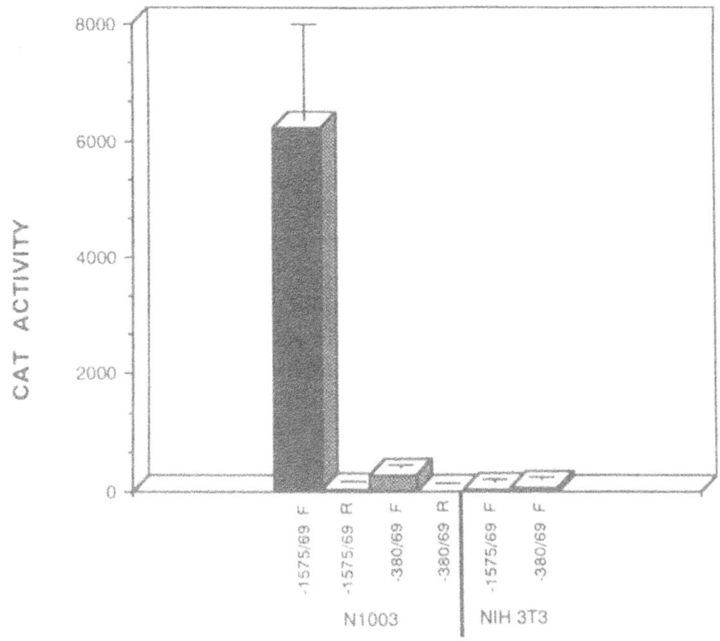

Figure 3. A lens-cell preferred promoter in the guinea pig ζ-crystallin gene. Two constructions of the putative lens promoter were tested in forward and reverse orientations by transient transfection into rabbit lens N1003 cells. Both constructions were also tested in the forward orientation in NIH 3T3 mouse fibroblast cells. CAT activity was normalized to co-transfected control construct expressing β-galactosidase (Kim et al., 1991).

kidney or brain. This is a striking demonstration of lens-specificity, or marked lens-preference, by this enzyme crystallin gene promoter.

The clear implication of these results is that ζ-crystallin achieved high expression in lens and a new role as a structural protein through the acquisition of a new promoter with lens specificity. The lens promoter may have appeared *de novo* in the first intron of the pre-existing gene for ζ-crystallin. Alternatively, ζ-crystallin, as an enzyme in mammals may already have possessed two promoters with differential activity. High expression in lens may then have occurred by lens-specific enhancement of the existing downstream promoter. If the latter proposition is true, alternative transcripts of ζ-crystallin should be detectable even in species, such as mouse and human, which do not use ζ-crystallin as an abundant lens structural protein.

This is now being addressed by both cDNA and genomic cloning of ζ-crystallin from mouse and human (Gonzalez et al in preparation). In both species, transcripts from liver have been cloned and sequenced, showing some similarity to the homologous sequences in the guinea pig. An identical mouse ζ-crystallin was also cloned from αTN4-1 cells, a transformed mouse lens cell line, however, although these cells share some properties consistent with their lens derivation they also have major differences and may not reflect authentic expression of lens genes (Shaughnessy and Wistow, 1992).

IMPLICATIONS

The direct recruitment of functional enzymes as crystallins demonstrates the pragmatism of evolutionary processes. Enzyme crystallins have two quite separate functions, first as enzymes, expressed in many tissues where they mediate important metabolic reactions, and second as crystallins, expressed at extremely high levels in the lens where they contribute to supramolecular structure and light refraction. The latter function seems to be independent of enzyme activity since there are examples of recruited enzymes which lose activity in lens [τ-crystallin (Wistow and Piatigorsky, 1987; Wistow et al., 1988), η-crystallin (Wistow and Kim, 1991) and possibly μ-crystallin (Kim et al., 1992)]. There is also one example in which gene duplication has produced a non-enzymatic, specialized crystallin (δ1-crystallin) in addition to a functional enzyme (argininosuccinate lyase/δ2-crystallin) (Wistow and Piatigorsky, 1990; Kondoh et al., 1991; Barbosa et al., 1991). This duplication was probably selected for by an adaptive conflict resulting from the dual selective pressures of two roles on a single gene.

However, even if activity is not required for all enzyme crystallins it could still have important effects in the lens. It might be expected that very high levels of functional enzymes would perturb lens metabolism. Even without full activity, enzymes may serve as binding proteins, sequestering cofactors and small molecules, including sugars and amino-acids. Indeed, elevated levels of enzymes serving as crystallins are associated with elevated levels of the appropriate cofactor in the lens (Rao and Zigler, 1990). This could have a direct functional role in maintaining the redox state of the lens or even in providing filters for UV radiation (Wistow et al., 1987). However it must also have an effect on the pools of cofactors, substrates and products in lens metabolism. Nevertheless, lenses seem to be capable of coping with any such effects. In a first attempt at mimicking the recruitment process by genetic engineering, the abundance of α-enolase has been increased 5-10 fold in a transgenic mouse lens, without evident ill-effect (R. Kim and G. Wistow, submitted).

The primary role of the recruited enzymes in lens must relate to the function of the lens as a refractive tissue. Rather than acting enzymatically, they may instead serve to dilute the effects of the γ-crystallins which are associated with the hard, high refractive index, non-accommodating lenses of fish and nocturnal rodents (Wistow and Piatigorsky,

1988). This may be one mechanism by which lenses of terrestrial species have adjusted their optical properties. Even so, it has been suggested that all the enzymes recruited as crystallins were expressed in lens prior to recruitment and continue normal expression in most species, perhaps for metabolic or developmental purposes (Wistow and Kim, 1991). For instance, some enzyme or stress protein crystallins could have a role in control of or response to the osmotic swelling which must occur during the enormous increase in volume in differentiated lens fiber cells. In any case the genes which acquire dual functions were active or easy to activate in the ancestral lens. By a variety of mechanisms the expression of these genes has been boosted to crystallin levels without compromising their primary role in other tissues. ζ-crystallin is the first example in which it is clear that different promoters are responsible for differential expression of an enzyme crystallin.

REFERENCES

Abedinia, M., T. Pain, E.M. Algar, and R.S. Holmes 1990, Bovine corneal aldehyde dehydrogenase: the major soluble corneal protein with a possible dual protective role for the eye, *Exp. Eye Res.* 51:419.

Barbosa, P., G.J. Wistow, M. Cialkowski, J. Piatigorsky, and W.E. O'Brien 1991, Expression of duck lens δ-crystallin cDNAs in yeast and bacterial hosts: δ2-crystallin is an active argininosuccinate lyase, *J. Biol. Chem.* 266:22319.

Berman, E.R. 1991, Perspectives in Vision Research: "Biochemistry of the eye", 1st ed. New York: Plenum Press.

Bloemendal, H. (ed). 1981, "Molecular and Cellular Biology of the Eye Lens". New York: Wiley.

Borras, T., B. Persson, and H. Jornvall 1989, Eye lens zeta-crystallin relationships to the family of long-chain alcohol/polyol dehydrogenases, *Biochemistry.* 28:6133.

Carper, D., C. Nishimura, T. Shinohara, B. Dietzschold, G. Wistow, C. Craft, P. Kador, and J.H. Kinoshita 1987, Aldose reductase and rho-crystallin belong to the same protein superfamily as aldehyde reductase, *FEBS Lett.* 220:209.

Carper, D.A., G. Wistow, C. Nishimura, C. Graham, K. Watanabe, Y. Fujii, H. Hayashi, and O. Hayaishi 1989, A superfamily of NADPH-dependent reductases in eukaryotes and prokaryotes, *Exp. Eye Res.* 49:377.

Chiou, S.H. 1988, A novel crystallin from octopus lens, *FEBS Lett.* 241:261.

Cooper, D.L., E.W. Baptist, J.J. Enghild, N. Isola, and G.K. Klintworth 1991, Mixed oligonucleotide primed amplification cDNA cloning (MOPAC) of bovine corneal protein 54K (BCP54) reveals it is a homologue of the tumor inducible (class 3) rat aldehydrogenase gene, *Gene.* 98:201.

Eklund, H., J.P. Samama, and T.A. Jones 1984, Crystallographic investigations of nicotinamide adenine dinucleotide binding to horse liver alcohol dehydrogenase, *Biochemistry.* 23:5982.

Favreau, L.V., and C.B. Pickett 1991, Transcriptional regulation of the rat NAD(P)H:quinone reductase gene. Identification of regulatory elements controlling basal level expression and inducible expression by planar aromatic compounds and phenolic antioxidants, *J. Biol. Chem.* 266:4556.

Garland, D., P.V. Rao, A. Del Corso, U. Mura, and J.S. Zigler Jr. 1991, zeta-Crystallin is a major protein in the lens of Camelus dromedarius, *Arch. Biochem. Biophys.* 285:134.

Graur, D., W.A. Hide, and W.H. Li 1991, Is the guinea-pig a rodent?, *Nature.* 351:649.

Harding, J.J., and M.J.C. Crabbe. 1984, The lens: Development, proteins, metabolism and cataract. *in* "The Eye", H. Davson, ed: The Eye, Vol. 1B. Academic Press, New York, 207.

Huang, Q.L., P. Russell, S. Stone, and J.S. Zigler 1987, zeta-crystallin, a novel lens protein from the guinea pig, *Curr. Eye. Res.* 6:725.

Huang, Q.L., X.Y. Du, S.H. Stone, D.F. Amsbaugh, M. Datilies, T.S. Hu, and J.S. Zigler Jr. 1990, Association of hereditary cataracts in strain 13/N guinea pigs with mutation of the gene for zeta-crystallin, *Exp. Eye Res.* 50:317.

de Jong, W.W., W. Hendriks, J.W. Mulders, and H. Bloemendal 1989, Evolution of eye lens crystallins: the stress connection, *Trends Biochem. Sci.*. 14:365.

Kim, R.Y., T. Lietman, J. Piatigorsky, and G.J. Wistow 1991, Structure and expression of the duck alpha-enolase/tau-crystallin encoding gene, *Gene.* 103:193.

Kim, R.Y., R. Gasser, and G.J. Wistow 1992, Mu-crystallin is a mammalian homologue of Agrobacterium ornithine cyclodeaminase and is expressed in retina and brain, *Proc. Natl. Acad. Sci. USA.* in press.

Kondoh, H., I. Araki, K. Yasuda, T. Matsubasa, and M. Mori 1991, Expression of the chicken 'delta 2-crystallin' gene in mouse cells: evidence for encoding of argininosuccinate lyase, *Gene.* 99:267.

Linial, M., K. Miller, and R.H. Scheller 1989, VAT-1: an abundant membrane protein from Torpedo cholinergic synaptic vesicles, *Neuron.* 2:1265.

Montgomery, M.K., and M.J. McFall-Ngai 1992, The muscle-derived lens of a squid bioluminescent organ is biochemically convergent with the ocular lens. Evidence for recruitment of aldehyde dehydrogenase as a predominant structural protein, *J. Biol. Chem.*:in press.

Mulders, J.W., W. Hendriks, W.M. Blankesteijn, H. Bloemendal, and W.W. de Jong 1988, Lambda-crystallin, a major rabbit lens protein, is related to hydroxyacyl-coenzyme A dehydrogenases, *J. Biol. Chem.* 263:15462.

Pailhoux, E.A., A. Martinez, G.M. Veyssiere, and C.G. Jean 1990, Androgen-dependent protein from mouse vas deferens. cDNA cloning and protein homology with the aldo-keto reductase superfamily, *J. Biol. Chem.*:19932.

Piatigorsky, J., and G. Wistow 1991, The recruitment of crystallins: new functions precede gene duplication, *Science.* 252:1078.

Rao, P.V., and J.S. Zigler Jr. 1990, Extremely high levels of NADPH in guinea pig lens: correlation with zeta-crystallin concentration, *Biochem. Biophys. Res. Commun.* 167:1221.

Rao, P.V., and J.S. Zigler Jr. 1991, Zeta-crystallin from guinea pig lens is capable of functioning catalytically as an oxidoreductase, *Arch. Biochem. Biophys.* 284:181.

Rao, P.V., C.M. Krishna, and J.S. Zigler 1992, Identification and characterization of the enzymatic activity of zeta-crystallin from guinea pig lens. A novel NADPH:quinone oxidoreductase, *J. Biol. Chem.* 267:96.

Reddan, J.R., A.B. Chepelinsky, D.C. Dziedzic, J. Piatigorsky, and E.M. Goldenberg 1986, Retention of lens specificity in long-term cultures of diploid rabbit lens epithelial cells, *Differentiation.* 33:168.

Rodokanaki, A., R.K. Holmes, and T. Borras 1989, Zeta-crystallin, a novel protein from the guinea pig lens is related to alcohol dehydrogenases, *Gene.* 78:215.

Rondeau, J.M., F. Tete-Favier, A. Podjarny, J.M. Reymann, P. Barth, J.F. Biellmann, and D. Moras 1992, Novel NADPH-binding domain revealed by the crystal structure of aldose reductase, *Nature.* 355:469.

Shaughnessy, M., and G. Wistow 1992, Absence of MHC gene expression in lens and cloning of dbpB/YB-1, a DNA-binding protein expressed in mouse lens, *Curr. Eye. Res.* 11:175.

Tomarev, S.I., R.D. Zinovieva, S.M. Dolgilevich, S.V. Luchin, A.S. Krayev, K.G. Skryabin, and G.G. Gause Jr. 1984, A novel type of crystallin in the frog eye lens. 35-kDa polypeptide is not homologous to any of the major classes of lens crystallins, *FEBS. Lett.* 171:297.

Tomarev, S.I., R.D. Zinovieva, and J. Piatigorsky 1991, Crystallins of the octopus lens. Recruitment from detoxification enzymes, *J. Biol. Chem.* 266:24226.

Winters, C.J., D.T. Molowa, and P.S. Guzelian 1990, Isolation and characterization of cloned cDNAs encoding human liver chlordecone reductase, *Biochemistry.* 29:1080.

Wistow, G. 1990, Evolution of a protein superfamily: relationships between vertebrate lens crystallins and micro-organism dormancy proteins, *J. Mol. Evol.* 30:140.

Wistow, G., and H. Kim 1991, Lens protein expression in mammals: taxon-specificity and the recruitment of crystallins, *J. Mol. Evol.* 32:262.

Wistow, G., and J. Piatigorsky 1987, Recruitment of enzymes as lens structural proteins, *Science.* 236:1554.

Wistow, G., and J. Piatigorsky 1988, Lens crystallins: evolution and expression of proteins for a highly specialized tissue, *Ann. Rev. Biochem.* 57:479.

Wistow, G., and J. Piatigorsky 1990, Gene conversion and splice-site slippage in the argininosuccinate lyases/delta crystallins of the duck lens: members of an enzyme superfamily, *Gene.* 96:263.

Wistow, G., A. Anderson, and J. Piatigorsky 1990, Evidence for neutral and selective processes in the recruitment of enzyme-crystallins in avian lenses, *Proc. Natl. Acad. Sci. USA.* 87:6277.

Wistow, G.J., J.W. Mulders, and W.W. de Jong 1987, The enzyme lactate dehydrogenase as a structural protein in avian and crocodilian lenses, *Nature.* 326:622.

Wistow, G.J., T. Lietman, L.A. Williams, S.O. Stapel, W.W. de Jong, J. Horwitz, and J. Piatigorsky 1988, Tau-crystallin/alpha-enolase: One gene encodes both an enzyme and a lens structural protein, *J. Cell. Biol.* 107:2729.

Zigler, J.S., Jr., and J.B. Sidbury Jr. 1976, A comparative Study of the Beta-crystallins of Four Sub-mammalian species, *Comp. Biochem. Physiol.* 55B:19.

MEMBERS OF THE ALDH GENE FAMILY ARE LENS AND *CORNEAL* CRYSTALLINS

[1]David L. Cooper, [1]Narayana R. Isola, [1]Karen Stevenson, and [2]Edward W. Baptist

[1]Department of Pathology
University of Pittsburgh School of Medicine
723B Scaife Hall
Pittsburgh, PA 15261

[2]Department of Medicine
Lineberger Comprehensive Cancer Center
University of North Carolina
Chapel Hill, NC 27599-7295

INTRODUCTION

The parallel demands placed on both the lens and the cornea and on the molecules involved in fulfilling the visual functions of these tissues, require that these proteins be stable and resistant to deleterious influences found in our environment (e.g. radiant light, free radicals, heat). To maintain transparency both the lens and the cornea must of necessity avoid structural disintegration. Embryologically, the cornea and the lens share a somatic ectodermal origin. It should not be surprising that these tissues share many molecular components or undergo similar molecular events.

The crystallins, the abundant proteins of the eye lens, belong to one of three distinct families of 'ubiquitious' genes represented in all vertebrates or are members of a diverse group of proteins that are expressed in a 'taxon-specific' fashion. Sequence comparisons have shown that the ubiquitous α, β, and γ crystallins belong to superfamilies extending outside the lens, indicating that evolution has recruited these ubiquitous crystallins from among already existing heat-shock proteins and microorganismal dormancy proteins (Ingolia and Craig, 1990). Marked sequence relationships have been found between many taxon-specific crystallins and specific enzymes expressed at significantly lower levels in other tissues. It has been suggested that other properties, such as thermodynamic stability or high intracellular solubility, may be the basis for their selection as crystallins. The recruitment of enzymes to serve the lens in a new structural role has been called 'gene sharing' (Piatiagorsky et al., 1988).

In the cow cornea, the major soluble corneal protein, bovine corneal protein molecular weight 54,000 D (BCP54), makes up 40% of the soluble epithelial cell protein (Alexander et al., 1981). Recently, we have utilized mixed oligodeoxyribonucleotide primers complementary to the reverse translation products of amino acid sequence obtained from *Staphylococcus aureus* V8 digested BCP54 fragments and PCR to generate the first reported cDNA probe to BCP54 (Cooper et al., 1991). The BCP54 cDNA probe generated by this mixed oligonucleotide primed amplification of cDNA (MOPAC) technique was cloned and dideoxy sequenced. A GenBank library search (version 63.0) revealed a strong similarity to the previously cloned cDNA of rat liver (class 3) tumor-associated aldehyde dehydrogenase (RATALD) (Jones et al., 1988). Nucleotide and amino acid sequence alignment of the BCP54 translation product revealed it as 81% and 84% identical with ALDH 3 at the nucleotide and amino acid levels, respectively.

Similar to the lens enzyme crystallins, ALDH class 3 (ALDH 3) is present in the cornea in amounts that exceed any likely catalytic role. It has been suggested the corneal function of BCP54/ALDH 3 is derived from its retained enzymatic activity, or its putative ability to absorb ultraviolet light (Holmes and VandeBerg, 1990; Abedinia et al., 1990). We have suggested that the presence of this enzymatically active protein at levels that exceed levels sufficient for enzymatic function, but consistent with fulfilling a structural role, extends the phenomena of 'gene sharing' to the cornea (Cooper et al., 1991). The crystallins may or may not maintain catalytic activity in the lens and their recruitment by altered gene expression allows for some crystallins to maintain a different role in another tissue or cell type. Many of the taxon specific crystallins are oxido-reductases, as are the ALDH enzymes, that bind pyridine nucleotides as cofactors and thus may be useful in sequestering reduced cofactors as ultraviolet filters or as reducing potential (Zigler and Rao, 1991). Most aldehydes are toxic to the cell and aldehyde scavenging has a definite selective advantage in preventing the formation of Schiff's base adducts of proteins. Such adducts would almost certainly be detrimental to maintaining the transparent properties of the visual organ. To fullfill such a cellular function does not warrant that 40% of corneal soluble protein be aldehyde dehydrogenase. The high water solubility, and structural properties both at the protein and gene levels of this multifunctional enzyme make it an ideal choice for recruitment in the lens and cornea. The recognition of gene sharing in the cornea is a novel concept which has implications on the evolution of gene sharing *per se* and the utility of such recruitment in the evolution of a complex organ such as the vertebrate eye. Based on analysis of the thiol proteinesterases, aldehyde dehydrogenases and crystallins we note a supra family relationship between these members many of which could have evolved by exon shuffling and gene duplication from a limited number of ancestral genes.

EXPERIMENTAL

We have assayed for BCP54/ALDH 3 in the corneas of various vertebrate classes **(Table 1)**. All mammals (dog, mouse, cow, pig), including two classes of marsupials

TABLE 1 ALDH Activity in Corneal Extracts of Several Vertebrate Classes

		Enzymatic Activity	
Mammal	MW of MCP(kD)	ALDH 1/2	ALDH 3
Dog	54	nt	nt
Pig	54	-	+
Cow	54	-	+
Mouse	54	-	+
Opossum	nt	-	+
Fish			
Bowfin	60	-	-
Chain Pickerel	56	+	-
Gizzard Shad	27,58	-	-
Redhorse	60	+	-
Striped Bass	60,58,27	-	-
Non-mammal other classes			
Chicken	27,22	-	-
Garter Snake	64	-	-
Xenopus	12,66	-	-

Whole corneas were homogenized in ice-cold extraction buffer (0.1M NaCl, 40mM Tris-HCl, pH 7.8) followed by centrifugation of the homogenate at 30,000 x g at 4°C for 30 minutes (Piatiagorsky et al., 1988; Alexander et al., 1981). Aldehyde dehydrogenase activity was determined spectrophotometrically by monitoring the change in A_{340} caused by NADH or NADPH production during the oxidation of aldehyde substrate (Lindahl and Evces, 1984). The reaction mixture contained 1.0 ml of 60 mM sodium phosphate buffer, pH 8.5, containing 1 mM EDTA and 1 mM mercaptoethanol, 1.0ml of 7.5 mM NAD or NADP in buffer, 0.25 ml of 75 mM propionaldehyde or a saturated benzaldehyde solution, 100ul of sample, and water to 3.0 ml. (NAD, nicotinamide adenine dinucleotide. NADP, nicotinamide adenine dinucleotide phosphate).

(opossum, kangaroo), examined for the presence of the BCP54/ALDH 3 were positive and this protein was predominant in corneal epithelial cell scrapings (cow, 40%) or among the major corneal proteins when total cornea was extracted (pig, dog,).

Preliminary survey of members of several other non-mammalian vertebrate classes (bird, snake, amphibia, bony fish) by Western blotting, enzymatic assay, and direct amino acid sequencing, has not detected the expression of BCP54/ALDH 3 (Cooper et al., 1991). Protein preparations from these corneal samples were tested with substrates capable of discerning the ALDH class 1/ALDH class 2 (ALDH 1/2) enzymes from ALDH 3. In our experiments, as indicated above, only mammalian corneal extracts were found capable of utilizing as substrate/coenzyme both the respective pairs of propionaldehyde-NAD and benzaldehyde-NADP, indicative of ALDH3 activity (Lindahl and Evces, 1984).

No such ALDH 3 activity was found in any other vertebrate class. Unexpectedly however, ALDH 1/2 activity utilizing propionaldehyde-NAD as substrate/coenzyme was detected in the cornea of two species of bony fish (e.g. chain pickerel and redhorse). Indicative of a probable structural role, the level of total ALDH 1/2 activity in milliunits/mg of fish cornea of the two bony fish species was comparable to or exceeded the level of ALDH3 activity determined to be present in mouse cornea (Table 2).

TABLE 2 ALDH Activity in the Corneas of Several Vertebrate Classes

Species	Tissue	Propionaldehyde/NAD mUnits/mg protein	Benzaldehyde/NADP mUnits/mg protein
Mouse	Eye	2.03	2.14
Mouse	Stomach	0.71	0.85
Mouse	Liver	0.48	0.13
Chicken	Cornea	0	0
Xenopus	Cornea	0	0
Redhorse	Cornea	3.2	0
Striped Bass	Cornea	0	0
Chain Pickerel	Cornea	2.4	0
Gizzard Shad	Cornea	0	0
Bowfin	Cornea	0	0

Corneal extracts were tested for ALDH activity by their ability to utilize propionaldehyde/NAD or benzaldehyde/NADP as substrate /coenzyme combinations. Protein concentrations were determined by the method of using bovine serum albumin as standard. Activities are expressed as milliunits (1 milliunit = 1 nmole substrate converted/min.) per mg of liver.

DISCUSSION

Corneal Crystallins

ALDH 1/2 and ALDH 3 gene activity has now been confirmed in the soluble protein fractions extracted from the corneas of two different vertebrate classes. Additionally, it has been recently shown that the most abundant mammalian taxon-specific crystallin identified to date is the cytoplasmic aldehyde dehydrogenase (class 1, ALDH 1)/η-crystallin (Wistow and Kim, 1991), which accounts for up to 25% of the soluble protein in the lenses of the elephant shrew. The ALDH 1 and 3 members of the ALDH gene family appear evolutionarily to represent enzymes recruited to both lens and cornea to fulfill structural roles, as taxon-specific lens and *corneal crystallins*. The ALDH gene family is the first intriguing example of 'gene sharing' that includes the utilization of alternative members of the same gene family (e.g. ALDH 1, ALDH 1/2, ALDH 3), in different tissues (e.g. cornea, lens) across species and class boundaries (e.g. mammals, reptiles) . The taxon distribution may be exceedingly wide, as is the case for the presence of ALDH 3 in the corneas of all mammals or as has been found for ALDH 1 and ALDH 1/2 limited to the lens of the elephant shrew, or some but not all bony fish species, respectively (Figure 1).

The ALDH gene family is unique among the multifunctional crystallins and genes known to be examples of 'gene sharing'. It is the only gene family that as shown above, contains at least two members who function as crystallins (defined here as major soluble structural proteins), in both lens (ALDH 1) and cornea (ALDH 3, and ALDH 1/2). Interestingly, ALDH 1 has most recently been ascribed another unrelated function, that being the role of androgen receptor in genital fibroblasts (Pereira et al., 1991). Thus the ALDH gene family has at least five known functional roles: (1) η-crystallin/ALDH 1 functions as a mammalian lens structural protein in the elephant shrew; (2) BCP54/ALDH 3 functions as the major soluble mammalian corneal protein; (3) either ALDH 1 or ALDH 2, or the predecessor form of ALDH 1/2 function in some fish species as a corneal protein; (4) ALDH 1 functions in genital fibroblasts as an androgen receptor; and finally (5) all retain their thiol esterase enzymatic function and are members of the oxido-reductases family of enzymes .

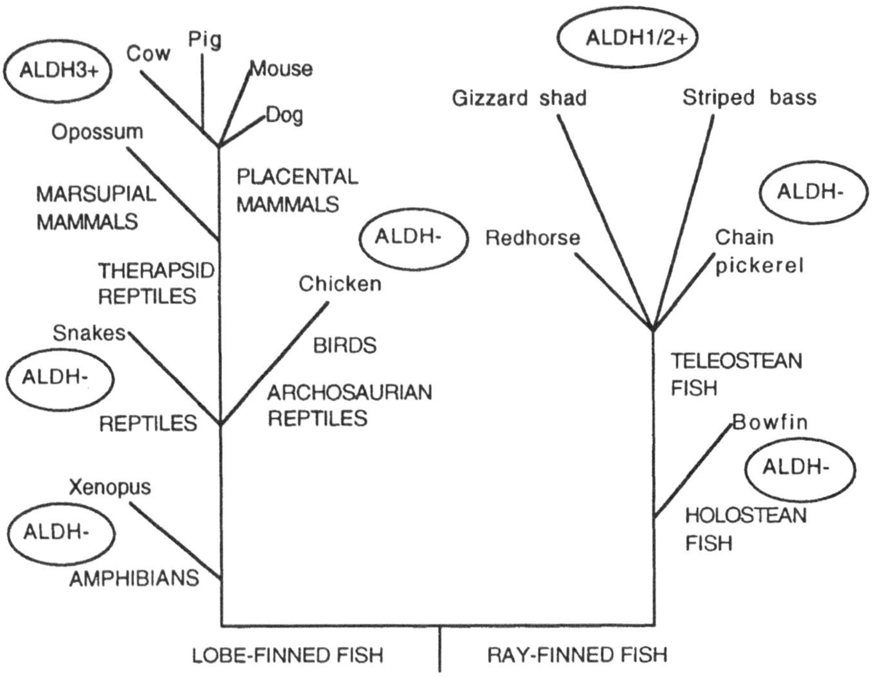

Figure 1. Taxon specificity of ALDH3 corneal expression. Aldehyde dehydrogenase activity in the cornea of various species is noted on a rooted phylogenetic tree.

A number of examples of genes with different roles exhibited elsewhere have now been described. Interestingly, glycosylation site binding protein (GSBP), a component of oligosaccharide transferase also exhibits three other functions (Geetha-Habib et al., 1988) which include: (1) protein disulfide isomerase, (2) thryoid hormone binding protein, and (3) the β-subunit of prolylhydroxylase. Although there may be a limited number of examples, both GSBP and ALDH appear to provide examples of superfamilies whose members are associated with a predisposition for acquiring new functions. Recruitment of such genes is dependent on the ability to modify the recruited genes level of expression, and the structural qualities of the gene product.

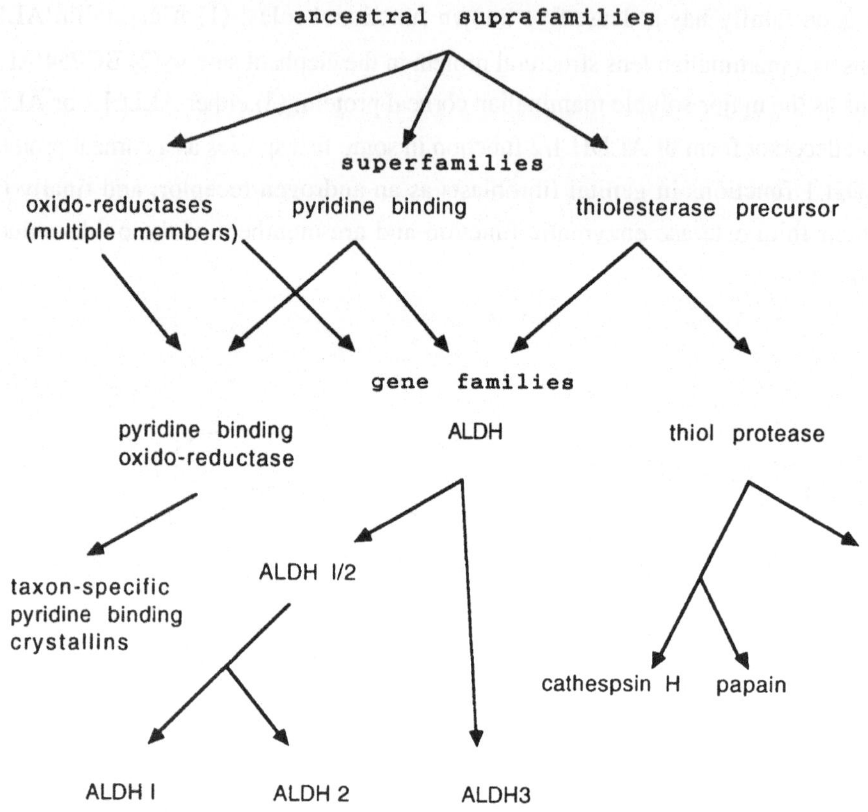

Figure 2. Evolutionary relationship of lens and corneal crystallins. Ancestral suprafamilies duplicated and underwent exon shuffling to produce the superfamilies of the oxido-reductases, the pyridine binding proteins, and the thiolesterase precursor. Subsequent exon shuffling and gene duplication produced the ALDH, pyridine oxido-reductase, and thiol protease gene families whose current day members arose by specialization following another round of gene duplication.

Structural characteristics and sequence alignment of the ALDH 3 with ALDH 1 and ALDH 2 suggest a distant relationship at the level of the superfamily (Hempel, J. et al., 1989). ALDH 1 is marginally higher in number of matching residues to ALDH 3 than ALDH 2. Structurally ALDH 3 appears to have lost the first two protein encoding exons that are still found to be shared by ALDH 1 and 2. A relationship of ALDH 3 to ALDH 2 has apparently been maintained since their 3' ends are slightly more homologous in sequence than to ALDH 1. The nearly equal degree of relatedness of ALDH 1/2 to the ALDH 3 structure, indicates that the divergence from the common ancestral gene of the these three enzymes occurred prior to the divergence of the ALDH 1/2 families.

The ALDH Super Family and 'Gene Sharing'

Most recently, the aldehyde dehydrogenases and thiol proteases have been hypothesized to have evolved from a common thiolesterase precursor (Hempel et al.,

1991). Based on the sequence comparison and structural/functional similarities in the phenotypes which might be responsible for these members to be chosen for gene sharing a phylogenetic tree implicating a supra family relationship is constructed (Figure 1). In the construction of our model we also assume a supra family functional/structural relationship to the taxon-specific oxido-reductase and pyridine binding proteins. The different functional phenotypes can be listed as: (1) structural protein, (2) pyridine nucleotide binding, and (3) generalized oxidoreductase activity, of which thiol esterase activity is a specific oxidoreductase activity (Table 3).

Ancestral suprafamilies encoding the components of multifunctional proteins gave rise to the superfamilies such as thiol proteinases, oxidoreductases, and pyridine nucleotide binding. Further divergence through exon shuffling gave rise to different combinations of these phenotypes which ultimately led to the taxon-specific pyridine nucleotide binding oxido reductases (i.e. the taxon-specific crystallins), the ALDH gene family, and the thiol proteases. All three phenotypes are present in the taxon-specific crystallins and the ALDH gene family members while the thiol protease family possess only one (thiol esterase activity) phenotype. According to the proposed model a 'ALDH 1 like' gene duplication gave rise to the ancestor of the ALDH 1/2 and ALDH 3 genes. Acquisition of multifunctionality may have preceeded this duplication and the loss of function led to specialization in the three genes as evidenced by the loss of the 5' exons of ALDH 3. Alternatively this duplication could have been driven by the acquistion of multifunctionality as has been proposed as a possible mechanism involved in 'gene sharing' (Piatigorsky and Wistow, 1991). The function that gives a selective advantage to such a duplication could have been either the androgen receptor function or the structural gene function of both lens and corneal crystallins which led to specialization of class 1 from 2 following gene duplication. Alternatively, this duplication could have been a classical gene duplication that led to divergence of enzymatic function and estabishment of the ALDH 1 and ALDH 2 genes. The associated functions of ALDH 1 could then result from modification of the expression of this gene in a number of different tissues (e.g. cornea, lens, genital fibroblasts). The arisal of ALDH 3 gene may reflect modification of gene expression in the cornea or specialization by the addition of aromatic substrate activity to this form of the enzyme.

Although multifunctionality, which is a prerequisite for 'gene sharing' may be limited to some gene families, 'gene sharing' is further limited to gene families whose members are amenable to gene regulation by tissue specific factors and have the required inherent physical properties to play the role required of specialized structural proteins. It is tempting to speculate a supra gene family relationship between the thiol proteinesterases, aldehyde dehydrogenases, and the taxon-specific crystallins (Figure 2). The ALDH family represents a process of specialization through gene duplication in an intermediate stage where the duplicated gene products have specialized sufficiently enough to be recruited as structural proteins while still retaining enzymatic activity.

TABLE 3 Structural/Functional Phenotypes Exhibited by the Thiol Proteases, Lens and Corneal Crystallins

	PHENOTYPES			
	Pyridine binding	Generalized Oxidoreductase activity	Specialized oxidoreductase, Thiol esterase activity	Structural protein
Thiol proteases				
Papain	-	-	+	-
Cathepsin	-	-	+	-
Lens crystallins, all vertebrates				
α, β, γ	-	-	-	+
taxon specific				
ε, lactate dehydrogenase, some birds and reptiles	+	+	-	+
λ, hydroxyacyl-CoA dehydrogenase, some mammals	+	+	-	+
ζ, alcohol dehydrogenase, some mammals	+	+	-	+
ρ, NADPH-dependent reductases, frogs	+	+	-	+
η, aldehyde dehydrogenase (class 1), some mammals	+	+	+	+
δ, argininosuccinate lyase, some birds and reptiles	-	-	-	+
Corneal crystallins, taxon-specific				
aldehyde dehyddrogenase (class 1/2), some fish	+	+	+	+
aldehyde dehydrogenase (class 3), all mammals	+	+	+	+

More detail for most of these gene products can be found elsewhere (Jones et al., 1988; Wistow and Kim, 1991; Hempel et al, 1991).

Analysis of the ALDH gene family and comparison with the crystallins and similar multifunctional proteins leads us to propose that: (1) specific functional/structural motifs may be more predisposed to multifunctionality, (2) suprafamily relationships may be more complex than previously imagined, (3) reptilian ancestors who maintain ALDH 1/2 expression may exist and link these earlier species to the later evolved mammals, and (4) a closer look for other possible cellular functions of ALDH 2 is likely to be rewarding.

SUMMARY

Many of the major lens proteins, known as crystallins, responsible for the structural integrity and functional utility of this visual tissue have been previously shown to be recruited proteins. This phenomena of a protein that is expressed and functions elsewhere acquiring a new function in another tissue has been termed 'gene sharing'. It is now becoming obvious that the cornea of vertebrates has similarily acquired proteins, and that at least one corneal protein, ALDH3 belongs to a gene family that has been previously identified as a lens crystallin. The recognition that both lens and corneal crystallins exist is a novel concept that has implications that involve the process by which multifunctional gene products have evolved.

Members of the ALDH gene family function in both the cornea and lens as crystallins and the acquistion of multifunctionality by this gene family is unique. Based on our analysis we have deduced a supragene family relationship between the thiol proteinesterases, aldehyde dehydrogenases, and the taxon-specific crystallins. Evolution of a complex organ such as the vertebrate eye is not a sequential and gradual process such as the Darwinian Giraffe's neck, since the eye can provide selective advantage only as a complete organ. Catastrophic theory proposes that the complex vertebrate eye with its lens, and focussing mechanism arose from the primitive eye spot which contained originally only the photoreceptor system by a one step event . In the evolution of the vertebrate eye it is evolutionarily plausible that several pre-existing proteins have been recruited to perform a structural role for this complex organ. It is also incumbent in evolutionary thought that any inherent enzymatic activity associated with this protein would be purely an incidental addition to the organ. However, the fact that most of these have pyridine nucleotide binding capacity, which is presumed important in giving protection from UV exposure, is noteworthy. Finally, to construct the vertebrate eye in one step from the existing visual pigment system such as the eyespot of unicellular organisms the following criteria would apparently be advantageous: (1) high water solubility; (2) transparency; and (3) common genetic regulatory elements (e.g. promoters/enhancers).

Although it is an important observation that certain members of the aldehyde dehydrogenase gene family are present as structural proteins in the cornea and lens, it is not surprising that the phenomenon of gene sharing extends to another occular tissue such as the cornea. In this context, it will be interesting to note if similar multifunctional gene products will be found as frequently in organs other than the eye.

REFERENCES

Abedinia M, Pain T, Alabar EM, Holmes RS. Bovine corneal aldehyde dehydrogenase: the major soluble corneal protein with a possible dual protective role for the eye. Exp. Eye Res. 51, 419 (1990).

Alexander RJ, Silverman B, Henley WL. Isolation and characterization of BCP54, the major soluble protein of bovine cornea. Exp. Eye Res. 32, 205 (1981).

Cooper DL, Baptist EW, Klintworth G, Isola NR. Bovine corneal protein 54K (BCP54) is a homologue of the tumor-associated class 3 rat aldehyde dehydrogenase (RATALD). Gene 98, 201 (1991).

Cooper DL and Baptist EW. Degenerate oligonucleotide sequence-directed cross-species PCR cloning of the BCP 54/ALDH 3 cDNA: Priming from inverted repeats and formation of tandem primer arrays. PCR Method and Applications 1 (1991).

Cooper DL, Baptist EW, Isola NR, Klintworth GK, Gottsman M. Bovine corneal protein 54K is a homologue of rat tumor-associated ALDH: Another example of gene sharing? Invest. Opthalmol. Vis. Sci. 32 (Suppl): 560 (1991).

Cooper DL, Baptist EW, Enghild J, Lee H, Isola N, Klintworth GK. Partial amino acid sequence of bovine corneal protein 54K (BCP54). Curr. Eye Res. 9, 781 (1990).

De Jong WW, Hendriks W, Mulders JWM, Bloemendal H. Evolution of eye lens crystallins: the stress connection. Trends Biochem. Sci. 14, 365 (1989).

Geetha-Habib M, Noiva R, Kaplan HA, Lennarz WJ. Glycosylation site binding protein, a component of oligosaccharyl transferase, is highly similar to three other 57 kd luminal proteins of the ER. Cell 54, 1053 (1988).

Hempel J, Nicholas H, Jornvall H. Thiol proteases and aldehyde dehydrogenases: evolution from a common thiolesterase precursor? Proteins 11, 176 (1991).

Hempel J, Harper K, Lindahl R. Inducible(Class III) aldehyde dehydrogenase from rat hepatocellular carcinoma I and II enzymes. Biochemistry 28, 1160 (1989).

Hempel J and Lindahl R. Class III aldehyde dehydrogenases from rat liver: superfamily relationship to classes I and II and functional interpretations. In Weiner H and Flynn TG, Eds, Enzymology and Molecular Biology of Carbonyl Metabolism 2. Liss, New York, 1989, pp 3.

Holmes RS, VandeBerg JL. Alcohol dehydrogenase and aldehyde dehydrogenase in the baboon. Exp. Eye Res. 13, 383 (1986)

Ingolia TD and Craig EA. Four small drosophila heat shock proteins are related to each other and to mammalian α-crystallin. Proc. Natl. Acad. Sci. U.S.A. 79, 2360 (1982).

Jones DE, Brennan MD, Hempel J, Lindahl R. Cloning and complete nucleotide sequences of a full length cDNA encoding a catalyticaly functional tumor-associated aldehyde dehydrogenase. Proc. Natl. Acad. Sci. U.S.A. 85, 1782 (1988).

Lindahl R, Evces S. Comparative subcellular distribution of aldehyde dehydrogenase in rat, mouse and rabbit liver. Biochem. Pharmac. 33, 3383 (1984).

Pereira F, Rosenmann E, Nylen E, Kaufman M, Pinsky L, Wrogemann K. The 56 kDa androgen binding protein is an aldehyde dehydrogenase. Biochem. Biophys. Res. Commun. 175, 831 (1991).

Piatigorsky J and Wistow G. The recruitment of crystallins: new functions precede gene duplication. Science 252, 1078 (1991).

Piatigorsky J and Wistow G. Enzyme-crystallins: gene sharing as evolutionary strategy. Minireview. Cell 57, 197 (1989).

Piatiagorsky J, O'Brien WE, Norman BL, Kalumuck K, Wistow GJ, Borras T, Nickerson JM, Wawrousek EF. Gene sharing by δ -crystallin and argininosuccinate lyase. Proc. Natl. Acad. Sci. U.S.A. 85, 3479 (1988).

Silverman B, Alexander RJ, Henley WL. Tissue and species specificity of BCP54, the major soluble protein of bovine cornea. Exp. Eye Res. 33, 19 (1981).

Wistow G and Kim HJ. Lens protein expression in mammals: taxon-specificity and the recruitment of crystallins. J. Mol. Evol. 32, 262 (1991).

Wistow GJ. Evolution of a protein superfamily: relationships between vertebrate lens crystallins and microorganism dormancy proteins. J. Mol. Evol. 30, 140 (1990).

Wistow G and Piatigorsky J. Lens crystallins: the evolution and expression of proteins for a highly specialized tissue. Ann. Rev. Biochem. 57, 479 (1988).

Zigler JS, Jr. and Rao PV. Enzyme/crystallins and extremely high pyridine nucleotide levels in the eye lens. FASEB J. 5, 223 (1991).

RETINOIC ACID SYNTHESIS IN THE DEVELOPING RETINA

Peter McCaffery and Ursula C. Dräger

Harvard Medical School
Department of Neurobiology
Boston MA 02115

MAP FORMATION IN THE CENTRAL NERVOUS SYSTEM

Neuronal information in the vertebrate brain is represented in the form of topographical maps: there are multiple topographical representations in the brain of the visual world, of auditory space, of the tactile body surface and of movement directions. Information processing in the framework of maps is a general feature of the nervous system throughout vertebrates and even in lower species. The main advance through evolution is not in the principle of mapping, but in the number and specialization of maps. The maps are formed in the embryonic brain due to a capacity of neuronal processes from one brain area to search out precisely their address in the next area. The molecular basis of this neuronal specificity is still largely unknown. The main working hypothesis, formulated for the example of the visual system, postulates two sets of cell-surface markers, one graded in the antero-posterior, the other in the dorso-ventral axis of the retinal maps, which enable optic growth cones to match up with corresponding markers in their targets (Sperry, 1963). The expression of such graded markers needs to be preceded by an asymmetry in gene transcription. The enzymes described here provide a direct access to this early step of neuronal specification in the retina, as their product, retinoic acid, regulates gene transcription through nuclear receptors (Chambon et al., 1991).

CLASS-1 ALDEHYDE DEHYDROGENASE IS A MARKER FOR THE DORSAL RETINA

In a search for factors involved in the axial polarization of the retina in the embryonic mouse, we compared protein fractions isolated from dorsal and ventral retina halves (McCaffery et al., 1991). Several hundred protein spots, resolved by two-dimensional gel electrophoresis, are identical in the two samples, except for a 53kD protein that is much more abundant in dorsal than

Enzymology and Molecular Biology of Carbonyl Metabolism 4
Edited by H. Weiner, Plenum Press, New York, 1993

ventral retina. Microsequencing identified it as most similar to the major cytosolic class 1 aldehyde dehydrogenase isoform in rat and human, also termed E1 or ALDH-1 (Dunn et al., 1989; Lindahl and Hempel, 1990). Immunohistochemistry with antisera to human and rat class 1 aldehyde dehydrogenase on embryonic mouse sections (Figure 1) shows bright labeling of the dorsal retina (Lindahl and Evces, 1984a; Lindahl and Evces, 1984b; Russo and Hilton, 1988). The enzyme becomes first detectable in the dorsal region of the early eye vesicle at embryonic day 9 (E9), and expression in dorsal retina persists into the adult retina. The identity with the murine class 1 isoform AHD-2 was confirmed through direct comparisons with purified enzyme prepared from adult mouse liver (Manthey et al., 1990). A plausible role for the function of the enzyme in the embryonic retina was indicated by the report that AHD-2 is the major aldehyde dehydrogenase in the adult mouse which can oxidize retinaldehyde to retinoic acid (Lee et al., 1991).

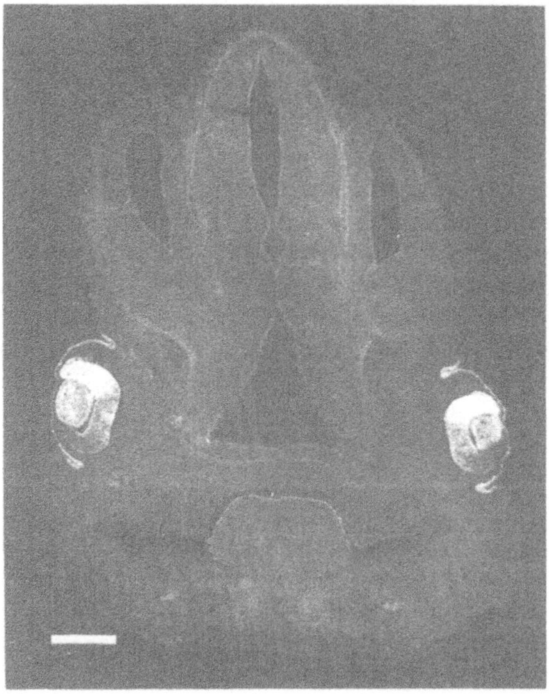

Figure 1. Coronal section through forebrain of an E12 mouse embryo, labeled with antisera to class 1 aldehyde dehydrogenase. Note the bright labeling of the dorsal retina and of the surface ectoderm overlying the eye, which will give rise to the cornea. Scale 300μm.

MEASUREMENTS OF RETINOIC ACID SYNTHESIS IN THE EMBRYONIC RETINA

Determinations of retinoic acid synthesizing capacity in extracts of embryonic retinas showed an overall high rate of synthesis, and under the experimental conditions retinoic acid production in dorsal retina halves exceeded ventral production about tenfold (McCaffery et al., 1992a); as synthesis was measured with a photospectrometric high-pressure liquid chromatography

(HPLC) system, the experimental conditions required rather high (25μM) retinaldehyde levels. Because the literature on retinoic acid synthesis is rich in discrepancies, we sought to confirm the HPLC measurements with a different method for retinoic acid detection: a reporter cell line which responds to retinoic acid, diffusing from co-cultured tissue pieces, by synthesizing beta-galactosidase, a compound that can be easily visualized with a histochemical reaction (Wagner et al., 1992). Explants of embryonic tissues onto the reporter cells generated a further example of a discrepancy in retinoic acid measurements: like the HPLC method the reporter cells reveal overall very high levels of retinoic acid synthesis in the embryonic retina, much higher than for instance in the embryonic cortex; in contrast to the HPLC method, however, the reporter cells show a two-to-three fold higher synthesis in ventral than dorsal retina.

In order to solve this paradox, we developed enzymatic micro-assays based on the reporter cells (McCaffery et al., 1992a). It turns out that the cells do not require explants of live tissue for the detection of retinoic acid, but they react in a very similar way to tissue homogenized in cell culture medium. These homogenates then allow enzymatic experiments: testing of different retinaldehyde levels, coenzymes and inhibitors; in addition, the beta-galactosidase synthesis can be quantified by performing the tests in 96-well plates and evaluating them colorimetrically in an ELISA reader. When such tests are done on dorsal and ventral embryonic retina halves, they provide evidence for a retinoic acid synthetic pathway in the ventral retina which differs from the AHD-2 mediated synthesis in dorsal retina. Tests with different retinaldehyde concentrations give an explanation for the discrepancy between the HPLC and reporter cell results: the dorsal pathway is much more powerful at high retinaldehyde levels as used in the HPLC assay, and ventral synthesis exceeds dorsal synthesis at low and presumably physiological retinaldehyde levels. This points to a lower Km of the ventral pathway as compared to AHD-2. The two pathways are differently affected by enzyme inhibitors: dorsal synthesis is preferentially depressed by very low concentrations of disulfiram, a specific inhibitor of the class 1/AHD-2/E1 isoform (Vallari and Pietruszko, 1982), and ventral synthesis is relatively more affected by para-hydroxymercuribenzoate and citral. Like AHD-2, the ventral pathway is NAD-dependent, indicating that it is also a dehydrogenase.

A ZYMOGRAPHY-BIOASSAY FOR RETINOIC-ACID SYNTHESIZING ENZYMES

Attempts to visualize the two enzymes by standard zymography in isoelectric focusing gels showed a strong AHD-2 signal, located in the basic portion of the gel, in samples from dorsal embryonic retina. However, we could never detect a ventral-specific signal with this method. It is likely that the ventrally localized enzyme is relatively specific for retinaldehyde, a substrate incompatible with standard zymography. To test specifically for retinaldehyde oxidation, we applied the reporter cell assay to the analysis of the isoelectric focusing gels (McCaffery et al., 1992a). Several neighboring lanes are cut with a multi-blade device into 30-40 slices, which are transferred into 96-well plates, and the eluted proteins are assayed for retinoic acid synthetic capacity. This zymography-bioassay works very reliably even with minute amounts of tissue. In addition to AHD-2 in dorsal retina, it shows the enzymatic activity in ventral retina: a strong signal at pH 5.6 (labeled V1 in Figure 2) and a signal precipitated at the origin of the sample (V3 in Figure 2). The activity at the origin is not an integral membrane protein, as it can be brought into solution by repeated washing; probably it represents an aggregate form of the pI 5.6 (V1) enzyme. The small AHD-2 peak in the ventral sample, as well as the small V1 peak in the dorsal sample, probably represent mainly imprecisions in the dissections of the tiny embryonic retinas into dorsal and ventral halves. From tests on retinas dissected into smaller fractions, we have the impression that the two activities are segregated into a dorsal and a ventral compartment with very little overlap between the two territories.

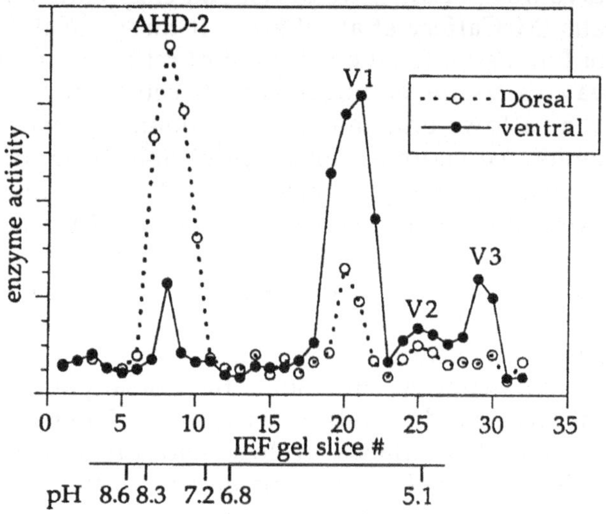

Figure 2. Zymography-bioassays showing retinoic-acid synthesizing enzyme activities in dorsal and ventral halves of retinas from a litter of E13.5 mouse embryos. The enzyme activity measurements represent arbitrary units based on colorimetric readings of the beta-galactosidase activity induced in the reporter cells.

DEVELOPMENTAL CHANGES IN PATTERNS OF RETINOIC-ACID SYNTHESIZING ENZYMES

Between the V1 and V3 peaks in the zymograph of Figure 2 a rather insignificant third peak, labeled V2, is visible, which we initially ignored because its detectability seemed variable. It turns out now that this variability is due to age differences. In younger retinas (Figure 3), this V2 peak is very prominent. It represents, in fact, the earliest form of retinoic-acid synthesizing enzyme detectable in the forebrain-fold region of the E8.0 embryo, and it is first expressed when the future eyes are just beginning to be demarcated as a pit in the anterior part of the neural plate. We do not know its significance. It differs from the V1 activity in being less efficient: short incubation times of E9 retinas with retinaldehyde give a much higher V1 peak and longer incubation times a relatively stronger V2 peak. Because V2 seems to co-localize in the retina with the V1 and V3 activities, it might represent an immature form of V1.

Very high activities of the ventral-type enzymes can be detected in the general regions of the optic pit (E8.0) and optic groove (E8.5-9) at early developmental stages, and no activity is detectable in the neural-plate regions destined to develop into most of the brain. It is impossible to tell, whether the activity within the optic groove is restricted to the ventral edge, because the determination of the spatial localization is limited by the size of tissue fractions that can be reliably dissected. In view of the very high enzyme activities, however, it seems likely that the entire optic groove, in addition to the surrounding tissue, express the enzymes. When AHD-2 becomes first detectable at E9.0, a day later than the ventral activity, it is already restricted to the dorsal portion of the future retina. As soon as dissections into dorsal and ventral retina become practicable (around E11), dorsal and ventral enzymes seem to be spatially segregated.

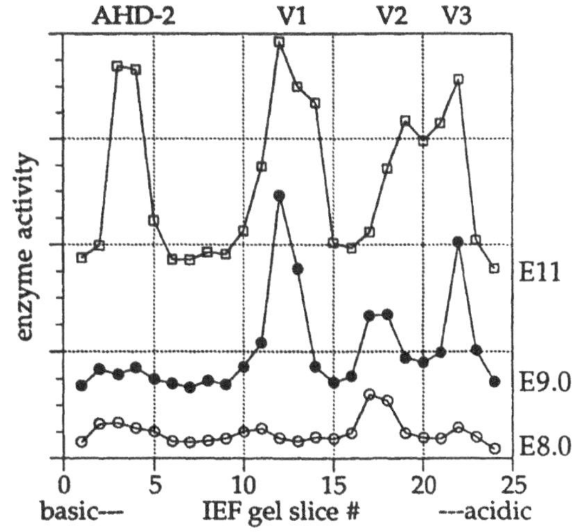

Figure 3. Zymography-bioassays showing retinoic-acid synthesizing enzyme activities in E8.0, E9.0 and E11 mouse embryos. In the E8 embryos the rostral parts of the neural plate including the forebrain folds with beginning optic pits were dissected, in the E9 embryos the forebrain regions with attached optic vesicles, and in the E11 embryos the retinas only. Comparative measurements of different dissected body parts of the early embryo show that within the head practically all synthetic activity is localized in the eye regions. The E11 and E9 samples represent one litter of embryos and the E8 sample two litters. The samples were not normalized for protein content.

In the later embryo, after E14, V2 is no longer detectable, and all ventral activity consists of V1 and V3. Levels of both AHD-2 and the ventral activities decrease later in development, with the decline being moderate during later fetal stages, and pronounced during the first two postnatal weeks (McCaffery et al., 1992b). The retina samples shown in Figure 4 were normalized for protein content and processed in parallel in order to illustrate this decline. Detectability of the ventral dehydrogenase reaches noise levels around the third postnatal week. A sizeable amount of AHD-2 persists in the adult retina, where its presence in dorsal retina is still very striking in immunohistological preparations due to its very high protein levels (McCaffery et al., 1991), but it is substantially reduced compared to the embryo.

ATTEMPTS AT DETERMINATION OF THE MOLECULAR WEIGHT OF THE V1 DEHYDROGENASE

In a screen for retinoic-acid generating dehydrogenases in the adult mouse, Lee et al. (1992) found such activity only in the class-1 aldehyde dehydrogenases, with AHD-2 providing the major and AHD-7 a minor contribution; both of these isoforms have a basic pI. In order to determine the molecular weight of the ventral enzymes, protein fractions from dorsal and ventral embryonic retina halves were eluted from the slices of the isoelectric focusing gels. Some of the protein was tested for retinoic-acid synthesizing capacity, and the remaining part was processed by SDS-PAGE combined with a

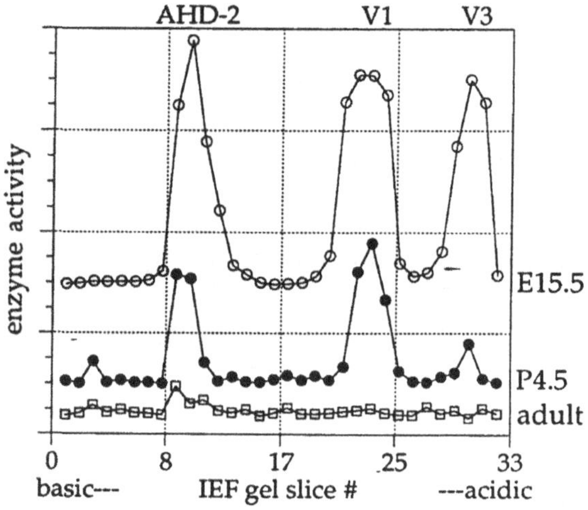

Figure 4. Zymography-bioassays showing retinoic-acid synthesizing enzyme activities in retinas from E15, P4.5 and adult mouse.

sensitive silver stain. The preferential localization of AHD-2 in the dorsal sample is very obvious due to its very high protein levels: we estimate that AHD-2 accounts for about 1% of soluble protein in the dorsal embryonic retina. Protein levels in the region containing the V1 activity peak, by comparison, are rather low, and only with a pre-purification step on Sepharose-blue beads does a band become visible whose presence and intensity correlates with the V1 activity; see Figure 5. In protein levels this band is at least two orders of magnitude less abundant than AHD-2. As the ventral retina contains an even higher endogenous retinoic acid concentration than the dorsal retina, at the protein level the ventral dehydrogenase is several hundredfold more powerful than AHD-2. The candidate signal for V1 has a molecular weight of 60kD, a value slightly higher than that described for the group of nonspecific aldehyde dehydrogenases. It might represent a novel dehydrogenase or, more likely, a

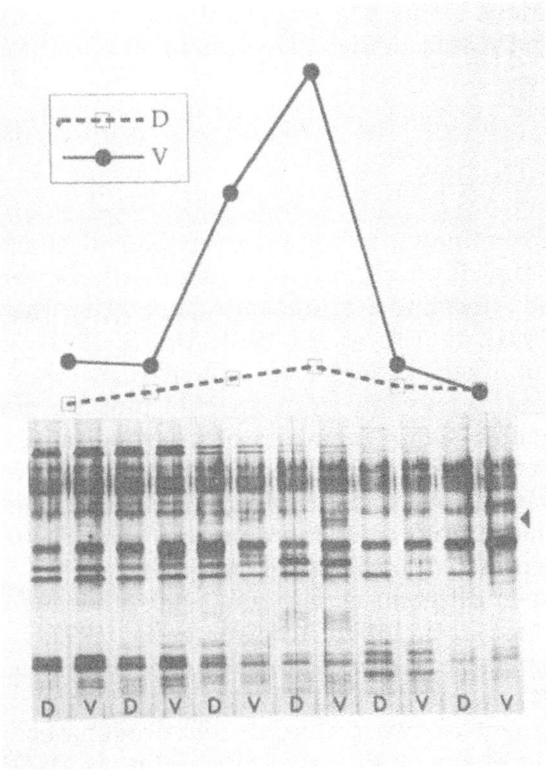

Figure 5. Zymography-bioassay traces showing the V1 activity in ventral halves of embryonic retinas (solid curve) and the corresponding gel region from the dorsal sample (broken curve). Proteins from single gel slices of the two samples, separated by SDS-PAGE and visualized by silver staining are shown underneath, with pairs of dorsal (D) and ventral (V) lanes lined up with the corresponding pairs of activity measurements. Note the faint band in the ventral samples, pointed out by the triangle, which correlates with the V1 activity. Its molecular weight is about 60kD.

dehydrogenase that is being studied in a different context and whose retinoic-acid synthesizing capacity has not been recognized.

ENZYME DIVERSITY AND LOCAL RETINOIC ACID SYNTHESIS

In screens of different tissues of the embryonic and adult mouse with the zymography-bioassay, we can identify at least half a dozen retinoic-acid synthesizing dehydrogenases in addition to the class-1 enzymes. We also find two different oxidases, which, however, make only a minor contribution (McCaffery & Dräger, unpublished observations). Comparisons of endogenous retinoic-acid levels with levels of retinoic-acid synthesizing enzymes show a good correlation for all organs and tissues we have tested. This indicates that in contrast to the related endocrine steroid- and thyroid-hormone systems, where the ligands are supplied by the blood circulation, the retinoic acid system seems to depend on local synthesis of the ligand, i.e. it operates predominantly as a paracrine system (Duester et al., 1991). This paracrine nature renders the local expression and regulation of synthetic enzymes an important component in retinoic acid actions. This is a neglected aspect by comparison to the main focus of retinoic acid research, the characterization of retinoic acid receptors and binding proteins.

RETINOIC ACID ACTIONS

Retinoic acid is known to have very powerful effects in many different systems, ranging from duplication and transformation of embryonic structures, to anti-carcinogenic effects, to a rejuvenating action on the skin (Brockes, 1989; Chambon et al., 1991; de Thé et al., 1991; Tabin, 1991). It exerts its actions through binding to nuclear receptors, which turns these into transcription factors that regulate expression from specific binding sites in the promotor regions of many genes. Other members of the superfamily of ligand-activated transcription factors include the steroid- and thyroid-hormone and vitamin-D receptors. The diversity and specificity of retinoic acid actions is contributed to three features of the receptors: (1) there are different classes and subclasses of retinoic acid receptors; (2) the receptors can act as homodimers or as heterodimers between different classes and with other members of the nuclear receptor family (Glass et al., 1990; Kliewer et al., 1992; Leid et al., 1992); (3) different stereo-isomers of retinoic acid, such as 9-cis retinoic acid, have specific affinity for particular receptor types (Heyman et al., 1992; Levin et al., 1992).

The arrangement of two different dehydrogenases spatially segregated along the dorso-ventral axis of the embryonic retina is identical in species as far apart as mice, chick and zebrafish. The two enzymes differ in several characteristics, such as substrate affinity, catalytic efficiency and inhibitor susceptibility (McCaffery et al., 1992a). We found several inhibitors that selectively depress activity of one or the other enzyme, and even this selectivity is conserved (McCaffery et al., 1992b; Marsh-Armstrong et al., 1992). An explanation for the use of different enzymes might be tied to the mechanisms that generate the diversity and specificity of retinoic acid actions: the enzymes may serve specific functions in addition to retinoic acid synthesis, and/or they may differ in their catalysis of retinoid isomers. An example of such an additional function is known for AHD-2: a steroid-binding protein that is missing in people with testicular feminization has recently been identified as the human homolog of AHD-2 (Pereira et al., 1991). This property could create the conditions for the heterodimeric action between retinoic-acid receptors and different members of the nuclear receptor family. In addition, the steroid-binding capacity of the class-1 enzyme could play a role in the dorso-ventral segregation of the enzymes in the embryonic retina: it might represent a mechanism to suppress expression of the ventral enzymes in cells that contain AHD-2.

POSSIBLE ACTIONS OF THE ENZYMES IN THE EMBRYONIC EYE

The novel enzyme becomes first detectable in the optic pit region of the neural plate, and it might be a factor essential for the determination of the eye as an organ in the early embryo. Its product could direct expression of Pax-6 transcripts in the eye region. The Pax-6 gene is a member of a multigene family of transcription factors which were first identified through homology to

Drosophila segmentation genes and which are being implicated in pattern formation in the early embryo (Walther and Gruss, 1991). The expression of the Pax genes is known to be regulated by retinoic acid. AHD-2 is expressed a day later in the dorsal part of the eye vesicle, which might signal the determination of the dorso-ventral axis of the retina, an event possibly mediated through the expression of Pax-2 in ventral retina (Nornes et al., 1990). The final goal of this cascade of transcriptional regulation might be the graded expression of a cell-surface marker such as the TOP molecule which shows a dorso-ventral gradient in the embryonic retina (Trisler et al., 1981), and which may guide optic axons to their target location in the tectum (Trisler and Collins, 1987). The effects of the asymmetry in retinoic acid synthesizing enzymes is probably not restricted to the retina, as the spatial asymmetry might be propagated to the optic tectum via differences in retinoic acid levels in optic growth cones. Newly outgrowing optic axons from dorsal retinal regions, in particular their growth cones, are rich in AHD-2 (McCaffery et al., 1991) and axons emerging from central and later ventral retinal regions express cellular retinoic acid binding protein I (McCaffery et al., 1992b), a compound now linked to retinoic acid breakdown (Boylan and Gudas, 1991).

Acknowledgements

We thank Dr. Michael Wagner for the retinoic acid reporter cell line, and Dr. Ronald Lindahl and Dr. John Hilton for gifts of the aldehyde dehydrogenase antisera. This work was supported by NIH grant EY03819.

REFERENCES

Boylan, J.F., and Gudas, L.J., 1991, Overexpression of the cellular retinoic acid binding protein-I (CRABP-I) results in a reduction in differentiation-specific gene expression in F9 teratocarcinoma cells, *J. Cell Biol.* 112:965.

Brockes, J.P., 1989, Retinoids, homeobox genes, and limb morphogenesis, *Neuron* 2:1285.

Chambon, P., Zelent, A., Petkovich, M., Mendelsohn, C., Leroy, P., Krust, A., Kastner, P., and Brand, N., 1991, The family of retinoic acid nuclear receptors, *in:* Retinoids: 10 Years on. Edited by J.-H. Saurat, Karger, Basel.

de Thé, H., Lavau, C., Marchio, A., Chomienne, C., Degos, L., and Dejean, A., 1991, The PML-RARα fusion mRNA generated by the t(15;17) translocation in acute promyelocytic leukemia encodes a functionally altered RAR, *Cell* 66:675.

Duester, G., Shean, M.L., McBridge, M.S., and Steward, M.J., 1991, Retinoic acid response element in the human alcohol dehydrogenase gene ADH3: implications for regulation of retinoic acid synthesis, *Molec. Cell. Biol.* 11:1638-1646.

Dunn, T.J., Koleske, A.J., Lindahl, R., and Pitot, H.C., 1989, Phenobarbital-inducible aldehyde dehydrogenase in the rat. cDNA sequence and regulation of the mRNA by phenobarbital in responsive rats, *J. Biol. Chem.* 264:13057.

Glass, C.K., Devary, O.V., and Rosenfeld, M.G., 1990, Multiple cell type specific proteins differentially regulate target gene sequence recognition by the α retinoic acid receptor, *Cell* 63:729.

Heyman, R.A., Mangelsdorf, D.J., Dyck, J.A., Stein, R.B., Eichele, G., Evans, R.M., and Thaller, C., 1992, 9-Cis retinoic acid is a high affinity ligand for the retinoid X receptor, *Cell* 68:397.

Kliewer, S.A., Umesono, K., Mangelsdorf, D.J., and Evans, R.M., 1992, The retinoid X receptor interacts directly with nuclear receptors involved in retinoic acid, thyroid hormone, and vitamin D3 signaling, *Nature* 355:446.

Lee, M.-O., Manthey, C.L., and Sladek, N.E., 1991, Identification of mouse liver aldehyde dehydrogenases that catalyze the oxidation of retinaldehyde to retinoic acid, *Biochem. Pharmacol.* 42:1279.

Leid, M., Kastner, P., Lyons, R., Nakshatri, H., Saunders, M., Zacharewski, T., Chen, J.Y., Staub, A., Garnier, J.M., Mader, S., and Chambon, P., 1992, Purification, cloning and RXR identity of the HeLa cell factor with which RAR or TR heterodimerizes to bind target sequences efficiently, *Cell* 68:377.

Levin, A.A., Sturzenbecker, L.J., Kazmer, S., Bosakowski,T., Huselton, C., Allenby, G., Speck, J., Kratzeisen, C., Rosenberger, M., Lovey, A., and Grippo, J.F., 1992, 9-Cis retinoic acid tereoisomer binds and activates the nuclear receptor RXRα, *Nature* 355:359.

Lindahl, R., and Evces S., 1984a, Rat liver aldehyde dehydrogenase. I. Isolation and characterization of four high Km normal liver isozymes, *J. Biol. Chem.* 259:11986.

Lindahl, R., and Evces, S., 1984b, Rat liver aldehyde dehydrogenase. I. Isolation and characterization of four inducible isozymes, *J. Biol. Chem.* 259:11991.

Lindahl, R., and Hempel, J., 1990, *Aldehyde dehydrogenases: what can be learned from a baker's dozen sequences?* In:*Enzymology and Molecular Biology of Carbonyl Metabolism III*. Edited by H. Weiner, B. Wermuth and D. W. Crabb. Plenum Press, New York.

Manthey, C.L., Landkamer, G.J., and Sladek, N.E., 1990, Identification of the mouse aldehyde dehydrogenases important in aldophosphamide detoxification, *Cancer Res.* 50:4991.

Marsh-Armstrong, N.R., McCaffery, P., Dowling, J.E., Gilbert, W., and Dräger, U.C., 1992, Effects of inhibitors of retinoic acid synthesis in vertebrate embryos., 22th Ann. Mtg. Soc. Neurosci., abstracts.

McCaffery, P., Lee, M.-O., Wagner, M.A., Sladek, N.E., and Dräger, U.C., 1992a, Asymmetrical retinoic acid synthesis in the dorso-ventral axis of the retina, *Development* 115:371.

McCaffery, P., Posch, K.C., Napoli, J.L., Gudas, L., and Dräger, U.C, 1992b, Changing patterns of the retinoic acid system in the developing retina, *Submitted for publication.*

McCaffery, P., Tempst, P., Lara, G., and Dräger, U.C., 1991, Aldehyde dehydrogenase is a positional marker in the retina, *Development* 112:693.

Nornes, H.O., Dressler, G.R., Knapik, E.W., Deutsch, U., and Gruss, P., 1990, Spatially and temporally restricted expression of Pax2 during murine neurogenesis, *Development* 109:797.

Pereira, F., Rosenmann, E., Nylen, E., Kaufman, M., Pinsky, L., and Wrogemann, K., 1991, The 56kDa androgen binding protein is an aldehyde dehydrogenase, *Biochem. Biophys. Res. Commun.* 175:831.

Russo, J.E., and Hilton, J., 1988, Characterization of cytosolic aldehyde dehydrogenase from cyclophosphamide resistant L1210 cells, *Cancer Res.* 48:2963.

Sperry, R.W., 1963, Chemoaffinity in the orderly growth of nerve fiber patterns and connections, *Proc. Natl. Acad. Sci. USA* 50:703.

Tabin, C.J., 1991, Retinoids, homeoboxes, and growth factors: toward molecular models for limb development, *Cell* 66:199.

Trisler, D., and Collins, F., 1987, Corresponding spatial gradients of TOP molecules in the developing retina and optic tectum, *Science* 237:1208.

Trisler, G., Schneider, M.D., and Nirenberg, M., 1981, A topographic gradient of molecules in retina can be used to identify neuron position, *Proc. Natl. Acad. Sci. USA* 78:2145.

Vallari, R.C., and Pietruszko, R., 1982, Human aldehyde dehydrogenase: mechanism of inhibition by disulfiram, *Science* 216:637.

Wagner, M., Han, B., and Jessell, T.M., 1992, Regional differences in retinoid release from embryonic neural tissue detected by an in vitro reporter assay. *Development* 116: in press.

Walther, C., and Gruss, P., 1991, Pax-6, a murine paired box gene, is expressed in the developing CNS, *Development* 113:1435.

HUMAN LIVER HIGH KM ALDEHYDE DEHYDROGENASE (ALDH4): PROPERTIES AND STRUCTURAL RELATIONSHIP TO THE GLUTAMIC γ-SEMIALDEHYDE DEHYDROGENASE

Dharam P. Agarwal,[1] Rolf Eckey,[1] John Hempel,[2] and H. Werner Goedde [1]

[1]Institute of Human Genetics, University of Hamburg, Hamburg
[2]Department of Molecular Genetics and Biochemistry,
University of Pittsburgh School of Medicine, Pittsburgh

INTRODUCTION

In humans aldehyde dehydrogenase (ALDH) represents an important metabolic system for the detoxification of aliphatic and aromatic aldehydes. Moreover, ALDH exhibits genetic heterogeneity and the polymorphic forms seem to contribute to differences in acute and chronic outcome of excessive alcohol drinking. Human liver contains a number of ALDH enzyme forms (isozymes) which differ in their kinetic, electrophoretic and structural properties (Harada et al., 1980; Pietruszko et al., 1987; Agarwal et al., 1990). The hepatic ALDH1 (cytosolic) and ALDH2 (mitochondrial) isozymes have a very low Km for acetaldehyde, and are supposed to play a major role in the oxidation of toxic acetaldehyde produced from ethanol. In stomach a high Km isozyme (ALDH3) form is present which shows a very high Km for acetaldehyde and propionaldehyde but a relatively low Km for aromatic aldehydes such as 3-nitrobenzaldehyde (Eckey et al., 1991). A tumor-associated ALDH form, detected in a human hepatocarcinoma (Agarwal et al., 1989), shares many physico-chemical properties with the constitutive form of human stomach. In addition, a cathodically migrating mitochondrial ALDH enzyme, hitherto designated as E4 or ALDH4, has been characterized in human liver (Forte-McRobbie and Pietruszko, 1986).

On the basis of its substrate specificity, this enzyme has been identified as glutamic γ-semialdehyde dehydrogenase (GSDH, EC 1.5.1.12) or 1-pyrroline-5-carboxylate (P5C) dehydrogenase. ALDH4 or GSDH catalyzes the reduction of glutamate to an intermediate, glutamic γ-semialdehyde, which is spontaneously converted to the cyclic compound 1-pyrroline-5-carboxylate (P5C). The latter product is an important intermediate in the synthesis of proline and ornithine, and in the interconversion of glutamic acid and proline.

Although human GSDH has been purified and characterized with respect to substrate specificity and other physico-chemical as well as molecular properties, no primary structure data have been so far reported for the human liver ALDH4.

Enzymology and Molecular Biology of Carbonyl Metabolism 4
Edited by H. Weiner, Plenum Press, New York, 1993

In the present study, we have purified ALDH4 from autopsy human livers and have studied its kinetic and immunological properties. In addition, the amino acid sequences of some selected peptides from the human liver ALDH4 are compared with the known sequence of 1-pyrroline-5-carboxylate dehydrogenase from the yeast *Saccharomyces cerevisiae*.

METHODS

Tissue extracts were prepared by homogenizing the autopsy liver samples in 30 mM sodium phosphate buffer, pH 6.0 (1 ml buffer/g tissue), containing 1 mM EDTA + 1 mM dithioerythritol.

Enzyme Purification

The extract was centrifuged for 15 min at 27,000g and the supernatant dialyzed overnight against a 20 fold volume of extraction buffer. The supernatant was applied to a 5 x 45 cm column of DEAE-Sephadex A50 equilibrated with the extraction buffer. The eluate containing GSDH activity (substrate: glutamic γ-semialdehyde) was concentrated and subjected to gel filtration on a 5 x 70 cm column of Sephadex G100. Fractions showing GSDH activity were pooled and applied to a 2.5 x 15 cm column of 5'AMP-Sepharose 4B equilibrated with the extraction buffer. The bound enzyme was eluted with a 100 mM sodium phosphate buffer, pH 8.0 containing 1 mM EDTA, 1 mM dithioerythritol and 0.5 mg NADH/ml. The enzyme was further purified by preparative isoelectric focusing on agarose gel in a pH-range of 3.5-9.5 (Eckey et al., 1988). Remaining impurities were removed by chromatography on a 1 x 50 cm column of Red Sepharose CL-6B equilibrated with the extraction buffer. GSDH was eluted with a linear salt gradient (0-1M NaCl).

Preparation of Antobodies and Double Diffusion Test

Specific antibodies against human liver ALDH4 were raised in rabbits by immunizing the animals with the purified enzyme protein. The IgG fraction of the antisera were salted out and dissolved in distilled water. The purified antibodies were used in double diffusion test to study immunological cross-reaction of ALDH4 antibodies with liver homogenates of man, rat, cow and sheep.

Determination of Molecular Weight and Subunit Composition

Polyacrylamide gradient gel electrophoresis and SDS-polyacrylamide slab gel electrophoresis were carried out under standard conditions for the determination of native molecular weight and subunit composition.

Peptide and Amino Acid Sequence Analysis

The purified ALDH4 protein was S-(^{14}C) carboxymethylated and cleaved with trypsin as described for the human liver ALDH1 (Hempel et al., 1984). The tryptic digest was prefractionated by ion exchange HPLC on a 4 x 250 mm poly (2-sulfoethyl aspartamide)-silica column (Nest Group, Southboro, MA. USA) in 5 mM potassium phosphate/25% CH_3CN, pH 3.0 at 1 ml/min. A gradient of the same solvent containing 0.4 M KCl was applied over 1 h. The eluate was pooled according to UV absorbing peaks and selected pools were separately applied to a C-18 reversed phase HPLC column in 0.1% trifluoroacetic acid and eluted with a linear gradient of acetonitrile, with UV-monitoring at 214 and 280 nm. Aliquots from UV-absorbing peaks were taken for hydrolysis (6N HCl, 0.5% phenol, 110°C in vacuo, 24 h) and compositional analysis (Beckman 6300 amino acid analyzer) to judge peptide purity. Material from selected peaks was submitted to automated Edman degradation in either a Beckman 890M or Porton 2090E sequenator with identification of the derivatives by HPLC.

RESULTS

Table 1 summarizes the results of a typical purification protocol. The purified preparation was found to be homogeneous by SDS-electrophoresis, and by a lack of cross-reaction with antisera raised against other proteins.

Some of the major physico-chemical properties of human liver ALDH4 (GSDH) as compared to other human ALDH isozymes are summarized in Table 1.

Table 1. Purification of ALDH4 (GSDH) from human liver

Step	Total protein mg	Total activity mUnits	Specific activity mUnits/mg	Yield %	Purification -fold
Homogenate	2768	117,359	42.4*	-	-
CM-Sephadex	1776	733	0.41	100	1
5'-AMP Sepharose	59.4	71.4	1.2	9.7	2.9
Preperative IEF	3.7	63.4	17.22	8.6	42
Preparative Electrophoresis	0.33	9.5	28.85	1.3	70

*Yield and fold purification was calculated from the initial specific activity after CM-Sephadex chromatography.

Table 2. Comparison of low and high Km isozymes of ALDH

Designation	Organ	Substrate	Cosubstrate	Km	EC Number
Low Km ALDHs					
ALDH1 (E1)	Liver	Acetaldehyde	NAD	<100 M	1.2.1.3
ALDH2 (E2)	Liver	Acetaldehyde	NAD	< 20 M	1.2.1.3
High Km ALDHs					
ALDH3 (?)	Stomach	Acetaldehyde	NAD	> 10 mM	1.2.1.3
		Benzaldehyde	NADP	> 53 M	
ALDH4 (E4)	Liver	Acetaldehyde	NAD	> 5 mM	1.5.1.12
(GSDH)		Glutamic γ-semialdehyde	NADP	>200 μM	

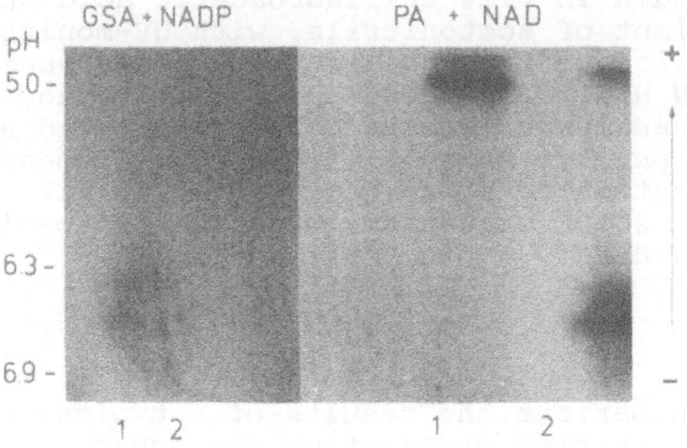

Fig. 1. Staining intensity of ALDH isozymes with various aldehyde substrates and coenzymes after separation by isoelectric focusing on agarose gel. L: liver extract; S: stomach extract. Substrates and coenzymes: A, propionaldehyde and NAD; B: 3-nitrobenzaldehyde and NADP; C: glutamic γ-semialdehyde and NADP.

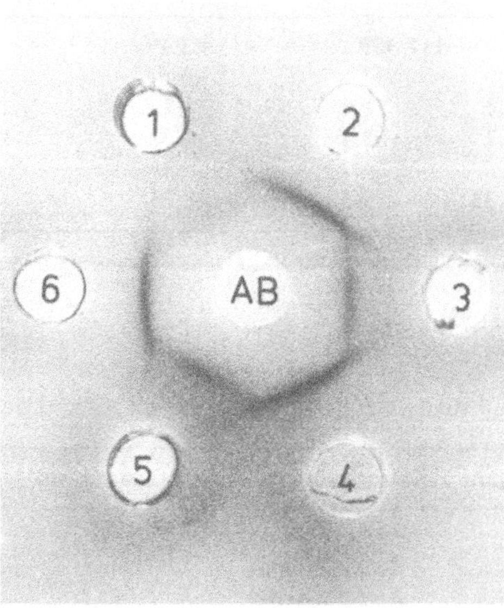

Fig. 2. Immunological cross-reaction of antibodies to human ALDH4 isozyme with liver homogenates from different animal species. Antibody to ALDH4 was placed in the central well and liver homogenates were placed in the outer wells; 2, 4, and 6: human liver homogenate; 1, 3, and 5: liver homogenates from rat, cow, and sheep, respectively.

In contrast to other ALDH isozymes, ALDH4 (GSDH) shows a significantly higher affinity for glutamic γ-semialdehyde than for propionaldehyde.

In double diffusion tests using liver and stomach extracts of horse, pig, cow, sheep and rat, the liver extracts of these animals reacted only with antibodies to human ALDH2 and ALDH4 (Fig. 2).

The purified ALDH4 migrated as a single band in SDS-polyacrylamide electrophoresis. The apparent subunit molecular weight of the subunits was about 60 KD as judged on the basis of calibration kit standards (Fig. 3).

Peptide Structure

Material from 11 HPLC peaks was analysed, yielding 11 sequences of six or more residues covering over 120 residues. Of these, nine sequences were placed unambiguously versus the sequence of the yeast P5C dehydrogenase with matching residues (Kryzwicki and Brandriss, 1984), ranging from 33 to 66% (Fig. 4).

DISCUSSION

A structure has been known for an enzyme from yeast (Kryzwicki and Brandriss, 1984), which catalyzes the same reaction as the present GSDH or ALDH4 from human liver. The ability to readily place several peptides vs. the yeast P5C dehydrogenase deduced sequence suggests a common ancestor for the enzymes from these two sources. The range in percent matching residues, and the presence of three sequences not yet alignable with the yeast enzyme suggests that some regions are substantially more conserved than others. Some of these sequences are being used as the basis for design of oligo-nucleotide probes for PCR amplification of the GSDH (ALDH4) gene. We expect that when the entire sequences becomes known, the three sequences for which no yeast enzyme counterpart was found will be incorporated into the entire sequence.

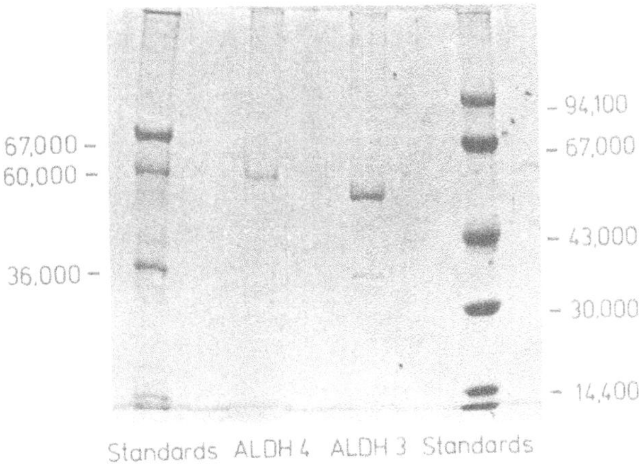

Fig. 3. SDS-polyacrylamide electrophoresis of purified ALDH3 and ALDH4.

Peptide	Sequence	% Identity
E-23	Tyr Ala Val Glu Leu Glu Gly Gln Gln Pro Ile Ser Val Leu Xxx Tyr	
176-191	Tyr Ala Ser Asp Leu Tyr Ala Gln Gln Pro Val Ser Arg Ala Asp Gly	44
F-19	Phe Cys Tyr Ala Asp Lys	
204-209	Phe Val Tyr Ala Val Ser	50
G-17	Ser Ala Phe Glu Tyr Gly Gly Gln Lys	
342-350	Gly Thr Phe Glu Phe Gln Gly Gln Lys	55
G-19	Ala Ala Asp Met Leu Ser Gly Pro Arg	
131-139	Ala Ala Asp Leu Ile Ser Thr Lys Tyr	45
G-23	Ala Ile Glu Ala Ala Leu Ala Ala Arg	
102-110	Val Met Asn Ala Val Lys Ala Ala Lys	33
G-26	Ser Thr Gly Ser Ile Val Gly Gln Gln Pro Phe Gly Gly Ala Gly Ser Ala	
521-537	Cys Thr Gly Ala Val Val Ser Gln Gln Pro Trp Gly Gly Ala Pro Met Ser	53
G-27	Ala Gln Ile Phe Leu Lys	
125-130	Ser Ala Ile Phe Leu Lys	66
G-48	Glu Ala Gly Leu Pro Pro Asn Ile Ile Gln Phe Val Pro Ala Asp ..	
252-274	Glu Ala Gly Leu Pro Lys Gly Val Ile Asn Phe Ile Leu Gly Asp ..	
	Gly Pro Leu Phe Gly Glu Xxx Val	
	Pro Val Gln Val Thr Asp Gln Val	39
J-32	Lys Tyr Val Pro His Ser Leu Trp Pro Gln Ile Lys	
357-368	Lys Tyr Lys Pro Glu Ser Lys Ser Glu Glu Phe Leu	33

Fig. 4. Comparison of amino acid sequences of tryptic peptides from human liver glutamic γ-semialdehyde dehydrogenase (this work) and yeast 1-pyrroline-5-carboxylate dehydrogenase (Kryzwicki and Brandriss, 1984). The upper sequence from each pair is the experimentally determined sequence of a tryptic peptide from the human enzyme, and is designated according to chromatographic isolation. The lower sequences are segments from the yeast enzyme, designated according to their position number in the complete structure. Identical residues are boxed.

REFERENCES

Agarwal, D.P., Eckey, R., Rudnay, A.C., Volkens, T., and Goedde, H.W., 1989, High Km aldehyde dehydrogenase isozymes in human tissues: constitutive and tumor-associated forms, *in*: Progress in Clinical and Biological Research, Vol. 290: "Enzymology and Molecular Biology of Carbonyl Metabolism 2", H. Weiner and T.G. Flynn, eds., Alan R. Liss, New York, p. 119.

Harada, S., Agarwal, D.P., and Goedde, H.W., 1980, Electrophoretic and biochemical studies of human aldehyde dehydrogenase isozymes in various tissues. *Life Sci.* 26:1771.

Eckey, R., Agarwal, D.P., Hempel, J., Timmann, R., and Goedde, H.W., 1991, Biochemical, immunological, and molecular characterization of a "high Km" aldehyde dehydrogenase, *in*: Advances in Experimental Medicine and Biology, Vol. 284: "Enzymology and Molecular Biology of Carbonyl Metabolism 3", H. Weiner, B. Wermuth and D.W. Crabb, eds., Plenum Press, New York, p. 43.

Forte-McRobbie, C.M. and Pietruszko, R., 1986, Purification and characterization of human liver high Km aldehyde dehydrogenase and its identification as glutamic γ-semialdehyde dehydrogenase. *J. Biol. Chem.* 261:2154.

Krzywicki, K.A. and Brandriss, M.C., 1984, Primary structure of the nuclear PUT2 gene involved in the mitochondrial pathway for proline utilization in *Saccharomyces cerevisiae*. *Mol. Cell. Biol.* 4:2837.

Eckey, R., Agarwal, D.P., Volkens, T., and Goedde, H.W., 1988, A simple and rapid method for the purification of aldehyde dehydrogenase isozymes from human liver, *in*: "Alcohol Toxicity and Free Radicals Mechanisms", R. Nordmann, C. Ribiere and H. Rouach, eds., Pergamon Press, New York, p. 171.

Hempel, J., von Bahr-Lindstrom, H., and Jornvall, H., 1984, Aldehyde dehydrogenase from human liver. Primary structure of the cytoplasmic isoenzyme. *Eur. J. Biochem.* 141:21.

Pietruszko, R., Ryzlak, M.T., and Forte-McRobbie, C.M., 1987, Multiplicity and identity of human aldehyde dehydrogenases. *Alcohol Alcoholism*, Suppl.1:175.

EFFECT OF SOME COMPOUNDS RELATED TO DISULFIRAM ON MITOCHONDRIAL ALDEHYDE DEHYDROGENASE *IN VITRO* AND *IN VIVO*

Trevor M. Kitson, Kathryn E. Kitson, and Leong Goh

Department of Chemistry and Biochemistry
Massey University
Palmerston North
New Zealand

INTRODUCTION

Since 1948, disulfiram (Et_2NCS-SS-$CSNEt_2$) has been widely used in the treatment of alcoholism. This is because it lowers the activity of aldehyde dehydrogenase (AldDH) *in vivo* such that if alcohol is drunk, a build-up of acetaldehyde ensues, resulting in an unpleasant reaction (Kitson, 1977).

It has been well established that cytoplasmic AldDH undergoes the following reaction with disulfiram *in vitro* (Vallari and Pietruszko, 1982; Kitson, 1983). First there is a very rapid reaction between the modifier and a particular cysteine residue, probably Cys-302 (Hempel *et al.*, 1985), leading to inactivation. Other modification studies have shown that Cys-302 is probably the essential nucleophile in the enzyme's catalytic mechanism (Kitson *et al.*, 1991). Secondly, the diethyldithiocarbamyl label is displaced from Cys-302 by a neighbouring but so far unidentified cysteine leading to the generation of a cystine bridge and the liberation of a second equivalent of diethyldithiocarbamate.

Whether or not the chemistry just described is relevant to the physiological situation, however, is a moot point. Upon absorption into the bloodstream, disulfiram is reduced, essentially quantitatively, to the diethyldithiocarbamate ion (Cobby *et al.*, 1977) which then undergoes further metabolism. Disulfiram itself is virtually undetectable *in vivo*. These observations have led some workers to propose various metabolites of disulfiram as the active modifier of AldDH *in vivo*, including S-methyl N,N-diethyldithiocarbamate ($Et_2NCSSMe$) (Yourick and Faiman, 1989) and S-methyl N,N-diethylmonothiocarbamate ($Et_2NCOSMe$) (Johansson *et al.*, 1989). In a previous study (Kitson, 1991) it was shown that these compounds have little effect except at very high concentration on sheep liver cytoplasmic AldDH *in vitro*. However, before they can be completely discounted, it was

Enzymology and Molecular Biology of Carbonyl Metabolism 4
Edited by H. Weiner, Plenum Press, New York, 1993

felt important to examine their effect on the mitochondrial enzyme, particularly in view of the reported importance of mitochondrial AldDH in the metabolism of acetaldehyde (Parilla *et al.*, 1974), and this was the main aim of the present work. We have also investigated the effect of dioxiram on mitochondrial AldDH *in vitro* and *in vivo*. Dioxiram is the name given to the analogue (and conceivable metabolite) of disulfiram of formula $Et_2NCO\text{-}SS\text{-}CONEt_2$, a compound that interestingly is far more water-soluble than disulfiram.

EXPERIMENTAL

Mitochondrial AldDH was isolated from sheep liver as described by Hart and Dickinson (1977) and purified from the cytoplasmic isoenzyme by pH-gradient ion-exchange chromatography (Dickinson *et al.*, 1981).

S-Methyl N,N-diethyldithiocarbamate, S-methyl N,N-diethylmonothiocarbamate and dioxiram were synthesised as previously described (Kitson, 1991).

The activity of AldDH in 50 mM sodium phosphate buffer, pH 7.4, was measured using an Aminco DW-2a spectrophotometer. Unless otherwise stated assays were carried out at 25 °C in a final volume of 3 ml using 1 mM NAD^+ and 1 mM acetaldehyde, or 100 µM p-nitrophenyl acetate. Modifiers were added as 15 µl of a solution in ethanol; the same volume of ethanol was added to control assays. In all cases, at least two or three identical assays were performed and the results (which agreed closely) were averaged to give the points in the Figures.

For the experiments *in vivo*, female Sprague-Dawley rats weighing 180-220 g were used and were starved overnight before drug administration. Prior to that they had access to standard laboratory rat food and water and were kept in a 12 hour light/dark cycle at 25 °C. Disulfiram or dioxiram was administered by intubation into the stomach as a suspension in 5% gum arabicum in water; the suspension was sonicated to ensure complete homogeneity. Control rats received 5% gum arabicum only. The concentration of the suspension was calculated such that each animal received a volume of about 4 ml with a drug dosage of either 200 or 500 mg per kg body weight. After various times the rats were killed by cervical dislocation. The liver was homogenised in 10 volumes of ice-cold 10 mM sodium phosphate buffer, pH 7.4, containing 0.25 M sucrose. After a preliminary low-speed spin (500 g for 5 min), mitochondria were sedimented at 20,000 g for 30 min and then solubilised using 0.25% sodium deoxycholate. Samples were assayed for low K_M-AldDH activity as described by Tottmar and Marchner (1976). Protein concentrations were determined by the method of Bradford (1976).

RESULTS

The preparation of mitochondrial AldDH used here retained 98.5% of its activity when assayed (over a short period of time) in the presence of 10 µM disulfiram, confirming that it was essentially free from contamination with the disulfiram-sensitive cytoplasmic enzyme. In the presence of a high concentration of $Et_2NCSSMe$ (1 mM) the rate was 101% of the control and this was not significantly different if there was an incubation period of 15 min before addition of acetaldehyde.

The compound $Et_2NCOSMe$ was found to be an inhibitor of sheep liver mitochondrial AldDH. Fig. 1 shows the effect of adding $Et_2NCOSMe$ to an on-going assay. Typically under these conditions it takes a minute or two for the inhibition to become maximal, but this is followed by an upwardly curving rate of enzyme-catalysed reaction which was found to be characteristic of all assays carried out in the presence of $Et_2NCOSMe$. The Figure also shows that in the absence of acetaldehyde, the modifier gives no increase in A_{340} whatsoever, showing that its effects on the enzyme reaction are not due to traces of aldehydes as impurities in the $Et_2NCOSMe$ preparation (which anyway was triply distilled).

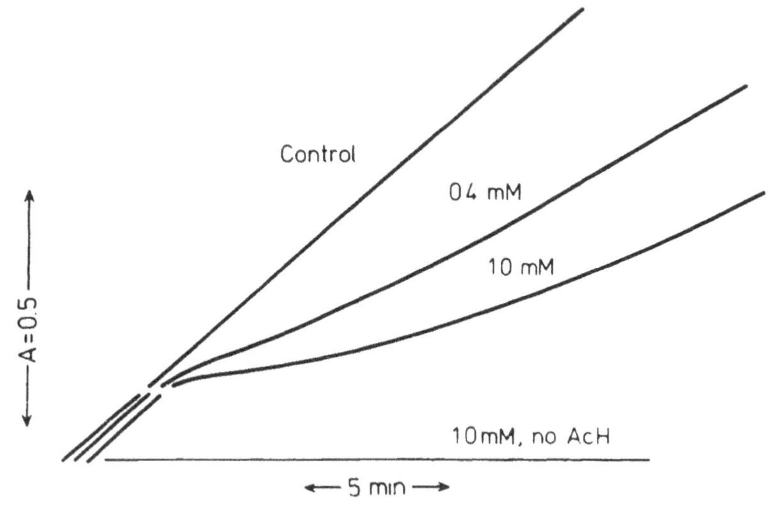

Figure 1. $Et_2NCOSMe$ (0.4 or 1.0 mM) was added to an on-going assay of mitochondrial AldDH (1.1 μM) at the break in the trace. Also shown is the lack of increase in A_{340} when enzyme is incubated with NAD^+ and $Et_2NCOSMe$ (1 mM) only (no AcH).

In Fig. 2 we see the effect of adding modifier to enzyme in the presence of NAD^+, before acetaldehyde is added last to initiate the assay. The assays were monitored for several minutes and tangents drawn after various times to obtain the data shown in the Figure. This again illustrates how the initial level of inhibition tends to be overcome as the enzyme turns over. For pronounced loss of activity, quite a high concentration of $Et_2NCOSMe$ is necessary. When the modifier was added to enzyme in the absence of NAD^+, with NAD^+ and acetaldehyde added subsequently, the results were similar (Fig. 3) except that somewhat lower concentrations of $Et_2NCOSMe$ were required for a given level of inhibition.

The enzyme was found to be more sensitive to $Et_2NCOSMe$ when assayed

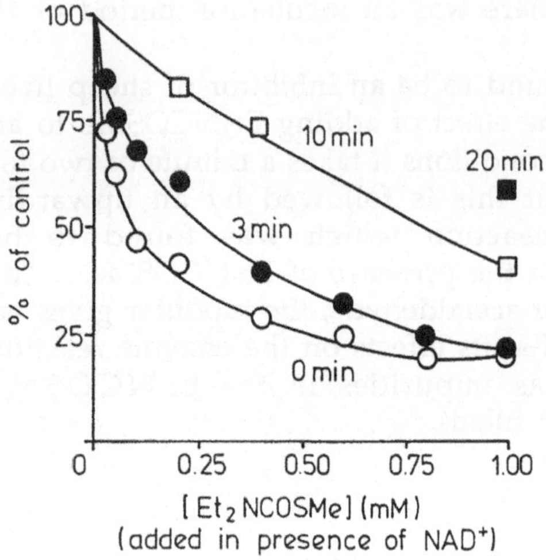

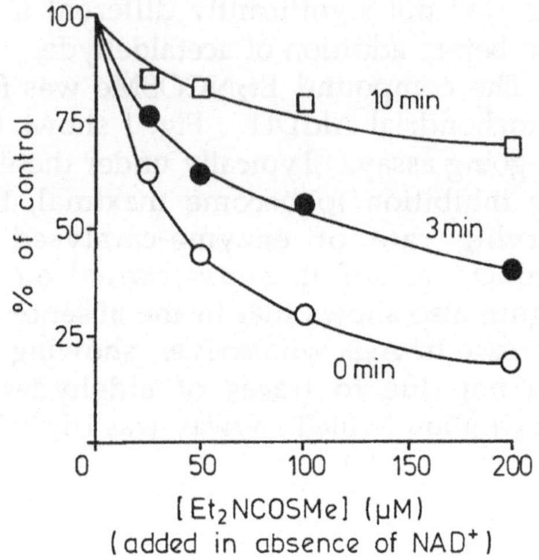

Figure 2. (LEFT) $Et_2NCOSMe$ **was added to mitochondrial AldDH (1.1 μM) in the presence of** NAD^+ **at 25 °C and then acetaldehyde (1 mM) was added. The rate of the enzyme-catalysed reaction was then measured at various times as described under Experimental and expressed as a percentage of the corresponding control value to give the data in the Figure.**

Figure 3. (RIGHT) $Et_2NCOSMe$ **was added to mitochondrial AldDH (1.1 μM) at 25 °C and then** NAD^+ **(1 mM) and acetaldehyde (1 mM) were added. The data are expressed as for Figure 2.**

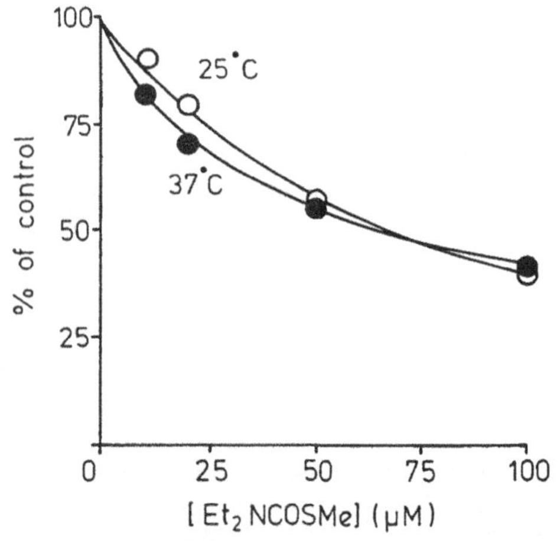

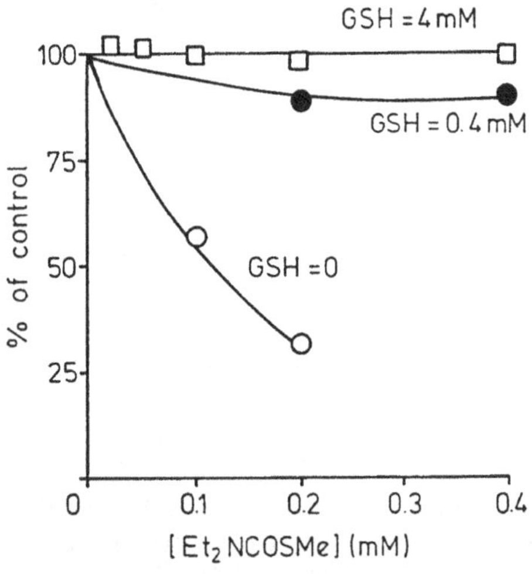

Figure 4. (LEFT) $Et_2NCOSMe$ **was added to mitochondrial AldDH (1.1 μM at 25 °C or 0.31 μM at 37 °C) in the presence of** NAD^+ **and then acetaldehyde (20 μM) was added. The maximal rate before the substrate ran out is expressed as a percentage of the corresponding control rate.**

Figure 5. (RIGHT) $Et_2NCOSMe$ **was added to mitochondrial AldDH (0.44 μM) at 37 °C in the presence of** NAD^+ **and the concentration of glutathione indicated and then acetaldehyde (20 μM) was added. The maximal rate before the substrate ran out is expressed as a percentage of the corresponding control rate.**

using the more physiologically reasonable concentration of acetaldehyde of 20 μM rather than 1 mM (Fig. 4). Again the assays curved upwards, but then fairly soon reached the point where the substrate was all used up. The data in the Figure were obtained by measuring the maximum rate that the enzyme-catalysed reaction had attained before substrate depletion became evident. The Figure also shows that the enzyme is marginally more sensitive still to $Et_2NCOSMe$ if assayed at body temperature rather than 25 °C.

Using conditions in which $Et_2NCOSMe$ achieves most inhibition (37 °C, 20 μM acetaldehyde) we looked at the effect of glutathione. Fig. 5 shows that glutathione provides very effective protection of the enzyme against $Et_2NCOSMe$. (The concentration of glutathione in liver is reported to be about 4 mM; Boyland and Chasseaud, 1969).

The data presented in Table 1 show that removal of $Et_2NCOSMe$ from the enzyme by prolonged dialysis does not remove the inhibitory effect and that the upward curvature of the modifier-treated assays is still present. The rate of the $Et_2NCOSMe$-treated samples was 48% of the control at the start of the assay and 69% after 5 min. The corresponding figures obtained without any dialysis were very similar, 42% and 61%.

Table 1. Lack of reversibility of effect of $Et_2NCOSMe$[1]

Sample		Specific activity (arbitrary units)	
		t=0	t=5 min
Modifier treated	#1	0.511	0.759
	#2	0.461	0.717
	#3	0.504	0.735
Control	#1	1.04	1.06
	#2	1.04	1.09
	#3	1.01	1.07

[1]Three different samples of mitochondrial AldDH (13 μM) were treated with $Et_2NCOSMe$ (1.2 mM) in a volume of 1 ml of 50 mM sodium phosphate buffer, pH 7.4, and then dialysed overnight in the cold against two changes of 1 l of the same buffer. Samples were then taken for assay in the usual way and for protein determination; the initial specific rate is recorded along with the rate after the assay had proceeded for 5 min.

Fig. 6 shows the effect of $Et_2NCOSMe$ and $Et_2NCSSMe$ on the esterase activity of mitochondrial AldDH using p-nitrophenyl acetate as substrate. All esterase rates, including controls, were found to curve downwards; the data in the Figure represent the initial rates of the assays.

The effect of disulfiram and dioxiram on mitochondrial AldDH *in vitro* is shown in Table 2. The modifiers were added to enzyme in the presence of NAD^+ and assays were initiated by the addition of acetaldehyde generally after half an hour. Disulfiram and dioxiram are approximately equally effective inactivators under these conditions.

When rats were dosed with disulfiram or dioxiram and subsequently their liver mitochondrial low-K_M AldDH was assayed, the results in Table 3 were obtained. The absolute value of the control specific activity was 20-30 nmol/min/mg of protein, in excellent agreement with the results of Tottmar

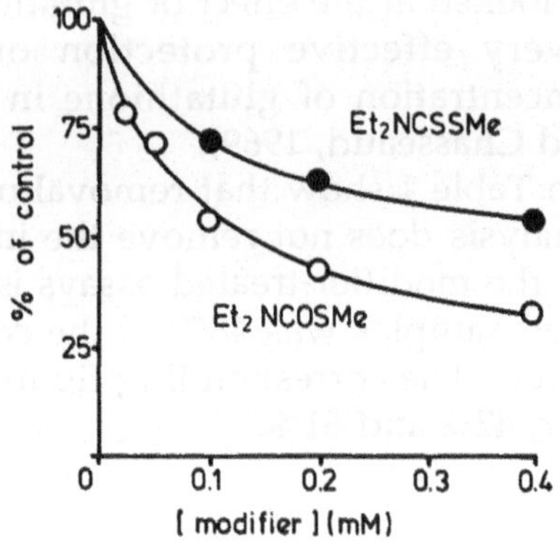

Figure 6. $Et_2NCOSMe$ or $Et_2NCSSMe$ was added to mitochondrial AldDH (0.44 μM) at 25 °C and then p-nitrophenyl acetate (0.1 mM) was added. The initial rate is expressed as a percentage of the corresponding control rate.

Table 2. Effect of disulfiram and dioxiram on sheep liver mitochondrial AldDH *in vitro*.[1]

Modifier	Conc. (μM)	Time (min)	Temperature (°C)	Rate (% of control)
Disulfiram	10	30	25	83
	50	30	25	54
	10	30	37	45
Dioxiram	50	~1	25	94
	10	30	25	86
	50	30	25	48
	10	30	37	52

[1]Mitochondrial AldDH (1.1 μM) was mixed with modifier in 50 mM sodium phosphate buffer, pH 7.4, in the presence of NAD^+ (1 mM), and after the time indicated, acetaldehyde (1 mM) was added to initiate the assay.

Table 3. Effect of disulfiram and dioxiram on rat liver mitochondrial AldDH *in vivo*.[1]

Compound	Dosage (mg/kg)	No of rats	Time before death (h)	Enzyme activity (% of control)
Dioxiram	200	5	2	84
	200	4	12	106
	200	4	24	71
	500	4	24	54
Disulfiram	200	6	24	30

[1]Drugs were administered as described under Experimental. After the time indicated the animals were killed and the activity of crude liver mitochondrial low K_M-AldDH was measured by the method of Tottmar and Marchner (1976).

and Marchner (1976), as was our figure for the extent of inactivation produced by disulfiram. Clearly however, dioxiram only causes pronounced loss of enzyme activity when administered in the very high dosage of 500 mg/kg body weight.

DISCUSSION

A major goal of many workers interested in the broad area of alcohol metabolism is to understand the precise chemical details of how AldDH becomes inactivated by disulfiram *in vivo*. Armed with this knowledge, more effective 'anti-alcohol drugs' may possibly be designed. If it is a metabolite, rather than disulfiram *per se,* that is the active species *in vivo,* then it must be possible to demonstrate that the species in question can affect the enzyme in a simple assay system *in vitro* and at a physiologically reasonable concentration. The first metabolite of disulfiram, the diethyldithiocarbamate ion, clearly does not fulfil this criterion; although administration of diethyldithiocarbamate provokes an anti-alcohol reaction (Deitrich and Erwin, 1971), it has no effect on AldDH *in vitro* (Kitson, 1975). The next metabolite, the methylated derivative of diethyldithiocarbamate, $Et_2NCSSMe$, can also be completely discounted, as shown by the work discussed above and previously (Kitson, 1991). The oxygen-containing further metabolite of disulfiram, $Et_2NCOSMe$, cannot be so easily dismissed, however. As shown in the Results section, this compound is a moderately effective inhibitor of sheep liver mitochondrial AldDH *in vitro* . The inhibition appears to be irreversible, in the sense that it is not abolished simply by dialysing away the modifier (Table 1), but on the other hand it is not constant, in the sense that the level of inhibition gets progressively less as the enzyme activity is being monitored (Figs. 1, 2 and 3). It appears that $Et_2NCOSMe$ chemically modifies AldDH to give an enzyme form with the characteristic of upwardly curving assays, but the explanation of this is not known. The modifier is more effective when the enzyme is assayed at 37 ^oC than at 25 ^oC and using 20 µM rather than 1 mM acetaldehyde. However, even under optimum conditions for reduction of enzyme activity, a concentration of $Et_2NCOSMe$ of more than 50 µM is required for 50%

inhibition. The plasma concentration of $Et_2NCOSMe$ detected in humans undergoing disulfiram treatment is only about 1 μM or less (Johansson and Stankiewicz, 1989). Furthermore, glutathione at high (but physiologically not unreasonable) concentrations protects mitochondrial AldDH from the effect of $Et_2NCOSMe$. Thus we are led to the conclusion that this compound cannot be the active species involved in the action of disulfiram *in vivo* and that the suggestions to this effect (Johansson et al., 1989; Petersen, 1989) are mistaken.

Unlike $Et_2NCSSMe$ and $Et_2NCOSMe$, the compound dioxiram (Et_2NCO-SS-$CONEt_2$) has not been detected as a metabolite of disulfiram, though it is conceivable it could arise from demethylation of $Et_2NCOSMe$ followed by oxidation of Et_2NCOS^-. Dioxiram was previously shown to be a very potent inactivator of cytoplasmic AldDH *in vitro,* though somewhat slower to react than disulfiram (Kitson, 1991). In the present work, we found that dioxiram was equally effective as disulfiram as a modifier of the mitochondrial enzyme (see Table 2). The mitochondrial form of sheep, horse or human AldDH is generally thought of as 'disulfiram-insensitive'. However, this is only relative to the highly susceptible cytoplasmic form. In Table 2 we see that in time a substantial inactivation of mitochondrial AldDH is caused by quite a low concentration of either disulfiram or dioxiram. Thus even mitochondrial AldDH is considerably more sensitive to disulfiram as such than it is to its metabolite, $Et_2NCOSMe$.

Since dioxiram is 500-fold more water-soluble than disulfiram, we had hopes that it might be a useful anti-alcohol drug. Perhaps it would be more readily absorbed from the gut enabling the use of lower dosages; perhaps it could be administered as a sub-cutaneous implant. The latter approach has been tried with disulfiram but is ineffective because of the compound's insolubility (Kitson, 1978). Our results, however, were disappointing (see Table 3). Dioxiram is considerably poorer than disulfiram as an inactivator of rat liver low K_M-AldDH *in vivo,* notwithstanding the fact that it is equally efficacious with the sheep liver mitochondrial enzyme *in vitro.* Presumably some unknown factor in the absorption, distribution, metabolism or elimination of dioxiram renders it less effective physiologically than disulfiram.

Returning to the original question of what is the active species that modifies the activity of AldDH *in vivo,* we are left with two alternatives. Either it is disulfiram itself or it is some metabolite of disulfiram other than those already discussed and discounted. Of course, if upon absorption, disulfiram were rapidly, totally and irreversibly reduced to diethyldithiocarbamate, then it absolutely could not react with AldDH *in vivo,* and the extraordinarily fast, specific and stoicheiometric reaction of disulfiram with cytoplasmic AldDH *in vitro* (Kitson, 1982) would be nothing more than a laboratory curiosity. This seems to us hard to believe. There are pathways for re-oxidation of diethyldithiocarbamate to disulfiram, and as argued before, perhaps only a very low concentration of the latter compound would be necessary to bring about loss of enzyme activity (Kitson, 1983). It could be argued that since acetaldehyde metabolism occurs in the mitochondria of rat liver (Parilla et al., 1974), the highly disulfiram-sensitive cytoplasmic enzyme is irrelevant. However, it is not firmly established for the human whether it is cytoplasmic or mitochondrial AldDH (or maybe a combination of both) that is important in the oxidation of ethanol-derived acetaldehyde (Bosron and Li, 1986). Recently, for example, it has become evident that some alcohol-intolerant individuals have a mutated form of cytoplasmic AldDH

(Yoshida *et al.*, 1989). (It has long been known that the Oriental flushing-response is caused by a malfunctional mutation of the mitochondrial enzyme; Goedde *et al.*, 1985.) Even if acetaldehyde metabolism is predominantly mitochondrial, then as discussed above, mitochondrial AldDH is by no means completely insensitive to low levels of disulfiram. Furthermore, the mitochondrial enzyme is extremely sensitive to certain mixed disulphides (Et_2NCS-SS-R), as originally shown by Kitson (1975) and further elaborated by MacKerell *et al.* (1985), and it may be that such mixed disulphides are produced *in vivo*.

Recently, in a very interesting paper, Hart and Faiman (1992) have shown that S-methyl N,N-diethylthiolcarbamate sulphoxide ($Et_2NCOSOMe$) is a potent inactivator of rat liver mitochondrial low K_M AldDH *in vitro* and *in vivo* and is a metabolite of disulfiram. Previously, Casida *et al.* (1974) showed that thiocarbamates very similar to $Et_2NCOSMe$ are converted to the sulphoxides by the action of the microsomal-NADPH system, and then rapidly metabolised by reaction with glutathione in the presence of liver soluble enzymes. Hart and Faiman's studies *in vitro* involved an incubation of the sulphoxide with intact mitochondria for 1 hour at 37 °C and thus it is not firmly established whether the sulphoxide itself or some further metabolite or breakdown product actually reacts with AldDH under the conditions of the experiment. The sulphoxide group by itself is certainly insufficient to explain the effect of $Et_2NCOSOMe$ on AldDH; enzyme assays can be carried out in 5% dimethyl sulphoxide without any effect on the rate (unpublished results). Possibly $Et_2NCOSOMe$ reacts with AldDH as follows:

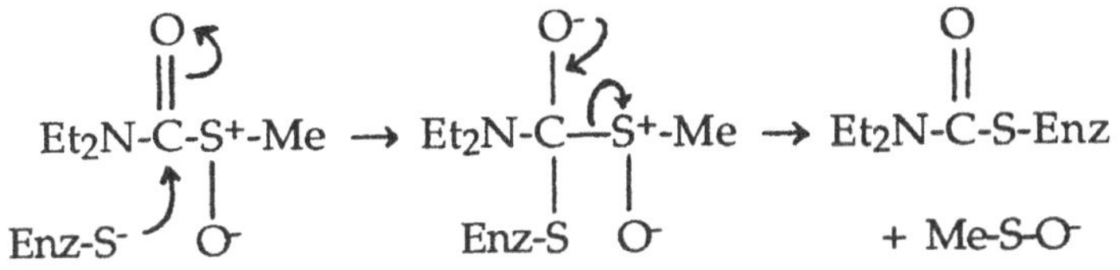

A similar thiocarbamate sulphoxide reacts in this way when heated with ethanethiol (Gozzo *et al.*, 1975). However, whether the methanesulphenate ion should be any better as a leaving group than the methanethiolate ion is debatable (and as we have seen $Et_2NCOSMe$ is not itself a very potent inactivator of AldDH). The compound *p*-nitrophenyl dimethylcarbamate inactivates AldDH by a similar mechanism, but the reaction is extremely slow (Kitson, 1989) even though *p*-nitrophenoxide is a fairly good leaving group. As an alternative, we speculate that the sulphoxide may possibly rearrange as follows:

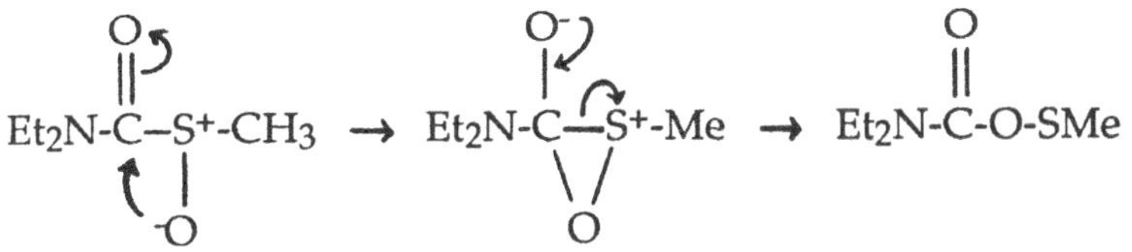

This would result in a compound expected to react rapidly with AldDH, donating an -SMe group to the enzyme's essential thiol group, with expulsion of the $Et_2NCO_2^-$ ion; this is essentially the same chemistry as previously studied with Et_2NCS-SSMe (MacKerell *et al.*, 1985; Kitson and Loomes, 1985). It should be easy to establish which of the two possibilities (if either) is correct; the second, but not the first, mechanism results in a modified form of the enzyme (Enz-S-S-Me) that could be reactivated by β-mercaptoethanol.

In conclusion, although the last few years have seen a vigorously growing interest in the chemistry behind disulfiram therapy, it is clear that as yet we still do not know exactly how AldDH becomes inactivated in the liver.

REFERENCES

Bosron, W.F., and Li, T.K., 1986, Genetic polymorphism of human liver alcohol and aldehyde dehydrogenases and their relationship to alcohol metabolism and alcoholism, *Hepatology*, 6:502.

Boyland, E., and Chasseaud, L.F., 1969, The role of glutathione and glutathione S-transferases in mercapturic acid biosynthesis, *Adv. Enzymol. Relat. Areas Mol. Biol.*, 32:173.

Bradford, M.M., 1976, A rapid and sensitive method for the quantitation of microgram quantities of protein utilizing the principle of protein-dye binding, *Anal. Biochem.*, 72:248.

Casida, J.E., Gray, R.A., and Tilles, H., 1974, Thiocarbamate sulphoxides: potent, selective, and biodegradable herbicides, *Science*, 184:573.

Cobby J., Mayersohn, M., and Selliah, S., 1977, The rapid reduction of disulfiram in blood and plasma, *J. Pharmacol. Exp. Ther.*, 202:724.

Deitrich, R.A., and Erwin, V.G., 1971, Mechanism of the inhibition of aldehyde dehydrogenase in vivo by disulfiram and diethyldithiocarbamate, *Mol. Pharmacol.*, 7:301.

Dickinson, F.M., Hart, G.J., and Kitson, T.M., 1981, The use of pH-gradient ion-exchange chromatography to separate sheep liver cytoplasmic aldehyde dehydrogenase from mitochondrial enzyme contamination, and observations on the interaction between the pure cytoplasmic enzyme and disulfiram, *Biochem. J.*, 199:573.

Goedde, H.W., Agarwal, D.P., Eckey, R., and Harada, S., 1985, Population genetic and family studies on aldehyde dehydrogenase deficiency and alcohol sensitivity, *Alcohol*, 2:383.

Gozzo, F., Masoero, M., Santi, R., Galluzzi, G., and Barton, D.H.R., 1975, On the thermal and chemical stability of carbamoyl sulphoxides, *Chem. Ind. (London)* , 221

Hart, G.J., and Dickinson, F.M., 1977, Some properties of aldehyde dehydrogenase from sheep liver mitochondria, *Biochem.J.*, 163:261.

Hempel, J., Kaiser, R., and Jornvall, H., 1985, Mitochondrial aldehyde dehydrogenase from human liver; primary structure, differences in relation to the cytosolic enzyme, and functional correlations, *Eur. J. Biochem.*, 153:13.

Johansson, B., Petersen, E.N., and Arnold, E., 1989, Diethylthiocarbamic acid methyl ester; a potent inhibitor of aldehyde dehydrogenase found in rats treated with disulfiram or diethyldithiocarbamic acid methyl ester, *Biochem. Pharmacol.*, 38:1053.

Johansson, B., and Stankiewicz, Z., 1989, Inhibition of erythrocyte aldehyde dehydrogenase activity and elimination kinetics of diethyldithiocarbamic acid methyl ester and its monothio analogue after administration of single and repeated doses of disulfiram to man, *Eur. J. Clin. Pharmacol.*, 37:133.

Kitson, T.M., 1975, The effect of disulfiram on the aldehyde dehydrogenases of sheep liver, *Biochem. J.*, 151:407.

Kitson, T.M., 1977, The disulfiram-ethanol reaction; a review, *J. Stud. Alcohol*, 38:96.

Kitson, T.M., 1978, On the probability of implanted disulfiram's causing a reaction to alcohol, *J. Stud. Alcohol*, 39:183.

Kitson, T.M., 1982, Further studies of the action of disulfiram and 2,2'-dithiodipyridine on the dehydrogenase and esterase activities of sheep liver cytoplasmic aldehyde dehydrogenase, *Biochem. J.*, 203:743.

Kitson, T.M., 1989, The action of cytoplasmic aldehyde dehydrogenase on methyl p-nitrophenyl carbonate and p-nitrophenyl dimethylcarbamate, *Biochem.J.*, 257:579.

Kitson, T.M., 1983, Mechanism of inactivation of sheep liver cytoplasmic aldehyde dehydrogenase by disulfiram, *Biochem J.*, 213:551.

Kitson, T.M., Hill, J.P., and Midwinter, G.G., 1991, Identification of a catalytically essential residue in sheep liver cytoplasmic aldehyde dehydrogenase, *Biochem.J.*, 275:207.

Kitson, T.M., and Loomes, K.M., 1985, Modification of thiol groups in cytoplasmic aldehyde dehydrogenase, *Alcohol* 2:97.

MacKerell, A.D., Vallari, R.C., and Pietruszko, R., 1985, Human mitochondrial aldehyde dehydrogenase inhibition by diethyldithiocarbamic acid methanethiol mixed disulfide: a derivative of disulfiram, *FEBS Lett.*, 179:77.

Parilla, R., Ohkawa, K., Lindros, K.O., Zimmerman, U. I.-P., Kobayashi, K., and Williamson, J.R., 1974, Functional compartmentation of acetaldehyde oxidation in rat liver, *J. Biol. Chem.*, 249:4926.

Petersen, E.N., 1989, Pharmacological effects of diethylthiocarbamic acid methyl ester, the active metabolite of disulfiram? *Eur. J. Pharmacol.*, 166:419.

Tottmar, O., and Marchner, H., 1976, Disulfiram as a tool in the studies on the metabolism of acetaldehyde in rats, *Acta Pharmacol. et Toxicol.*, 38:366.

Vallari, R.C., and Pietruszko, R., 1982, Human aldehyde dehydrogenase: mechanism of inhibition by disulfiram, *Science*, 216:637.

Yoshida, A., Dave, V., Ward, R.J., and Peters, T.J., 1989, Cytosolic aldehyde dehydrogenase (ALDH1) variants found in alcohol flushers, *Ann. Hum. Genet.*, 53: 1

Yourick, J.J., and Faiman, M.D., 1989, Comparative aspects of disulfiram and its metabolites in the disulfiram-ethanol reaction in the rat, *Biochem. Pharmacol.*, 38:413.

PHOTOAFFINITY LABELING OF ALDEHYDE DEHYDROGENASE FROM HORSE LIVER BY P^1-N^6-(4-AZIDOPHENYLETHYL)ADENOSINE-P^2-[4-(3-AZIDOPYRIDINIO)BUTYL] DIPHOSPHATE

Christoph Woenckhaus, Hanno Leibrock and Susanne Becker

Klinikum der J.W. Goethe Universität, Gustav Embden-Zentrum der Biologischen Chemie, Enzymologie, D-6000 Frankfurt/Main, Germany

INTRODUCTION

Affinity labeling of dehydrogenases has been achieved with various reactive groups bound to nucleotide pyrophosphates. Bromoketones react preferentially with sulfhydryl-groups of the protein (Jörnvall, et al., 1975; von Bahr-Lindström, et al., 1985). In the case of lactate dehydrogenase modification, the small molecule 3-bromoacetylpyridine is attached to the imidazole part of an essential histidine (Berghäuser, et al., 1971).

Diazonium groups react with several amino-acid side chains by forming azobridges, which are not stable enough under conditions of protein analysis (Burkhard, et al., 1981). The synthesis of some azido analogs of NAD^+ and $NADP^+$ have been described (Chen and Guillory, 1977; Koberstein, 1976). These coenzyme analogs are stable compounds, but they decompose upon irradiation with light of certain wavelengths forming highly reactive nitrenes which react with several amino-acid side chains of dehydrogenases. The inactivation reaction of dehydrogenases with azido analogs of NAD^+ is often described, but only in a few cases the modified amino-acid side-residue of the protein has been identified (Vogel, et al., 1992). Low yields and difficult synthesis impede the radioactive labeling of these coenzyme analogs, needed to facilitate the identification of the site of attack.

In the case of non-radioactive labels, the site of attack can only be analysed on the basis of coenzyme fragments of the coenzyme analog with distinguishable protein components (Woenckhaus and Filbrich, 1990).

Enzymology and Molecular Biology of Carbonyl Metabolism 4
Edited by H. Weiner, Plenum Press, New York, 1993

Preparation and properties of the inactivator

For the preparation of ^{14}C-labeled bis-azido analog we used uniformly labeled ^{14}C adenosine, or 8-^{14}C inosine. 250 μCi of the labeled adenosine was diluted with 1 g non-radioactive nucleoside. After acetylation of the ribose moiety the amino-group in position 6 was replaced by treatment with nitrous acid. The formed O-triacetyladenosine was transformed to O-triacetylinosine and the 6-chloro derivative was prepared by treatment with thionyl chloride. Upon hydrolysis of the acetyl-groups the chloro-atom in position 6 of the purine ring was exchanged by 4-aminophenylethylamine. The phosphorylation of the nucleoside was performed by the Yoshikawa method (Yoshikawa, et al., 1967). N^6-(4-aminophenylethyl)adenosine-5' phosphate and 4-(3-

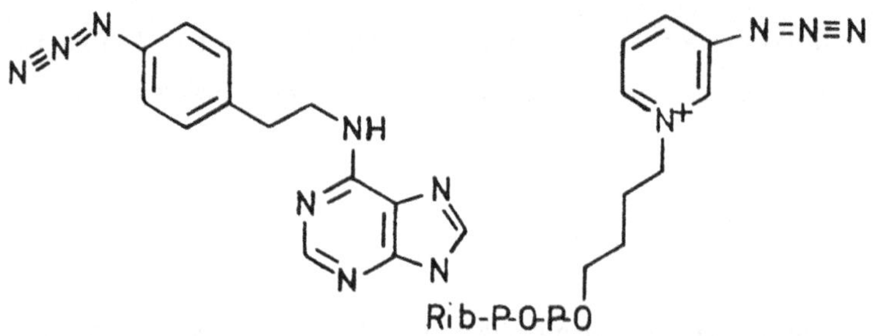

Figure 1. P^1-N^6-(4-azidophenylethyl)adenosine-P^2-[4-(3-azidopyridinio)butyl]diphosphate

aminopyridinio)-butyl phosphate react with dicyclohexyl-carbodiimide in aqueous pyridine to form the mixed anhydride, which was isolated by ion exchange chromatography on DEAE Sephadex A-25. The bis-amino compound was treated with nitrous acid and the resulting diazonium groups were then reacted with sodium azide to form the bis-azido analog P^1-N^6-(4-azidophenylethyl)adenosine-P^2-[4-(3-azido-pyridinio)butyl] diphosphate (Woenckhaus and Filbrich, 1990). The yield was 141 mg, 5% related to adenosine as starting material. The specific activity was 0.045 Ci/mole.

250 μCi of 8-^{14}C labeled inosine were mixed with 0.5 g non-radioactive inosine This mixture was used for the synthesis, as described above. The yield was 28 mg, 2.0% related to the starting material. The specific activity was 0.11 Ci/mole (Figure 1).

P^1-N^6-(4-aminophenylethyl)adenosine-P^2-[4-(3-aminopyridinio)butyl]diphosphate or the bis-azido derivative have no coenzymic activity; however, both compounds are competitive inhibitors with respect to NAD^+ in tests with aldehyde dehydrogenase from horse liver. The inhibition constant (K_I) of the bis-azido compound is 0.6 mM and the Michaelis-Menten constant of NAD+, under the same conditions, is 0.05 mM (Figure 2).

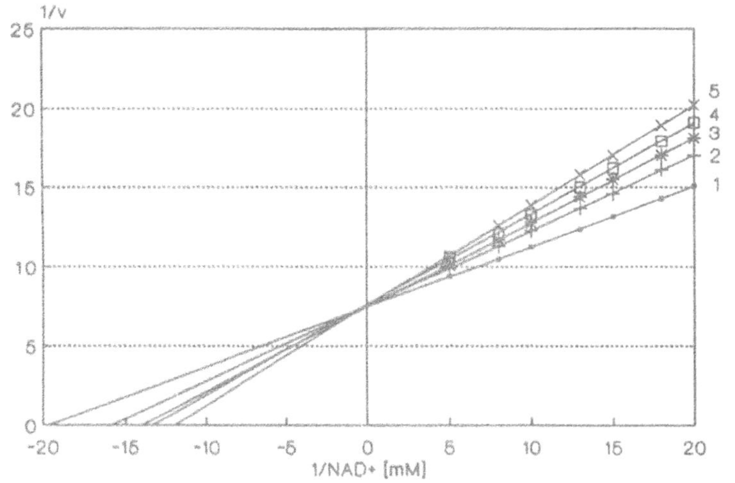

Figure 2. Competitive inhibition of ALDH from horse liver mitochondria by P^1-N^6-(4-azidophenylethyl) adenosine- P^2-[4-(3-azidopyridinio)butyl] diphosphate in 0.12 M pyrophosphate buffer pH 9, inhibitor concentrations: 1) 0.0 2) 0.05 mM, 3) 0.1 mM, 4) 0.2 mM, 5) 0.4 mM; K_I 0.6 mM.

Upon irradiation at 300-380 nm the azido groups are cleaved to form nitrenes which react to inactive compounds in aqueous medium. The optical properties change (Woenckhaus and Filbrich, 1990). Furthermore the IR-spectrum of the irradiated compound shows no band at 2080 cm^{-1}. This band is characteristic for the azido groups.

Modification of Aldehyde Dehydrogenase

The bis-azido compound forms a complex with aldehyde dehydrogenase. Upon irradiation at 300 - 380 nm the enzymic activity decreases. The loss of the catalytic function is caused by the light induced formation of a high reactive nitrene, which reacts with side-residues of the protein forming covalent bonds.

The rate of inactivation is dependent on temperature, pH, and the inactivator concentration. In the presence of NAD^+ only a slight decrease of inactivation is observed. The presence of NAD^+ and the pseudosubstrate acetophenone protects the catalytic properties of aldehyde dehydrogenase (Figure 3). The inhibition of catalytic activity was directly related to the incorporation of the inactivator (Figure 4). The incorporation was measured by precipitating the protein with trichloro acetic acid and scintillation counting (Berghäuser, 1975). Complete loss of the catalytic function occurs after incorporation of 2 moles of inactivator into 1 mole subunit (MW 55,000). Control measurements were performed with lactate dehydrogenase from pig heart (EC 1.1.1.27). Lactate dehydrogenase has only one binding domain for NAD^+ corresponding to the loss of enzymic activity caused by the incorporation of 1 mole inactivator into 1 mole subunit of the enzyme (MW 40,000).

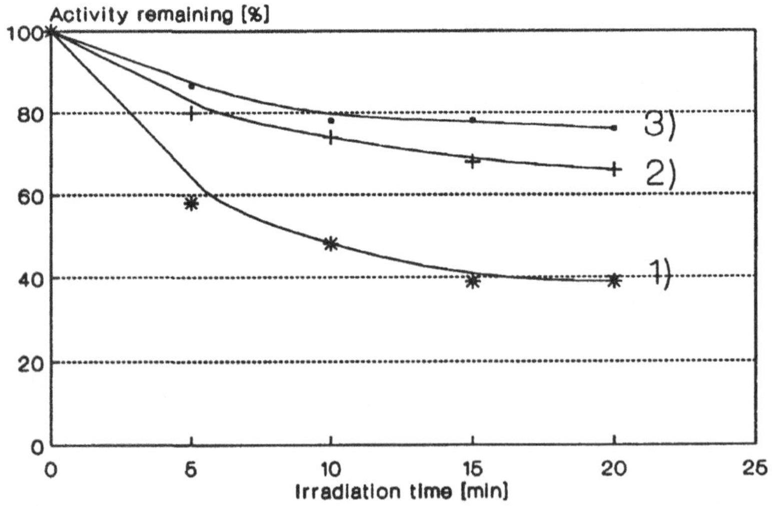

Figure 3. Photoinactivation of ALDH from horse liver mitochondrial enzyme by the bis-azido-inactivator in 0.12 M pyrophosphate pH 9.0, enzyme concentration $7x10^{-6}$ M, 1) inactivator $3.6x10^{-4}$ M 2) inactivator $3.6x10^{-4}$ M and NAD^+ $2x10^{-3}$ M , 3) inactivator $3.6x10^{-4}$ M, NAD^+ $3x10^{-3}$ M and acetophenone 10^{-3} M.

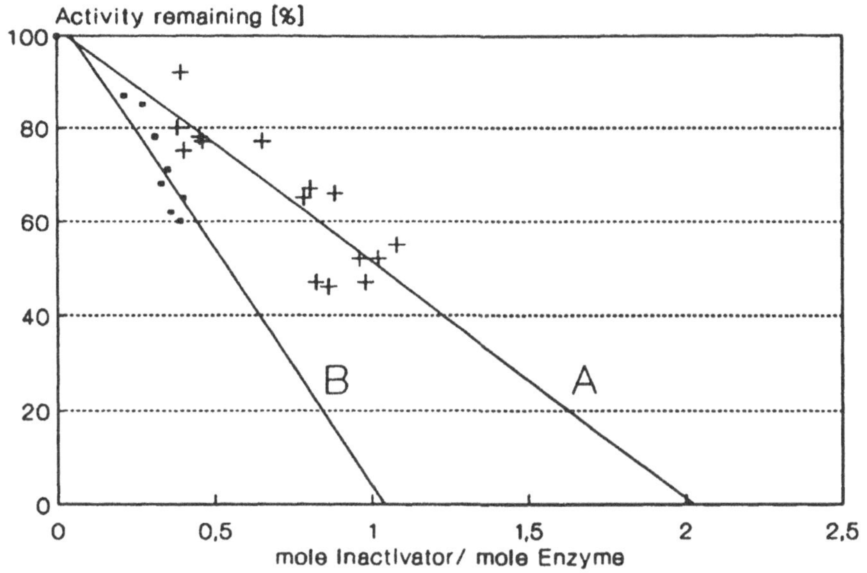

Figure 4. Incorporation of the ^{14}C labeled bis-azido derivative into ALDH from horse liver (A) and LDH from pig heart (B). MW 55,000 for ALDH and 40,000 for LDH.

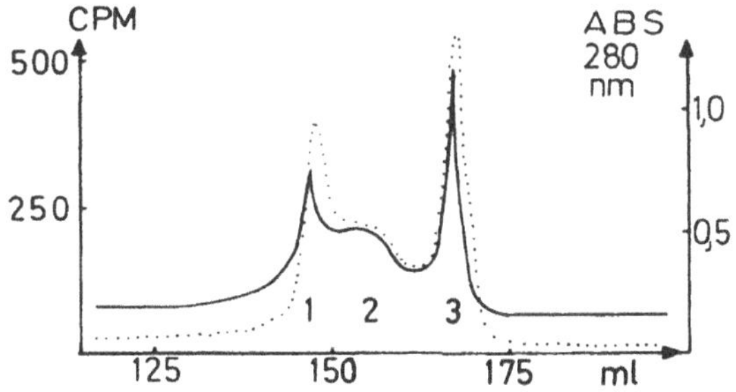

Figure 5. Elution pattern of a tryptic-digest of carboxymethylated ALDH, modified with ^{14}C labeled inactivator. Column 120 X 2 cm, eluted with 0.01 M ammonium hydrogencarbonate pH 8. Radioactivity (...) determined by scintillation counting of 0.2 ml of 2.5 ml fractions, absorption at 280 nm (---).

For the identification of the modified peptides and for the identification of the labeled side-residue, the modified enzyme was denaturated by dialysis in 9 M urea which contained 15 mM iodoacetamide. This enables the transfer of hydrogen sulfide residues to carboxymethylthio-derivatives avoiding oxydation reactions. Urea was removed by dialysis against water and the protein digested by treatment with trypsin (Heiland, et al., 1976).

The peptide mixture was first separated by chromatography on a Sephadex G 50 column (Figure 5). We obtained three labeled fractions. The radioactive labeled fractions 1 and 2 contained peptides, whereas fraction 3 was free of peptides and contained mostly fragments of the inactivator. The separation of the labeled peptides by reversed-phase HPLC-chromatography was not successful (King, et al., 1991). The elution pattern showed a nearly complete distribution of the radioactivity. Only one peptide fraction (P) contained some radioactivity. But rechromatography of this fraction (P) results in a complete loss of the label (Figure 6,7).

In contrast to reversed-phase chromatography, ion-exchange chromatography on DEAE-3SW minimizes the decomposition of the labeled peptides. The column was eluted by a linear salt gradient (Tris/HCL buffer pH 7, 0.05 M ; 0-0.5 M KCL). Most of the radioactivity was eluted in one major peak. The rest of radioactivity was scattered in smaller peptides. Even after two or three rechromatography steps, the loss of radioactive labeled peptides was small.

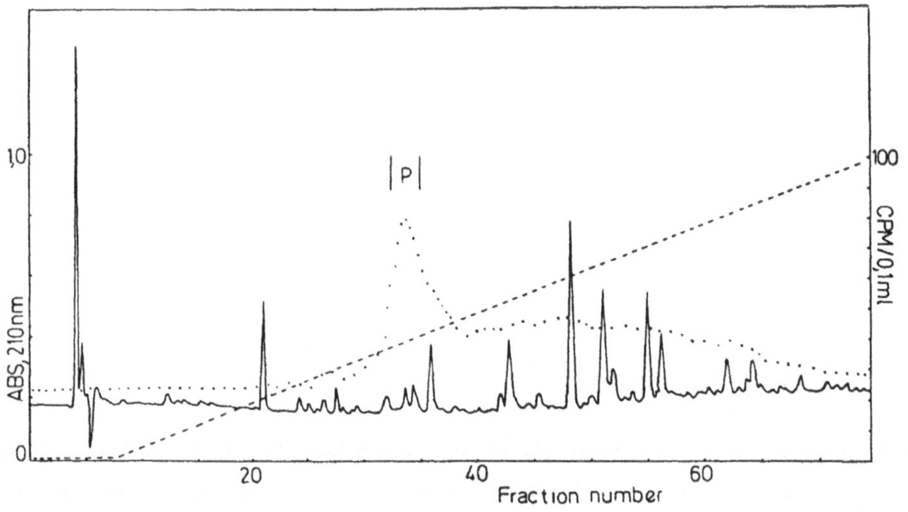

Figure 6. Peptide map of tryptic digest from ^{14}C labeled ALDH by HPLC. The peptide mixture was applied to a C18 μ Bondapak column initially in 0.1 % trifluoro acetic acid. The column was eluted with a linear gradient (0-100 %) of methanol (- - -) at a flow rate of 1 ml/min. Peptides were detected by the absorbance at 210 nm (---) and the radioactivity (. . .) by scintillation counting of 0.1 ml of 1.0 ml fractions.

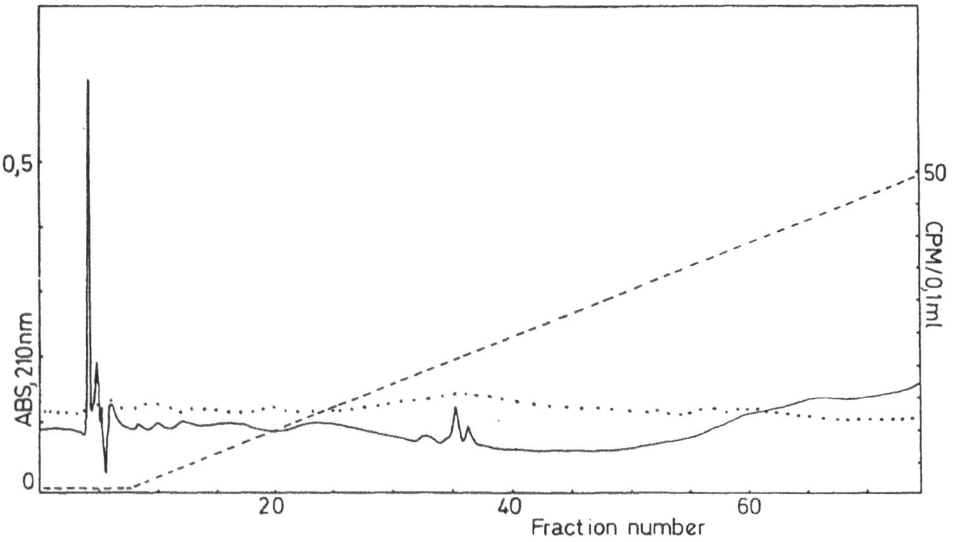

Figure 7. Rechromatography of the radioactive fraction (P) by the method described in (Figure 6).

DISCUSSION

P^1-N^6-(4-azidophenylethyl) adenosine-P^2-[4-(3-azidopyridinio) butyl] diphosphate resembles the coenzyme NAD^+. The compound can not catalyse any hydrogen transfer in presence of a dehydrogenase. But the competitive inhibition indicates, that the coenzyme analog is bound to the NAD^+ binding site of the enzyme. The compound has two different ring-systems: a purine and a pyridinium. Both are fixed to the different domains: the hydrophobic adenine will be bound to a hydrophobic binding pocket of the enzyme whereas, the hydrophilic azidopyridiniobutyl residue is attached to the hydrophilic pocket, which binds the nicotinamide ringsystem in the natural NAD^+-aldehyde dehydrogenase complex. In spite of the great structural alterations the ribose moiety is exchanged by a hydrophobic butyl hydrocarbon chain, and the nicotinamide ring by an azidopyridinium ring, it forms a complex similar to the binary NAD^+-ALDH complex. Investigations with the analog [4-(3-acetylpyridinio)butyl]-adenosine pyrophosphate reveal that the affinity of these analogs to dehydrogenases is decreased, but the complex-formation is efficient enough to make it function as hydrogen acceptor in several enzyme tests (Jeck, 1977). The azidophenylethyl residue bound to the adenine moiety is not involved in the complex-formation of the dehydrogenase, but may be able to react with superficial side-residues of the protein when irradiated. In a recent paper we described the inactivation of LDH by this analog (Woenckhaus and Filbrich, 1990). The analog is bound to two different peptides of LDH involving different azido-residues which are combined by the pyrophosphate bridge. The binding between the nucleotide pyrophosphate and the protein is unstable and can be cleaved in acidic medium. The separation of the tryptic-digest by ion exchange chromatography may be suitable to isolate the labeled peptides. We obtained unique fractions which lost small amounts of radioactivity and we intend to determine the position of label with this method. The enzymic activity of aldehyde dehydrogenase apoenzyme decreases gradually when irradiated at 300-380 nm. The presence of NAD^+ or P^1-N^6-(4-aminophenylethyl) adenosine-P^2-[4-(3-aminopyridinio)butyl] diphosphate protects the catalytic properties. In the presence of the bis-azido compound the inactivation rate increases. On the other hand, in the presence of NAD^+ and the pseudosubstrate acetophenone, the enzymic activity remains almost unchanged even in presence of the bis-azido compound.

The pseudosubstrate in absence of NAD^+ shows only a slight protection of the enzyme against the inactivation by the bis-azido analog. Acetophenone forms with NAD^+ and aldehyde dehydrogenase a less dissociable complex (Woenckhaus, et al., 1987). The bis-azido analog is not able to displace the natural coenzyme from this complex. Therefore, the amino acid residues of the coenzyme binding site remain unaffected. The substrate binding domain is reversibly blocked by acetophenone. Under these conditions the photo-activated analog can only react with solvent molecules forming an inactive compound which is unable to modify the enzyme protein. The hyperreactive sulfhydryl residue of cysteine 302 is also protected by acetophenone against chemical modification (Hempel and Pietruszko, 1981; Hempel, et al., 1985). The incomplete and slow inactivation of the enzyme by the bis-azido analog in the presence of acetophenone may be explained by the protection of this cysteine residue.

The incorporation of two moles inactivator into one mole subunit of aldehyde dehydrogenase may be based on two different binding centers. According to Weiner the two steps of the enzymatic reaction occur at two different centers (Tu and Weiner, 1988). In the first step the substrate is dehydrogenated and in the second step the formed acetyl-thio-ester is hydrolysed. On the other side, the inactivator may react with two different side-residues of the active center of the enzyme. Lactate dehydrogenase from pig heart has only one binding site. Corresponding to this, we found an incorporation of one mole bis-azido inactivator per mole subunit of the protein (Figure 3 B).

REFERENCES

Berghäuser, J., Falderbaum, I., and Woenckhaus, C., 1971, Zuordnung eines essentiellen Histidinrestes der Lactat-Dehydrogenase zur Substratbindungsstelle, Hoppe-Seyler's Z. Physiol. Chem., 352:52.

Berghäuser, J., 1975, A Reactive Arginine in Adenylate Kinase , Biochim. Biophys. Acta, 397:370.

Burkhard, A., Dworsky, A., Jeck, R., Pfeiffer, M., Pundak, S., and Woenckhaus, C., 1981, Diazoniumderivate des ADP zur Identifizierung essentieller Aminosäureseitenreste im aktiven Zentrum von Dehydrogenasen, Hoppe Seyler's Z. Physiol. Chem., 362:1079.

Chen, S., and Guillory, R., 1977, Arylazido-ß-alanine NAD^+, a NAD^+ Photoaffinity Analogue, J. Biol. Chem., 252:8990.

Heiland, I., and Brauer, D., 1976, Primary Structure of Protein L 10 from the Large Subunit of Escherichia Coli Ribosomes, Hoppe-Seyler's Z. Physiol. Chem., 357:1751.

Hempel, J., and Pietruszko, R., 1981, Selective Chemical Modification of Human Liver Aldehyde Dehydrogenase E_1 and E_2 by Iodoacetamide, J. Biol. Chem., 256:10889.

Hempel, J., Kaiser, R., and Jörnvall, H., 1985, Mitochondrial aldehyde dehydrogenase from human liver, Primary structure, differences in relation to the cytosolic enzyme, and functional correlations, Eur. J. Biochem. 153:13.

Jeck, R., 1977, The Properties of [(3-Acetylpyridinium)-n-alkyl] Adenosine Pyrophosphates, Structural Analogs of the Coenzyme NAD^+, Z. Naturforsch., 32c:550.

Jörnvall, H., Woenckhaus, C., and Johnscher, G., 1975, Modification of Alcohol Dehydrogenase with a Reactive Coenzyme Analogue, Eur. J. Biochem., 53:71.

King, S., Kim, H., and Haley, B., 1991, Strategies and Reagents for Photoaffinity Labeling of Mechanochemical Proteins, "Methods in Enzymology", Vol. 196, R. B. Vallee, ed., Academic Press Inc. p. 449.

Koberstein, R., 1976, 8-Azidoadenine Analogs of NAD^+ and FAD; Synthesis and Coenzyme Properties with NAD^+-Dependent and FAD-Dependent Enzymes, Eur. J. Biochem., 67:223.

Tu, G., C., and Weiner, H., 1988, Evidence for Two Distinct Active Sites on Aldehyde Dehydrogenase, J. Biol. Chem., 263:1218.

Vogel, P.,D., Nett, J.,H., Sauer, H., E., Schmadel, K., Cross, R., L., and Trommer, W., E., 1992, Nucleotide-binding Sites on Mitochondrial F_1-ATPase, in press J. Biol. Chem., 267.

von Bahr-Lindström, H., Jeck, R., Woenckhaus, C., Sohn, S., Hempel, J., and Jörnvall, H., 1985, Characterization of the Coenzyme Binding Site of Liver Aldehyde Dehydrogenase : Differential Reactivity of Coenzyme Analogues, Biochemistry, 24:5847.

Woenckhaus, C., Bieber, E., and Jeck, R., 1987, Studies on the Inactivation of Aldehyde Dehydrogenase (ALDH), in "Enzymology and Molecular Biology of Carbonyl Metabolism", H. Weiner and T. G. Flynn, eds., Alan R. Liss, New York, p. 53.

Woenckhaus, C., and Filbrich, R., 1990, Photoaffinity Labeling of Lactate Dehydrogenase by the Bis-azido Analog of NAD^+: P^1-N^6(4-azidophenylethyl)adenosine-P^2-[4-(3-azidopyridinio)butyl]di-phosphate, Z. Naturforsch., 45C:1044.

Yoshikawa, M., Kato, T., and Takenishi, T., 1967, A Novel Method For Phosphorylation of Nucleosides To 5`-Nucleotides, Tetrahedron Letters, 50:5065.

ALDEHYDE DEHYDROGENASE: ALDEHYDE DEHYDROGENATION AND ESTER HYDROLYSIS

Regina Pietruszko, Darryl P. Abriola, Erich E. Blatter
and Neeta Mukerjee

Center of Alcohol Studies
Rutgers University
Piscataway
New Jersey 08855-0969, U.S.A.

INTRODUCTION

Aldehyde dehydrogenase catalyses dehydrogenation of aldehydes as well as hydrolysis of esters. Because the aldehydic carbon is electron depleted by the carbonyl oxygen, it must be induced to give up the hydride ion. One of the ways in which hydride transfer can be induced is conversion of an aldehyde into a hemiacetal. The hemiacetal is then converted into an acyl intermediate during hydride transfer to NAD; only the acyl intermediate is formed from ester substrates. A similar mechanism is known to operate in another aldehyde dehydrogenase, glyceraldehyde-3-phosphate dehydrogenase (see review by Harris and Waters, 1976).

Visualization of the acyl intermediate from 4-*trans*-(N,N-dimethylamino)-cinnamaldehyde (a substrate for dehydrogenase reaction, DACA) and sheep liver cytoplasmic aldehyde dehydrogenase via spectroscopy was first reported by Buckley and Dunn (1982 and 1985). Dickinson (1985), employing *trans*-cinnamaldehyde, confirmed the findings of Buckley and Dunn (1982 and 1985). The covalent intermediate from DACA was also used for residue identification by Loomes et al. (1990). A substrate for the esterase reaction, 4-*trans*-(N,N-dimethylcinnamoyl)-imidazole, (DACI) was previously employed by Breaux and Bender (1976) to observe spectrophotometrically the acyl-enzyme intermediate in elastase catalysis. The structures of DACA and DACI are shown in Figure 1; both will form the same enzyme-acyl intermediate if the same amino acid residue of the enzyme is involved in its formation. Blatter et al., (1992) utilized both DACA and DACI and human cytoplasmic, E1, and mitochondrial, E2, isozymes for acyl intermediate formation, and identification of amino acid residues with which DACA and DACI reacted. Experiments described by Blatter et al., (1992) identify Cysteine 302 as the catalytic residue in both aldehyde dehydrogenation and ester hydrolysis. Participation of cysteine 302 in catalysis of ester hydrolysis is also supported by recent experiments of Kitson et al, (1991) who employed p-nitrophenyl ester of dimethyl carbamate for formation of the acyl intermediate.

However, there have been several reports (McGibbon et al., 1978; Motion et al., 1984; Deady et al., 1985; Tu and Weiner, 1988 a, b) arguing that active sites for aldehyde dehydrogenation and ester hydrolysis are separate. In the present communication, data from our laboratory are presented (Mukerjee and Pietruszko, 1992; Abriola and Pietruszko, 1992) which could also be interpreted in terms of distinct sites for aldehyde dehydrogenation and ester hydrolysis. This report will demonstrate how our (Blatter et al., 1992; Mukerjee and Pietruszko, 1992; Abriola and Pietruszko, 1992) and other McGibbon et al., 1978; Motion et al., 1984; Deady et al., 1985) apparently discrepant results are consistent with the same cysteine 302 participation in both aldehyde dehydrogenation and ester hydrolysis. Some arguments are also presented which help to assign a possible role for the glutamate 268 residue in aldehyde dehydrogenase catalysis.

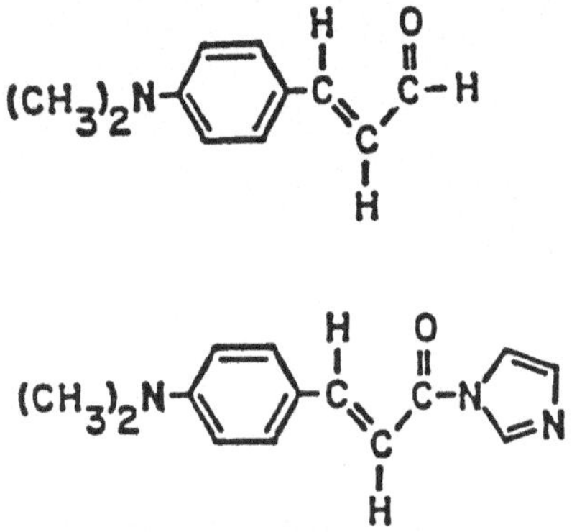

Figure 1. Structures of 4-*trans*-(N,N-dimethylamino)-cinnamaldehyde, DACA; and 4-*trans*-(N,N-dimethylamino)-cinnamoyl imidazole, DACI.

MATERIALS AND METHODS

Both DACA and DACI have pronounced spectra (Buckley and Dunn, 1982; Breaux and Bender, 1976) with extinction coefficients of ca. 30 $cm^{-1}mM^{-1}$. In addition both were tritium labelled to permit quantitation of spectral properties and detection at levels at which spectra were not easily visualized. The experimental procedures discussed in this paper were previously described in detail by Blatter et al, (1992), Abriola and Pietruszko (1992) and Mukerjee and Pietruszko (1992). Specificity of iodoacetamide modification of El isozyme was previously described (Hempel, 1981; Hempel and Pietruszko, 1981; Hempel et al., 1982) and more recently confirmed (Blatter, 1990). Experiments relating to the glutamate 268 residue are described by MacKerell et al, (1986) and by Abriola (1991) or Abriola et al., (1987). Dehydrogenase activity was measured in 0.1M pyrophosphate buffer, pH 9.0 containing 1mM EDTA, 500 μM NAD, 1mM propionaldehyde at 25°C in 3ml volume. Esterase activity was measured in 50 mM sodium phosphate buffer, pH 7.0, containing 1mM EDTA, 500 μM NAD and 150 μM p-nitrophenyl acetate. Results of the esterase assay were corrected by subtracting the rate of non enzymic hydrolysis of p-nitrophenyl acetate. To trap and quantitate the covalent intermediate, El or E2 isozymes

(1mg/ml) were incubated at 25°C for 60-100 sec in 50mM sodium phosphate buffer, pH 7.0, containing EDTA (1 mM) and DACA or DACI (50-200 μM) in the presence and absence of NAD (0.5-1 mM). The reaction was stopped by precipitation of enzyme with trichloroacetic acid (5% w/v, final concentration). After washing, the precipitate was dissolved in guanidine HCl or in 50mM sodium acetate buffer, pH 5.2, containing 2% w/v sodium dodecyl sulfate (SDS), for determination of spectra, scintillation counting and protein determination. Controls went through the same procedure except that DACA or DACI were omitted from the incubation.

RESULTS AND DISCUSSION

DACA and DACI as Substrates

The results presented in Table 1 demonstrate that both DACA and DACI are good substrates for E1 and E2 isozymes of human liver aldehyde dehydrogenase. Their kinetic constants are comparable in magnitude to those of substrates routinely employed in enzyme assays eg. propionaldehyde and p-nitrophenyl acetate. The Km values for both DACA and DACI are low micromolar values (see E1 isozyme and DACA) or are less than 1 μM. Thus, with E2 isozyme, substrate saturation (concentration = ca 100 x Km) can be achieved at 30 μM DACA and 40 μM DACI, respectively. These values are somewhat higher for E1 isozyme: 310 μM DACA and 80 μM DACI. According to Michaelis

Table 1. Kinetic constants of E1 and E2 isozymes with DACA and DACI.

Isozyme	Substrate	Km (μM)	kcat (IU)	kcat/Km
E1	DACA	3.1	0.04	0.01
"	Propanal	4.6	0.5	0.11
"	DACI	0.8	0.003	0.004
"	pNPA	1.1	0.1	0.09
E2	DACA	0.3	0.03	0.1
"	Propanal	0.9	0.25	0.28
"	DACI	0.4	0.02	0.05
"	pNPA	1.3	0.22	0.17

IU = μmol/min/mg; pNPA = p-nitrophenyl acetate. All kinetic constants were determined in sodium phosphate buffer, pH 7.0 and are averages of duplicate determinations.

Menten, at substrate [S] saturation, all enzyme [E] is present as enzyme:substrate [ES] complex:

[E] + [S] → [ES] → [E] + [P]
at saturation: [E] = [ES] eg. TOTAL enzyme is present as [ES]

Thus, both DACA and DACI could be easily employed for determination of stoichiometry of interaction with both isozymes.

Formation and Visualization of Covalent Intermediates

Covalent intermediates from both E1 and E2 isozymes with DACA and DACI were obtained by reacting with DACA + NAD, or with DACI (at saturating concentrations) and precipitating the enzyme from solution before the substrate was depleted. When the covalent intermediate was formed, the precipitated protein was yellow in color and acquired a characteristic spectrum (Figure 2). The amount of DACA or DACI covalently bound to aldehyde dehydrogenase could be easily quantitated from the amount of tritium label associated with DACA and DACI.

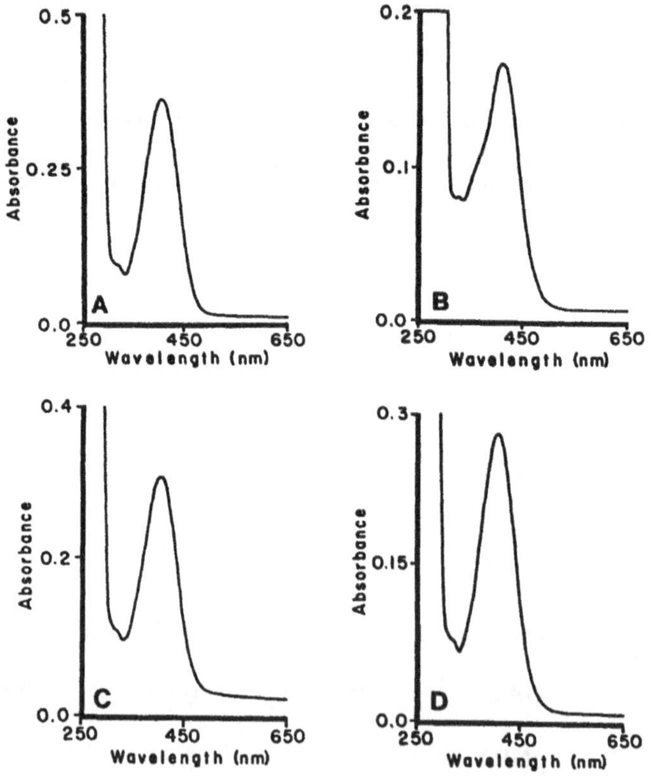

Figure 2. Spectra of covalent intermediates of E1 and E2 isozymes with DACA and DACI. E1 and E2 isozymes (2.25mg) were incubated with DACA (180 μM) and NAD (0.5mM) or DACI (100 μM) for 100 sec at 24°C in 30 mM phosphate buffer, pH 7.0, containing EDTA (1mM) and NAD (1mM). Following precipitation and washing, the precipitates were dissolved in 50 mM sodium acetate buffer, pH 5.2, containing 2% w/v SDS. UV-visible scans were obtained vs. sodium acetate/SDS buffer in the reference cell. A = E1 isozyme with DACA; B = E1 isozyme with DACI; C = E2 isozyme with DACA; D = E2 isozyme with DACI.

Identification of Cysteine 302 as the Amino Acid Residue that forms the Covalent Intermediate

The spectral properties of covalent intermediates from both DACA and DACI (with maxima at 408 nm) suggested involvement of a cysteine residue in the formation of the covalent intermediate. This was in agreement with previous reports of spectral properties (Buckley and Dunn, 1982 and 1985; Breaux and Bender, 1976) as well as the spectra obtained with standards during our experiments (Blatter et al., 1992). Pronase digestion of E2 isozyme, derivatized with both DACA and DACI, demonstrated that the label was indeed associated with a cysteine residue (Blatter et al., 1992; Abriola, 1991). In both cases tryptic digestion and peptide mapping resulted in isolation of the tryptic peptide comprising residues 273-307 of the primary structures of both E1 and E2 isozymes. This peptide contains cysteine residues in positions 275, 301 and 302 in E1 isozyme and in positions 301-303 in E2 isozyme. In the case of DACA, the label position could be unequivocally assigned to cysteine residues 301-303 of E2 isozyme via amino acid sequence analysis (Blatter et al., 1992; Blatter, 1990), and scintillation counting. Since the same covalent intermediate was formed from E1 and E2 isozymes, participation of cysteines 275 and 303 in covalent intermediate formation could be excluded. This left only cysteines 301 and 302 as likely participants. Assignment of label from both DACA and DACI to cysteine 302 was made on the basis that iodoacetamide, which reacts specifically with cysteine 302 of E1 isozyme (Hempel, 1981; Hempel and Pietruszko, 1981; Hempel et al., 1982) prevents formation of covalent intermediate, and also that cysteine 302 is the only cysteine residue that is conserved in all aldehyde dehydrogenases whose primary structure is known.

Quantitation of Covalent Intermediates and Stoichiometry

Since both DACA and DACI were tritium labelled, the amount of covalent intermediate formed could be readily calculated from specific activities of DACA and DACI. The stoichiometry of incorporation was found to depend on the specific activity of the enzymes employed in experiments. When specific activity was poor, label incorporation was low. When the isozymes had good specific activity, the amount of label incorporated increased proportionately to specific activity. If the incubation time was long and the substrate was exhausted, the amount of incorporated label was also low. When E1 and E2 isozymes with high specific activity (0.6 μmol NADH/min/mg for E1 and 1.6 μmol NADH/min/mg for E2) were used, and the reaction was stopped when the substrates (DACA or DACI) were still saturating, the highest incorporation stoichiometry was two molecules of either DACA or DACI per molecule of either E1 or E2 isozyme. This demonstrated that only 2 of the 4 cysteine 302 residues present in aldehyde dehydrogenase molecule participate in the acyl intermediate formation from either DACA or DACI (Blatter et al., 1992).

Relationship of Dehydrogenase and Esterase Activities in E1 and E2 Isozymes

Comparison of the esterase and dehydrogenase activities of E2 isozyme has been obtained recently in our laboratory (Mukerjee and Pietruszko, 1992) by studying substrate specificity of saturated straight chain aldehydes and p-nitrophenyl esters with the same acyl groups. Comparison of kcat/Km ratios for aldehyde and ester substrates in the absence of coenzyme and in the presence of NAD (1mM) and NADH (160μM, Figure 3) demonstrates that the substrate specificity profile for aldehydes and esters is totally distinct. This result, as well as those of other investigators (McGibbon et al., 1978; Motion et al., 1984; Deady et al., 1985), could be only interpreted in terms of distinct catalytic and binding sites for aldehydes and esters. Other experiments from our laboratory have shown that dehydrogenase but not esterase activity of E1 isozyme could be completely abolished by dicyclohexylcarbodiimide incorporation (Abriola and Pietruszko, 1992). The esterase activity remained essentially intact, but exhibited an initial "burst" before steady state velocity was established. This result could be interpreted in terms of distinct sites for aldehydes and esters. It could also be interpreted in terms of a single identical catalytic and binding site for aldehydes and esters with an assumption that the dehydrogenase reaction is more complex than the esterase reaction and as such would involve more amino acid residues at the active site than the esterase reaction. If latter is the case, the residue derivatized by dicyclohexylcarbodiimide would be essential for the dehydrogenase reaction but not important for the esterase reaction.

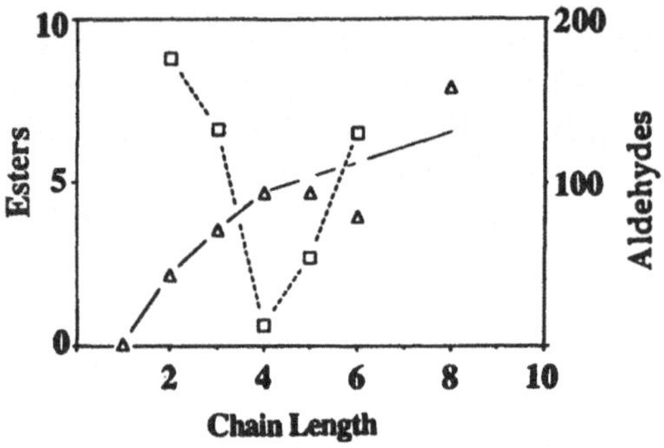

Figure 3. Comparison of chain length dependence of kcat/Km values of E2 isozyme with aldehydes and esters of the same acyl chain length. Aldehydes - △ = formaldehyde to n-octanaldehyde; esters - □ = p-nitrophenylacetate - p-nitrophenylcaproate in the presence of 160 μM NADH.

Inactivation of E1 Isozyme with Iodoacetamide

Iodoacetamide modifies the E1 isozyme with a stoichiometry of two iodoacetamide molecules per molecule of E1 isozyme (Hempel and Pietruszko, 1981). Almost complete loss of dehydrogenase activity and almost absolute specificity of label incorporation into cysteine 302 is observed (Hempel, 1981; Hempel and Pietruszko, 1981; Hempel et al., 1982). In Figure 4 the time course of inactivation of E1 isozyme with iodoacetamide (E1 = 10 μM; iodoacetamide = 100 μM) is presented. The reaction is followed by measuring both dehydrogenase and esterase activity. It can be seen that both activities are lost simultaneously. The ca 10% of esterase activity remaining, after dehydrogenase activity is completely lost, is due to non-specific protein catalysis of p-nitrophenyl acetate hydrolysis, for which these data were not corrected. The time course of activity loss shows a parallel disappearance of both dehydrogenase and esterase activity. At the end of this experiment, both iodoacetamide modified E1 isozyme and control, which was not modified by iodoacetamide, were incubated with either DACA or DACI and precipitated out of solution as described in Materials and Methods. The E1 isozyme which had been inactivated by iodoacetamide did not form a covalent intermediate with either DACA or

DACI. The covalent intermediate was formed from the control E1 and could be visualized via its yellow color and spectral properties.

The Role of Cysteine 302 in Aldehyde Dehydrogenase Catalysis

In these experiments, DACA was employed as an aldehyde substrate while DACI was used as an ester substrate. In both cases the same cysteine residue (cysteine 302) was shown to form the covalent acyl intermediate. Thus, cysteine 302 is the catalytic residue that participates in both aldehyde dehydrogenation and ester hydrolysis. A molecule of aldehyde dehydrogenase contains 4 Cysteine 302 residues. The stoichiometry of covalent intermediate formation from either DACA or DACI is only **two** per aldehyde dehydrogenase molecule. Could it be that two cysteine 302 residues are concerned with aldehyde dehydrogenation while the other two are concerned with ester hydrolysis?

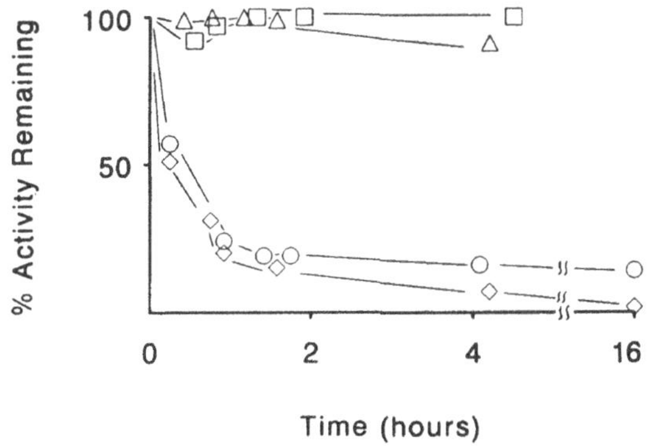

Figure 4. Inactivation of E1 isozyme with iodoacetamide. E1 isozyme (10 μM) was incubated with iodoacetamide (100 μM) in 50 mM phosphate buffer, pH 7.0, containing EDTA (1 mM), and NAD (0.6 mM) at 25°C. Samples of 30 μl were taken out for assay of dehydrogenase and 50 μl samples were used for assay of esterase activity. Esterase activity was measured in the presence of NAD (1 mM). Dehydrogenase standard assay was at pH 9.0 while esterase was assayed at pH 7.0 (Mukerjee and Pietruszko, 1992). Control dehydrogenase = ▲; control esterase = □; dehydrogenase activity loss with idoacetamide = ◊; esterase actvity loss with iodoacetamide = O.

The answer to this question could be unexpectedly obtained from our experiments with E1 isozyme derivatized by iodoacetamide. The specificity of iodoacetamide incorporation into E1 isozyme is almost absolute and occurs almost exclusively into cysteine 302 (Hempel, 1981; Hempel and Pietruszko, 1981; Hempel et al., 1982). Iodoacetamide is a small molecule, which would most likely not disturb the enzyme's tertiary structure, and it forms a stable covalent bond with the enzyme. As shown above (Figure 4), iodoacetamide incorporation abolishes both dehydrogenase and esterase activity of E1 isozyme. At the end of incubation, the control incubated without iodoacetamide readily formed covalent intermediates following short incubation with either DACA or DACI. The E1 isozyme, incubated with iodoacetamide, did not form covalent intermediates with either DACA or DACI. Thus, the same two cysteine 302 residues of aldehyde dehydrogenase molecule catalyse aldehyde dehydrogenation and ester hydrolysis.

In addition to published experimental work that argued for different sites of aldehyde dehydrogenation and ester hydrolysis, there were reports that argued for the same active site. It was previously demonstrated (Duncan, 1979; Loomes and Kitson, 1986) that during the esterase reaction, with p-nitrophenyl acetate as substrate in the presence of NADH, net formation of acetaldehyde occurred suggesting that the same active site was involved in catalysis of both reactions. In addition, Duncan (1985) published a paper using kinetic data of McGibbon et al., (1978) and Motion et al., (1984), which were used by the above authors as evidence for distinct active sites, to support an argument that the same active site is involved in both the esterase and the dehydrogenase activity. Our experiments (Blatter et al., 1992) with DACA and DACI demonstrate that cysteine 302 constitutes a central part of the catalytic area of aldehyde dehydrogenase active site as the residue which forms an enzyme acyl intermediate with both aldehyde and ester substrates, and are in agreement with Duncan's (1985) conclusions. Our more recent experiments (Mukerjee and Pietruszko, 1992), however, cannot be fitted into a model of the same catalytic and binding sites for aldehydes and esters. To reconcile our (Mukerjee and Pietruszko, 1992; Abriola and Pietruszko, 1992) and other (McGibbon et al., 1978; Motion et al., 1984; Deady et al., 1985) results with catalysis occurring at the same cysteine 302 residue it is necessary to postulate that the binding sites for aldehydes and esters are distinct. Both kinds of binding sites must be localized at the same active site, and both must overlap at the catalytic area, at cysteine 302.

The stoichiometry of acyl intermediate formation is always two, despite that the molecules of E1 and E2 isozyme contain four cysteine 302 residues. It is, therefore, proposed that only two of the four cysteine 302 residues of aldehyde dehydrogenase molecule are concerned with both aldehyde dehydrogenation and ester hydrolysis; the function of the other two cysteine 302 residues is unknown.

The Role of Glutamate 268 in Aldehyde Dehydrogenase Catalysis

The other conserved residue on the aldehyde dehydrogenase molecule which has been demonstrated to be localized at the active site (Abriola et al., 1987 and 1990) is glutamate 268. Its proximity to cysteine 302 is evident from the fact that glutamate 268 and cysteine 302 are the only two residues derivatized (Abriola et al., 1990) by bromoacetophenone - a well characterized affinity reagent (MacKerell et al., 1986). The other important information about this residue is that it occurs in the aldehyde dehydrogenase molecule as the "naked anion". This was deduced from the fact that it reacted with bromoacetophenone (Durst et al., 1975), which would normally react with amino, sulfhydryl and hydroxyl groups but not with carboxyls. In model systems, produced experimentally by employing crown ether catalysts (Pedersen, 1968), bromoacetophenone would not react with a carboxyl group unless the carboxyl was present as a "naked anion". Examination of literature on glutamate and aspartate residues with similar reactivity characteristics, indicated that their pKa values were considerably elevated over normal pKa values and occurred in the region normally observed with histidine (pKa 6-8). Takahashi et al. (1981) observed involvement in catalysis of a group with a pKa of 7.2 and another group with a pKa of 9.2 and suggested that the group with pKa 7.2 may be a histidine residue. Amino acid sequences of aldehyde dehydrogenases, however, show no conserved histidine in the molecule. Thus, it appears likely that glutamate 268 plays the same role in aldehyde dehydrogenase catalysis that histidine does in proteases (Fersht, 1977). The proposed role for glutamate 268 residue (Figure 5) is that of ionization of cysteine 302 and consequent facilitation of formation of a thiohemiacetal from an aldehyde. Thus, presence of the glutamate 268 "naked anion" would make it possible for an aldehyde to acquire the suitable structure for hydride transfer to NAD to occur. This role is also supported by the fact that the only aldehyde dehydrogenase sequenced (see D.W. Crabb et al., in this book), that does not contain glutamate 268 residue (Glu 268 = Arg), requires Coenzyme A in addition to NAD for catalysis. Although the precise mechanism of catalysis of Coenzyme A linked aldehyde dehydrogenase is also unknown, it has been

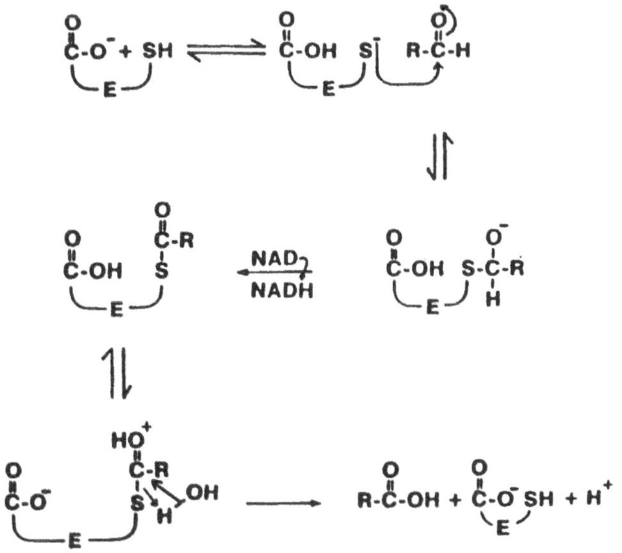

Figure 5. The role of Glutamate 268 in aldehyde dehydrogenase catalysis. E = aldehyde dehydrogenase; SH = cysteine 302 residue; C00H = glutamate 268 residue.

proposed that semimercaptals derived from Coenzyme A and aldehydes are true substrates (Burton and Stadtman, 1953).

Another function of a "naked anion" might possibly be in the repelling of the acid products of aldehyde dehydrogenation from the active site to cause their dissociation. It is well known that acids formed by aldehyde dehydrogenation do not bind to the enzyme, and also that the enzyme cannot be inactivated by iodoacetic acid, although it is readily inactivated by iodoacetamide. The unusual conservation of other residues surrounding glutamate 268 may also indicate other functions such as substrate recognition.

ACKNOWLEDGEMENTS

Financial support of NIAAA Research Scientist Award (K05 AA00046), Charles and Johanna Busch Memorial Fund, and R. Brinkley Smithers Institute for Alcoholism Prevention is acknowledged.

REFERENCES

Abriola, D.P., 1991, Active Site Studies of Aldehyde Dehydrogenase, PhD. Dissertation, Rutgers University.

Abriola, D.P., Fields, R., Stein, S., MacKerell, A.D., Jr., and Pietruszko, R,. 1987, Active site of human aldehyde dehydrogenase, Biochemistry 26:5679.

Abriola, D.P., MacKerell, A.D., Jr., and Pietruszko, R., 1990, Correlation of loss of activity of human aldehyde dehydrogenase with reaction of bromoacetophenone with glutamic acid-268 and cysteine-302 residues. Partial-sites reactivity of aldehyde dehydrogenase, Biochem. J. 266:179.

Abriola, D.P., and Pietruszko, R., 1992, Modification of aldehyde dehydrogenase with dicyclohexylcarbodiimide: separation of dehydrogenase from esterase activity, J. Prot. Chem. 11:59.

Blatter, E.E., 1991, The Role of Cysteine 302 in the Catalytic Mechanism of Aldehyde Dehydrogenase, PhD. Dissertation, Rutgers University (Dissert. Abstr. Int. B51:3358; University Microfilms no. DA9034880).

Blatter, E.E., Abriola, D.P., and Pietruszko, R., 1992, Aldehyde dehydrogenase: covalent intermediate in aldehyde dehydrogenation and ester hydrolysis, Biochem. J. 282:353.

Breaux, E.J., and Bender, M.L., 1976, Direct spectrophotometric observation of an acyl-enzyme intermediate in elastase catalysis, Biochem. Biophys. Res. Commun. 70:235.

Buckley, P.D., and Dunn, M.F., 1982, Observation of acyl-enzyme intermediate in the sheep liver aldehyde dehydrogenase catalytic mechanism via rapid scanning UV-visible spectroscopy, Enzymology of Carbonyl Metabolism, eds. H. Weiner and B. Wermuth, Alan Liss Inc. New York, pp. 23-35.

Burton, R.M., and Stadtman, E.R., 1953, The oxidation of acetaldehyde to acetyl coenzyme A, J.Biol. Chem. 202:873.

Deady, L.W., Buckley, P.D., Bennett, A.F., and Blackwell, 1985, Kinetics of inhibition and hysteresis of sheep liver cytoplasmic aldehyde dehydrogenase with glyoxylic acid: further evidence relating to the two site model for aldehyde oxidation, Arch. Biochem. Biophys. 243:586.

Dickinson, F.M., 1985, Studies on the mechanism of sheep liver cytoplasmic aldehyde dehydrogenase, Biochem. J. 225:159.

Duncan, J.R.S., 1979, Reversal of part of the aldehyde dehydrogenase reaction pathway during the hydrolysis of an ester, Biochem. J. 183:459.

Duncan, R.J.S., 1985, Aldehyde dehydrogenase. An enzyme with two distinct catalytic activities at a single type of active site, Biochem. J. 230:261.

Dunn, M.F., and Buckley, P.D., 1985, Kinetic and spectroscopic characterization of the sheep liver aldehyde dehydrogenase acyl-enzyme, Enzymology of Carbonyl Metabolism, eds. T.G. Flynn and H. Weiner, Alan Liss Inc. New York, pp 15-27.

Durst, H.D., Milano, M., Kikta, E.J., Jr., Connelly, S.A., and Grushka, E., 1975, Phenacyl esters of fatty acids via crown ether catalysts for enhanced ultraviolet detection in liquid chromatography, Anal. Chem. 47:1797.

Fersht, A., 1977, Enzyme Structure and Mechanism, W.H Freeman and Company, New York, San Francisco.

Harris, J.I., and Waters, M., 1976, Glyceraldehyde-3-phosphate dehydrogenase, The Enzymes, vol 13, 3rd ed., ed. P.D. Boyer, pp 1-49, Academic Press, New York.

Hempel, J.D., 1981, Chemical Modification of Human Liver Aldehyde Dehydrogenase Isoenzymes E1 and E2, PhD. Dissertation, Rutgers University (Dissert. Abstr. Int. B42:3664; University Microfilms no. DA80204216).

Hempel, J.D., and Pietruszko, R., 1981, Selective chemical modification of human liver aldehyde dehydrogenases E1 and E2 by iodoacetamide, J. Biol. Chem. 256:10889.

Hempel, J.D., Pietruszko, R., Fietzek, P., and Jornvall, H., 1982, Identification of a segment containing a reactive cysteine residue in human liver cystoplasmic aldehyde dehydrogenase (isoenzyme E1), Biochemistry 21:6834.

Kitson, T.M., Hill, J.P., and Midwinter, G.G., 1991, Identification of catalytically essential nucleophilic residue in sheep liver cytoplasmic aldehyde dehydrogenase, Biochem. J. 275:207.

Loomes, K.M., and Kitson, T.M., 1986, Aldehyde dehydrogenase catalyses acetaldehyde formation from 4-nitrophenyl acetate and NADH, Biochem. J. 238:617.

Loomes, K.M., Midwinter, G.G., Blackwell, L.F., and Buckley, P.D., 1990, Evidence for reactivity of serine-74 with *trans*-4-(N,N-dimethylamino)cinnamaldehyde during oxidation by the cytoplasmic aldehyde dehydrogenase from sheep liver, Biochemistry 29:2070.

MacKerell, A.D., Jr., MacWright, R.S., and Pietruszko, R., 1986, Bromoacetophenone as an affinity reagent for human liver aldehyde dehydrogenase, Biochemistry 25:5182.

McGibbon, A.K.H., Haylock, S.J., Buckley, P.D., and Blackwell, L.F., 1978, Kinetic studies on the esterase activity of cytoplasmic sheep liver aldehyde dehydrogenase, Biochem. J. 171:533.

Motion, R.L., Blackwell, L.F., and Buckley, P.D., 1984, Activating effect of p-(chloromercuri)benzoate on the cytoplasmic aldehyde dehydrogenase from sheep liver, Biochemistry 23:6851.

Mukerjee, N., and Pietruszko, R., 1992, Human mitochondrial aldehyde dehydrogenase: substrate specificity. Comparison of dehydrogenase with esterase reaction, Arch. Biochem. Biophys. *in press*.

Pedersen, C.J., 1968, Ionic complexes of macrocyclic polyethers, Fed. Proc. 27:1305.

Takahashi, K., Filmer, D.L., and Weiner, H., 1981, Effects of pH on horse liver aldehyde dehydrogenase: alterations in metal ion activation, number of functioning active sites and hydrolysis of the acyl intermediates, Biochemistry 21:6225.

Tu, G.-C., and Weiner, H., 1988a, Identification of the cysteine residue in the active site of horse liver mitochondrial aldehyde dehydrogenase, J. Biol. Chem. 263:1212.
Tu, G.-C and Weiner, H., 1988b, Evidence for two distinct active sites on aldehyde dehydrogenase, J. Biol. Chem. 263:1218.

IS THE SINGLE SITE BINDING MODEL FOR ALDEHYDE DEHYDROGENASE AN OVERSIMPLIFICATION? THE ONE-SITE, TWO-SITE DEBATE REVISITED

Jeremy P. Hill*, Adrian F. Bennett, Paul D. Buckley, Rosemary L. Motion* and Leonard F. Blackwell

Department of Chemistry and Biochemistry, Massey University Palmerston North, New Zealand. * New Zealand Dairy Research Institute, Palmerston North, New Zealand

INTRODUCTION

The kinetics of the cytosolic aldehyde dehydrogenase (ALDH) catalysed oxidation of propionaldehyde are complex. For example, at concentrations of propionaldehyde below about 200 μM propionaldehyde (low propionaldehyde concentration) the Lineweaver-Burk plot is linear, but at higher propionaldehyde concentration (20 mM) substrate activation occurs (MacGibbon *et al.*, 1977a). At low propionaldehyde concentrations it is generally agreed that aldehyde dehydrogenase oxidises propionaldehyde by an ordered bi bi mechanism (Hill *et al.*, 1991) as shown in scheme I in which isomerisation of binary enzyme.NADH complexes (k_5) and subsequent NADH release (k_6) are together rate limiting in the steady-state phase of the reaction. In scheme I *E represents a conformationally-rearranged form of the enzyme.

$$E + NAD^+ \rightleftharpoons E.NAD^+ + ald \rightleftharpoons E.NAD^+.ald \rightleftharpoons E.NADH.acyl \rightarrow {}^*E.NADH$$

$$\overset{k_5}{\rightleftharpoons} E.NADH \overset{k_6}{\rightleftharpoons} E + NADH$$

SCHEME I

The downward curvature of the Lineweaver-Burk plot for ALDH at high propionaldehyde concentrations (see Fig. 1a) is believed to be due to the aldehyde binding at a low affinity site on the enzyme. As a consequence of this an enzyme.NADH.aldehyde complex accumulates at high aldehyde concentrations for which the steps controlling NADH release are faster than from the enzyme.NADH complexes of scheme I (Dickinson, 1985). Blackwell *et al.* (1983) suggested that this complex arose from high concentrations of propionaldehyde binding in a second site (designated P2) not identical to the normal aldehyde binding site (two site model). On the other hand, Dickinson (1985), Duncan (1985), and Kitson (1987) identify this low affinity propionaldehyde binding site with the usual active site functional group (presumed to be a thiol) but in the enzyme.NADH complexes of scheme I (single site model). The models differ essentially in that the two site model allows the simultaneous binding of two propionaldehyde molecules per active site for all enzyme complexes of scheme I, while the one site model allows only one.

Since both models postulate the accumulation of enzyme.NADH.aldehyde complexes with the same overall composition, steady-state measurements cannot be used to distinguish between them. However pre-steady state kinetic measurements do allow the possibility of resolving the controversy. Under the two site model, where a second aldehyde molecule may bind in parallel even when the active site is acylated or occupied by aldehyde, differences between the burst rate constants or amplitudes at high and low propionaldehyde concentrations can occur. However the single site model does not allow this, since according to this model the low affinity aldehyde binding site is not available until after acyl-enzyme hydrolysis and the completion of the absorbance and fluorescence bursts. We now examine the effect of aldehyde concentration on the steady-state and pre-steady-state kinetics of cytoplasmic aldehyde dehydrogenase from sheep liver at pH 7.6 in the presence and absence of p-(chloromercuri)benzoate (PCMB) and in the presence and absence of disulphiram. These studies indicate the presence of more than one type of binding site on aldehyde dehydrogenase, and call into question the simple one-site model for this enzyme.

EXPERIMENTAL PROCEDURES

Sheep liver cytosolic aldehyde dehydrogenase (ALDH) was prepared essentially as described by MacGibbon *et al.* (1979) with the inclusion of a pH gradient step to remove mitochondrial contamination as described by Dickinson *et al.*, 1981. NAD^+ (grade III), NADH (grade III), p-(chloromercuri)benzoate (PCMB) and disulphiram were purchased from Sigma Chemicals, St. Louis, MO. USA. Propionaldehyde was purchased from Koch-Light Laboratories, Colnbrook, Bucks, U.K. All solutions were prepared in distilled water.

METHODS

ALDH was assayed as described by Blackwell *et al.*, (1987) and was dialysed for at least 4 hours (usually overnight) against 2 x 4 litre changes of 25 mM pH 7.3 phosphate buffer to remove 2-mercaptoethanol before experiments with PCMB and disulphiram. NADH burst experiments were carried out on a Durrum Gibson D110 stopped flow spectrophotometer also as described by Blackwell *et al.*, (1987) and proton burst experiments were carried out according to Bennett *et al.*, (1982). All solutions for proton release experiments contained phenol red (20 μM), 0.1 M Na_2SO_4, and 0.1 M $NaNO_3$ and were adjusted to pH 7.6 with 0.01 M HCl or NaOH immediately before use. Solutions containing 7.0 μM enzyme with 2 mM NAD^+ (with and without 60 μM PCMB) were mixed with propionaldehyde (400 μM or 40 mM) in the stopped flow spectrophotometer as indicated in Table 2. The reactions were followed by monitoring the decrease in absorbance of phenol red at 560 nm. Unless stated otherwise all concentrations are those after mixing. Enzyme concentrations are NADH binding site concentrations as determined from a V_{max} assay at 20 mM concentrations of propionaldehyde (Blackwell *et al* ., 1987). The maximum activating level of PCMB was determined separately for each set of experiments by titrating the steady-state activity until a maximum value was obtained. This level varied for different enzyme preparations but was usually in the range [PCMB]:[ALDH] 2:1 to 4:1. The maximum activating levels of PCMB will be referred to as activating-PCMB throughout this paper. Disulphiram was added as an ethanol solution so that the final concentration of ethanol did not exceed 3%.

RESULTS AND DISCUSSION

Effect of propionaldehyde concentration on the PCMB-modified enzyme

The effect of the propionaldehyde concentration on the steady-state rate of oxidation of propionaldehyde by PCMB-activated ALDH is shown in Fig. 1. The enzyme was clearly activated at low propionaldehyde concentrations irrespective of the order of mixing (lines b & c). Activation by PCMB arises from a 2-3 fold increase in the rate-controlling steps involving the release of NADH from the *E.PMB.NADH species (Motion R.L., 1986; where PMB represents the p-(mercuri)benzoate moiety) hence modification by PCMB is not detrimental *per se*.. However when the propionaldehyde concentration was increased above 200 μM (Fig. 1) marked inhibition occurred. The onset of the inhibition appeared at about the same propionaldehyde concentration as substrate activation began in the unmodified control reaction (Fig. 1a) suggesting the same low affinity binding site is involved. The predominant steady-state complexes will in this case be *E.PMB.NADH.Ald and E.PMB.NADH.Ald. While either only activating-PCMB is bound to the enzyme, or only propionaldehyde is bound to the low affinity aldehyde binding site, the effect is a net activation of the enzymic activity. However, if both compounds are bound simultaneously inhibition arises. The location of the low affinity propionaldehyde binding site is unknown but it clearly cannot involve the activating thiol group in the PCMB-modified enzyme.

Wherever the propionaldehyde binds, the formation of the quaternary complexes leads rapidly to a loss of functional enzyme as is shown by the rapid decay in nucleotide fluorescence following the burst when ALDH is mixed with activating-PCMB and 40 mM propionaldehyde (Fig.2). The fluorescence decay arises from dissociation of NADH from an enzyme species which probably contains both PCMB and propionaldehyde.

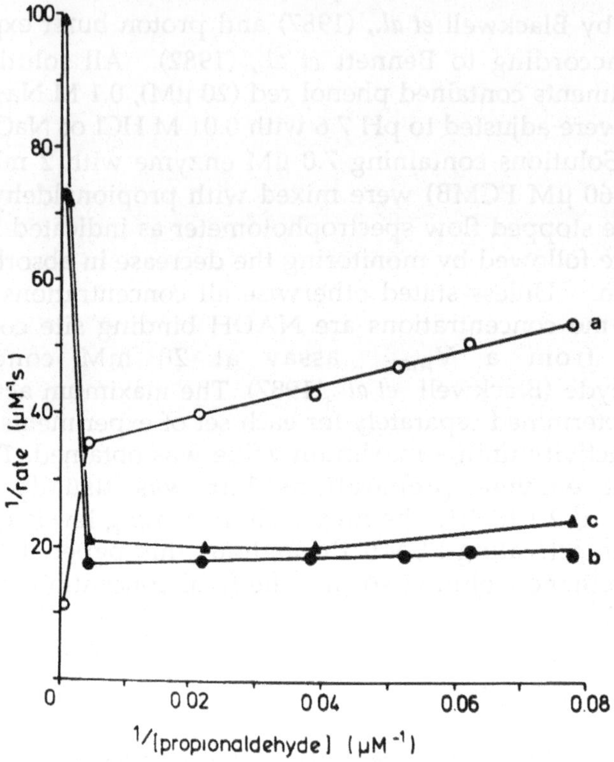

Fig. 1. Double reciprocal plots showing the effect of PCMB on the activity of sheep liver cytosolic ALDH. (a) Enzyme (0.16 μM) was added to a mixture of NAD^+ (1 mM) and propionaldehyde in 25 mM sodium phosphate buffer, pH 7.6 at 25°C and the reaction followed by changes in NADH absorbance at 340 nm (o). (b) PCMB (0.3 μM) was then added to each on going assay and the production of NADH followed as above (•). (c) Data for the final double reciprocal plot were obtained by mixing enzyme (0.16 μM), NAD^+ (1 mM) and PCMB (0.3 μM) prior to the addition to the other assay reagents as described above (▲).

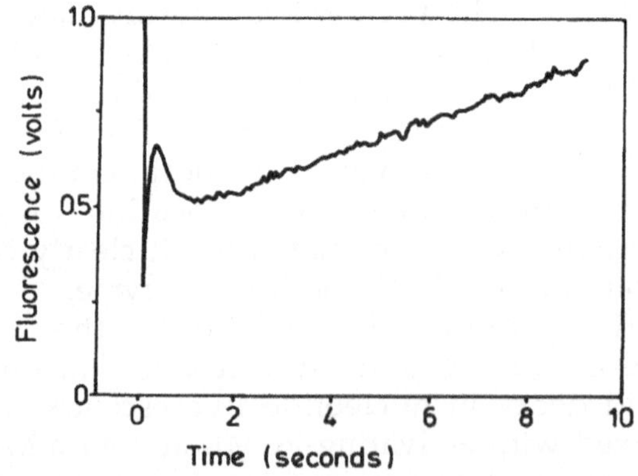

Fig. 2. Fluorescence decay following burst at high propionaldehyde concentrations. ALDH (6.8 μM), NAD^+ (2 mM) and PCMB (13.0 μM) were rapidly mixed with propionaldehyde (40 mM) at pH 7.6. The decay constant was calculated as described in methods.

The decay rate constant for the loss of NADH (3.5 s^{-1}) is significantly larger than that obtained for the displacement of NADH from either E.NADH.PMB or E.NADH (MacGibbon *et al.*, 1977b). Hence, the combination of activating levels of PCMB and high levels of propionaldehyde results in an additive effect on the NADH dissociation rates. In the absence of any other effects this would be expected to lead to a large activation of the steady-state rate. However, after NADH dissociates from the enzyme, a significant fraction of the enzyme must remain non-functional, since inhibition results. Evidence has recently been presented (Henehan & Tipton, 1991) for the existence of a dead-end E.Ald complex for the native enzyme in the presence of high concentrations of propionaldehyde and its dissociation constant has been measured. The effect of PCMB modification as in the present study may simply be to tighten the binding of aldehyde to the low affinity binding site on the free enzyme. This would allow the inhibition effects to be observed at lower substrate concentrations than with the native enzyme. It is significant that a similar marked decay in fluorescence is observed following the burst in nucleotide fluorescence even for the native enzyme at still higher propionaldehyde concentrations (33 mM after mixing) (Hart and Dickinson, 1982). It may be that whenever high concentrations of propionaldehyde are bound to the enzyme (at the low affinity aldehyde binding site), simultaneously with another modifying species, inhibition results.

Effect of PCMB on the pre-steady-state bursts in NADH fluorescence and proton release.

The effect of activating-PCMB on the NADH burst in nucleotide fluorescence is shown in Table 1. At low concentrations of propionaldehyde (200 μM after mixing) pre-incubating the E.NAD+ solution with activating-PCMB had no effect on the rate constant for the burst (Table 1) but the steady-state rate immediately following the burst was activated as seen in the steady-state experiments. However the amplitude of the burst process decreased by 38%. When the experiments were repeated at high

Table 1 The effect of activating levels of PCMB on nucleotide fluorescence burst parameters [a]

[propionaldehyde]	mixing conditions	k_b s^{-1}	V_{max}/V_{max_0}	A_b volts
200 μM	E, NAD+/prop	9.8 ± .8	1.00	0.64 ± 0.06
200 μM	E,NAD+PCMB/prop	9.0 ± 2	2.38	0.40 ± 0.08
20 mM	E,NAD+/prop	16 ± 1	2.02	0.68 ± 0.02
20 mM	E,NAD+,PCMB/prop	13 ± 1[b,c]	1.09[b,c]	0.42 ± 0.07[b]

[a] $[AlDH]_0$ = 3.0 μM; $[NAD^+]_0$ = 2 mM; $[PCMB]_0$ = 13 μM

[b] Burst followed by displacement; see Fig.2

[c] Displacement amplitude = 0.17 volts; λ = 3.5 s^{-1}.

concentrations of propionaldehyde (20 mM after mixing) several interesting differences were apparent. The steady-state rate following the burst for the control reaction (in the absence of PCMB) was activated (202%, Table 1) by the higher propionaldehyde concentrations as expected and the amplitude was the same as for the control burst at low (200 μM) concentrations of propionaldehyde. When activating-PCMB was pre-incubated with $E.NAD^+$ the amplitude of the burst process was again reduced by about 38% and the final steady-state rate was reduced to a value close to the low propionaldehyde control (Table 1). However, there was now a decay in the nucleotide fluorescence immediately following the burst (see Fig. 2) as discussed above corresponding to an apparent displacement of NADH equal to 40% of the burst amplitude. Gel filtration experiments suggest that the loss in burst amplitude is due to partial dissociation of the enzyme tetramers after modification with PCMB.

The fact that there was no significant effect on the burst rate constants could be taken as evidence in favour of a single-site model. However, propionaldehyde is clearly able to bind to the species prior to the *E.NADH complex of scheme I as indicated by the following data. The results of pre-incubating $E.NAD^+$ with activating-PCMB before rapidly mixing with propionaldehyde and monitoring in absorbance at 560 nm are given in Table 2. As found with the nucleotide fluorescence burst data there was an activation (170%) of the steady-state rate in the presence of PCMB at low (100 μM) propionaldehyde concentrations and the burst amplitude was again reduced (by about 35%) compared with the control. Again, there was no significant effect on the burst rate constant. However, when the experiments were repeated with 20 mM propionaldehyde although there was no significant effect of activating-PCMB on the burst rate constant, the burst amplitude was increased nearly 2 fold (Table 2). The pre-steady-state experiments clearly demonstrate that sufficiently high concentrations of propionaldehyde can bind to the PCMB-modified enzyme even though the active site is already occupied. Since during the burst experiments the enzyme is in the form of either $E.NAD^+.Ald$ or E.NADH. Acyl (see scheme I) simultaneous binding of propionaldehyde to two sites with different

Table 2 Effect of activating levels of PCMB on proton burst at 20 mM concentrations of propionaldehyde[a]

[propionaldehyde]	mixing conditions	k_b s^{-1}	V_{max}/V_{max_0}	A_b volts
100 μM	E,NAD^+/prop	6.96 ± .5	1.00	0.107 ± .003
100 μM	$E.NAD^+$,PCMB/prop	5.97 ±.3	1.69	0.066 ± .007
20 mM	E,NAD^+/prop	6.40 ± 1.3	3.15	0.100 ± .013
20 mM	E,NAD^+,PCMB/prop	4.69 ±.4	1.88	0.191 ± .009

[a] [AlDH] = 15 μM; [NAD^+] = 2 mM; [PCMB] = 60 μM; [phenol red] = 20 μM.

affinities is clearly required to explain this result. If this were not the case no effect on the burst amplitude could be observed. This result provides support for a two site model as originally proposed by Blackwell *et al.*, (1983).

Effect of Disulphiram on the burst in NADH fluorescence

The general concensus of opinion is that disulphiram exerts its effect by reacting with cys-302, thought to be the active-site acylation centre, thus inactivating the enzyme. Since on the one-site model this would also be the low affinity aldehyde binding site experiments in which aldehyde and disulphiram can compete for the same enzyme species might be revealing. Hence we examined two pre-steady-state situations: i) E.NAD$^+$ rapidly mixed with a solution containing 120 μM propionaldehyde and varying concentrations of disulphiram. Depending on the relative rates of binding it would be anticipated that competition for E.NAD$^+$ would occur but there would be free access for disulphiram to *E.NADH, ii) If the same E.NAD$^+$ mixture is rapidly mixed with 48 mM propionaldehyde containing varying concentrations of disulphiram competition for *E.NADH alone would be expected.

When disulphiram was premixed with enzyme and NAD$^+$ before mixing with propionaldehyde, there was a steady decrease in the burst amplitude as the ratio of the concentration of disulphiram to enzyme was increased, (Fig. 3) although a very low amplitude burst was still present even at very high disulphiram to enzyme ratios. The results of competition experiments in which enzyme and NAD$^+$ were mixed with disulphiram and propionaldehyde are shown in Tables 3 and 4. At low propionaldehyde concentrations (60 μM), as the [disulphiram]:[enzyme] increased from 0 to about 2.5, the burst rate constant remained unchanged (Table 3) but the amplitude of the burst decreased (Fig. 3). This result was similar to that obtained when enzyme and disulphiram were pre-mixed (Fig. 3) However a lag was observed after the burst as would be expected if some competition was taking place between propionaldehyde and disulphiram.

At high propionaldehyde concentrations (24 mM) however the results were completely different. There was no change in the burst amplitude as the disulphiram concentration was increased (Fig. 4), a result which suggests that at these concentrations propionaldehyde competes better for E.NAD$^+$ than the relatively small concentrations of disulphiram. The significant result is that despite this there was a three-fold increase in the burst rate constant as the [disulphiram]:[enzyme] ratio increased (Table 4). After the burst there was a loss of amplitude, the extent of which increased as the concentration of disulphiram increased (Fig. 4), to a maximum of about 86 per cent of the burst amplidude at the highest disulphiram concentration. (Fig 5). At the highest disulphiram concentration the biphasic decay in the fluorescence was fitted to give decay constants which correspond to a k_{cat} of 0.5 s^{-1} (Blackwell *et al.*, 1987). This result is consistent with the activation of the off rates expected at high concentrations of propionaldehyde and the lack of an effect of disulphiram on the dissociation constant for NADH.

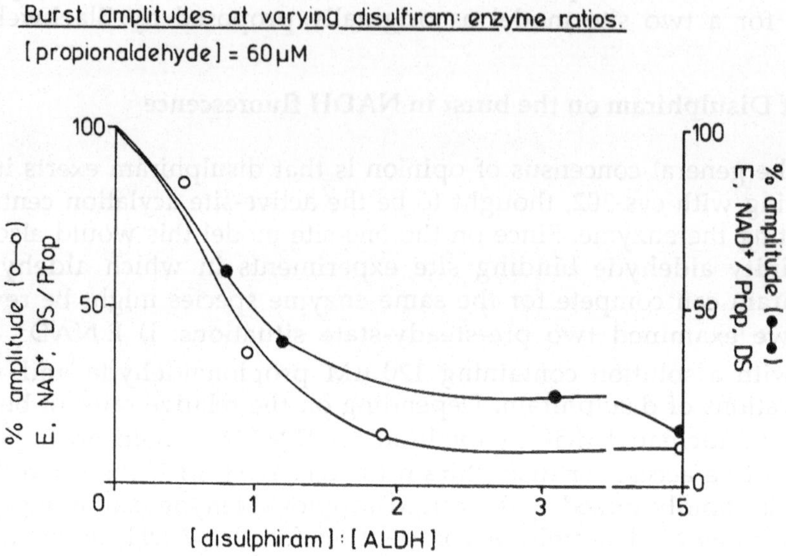

Fig. 3. Burst amplitudes at varying disulphiram:enzyme ratios. ALDH (10 μM) and NAD^+ (2 mM) were rapidly mixed with propionaldehyde at pH 7.6 (o) disulphiram premixed with the ALDH and NAD^+ solution; (●) disulphiram added to the propionaldehyde (120 μM) syringe.

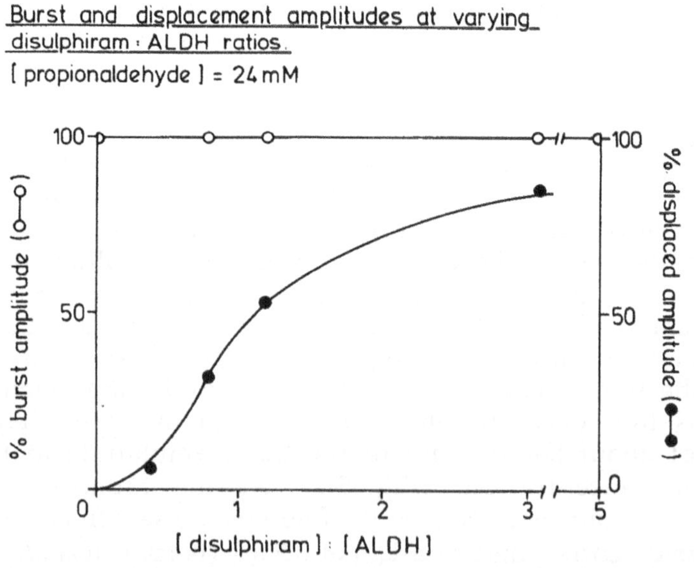

Fig. 4. ALDH (10 μM) and NAD^+ (2 mM) were rapidly mixed with propionaldehyde (48 mM) containing varying concentrations of disulphiram (0 to 16 μM) at pH 7.6.

Table 3 The effect of disulphiram on the burst in nucleotide fluorescence at low propionaldehyde concentrations

[ALDH] μM	[Disulphiram] μM	burst rate constant s^{-1}	burst amplitude volts	displacement amplitude volts
10	0	8.6± 0.4	9.2	
10	4	9.2 ± 0.4	5.8	0.8
10	6	-	4.1	1.0
10	16	8.0 ± 0.25	2.5	0.6
10	25	-	1.5	0.2

ALDH and NAD+ (2 mM) were mixed with propionaldehyde (120 μM) and a variable concentration of disulphiram. pH 7.6 and 25°C.

Table 4 The effect of disulphiram on the burst in nucleotide fluorescence at high propionaldehyde concentrations.

[ALDH] μM	[Disulphiram] μM	burst rate constant s^{-1}	burst amplitude volts	displacement amplitude volts
10	0	10.5 ± 0.8	9.3	-
10	2	14.2 ± 0.08	8.7	0.5
10	4	-	9.0	3.0
10	6	31 ± 0.4	9.4	5.3
10	16	27 ± 2	10.5	8.0*

* Under these conditions the "rate" constants for biphasic displacement were $\lambda_f = 2 \pm 0.11\ s^{-1}$ and $\lambda_s = 0.66 \pm 0.4\ s^{-1}$.

ALDH and NAD+ (2 mM) were rapidly mixed with propionaldehyde (48 mm) and a variable concentration of disulfiram as shown in the Table. pH 7.6 and 25°C.

The activation of the burst rate constant, without any loss in amplitude of the burst requires the simultaneous binding of disulphiram and aldehyde. Since during the burst process the active site nucleophile is either blocked by aldehyde (during the early steps in the pre-steady-state phase of the reaction) or is acylated (after hydride transfer occurs) disulphiram during the period of the burst cannot be binding at the active site nucleophile. If the active site nucleophile is cysteine-302 (Abriola and Pietrusczko 1992) then the disulphiram must be reacting with some other thiol group under the conditions used for the results shown in Table 4. It is interesting that a stoichiometry of 1.5 to 2.0 disulphiram molecules to enzyme is required to cause enzyme activity to fall to near the minimum observed using this reagent (Dickinson *et al.*, 1981). This modification still results in the loss of most (but even under the most stringent conditions not all) of the activity. Thus, reaction with more than one site can occur

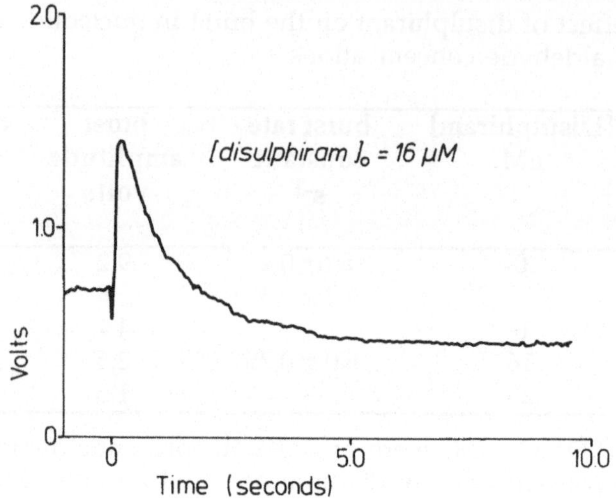

Fig. 5. **Displacement of NADH following the burst at high concentrations of propionaldehyde and disulphiram. ALDH (10 µM) and NAD$^+$ (2 mM) were rapidly mixed with propionaldehyde (48 mM) containing disulphiram (16 µM) at pH 7.6.**

even with disulphiram and this reaction causes loss of functional enzyme in the presence of high concentrations of propionaldehyde. Inhibition of the steady-state enzymic activity can therefore be caused by disulphiram through an indirect effect and hence caution must be exercised when interpreting steady-state inhibition data.

The Single-Site versus Two-Site Debate Revisited

The pre-steady state results described in this paper are difficult to rationalise on the basis of the single site model as commonly perceived. If a second propionaldehyde binding domain exists in the presence of PCMB, its involvement also in the activation effects observed with high concentrations of propionaldehyde for the native enzyme cannot be discounted. Perhaps propionaldehyde at high concentrations, PCMB and 2,2-dithiodipyridine (Kitson, 1982) all activate the enzyme, through modification of the same activating thiol group in a second modifier site as originally envisaged by Blackwell *et al.*, (1983). Although disulphiram, in the absence of the competition provided by high concentrations of propionaldehyde, does appear to exert its inactivating effects by reaction with a specific thiol, rapid reaction with a second thiol can also take place. The increase in the burst rate constant which results with high concentrations of propionaldehyde and higher levels of disulphiram, requires that reaction with disulphiram is not taking place at the active site nucleophile. The modification does not adversely affect the steps up to at least acyl-enzyme formation and the inhibition presumably results from the combined effects of disulphiram and propionaldehyde.

There are many implications arising from these results which remain to be explored. However, it is clear that the results reported here do not fit the simple one-site model. The enzyme appears to possess two binding sites for a number of substrates and modifying agents and simultaneous occupation of any two leads to inhibition The existence of other binding sites and their possible effects on enzyme catalysed reactions must be kept in mind as X-ray crystallographic studies and mutagenesis studies continue on this important but complex enzyme.

Acknowledgements

We gratefully acknowledge the receipt of a post-doctoral fellowship (JHP).

REFERENCES

Abriola, D.P., and Pietruszko, R., 1992, Modification of aldehyde dehydrogenase with dicyclohexylcarbodiimide: separation of dehydrogenase from esterase activity, J. Protein Res., **11**: 59-70.

Bennett, A.F., Buckley, P.D., and Blackwell, L.F., 1982, Proton release during the pre-steady-state oxidation of aldehydes by aldehyde dehydrogenase. Evidence for a rate-limiting conformational change, Biochemistry, **21**: 4407-4413.

Blackwell, L.F., Bennett, A.F., and Buckley, P.D., 1983, Relationship between the mechanisms of the esterase and dehydrogenase activities of the cytoplasmic aldehyde dehydrogenase from sheep liver. An alternative view, Biochemistry, **22**:3784-3791.

Blackwell, L.F., Motion, R.L., MacGibbon, A.K.H., Hardman, M.J., and Buckley, P.D., 1987, Evidence that the slow conformation change controlling NADH release from the enzyme is rate-limiting during the oxidation of propionaldehyde by aldehyde dehydrogenase, Biochem. J., **242**:803-808.

Dickinson, F.M., 1985, Studies on the mechanism of sheep liver cytosolic aldehyde dehydrogenase, Biochem. J., **225**:159-165.

Dickinson, F.M., Hart G.J., and Kitson, T.M., 1981, The use of pH gradient ion-exchange chromatography to separate sheep liver cytoplasmic aldehyde dehydrogenase from mitochondrial contamination and observations on the interaction between the pure cytoplasmic enzyme and disulphiram, Biochem. J., **199**:573-579.

Duncan, R.J.S., 1985, Aldehyde dehydrogenase, An enzyme with two distinct catalytic activities at a single type of active site, Biochem. J., **230**:261-267.

Hart, G.J., and Dickinson, F.M., 1982, Kinetic properties of highly purified preparations of sheep liver cytoplasmic aldehyde dehydrogenase, Biochem. J., **203**:617-627.

Henehan, G.T.M., and Tipton, K.F., 1991, The effects of assay temperature on the complex kinetics of aldehyde oxidation by aldehyde dehydrogenase from human erthrocytes, Biochem. Pharmacol., **42**:979-984.

Hill J.P., Blackwell, L.F., Buckley, P.D., and Motion, R.L., 1991, Steady-state and pre-steady-state kinetics of propionaldehyde oxidation by sheep liver cytosolic aldehyde dehydrogenase at pH 5.2. Evidence that the release of NADH remains rate-limiting in the enzyme mechanism at acid pH values. Biochemistry., **30**:1390-1394.

Kitson, T.M., 1982, The activation of aldehyde dehydrogenase by diethylstilboestrol and 2,2'-dithiodipyridine, Biochem. J., **207**:81-89.

Kitson, T.M., 1987, Effect of disulphiram on the pre-steady-state burst in the reactions of sheep liver cytoplasmic aldehyde dehydrogenase, Biochem. J., 248, 989-991.
MacGibbon, A.K.H., Buckley, P.D., and Blackwell, L.F., 1977a, Kinetics of sheep liver cytoplasmic aldehyde dehydrogenase, Eur. J. Biochem., 77:93-100.
MacGibbon, A.K.H., Buckley, P.D., and Blackwell, L.F., 1977b, Evidence for two-step binding of reduced nicotinamide adenine dinucleotide to aldehyde dehydrogenase, Biochem. J., 165:455-462.
MacGibbon, A.K.H., Motion, R.L., Crow, K.E., Buckley, P.D., and Blackwell, L.F., 1979, Purification and properties of sheep-liver aldehyde dehydrogenases, Eur. J. Biochem., 96:585-595.
Means, G.E. and Feeney, R.E. (1971). in *Chemical Modification of Proteins*. Holden-day, San Francisco, CA, pp 69-76, 81-83.
Motion R.L., 1986, Structural and mechanistic studies of sheep liver aldehyde dehydrogenase, Ph.D. Thesis, Massey University, New Zealand.

CRYSTALLIZATION AND PRELIMINARY X-RAY ANALYSIS OF BOVINE MITOCHONDRIAL ALDEHYDE DEHYDROGENASE AND HUMAN GLUTATHIONE-DEPENDENT FORMALDEHYDE DEHYDROGENASE

Thomas D. Hurley[1], Zhongning Yang[1], William F. Bosron[1], and Henry Weiner[2]

[1]Department of Biochemistry and Molecular Biology
Indiana University School of Medicine
Indianapolis, IN 46202

[2]Department of Biochemistry
Purdue University
West Lafayette, IN 47907

INTRODUCTION

The metabolism of acetaldehyde and formaldehyde is catalyzed by two distinct enzyme systems. The oxidative metabolism of acetaldehyde to acetate is performed, primarily, by the mitochondrial form of aldehyde dehydrogenase (ALDH2) (Cao, et al., 1988; Svanas & Weiner, 1985). While the oxidative metabolism of formaldehyde to formate can be accomplished through a hydrated glutathione adduct by the glutathione-dependent formaldehyde dehydrogenase located in the cytosol (Uotila & Koivusalo, 1989). The glutathione-dependent formaldehyde dehydrogenase has recently been demonstrated to be identical to the Class 3 form of alcohol dehydrogenase (Koivusalo, et al., 1989).

The multiple molecular forms of aldehyde dehydrogenase have been grouped into three classes of enzymes based on their cellular localization and enzymatic properties (Lindahl & Hempel, 1991). The Class I enzymes (ALDH1) include most of the cytosolic forms, Class II (ALDH2) includes most of the mitochondrial forms, while the Class III enzymes (ALDH3) include a variety of tumor specific, inducible, and those enzymes located within the endoplasmic reticulum. The Class II enzymes, like the Class I enzymes, are tetramers of identical 55 kD subunits. The Class I and Class II enzymes share approximately 70% sequence identity. The Class II enzymes share approximately 95% sequence identity across animal species. The Class III enzymes tend to be smaller in size and form dimers instead of tetramers. Additionally, the Class III enzymes share only 25-

30% sequence identity with either the Class I or the Class II enzymes (Hempel & Lindahl, 1989). Recently, preliminary crystallographic data for one of these Class III enzymes was reported (Rose, et al., 1990). The physiological role of each aldehyde dehydrogenase is not known with certainty. Aside from its apparent importance in the metabolism of acetaldehyde derived from the metabolism of ethanol, ALDH2 has also been implicated in the metabolism of biogenic amimes, polyamines and the products of lipid peroxidation (Tank, et al., 1981; Ambroziak & Pietrusko, 1991).

Alcohol dehydrogenases, like the aldehyde dehydrogenases, have been grouped into distinct classes of enzymes dependent on their sequence identity and catalytic properties (Burnell & Bosron, 1989). In humans, the Class I enzymes (α, β, and γ) are dimeric enzymes comprised of 40 kD subunits. These isoenzymes share greater than 93% sequence identity and form active heterodimers (eg. $\alpha\beta$). The human Class II (π), Class III (χ), and Class IV (σ) are more distantly related to one another and share approximately 60% sequence identity among each other and the Class I enzymes. The π, χ, and σ enzymes do not form heterodimers with each other or with the Class I enzymes.

The χ enzyme has recently been shown to be identical to the glutathione-dependent formaldehyde dehydrogenase (Koivusalo, et al., 1989). Thus, the same polypeptide is capable of performing two seemingly diverse enzymatic activities, the oxidation of aliphatic alcohols to their corresponding aldehydes and the oxidation of the formaldehyde-glutathione adduct S-hydroxymethylglutathione to S-formylglutathione. The S-formylglutathione is subsequently cleaved by a separate hydrolase to yield formic acid and glutathione (Uotila & Koivusalo, 1989). The human glutathione-dependent formaldehyde dehydrogenase enzyme is a dimer of 40 kD subunits. Each subunit shares 58% sequence identity to the human $\beta_1\beta_1$ alcohol dehydrogenase enzyme, whose structure has recently been determined (Hurley, et al., 1991).

We report here the crystallization and preliminary investigation of the X-ray diffraction properties of two distinct enzymes involved in the metabolism of C1 and C2 aldehydes. The bovine liver mitochondrial aldehyde dehydrogenase (ALDH2) was crystallized from solutions containing NAD^+. In an effort to investigate the seemingly diverse protein substrate interactions necessary to catalyze the oxidation of alcohols and the formaldehyde-glutathione adduct, the human $\chi\chi$ alcohol dehydrogenase, or glutathione-dependent formaldehyde dehydrogenase, was crystallized from solutions containing both NAD^+ and cinnamyl alcohol.

METHODS

Crystallization and Data Collection on Bovine ALDH2

All crystals were obtained using the sitting drop method of vapour diffusion and the CRYSCHEM plates (Charles Supper Company, Inc.). The crystals were grown from solutions containing 40 mM sodium citrate, 1 mM NAD^+ (grade I, Boehringer Mannheim), and 21-24% (w/v) polyethylene glycol 3400. Crystals suitable for X-ray diffraction analysis could be obtained between pH 5.3 and pH 5.5 with a protein concentration of 5-9 mg/ml. The average size of the crystals was 0.5 x 0.05 x 0.05 mm and their external morphology resembled hexagonal rods. Crystals greater than 0.1 mm in their two smaller dimensions were examined for their ability to diffract X-rays.

Small angle precession photographs obtained from these crystals were consistant with an orthorhombic space group with approximate cell dimensions of 150 x 160 x 100 Å and exhibited diffraction to at least 3.0 Å. High-resolution X-ray diffraction data was collected using a Nicolet multiwire area detector equipped with a helium path and a Rigaku Rotaflex

RU-200B rotating anode generator. Three crystals were used to collect a data set to 2.85 Å with a crystal to detector distance of 20 cm. The individual data sets were processed, scaled, and merged using the XENGEN program package (Howard, et al., 1987).

Crystallization and Data Collection on Human $\chi\chi$ ADH

Crystals suitable for X-ray analysis were grown using the sitting drop method of vapour diffusion from solutions containing 240 mM MES, pH 6.6, 1 mM NAD^+, 5 mM cinnamyl alcohol, and 8.5% PEG 8000 at a protein concentration of 15-20 mg/ml. The crystals grew as flat plates within one week and exhibited an average size of 0.3 x 0.3 x 0.05 mm. Crystals with their smallest dimension greater than 0.1 mm were examined for their ability to diffract X-rays. Three crystals were used to collect high-resolution X-ray diffraction data to 2.88 Å on a Nicolet multiwire area detector equipped with a helium path and a Rigaku RU-200B rotating anode generator. The data sets were processed, scaled, and merged using the program package XENGEN (Howard, et al., 1987).

RESULTS AND DISCUSSION

Bovine ALDH2

A total of 48,772 reflections were collected to 2.85 Å (83% complete, 91% complete to 3.0 Å). The data set was processed as the space group P222. However, systematic absences of the h00, 0k0, and 00l = 2n+1 reflections indicate that the correct space group is $P2_12_12_1$. The cell dimensions of these crystals are a = 153.74 Å, b = 159.37 Å, c = 101.45 Å. The volume of the asymmetric unit is 621,400 Å^3. If a tetramer comprises the asymmetric unit this yields a V_m of 2.82 and a probable solvent content of 56% which is within the normal range for protein crystals (Matthews, 1968). This crystal form for ALDH2 does not seem to be related in any simple way to the crystal form determined for the rat tumor ALDH3 enzyme. This is not too surprising considering the low sequence identity between the two enzymes (25%) and the smaller size of the ALDH3 enzyme (50 kD verus 55 kD for ALDH2). The ALDH3 enzyme crystallized in the space group $P2_1$ with cell dimensions of a = 65.1 Å, b = 170.7 Å, c = 47.1 Å, β = 110.5° and a dimer apparently comprises the asymmetric unit (Rose, et al., 1990).

In an effort to determine the arrangement of the subunits within the asymmetric unit of this crystal form, a self-rotation function using the data between 10 and 5 Å was calculated with the Crowther rotation function (Crowther, 1972) as implemented in the program package MERLOT (Fitzgerald, 1988). The self-rotation function exhibited no peaks at κ = 90° and two strong peaks at κ = 180°, one at ϕ = 30°, ψ = 90° and the other at ϕ = 120°, ψ = 90°. These calculations suggest that the subunits of the tetramer are arranged according to the point group 222 (the subunit arrangement observed in hemoglobin), with one of the two-fold axes nearly parallel to the c axis and the other two 2-fold axes lying in the plane defined by the a and b axes (**Figure 1**). A search for isomorphous heavy atom derivatives is currently underway.

Human $\chi\chi$ ADH

A total of 18,666 reflections were collected to 2.88 Å (80% complete, 87% complete to 3.0 Å). The data set was originally indexed as the space group P222. Inspection of the merged and scaled data set revealed that all h + k = 2n+1 reflections were either absent or weak with $F/\sigma_F < 2.0$. The data was subsequently indexed and processed as C222. Systematic absences of the 00l = 2n+1 reflections, in the final data set, indicated that the correct space group is $C222_1$. The cell dimensions of these crystals are a = 142.10 Å, b = 202.34 Å, c = 69.58 Å. The volume of the asymmetric unit is 250,060 Å^3. If a dimer comprises the asymmetric unit this yields a V_m of 3.12 and a probable solvent content of 60% (Matthews, 1968).

Alcohol dehydrogenase isoenzymes have been observed to crystallize and diffract X-rays to high resolution in three different crystal forms, orthorhombic $C222_1$, monoclinic C2, and triclinic P1. The orthorhombic and monoclinic forms have, generally, been observed for apo-enzymes and for complexes with substrates which cause the enzyme to

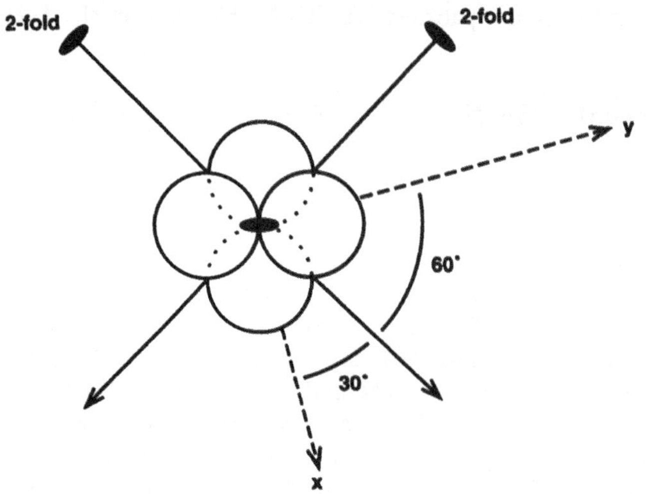

Figure 1. A representation of the subunit arrangement (point group 222) for bovine ALDH2 which is consistent with the results of the self-rotation function. Each circle represents a subunit of the enzyme. The 2-fold axes identified in the self-rotation function are represented as the two solid lines running diagonally through the subunits. The dashed axes, x and y, represent the crystallographic axes a and b, respectively. As represented in this figure these two 2-fold axes lie approximately 30° and 60° from the x (a) or y (b) axes. The third 2-fold axis lies parallel to the crystallographic c axis and is represented here as the ellipse at the center of the tetrameric subunit arrangement.

remain in the so called "open" conformation (eg. horse EE complexed with ADP-ribose) and the triclinic forms have been observed for enzyme substrate complexes in the so called "closed" conformation (eg. horse EE complexed with NADH and DMSO; and human $\beta_1\beta_1$ complexed with NAD^+) (Eklund, et al., 1976; Eklund, et al., 1981; Hurley, et al., 1991). The human Class III ADH does crystallize in one of the commonly observed space groups, but it does not follow the pattern observed for the Class I enzymes in which crystals grown in the presence of coenzyme and substrate/inhibitor are of the triclinic class. This result may suggest that the human $\chi\chi$ enzyme possesses an overall molecular shape which retains some of the same crystal packing forces as the Class I enzymes. In contrast, the way in which the enzyme - substrate interactions affect the crystal packing is quite different. We are currently trying to solve the structure of this enzyme by molecular replacement with the human $\beta_1\beta_1$ ADH structure as the model.

ACKNOWLEDGEMENTS

T.D.H. is a recipient of a Scientist Development Award (K21-AA00150) and H.W. is a recipient of a Senior Scientist Award (K05-AA00028) from the National Institute of Alcohol Abuse and Alcoholism. This work was supported in part by grants from the National Institute of Alcohol Abuse and Alcoholism (R01-AA07112, R01-AA05812 and P50-AA07611). The authors gratefully acknowledge Dr. Mario Amzel for his insightful discussions during the collection and processing of the data and for providing access to the X-ray equipment within the Department of Biophysics and Biophysical Chemistry at the Johns Hopkins University Medical School. We thank Laura Robinson for purifying the bovine mitochondrial aldehyde dehydrogenase.

REFERENCES

Ambroziak, W., and Pietrusko, R., 1991, Human aldehyde dehydrogenase: Activity with aldehyde metabolites of monoamines, diamines, and polyamines. J. Biol. Chem., **266**:13011

Burnell, J.C., and Bosron, W.F., 1989, Genetic polymorphism of human liver alcohol dehydrogenase and kinetic properties of the isoenzymes, in "Human metabolism of Alcohol," Volume II, Crow, K.E. and Batt, R.D., eds., CRC Press, Florida, p. 65.

Cao, Q.-N., Tu, G.-C., and Weiner, H., 1988, Mitochondria as the primary site of acetaldehyde metabolism in beef and pig liver slices. Alcoholism: Clin. Exp. Res., **12**:720.

Crowther, R.A., 1972, The fast rotation function, in "The molecular replacement method," Rossmann, M.G., ed., Gordon and Breach, New York, p. 174.

Eklund, H., Nordström, B., Zeppezauer, E., Söderlund, G., Ohlsson, I., Boiwe, T., Söderberg, B.-O., Tapia, O., Brändén, C.-I., and Åkeson, Å., 1976, Three-dimensional structure of horse liver alcohol dehydrogenase at 2.4 Å resolution. J. Mol. Biol., **102**:27.

Eklund, H., Samama, J.-P., Wallén, L., Brändén, C.-I., Åkeson, Å., and Jones, T.A., 1981, Structure of a triclinic ternary complex of horse liver alcohol dehydrogenase at 2.9 Å resolution. J. Mol. Biol., **146**:561.

Fitzgerald, P.M.D., 1988, MERLOT, an integrated package of computer programs for the determination of crystal structures by molecular replacement. J. Appl. Cryst., **21**:273.

Hempel, J., and Lindahl, R., 1989, Class III aldehyde dehydrogenase from rat liver: Super family relationship to classes I and II and functional interpretation, in "Enzymology and molecular biology of carbonyl metabolism 2," Weiner, H. and Flynn, T.G., eds., Alan R. Liss, Inc., New York, p. 3.

Howard, A.J., Gilliland, G.L., Finzel, B.C., Poulos, T.L., Ohlendorf, D.H., and Salemme, F.R., 1987, The use of an imaging proportional counter in macromolecular crystallography. J. Appl. Crystallogr., **20**:383.

Hurley, T.D., Bosron, W.F., Hamilton, J.A., and Amzel, L.M., 1991, Structure of human $\beta_1\beta_1$ alcohol dehydrogenase: catalytic effects of non-active site substitutions. Proc. Nat. Acad. Sci. USA, **88**:8149.

Lindahl, R., and Hempel, J., 1991, Aldehyde dehydrogenase: What can be learned from a baker's dozen sequences?, in "Enzymology and molecular biology of carbonyl metabolism 3," Weiner, H., Wermuth, B., and Crabb, D.W., eds., Plenum Press, New York, p. 1.

Koivusalo, M., Baumann, M., and Uotila, L., 1989, Evidence for the identity of glutathione-dependent formaldehyde dehydrogenase and class III dehydrogenase. FEBS Lett., **257**:105

Matthews, B.M., 1968, Solvent content of protein crystals. J. Mol. Biol., **33**:491.

Rose, J.P., Hempel, J., Kuo, I., Lindahl, R., and Wang, B.-C., 1990, Preliminary crystallographic analysis of class 3 rat liver aldehyde dehydrogenase. Proteins: Structure, Function, and Genetics, **8**:305.

Svanas, G.W., and Weiner, H., 1985, Aldehyde dehydrogenase activity as the rate-limiting factor of acetaldehyde metabolism in rat liver. Arch. Biochem. Biophys., **236**:36.

Tank, A.W., Weiner, H., and Thurman, J.A., 1981, Enzymology and subcellular localization of aldehyde oxidation in rat liver. Biochem. Pharm., **30**:3265.

Utiola, L., and Koivusalo, M., 1989, Glutathione-Dependent Oxidoreductases: Formaldehyde Dehydrogenase, in "Coenzymes and cofactors, Vol. III, glutathione. chemical, biochemical and medical aspects, part A," Dolphin, D., Poulson, R., and Abramovic, O., eds., John Wiley and Sons, Inc., p. 517.

THE ALDO-KETO REDUCTASES: AN OVERVIEW

T. Geoffrey Flynn and Nancy C. Green

Department of Biochemistry
Queen's University
Kingston, Ontario, Canada K7L 3N6

INTRODUCTION

The past few years have seen real and rapid advances in our knowledge of the structure and function of the aldo-keto reductases. It is now evident that these enzymes constitute a superfamily of monomeric oxidoreductases (Carper *et al.*,1987; Bohren *et al.*,1989; Carper *et al.*,1989). The three major enzymes comprising this group are aldehyde reductase, aldose reductase and carbonyl reductase. These enzymes are functionally related through a broad and overlapping substrate specificity and an attempt to classify them on this basis was made at the Bern meeting in 1982 (Turner & Flynn, 1982). This classification has not been accepted universally and many workers retain and use a varied nomenclature. This of course serves only to confuse and it is incumbent on workers in the field to agree on a sensible nomenclature. This should be one aim for the next meeting to be held in New Zealand in 1994.

ALDEHYDE REDUCTASE (HIGH K_M ALDEHYDE REDUCTASE: GLUCURONATE REDUCTASE: HEXONATE DEHYDROGENASE: ALR1)

This is the flagship enzyme of the aldo-keto reductase fleet yet the least progress has been made with this enzyme. This is very evident from the fact that not a single paper in the present volume is concerned with aldehyde reductase. It was established early on that enzymes known as the high-K_M aldehyde reductase (Turner & Tipton, 1972), glucuronate reductase (Bosron & Prairie, 1972) and hexonate dehydrogenase (Mano *et al.*,1961) were one and the same and this enzyme was given the designation aldehyde reductase 1 or ALR1 (EC 1.1.1.2) by Turner and Flynn (1982). Aldehyde reductase uses NADPH exclusively as coenzyme and has a broad substrate specificity for aromatic and aliphatic aldehydes (Flynn, 1982). ALR1 does not reduce glucose and other aldose sugars but it does reduce D-glucuronate to L-gulonate and it has been proposed that this is its primary physiological role (Tulsiani & Touster, 1979). The enzyme also reduces isocorticosteroids in the long pathway of steroid catabolism (Wermuth & Monder, 1983) and also is a very good catalyst for the reduction of aldehydes from biogenic amines (Turner & Tipton, 1972). Aldehyde reductase therefore appears to have a variety of functions and it might be supposed that it is a general housekeeping enzyme for the detoxication of a variety of different aldehydes. However, there are other enzymes available for the reduction of toxic aldehydes (e.g. aldehyde dehydrogenase and aldose reductase) and a defined metabolic role for ALR1 has not been established.

From a mechanistic and structural point of view the least progress has been made with ALR1. Its kinetic mechanism has been investigated by a number of workers (e.g. Davidson & Flynn, 1979a; Wermuth & von Wartburg, 1980; Morpeth & Dickinson, 1981). With simple substrates like D-glyceraldehyde, the enzyme displays an ordered sequential mechanism with NADPH binding first (Davidson & Flynn, 1979a). With other substrates, the mechanism may change depending on the aldehyde used (Magnien & Branlant, 1983). By and large though, the accepted mechanism for ALR1 is an ordered bi-bi mechanism. The rate limiting step has not been determined but it is likely to be the rate of isomerization of the binary complex $E^{*}.NADP \rightleftarrows E.NADP^{+}$ as is the case for aldose reductase (Kubiseski *et al.*,1992). Aldehyde reductase is a monomeric oxidoreductase with a molecular weight of about 35,000. The amino acid sequence of ALR1, reported by Wermuth *et al.* (1987) using protein sequencing techniques, was the first complete sequence of any aldo-keto reductase. At the time, the sequence of ALR1 exhibited no significant homology with any other enzyme. However, it is now known that is has significant sequence identity with aldose reductase (Bohren *et al.*,1989), and the enzymes are clearly closely related.

The three dimensional structure of aldehyde reductase has not yet been established although at least two groups are working on it. Given the sequence similarity to aldose reductase the enzyme is almost certainly also a TIM β-barrel enzyme (Farber & Petsko, 1990). Eagerly awaited features of a comparison of the crystal structures of ALR1 and aldose reductase are the structural characteristics that determine the ability of aldose reductase to bind both NADPH and NADH. Chemical modification of active site residues of ALR1 has suggested that both lysine (Flynn *et al.*, 1980) and arginine (Davidson & Flynn, 1979b) play critical roles in the binding of coenzyme. However, similar roles have been postulated for the same residues in aldose reductase (Morjana *et al.*,1989; Flynn *et al.*,1989) and the precise positioning of these amino acids and others to allow NADH binding in ALR2 and to exclude it in ALR1 remains to be elucidated.

Despite cloned cDNA of ALR1 being available for several years little work has been done on expressing the enzyme and studying its activity by site-directed mutagenesis. This approach has been confined largely to aldose reductase. However, one suspects that increased attention will be focused on ALR1 when the three-dimensional structure is known.

ALDOSE REDUCTASE (LOW-K_M ALDEHYDE REDUCTASE: ALR2)

While aldehyde reductase may be the flagship enzyme, ALR2 is the star-ship of the enterprise. More attention has been focused on this enzyme because of its suspected role in the etiology of diabetic complications and because of the potential this enzyme has for rational drug design. Early in the 1980's aldose reductase was shown to be identical with the low-K_M aldehyde reductase (Cromlish & Flynn, 1983) an enzyme first characterized by Turner and Tipton (1972). Workers in the aldose reductase field fall into two distinct camps depending on the particular perspective. Most of the early work on aldose reductase was done by workers interested in the role of the enzyme in diabetic complications (for review see Gabbay, 1975). The involvement of ALR2 in the polyol pathway and diabetic complications has been extensively investigated by Jin Kinoshita and his colleagues at the NIH and the fact that inhibition of the enzyme prevented the onset of cataract in experimental diabetes was also shown by the Kinoshita group (Fukushi *et al.*,1980). Much of the work on aldose reductase has been concerned with the development of specific aldose reductase inhibitors (Kador *et al.*,1985) and little attention was paid to this enzyme by enzymologists until it was shown that aldose reductase was identical with the low-K_M aldehyde reductase.

From the mid 1980's on significant progress has been made on the enzymology and protein chemistry of aldose reductase. The amino acid sequence of the enzyme has been determined from a number of sources both by cloning and sequencing cDNA (Carper *et al.*,1987; Bohren *et al.*,1989; Garcia-Perez *et al.*,1989; Nishimura *et al.*,1990; Pailhoux *et al.*,1990) and by conventional protein sequencing techniques (Schade *et al.*,1990). Although, apparently, there is only one gene for aldose reductase (Graham *et al.*,1991a; Graham *et al.*,1991b) there are at least three pseudogenes (Graham *et al.*,1991b) and Southern blotting of genomic DNA shows the presence of a number of related genes (Nishimura *et al.*,1988; Bohren *et al.*,1989). These include prostaglandin F synthase (Watanabe *et al.*,1988), frog lens ρ-crystallin (Tomarev *et al.*,1984) and others. Aldose reductase seems to belong to its own subfamily of enzymes thus making the aldo-keto reductases a widely encompassing super-family as suggested by (Carper *et al.*,1987; Bohren *et al.*,1989). It also appears that the primary structure of aldose reductase has been adapted for other purposes which makes the role of this enzyme quite intriguing. For example, the microorganism *Leishmania major* expresses a developmentally regulated protein that has significant sequence similarity to aldose reductase (Samaras & Spithill, 1989). The ubiquity of the aldose reductase sequence in several diverse proteins makes one suspect that its primary role is not that of the detoxication of aldehydes.

Because the K_M of aldose reductase for biogenic aldehydes is lower than that of aldehyde reductase for the same substrates, the primary physiological role of aldose reductase was postulated to be that of detoxifying these compounds, particularly in the brain (Turner *et al.*,1974). To workers more interested in the enzyme's role in diabetic complications it had to be supposed that the prime physiological substrate for the enzyme is glucose. Since the K_M for glucose ranges between 100-600 mM for the enzyme from most species there has been much scepticism regarding glucose being the physiological substrate (Cromlish & Flynn, 1983). However, there is strong evidence that it is the acyclic form of glucose, not the cyclic, which is used as a substrate by ALR2. Indeed a K_M of 0.66 μM has been calculated for aldose reductase with glucose in this form (Inagaki *et al.*,1982). In addition, Grimshaw (1986) has shown that in the reduction of D-glucose by ALR2 the rate of formation of the free carbonyl (i.e. the acyclic form) is rate limiting for the overall reaction.

Searches for a physiological role notwithstanding, the fact still remains that aldose reductase and aldehyde reductase are capable of reducing a very large variety of aldehydes. The reason ALR2 is able to do this is now becoming apparent from the three-dimensional structure recently elucidated independently by Biellmann and his colleagues (Rondeau *et al.*, 1992) and by Gabbay's group in collaboration with Quiocho (Wilson *et al.*,1992). The likely substrate binding site is a very large shallow depression formed in the main by aromatic and aliphatic amino acids. This apolarity presumably allows the binding of molecules as diverse as p-nitrobenzaldehyde and isocorticosteroids. It begs the question somewhat as to how it allows binding and correct positioning of molecules such as D-glyceraldehyde and the straight-chain form of glucose for catalysis. Precise definition of the substrate binding site will entail some difficulty for the crystallographer since substrate cannot bind in the absence of NADPH, and crystals of the ternary rather than binary complex will be required.

The binding of coenzyme to aldose reductase causes quite a large conformational change in the enzyme. This has been suggested by the kinetic analyses of Grimshaw *et al.*, (1990), and Kubiseski *et al.*, (1992), from changes in intrinsic fluorescence upon coenzyme binding (Kubiseski *et al.*, 1992) and by superimposition of the α-carbon backbone of pig ADP-ribose phosphate (ADPRP) bound enzyme (which is essentially the structure of the free enzyme since no fluorescence changes are shown upon ADPRP binding) on the α-carbon backbone of human NADPH-bound aldose reductase (Borhani, D. and Flynn, T.G., unpublished results). The postulated movement in a C-terminal loop of the enzyme is large and serves to anchor the NADPH tightly in position.

The recent work on the elucidation of the three-dimensional structure of aldose reductase together with the successful expression of aldose reductase cDNA in a number of systems will undoubtedly produce a spate of site-directed mutagenesis studies designed to elucidate the role of specific residues in the mechanism of action of the enzyme. Some studies have already appeared (Old *et al.*,1990; Bohren *et al.*,1991; Yamaoka *et al.*,1992) but knowledge of the three-dimensional structure will give these studies greater meaning.

CARBONYL REDUCTASE (ALR3)

Carbonyl reductase, like ALR1 and ALR2, is a cytosolic monomeric NADPH-dependent oxidoreductase and because of a broad, similar overlapping substrate specificity has been classified as one of the aldo-keto reductases (Turner & Flynn, 1982). However, there are significant functional differences between ALR3 and the other two enzymes. ALR3 has a distinct preference for ketones (e.g. quinones rather than aldehydes) and its stereochemistry is different in that it catalyzes the transfer of the pro-4S hydrogen rather than the pro-4R (Wermuth, 1981). The determination of the primary structure of carbonyl reductase has shown, however, that the difference between ALR3 and ALR1 and ALR2 is even more profound (Wermuth *et al.*,1988). The amino acid sequence of carbonyl reductase bears no significant homology with either aldose or aldehyde reductase. Despite the similarity in substrate specificity the lack of structural similarity (the chain length of 277 amino acids is also shorter than that of aldehyde and aldose reductase) opens the question as to whether this enzyme is properly classified with the other two. Indeed, the amino acid sequence of carbonyl reductase shows some relationship to the short chain alcohol dehydrogenase/polyol dehydrogenases e.g. ribitol dehydrogenase and Wermuth *et al.* (1988) have raised the interesting possibility that a similar relationship exists between the aldo-keto reductases as does between the alcohol dehydrogenases. The latter functionally similar enzymes are grouped into two families with no apparent structural homologies and with differing stereospecificities. A more recent finding emphasizes the need for a complete re-evaluation of the role and classification of ALR3. Tanaka *et al.* (1992) have shown that subject to species variation, 20β-hydroxysteroid dehydrogenase is identical both functionally and structurally with carbonyl reductase. The role of the aldo-keto reductases in steroid metabolism becomes more and more evident.

The difference in structure between carbonyl reductase and aldehyde and aldose reductase raises the intriguing possibility that the three-dimensional structure of carbonyl reductase might be more typical of dehydrogenases in general i.e. will contain a dinucleotide binding domain or Rossmann fold (Rossmann *et al.*,1974) as opposed to a TIM β-barrel. Determination of the three-dimensional structure of carbonyl reductase will not only resolve this question but will also provide the means of examining contrasting mechanisms of hydride transfer in functionally similar oxidoreductases.

Much of the work on carbonyl reductase has been concerned with its role as a drug detoxifying enzyme and it has been identified with daunorubicin reductase and quinone reductase. The cloning and expression of the gene for carbonyl reductase leaves no doubt that carbonyl and daunorubicin reductases are identical enzymes (Forrest *et al.*,1991). The expression of this enzyme allows a detailed examination of the role of carbonyl reductase in the metabolism of anticancer drugs such as daunorubicin.

CONCLUSION

It is clear that we are at a very interesting point with respect to research on the aldo-keto reductases. Despite significant advancement in our knowledge of the structure and mechanism of action of these enzymes we are still no nearer to the vexed question of their physiological roles. Indeed, as time progresses more and more possibilities are uncovered. Recently, for example, it has been suggested that a primary metabolic role for ALR2 may be the reduction of the highly toxic metabolite, methylglyoxal (Vander Jagt *et al.*,1992). Given the wide variety of possible physiological substrates and metabolites, it is understandable that a variety of names have been applied to the various aldo-keto reductases. However, now that structural relationships between the various enzymes have been established, it should be possible to decide on a mutually acceptable nomenclature. It may well be that the designations ALR1, ALR2 and ALR3 are not appropriate given the structural dissimilarity of ALR3 to the other two. However, this nomenclature was at least arrived at by a logical consideration of the functional characteristics of the enzymes in a similar fashion to the alcohol dehydrogenases. Indeed, the abbreviation ALR was felt to be consistent with that of the other carbonyl metabolizing enzymes particularly aldehyde dehydrogenase which has the designation ALDH.

The papers presented in this section of the book are broadly representative of current research in the aldo-keto reductase field. The determination of the three-dimensional structure of aldose reductase has been the outstanding feature of research in the field over the past few years. It has paved the way for a large number of exciting new experiments designed to explore the kinetic mechanism and in particular to elucidate the salient features of the substrate binding site. It is to be expected that the three-dimensional structures of both aldehyde and carbonyl reductase will be presented in the not too distant future.

We look forward with great anticipation to the latest findings in this rapidly moving field when we next meet in the land of the silver fern in 1994.

REFERENCES

Bohren, K.M., Bullock, B., Wermuth, B. & Gabbay, K.H.,1989, The aldo-keto reductase superfamily. cDNAs and deduced amino acid sequences of human aldehyde and aldose reductases, *J. Biol. Chem.* 264:9547.

Bohren, K.M., Page, J.L., Shankar, R., Henry, S.P. & Gabbay, K.H.,1991, Expression of human aldose and aldehyde reductases. Site- directed mutagenesis of a critical lysine 262, *J. Biol. Chem.* 266:24031.

Bosron, W.F. & Prairie, R.L.,1972, Triphosphopyridine Nucleotide-linked Aldehyde Reductase, *J. Biol. Chem.* 247:4480.

Carper, D., Nishimura, C., Shinohara, T., Dietzchold, B., Wistow, G., Craft, C., Kador, P. & Kinoshita, J.H.,1987, Aldose reductase and p-crystallin belong to the same protein superfamily as aldehyde reductase, *FEBS. Lett.* 220:209.

Carper, D.A., Wistow, G., Nishimura, C., Graham, C., Watanabe, K., Fujii, Y., Hayashi, H. & Hayaishi, O.,1989, A superfamily of NADPH-dependent reductases in eukaryotes and prokaryotes, *Exp. Eye Res.* 49:377.

Cromlish, J.A. & Flynn, T.G.,1983, Identity of Pig Muscle Aldehyde Reductase with Pig Lens Aldose Reductase and with the Low Km Aldehyde Reductase of Pig Brain and Pig Kidney, *J. Biol. Chem.* 258:3583.

Davidson, W.S. & Flynn, T.G.,1979, Kinetics and Mechanism of Action of Aldehyde Reductase From Pig Kidney, *Biochem. J.* 177:595.

Davidson, W.S. & Flynn, T.G., 1979b, A functional arginine residue in NADPH-dependent aldehyde reductase from pig kidney. *J. Biol. Chem.* 254:3724.

Farber, G.K. & Petsko, G.A.,1990, The evolution of alpha/beta barrel enzymes, *TIBS* 15:228.

Flynn, T.G., 1982. Aldehyde reductases: monomeni NADPH-dependent oxidoreductases with multi-functional potential, *Biochem. Pharmacol.* 31:

Flynn, T.G., Charingtion, B., Lyons, C., Chao, H., Hyndman, D.J. & Morjana, N.A.,1989, Chemical Modification of Aldehyde and Aldose Reductase by Pyridoxal 5'-Phosphate, *Prog. Clin. Biol. Res.* 290:251.

Flynn, T.G., Gallerneault, C., Ferguson, D., Cromlish, J.A., Davidson, W.S., 1980, Studies on the active site of pig kidney aldehyde reductases, *Biochem Soc. Transac.* 9:273.

Forrest, G.L., Akman, S., Doroshow, J., Rivera, H. & Kaplan, W.D.,1991, Genomic sequence and expression of a cloned human carbonyl reductase gene with daunorubicin reductase activity, *Mol. Pharmacol.* 40:502.

Fukushi, S., Merola, L. & Kinoshita, J.H.,1980, Altering Cataract in Diabetic Rats, *Invest. Ophthalmol. Vis. Sci.* 19:313.

Gabbay, K.H.,1975, Hyperglyceamia, Polyol Metabolism and Complications of Diabetes Mellitus, *Annu. Rev. Med.* 26:521.

Garcia-Perez, A., Martin, B., Murphy, H.R., Uchida, S., Murer, H., Cowley, B.D.,Jr., Handler, J.S. & Burg, M.B.,1989, Molecular cloning of cDNA coding for kidney aldose reductase. Regulation of specific mRNA accumulation by NaCl-mediated osmotic stress, *J. Biol. Chem.* 264:16815.

Graham, A., Brown, L., Hedge, P.J., Gammack, A.J. & Markham, A.F.,1991a, Structure of the Human Aldose Reductase Gene, *J. Biol. Chem.* 266:6872.

Graham, C., Szpirer, C., Levan, G. & Carper, D.,1991b, Characterization of the aldose reductase-encoding gene family in rat, *Gene* 107:259.

Grimshaw, C.E.,1986, Direct measurement of the rate of ring opening of D-glucose by enzyme-catalyzed reduction, *Carbohydr. Res.* 148:345.

Grimshaw, C.E., Shahbaz, M. & Putney, C.G.,1990, Mechanistic Basis for Nonlinear Kinetics of Aldehyde Reduction Catalyzed by Aldose Reductase, *Biochemistry* 29:9947.

Inagaki, K., Miwa, I. & Okuda, J.,1982, Affinity Purification anf Glucose Specificity of Aldose Reductase from Bovine Lens, *Arch. Biochem. Biophys.* 216:337.

Kador, P.F., Robison, W.G.,Jr & Kinoshita, J.H.,1985, The Pharmacology of Aldose Reductase Inhibitors, *Annu. Rev. Pharmacol. Toxicol.* 25:691.

Kubiseski, T.J., Hyndman, D.J., Morjana, N.A. & Flynn, T.G.,1992, Studies on pig muscle aldose reductase. Kinetic mechanism and evidence for a slow conformational change upon coenzyme binding, *J. Biol. Chem.* 267:6510.

Magnien, A. & Branlant, G.,1983, The Kinetics and Mechanism of Pig Liver Aldehyde Reductase, *Eur. J. Biochem.* 131:375.

Mano, Y., Suzudi, K., Yamada, K. & Shinazono, N.,1961, Enzymic Studies on TPN L-hexonate Dehydrogenase from Rat Liver, *J. Biochem. (Tokyo)* 49:618.

Morjana, N.A., Lyons, C. & Flynn, T.G.,1989, Aldose reductase from human psoas muscle. Affinity labeling of an active site lysine by pyridoxal 5'-phosphate and pyridoxal 5'-diphospho-5'-adenosine, *J. Biol. Chem.* 264:2912.

Morpeth, F.F. & Dickinson, F.M.,1981, Kinetic Studies of the Mechanism of Pig Kidney Aldeyde Reductase, *Biochem. J.* 193:485.

Nishimura, C., Graham, C., Hohman, T.C., Nagata, M., Robison WG, Jr. & Carper, D.,1988, Characterization of mRNA and genes for aldose reductase in rat, *Biochem. Biophys. Res. Commun.* 153:1051.

Nishimura, C., Matsuura, Y., Kokai, Y., Akera, T., Carper, D., Morjana, N., Lyons, C. & Flynn, T.G.,1990, Cloning and expression of human aldose reductase, *J. Biol. Chem.* 265:9788.

Old, S.E., Sato, S., Kador, P.F. & Carper, D.A.,1990, *In vitro* expression of rat lens aldose reductase in *Escherichia coli*, *Proc. Natl. Acad. Sci. USA* 87:4942.

Pailhoux, E.A., Martinez, A., Veyssiere, G.M. & Jean, C.G.,1990, Androgen-dependent protein from mouse vas deferens. cDNA cloning and protein homology with the aldo-keto reductase superfamily, *J. Biol. Chem.* 265:19932.

Rondeau, J-M., Tête-Favier, F., Podjarny, A., Reymann, J.-M., Barth, P., Biellmann, J.-F. and Moras, D. 1992. Novel NADPH-binding domain revealed by the crystal structure of aldose reductase. *Nature* 355:469.

Rossmann, M.G., Moras, D. & Olsen, K.W.,1974, Chemical and Biological Evolution of a Nucleotide-binding Protein, *Nature* 250:194.

Samaras, N. & Spithill, T.W.,1989, The Developmentally Regulated P100/11E Gene of Leishmanina major Shows Homology to a Superfamily of Reductases Genes, *J. Biol. Chem.* 264:4251.

Schade, S.Z., Early, S.L., Williams, T.R., Kezdy, F.J., Heinrikson, R.L., Grimshaw, C.E. & Doughty, C.C.,1990, Sequence analysis of bovine lens aldose reductase, *J. Biol. Chem.* 265:3628.

Tanaka, M., Ohno, S., Adachi, S., Nakajin, S., Shinoda, M. & Nagahama, Y.,1992, Pig Testicular 20beta-hydroxysteroid Dehydrogenase Exhibits Carbonyl Reductase-like Structure and Activity, *J. Biol. Chem.* 267:13451.

Tomarev, S.I., Zinovieva, R.D., Dolgilevich, S.M., Luchin, S.V., Krayev, A.S., Skryabin, D.G. & Gause, G.G.,1984, A Novel Type of Crystallin in the Frog Eye Lens. 35-kDa Polypeptide is not Homologous to any of the Major Classes of Lens Crystallins, *FEBS Lett.* 171:297.

Tulsiani, D.R.P. & Touster, O.,1979, Studies on Dehydrogenases of the Glucuronate-xylulose Cycle in the Livers of Diabetic Mice and Rats, *Diabetes* 28:793.

Turner, A.J. & Flynn, T.G. (1982) in Enzymology of Carbonyl Metabolism: Aldehyde Dehydrogenase and Aldo/Keto Reductase (Weiner, H. & Wermuth, B., eds.), pp. 401-402, Alan R. Liss, New York

Turner, A.J., Illingworth, J.A. & Tipton, K.F.,1974, Simulation of Biogenic Amine Metabolism in Brain, *Biochem. J.* 144:353.

Turner, A.J. & Tipton, K.F.,1972, The Characterization of Two Reduced Nicotinamide-Adenine Dinucleotide Phosphate-Linked Aldehyde Reductases from Pig Brain, *Biochem. J.* 130:765.

Vander Jagt, D.L., Robinson, B., Taylor, K.K. & Hunsaker, L.A.,1992, Reduction of trioses by NADPH-dependent aldo-keto reductases. Aldose reductase, methylglyoxal, and diabetic complications, *J. Biol. Chem.* 267:4364.

Watanabe, K., Fujii, Y., Nakayama, K., Ohkubo, H., Kuramitsu, S., Kagamiyama, H., Nakanishi, S. & Hayaishi, O.,1988, Structural similarity of bovine lung prostaglandin F synthase to lens epsilon-crystallin of the European common frog, *Proc. Natl. Acad. Sci. USA* 85:11.

Wermuth, B., 1981, Purification and Properties of an NADPH-dependent Carbonyl Reductase, *J. Biol. Chem.* 256:1206.

Wermuth, B., Bohren, K.M., Heinmann, G., von Wartburg, J.P. & Gabbay, K.H.,1988, Human Carbonyl Reductase: Nucleotide Sequence Analysis of a cDNA and amino acid Sequence of the Encoded Protein, *J. Biol. Chem.* 263:16185.

Wermuth, B. & Monder, C.,1983, Aldose and Aldehyde Reductase Exhibit Isocorticosteroid Reductase Activity, *Eur. J. Biochem.* 131:423.

Wermuth, B., Omar, A., Forster, A., di Francesco, C., Wolf, M., von Wartburg, J.P., Bullock, B. & Gabbay, K.H.,1987, Primary structure of aldehyde reductase from human liver, *Prog. Clin. Biol. Res.* 232:297.

Wermuth, B. & von Wartburg, J.P.,1980, Kinetic Studies on NADPH-linked Aldehyde Reductase from Human Liver, *Adv. Exp. Med. Biol.* 132:189.

Wilson, D.K., Bohren, K.M., Gabbay, K.H. & Quiocho, F.A.,1992, An unlikely sugar substrate site in the 1.65 Å structure of the human aldose reductase holoenzyme implicated in diabetic complications, *Science* 257:81.

Yamaoka, T., Matsuura, Y., Yamashita, K., Tanimoto, T. & Nishimura, C.,1992, Site-directed mutagenesis of His-42, His-188 and Lys-263 of human aldose reductase, *Biochem. Biophys. Res. Commun.* 183:327.

LOCATION OF AN ESSENTIAL ARGININE RESIDUE IN THE PRIMARY STRUCTURE OF PIG ALDOSE REDUCTASE

Terrance J. Kubiseski, Nancy C. Green and T. Geoffrey Flynn

Department of Biochemistry
Queen's University
Kingston, Ontario, Canada K7L 3N6

INTRODUCTION

There have been several recent advances in the elucidation of the structure and mechanism of action of aldose reductase (ALR2; EC 1.1.1.21). The enzyme exhibits ordered sequential kinetics and during the kinetic mechanism a conformational change occurs upon coenzyme binding (Grimshaw et al.,1990; Kubiseski et al.,1992). Unlike most oxidoreductases the rate limiting step is not the release of $NADP^+$ but rather, in the forward direction at least, the rate of isomerization of the binary enzyme-coenzyme complex, $E^{*}\cdot NADP \leftrightarrows E\cdot NADP$ (Kubiseski et al.,1992). Chemical modification studies previously have suggested several amino acid residues essential for the function of the enzyme e.g. lysine (Morjana et al.,1989), arginine (Halder et al.,1985) and cysteine (Liu et al.,1989). The cloning and sequencing of ALR2 from several species has enabled the precise location of such residues in the primary structure (Carper et al.,1987; Garcia-Perez et al.,1989; Nishimura et al.,1990; Schade et al.,1990). Moreover, the development of expression systems for ALR2 has allowed an examination of critical residues by site-directed mutagenesis (Bohren et al.,1991; Yamaoka et al.,1992).

A major breakthrough in the study of ALR2 was made recently by the elucidation of the three-dimensional structure of the enzyme by X-ray crystallography (Rondeau et al.,1992; Wilson et al.,1992). Thus, a combination of site-directed mutagenesis, enzyme expression and X-ray crystallography can now be used to examine residues involved in the binding of coenzyme and substrates, in the binding of aldose reductose inhibitors and in the mechanism of hydride transfer.

A pre-requisite to the study of a functional amino acid residue in any enzyme is the determination of the enzyme's primary structure. In the present paper we describe the determination of the amino acid sequence of pig aldose reductase from the sequence of cDNA for ALR2 and some preliminary studies on the location of an essential arginine.

Enzymology and Molecular Biology of Carbonyl Metabolism 4
Edited by H. Weiner, Plenum Press, New York, 1993

EXPERIMENTAL PROCEDURES

Materials

Aldose reductase was prepared from pig muscle obtained at the Ross McFedridge Abattoir of Glenburnie, Ontario, by the procedure described previously (Cromlish and Flynn, 1983). DL-glyceraldehyde, phenylglyoxal, and trypsin (Grade III) were purchased from the Sigma Chemical Co. Sephadex G-50 was obtained from Pharmacia. [7-^{14}C]-Phenylglyoxal was purchased from Amersham Corp. Trichloroacetic acid was purchased from Pierce. HPLC grade acetonitrile and sodium phosphate were obtained from BDH (Canada) Ltd., Dorval, Quebec. NADPH and NADP$^+$ were obtained from Boehringer Mannheim Canada Ltd., Laval, Quebec.

Standard Enzyme Steady State Activity Assay

Enzyme assays were performed at 25°C using a Hewlett-Packard HP8452A diode array spectrophotometer linked to a Hewlett-Packard 9000-300 (model 98561A) desktop computer as previously described (Kubiseski et al.,1992).

Modification of ALR2 with Phenylglyoxal

Purified ALR2 (0.20 mg/ml) was incubated with varying concentrations of phenylglyoxal at 25°C in the dark in 75 mM sodium bicarbonate buffer (pH 8.0). The reaction was initiated by adding phenylglyoxal. Ten μL aliquots were removed at various time intervals and the enzyme activities were then determined. The initial activity of ALR2 was measured in triplicate immediately before adding reagent and the enzyme activity during the time course of reaction was expressed as a fraction of the control.

Enzymatic Digestion of Modified ALR2

ALR2 (0.170 mg) was incubated with 5 mM [7-^{14}C]-phenylglyoxal in 75 mM sodium bicarbonate buffer, pH 8.0 (in a total volume of 100 μL) at 25°C in the dark for 60 min. The samples were then desalted by the rapid microcentrifuge desalting techniques described by Penefsky (1971). One mL syringes were plugged with a small amount of glass wool. Preswollen Sephadex G-50 in H_2O was loaded into the column to within one cm of the top. The syringe was centrifuged in a 15 mL Corex tube in a clinical bench top centrifuge for 4 min. at 1800 rpm to remove excess buffer. The column was then equilibrated with 4 washes of 100 μL of H_2O and recentrifuged. The sample in 100 μL was applied to the column and recentrifuged with collection into Eppendorf tubes. The samples were then dried down and digested with trypsin by a method described by Stone et al. (1989) with a few modifications. To the dry protein was added 50 μL of 8 M urea, 0.4 M ammonium bicarbonate, 4 mM calcium chloride buffer and 10 μL of 45 mM dithiothreitol. This was then incubated at 50°C for 10 min. and, after cooling to room temperature, 10 μL of 120 mM iodoacetamide was added, and the sample was allowed to incubate at room temperature for 10 min. The denatured enzyme was then digested by adding 140 μL of H_2O and 5.0 μL of trypsin (1.6 mg/mL). The reaction was incubated at 37°C for 8 hrs., and the reaction was stopped by acidifying the digest with 0.1% trifluoracetic acid.

Peptide Purification and Isolation

The enzyme digests were subjected to HPLC using a reverse phase column (0.46 x 25 cm Vydac C_{18}) pre-equilibrated with 0.1% aqueous trifluoracetic acid. The peptide peaks were eluted with a linear gradient of 0-60% acetonitrile containing 0.1% triflouroacetic acid at a flow rate of 1 mL/min. The peptides were monitored at 217 nm and the radioactivity in individual fractions was determined by scintillation counting using Beckman Ready Cap solvent free scintillation medium. The [7-^{14}C]-phenylglyoxal labelled peptides were purified by reapplying them to the HPLC column and the peptides were eluted by a shallower linear gradient of acetonitrile at a flow rate of 1 mL/min. Fractions containing labelled, purified peptides were dried down to 50 μL and kept at -20°C for subsequent identification by amino acid sequencing.

Peptide Sequencing

Peptides were subjected to sequencing by automatic Edman degradation in a Model 470A gas phase sequencer (Applied Biosystems) with on-line phenylthiohydantoin-amino acid analysis.

Cloning and Sequencing of Aldose Reductase cDNA

Full length human aldose reductase cDNA (Nishimura et al.,1990) was labelled using the random primer method (Feinberg et al.,1983) and [α^{32}P] dCTP (ICN). This was used to screen 8.0 x 10^5 clones of a λgt10 porcine brain cDNA library (Clonetech) following essentially the protocols outlined in Maniatis et al. (1982). Positive clones were rescreened until plaque purified. Insert sizes were determined using the polymerase chain reaction (PCR) (Saiki et al.,1985) primed by λgt10 rightward and leftward oligonucleotides primers (Oligonucleotide Synthesis Laboratory, Queen's University) followed by electrophoresis in 1% agarose. Clones with the largest inserts were subcloned into pGEM 3Z at the EcoRI site. One subclone was sequenced on both strands using dideoxy sequencing kits (Promega and Pharmacia) following the manufacturers' protocols. Successive sequencing primers (17-mers) were synthesized as DNA sequence was determined. All enzymes used were from either Pharmacia, BRL or Promega. As well, all chemicals used were reagent grade or better.

RESULTS

Cloning and Sequencing of cDNA for ALR2

Several putative clones were obtained from the pig brain cDNA library and their insert sizes were determined via PCR using λgt10 sites found in the vector. One of these clones contained the complete coding sequence for ALR2 (Fig. 1) together with 19 bp of 5′ untranslated sequence. Comparison of the deduced amino acid sequence (Fig. 1) with the amino acid sequence of ALR2 from other species revealed an amino acid sequence identity of between 84 and 89%. Thus, as pointed out previously (Bohren et al.,1989) there is a high degree of sequence homology between aldose reductases from different species.

Reaction of ALR2 with Phenylglyoxal

Reaction of ALR2 with different concentrations of phenylglyoxal resulted in a time dependent inactivation of the enzyme activity. The inactivation curve obtained with 1 mM reagent is shown in Fig. 2. When the enzyme (0.20 mg/mL) was incubated with 5 mM of phenylglyoxal for 40 minutes, no activity remained. The inactivation of the enzyme follow pseudo-first order kinetics over a concentration range of 0.25-5.0 mM phenylglyoxal. The plot of the logarithm of residual activity versus the time of reaction with phenylglyoxal was

```
-19       -10                        15                  30
CGGGAGGAG CTCCGCAGTC ATG GCC AGC CAC CTC GTG CTC TAC ACC GGT GCC AAG
                     Met Ala Ser His Leu Val Leu Tyr Thr Gly Ala Lys

         45                  60                  75                  90
ATG CCC ATC TTG GGG CTG GGC ACC TGG AAG TCC CCT CCA GGC AAA GTG ACA GAG
Met Pro Ile Leu Gly Leu Gly Thr Trp Lys Ser Pro Pro Gly Lys Val Thr Glu

                105                 120                 135
GCT GTG AAG GTG GCC ATT GAC CTT GGG TAC CGC CAC ATT GAC TGT GCC CAT GTG
Ala Val Lys Val Ala Ile Asp Leu Gly Tyr Arg His Ile Asp Cys Ala His Val

    150                 165                 180                 195
TAC CAG AAC GAG AAC GAG GTG GGG CTG GGC CTC CAG GAG AAG CTC CAA GGG CAG
Tyr Gln Asn Glu Asn Glu Val Gly Leu Gly Leu Gln Glu Lys Leu Gln Gly Gln

            210                 225                 240
GTG GTG AAG CGT GAG GAC CTC TTC ATC GTC AGT AAG CTG TGG TGC ACA GAC CAT
Val Val Lys Arg Glu Asp Leu Phe Ile Val Ser Lys Leu Trp Cys Thr Asp His

255                 270                 285                 300
GAG AAG AAC CTG GTG AAA GGG GCC TGC CAG ACG ACC CTC CGC AAC CTG AAG CTG
Glu Lys Asn Leu Val Lys Gly Ala Cys Gln Thr Thr Leu Arg Asn Leu Lys Leu

        315                 330                 345                 360
GAC TAC CTG GAC CTC TAC CTT ATC CAC TGG CCC ACT GGC TTC AAG CCT GGC AAG
Asp Tyr Leu Asp Leu Tyr Leu Ile His Trp Pro Thr Gly Phe Lys Pro Gly Lys

                375                 390                 405
GAC CCT TTC CCG CTG GAT GGC GAC GGC AAC GTG GTT CCT GAT GAG AGC GAT TTT
Asp Pro Phe Pro Leu Asp Gly Asp Gly Asn Val Val Pro Asp Glu Ser Asp Phe

    420                 435                 450                 465
GTG GAG ACG TGG GAG GCC ATG GAG GAG CTG GTG GAC GAA GGG CTG GTG AAA GCC
Val Glu Thr Trp Glu Ala Met Glu Glu Leu Val Asp Glu Gly Leu Val Lys Ala

            480                 495                 510
ATT GGC GTC TCC AAC TTC AAC CAT CTG CAA GTG GAG AAG ATC TTA AAC AAA CCT
Ile Gly Val Ser Asn Phe Asn His Leu Gln Val Glu Lys Ile Leu Asn Lys Pro

525                 540                 555                 570
GGC CTG AAA TAC AAG CCA GCC GTT AAC CAG ATC GAG GTC CAC CCA TAC CTC ACT
Gly Leu Lys Tyr Lys Pro Ala Val Asn Gln Ile Glu Val His Pro Tyr Leu Thr

        585                 600                 615                 630
CAG GAG AAG TTA ATC GAA TAC TGC AAG TCC AAA GGC ATC GTG GTG ACT GCC TAC
Gln Glu Lys Leu Ile Glu Tyr Cys Lys Ser Lys Gly Ile Val Val Thr Ala Tyr

                645                 660                 675
AGC CCC CTC GGC TCT CCC GAC AGG CCC TGG GCC AAG CCT GAG GAC CCT TCC CTC
Ser Pro Leu Gly Ser Pro Asp Arg Pro Trp Ala Lys Pro Glu Asp Pro Ser Leu

    690                 705                 720                 735
CTG GAG GAC CCC AGG ATC AAA GCG ATT GCA GCC AAG TAC AAT AAA ACC ACA GCC
Leu Glu Asp Pro Arg Ile Lys Ala Ile Ala Ala Lys Tyr Asn Lys Thr Thr Ala

            750                 765                 780
CAG GTT CTG ATC CGG TTC CCC ATG CAG AGG AAC TTG ATC GTC ATC CCC AAG TCC
Gln Val Leu Ile Arg Phe Pro Met Gln Arg Asn Leu Ile Val Ile Pro Lys Ser

795                 810                 825                 840
GTG ACG CCT GAA CGC ATT GCC GAG AAC TTC CAG GTC TTT GAC TTT GAA CTG AGC
Val Thr Pro Glu Arg Ile Ala Glu Asn Phe Gln Val Phe Asp Phe Glu Leu Ser

        855                 870                 885                 900
CCT GAG GAT ATG AAC ACC TTA CTG AGC TAC AAC AGG AAC TGG AGG GTC TGT GCC
Pro Glu Asp Met Asn Thr Leu Leu Ser Tyr Asn Arg Asn Trp Arg Val Cys Ala

                915                 930                 945
TTG ATG AGC TGT GCC TCC CAC AAG GAT TAT CCC TTT CAC GAG GAA TAT TGA
Leu Met Ser Cys Ala Ser His Lys Asp Tyr Pro Phe His Glu Glu Tyr  *
```

Figure 1. cDNA and derived amino acid sequence of pig brain aldose reductase.

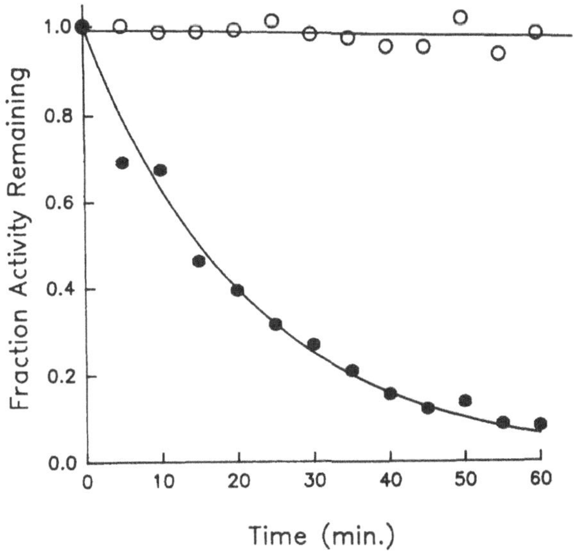

Figure 2. Modification of ALR2 in the presence of phenylglyoxal (●, 1 mM), and absence of phenylglyoxal (○). Experimental procedure as described in text.

linear as predicted by equation 1:

$$-\ln (E/Eo) = k_{obs}\, t \quad (1)$$

where E and Eo are the activity at time t and zero time, respectively, and k_{obs} is the observed first-order rate constant for the loss of enzymatic activity.

Stoichiometry of Reaction of ALR2 with Phenylglyoxal

The number of amino acid residues modified by phenylglyoxal which affect the enzymatic activity was determined from the inactivation kinetics as described by (Nihira et al.,1981). The measurement of the apparent velocity constant of inactivation (k_{obs}) versus inhibitor concentration can be described by equation 2:

$$E + nI \rightarrow EI_n \quad (2)$$

where n is the number of molecules of reagent bound per enzyme which affects enzymatic activity. The rate of ALR2 inactivation is:

$$d[E]/dt = -k[E]\,[I]^n \quad (3)$$

where k is the second order rate constant for the reaction. The pseudo-first order rate constant is described by $k_{obs} = k[I]^n$ so that $d[E]/dt = -k_{obs}[E]$. By taking logarithm of k_{obs} the following equation is obtained

$$\log k_{obs} = n \log [I] + \log k. \quad (4)$$

Plotting log k_{obs} versus log of reagent concentration gave a straight line with slope = n. The slope of this line was 0.81, indicating that one molecule of reagent was bound per molecule of the enzyme.

Identification of Phenylglyoxal Modified Peptides

The [7-^{14}C]-phenylglyoxal labelled peptide was isolated by applying the tryptic digest of [7-^{14}C]-phenylglyoxal modified ALR2 in the absence of any substrate to a HPLC C_{18} reverse phase column and the eluent was monitored at 217 nm. Aliquots of all fractions were withdrawn to determine the presence of ^{14}C-labelled materials by scintillation counting. The elution profile of the digest showed three major peaks and several minor peaks. Peaks I and II routinely appeared in all mock digestions carried out in the absence of trypsin demonstrating that these two peaks were not peptides generated by trypsin. The radioactive peak III was eluted by 26.3% acentonitrile after 35 minutes. The labelled peptide was further purified by HPLC on a C_{18} reverse phase column. The amino acid sequence of the purified peptide was then determined. Sequence analysis showed that the amino acid sequence of this peptide correspond to residues 263-269 in which Arg-268 showed a very low yield (< 1%) of phenylthiohydantoin derivative. Therefore, the arginine at position 268 was assumed to be the essential residue modified by phenylglyoxal.

DISCUSSION

As expected the amino acid sequence of pig ALR2 shows a high degree of identity with ALR2 from other species (84-89%). Since the three-dimensional structure of pig ALR2 (Rondeau et al.,1992) is also virtually identical with that of human ALR2 (Wilson et al.,1992) then mechanistic and structural studies conducted with the pig enzyme, particularly those directed at rational drug design, should be valid for the human enzyme. Previous work in our laboratory (Cromlish et al.,1983) and in Doughty's (Schade et al.,1990) has shown that there is only a single post-translational modification in ALR2, namely, acetylation of the N-terminal alanine. There has been some speculation as to the presence of disulfide bonds in ALR2 (Schade et al.,1990) but the presence of Val 186 in the pig enzyme in place of Cys precludes a disulfide bond with Cys 199. Disulfide bonds involving Cys 298 and Cys 303 have been clearly precluded by the crystal structure (Wilson et al.,1992).

There have been many previous studies designed to elucidate amino acids involved in binding and catalysis at the active site of ALR2. Studies from this laboratory have shown Lys 262 to be present at the coenzyme binding site (Morjana et al.,1989). This residue is part of a motif, -Ile-Pro-Lys-Ser-, which occurs in all aldo-keto reductases (Bohren et al.,1989). Other studies have implicated arginine (Halder et al.,1985) histidine (Bhatnagar et al.,1987) and cysteine (Liu et al.,1989) in catalysis. These studies have now been extended in many laboratories using the technique of site directed mutagenesis (Bohren et al.,1991; Yamaoka et al.,1992). It is clear from these previous studies that modification (Morjana et al.,1989) and mutagenesis (Bohren et al.,1991) of Lys 262, while affecting both K_M and k_{cat}, does not prevent the binding of coenzyme. Chemical modification of ALR2 with the arginine specific reagent phenylglyoxal on the other hand causes an inactivation of the enzyme with the modification of arginine 268 a residue which like Lys 262 is conserved in all aldose reductases. This arginine is located in the coenzyme binding site (Wilson et al.,1992) and is ideally placed to form hydrogen bonds and a salt-linkage with the 2′-OH phosphate on the ribose of the adenine moiety of NADPH. Site-directed mutagenesis of this residue should confirm its importance in the coenzyme binding site.

Acknowledgment

This work was supported by a grant from the Medical Research Council of Canada and by ICI-Pharma. The authors are grateful to Ms. Dianne Hyde-Kelcey and Mr. John Watson for their help in preparing the manuscript.

REFERENCES

Bhatnagar, A., Das, B. & Srivastava, S.K.,1987, Diethyl pyrocarbonate inactivation of human placental aldehyde reductase II of Texas Medical Branch, Galveston 77550, *Biochim. Biophys. Acta* 916:179.

Bohren, K.M., Bullock, B., Wermuth, B. & Gabbay, K.H.,1989, The aldo-keto reductase superfamily. cDNAs and deduced amino acid sequences of human aldehyde and aldose reductases, *J. Biol. Chem.* 264:9547.

Bohren, K.M., Page, J.L., Shankar, R., Henry, S.P. & Gabbay, K.H.,1991, Expression of human aldose and aldehyde reductases. Site- directed mutagenesis of a critical lysine 262, *J. Biol. Chem.* 266:24031.

Carper, D., Nishimura, C., Shinohara, T., Dietzchold, B., Wistow, G., Craft, C., Kador, P. & Kinoshita, J.H.,1987, Aldose reductase and ρ-crystallin belong to the same protein superfamily as aldehyde reductase, *FEBS. Lett.* 220:209.

Cromlish, J.A. & Flynn, T.G.,1983, Identity of pig muscle aldehyde reductase with pig lens aldose reductase and with the low K_M aldehyde reductase of pig brain and pig kidney, *J. Biol. Chem.* 258:3583.

Feinberg, A.P. & Vogelstein, B.,1983, A technique for radiolabelling DNA restriction endonuclease fragments to high specific activity, *Anal. Biochem.* 132:6.

Garcia-Perez, A., Martin, B., Murphy, H.R., Uchida, S., Murer, H., Cowley, B.D.,Jr., Handler, J.S. & Burg, M.B.,1989, Molecular cloning of cDNA coding for kidney aldose reductase. Regulation of specific mRNA accumulation by NaCl-mediated osmotic stress, *J. Biol. Chem.* 264:16815.

Grimshaw, C.E., Shahbaz, M. & Putney, C.G.,1990, Mechanistic basis for nonlinear kinetics of aldehyde reduction catalyzed by aldose reductase, *Biochemistry* 29:9947.

Halder, A.B., James, M. & Crabbe, C.,1985, Chemical modification studies on purified bovine lens aldose reductase, *Ophthalmic. Res.* 17:185.

Kubiseski, T.J., Hyndman, D.J., Morjana, N.A. & Flynn, T.G.,1992, Studies on pig muscle aldose reductase. Kinetic mechanism and evidence for a slow conformational change upon coenzyme binding, *J. Biol. Chem.* 267:6510.

Liu, S., Bhatnagar, A., Das, B. & Srivastava, S.K.,1989, Functional cysteinyl residues in human placental aldose reductase, *Arch. Biochem. Biophys.* 275:112.

Maniatis, T., Fritsch, E.F. & Sambrook, J. (1982) Molecular Cloning; A Laboratory Manual, Cold Spring Harbor Laboratory, Cold Spring Harbor, N.Y.

Morjana, N.A., Lyons, C. & Flynn, T.G.,1989, Aldose reductase from human psoas muscle. Affinity labeling of an active site lysine by pyridoxal 5'-phosphate and pyridoxal 5'-diphospho-5'-adenosine, *J. Biol. Chem.* 264:2912.

Nihira, T., Toraya, T. & Fukui, S.,1981, Modification of tryptophanase with tetranitromethane, *Eur. J. Biochem.* 119:273.

Nishimura, C., Matsuura, Y., Kokai, Y., Akera, T., Carper, D., Morjana, N., Lyons, C. & Flynn, T.G.,1990, Cloning and expression of human aldose reductase, *J. Biol. Chem.* 265:9788.

Penefsky, S.H.,1971, A centrifuged-column procedure for the measurement of ligand binding by beef heart F1, *Methods Enzymol.* 56:527.

Rondeau, J.-M., Tête-Favier, F., Podjarny, A., Reymann, J.-M., Barth, P., Biellmann, J.-F. & Moras, D.,1992, Novel NADPH-binding domain revealed by the crystal structure of aldose reductase, *Nature* 355:469.

Saiki, R.K., Scharf, S., Faloona, F., Mullis, D.B., Horn, G.T., Erlich, H.A. & Arnhein, N.,1985, Enzymatic amplification of β-globin genomic sequences and restriction site analysis for diagnosis of sickle cell anemia, *Science* 230:1350.

Schade, S.Z., Early, S.L., Williams, T.R., Kézdy, F.J., Heinrikson, R.L., Grimshaw, C.E. & Doughty, C.C.,1990, Sequence analysis of bovine lens aldose reductase, *J. Biol. Chem.* 265:3628.

Stone, C.L., Li, T.-K. & Bosron, W.F.,1989, Stereospecific oxidation of secondary alcohols by human alcohol dehydrogenases, *J. Biol. Chem.* 264:11112.

Wilson, D.K., Bohren, K.M., Gabbay, K.H. & Quiocho, F.A.,1992, An unlikely sugar substrate site in the 1.65 Å structure of the human aldose reductase holoenzyme implicated in diabetic complications, *Science* 257:81.

Yamaoka, T., Matsuura, Y., Yamashita, K., Tanimoto, T. & Nishimura, C.,1992, Site-directed mutagenesis of His-42, His-188 and Lys-263 of human aldose reductase, *Biochem. Biophys. Res. Commun.* 183:327.

CYS^{298} IS RESPONSIBLE FOR REVERSIBLE THIOL-INDUCED VARIATION IN ALDOSE REDUCTASE ACTIVITY

Kurt M. Bohren[1], and Kenneth H. Gabbay[1,2]

[1]Department of Pediatrics and [2]Cell Biology
Baylor College of Medicine
Houston, Texas 77030

INTRODUCTION

Aldose reductase (EC 1.1.1.21) has been purified to apparent homogeneity from a variety of tissues including placenta, brain, nerves, kidney, muscle and lens. Multiple molecular forms of aldose reductase have been claimed to be isolated from bovine lens (Jedziniak et al, 1971) and bovine kidney (Gabbay et al 1974). These forms were subsequently described by some authors whereas others found a single form only (for a review see Wermuth, 1985). Conversion by a reducing agent (ß-mercaptoethanol) of one form to a more acidic but activity-retaining form was reported by Wermuth et al (1982), and differential susceptibility to inhibition of different enzyme forms was first described by Maragoudakis et al (1984). Nonlinear kinetics were often attributed to the presence of multiple forms. Thus, two kinetically distinct forms of human erythrocyte aldose reductase were postulated (Srivastava et al 1985) and the presence of bovine aldose reductase oxidized by oxygen radical generating systems was suggested as a possible cause for the nonlinear kinetics (Del Corso et al, 1987). More recently, data seem to firmly establish the existence of different forms of aldose reductase: "activated" and "unactivated" forms were isolated from bovine kidney (Grimshaw (1990) and their persistent peculiar kinetic behavior was then essentially rationalized (Grimshaw, 1991, Grimshaw et al 1990, Kubiseski et al, 1992). Recent advances in molecular biology led to the *in vitro* expression of rat lens aldose reductase (Old et al 1990) and human aldose reductase (Grundmann et al 1990, Carper et al 1990, Nishimura et al 1990, Bohren et al 1991). Multiple molecular forms of recombinant aldose reductase have so far not been reported with the exception of

Enzymology and Molecular Biology of Carbonyl Metabolism 4
Edited by H. Weiner, Plenum Press, New York, 1993

some charge heterogeneity that is evident upon isoelectric focusing (Bohren et al. 1991).

To investigate ß-mercaptoethanol-induced molecular forms of human aldose reductase, we mutated residues Cys^{80}, Cys^{298}, and Cys^{303} to serine residues. In addition, since a $Lys^{262} \rightarrow Met^{262}$ mutant showed enhanced kinetic responses to ß-mercaptoethanol, a double mutant (Lys262Met-Cys298Ser) was constructed to investigate these exaggerated variations. We report some of the properties of the modified enzymes and show the formation of interchangeable molecular forms of recombinant wild-type aldose reductase from a single coding source. The distinct kinetic properties can be attributed solely to different oxidation states of the Cys^{298} residue, a residue which is an integral part of the active site pocket of the enzyme (Wilson et al, 1992).

MATERIALS AND METHODS

Construction of wild-type and a $Lys^{262} \rightarrow Met^{262}$ (AR_{K262M}) mutant human aldose expression plasmid in the pET system was described earlier (Bohren et al, 1991). The $Cys^{80} \rightarrow Ser^{80}$ (AR_{C80S}), $Cys^{298} \rightarrow Ser^{298}$ (AR_{C298S}), and $Cys^{303} \rightarrow Ser^{303}$ (AR_{C303S}) mutation in aldose reductase were created by oligo-directed mutagenesis. The aldose reductase cDNA was subcloned into M13mp18 using the EcoRI restriction sites. Mutagenesis was done by using the Amersham Corp. *in vitro* mutagenesis system (version 2.1) and appropriate primers. The double mutation, $Lys^{262} \rightarrow Met\text{-}Cys^{298} \rightarrow Ser$ ($AR_{K262M\text{-}C298S}$) was made using the polymerase chain reaction as described earlier (Bohren et al, 1991) and cDNA of AR_{C298S} as template. cDNA bearing the various mutations was introduced into pET11a using the polymerase chain reaction as previously described (Bohren et al , 1991). To minimize PCR copying errors of the templates, 15 cycles or less were carried out using Vent polymerase (New England Biolabs) which has proofreading activity. The oligonucleotide primers were synthesized by National Biosciences (Plymouth, MN), and all constructs were completely sequenced (Sequenase kit, United States Biochemical) to verify the desired mutations and insure that no other mutations had occurred. The recombinant plasmids were overexpressed and purified as described in detail elsewhere (Bohren et al, 1991)

SDS-Polyacrylamide Gel Electrophoresis and Isoelectric Focusing : SDS-polyacrylamide gels were run on precast 10-15% gradient gels on the Phastsystem (Pharmacia) in the presence of 2% ß-mercaptoethanol as described by the manufacturer. Isoelectric focusing was carried out using precast isoelectric focusing gels (pH 3-9), according to the manufacturer's (Pharmacia) instructions. The protein concentration of the purified enzymes was determined with the Bio-Rad Protein Assay Kit, using bovine gamma-globulin as the standard.

Enzyme Assays and Kinetic Analysis: Enzymatic activities were assayed during purification by measuring the rate of enzyme-dependent decrease of NADPH absorption at 340 nm in a Gilford Response spectrophotometer at 25 °C. The standard reaction mixture contained 0.1 M sodium phosphate buffer, pH 7.0, 0.2 mM NADPH, and 1 mM DL-glyceraldehyde or 10 mM DL-glyceraldehyde (to assay mutants carrying the K262M mutation) in a total volume of 1 ml. Kinetic constants were determined in the same way except for varying the substrate or cofactor concentrations. Each data point (initial velocity) was determined in duplicate over at least six different substrate concentrations. Control assays, lacking either substrate or enzyme, were routinely included and the rates, if any, were subtracted from the reaction rates. Kinetic constants were calculated by fitting the Michaelis-Menten function directly in the hyperbolic form to the data with an unweighted least-squares analysis using the Marquardt-Levenberg algorithm provided with SigmaPlot, version 4.1. In cases where substrate inhibition occurred, the Michaelis-Menten function was modified to the following equation, $v_i = VA/(Km + A + A^2/K_i)$, where v_i is the initial velocity, A is the substrate concentration, V is the apparent maximal velocity, Km is the apparent Michaelis constant, and K_i is the apparent substrate inhibition constant. To measure the reverse reaction, 0.2 mM NADP+ was substituted for NADPH in the standard reaction buffer, and various concentrations of xylitol were used.

RESULTS

All constructs were readily expressed and purified with a yield of 10-20 mg per 3-liter culture. The purified enzymes show single bands on SDS-polycacrylamide gel electrophoresis. Upon isoelectric focusing, the wild-type and all mutant enzymes showed a major band at pH 6.2, with the exception of the double mutant ($AR_{K262M-C298S}$) that had a more acidic isoelectric point of 5.9 which is due to the $Lys^{262} \rightarrow Met$ mutation (Bohren et al 1991). Due to the unmodified N-terminal, the isoelectric points of these recombinant aldose reductases are all 0.3 pH units higher than the native protein. One minor band at pH 5.3 could be enhanced in the presence of 1% ß-mercaptoethanol (Fig. 1), a phenomenon first described for native human brain aldose reductase (Wermuth et al 1982). As can be seen in Figure 1, the minor band seems to be present before treatment with excess ß-mercaptoethanol and becomes more prominent after its addition. A possible reason is the inclusion of 7 mM ß-mercaptoethanol in all buffers used during the purification procedure. Interestingly, the acidic band is most prominent with the protein carrying the C80S mutation (AR_{C80S}).

As summarized in Table 1, only the AR_{C298S} mutant exhibits marked differences in apparent kinetic constants as compared to the wild-type with an average

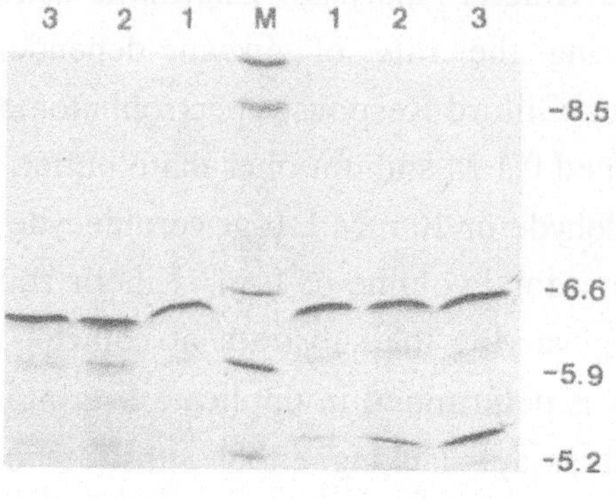

Fig 1. Polyacrylamide gel isoelectric focusing of Cys→Ser mutant aldose reductases. Marker proteins with characteristic isolectric points are in the middle lane (M). Mutant proteins were dialyzed aginst 5 mM sodium phosphate, pH 7.4, containing 0.1 mM EDTA and 7 mM ß-mercaptoethanol. Proteins on the right of the markers were additionally incubated in 1 % β-mercaptoethanol prior to loading. AR_{C298S} (1), AR_{C303S} (2), AR_{C80S} (3).

4.2-fold increase in k_{cat}. Since Km values increased even more for all substrates tested (average: 43-fold) the net effect is a marked decrease in catalytic effectiveness (k_{cat}/Km) of this mutant for the reduction of aldehydes. The pH dependency of k_{cat}/Km for the reduction of DL-glyceraldehyde by ARwild and AR_{C298S} is shown in Figure 2. Both bell-shaped curves have a maximum around pH 7 and result from a concurrent decrease in k_{cat} and Km with pH.

To study the effect of ß-mercaptoethanol on aldose reductase, all available mutants were dialyzed for two days against 5 mM sodium phosphate, pH 7.4 containing 0.1 mM EDTA and 10 mM or 0 mM ß-mercaptoethanol, and Km and k_{cat} were determined using DL-glyceraldehyde as substrate. All but the AR_{C298S} mutant showed significant increases in Km in the absence of ß-mercaptoethanol, whereas the k_{cat} did not change significantly (Table 2). This "redox behavior" is experimentally easiest to follow if the difference in Km is reasonably big and in the millimolar range as it is the case with the AR_{K262M} mutant (Figure 3A) where the $Km_{DL\text{-}glyceraldehyde}$ changed from 0.6 $\pm$ 0.1 mM to 2.4 $\pm$ 0.2 mM. These changes in Km were completely reversible, and the results of an extensive experiment with AR_{K262M} in which the buffers were changed three times for two samples are shown in Figure 4. This thiol-induced change of $Km_{DL\text{-}glyceraldehyde}$ was also caused by 5 mM DTT in AR_{K262M}, however other mutants were not tested. Since there was no observed ß-mercaptoethanol-induced change in $Km_{DL\text{-}glyceraldehyde}$ in the AR_{C298S} mutant, it was concluded that cysteine^{298} must be responsible for the redox behavior. An additional Cys^{298}→Ser mutation in AR_{K262M}, which exhibits enhanced thiol sensitivity with respect to Km, should also eliminate its kinetic sensitivity towards ß-mercaptoethanol. That this is indeed so is shown by the results of the appropriate experiment in Figure 3B: The $Km_{DL\text{-}glyceraldehyde}$ (27 $\pm$ 3 mM) of $AR_{K262M\text{-}C298S}$ is insensitive to ß-mercaptoethanol just as is the $Km_{DL\text{-}glyceraldehyde}$ (0.9 $\pm$ 0.1 mM) of AR_{C298S}.

Table 1. Apparent Kinetic Constants of AR_{wild}, AR_{C80S}, AR_{C298S}, AR_{C303S}

k_{cat} [s^{-1}] and k_{cat}/Km [$s^{-1}M^{-1}$] values for AR_{wild} and AR_{C303S} from Bohren et al (1991, 1992); standard deviation < 15 %

	AR_{wild}	AR_{C80S}	AR_{C298S}	AR_{C303S}
DL-Glyceraldehyde				
Km (mM)	0.02	0.05	0.9	0.03
k_{cat}	0.45	0.46	1.60	0.34
k_{cat}/Km	$22.7x10^3$	$9.5x10^3$	$1.8x10^3$	$12.8x10^3$
D-Xylose				
Km (M)	0.06		0.58	0.07
k_{cat}	0.3		1.51	0.26
k_{cat}/Km	49.0		2.6	40.0
D-Glucose				
Km (M)	0.11	0.14	3.30#	0.14
k_{cat}	0.15	0.28	0.55	0.16
k_{cat}/Km	1.3	1.9	0.2	1.2
D-Glucuronate				
Km (mM)	4.9	8.5	437	4.7
k_{cat}	0.28	0.24	1.23	0.20
k_{cat}/Km	58.0	29.0	2.8	43.0
NADPH#$				
Km (μM)	1.9		4.0	1.7
Xylitol				
Km (M)	0.19	0.29	0.52	
k_{cat}	0.03	0.04	0.04	
k_{cat}/Km	0.2	0.1	0.1	

#) 30% > standard deviation > 15 %
$) DL-glyceraldehyde as substrate

With AR_{C298S} being insensitive towards thiols, the question arose whether differences in inhibitor susceptibility also exist between this mutant and the wild-type enzyme. Since Bhatnagar et al (1989) demonstrated that ß-mercaptoethanol containing buffers reduce inhibition by Tolrestat and Sorbinil, these two inhibitors which are thought to bind to different sites (Bhatnagar, 1990) were tested: the concentration of Sorbinil necessary to inhibit AR_{C298S} by 50% was around 1 μM, equal to a 100-fold decrease over the concentration needed to inhibit the wild type enzyme by the same amount. No significant difference in the inhibition of AR_{wild} and AR_{C298S} by Tolrestat was observed (Fig5.).

TABLE 2. ß-mercaptoethanol dependency of apparent $Km_{DL\text{-}glyceraldehyde}$ of aldose reductases

βME	AR_{wild}	AR_{C80S}	AR_{C303S}
0 mM	50±5 μM	80±9 μM	41±5 μM
10 mM	24±2 μM	48±6 μM	26±3 μM
βME	**AR_{C298S}**	**AR_{K262M}**	**$AR_{K262M\text{-}C298S}$**
0 mM	0.92±0.05 mM	2.4±0.2mM	27±3mM
10 mM	0.87±0.05 mM	0.6±0.2mM	27±3mM

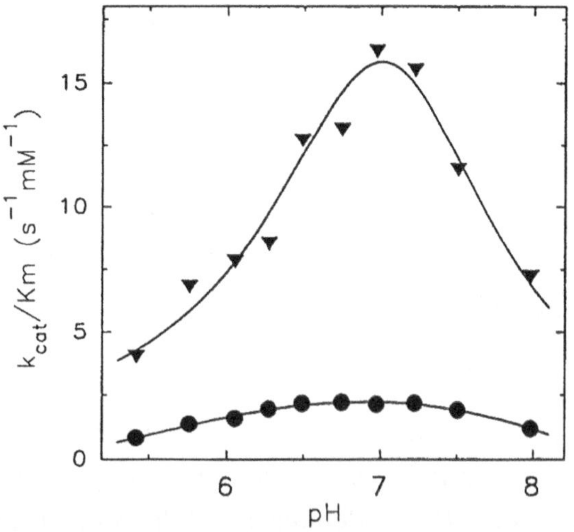

Fig.2. pH dependency of k_{cat}/Km for the reduction of DL-glyceraldehyde by aldose reductases. Initial velocities were measured as described under "Materials and Methods" in 0.1 M sodium phosphate at various pHs. AR_{wild} (▲), AR_{C298S} (●).

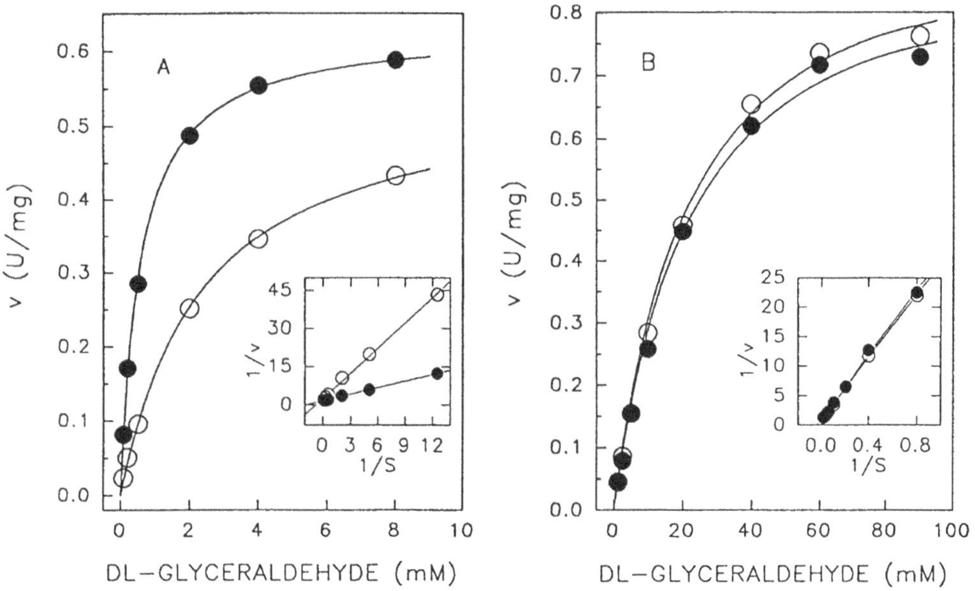

Fig.3. ß-mercaptoethanol dependency of initial velocity *versus* DL-glyceraldehyde concentration curves of mutant aldose reductases. Enzymes were dialyzed for two days against 5 mM sodium phosphate,pH 7.4, containing 0.1 mM EDTA and 0 mM (○) or 10 mM (●) ß-mercaptoethanol. NADPH concentration is 0.25 mM. Insets show double-reciprocal plots of the same data that were used for the direct plots. AR_{K262M} (A), $AR_{K262M\text{-}C298S}$ (B).

DISCUSSION

The existence of different molecular forms of aldose reductase has been the basis of many speculations and controversies. Rationalization of the existence of multiple molecular forms includes microheterogeneity, existence of isoenzyme as products of a multigene family, change of one form to another by glucose 6-phosphate, and conversion by oxidative stress and thiol-disulfide exchange.

In an effort to analyze the thiol-induced forms, we investigated the thiol-induced modifications of recombinant human aldose reductase and mutants of cysteine residues (Cys^{80}, Cys^{92}, Cys^{298}, Cys^{303}) that are not conserved between aldose and aldehyde reductase (Bohren et al, 1989). These residues were individually mutated to serine residues. While this work was in progress, ongoing x-ray crystallographic work in our laboratory as well as a low resolution structure of pig aldose reductase (Rondeau et al, 1992; Wilson et al 1992) showed that Cys^{92} is not exposed to solvent. Only three Cys→Ser mutants were therefore investigated in detail (AR_{C80S}, AR_{C298S}, and AR_{C303S}). They appear as single bands on SDS-polyacrylamide electrophoresis and show charge heterogeneity on isoelectrofocusing gels (Figure 1). As shown earlier (Wermuth et al 1982) for the native human brain aldose reductase, one particular form with a distinctly more acidic IEP could be induced by ß-mercaptoethanol in all mutants, albeit to different degrees. Kinetic analysis of native and mutant enzymes that include a Lys^{262}→Met mutant (Bohren et al, 1991) demonstrates that dialysis against 10 mM ß-mercaptoethanol significantly decreases the $Km_{DL\text{-}glyceraldehyde}$ of the wild type and all Cys→Ser mutants with the exception of AR_{C298S}, where no thiol-induced change is observed. This drop in the value of Km was not accompanied with a significant change in k_{cat}, an observation also made with the human muscle aldose reductase (Vander Jagt et al, 1990). The reversibility of this phenomenon, which might be called "redox behavior" of aldose reductase, was unquestionably shown with the Lys^{262}→Met mutation which exhibits, for reasons not yet clear, an exaggerated response to the presence or absence of thiols.

The absence of thiol-induced, kinetically observable effects on Km in the Cys_{298}→Ser mutant and the Lys^{262}→Met-Cys^{298}→Ser double mutant indicates that Cys^{298} is the critical residue responsible for the thiol-induced kinetic changes in aldose reductase. The persistant presence of this behavior in the Cys^{303}→Ser mutant precludes any disulfide formation between this residue and residue Cys^{298}. Our 1.65 Å refined atomic structure of the aldose reductase holoenzyme (Wilson et al, 1992) shows that the sulfhydryl of Cys^{298} cannot form any disulfide bridges with any of the remaining cysteines. With the critical role of Cys^{298} established, the remaining problem is the chemical and kinetic mechanism of the thiol-induced modification. At this time we can only speculate. A possible chemical mechanism involves the oxidation of ß-mercaptoethanol by oxygen to its disulfide and subsequent disulfide exchange reaction with the sulfhydryl of exposed cysteine residues:

$$2\,HOCH_2CH_2\text{-}SH + \tfrac{1}{2}O_2 \rightarrow HOCH_2CH_2\text{-}S\text{-}S\text{-}CH_2CH_2OH + H_2O$$

$$HOCH_2CH_2\text{-}S\text{-}S\text{-}CH_2CH_2OH + HS\text{-protein} \rightarrow HOCH_2CH_2\text{-}SH + HOCH_2CH_2\text{-}S\text{-}S\text{-protein}$$

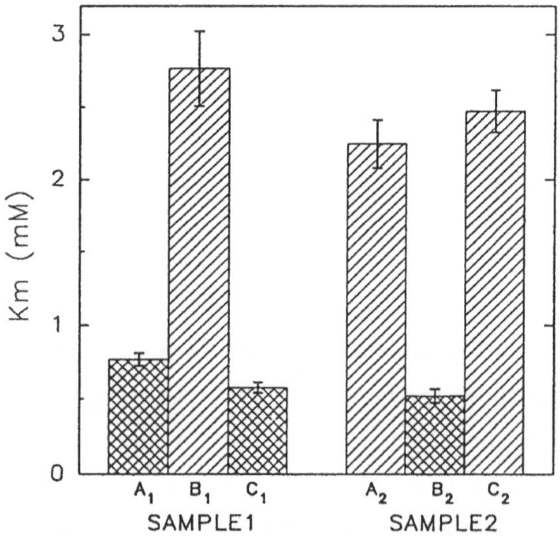

Fig.4. Reversibility of the ß-mercaptoethanol effect on $Km_{DL\text{-}glyceraldehyde}$ of AR_{K262M}. Two samples were dialyzed for two days against 5 mM sodium phosphate, pH 7.4, containing 0.1 mM EDTA and 0 mM (hatched bars) or 10 mM (crosshatched bars) ß-mercaptoethanol. Apparent Km values were determined as described under "Material and Methods". This procedure was repeated twice, yet, the dialysis buffers were switched each time (A_n, B_n, C_n).

It is generally known that the initially high redox potential of an aerobic solution quickly converts ß-mercaptoethanol to its disulfide. However, thiols do not oxidize without a catalyst which can be free metals or traces of metal ions, metal complexes or selenium as selenite. The rate of oxidation is controlled by various factors which include the catalyst, the temperature, pH, and buffer (for a review see Jocelyn, 1972). EDTA, which we include in all buffers used to purify aldose reductase, inhibit Cu^{++} or Fe^{++} catalysis. Similarly, disulfide-bond cleavage, re-formation, and the chemical modification of proteins by attachments of side chains according to the scheme above is also determined by the reaction conditions (Smithis, 1965). Atomic structure analysis of aldose reductase revealed a region of electron density not associated with the protein, the NADPH or water molecules (Wilson et al 1992). This unidentified electron density is in extremely close proximity to Cys^{298} but is not merged to the sulfur atom electron density. The fact that we cannot see with certainty any extra electron density coupled to the sulfur atom of Cys^{298}, however, does not contradict the proposed explanation since conditions for crystal growth were very different from the conditions used to induce the thiol effect on $Km_{DL\text{-}Glyceraldehyde}$ of aldose reductase.

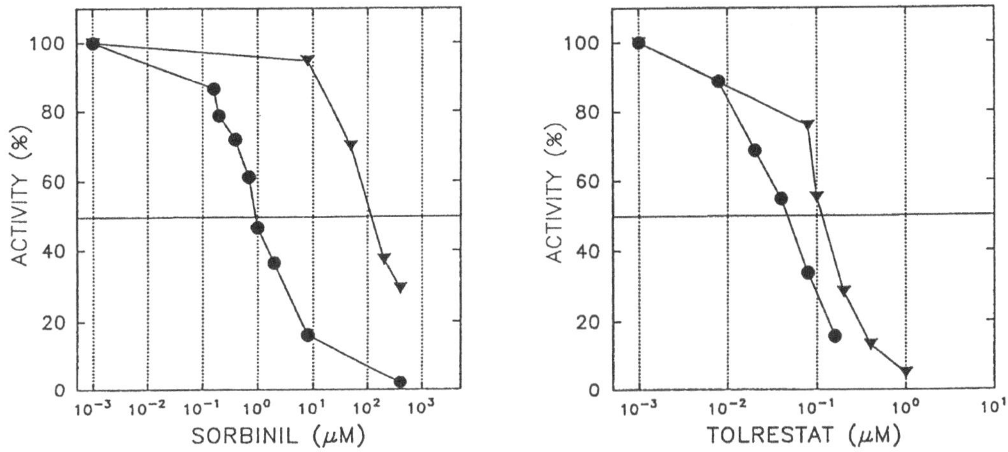

Fig.5. Inhibition by Sorbinil and Tolrestat. Wild type aldose reductase (▼) and a Cys^{298}→Ser mutant aldose reductase (●) were assayed in the presence of increasing concentration of inhibitors as indicated in a logarithmic scale. DL-glyceraldehyde at 3 mM and NADPH at 0.2 mM were used as substrate and cofactor, respectively.

In what way does modification of Cys^{298} influence the kinetics of aldose reductase? An appropriate way to look at the problem is to compare the chemistry of the sulfhydryl of the wild-type and the hydroxyl of the AR_{C298S} mutant enzyme. The replacement of the sulfur atom with the oxygen atom in a Cys→Ser mutation is the most conservative mutation between any pair of amino acid residues, and yet interchange of one amino acid with the other has never produced an equivalent functionality in any protein (Gosh, 1986). As in the case of aldose reductase, this interconversion leads to an average 4-fold increase in k_{cat} and an average 40-fold increase in Km. One reason may be the different geometry of the Cß-Sτ-H bonds in cysteine and Cß-Oτ-H bonds in serine: the atomic radius of sulfur is 0.4 Å larger than oxygen and the C-O-H angle is 110° vs 96° in C-S-H (He et al, 1991). Cysteine is also weaker at hydrogen bonding than serine. Inspection of the vicinity of Cys^{298} in the atomic structure of aldose reductase shows that the sulfhydryl of Cys^{298} is 5.38 Å away from the carbonyl oxygen of the nicotinamide and 4.24 Å from the 4-*pro*-R hydrogen on C4 of the nicotinamide. It's closest neighbor is the hydroxyl of Tyr^{209} (4.03 Å) which stacks against the α face (Rose et al, 1980) of the nicotinamide ring. Thus, a change in the hydrogen bonding interaction between the proton of the sulfhydryl and a close neighbor could easily perturb the active site and alter catalysis, k_{cat} and Km. It remains to be seen whether the perturbance brought about by the serine mutation or the modified cysteine causes stabilization or destabilization of the transition state, or whether the nicotinamide interaction with the substrate is disturbed.

Acknowledgements: This work was supported by grants from the National Institutes of Health DK-39,044, the Juvenile Diabetes Foundation (1891113 and 1901111), the Harry B. and Aileen B. Gordon Foundation, and the Retina Research Foundation.

REFERENCES

Bohren, K.M., Bullock,B., Wermuth,B., and Gabbay, K.H., 1989, The aldo-keto reductase superfamily, *J.Biol.Chem.* 264:9547-9551

Bohren, K.M., Page, J.L., Shankar, R., Henry, S.P.,and Gabbay, K.H. ,1991, Expression of human aldose and aldehyde reductases: site-directed mutagenesis of a critical lysine^{262}, *J.Biol.Chem*, 266: 24031-24037

Bohren, K.M., Grimshaw, C.E., and Gabbay, K.H., 1992, Catalytic effectiveness of human aldose reductase: Critical role of C-terminal domain, *J.Biol.Chem.*, in press

Bhatnagar, A., Liu, S., and Srivastava, S.K., 1989, Involvment of sulfhydryl residues in aldose reductase-inhibitor interaction, *Mol.Pharmacol.* 36: 825-830

Bhatnagar, A., Liu, S., Das, B., Ansari, N.H., and Srivastava, S.K., 1990, Inhibition kinetics of human kidney aldose reductase inhibitors, *Biochem. Pharmacol.* 30:1115-1124

Carper, D.,Sato S., Old, S., Chung, S. Kador, P.F. 1991, In vitro expression of human placental aldose reductase in Escherichia coli, *Adv.Exp.Med.Biol.* 284: 129-138

Del Corso, A., Camici, M., and Mura,U. 1987, *In vitro* modification of bovine aldose reductase activity, *Biochem. Biophys. Res. Commun.* 148: 369-375

Gabbay, K.H.,and Cathcart, E.S. ,1974, Purification and immunological identification of aldose reductase, *Diabetes* 23: 460-468

Gosh, S., Bock, S., Rokita, S., and Kaiser, E., 1986, Modification of the active site of alkaline phosphatase by site-directed mutagenesis, *Science* 231: 145-148

Grimshaw, C.E., 1990, Chromatographic separation of activated and unactivated forms of aldose reductase, *Arch. Biochem. Biophys*. 278: 273-6

Grimshaw, C.E.,1991, A kinetic perspective on the peculiarity of aldose reductase, *Adv.Exp.Med.Biol.* 284: 217-228

Grimshaw, C.E., Shahbaz,M., and Putney, C.G., 1990, Mechanistic basis for nonlinear kinetics of aldehyde reduction catalyzed by aldose reductase, *Biochemistry* 29: 9947-955

Grimshaw,C.E., Shahbaz,M., and Putney, C.G., 1990, Spectroscopic and kinetic characterization of nonenzymic and aldose reductase mediated covalent NADP-glycolaldehyde adduct formation, *Biochemistry* 29: 9936-46

Grundmann, U., Bohn, H., Obermeier, R.,and Amann E., 1990, Cloning and prokaryotic expression of a biologically active human placental aldose reductase, *DNA. Cell.Biol.* 9: 149-157

He, J.J., Quiocho,F.A., 1991, A nonconservative Serine to Cysteine mutation in the sulfate-binding protein, a transport receptor, *Science*, 251: 1479-1481

Jedziniak, J.A., and Kinoshita, J.H., 1971, Activators and inhibitors of of lens aldose reductase, *Invest. Ophtalmol.* 10: 357-366

Kubiseski, T.J, Hyndman, D.J., Morjana, N.A, and Flynn,T.G. 1992, Studies on pig muscle aldose reductase - kinetic mechanism and evidence for slow conformational change upon coenzyme binding, *J.Biol.Chem.* 267: 6510-6517

Maragoudakis, M.E., Wasvary, J., Hankin,H. and Gargiuolo, P., 1984, Human placenta aldose reductase: forms sensitive and insensitive to inhibition by Alrestatin, *Mol. Pharmacol.* 25:425

Nishimura, C., Matsuura Y., Kokai Y., Akera,T., Carper, D., Morjana,,N., Lyons,C.,and Flynn,T. G.,1990, Cloning and expression of human aldose reductase, *J.Biol.Chem.* 265: 9788-9792

Old, S.E., Sato, S. Kador, P.F.,and Carper,D.A., 1990, In vitro expression of rat lens aldose reductase in *E.coli*, *Proc.Natl.Acad.Sci.USA*. 87: 4942-4955

Rondeau, J.-M., Tête-Favier, F., Podjarny, A., Reyman. J.-M, Barth,P., Biellmann, J.-F. and Moras, D., 1992, Novel NADPH-binding domain revealed by the crystal structure of aldose reductase, *Nature* 355: 469-472

Rose, I.A., Hanson, K., Wilkinson, K.D., Wimmer, M.J.,1980, A suggestion for naming faces of ring compounds, *Proc.Natl.Acad.Sci.USA*, 77: 2439-2441

Smithies, O., 1965, Disulfide-bond cleavage and formation in proteins, *Science* 150: 1595-1598

Srivastava, S.K., Hair, G.A., and Das, B., 1985, Activated and unactivated forms of human erythrocyte aldose reductase, *Proc.Natl.Acad.Sci.USA* 82: 7222-78226

Wermuth, B., 1985, Aldo-keto reductases, *in* "Enzymology of Carbonyl Metabolism 2: Aldehyde Dehydrogenase, Aldo-Keto Reductase, and Alcohol Dehydrogenase", Flynn, T.G. and Weiner,H., eds., Alan R.Liss Inc., New York., 209-230

Wermuth, B. Bürgisser, H., Bohren, K.M.,and von Wartburg, J.-P., 1982, Purification and characterization of human brain aldose reductase, *Eur.J.Biochem.* 127: 279-284

Wilson, D.K., Bohren, K.M., Gabbay, K.H., and Quiocho, F.A.,1992, An unlikely sugar substrate site in the 1.65 Å structure of the human aldose reductase holoenzyme implicated in diabetic complications, *Science* 257: 81-84

SUBSTRATE SPECIFICITY OF REDUCED AND OXIDIZED FORMS OF HUMAN ALDOSE REDUCTASE

David L. Vander Jagt and Lucy A. Hunsaker

Department of Biochemistry
University of New Mexico
School of Medicine
Albuquerque, NM 87131

INTRODUCTION

NADPH-dependent oxidoreductases of the aldo-keto reductase family are widely distributed in man and animals (Wirth and Wermuth, 1985; Wermuth, 1985; Grimshaw and Mathur, 1989; Carper et al., 1987). The broad overlapping substrate specificities of these enzymes in the reduction of aldehydes and ketones suggests a role in detoxification of reactive carbonyls. Aldose reductase (EC 1.1.1.21; alditol:NAD(P) oxidoreductase) has received special attention because of its possible role in the development of diabetic complications (Gabbay, 1973; Kador and Kinoshita, 1985). Aldose reductase catalyzes the reduction of glucose to sorbitol in the polyol pathway which is normally a minor pathway for metabolism of glucose but may become important during hyperglycemia (Hers, 1956; Kador, 1988). Sorbitol accumulates in certain tissues that are prone to diabetic complications and may damage those tissues through a hyperosmotic mechanism (Kinoshita; 1974). Sorbitol may also interfere with the uptake and processing of myo-inositol, leading to impairment of cellular processes that are regulated by phosphatidylinositol metabolism (Greene et al., 1987; Finegold et al., 1983).

An alternative view of the development of diabetic complications is that hyperglycemia leads to nonenzymatic glycation of longlived proteins such as those found in basesment membranes (Pongor et al., 1984; Brownlee et al., 1987). A number of products from the reaction of basement membrane proteins with glucose or glucose-derived compounds have been described in studies of diabetic animals (Brownlee, 1989; Richard et al., 1991). Aminoguanidine, a relatively nontoxic nucleophile that interferes with the reactions by which glucose and related compounds produce advanced glycosylation endproducts including fluorescent crosslinks in basement membrane proteins, can prevent a number of diabetic complications (Brownlee et al., 1986).

Enzymology and Molecular Biology of Carbonyl Metabolism 4
Edited by H. Weiner, Plenum Press, New York, 1993

Numerous aldose reductase inhibitors have been developed, many of which are effective in preventing the development of a wide variety of diabetic complications in experimental animal models of diabetes (Mayer and Tomlinson, 1983; Robinson et al., 1986; Simard-Duquesne et al., 1985; Unakar et al., 1989; Williams and Odom, 1986; Beyer-Mears et al., 1984). Some of these aldose reductase inhibitors also prevent thickening of the basement membrane. This suggests that there may be a role for aldose reductase and the polyol pathway in the development of complications that have generally been ascribed to nonenzymatic glycation.

Recently, we reported that methylglyoxal, a toxic 2-oxo-aldehyde produced nonenzymatically from triose phosphates and enzymatically from metabolism of acetone, is a preferred physiological substrate for human aldose reductase (Vander Jagt et al., 1992). Formation of methylglyoxal is increased during hyperglycemia (Thornalley, 1988). In addition, methylglyoxal and its reduction product acetol were shown to modify proteins, leading to the formation of fluorescent products with spectral properties similar to those produced by glucose. We proposed an integrative model for diabetic complications that combines the polyol pathway theory and the nonenzymatic glycation theory but with emphasis both on methylglyoxal/acetol metabolism and glucose metabolism (Vander Jagt et al., 1992). This integrative model has a central role for aldose reductase.

In previous studies of human aldose reductase, we demonstrated that aldose reductase can easily be oxidized to a form with markedly different kinetic properties and altered sensitivity to inhibitors (Vander Jagt et al., 1990a,b). The reduced form appears to be predominant in normal tissues. It is not known whether the oxidized form exists in tissues under oxidative stress, such as diabetic tissues. Consequently, the significance of the oxidized form in not clear. In the present study, we have compared some of the kinetic properties of reduced and oxidized human aldose reductase with emphasis on methylglyoxal as substrate.

MATERIALS AND METHODS

Chemicals

Methylglyoxal was from Aldrich. D,L-Glyceraldehyde, acetol and NADPH were from Sigma. Acetol and methylglyoxal were purified by distillation. Red Sepharose CL-6B and chromatofocusing resin PBE 94 were from Pharmacia LKB. A Bio-Gel HPHT (hydroxylapatite) HPLC column was from Bio-Rad.

Enzyme Assays

Routine assays of aldose reductase during its purification were carried out in 1 ml volumes of 0.1 M sodium phosphate, pH 7, containing 0.1 mM NADPH and 10 mM D,L-glyceraldehyde. Reactions were monitored at 340 nm, 25°C, with a Perkin Elmer Lambda 6 spectrophotometer.

Enzyme Purification

Aldose reductase was purified from human skeletal muscle as described previously (Vander Jagt et al., 1990b). Reductase activity was extracted from the 100,000xg supernatant fraction with Red Sepharose, followed by chromatofocusing on Pharmacia PBE 94 and chromatography on a Bio-Gel HPHT HPLC column.

This procedure, which can be accomplished in 2 days, affords homogeneous aldose reductase with high recoveries of activity. In addition, this procedure provides enzyme that has not changed its kinetic properties during purification due to oxidation. Addition of DTT to the purified enzyme protects against oxidation and reduces any oxidized enzyme present.

Formation of Oxidized Aldose Reductase

Dialysis of aldose reductase to remove any DTT or NADP present results in the oxidation of aldose reductase. In reduced aldose reductase, all seven cysteine residues are reduced whereas, in oxidized aldose reductase, there are five free thiol groups (Vander Jagt et al., 1990b).

Enzyme Kinetics

Kinetic studies of human aldose reductase were carried out at pH 7, 25°C. K_m and k_{cat} values were determined by nonlinear regression analysis of initial rate data using the Enzfitter program. K_i values for aldose reductase inhibitors were generally determined by linear regression analysis of Dixon plots.

RESULTS AND DISCUSSION

Substrate Specificity of Reduced and Oxidized Aldose Reductase

Aldose reductase from human skeletal muscle was used for all of the experiments in this study. The reduced form of the enzyme was isolated from autopsy tissue. The oxidized form was prepared by dialysis of the reduced form (Vander Jagt et al., 1990b). For convenience, reduced and oxidized aldose reductase will be denoted as AR_R and AR_O, respectively. The substrate properties of AR_R and AR_O are summarized in Table 1. The broad specificity of aldose reductase extends from small water soluble substrates such as glyceraldehyde to fairly hydrophobic substrates such as p-nitrobenzaldehyde.

Of particular interest are the potential physiological substrates glucose, methylglyoxal and acetol. Glucose is a substrate for AR_R, K_m = 68 mM. Although this is a high K_m compared with some of the other substrates, glucose levels can reach 25-50 mM in uncontrolled diabetes. AR_O, by comparison, does not effectively catalyze the reduction of glucose due to very weak affinity for the substrate. Methylglyoxal is 10^4 times better than glucose as a sustrate for AR_R. This resides almost entirely in K_m since k_{cat} values are almost independent of substrate. For AR_R, this insensitivity of k_{cat} is consistent with a rate determining conformational change of the E-NADP complex prior to release of NADP, as has been suggested for bovine and pig aldose reductases (Grimshaw et al., 1990; Kubiseski et al., 1992). For AR_O, k_{cat} is more sensitive to changes in substrate, suggesting that a conformational change is not entirely rate determining.

Methylglyoxal: A Preferred Substrate for Aldose Reductase

The reduction of methylglyoxal by NADPH, catalyzed by AR_R, primarily results in reduction of the aldehyde functional group to form acetol; reduction of the ketone group to form D-lactaldehyde is a minor reaction. Furthur reduction of acetol leads selectively to formation of L-1,2-propanediol (figure 1; Vander Jagt et

Table 1. Substrate Specificities of Reduced and Oxidized Aldose Reductase

Substrate	AR_R		AR_O	
	k_{cat} (min^{-1})	K_m (mM)	k_{cat} (min^{-1})	K_m (mM)
Glucose	78	68	--	>1000
D,L-Glyceraldehyde	133	0.013	137	0.73
Methylglyoxal	142	0.008	159	2.4
Acetol	73	2.2	27	1.5
p-Nitrobenzaldehyde	160	0.003	107	0.032

All data at pH 7, 25°C (from Vander Jagt et al., 1990 b; 1992).

al., 1992). Methylglyoxal is 500 times better than acetol as a substrate for AR_R. Both acetol and propanediol can accumulate to near millimolar levels in uncontrolled diabetics (Reichard et al., 1986).

Kinetic Patterns for AR_R and AR_O

The kinetic patterns for the steady state analysis of AR_R and AR_O with methylglyoxal as substrate are shown in figures 2 and 3. For both forms of the enzyme, the data suggest an ordered sequential mechanism with NADPH the first reactant to bind and NADP the last product to leave. Ordered bi bi mechanisms have been proposed for bovine and pig aldose reductases (Grimshaw et al., 1990; Kubiseski et al., 1992).

The Michaelis constants for NADPH and the binding constants for NADP, determined from the secondary plots in figures 2 and 3, are summarized in Table 2. Cofactor binding is similar for AR_R and AR_O, unlike substrate binding affinities.

Inhibition of AR_R and AR_O by Aldose Reductase Inhibitors

The inhibition of AR_R and AR_O by the ARI sorbinil, tolrestat and statil was monitored at pH 7, 25°C, in 0.1 M sodium phosphate. The results are summarized in Table 3. In all cases, the binding of ARI is tighter to AR_R than to AR_O. In addition, inhibition patterns consistently show that the binding of ARI is noncompetitive with the binding of NADPH. Representative plots are shown in figure 4 for inhibition of AR_R by statil.

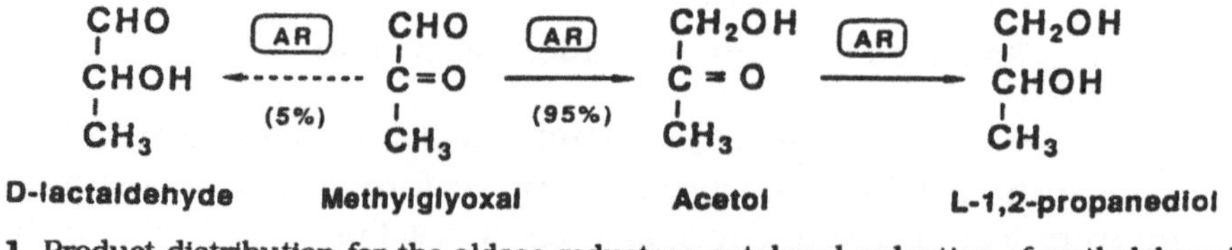

Figure 1. Product distribution for the aldose reductase-catalyzed reduction of methylglyoxal and acetol.

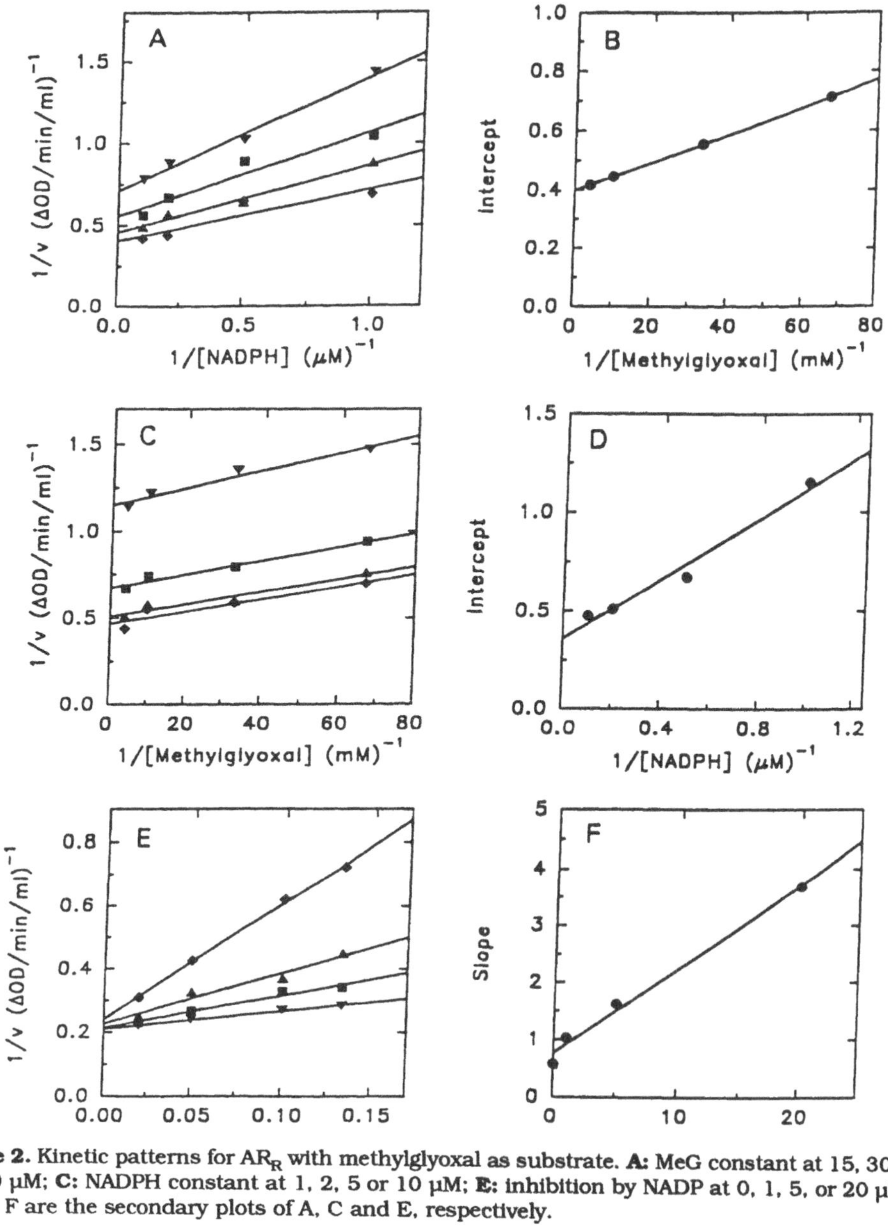

Figure 2. Kinetic patterns for AR_R with methylglyoxal as substrate. **A:** MeG constant at 15, 30, 100 or 250 μM; **C:** NADPH constant at 1, 2, 5 or 10 μM; **E:** inhibition by NADP at 0, 1, 5, or 20 μM; B, D and F are the secondary plots of A, C and E, respectively.

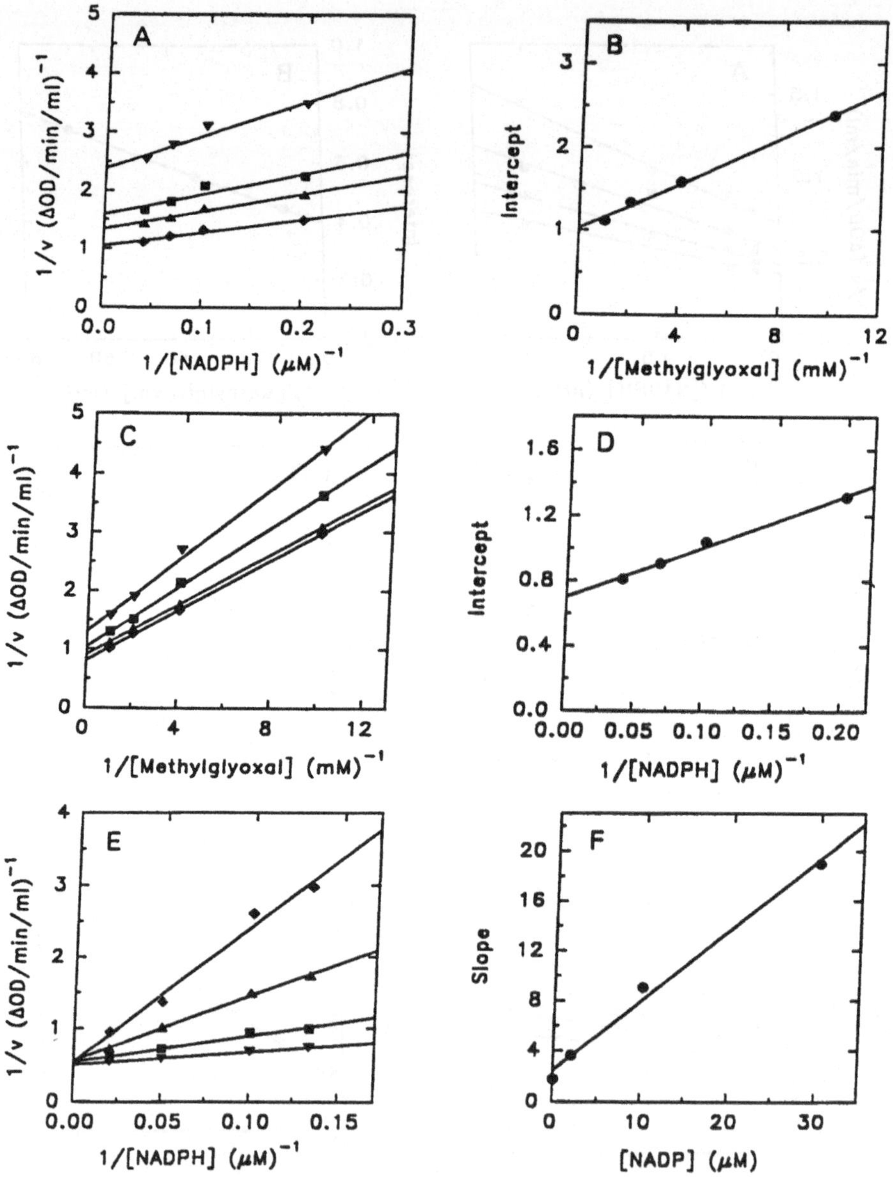

Figure 3. Kinetic patterns for AR_O with methylglyoxal as substrate. **A:** MeG constant at 100, 250, 500 or 10000 μM; **C:** NADPH constant at 5, 10, 30 or 50 μM; **E:** inhibition by NADP at 0, 2, 10 or 30 μM; B, D and F are the secondary plots of A, C and E, respectively.

Table 2. Cofactor Binding to AR_R and AR_O

Cofactor Binding to AR_R and AR_O

	AR_R	AR_O
NADPH	K_m = 2.2 μM	K_m = 4.4 μM
NADP	K_i = 0.5 μM	K_i = 0.2 μM

Table 3. Inhibition of AR_R and AR_O by ARI

Inhibition of AR_R and AR_O by ARI

Inhibitor	K_i (μM)	
	AR_R	AR_O
Sorbinil	0.34	416
Tolrestat	0.004	0.20
Statil	0.022	2.6

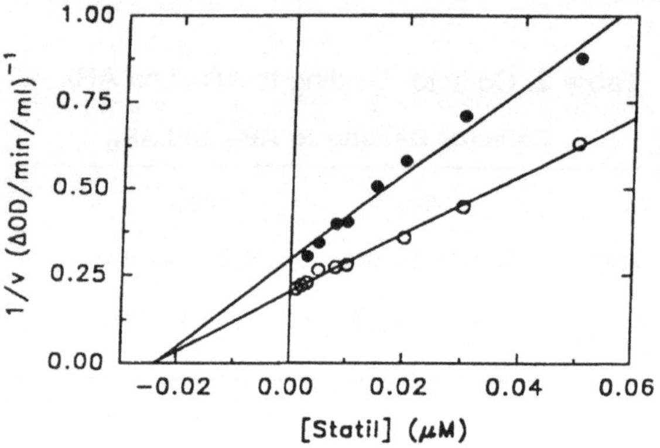

Figure 4. Dixon plot of the inhibition of AR_R by Statil shows noncompetitive inhibition with respect to the binding of NADPH. Concentrations of NADPH were constant at 10 and 40 μM with saturating methylglyoxal.

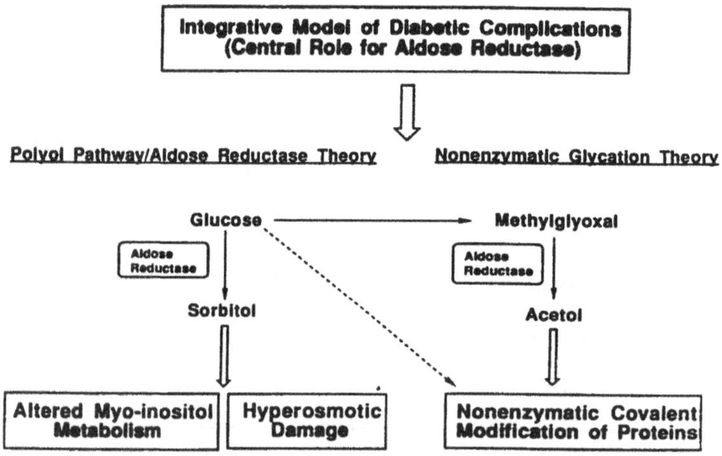

Figure 5

Integrative Model of Diabetic Complications

An integrative model of diabetic complications is presented in figure 5. This model, which has a central role for aldose reductase in the development of all diabetic complications, is based upon the concept that reduction of glucose to sorbitol and reduction of methylglyoxal to acetol are both important reactions in the etiology of complications.

ACKNOWLEDGEMENT

This work was supported by NIH grant DK43238

REFERENCES

Beyer-Mears, A., Ku, L. and Cohen, M.P., 1984, Glomerular polyol accumulation in diabetes and its prevention by oral sorbinil, *Diabetes* **33**: 604.

Brownlee, M. in: "Complications of Diabetes Mellitus," B. Draznin, S. Melmed and D. LeRoith, eds., A.R. Liss, New York (1989).

Brownlee, M., Vlassara, H., Kooney, A., Ulrich, P. and Cerami, A., 1986, Aminoguanidine prevents diabetes induced arterial wall proteins crosslinking, *Science* **232**: 1629.

Brownlee, M., Vlassara, H. and Cerami, A. in: Diabetic Complications: Scientific and Clinical Aspects," M.J.C. Crabbe, ed., Churchill-Livingstone, Edinburgh (1987).

Carper, D., Nishimura, C., Shinohara, T., Dietzchold, B., Wistow, G., Craft, C., Kador, P. and Kinoshita, J.H., 1987, Aldose reductase and p-crystalline belong to the same protein super family as aldehyde reductase, *FEBS Lett* **220**: 209.

Finegold, D., Lattimer, S.A., Nolle, S., Bernstein, M. and Greene, D.A., 1982, Polyol pathway activity and myo-inositol metabolism: a suggested relationship in the pathogenesis of diabetic neuropathy, *Diabetes* **32**: 988.

Gabbay, K.H., 1973, Role of sorbitol pathway in neuropathy, *Adv. Metab. Dis.* (Suppl. 2): 417.

Greene, D.A., Lattimer, S.A. and Sima, A.A.F., 1987, Sorbitol, phosphoinositides, and sodium potassium-ATPase in the pathogenesis of diabetic complications, *N.Engl. J. Med.* **316**: 599.

Grimshaw, C.E. and Mather, E.J., 1989, Immunoquantitation of aldose reductase in human tissues, *Anal. Biochem.* **176**: 66.

Grimshaw, C.E., Shahbaz, M. and Putney, C.G., 1990, Mechanistic basis for nonlinear kinetics of aldehyde reduction catalyzed by aldose reductase, *Biochem.* **29**: 9947.

Hers, H.G., 1956, Le mecanisme de la transformation de glucose en fructose par les vesicules seminals, *Biochim. Biophys. Acta* **22**: 202.

Kador, P.F., 1988, The role of aldose reductase in the development of diabetic complications, *Med. Res. Rev.* **8**: 325.

Kador, P.F. and Kinoshita, J.H., 1985, Role of aldose reductase in the development of diabetes-associated complications, Am. J. Med. **79**: 8.

Kinoshita, J.H., 1974, Mechanisms initiating cataract formation. Proctor Lecture, *Invest. Ophthalmol.* **13**: 713.

Kubiseski, T.J., Hyndman, D.J., Morjana, N.A. and Flynn, T.G., 1992, Studies on pig muscle aldose reductase, *J. Biol. Chem.* **267**: 6510.

Mayer, J.H. and Tomlinson, D.R., 1983, Prevention of defects of axonal transport and nerve conduction velocity by oral administration of myo-inositol or an aldose reductase inhibitor in streptozotocin-diabetic rats, *Diabetologia* **25**: 433.

Pongor,S., Ulrich, P.C., Bencsath, F.A. and Cerami, A., 1984, Aging of proteins: isolation and identification of a fluorescent chromophore from the reaction of polypeptides with glucose, *Proc. Natl. Acad. Sci. USA* **81**: 2684.

Reichard, G.A., Jr., Skutches, C.L., Hoeldtke, R.D. and Owen, O.E., 1986, Acetone metabolism in humans during diabetic ketoacidosis, *Diabetes* **35**: 668.

Richard, S., Tamas, C., Sell, D.R. and Monnier, V.M., 1991, Tissue-specific effects of aldose reductase inhibition on fluorescence and cross-linking of extracellular matrix in chronic galactosemia, *Diabetes* **40**: 1049.

Robinson, W.G., Jr., Kinoshita, J.H. and Kador, P.F., 1986, The polyol pathway in retinal microangiopathy, *Drugs* **32**: 19.

Simard-Duquense, N., Greselin, E., Gonzalez, R. and Dvornik, D., 1985, Prevention of cataract development in severely galactosemic rats by the aldose reductase inhibitor, tolrestat, *Proc. Soc. Exp. Biol. Med.* **178**: 599.

Thornalley, P.J., 1988, Modification of the glyoxalase system in human red blood cells by glucose in vitro, *Biochem. J.* **254**: 751.

Unakar, N., Tsui, J. and Johnson, M., 1989, Aldose reductase inhibitors and prevention of galactose cataracts in rats, *Invest. Ophthalmol. Vis. Sci.* **30**: 1623.

Vander Jagt, D.L., Hunsaker, L.A., Robinson, B., Stangebye, L.A. and Deck, L.M., 1990a, Aldehyde and aldose reductases from human placenta: heterogeneous expression of multiple enzyme forms, *J. Biol. Chem.* **265**: 10912.

Vander Jagt, D.L., Robinson, B., Taylor, K.K. and Hunsaker, L.A., 1990b, Aldose reductase from human skeletal and heart muscle: interconvertible forms related by thiol-disulfide exchange, *J. Biol. Chem.* **265**: 20982.

Vander Jagt, D.L., Robinson, B., Taylor, K.K. and Hunsaker, L.A., 1992, Reduction of trioses by NADPH-dependent aldo-keto reductase: aldose reductase, methylglyoxal and diabetic complications, *J. Biol. Chem.* **267**: 4364.

Wermuth, B., Aldo-keto reductases, in: "Enzymology of Carbonyl Metabolism 2: Aldehyde Dehydrogenase, Aldo-Keto Reductase, and Alcohol Dehydrogenase", T.G. Flynn and H. Weiner, eds., A.R. Liss, New York (1985).

Williams, W.F. and Odom, J.D., 1986, Study of aldose reductase inhibition in intact lenses by ^{13}C nuclear magnetic resonance spectroscopy, *Science* **233**: 233.

Wirth, H-P. and Wermuth, B., 1985, Immunochemical characterization of aldo-keto reductases from human tissues, *FEBS Lett.* **187**: 280.

KINETIC ALTERATION OF HUMAN ALDOSE REDUCTASE BY MUTAGENESIS OF CYSTEINE RESIDUES

J. Mark Petrash, Theresa Harter, Ivan Tarle and David Borhani*

Departments of Ophthalmology and Visual Sciences and of Genetics, Washington University School of Medicine, St. Louis, MO 63110 and *BioCryst Pharmaceuticals, Inc., Birmingham, AL 35244

INTRODUCTION

Aldose reductase (ALR2[1]: alditol:NADPH oxidoreductase: E.C. 1.1.1.21) catalyzes the NADPH-linked reduction of aldoses to their corresponding alcohols or polyols, the first step of the polyol pathway. Enhanced flux of glucose through the polyol pathway and consequent biochemical imbalances are thought to be crucial to the onset and progression of many complications of diabetes mellitus including cataract, retinopathy, neuropathy and nephropathy (Kinoshita and Nishimura, 1988). In light of its rate-limiting position in the polyol pathway as well as its apparent metabolic dispensability (Yancey et al., 1990), strategies to control or prevent the onset of diabetic complications through inhibition of aldose reductase are being aggressively pursued. While a structurally-diverse array of aldose reductase inhibitors (ARI) have yielded impressive results in animal studies, their effectiveness in arresting or preventing diabetic neuropathy (Boulton et al., 1990) and retinopathy (Sorbinil Retinopathy Trial Research Group, 1990) in human trials has been less encouraging (Frank, 1990).

Altered (so-called "activated") forms of aldose reductase, characterized by changes in kinetic parameters for aldose and aldehyde substrates and reduced susceptibility to inhibition by a variety of ARIs, have been described recently (Das and Srivastava, 1985; Del Corso et al., 1989a,b; Vander Jagt, 1990). Such isoforms have been extracted from tissues obtained from diabetic patients (Das and Srivastava, 1985), from cells cultured under hyperglycemic conditions (Srivastava et al., 1986), or produced by treatment of purified aldose reductase under conditions thought to induce protein oxidation (Del Corso et al., 1987; Vander Jagt et al., 1990). Conversion from native to the modified form is prevented or reversed by thiol reducing agents and occurs with loss of two titratable protein thiol groups, lending support to the idea that formation of an intramolecular disulfide bridge is the structural basis for the altered kinetic behavior (Vander Jagt et al., 1990). The rationale for studying this process takes on added importance if one considers that inhibition of altered forms of aldose reductase *in vivo* would be expected to be very poor considering the marked loss of sensitivity of these isoforms to ARIs.

Human aldose reductase contains 7 cysteine residues, 2 or 3 of which appear to be relatively solvent-accessible on the basis of their reaction kinetics with 5,5'-dithiobis-(2-nitrobenzoic acid) (Liu et al., 1989; Vander Jagt et al., 1990). The recently-solved structure of porcine lens aldose reductase revealed the presence of 3 solvent exposed cysteine residues located in a region of the protein relatively near the active site (Rondeau et al., 1992). These cysteine residues correspond to cysteines -80, -298 and -303 in human aldose reductase. In order to investigate the potential involvement of these cysteines in catalysis, we constructed mutants with serine substitutions at each individual cysteine site. Mutant ALR2:C298S displayed kinetic properties characteristic of the "activated" aldose reductase such as elevated apparent K_m and V_{max} for aldose and aldehyde substrates and reduced sensitivity to certain aldose reductase inhibitors. We have examined the structure of this enzyme by x-ray crystallography to evaluate the spatial distribution and potential interaction of the remaining cysteine residues.

EXPERIMENTAL METHODS

Complementary DNA sequences encoding aldose reductase were isolated from a human placenta cDNA library by screening with a bovine lens aldose reductase cDNA (Petrash et al., 1989). Plasmids containing cDNA sequences were recovered from bacteriophage vectors by excision with helper phage (Figure 1). Aldose reductase coding sequences were then transferred into the expression plasmid pMON5842 (see Olins and Rangwala, 1990) giving rise to pMON5997. For mutagenesis, single stranded wild type coding sequences were produced from phagemid stocks and reacted with synthetic oligonucleotides encoding one or more nucleotide substitutions using the site directed mutagenesis procedure developed by Kunkel and coworkers (1987). The mutated coding sequences were then exchanged for wild type sequences in pMON5997 for over-expression in *Escherichia coli* strain JM101. Nucleotide sequencing of mutant plasmids confirmed the mutations and assured that no unintended nucleotide sequence changes had been incorporated.

Expression of human aldose reductase in shaker flask cultures of *Escherichia coli* was conducted as described previously (Petrash et al., 1992). Recombinant aldose reductase was extracted from host cells by osmotic shock, concentrated by precipitation with 50-80% ammonium sulfate, and purified by chromatofocusing and hydroxylapatite column chromatography. All purification steps were conducted at 4°C in buffers containing 1 mM DTT.

Crystals of the ALR2:C298S·NADPH complex were grown by vapor diffusion in hanging drops containing 15 mg/ml ALR2:C298S, 9% polyethylene glycol (PEG) 6000, 25 mM MES, 1 mM DTT, 0.1 mM EDTA, 1 mM NADPH, pH 7. Drops were equilibrated at 4°C against 1 M NaCl; box-like crystals grew from precipitated protein in a few days. These crystals, which belong to space group *I222*, a=74.71 Å, b=85.32 Å, c=106.84 Å, diffract X-rays to at least 2.2 Å resolution. The crystal structure was solved by the method of molecular replacement, using porcine aldose reductase apoenzyme as the search model (Rondeau et al., 1992; J.-M Rondeau, F. Tête-Favier, A. Podjarny, D. Moras, personal communication). Rigid body refinement (CORELS; Sussman et al., 1977) of the initial model (*R*-factor 0.456, 20-3 Å, $F > \sigma_F$), deletion of three large loops, and further rigid body refinement reduced the *R*-factor to 0.305 (6-3 Å). Multiple cycles of least squares refinement of the model (TNT; Tronrud et al., 1987) and manual rebuilding using computer graphics (FRODO; Jones, 1978) allowed the re-introduction of most of the deleted residues, and gradual extension to higher resolution. Since clear difference density was present for the coenzyme, NADPH was added to the model at this stage. The current model contains 308 amino acid residues and NADPH (2366 atoms); the *R*-factor for all data from 6-2.4 Å is 0.286, with RMS deviations in bond lengths of 0.022 Å and in bond angles of 3.85°. Completion of the refinement at the diffraction limit of these crystals (2.2 Å) is in progress.

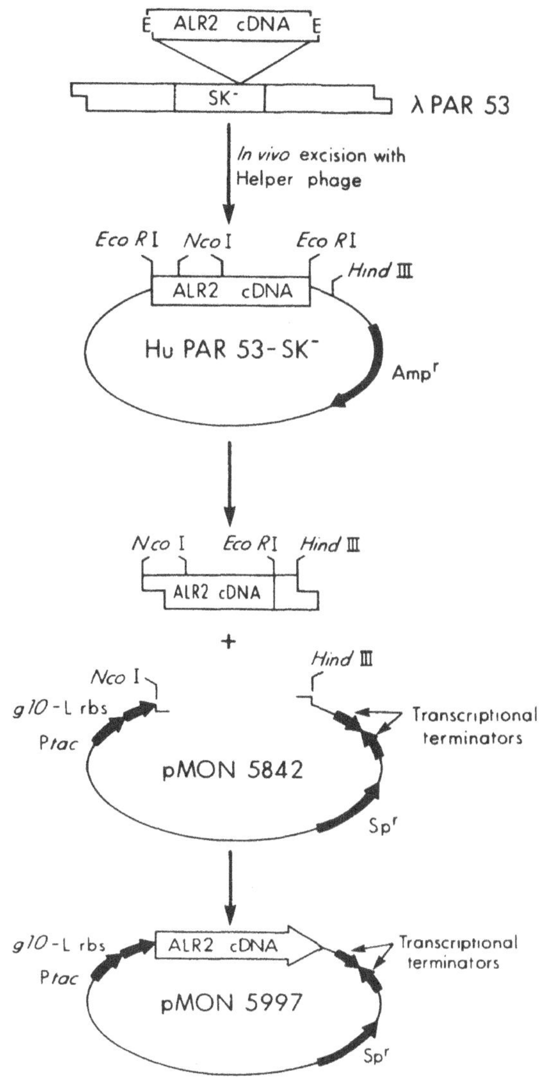

Figure 1. Construction of aldose reductase expression plasmid pMON5997. cDNA sequences contained in bacteriophage isolate λPAR 53 were recovered by *in vivo* excision of its phagemid vector (HuPAR53SK⁻). cDNA sequences were excised from the phagemid by digestion with *Hin*d III and *Nco*I; the resulting cDNA fragment was transferred to expression vector sequences derived from pMON5842 (a derivative of the vector described by Olins and Rangwala, 1990) giving rise to pMON5997. Functional elements in the expression plasmids include a *tac* promoter (P*tac*), a ribosome binding domain (*g10*-L rbs), transcriptional terminators derived from phage P22 *ant* gene, and a spectinomycin resistance gene (Sp^r).

RESULTS AND DISCUSSION

Cloning and Expression of Human Aldose Reductase

Marked variability observed in the properties of human aldose reductase extracted from different tissues or studied at various levels of purification (Poulsom, 1987) provide the rationale for establishing a stable and renewable source of aldose reductase for detailed chemical study. In light of the significant occupational hazards faced when using human donor tissue, animal tissues would seem to be an alternative source of starting material. However, species differences in the kinetic properties of aldose reductases also indicate that evaluation of the enzyme with regard to potential human therapy would best be carried out with enzyme derived from human sources (Kador et al., 1980). To satisfy these needs, we focused our initial efforts on establishing a recombinant source of the enzyme by cloning and expressing cDNA sequences encoding human aldose reductase.

Human aldose reductase cDNA clones were isolated from a placenta library by screening with a probe derived from bovine lens aldose reductase (Petrash and Favello, 1989). The coding regions of 6 out of 36 aldose reductase cDNA clones isolated from the library were substantially or completely analyzed by nucleotide sequencing and were found to contain identical sequences in their overlapping regions. The consensus sequence was also perfectly co-linear with human aldose reductase cDNA sequences published previously (Graham et al., 1989; Grundmann et al., 1990). These results indicate that a single aldose reductase gene is expressed in human placenta and that the coding sequence is identical to that expressed in human muscle and retina (Nishimura et al., 1990).

Steps involved in the transfer of aldose reductase coding sequences from the original bacteriophage cloning vector to the final expression vector construct are outlined in Figure 1. Abundant quantities of aldose reductase activity were obtained from *Escherichia coli* cultures transformed with the aldose reductase expression plasmid pMON5997. As noted previously (Grundmann et al., 1990), the enzyme could be readily extracted from host cells by osmotic shock, presumably reflecting its localization in the periplasmic space. Rapid purification by chromatofocusing (Figure 2) and hydroxylapatite chromatography provided an enzyme preparation virtually homogeneous judging by the appearance of a single protein

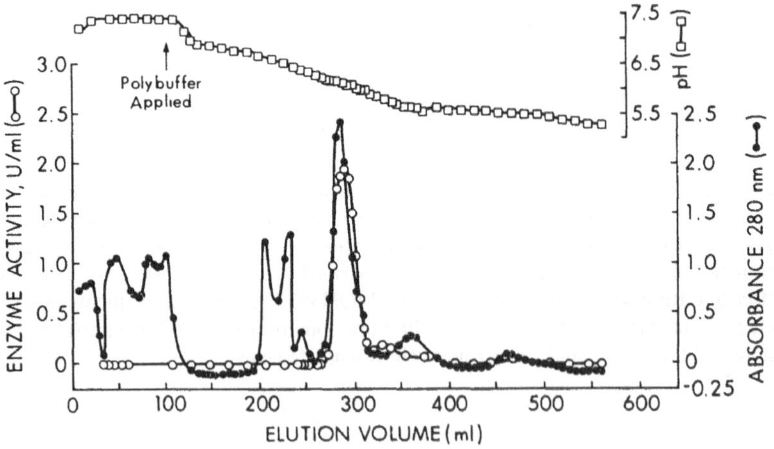

Figure 2. Purification of recombinant human aldose reductase by column chromatofocusing. This representative chromatogram was obtained by loading approximately 200 mg protein extracted from E. coli cells transformed with pMON5997. Each fraction was measured for absorbance at 280 nm (•–•), aldose reductase activity (o-o) and pH (□–□).

band following SDS-polyacrylamide gel electrophoresis (SDS-PAGE). Recombinant and native human aldose reductases co-migrated at a position corresponding to M_r~36,000 on SDS-PAGE and cross-reacted with antibodies directed against bovine lens aldose reductase. Ten cycles of Edman degradation of the recombinant aldose reductase revealed the following sequence: NH_2-Ala-Ser-Arg-Leu-Leu-Leu-Asn-Asn-Gly-Ala-COO^-. This structure agrees with the sequence predicted from the cDNA, and indicates that the amino terminal alanine residue is not covalently blocked (acetylated) as is the case for aldose reductase synthesized in eucaryotic cells (Schade et al., 1990).

Although other investigators have reported over-expression of recombinant human aldose reductase in procaryotic (Grundmann et al., 1990; Carper et al., 1990; Bohren et al., 1992) and eucaryotic (Nishimura et al, 1990) host cells, concerns about the usefulness and reliability of recombinant-derived aldose reductase as a substitute for enzyme purified from human tissues have been raised (Carper et al., 1991). Therefore, we devoted considerable effort in our initial studies to evaluate the relatedness of aldose reductases from recombinant and native sources. Michaelis constants (K_m) and turnover numbers (k_{cat}) measured with recombinant human aldose reductase using a variety of structurally-diverse substrates were strikingly similar to that reported previously for native human aldose reductase (Table 1). The sensitivity of recombinant aldose reductase to inhibition by a variety of ARIs was also similar to that reported for aldose reductase extracted from native human tissues. Tolrestat (*N*-[[5-(trifluoromethyl)-6-methoxy-1-naphthalenyl]thioxomethyl]-*N*-methylglycine) and zopolrestat (3,4-dihydro-4-oxo-3-[[5-(trifluoromethyl)-2-benzothiazolyl]methyl]-1-phthalazineacetic acid); Mylari et al, 1991) were potent inhibitors, with K_i values ~6 to 16-fold lower than that for sorbinil (*(S)*-6-fluoro-spiro-[chroman-4,4'-imidazolidine]-2',5'-dione).

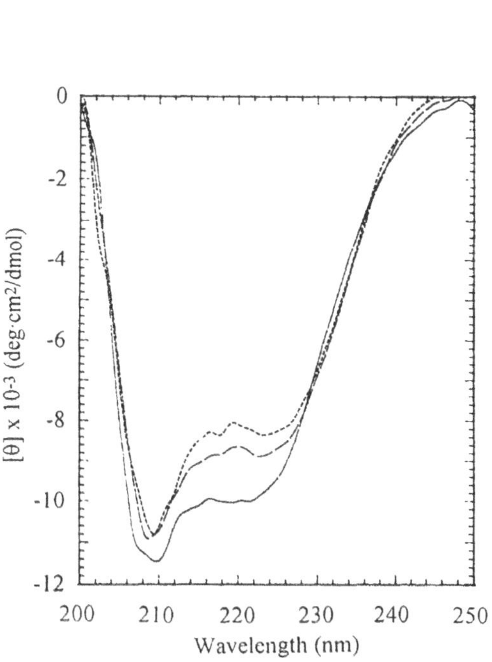

Figure 3. Circular dichroism spectra of human aldose reductase. Spectra of the enzyme (5 μM in 50 mM potassium phosphate, pH 7.0) and coenzyme or coenzyme analogs (≥200 μM) were recorded at 23°C using a 0.1 cm path length quartz cell under constant nitrogen flush. Values are corrected for buffer and coenzyme (or analog). ALR2 (——); ALR2 plus NADPH (— —); ALR2 plus $NADP^+$ (- - -). Spectra collected with ALR2 in presence of 2',5'-adenosine diphosphate were identical to that collected with the enzyme alone.

We also examined the interaction of the recombinant ALR2 with NADPH, $NADP^+$ and the coenzyme analog adenosine 2',5'-diphosphate (2',5'-ADP). As shown in Figure 3, binding of either NADPH or its oxidized form, $NADP^+$, induced a major conformational change as measured by circular dichroism spectroscopy. In contrast, binding of 2',5'-ADP failed to induce a measurable conformational change although this analog was a potent inhibitor (K_i=5 μM; competitive versus NADPH). Crystallographic studies have now provided the structural basis for this conformational change (*vide infra*).

Table 1. Comparison of Recombinant and Native Human Aldose Reductases.

	Recombinant ALR2	Native ALR2[a]
Specific Activity (U/mg)	3.5	2.4
DL-glyceraldehyde		
K_m (μM)	54	72
k_{cat} (sec^{-1})	1.0	1.2
k_{cat}/K_m ($sec^{-1}M^{-1}$)	1.9×10^4	2×10^4
D-glucose		
K_m (mM)	212	651
k_{cat} (sec^{-1})	0.09	0.61
k_{cat}/K_m ($sec^{-1}M^{-1}$)	0.41	0.94
Benzaldehyde		
K_m (μM)	73	224
k_{cat} (sec^{-1})	0.44	0.86
k_{cat}/K_m ($sec^{-1}M^{-1}$)	6.1×10^3	4×10^3
p-nitrobenzaldehyde		
K_m (μM)	16	31
k_{cat} (sec^{-1})	0.47	1.3
k_{cat}/K_m ($sec^{-1}M^{-1}$)	3×10^4	4×10^4
Effect of Inhibitors (K_i, μM)		
Sorbinil	0.098	0.22[b]
Tolrestat	0.015	0.017[b]
Zopolrestat	0.006	0.002[c]

Specific activity was determined at 23°C using 5 mM DL-glyceraldehyde, 150 μM NADPH, 0.4 M lithium sulfate, 10 mM sodium phosphate, pH 6.2, and 5 mM β-mercaptoethanol. Substrate kinetic parameters were determined under these conditions but with the indicated substrates. Inhibition constants (K_i) were determined at pH 7.1 in 50 mM potassium phosphate, 0.4 M ammonium sulfate, 150 μM NADPH with DL-glyceraldehyde as variable substrate. Comparative data are from: [a] (Morjana and Flynn, 1989); [b] (Vander Jagt et al., 1990); [c] (Mylari et al., 1991).

Mutagenesis of Cys-80, -298, and -303

Recent structure determination of porcine and human aldose reductases revealed that the enzyme consists of a β/α-barrel structural motif, with the coenzyme-binding domain located at the carboxy-terminal ends of the β-strands (Rondeau et al., 1992; Wilson et al., 1992; Borhani et al., 1992). Three cysteine residues, namely those corresponding to cysteine residues 80, 298 and 303 in human aldose reductase, are located in this region. Previous studies have implicated the involvement of cysteine residues in inhibitor and coenzyme binding and in post-translational modifications leading to enzyme isoforms with altered kinetic properties. As an initial attempt to evaluate the potential role of these cysteine residues in catalysis and inhibitor binding, we examined the kinetic properties of mutants containing serine substitutions at cysteine residues 80 (ALR2:C80S), 298 (ALR2:C298S) and 303 (ALR2:C303S).

Kinetic Effects of Cysteine Mutations. In comparison with ALR2:WT, all three cysteine→serine mutants were expressed with equal if not greater efficiency in *E. coli* and could be routinely isolated to apparent homogeneity without modification of the purification procedure. All mutants exhibited robust reductase activity with a variety of aldose and aromatic aldehyde substrates. Although detailed kinetic properties of these mutants will be published elsewhere (Petrash et al., 1992), three model substrates are presented here. These compounds were chosen as they represent three structural classes of potential aldose reductase substrates: glyceraldehyde as a model for short-chain "aldoses" or trioses (Vander Jagt, 1992), galactose as representative of potential hexose substrates utilized in the polyol pathway and linked to ocular disease (Kinoshita and Nishimura, 1988), and benzaldehyde as an aromatic aldehyde similar in structure and hydrophobicity to other substituted aromatic aldehydes that could be potential physiological substrates of the aldo-keto reductases (Flynn, 1982).

When quantified under standard assay conditions (5 mM DL-glyceraldehyde as substrate and 150 μM NADPH as coenzyme), the specific activity (units/mg) of the C80S and C303S mutants was approximately 50% that of wild type (Table 2). However, the specific activity of the C298S mutant was approximately 2.5-fold that of wild type, due in part to an almost 6-fold increase (relative to wild type) in k_{cat}. The apparent $K_{m\ (DL\text{-}glyceraldehyde)}$ for all mutants was elevated, particularly so for ALR2:C298S which was increased approximately 57-fold compared to wild type. Despite differences in their respective K_m and k_{cat} values, the catalytic efficiency (k_{cat}/K_m) measured with DL-glyceraldehyde was uniformly decreased among all the mutants. Similar changes in kinetic parameters were also apparent when reactivity with other substrates was examined. As with DL-glyceraldehyde, the K_m and k_{cat} (measured with galactose and benzaldehyde) observed with ALR2:C298S were markedly elevated relative to wild type. Mutants C80S and C303S showed higher (approximately 6 to 10-fold) apparent $K_{m\ (galactose)}$; utilization of this substrate by the C303S mutant was particularly poor (k_{cat}/K_m reduced >50-fold). In contrast, the C80S was notable in its enhanced reactivity with benzaldehyde (decreased K_m and increased k_{cat}; k_{cat}/K_m increased to a level 4.5-fold that of wild type).

The kinetic behavior of mutant C298S is strikingly similar to that attributed to "activated" forms of aldose reductase. This is exemplified by its elevated K_m and V_{max} for aldehyde substrates and its reduced sensitivity to inhibition by ARIs such as sorbinil. As conversion of native aldose reductase to the modified form is thought to occur concomitantly with formation of an intramolecular disulfide bridge, a structural model of C298S was examined to evaluate the spatial distribution and chemical environment of the cysteine residues.

STRUCTURE OF ALR2:C298S COMPLEXED TO NADPH

Recombinant human aldose reductase (C298S mutant) was crystallized in the presence of NADPH. These crystals, which diffract X-rays to at least 2.2 Å resolution, belong to a different space group (*I222*) than previously reported crystals of porcine aldose reductase apoenzyme (*P1*; Rondeau et al., 1992) or wild type recombinant human aldose reductase complexed with NADPH ($P2_12_12_1$; Wilson et al., 1992). The crystal structure of ALR2:C298S·NADPH was solved by molecular replacement; the search model was one monomer of porcine aldose reductase apoenzyme. The structure has been refined at 2.4 Å resolution to an *R*-factor of 0.286.

The structure of ALR2:C298S is for the most part identical to the previously described structures of wild type human (Wilson et al., 1992) and porcine (Rondeau et al., 1992) aldose reductases. As shown in Figure 4, ALR2:C298S consists predominantly of a $(\beta/\alpha)_8$-barrel; this regular fold is preceded by a β-hairpin, and is interrupted by three large *C*-terminal loops and two extra α-helices. Comparison of our model with that of the porcine apoenzyme reveals the nature of the conformational changes which occur upon coenzyme binding. Upon binding NADPH, loop 7 of ALR2:C298S rotates by 51° around Gly 213 and Ser 214. This movement brings residues 213-216 into contact with the pyrophosphate and ribose-2'-phosphate regions of NADPH, and locks the coenzyme into its binding cleft. The detailed interactions between aldose reductase and NADPH which we observe appear to be identical to those found for the wild type recombinant human enzyme (Wilson et al., 1992). When the localized changes which occur upon coenzyme binding are excluded from the analysis, our

Table 2. Kinetic constants for wild type and mutant aldose reductases

	Wild Type	C80S	C298S	C303S
Specific Activity	3.5	1.7	9.0	1.7
DL-glyceraldehyde				
K_m (μM)	54	195	3100	405
k_{cat} (sec^{-1})	1.02	0.38	5.82	0.82
k_{cat}/K_m ($M^{-1}sec^{-1}$)	18907	1931	1878	2035
D-Galactose				
K_m (mM)	94.5	578	2620	889
k_{cat} (sec^{-1})	0.32	0.83	1.93	0.05
k_{cat}/K_m ($M^{-1}sec^{-1}$)	3.4	1.4	.74	.06
Benzaldehyde				
K_m (μM)	73	28	1061	68
k_{cat} (sec^{-1})	0.44	0.76	4.62	0.30
k_{cat}/K_m ($M^{-1}sec^{-1}$)	6093	27473	4404	4420
K_i sorbinil (μM)*	0.37	0.84	3.81	0.55

Kinetics measurements were carried out with recombinant enzymes purified to apparent homogeneity. Specific activities were determined as described in Table 1. Kinetics constants were determined at 23°C in 50 mM potassium phosphate buffer, pH 7.0, containing 0.4 M ammonium sulfate and 150 μM NADPH. Rate measurements were analyzed using a software program originally described by Cleland (1979). Protein concentration was determined by the dye-binding method of Bradford (1976). *Prior to kinetics analysis to determine *Ki* values, the enzymes were treated with 0.1 M DTT, followed by rapid filtration through a desalting column (Petrash et al., 1992).

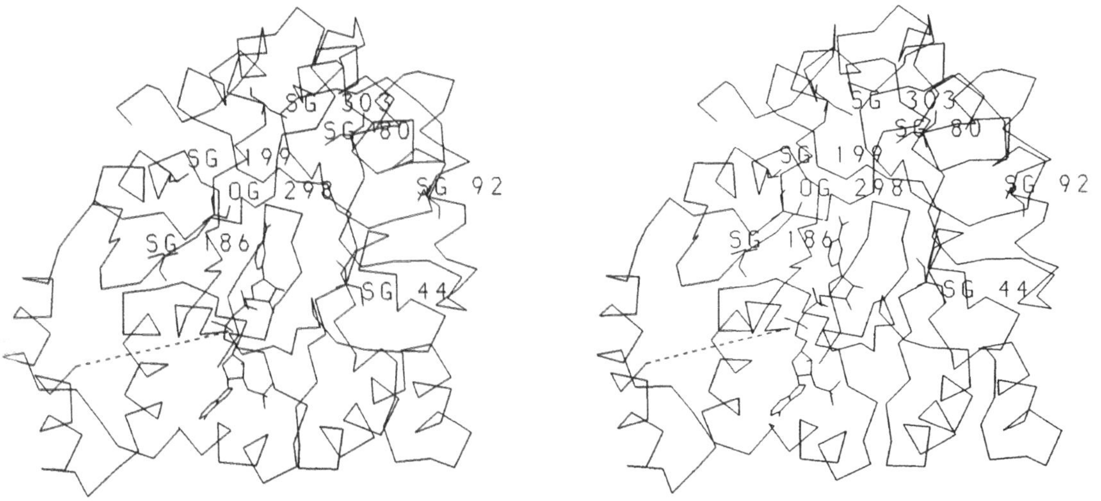

Figure 4. Stereoview of the α-carbon backbone of ALR2:C298S. The view is down the mouth of the barrel; the *N*-terminal β-hairpin is at the bottom center of the picture, and the *C*-terminus is at the upper left. NADPH is shown, as are the six cysteines and Ser 298, which are labelled at the γ-sulfur (oxygen) atoms. The dashed lines represent residues 217-229 of loop 7, which are not yet included in our model.

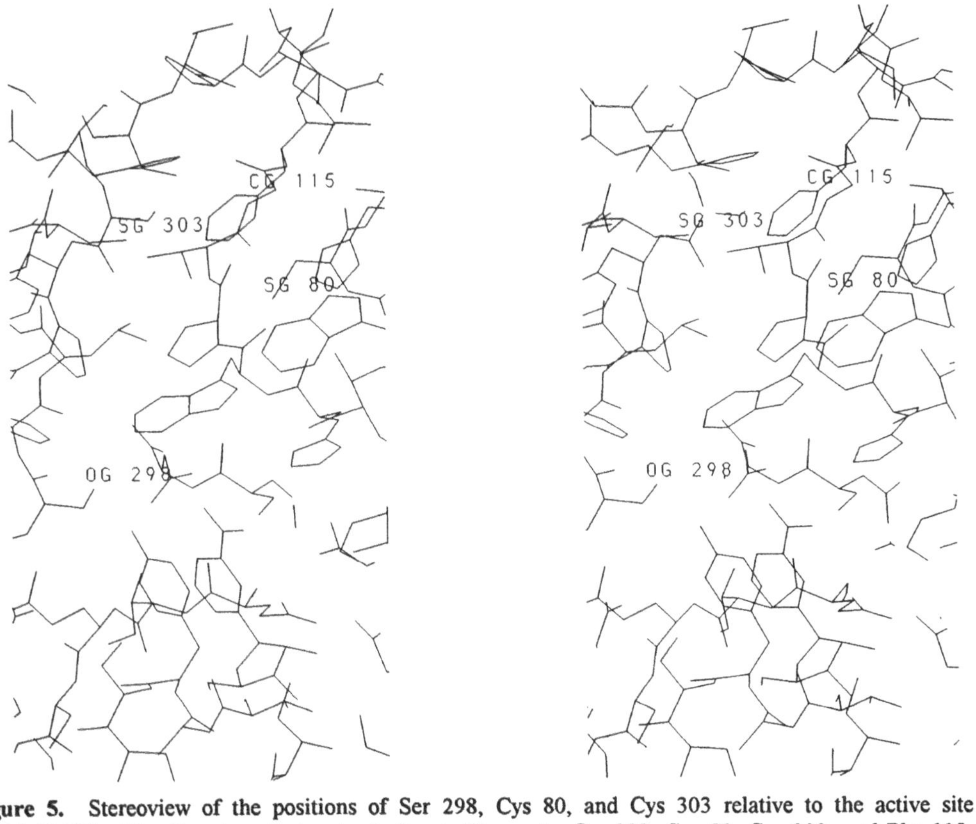

Figure 5. Stereoview of the positions of Ser 298, Cys 80, and Cys 303 relative to the active site of ALR2:C298S. The direction of view is similar to Figure 4. Ser 298, Cys 80, Cys 303, and Phe 115 are labelled. The nicotinamide ribose portion of NADPH is visible at the lower right.

model of ALR2:C298S superimposes on the porcine apoenzyme model with an RMS deviation of 0.6 Å (293 α-carbon atoms).

The proximity of cysteine residues 80, 298 and 303 to the active site of aldose reductase is clear from our structure. As shown in Figure 5, the nicotinamide ring of NADPH is tucked deeply in the center of the β-barrel, the walls of which are lined with hydrophobic residues. The active site of aldose reductase, as identified by the pro-*R* C4 hydrogen atom of NADPH, thus has the shape of an inverted cone, with the nicotinamide ring exposed to solvent at the point of the cone. The locations of cysteines 80 and 303, as well as the mutated residue, serine 298, are essentially identical in our structure and the porcine apoenzyme. These residues form the apices of an isosceles triangle; the sulfur atoms of Cys 80 and Cys 303 are separated by 6.6 Å, and both of these sulfur atoms are about 10.5 Å from the hydroxyl oxygen atom of Ser 298. Cysteines 80 and 303 are quite distant from the active site. In contrast, Ser 298, which forms part of the upper wall of the inside of the β-barrel, is in van der Waals contact with the nicotinamide ring of NADPH and thus is clearly "in" the active site.

In our structure, we find no evidence for the formation of any disulfide bridges. It is clear that Cys 80, Cys 298, and Cys 303 are exposed to solvent; these residues are susceptible to chemical modification (Rondeau et al., 1992). Formation of a disulfide bridge between the two most closely apposed of these three residues (Cys 80 and Cys 303) would require the movement of Phe 115, which physically separates Cys 80 and Cys 303. Movement of Phe 115 would be expected to be thermodynamically unfavorable, as it would require the disruption of the hydrophobic packing of numerous residues in this region of the enzyme (see Figure 5). As can be seen in Figure 4, the remaining cysteines, Cys 44, Cys 92, Cys 186, and Cys 199, are all distant from the active site (>12.5 Å), distant from each other (>11.0 Å), and relatively sequestered in the interior of the enzyme. Disulfide bridges among these residues would appear to be impossible to construct. On the basis of our model, one may conclude that the modified kinetic behavior of ALR2:C298S does not arise from the formation of a disulfide bridge between any two of the remaining 6 cysteine residues. Indeed, our structural model would predict that none would be capable of forming without dramatic reorganization of the hydrophobic core of the β/α barrel. Therefore, we question the hypothesis that oxidative modification of aldose reductase involves formation of a disulfide bridge (Bhatnagar et al., 1989; Vander Jagt et al., 1990).

It seems plausible to conclude that cysteine residues do not participate directly in catalysis, as the cysteine mutants containing serine substitutions nearest the active site displayed near-wild type activity levels under standard assay conditions. However, demonstration that any given residue is "unessential" to the catalytic reaction will require the construction of additional mutants with other structurally conservative substitutions at each given site. Further study with additional substrates will also be required to sort out the apparent structural and/or functional roles each of these residues may contribute to the enzyme. As pointed out by Plapp and coworkers in their study of alcohol dehydrogenases (1991), single mutations can potentially affect many steps in the catalytic mechanism and that is probably what we are measuring with these cysteine mutations. It is difficult to ascribe a definitive role to Cys 298 in the catalytic mechanism, and thereby understand why mutation of this residue to serine affects the enzyme's kinetic parameters. One possibility consistent with the kinetic and structural data we have presented is that Cys 298 helps to bind and position the nicotinamide ring of NADPH and/or the substrate, and that mutation of this residue to serine appears to modify these interactions. Alternatively, mutation of this residue may affect the rate at which enzyme isomerization occurs with the binding and release of NADP(H) (Borhani et al., 1992), a step thought to be rate-limiting for aldehyde reduction catalyzed by aldose reductase (Kubiseski et al., 1992).

ACKNOWLEDGMENTS

This work was supported by research grants from the National Eye Institute (EY05856 and 5T32 EY07108), a Diabetes Research and Training Center grant (P60 DK20579) and in part by grants to the Department of Ophthalmology and Visual Sciences from Research to Prevent Blindness, Inc. We would like to thank Dr. Jean-Michel Rondeau (Biostructure, S.A.), and Dr. Alberto Podjarny and Professor Dino Moras (University of Strasbourg) for providing the porcine aldose reductase apoenzyme and 2'-monophosphoadenosine-5'-diphosphoribose (ADPRP)-complex coordinates to BioCryst prior to publication.

REFERENCES

Bhatnagar, A., Liu, S., Das, B., Srivastava, S.K., 1989, Involvement of sulfhydryl residues in aldose reductase-inhibitor interaction. *Mol. Pharmacol.* 36,825-830.

Bohren, K.M., Page, J.L., Shankar, R., Henry, S.P., and Gabbay, K.H., 1992, Expression of human aldose and aldehyde reductases: Site-directed mutagenesis of a critical lysine 262. *J. Biol. Chem.* 266,24031-24037.

Borhani, D.W., Harter, T.M., and Petrash, J.M., 1992, The crystal structure of the aldose reductase·NADPH binary complex. *J. Biol. Chem.* (*submitted*).

Boulton, A.J.M., Levin, S., and Comstock, J., 1990, A multicentre trial of the aldose-reductase inhibitor, tolrestat, in patients with symptomatic diabetic neuropathy. *Diabetologia*, 33:431-437.

Bradford, M.M., 1976, A rapid and sensitive method for the quantitation of microgram quantities of protein using the principle of dye-binding. *Anal. Biochem.* 72,248-254.

Carper, D., Sato, S., Old, S., Chung, S., and Kador, P.F., 1990, In vitro expression of human placental aldose reductase in Escherichia coli. *In*,: "Enzymology and Molecular Biology of Carbonyl Metabolism 3," H. Weiner *et al.*, eds., Plenum Press, New York, NY.

Carper, D.A., Old, S.E., Sato, S. and Kador, P.F., 1991, Characterization of recombinant human placenta and rat lens aldose reductase expressed in *Escherichia coli*. ARVO abstracts. Supplement to *Invest. Ophthal. Vis. Sci.*, 1991, p.975, J.B. Lippincott, Philadelphia.

Cleland, W.W., 1979, Stastical analysis of enzyme kinetic data. *Methods Enzymol.* 63,103-138.

Das, B. and Srivastava, S.K., 1985, Activation of aldose reductase from human tissue. *Diabetes* 34,1145-1151.

Del Corso, A., Camici, M. and Mura, U., 1987, *In vitro* modification of bovine lens aldose reductase activity. *Biochem. Biophys. Res. Commun.* 148:369-375.

Del Corso, A., Barsacchi, D., Camici, M., Garland, D., and Mura, U. 1989a, Bovine lens aldose reductase: Identification of two enzyme forms. *Arch. Biochem. Biophys.* 270:604-610.

Del Corso, A., Barsacchi, D., Giannessi, M., Tozzi, M.G., Camici, M., and Mura, U., 1989b, Change in stereospecificity of bovine lens aldose reductase modified by oxidative stress. *J. Biol. Chem.* 264,17653-17655.

Del Corso, A., Barsacchi, D., Giannessi, M., Tozzi, M.G., Camici, M., Houben, J.L., Zandomeneghi, M., and Mura, U., 1990, Bovine lens aldose reductase: Tight binding of the pyridine coenzyme. *Arch. Biochem. Biophys.* 283:512-518.

Flynn, T.G., 1982, Aldehyde reductases: Monomeric NADPH-depencent oxidoreductases with multifunctional potential. *Biochem. Pharmacol.* 31,2705-2712.

Frank, R.N.,1990, Aldose reductase inhibition: The chemical key to the control of diabetic retinopathy? (editorial) *Arch. Ophthal.* 108, 1229-1231.

Graham, A., Hedge, P.J., Powell, S.J., Riley, J., Brown, L., Gammack, A., Carey, F., and Markham, A.F.,1989, Nucleotide sequence of cDNA for human aldose reductase. *Nucl. Acids Res.* 17, 8368.

Grundmann, Ulrich, Bohn, H., Obermeier, R., and Amann, E., 1990, Cloning and prokaryotic expression of a biologically active human placental aldose reductase. *DNA and Cell Biol.* 9, 149-157.

Jones, T.A., 1978, A graphics model building and refinement system for macromolecules. *J. Appl. Cryst.* 11,268-272.

Kador, P.F., Kinoshita, J.H., Tung, W.H., and Chylack, L.T.,Jr., 1980, Differences in the susceptibility of various aldose reductases to inhibition. *Invest. Ophthal. Vis. Sci.* 19:980-982

Kinoshita, J.H. and Nishimura, C., 1988, The involvement of aldose reductase in diabetic complications. *Diabetes/Metabolism Reviews*, 4:323-337.

Kubiseski, T.J., Hyndman, D.J., Morjana, N.A., Flynn, T.G., 1992, Studies on pig muscle aldose reductase: Kinetic mechanism and evidence for a slow conformational change upon coenzyme binding. *J. Biol. Chem.* 267,6510-6517.

Kunkel, T.A., Roberts, J.D. and Zakour, R.A., 1987, Rapid and efficient site-specific mutagenesis without phenotypic selection. *Methods Enzymol.* 154, 367-382.

Liu, S., Bhatnagar, A., Das, B., and Srivastava, S.K., 1989, Functional cysteinyl residues in human placental aldose reductase, *Arch. Biochem. Biophys.* 275:112-121.

Morjana, N.A. and Flynn, T.G., 1989, Aldose reductase from psoas muscle: Purification, substrate specificity, immunological characterization, and effect of drugs and inhibitors, *J. Biol. Chem.* 264,2906-2911.

Mylari, B.L., Larson, E.R., Beyer, T.A., Zembrowski, W.J., Aldinger, C.E., Dee, M.F., Siegel, T.W., and Singleton, D.H., 1991, Novel, potent aldose reductase inhibitors: 3,4-dihydro-4-oxo-3-[[5-(trifluoromethyl)-2-benzothiazolyl]methyl]-1-phthalazine-acetic acid (Zopolrestat) and congeners, *J. Med. Chem.* 34,108-122.

Nishimura, C., Matsuura, Y., Kokai, Y. Akera, T., Carper, D., Morjana, N., Lyons, C., and Flynn, T.G.,1990, Cloning and expression of human aldose reductase. *J. Biol. Chem.* 265, 9788-9792.

Olins, P.O. and Rangwala, S.H., 1990, Vector for enhanced translation of foreign genes in *Escherichia coli*. *Methods Enzymol.* 185,115-119.

Petrash, J.M., Harter, T.M., Devine, C., Olins, P., Bhatnagar, A., Liu, S., and Srivastava, S.K., 1992, Involvement of cysteine residues in catalysis and inhibition of human aldose reductase: Site-directed mutagenesis of cys-80, -298 and -303. *J. Biol. Chem.* (*submitted*).

Petrash, J.M. and Favello, A.D. ,1989, Isolation and characterization of cDNA clones encoding aldose reductase. *Curr. Eye Res.* 8, 1021-1027.

Plapp, B.V., Ganzhorn, A.J., Gould, R.M., Green, D.W., Warth, J.T., and Kratzer, D.A., 1991, Catalysis by yeast alcohol dehydrogenase. *Adv. Exp. Med. Biol.* 284,241-251.

Poulsom, R., 1987, Comparison of aldose reductase inhibitors *in vitro*: Effects of enzyme purification and substrate type. *Biochem. Pharm.* 36,1577-1581.

Rondeau, J.-M., Tête-Favier, F., Podjarny, A., Reymann, J.-M., Barth, P., Biellmann, J.-F., and Moras, D., 1992, Novel NADPH-binding domain revealed by the crystal struture of aldose reductase. *Nature* 355:469-472.

Schade, S.Z., Early, S.L., Williams, T.R., Kezdy, F.J., Heinrikson, R.L., Grimshaw, C.E. and Doughty, C.C., 1990, Sequence analysis of bovine lens aldose reductase. *J. Biol. Chem.* 265,3628-3635.

Sorbinil Retinopathy Trial Research Group, 1990, A randomized trial of sorbinil, an aldose reductase inhibitor, in diabetic retinopathy. *Arch. Ophthalmol.* 108, 1234-1244.

Srivastava, S.K., Ansari, N.H., Hair, G.A., Jaspan, J., Rao, M.B., and Das, B., 1986, Hyperglycemia-induced activation of human erythrocyte aldose reductase and alteration in kinetic properties. *Biochim. Biophys. Acta* 870:302-311.

Sussman, J.L., Holbrook, S.R., Church, G.M., Kim, S.-H., 1977, A structure-factor least-squares refinement procedure for macromolecular structures using constrained *and* restrained parameters. *Acta Cryst.* A23,800-804.

Tronrud, D.E., Ten Eyck, L.F., Matthews, B.W., 1987, An efficient general-purpose least-squares refinement program for macromolecular structures. *Acta Cryst.* A43,489-501.

Vander Jagt, D.L., Robinson, B., Taylor, K.K., and Hunsaker, L.A., 1990, Aldose reductase from human skeletal and heart muscle: Interconvertible forms related by thiol-disulfide exchange. *J. Biol. Chem.* 265: 20982-20987.

Vander Jagt, D.L., Robinson, B., Taylor, K.K., and L. Hunsaker, 1992, Reduction of trioses by NADPH-dependent aldo-keto reductases: Aldose reductase, methylglyoxal, and diabetic complications. *J. Biol. Chem.* 267,4364-4369.

Wilson, D.K., Bohren, K.M., Gabbay, K.H., and Quiocho, F.A., 1992, An unlikely sugar substrate site in the 1.65Å structure of the human aldose reductase holoenzyme implicated in diabetic complications. *Science* 257:81-84.

Yancey, P.H., Haner, R.G., and Freudenberger, T.H., 1990, Effects of an aldose reductase inhibitor on organic osmotic effectors in rat renal medulla. *Am. J. Physiol.* 259,F733-F738.

INHIBITION OF ALDOSE REDUCTASE BY (2,6-DIMETHYLPHENYLSULPHONYL)NITROMETHANE: POSSIBLE IMPLICATIONS FOR THE NATURE OF AN INHIBITOR BINDING SITE AND A CAUSE OF BIPHASIC KINETICS

Walter H. J. Ward*, Peter N. Cook, Donald J. Mirrlees,
David R. Brittain, John Preston, Frank Carey,
David P. Tuffin and Ralph Howe

ICI Pharmaceuticals, Mereside, Alderley Park, Macclesfield,
Cheshire, SK10 4TG, U. K.

ABSTRACT

Aldose reductase (aldehyde reductase 2, ALR2) is often isolated as a mixture of two forms which are sensitive (ALR2S), or insensitive (ALR2I), to inhibitors. We show that ICI 215918 ((2,6-dimethylphenylsulphonyl)-nitromethane) follows either noncompetitive, or uncompetitive kinetics with respect to aldehyde for ALR2S, or the closely related enzyme, aldehyde reductase (aldehyde reductase 1, ALR1). Similar behaviour is exhibited by two other structural types of aldose reductase inhibitor (ARI), spiro-hydantoins and acetic acids, when either aldehyde, or NADPH is varied. For ALR2S, we have demonstrated kinetic competition between a sulphonylnitro-methane, an acetic acid and a spirohydantoin. Thus, different ARIs probably have overlapping binding sites. Published studies imply that ALR2 follows an ordered mechanism where coenzyme binds first and induces a reversible conformation change (E.NADPH → E^*.NADPH). Reduction of aldehyde appears rate-limited by the step $E^*.NADP^+ \rightarrow E.NADP^+$. Spontaneous activation converts ALR2S into ALR2I and increases k_{cat}. This must be associated with acceleration of the rate-determining step. We now propose the following hypothesis to explain characteristics of ARIs. (1) Inhibitors preferentially bind to the E^* conformation. (2) The ARI binding site contains residues in common with that for aldehyde substrates. When aldehyde is varied, uncompetitive inhibition arises from association at the site for alcohol product in the $E^*.NADP^+$ complex which has little affinity for the substrate. Any competitive inhibition arises from use of the aldehyde site in the E^*.NADPH complex. (3) Acceleration of the $E^*.NADP^+ \rightarrow E.NADP^+$ step upon activation of ALR2 reduces steady state levels of E^* and so decreases sensitivity to ARIs.

INTRODUCTION

Aldose reductase (aldehyde reductase 2, or ALR2) (E. C. 1.1.1.21) is strongly associated with the pathogenesis of some symptoms of diabetes (see Kador, 1988). The enzyme catalyses reduction of the open-chain, aldehyde form of glucose to sorbitol (Grimshaw, 1986) and there is evidence suggesting that excessive flux leads directly, or indirectly, to damage in nerve, kidney, retina and lens. Accordingly, inhibition of ALR2 represents a potential therapeutic approach to reducing the development of diabetic complications.

ALR2 is closely related to other members of the aldo-keto reductase family, with aldehyde reductase (aldehyde reductase 1 or ALR1, E. C. 1.1.1.2) exhibiting most homology in structure and function (Flynn, 1982; Bohren, et al., 1989). Both ALR1 and ALR2 seem to follow a compulsory ordered mechanism where NADPH is the first substrate to bind and $NADP^+$ is the last product to dissociate (see Daly & Mantle, 1982; Worrall, et al., 1986; De Jongh, et al., 1987; Kubiseski, et al., 1992). Protein fluorescence and kinetic studies indicate that association with NADPH, or $NADP^+$ induces a conformation change in ALR2 (E to E^*) (Grimshaw, et al., 1989; Kubiseski, et al., 1992). There is also evidence for coenzyme inducing conformation changes in ALR1 (Daly & Mantle, 1982; Worrall, et al., 1986), so that a Di-Iso-Ordered mechanism (Fig. 1) may apply to both ALR1 and ALR2.

SYMBOLS AND ABBREVIATIONS

* author for correspondence. E, enzyme; E^*, E in modified conformation; I, inhibitor; S, substrate; A and B, first and second substrates to bind; P and Q, first and second products to dissociate. K_i and K_{ies}, dissociation constants for I at zero and saturating substrate concentrations respectively. k_{cat}, turnover number. ALR1, aldehyde reductase 1, aldehyde reductase; ALR2, aldehyde reductase 2, aldose reductase; ALR2I, ALR2 which is insensitive to inhibitors; ALR2S, ALR2 which is sensitive to inhibitors; ARI, aldose reductase inhibitor, specifically the following structural types: spirohydantoin, acetic acid, or sulphonylnitromethane; ICI 215918, (2,6-dimethylphenyl-sulphonyl)nitromethane.

Figure 1. Proposed Di-Iso-Ordered Bi Bi mechanism for ALR2. (See Kubiseski, et al., 1992). A, NADPH; B, aldehyde; P, alcohol; Q, $NADP^+$; E^*, modified conformation of ALR2.

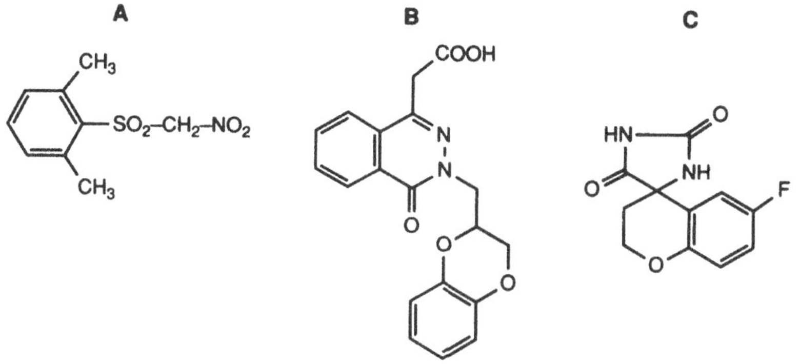

Figure 2. Structures of inhibitors used in this study. (A) ICI 215918, (2,6-dimethylphenylsulphonyl)nitromethane; (B) Compound I, 2-(2H,3H-benzodioxin-2-yl-methyl)-1H,2H-1-oxophthalazin-4-ylacetic acid; (C) sorbinil, d-6-fluoro-spiro-(chroman-4,4′-imidazolidine)-2′,5′-dione.

Spirohydantoins and acetic acids were the first aldose reductase inhibitors (ARIs) to be evaluated for clinical potential (see Kador, 1988). The inhibitor now under investigation, ICI 215918 (Fig. 2), is an example of a new structural type of ARI, the sulphonylnitromethanes.

ALR2 preparations often behave as a mixture of kinetic forms, one sensitive (ALR2S) to spirohydantoins and acetic acids, and the other insensitive (ALR2I) (see Maragoudakis, et al., 1984; Grimshaw, et al., 1989; Del Corso, et al., 1989; Vander Jagt, et al., 1990; Ward, et al., 1991). We have investigated the characteristics of ICI 215918 using ALR2 from bovine lens and ALR1 from bovine kidney. Our results and other published studies are discussed in terms of the possible nature of an ARI binding site and a putative relationship between ALR2S and ALR2I.

EXPERIMENTAL

Sources of reagents, inhibitors and enzyme preparations have been described previously (Ward, et al., 1991). Measured substrate specificity and inhibitor sensitivity demonstrate no detectable contamination of the

ALR2 preparation with ALR1 and vice-versa (see Ward, et al., 1990).

ALR1 and ALR2 activities were assayed spectrophotometrically (see Ward, et al., 1991). The various ARIs all behave as reversible inhibitors. Mechanism of inhibition was identified by analysing the measured initial velocities as described previously (Ward, et al., 1991). Rate equations were fitted to the velocities using unweighted multivariate non-linear regression. An F-test was used to estimate the probability, P, that the improvement in fit associated with inclusion of any individual parameter arose due to chance. Thus, the term was taken as justified by the data if P was low (< 1%), but not if it was high.

RESULTS AND DISCUSSION

Inhibition of aldose reductase follows biphasic kinetics

Biphasic kinetics are seen on a Dixon (1953) plot for ICI 215918 (Fig. 3). The sulphonylnitromethane, therefore, exhibits similar properties to other ARIs. Possible causes of biphasic kinetics include alternative reaction pathways, autoxidation of glucose, slow formation of an inhibitory enzyme-bound complex containing aldehyde and NADPH, contamination with ALR1, proteolysis, binding of a regulatory molecule, glycosylation, occurrence of isoenzymes, or markedly different enzymes with ALR2 activity (see Cromlish & Flynn, 1983; Maragoudakis, et al., 1984; De Jongh, et al., 1987; Grimshaw, 1990; Ward, et al., 1991). Grimshaw, et al., (1989) demonstrated that during storage homogeneous enzyme purified from bovine kidney is converted from a low K_m/low V_{max} form into a high K_m/high V_{max} form. This activation process reduces sensitivity to several ARIs, including spirohydantoins (sorbinil, AL 1576) and acetic acids (ponalrestat, tolrestat). Protein fluorescence, circular dichroism and chemical modification experiments show that activation involves changes in the secondary structure of ALR2 (Grimshaw, et al., 1989).

Mechanism of inhibition

The kinetics of inhibition have been characterized for a number of ARIs. Spirohydantoins (such as sorbinil, AL 1567 and AL 1576) and acetic acids (such as alrestatin, ponalrestat and tolrestat) inhibit both ALR1 and ALR2 by following either an uncompetitive, or noncompetitive, mechanism when either aldehyde, or NADPH, is varied (see Griffin & McNatt, 1986; De Jongh, et al., 1987; Ward, et al., 1990; Bhatnagar, et al., 1990 and Sato & Kador, 1990). ICI 215918 is a representative of a new structural type of ARIs, the sulphonylnitromethanes. This compound leads to linear mixed noncompetitive inhibition of bovine kidney ALR1 when D-glucuronate is varied

(Ward, et al., 1991). The values of K_i and K_{ies} are respectively 10 ± 2 µM and 1.8 ± 0.1 µM. The kinetics are uncompetitive when D-glucose is the varied substrate for ALR2S from bovine lens (K_{ies} = 0.10 ± 0.01 µM, $K_i >> K_{ies}$) (Ward, et al., 1991). ICI 215918, therefore, exhibits similar behaviour to other ARIs, suggesting that the sulphonylnitromethanes, acetic acids and spirohydantoins may share overlapping binding sites.

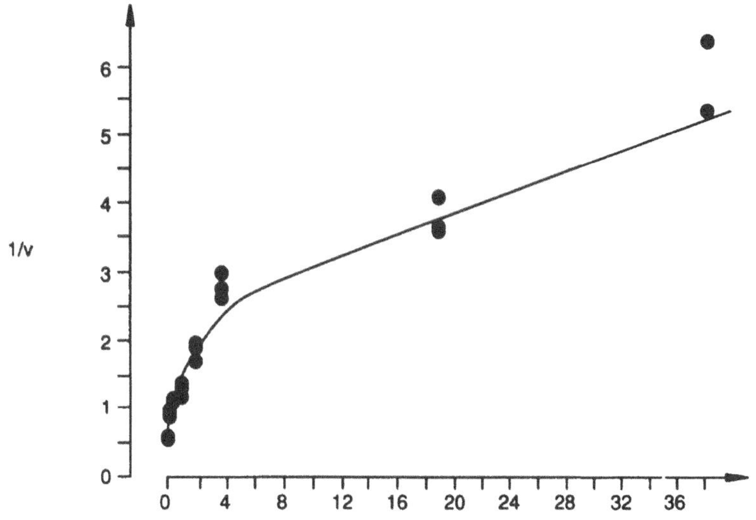

Figure 3. Biphasic inhibition of ALR2 by ICI 215918 (from Ward, et al., 1991). Assays were performed at 167 mM D-glucose (see **Experimental**). The IC_{50} values for ALR2S and ALR2I were 0.23 ± 0.04 and 30 ± 20 µM, respectively, with 77% of the total activity being due to ALR2S. Rate was measured in nmol NADPH utilized/min/mg protein.

Inhibition of ALR2S by mixtures of compounds

Rates were measured at 100 mM D-glucose when varying the levels of two inhibitors and then analysed using to the relationship

$$v = \{v_0/(1 + [I]/K_i' + [X]/K_x' + \beta[I][X]/K_i'K_x')\} + v_I$$

where v_0 is rate in the absence of inhibitors. K_i' and K_x' are the IC_{50} values for the two inhibitors, I and X, and v_I is the rate due to ALR2I. The independence term, β, is equal to zero when I and X compete for the same site. If I and X bind independently, then β = 1. An intermediate value of β indicates that binding of one inhibitor decreases affinity for the second. β is the inverse of the Yonetani-Theorell (1964) interaction constant, α. One advantage of using β is that, if its value tends to zero or one, it becomes a redundant parameter which can be rejected using an F-test.

Experiments have been performed using pairwise combinations of one representative from each of three structural types of ARI (Ward, et al., 1991) (see Fig. 2): the spirohydantoins (sorbinil), the acetic acids (Compound I) and the sulphonylnitromethanes (ICI 215918). The measured value of β is close to zero in each case, suggesting that the three inhibitors share overlapping binding sites on ALR2S (Table 1). These results are consistent with the different inhibitors following similar mechanisms, and with kinetic competition experiments between acetic acids and spirohydantoins (Kador, et al., 1986; Bhatnagar, et al., 1990; Sato & Kador, 1990). Molecular modelling and structure-activity relationships also suggest a single inhibitor binding site (Kador & Sharpless, 1983).

Table 1. Inhibition of ALR2S by mixtures of ARIs[1].

Inhibitors used		β
ICI 215918	Sorbinil	0.03 ± 0.12
ICI 215918	Compound I	0.04 ± 0.09
Compound I	Sorbinil	-0.08 ± 0.05

[1]From Ward, et al. (1991).

Implications for the nature of an ARI binding site and a cause of biphasic kinetics

We now propose a hypothesis (Fig. 4) which may explain several kinetic properties of ALR2 and its inhibitors:

1. ARIs preferentially bind to the E^* conformation.
2. The ARI binding site contains residues in common with that for aldehyde substrates. When aldehyde is varied, uncompetitive inhibition arises from association at the site for alcohol product in the $E^*.NADP^+$ complex which has little affinity for the substrate. Any competitive inhibition arises from utilisation of the aldehyde site in the E^*.NADPH complex.
3. Acceleration of the $E^*.NADP^+ \rightarrow E.NADP^+$ step on activation of ALR2 leads to a decrease in sensitivity to ARIs because steady state levels of E^* are reduced.

ARIs may bind at the site for aldehyde substrate and alcohol product. Uncompetitive kinetics indicate that saturation with the varied substrate cannot overcome inhibition. This may arise because:

1. The varied substrate and inhibitor use different binding sites, or

2. A single site preferentially binds the varied substrate at an early point in the reaction pathway, but subsequent changes in specificity favour association with the inhibitor.

Association of an enzyme-containing species with inhibitor is dependent, not only on binding affinity, but also on the steady state levels of the appropriate intermediate. For example, the velocity of the forward reaction of ALR2 in the absence of inhibitors is rate-limited by the step $E^*.Q \rightarrow E.Q$ (Grimshaw, *et al*., 1989; Kubiseski, *et al*., 1992) so that E.Q is not present at significant levels during an initial rate experiment. A

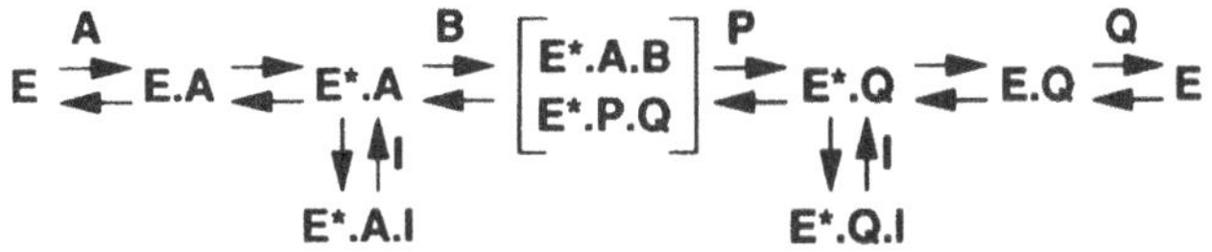

Figure 4. Proposed kinetic scheme for the action of ARIs. E, enzyme; E^*, E in modified conformation; I, inhibitor; A and B, first and second substrates to bind (NADPH and aldehyde respectively); P and Q, first and second products to dissociate (alcohol and $NADP^+$ respectively).

K_{ies} term when aldehyde is varied could, therefore, imply association with $E^*.A.B/E^*.P.Q$, or $E^*.Q$. Published studies on five ARIs when alcohol is the varied substrate (Table 2) appear to be inconsistent with binding to ternary complex and so imply association with $E^*.Q$, presumably at the site for alcohol product. These results, therefore, support our hypothesis provided that the conclusions reported for different ARIs against ALR1 are valid for ICI 215918 and ALR2. This is deduced from the failure to detect a K_{ies} term in the back reaction, suggesting that the inhibitors cannot associate with E.A, $E^*.A$, or $E^*.A.B/E^*.P.Q$. Lack of significant binding to E.A, or $E^*.A$ could be because these complexes are present only at low levels since the catalytic step may be rate-limiting in the back reaction (Kubiseski, *et al*., 1992). Alcohol and aldehyde probably use the same site, thus, it seems that any K_i term in the forward reaction is due to association with the aldehyde site in the $E^*.A$ complex. Similar mechanisms have been proposed for inhibition of sheep liver ALR1 by alrestatin (an acetic acid) and sorbinil (De Jongh, *et al*., 1987), and ox kidney ALR1 by carboxylic acids (benzoic acid and sodium valproate) (Daly & Mantle, 1982; Worrall, *et al*., 1986).

Aldehyde reductases employ a range of carbonyl compounds as substrates (see Flynn, 1982), with ALR2 utilising glucose only as its open-chain aldehyde form (Grimshaw, 1986). ARIs contain groups which could resemble the carbonyl moiety of aldehyde reductase substrates. This may explain the function of key moieties in ARIs such as the acetate residue of ponalrestat, the hydantoin moiety of sorbinil and the nitro group of ICI 215918. The proposal of a shared binding site for substrates, products and inhibitors is consistent with the results of chemical modification experiments using ALR1 from human liver (Bohren, et al., 1988; Wermuth, 1990). These studies show that a mixture of $NADP^+$ with L-gulonate, trimethylglutarate, or diphenylhydantoin can protect an Arg residue in ALR1 from covalent modification.

Table 2. Inhibition of aldehyde reductases by ARIs when alcohols are the varied substrates.

Inhibitor	Mechanism of inhibition	Varied substrate	Enzyme	Reference
Spirohydantoins				
Sorbinil	Competitive	Gulonate	Rat kidney ALR1	Sato & Kador, 1990
AL 1567	Competitive	Benzyl-OH	Rat lens ALR2	Griffin & McNatt, 1986
AL 1576	Competitive	Gulonate	Rat kidney ALR1	Sato & Kador, 1990
Acetic acids				
Tolrestat	Competitive	Gulonate	Rat kidney ALR1	Sato & Kador, 1990
Ponalrestat	Competitive	Gulonate	Rat kidney ALR1	Sato & Kador, 1990

Binding of compounds at a site which is distinct from that for substrates seems unlikely for several reasons:

1. Apparent competitive inhibition in the reverse reaction suggests that alcohol substrates can displace bound ARIs.
2. Such a model also is difficult to reconcile with the results of chemical modification experiments.
3. The observed conservation of a high affinity binding site in ALR1 and ALR2 implies that it has an important function in vivo. However, if one cannot assume that substrate, or product use the site, then no natural ligand seems to be known.
4. Conversely, it is possible to propose a biological function for the

binding specificity of the ARI binding site in our hypothesis. Inhibitors which follow uncompetitive kinetics would bind preferentially to $E^*.Q$, whereas aldehydes would favour $E^*.A$. Reduced affinity for aldehyde in the $E^*.Q$ complex could avoid substrate inhibition. This idea is supported by substrate inhibition occurring only at very high concentrations of aldehyde for ALR1 (Worrall, et al., 1986; De Jongh, et al., 1987) and ALR2 (Ward, et al., 1991).

5. ARI binding to the aldehyde/alcohol site is consistent with a simple model to explain biphasic kinetics (see below).

A possible cause of biphasic kinetics. The presence of ALR2S together with ALR2I leads to biphasic kinetics. Activation of ALR2S to ALR2I seems to be due to acceleration of a rate-determining conformation change ($E^*.Q \rightarrow E.Q$) (Grimshaw, et al., 1989). Thus, activation could decrease steady state levels of E^* relative to E. ALR2I has a lower affinity for aldehydes than does ALR2S (Grimshaw, et al., 1989; Ward, et al., 1991). The observed correlation between an increase in k_{cat}, decreased affinity for aldehydes and reduced affinity for inhibitors could, therefore, arise because substrates and ARIs associate preferentially with the E^* conformation. Activation is reported to cause only a small decrease in binding for some ARIs (eg tolrestat) (Grimshaw, et al., 1989). Such compounds could have similar affinities for the E and E^* conformations. An additional, or alternative, factor which may contribute to biphasic kinetics is that ALR2S and ALR2I may have an intrinsic difference in affinity for ARIs. This would be consistent with the evidence for a difference in conformation between ALR2S and ALR2I (Grimshaw, et al., 1989).

Possible significance of the hypothesis. Confidence in our model could be increased by studies on the 3-D structure of ALR2 when complexed to substrates, products and inhibitors. The hypothesis predicts that crystallisation of complexes containing ARIs is more likely to succeed if coenzyme is present, and ALR2S is used rather than ALR2I. If substrates, products and inhibitors do share overlapping sites on ALR2, then it could be possible to combine information on structure-activity relationships in order to assist in the design of ARIs.

The model may have complex implications for inhibition of ALR2 in vivo. Binding of inhibitors appears to be favoured by the presence of coenzyme. For compounds which tend towards uncompetitive with respect to aldehyde, increasing levels of substrate could favour inhibition, but accummulation of alcohol product would offset inhibition.

ACKNOWLEDGEMENTS

We are extremely grateful to Professor T. G. Flynn for providing information on conformation changes in ALR2 prior to publication, and to Dr C. E. Grimshaw for his comments on our work.

REFERENCES

Bhatnagar, A., Liu, S., Das, B., Ansari, N. H., and Srivastava, S. K., 1990, Inhibition kinetics of human kidney aldose and aldehyde reductases by aldose reductase inhibitors, Biochem. Pharmacol., 39:1115.

Bohren, K. M., von Wartburg, J. -P., and Wermuth, B., 1988, Inactivation of carbonyl reductase from human brain by phenylglyoxal and 2,3-butanedione: a comparison with aldehyde reductase and aldose reductase, Biochim. Biophys. Acta, 916:185.

Bohren, K. M., Bullock, B., Wermuth, B., and Gabbay, K. H., 1989, The aldo-keto reductase superfamily, J. Biol. Chem., 264:9547.

Cromlish, J. A., and Flynn, T. G., 1983, Purification and characterisation of two aldose reductase isoenzymes from rabbit muscle, J. Biol. Chem., 258:3416.

Daly, A. K., and Mantle, T. J., 1982, The kinetic mechanism of the major form of ox kidney aldehyde reductase with D-glucuronic acid, Biochem. J., 205:381.

De Jongh, K. S., Schofield, P. J., and Edwards, M. R., 1987, Kinetic mechanism of sheep liver NADPH-dependent aldehyde reductase, Biochem. J., 242:143.

Del Corso, A., Barsacchi, D., Camici, M., Garland, D., and Mura, U., 1989, Bovine lens aldose reductase: identification of two enzyme forms, Arch. Biochem. Biophys., 270:604.

Dixon, M., 1953, The determination of enzyme inhibitor constants, Biochem. J., 55:170.

Flynn, T. G., 1982, Aldehyde reductases: monomeric NADPH-dependent oxidoreductases with multifunctional potential, Biochem. Pharmacol., 31:2705.

Griffin, B. W., and McNatt, L. G., 1986, Characterization of the reduction of 3-acetylpyridine adenine dinucleotide phosphate by benzyl alcohol catalyzed by aldose reductase, Arch. Biochem. Biophys., 246:75.

Grimshaw, C. E., 1986, Direct measurement of the rate of ring opening of D-glucose by enzyme-catalyzed reduction, Carbohydrate Research, 148:345.

Grimshaw, C. E., 1990, A kinetic perspective on the peculiarity of aldose reductase, in "Enzymology and Molecular Biology of Carbonyl Metabolism 3", H. Weiner, et al., eds, Plenum Press, New York, p. 217.

Grimshaw, C. E., Shahbaz, M., Jahangiri, G., Putney, C. G., McKercher, S. R., and Mathur, E. J., 1989, Kinetic and structural effects of activation of bovine kidney aldose reductase, Biochemistry, 28:5343.

Kador, P. F., 1988, The role of aldose reductase in the development of diabetic complications, Medicinal Research Reviews, 8:325.

Kador, P. F., and Sharpless, N. E., 1983, Pharmacophor requirements of the aldose reductase inhibitor site, Mol. Pharmacol., 24:521.

Kador, P. F., Kinoshita, J. F., and Sharpless, N. E., 1986, The aldose reductase inhibitor site, Metabolism, 35 Supplement:109.

Kubiseski, T. J., Hyndman, D. J., Morjana, N. H., and Flynn, T. G., 1992, Studies on pig muscle aldose reductase. Kinetic mechanism and evidence for a slow conformation change on coenzyme binding, J. Biol. Chem., 267: 6510.

Maragoudakis, M. E., Wasvary, J., Hankin, H., and Gargiulo, P., 1984, Human placental aldose reductase. Forms sensitive and insensitive to inhibition by alrestatin, Mol. Pharmacol., 25:425.

Sato, S., and Kador, P. F., 1990, Inhibition of aldehyde reductase by aldose reductase inhibitors, Biochem. Pharmacol., 40:1033.

Vander Jagt, D. L., Robinson, B., Taylor, K. K., and Hunsaker, L. A., 1990, Aldose reductase from human skeletal and heart muscle, J. Biol. Chem., 265:20982.

Ward, W. H. J., Sennitt, C. M., Ross, H., Dingle, A., Timms, D., Mirrlees, D. J., and Tuffin, D. P., 1990, Ponalrestat: a potent and specific inhibitor of aldose reductase, Biochem. Pharmacol., 39:337.

Ward, W. H. J., Cook, P. N., Mirrlees, D. J., Brittain, D. R., Preston, J., Carey, F., Tuffin, D. P., and Howe, R., 1991, (2,6-dimethylphenylsulphonyl)nitromethane: a new structural type of aldose reductase inhibitor which follows biphasic kinetics and uses an allosteric binding site, Biochem. Pharmacol., 42:2115.

Wermuth, B., 1990, Inhibition of aldehyde reductase by carboxylic acids, in, "Enzymology and Molecular Biology of Carbonyl Metabolism 3", H. Weiner, et al., eds, Plenum Press, New York, p. 197.

Worrall, D. M., Daly, A. K., and Mantle, T. J., 1986, Kinetic studies on the major form of aldehyde reductase in ox kidney: a general kinetic mechanism to explain substrate-dependent mechanisms and the inhibition by anticonvulsants, J. Enzyme Inhibition, 1:163.

Yonetani, T., and Theorell, H., 1964, Studies on liver alcohol dehydrogenase complexes. III. Multiple inhibition kinetics in the presence of two competitive inhibitors, Arch. Biochem. Biophys., 106:243.

SEPIAPTERIN REDUCTASE AND ALR2 ("ALDOSE REDUCTASE") FROM BOVINE BRAIN

Thomas G. Dowling, John F. O' Rourke and Keith. F. Tipton

Department of Biochemistry
Trinity College Dublin
Dublin 2
Ireland

INTRODUCTION

Sepiapterin reductase (EC 1.1.1.153) catalyses the NADPH-dependent reduction of L-sepiapterin to 7,8-dihydrobiopterin as shown in Fig.1. It was first described in silkworms where it was shown to be a separate enzyme from Dihydrofolate reductase (Matsubara *et al.*, 1963). It has been purified from the livers of various species (Matsubara & Akino, 1964; Matsubara *et al.*, 1966; Katoh, 1971) and a homogeneous enzyme preparation has been obtained from rat erythrocytes (Sueoka and Katoh, 1982; Smith, 1987). It is a dimeric protein with a native relative molecular mass of about 56,000 (rat erythrocytes; Sueoka and Katoh, 1982) and is potently inhibited by N-acetyl derivatives of both serotonin and dopamine such as N-acetylserotonin, melatonin (Katoh *et al.*, 1982,1983), N-methoxyacetylserotonin and N-chloroacetydopamine (Smith *et al.*, 1992) with K_i values in the low micromolar range. Its probable function is in tetrahydrobiopterin (BH_4) biosynthesis where it may participate in the reduction of the diketo compound, 6-pyruvoyl-tetrahydropterin (6R-(1', 2'-dioxopropyl)-tetrahydropterin), to BH_4 (see Nichol *et al.*, 1985). It is is unclear whether Sepiapterin reductase alone catalyses the reduction of both keto groups or whether another reductase is also involved. (Milstien & Kaufman,1989a,b; Steinerstauch *et al.*, 1989). Sepiapterin itself, however, is no longer considered an intermediate in *de novo* BH_4 biosynthesis. Absence of the enzyme could give rise to an atypical form of phenylketonuria (see Niederwieser, 1987). Although Sepiapterin reductase was thought to be specific for the 6-lactoyl side chain of sepiapterin, it was recently found that it has a broad substrate specificity towards carbonyl compounds and can catalyse the NADPH-dependent reduction of many "typical" aldoketo reductase substrates such as *p*-nitrobenzaldehyde and 9,10-phenanthrenequinone (Katoh & Sueoka, 1984,1989; Sueoka & Katoh, 1985)

The multiple forms of soluble NADPH-dependent aldoketo reductases in mammalian

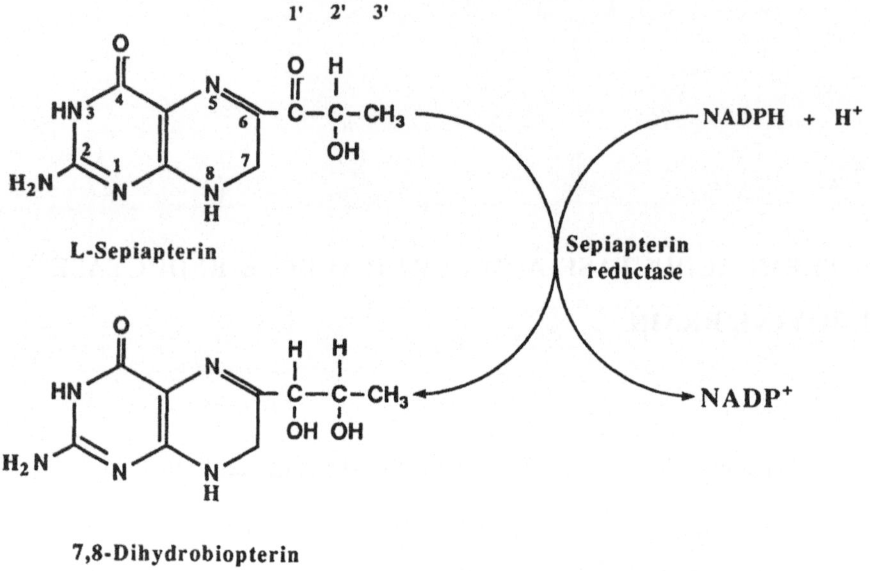

Figure 1. The reaction catalysed by Sepiapterin reductase. The enzyme catalyses the NADPH-dependent reduction of the yellow compound L-sepiapterin (6-(S)-lactoyl-7,8-dihydropterin) to 7,8-dihydrobiopterin (6-(L-*erythro*-1', 2'-dihydroxypropyl)-7,8-dihydropterin) which is colourless.

tissues have been extensively studied (see Flynn, 1982; Felsted & Bachur, 1980 for reviews). Although there has been confusion concerning the number, nomenclature and identity of reductases studied by various groups kinetic, sequence and immunological data on homogeneous proteins suggest that ALR1 (the "High K_m", valproate-sensitive Aldehyde reductase or L-Hexonate dehydrogenase); ALR2 ("Aldose reductase" or the "Low K_m" Aldehyde reductase, EC 1.1.1.21); ALR3 (Carbonyl reductase EC 1.1.1.184 or Prostaglandin 9-ketoreductase:15-hydroxy Prostaglandin dehydrogenase, Wermuth, 1981; Jarabak *et al.*, 1983, Schieber *et al.*, 1992); lung Prostaglandin F synthase (Watanabe *et al.* 1985,1988); Chlordecone reductase (Molowa *et al.* 1986; Winters *et al.* 1990); and rat liver 3α-Hydroxysteroid dehydrogenase/Dihydrodiol dehydrogenase (EC1.1.1.50 ; Penning *et al.*, 1984; Pawlowski *et al*, 1991) are distinct enzymes. The above proteins form a class of monomeric, NADPH-dependent reductases with relative molecular masses between 30,000 - 40,000. With the exception of ALR3 (Wermuth *et al.*,1988) they all share sequence homology and form part of a "superfamily" which also includes ρ-Crystallin from frog lens, Corynebacterium 2,5-Diketo-D-glucuronate reductase, mouse vas deferens androgen-dependent protein and the product of the yeast nuclear GCY gene (Carper *et al.*, 1989; Fujii *et al.*, 1990; Pailhoux *et al.*,1990). The crystal structure of ALR2 has revealed that, like Triose phosphate isomerase, it is an α/β barrel enzyme and that this structural motif is probably a general property of these homologous NADP-dependent oxidoreductases (Wilson *et al.*,1992; Rondeau *et al.*,1992). Non-monomeric aldoketo reductases have also been characterized. Both "specific" Succinic semialdehyde reductase (Cash *et al.*, 1979; Hearl & Churchich,1985) and lung pyrazole sensitive Carbonyl reductase (Nakayama *et al.*, 1982, 1986) also catalyse the NADPH-dependent reduction of carbonyl compounds but with different physico-chemical properties (see Table 1). Sepiapterin reductase, until the reports of Katoh & Sueoka mentioned above, had not been recognised as an aldoketo reductase. Furthermore, although one would expect the enzyme to occur in all tissues where BH_4 biosynthesis is required, an aldoketo reductase with similar properties has not been described by workers studying the number and type of aldehyde reductases in a given tissue (see, for example, Cromlish *et al.*, 1985; Hoffman *et al.*, 1980). This is particularly surprising in

Table 1. Some properties of soluble NADPH-dependent Aldoketo reductases. [1]The relative molecular masses were taken, where possible, from the amino acid sequence. [2]See text for references. [3]The enzyme from rat brain has been reported to be monomeric ([4]Rumigny *et al.*, 1980) but both the pig brain enzyme ([5]Hearl & Churchich, 1985) and the human brain enzyme ([6]Cash *et al.*, 1979) are dimeric.[7]ARI'S are not specific for ALR2 but also inhibit other enzymes including ALR1 (De Jongh *et al.*, 1987; Poulsom, 1986; Wilson *et al.*,1992). [8]3α-HSD from *Pseudomonas testosteroni* is B-stereospecific (Jarabak & Talalay,1960). [9]Nakayama *et al.*, (1982). [10]Sueoka & Katoh, (1982). [11]Penning *et al.*, (1984). ARI'S, aldose reductase inhibitors; CR, Carbonyl reductase; NSAID & SAID, Non-steroidal and steroidal anti-inflammatory drugs; NAS, N-acetyl serotonin; PG-F, Prostaglandin F; SSAR, "specific" Succinic semialdehyde reductase.

		[1] Approx. native M_r	[2] Stereo-specificity	[2] Homology with ALR1	Characteristic Inhibitors
ALR1	Monomeric	37,000	A	Yes	Valproate, Barbitone
ALR2	Monomeric	36,000	A	Yes	[7]ARI'S
ALR3	Monomeric	30,000	B	No	—
Rat liver 3α-HSD	Monomeric	37,000	[8]A	Yes	[11]NSAID & SAID
Chlordecone reductase	Monomeric	37,000	—	Yes	—
PG-F Synthase	Monomeric	37,000	—	Yes	—
SSAR	[3] Dimeric	110,000 [5] 90,000 [6] 45,000 [4]	B	—	—
Pyrazole sensitive CR	Tetrameric	86,000 [9]	B	—	Pyrazole
Sepiapterin reductase	Dimeric	55,000 [10]	B	No	[2]NAS, Melatonin

brain where BH_4 is essential for neurotransmitter biosynthesis. No procedure for purifying Sepiapterin reductase from brain has so far been published (but see Katoh *et al.*,1983 for inhibition studies on apparently homogeneous enzyme). This report describes some of the properties of homogeneous bovine brain Sepiapterin reductase and its relationship to other NADPH-dependent aldoketo reductases.

MATERIALS AND METHODS

L-sepiapterin and 7,8-dihydro-L-biopterin were obtained from Dr. B. Schircks Laboratories, Buechstrasse, Jona, Switzerland. *p*-Nitrobenzaldehyde was recrystallised twice from water and was dissolved in methanol before addition to the assay mixture. Other substrates (including sepiapterin and dihydrobiopterin) were dissolved in water. In all cases 50 µl of a concentrated solution was added to the reaction cocktail. The concentration of pyridine-3-aldehyde was estimated enzymically as described by Ryle & Tipton (1985). Protein was estimated by the method of Markwell *et al.* (1978) or by measuring the absorbance at 280nm.

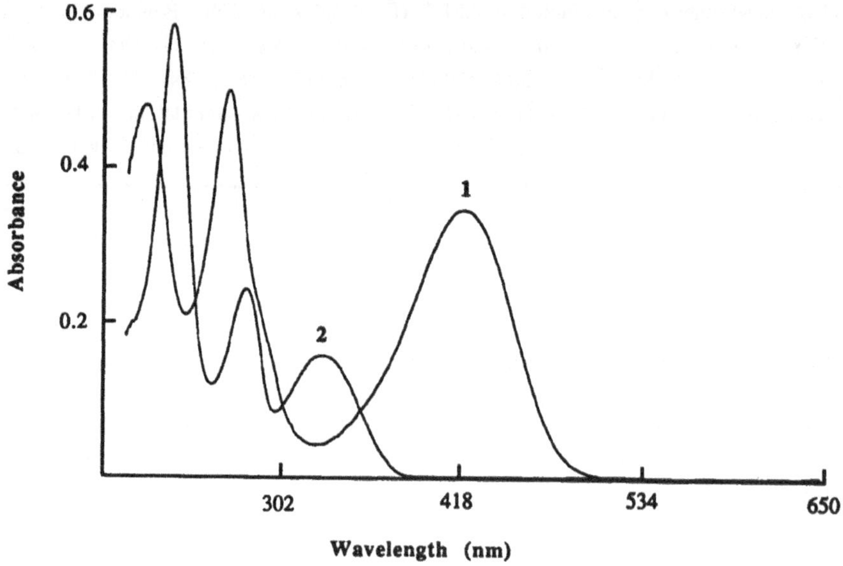

Figure 2. The absorbtion spectra of commercial L-sepiapterin (1) and 7,8-dihydrobiopterin (2) in 100mM sodium phosphate pH 7.0

Enzyme Assays

All assays were carried out spectrophotometrically using a Pye-Unicam SP8-200 double beam spectrophotometer or a Hewlett-Packard 8452A Diode Array spectrometer. Except where sepiapterin was substrate, the reaction was continuously monitored at 37° C by following the decrease in absorbance at 340nm due to NADPH oxidation and assuming an extinction coefficient of $6220M^{-1}cm^{-1}$. The reaction mixture contained 100mM sodium phosphate pH 7.0, 100 µM NADPH, either 1 mM *p*-nitrobenzaldehyde or 2 mM pyridine-3-aldehyde and enzyme in a final volume of 3ml. Blank rates, estimated in the presence of all components except either enzyme or aldehyde, were subtracted. Alternatively, when pure enzyme was available, blank rates were automatically subtracted using split-beam spectrophotometry, the assay being initiated by the addition of enzyme to the assay cuvette containing all other components and an equal volume of buffer to the reference cuvette.

Assays using Sepiapterin as Substrate

Sepiapterin reductase was assayed by a modification of the method of Katoh (1971). The reaction mixture was identical to that described above but with 50 µM sepiapterin replacing aldehyde. Because the product of the reaction, dihydrobiopterin, absorbs strongly at 340nm (E_m = 6220 $M^{-1}cm^{-1}$ at 330nm; Nagai (Matsubara), 1968) changes in optical density at this wavelength are small. Initial rates of reaction were measured at 37° C by

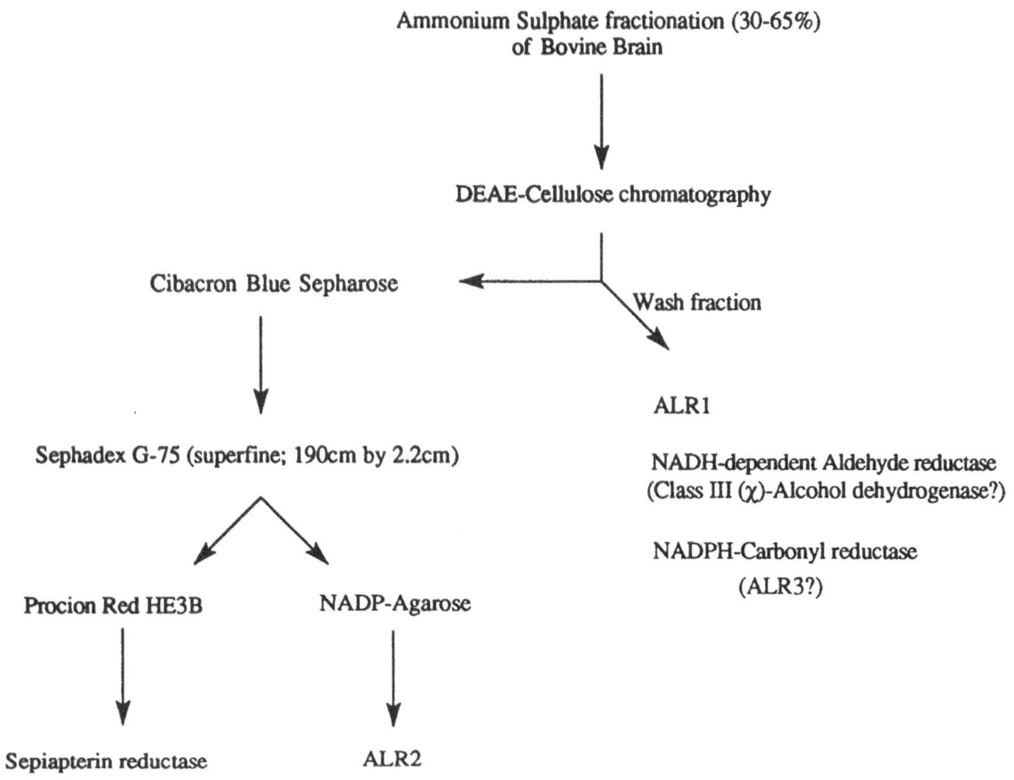

Figure 3. The scheme used for purification of ALR2 and Sepiapterin reductase. A dialysed ammmonium sulphate fraction from bovine brain was subjected to DEAE-Cellulose chromatography and the assymetrical peak obtained upon elution with KCl was adsorbed, after dialysis, onto Cibacron blue Sepharose. Gel filtration of the eluted material yielded two peaks of aldehyde reducing activity which were separately applied to either Procion Red HE3B Sepharose or NADP-agarose. Elution with NADPH yielded apparently homogeneous preparations of both Sepiapterin reductase and ALR2.

monitoring the decrease at 420nm where only sepiapterin absorbs (see Fig.2) and using an extinction coefficient of 10,400 $M^{-1}cm^{-1}$ (Katoh, 1971). Except in crude solutions of enzyme, no blank rates were observed in the absence of either cofactor or enzyme and assays were routinely started by the addition of one of these components. Stock solutions of sepiapterin in water (1.4 mg/ml or less) were stored protected from light. One unit of enzyme activity is defined as that amount which catalyses the transformation of 1 µmol of NADPH or sepiapterin per minute at 37°C.

Purification of Sepiapterin Reductase & ALR2 from Bovine Brain

Both Sepiapterin reductase and ALR2 were purified by a modification of the method of Ryle *et al.* (1984) for purification of ALR2. An outline of the scheme used is shown in Fig. 3 and details will be published elsewhere.

Determination of the Stereospecificity of Hydride Transfer

Both 4S-[4-^{3}H] NADPH (B-NADPH) and 4R-[4-^{3}H] NADPH (A-NADPH) were prepared by a modification of the method of Wermuth *et al.* (1979) and the scheme used is outlined in Fig.4. The stereospecificity of hydride transfer was then determined by incubating overnight homogeneous ALR2 or Sepiapterin reductase at 37° C in the presence of 100mM sodium phosphate pH 7.0, excess aldehyde or ketone substrate and either "A-labelled" or "B-labelled" NADPH. After passage of the reaction mixture through Centriflo membrane cones (Amicon Type CF25A, MW cutoff 25,000) the products of the reaction were separated either by gel-filtration on Sephadex G10 or by ion-exchange chromatography using QAE-Sephadex.The amount of radioactivity in the alcohol and NADP$^+$ products was determined by liquid scintillation counting.

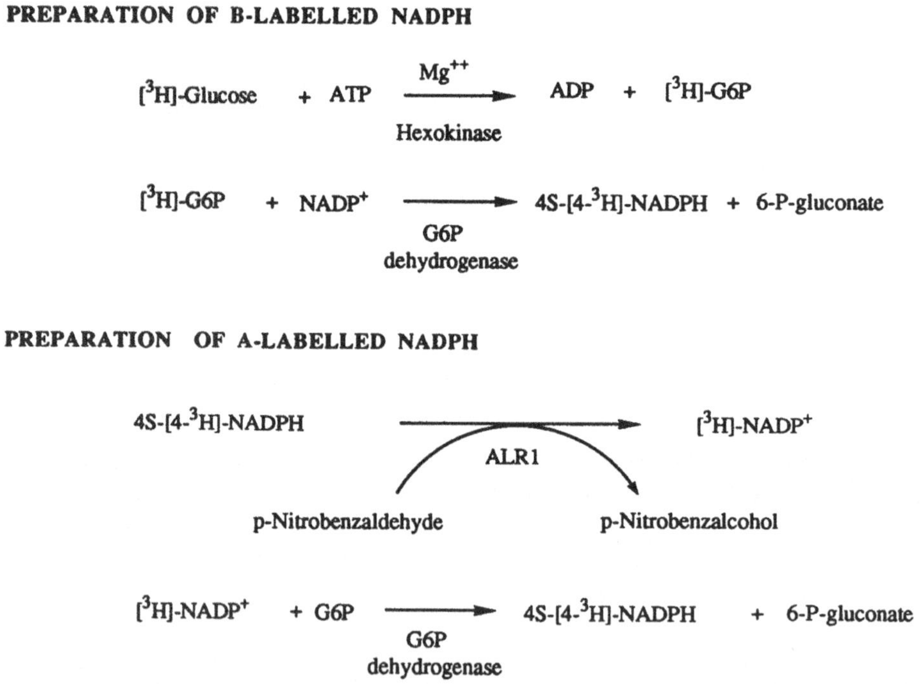

Figure 4. Synthesis of 4S-[4-^{3}H] NADPH and 4R-[4-^{3}H] NADPH. "B-labelled" NADPH was prepared by incubating tritiated glucose in the presence of both Hexokinase and Glucose-6-phosphate dehydrogenase (a "B-stereospecific" enzyme) and subsequent purification of the NADPH using QAE-Sephadex. "A-labelled" NADPH (4R-[4-^{3}H] NADPH) was prepared from "B-labelled" NADPH by first converting it to ^{3}H-NADP$^+$ in the presence of ALR1 (An "A-stereospecific" reductase). The ^{3}H-NADP$^+$ was then converted to *pro*-4R NADPH by incubation in the presence of non-radioactive glucose-6-phosphate and Glucose-6-phosphate dehydrogenase: G6P, Glucose-6-phosphate.

RESULTS

The purification scheme outlined in Fig.2 yields apparently homogeneous preparations of both Sepiapterin reductase and ALR2 as judged by gel electrophoresis in the presence and absence of sodium dodecyl sulphate followed by staining with Coomassie Blue. Although Sepiapterin reductase shows similar chromatographic properties to ALR2 on DEAE-cellulose and Cibacron Blue Sepharose, the two enzymes were completely separated by gel-filtration on Sephadex G-75 (Superfine; 190cm x 2.2cm) with Sepiapterin reductase eluting first.

The native relative masses of both Sepiapterin reductase and ALR2 were about 59,000 and 35,000, respectively, as determined by gel-filtration on a calibrated Sephadex G100 column. The relative molecular masses determined by SDS-gel electrophoresis were about 29,500 (Sepiapterin reductase) and 36,000 (ALR2). These results suggest that Sepiapterin reductase from bovine brain is dimeric. In some cases, however, a second band, of approximately double the relative mass, was visible after SDS-electrophoresis of Sepiapterin reductase. A typical purification, starting with four ox brains, yielded 6-8 mg of ALR2 and 300-700 μg of Sepiapterin reductase. The specific activities, determined in the presence of 2 mM p-nitrobenzaldehyde, were 0.68 Units/mg (ALR2) and 9.76 Units/mg (Sepiapterin reductase).The purification scheme described above separates both ALR2 and Sepiapterin reductase from ALR1, from an enzyme with similar properties to ALR3 and from glutathione-dependent formaldehyde dehydrogenase (Class III (χ)-Alcohol dehydrogenase). These latter three activities appear in the "unadsorbed fraction" on DEAE-Cellulose chromatography. Surprisingly, and in contrast to the known broad substrate specificity of the aldoketo reductases, sepiapterin was only reduced by the dimeric enzyme and did not act as a substrate for ALR1, the ALR3 fraction mentioned above or either NADH-dependent Class III (χ)-Alcohol dehydrogenase or horse liver Alcohol dehydrogenase. Neither glucose nor succinic semialdehyde, both of which are reduced by ALR2 (see Ryle *et al.*,1984), could act as substrates for Sepiapterin reductase.

Table 2. Determination of the stereospecificity of hydride transfer. Only the percentage radiolabel occurring in the alcohol product is shown. Where transfer did not take place the radioactivity was found in the $NADP^+$ fraction in each case.

	% Tritium in alcohol	
	4R-[4-^{3}H] NADPH	4S-[4-^{3}H] NADPH
ALR2	90-95%	< 5%
Sepiapterin reductase	< 5%	85-90%

Results investigating the stereospecificity of hydride transfer are shown in Table 2. It is seen that when Sepiapterin reductase was incubated in the presence of 4S-[4-^{3}H] NADPH, 85-90% of the tritium label appeared in the alcohol product. When incubations were carried out in the presence of 4R-[4-^{3}H] NADPH, less than 5% of the label was transferred. The opposite result was obtained when ALR2 was incubated under similar conditions. Thus it is clear that although ALR2 is an A-stereospecific enzyme, Sepiapterin reductase catalyses the transfer of the *pro*-4S hydrogen from NADPH.

DISCUSSION

The properties of Sepiapterin reductase distinguish it from other NADPH-dependent aldoketo reductases (see Table 1). No other purification procedure yielding homogeneous brain Sepiapterin reductase has so far been reported. The enzyme from bovine brain is apparently dimeric and in this respect it is similar to the enzyme purified from rat erythrocytes (Sueoka & Katoh, 1982). The native (59,000) and subunit (29,500) relative molecular masses are also similar to those reported by these workers (55,000 and 27,500, respectively). Sometimes SDS-gel electrophoresis of apparently homogeneous Sepiapterin reductase revealed two protein bands, the upper band being approximately double that of the lower one. It is possible that this upper band represents undissociated protein. Smith, (1987) have also reported two bands after SDS-gel electrophoresis of homogeneous rat erythrocyte Sepiapterin reductase.

Sepiapterin was not a substrate for any of the other bovine brain aldoketo reductases detected in this study and thus it may be used specifically to assay the enzyme in this tissue. This is in contrast to human brain where it is has been reported that the rates of reduction of sepiapterin by ALR3 and Sepiapterin reductase are comparable and that ALR2 can also reduce sepiapterin but with a very low specific activity (Park *et al.*, 1991). It has also been reported, however, that sepiapterin is not a substrate for either ALR3 or ALR2 from human brain or for ALR1 from human liver (Steinerstauch *et al.*, 1989). Whether these discrepancies are due to differences in enzyme preparations or to assay conditions or to other factors remains to be clarified.

Stereospecificity studies clearly distinguish both ALR2 and Sepiapterin reductase. ALR2 was found to transfer the *pro*-4R hydrogen from NADPH to substrate in agreement with previous results (Walton, 1973; Feldman *et al.*, 1977; Branlant & Biellmann, 1980). Sepiapterin reductase catalyses the transfer of the *pro*-4S hydrogen (B-stereospecific). It is similar in this respect to ALR3 (Wermuth,1981; Schieber *et al.*, 1992), "specific" Succinic semialdehyde reductase (Hoffman *et al.*,1980; Cromlish *et al.*,1985) and lung pyrazole sensitive Carbonyl reductase (Nakayama *et al.*,1986) but differs from ALR1 (Wermuth *et al.*,1979; Branlant & Biellmann, 1980), ALR2 and rat liver 3α-Hydroxysteroid dehydrogenase (Berséus & Björkhem, 1967; Askonas *et al.*,1991) which are A-stereospecific (see Table 1).

Although sepiapterin is no longer considered a *de novo* intermediate in BH_4 biosynthesis (see Nichol *et al.*,1985), Sepiapterin reductase has been shown to be essential for this process (Levine *et al.*,1990; Curtius *et al.*, 1985). The true substrate may be the diketo compound, 6-pyruvoyl tetrahydropterin (see Levine *et al.*, 1990; Smith & Nichol, 1986). There has been much controversy as to whether Sepiapterin reductase is the only reductase involved in BH_4 biosynthesis or whether it acts in conjunction with another NADPH-dependent enzyme (see Levine *et al.*,1990; Milstien & Kaufman, 1989a). It has been shown , for example, that the enzyme 6-pyruvoyl tetrahydropterin (2'-oxo) reductase, at one time thought to be a separate and distinct aldoketo reductase and to function in BH_4 biosynthesis (Milstien & Kaufman,1989a; Curtius *et al.*,1987), is identical with ALR2

(Steinerstauch *et al.*, 1989; Milstien & Kaufman, 1989b). It has also been shown that the reduction of both keto groups is catalysed by B-stereospecific reductase(s) (Curtius *et al.*, 1985). If this is true then the above results rule out ALR2 having a role in *de novo* BH_4 biosynthesis. The natural isomer of BH_4 is L-*erythro*-tetrahydrobiopterin (Kaufman,1963). Thus if two reductases are involved they must not only be B-stereospecific (Curtius *et al.*,1985) but must also transfer the hydride to the same face of each trigonal ketone. It is also possible that Sepiapterin reductase, like Histidinol dehydrogenase (EC 1.1.1.23) and 3-hydroxy-3-methylglutaryl CoA reductase (EC 1.1.1.34), catalyses a four electron transfer reaction without the release of the monoketo intermediate into free solution.

The properties of bovine brain Sepiapterin reductase distinguish it from other well-characterized aldoketo reductases. Moreover, sequences of the human and rat enzymes show no homology with members of the aldoketo reductase "superfamily" (Oyama *et al.*,1990; Ichinose *et al.*,1991). Variations in the metabolism of tetrahydrobiopterin have been associated with a variety of neurological disorders such as atypical phenylketonuria, Alzheimer's disease and Parkinson's disease. BH_4 may also be required for nitric oxide synthesis (Mayer *et al.*,1991). A study of the enzymes involved in its biosynthesis may lead to a better understanding of these processes.

REFERENCES

Askonas, L.J., Ricigliano, J.W. & Penning, T.M. (1991) The kinetic mechanism catalysed by homogeneous rat liver 3α-hydroxysteroid dehydrogenase. Evidence for binary and ternary dead-end complexes containing non-steroidal anti-inflammatory drugs. *Biochem. J.* 278 : 835-841.

Berséus, O. & Björkhem, I. (1967) Enzymatic conversion of a Δ^4-3-ketosteroid into a 3α-hydroxy-5β-steroid: Mechanism and stereochemistry of hydrogen transfer from NADPH. *Eur. J. Biochem.* 2 : 503-507.

Branlant, G. & Biellmann, J.-F. (1980) Purification and some properties of aldehyde reductases from pig liver. *Eur. J. Biochem.* 105 : 611-621.

Carper, D.A., Wistow, G., Nishimura, C., Graham, C., Watanabe, K., Fujii, Y., Hayashi, H. & Hayaishi, O. (1989) A superfamily of NADPH-dependent reductases in eukaryotes and prokaryotes. *Exp. Eye Res.* 49 : 377-388.

Cash, C.D., Maitre, M. & Mandel, P. (1979) Purification from human brain and some properties of two NADPH-linked aldehyde reductases which reduce succinic semialdehyde to 4-hydroxybutryate. *J. Neurochem.* 33 : 1169-1175.

Cromlish, J.A., Yoshimoto, C.K. & Flynn, T.G. (1985) Purification and characterization of four NADPH dependent aldehyde reductases from pig brain. *J. Neurochem.* 44 : 1477-1484.

Curtius, H.-Ch., Heintel, D., Ghisla, S., Kuster, T., Leimbacher, W. & Niederwieser, A. (1985) Tetrahydrobiopterin biosynthesis. Studies with specifically labelled (^{2}H)NAD(P)H and ^2H_2O and of the enzymes involved. *Eur. J. Biochem.* 148 : 413-419.

Curtius, H.-Ch., Steinerstauch, P., Leimbacher, W., Redweik, U., Takikawa, S.I. & Ghisla, S. (1987) Purification of 6-pyruvoyl tetrahydropterin 2'-keto reductase from human liver, *in*: "Unconjugated Pterins and Related Biogenic Amines", Curtius, H.-Ch., Blau, N. & Levine, R.A., eds., pp 89-98, de Gruyter, Berlin.

De Jongh, K.S., Schofield, P.J. & Edwards, M.R. (1987) Kinetic mechanism of sheep liver NADPH-dependent aldehyde reductase. *Biochem. J.* 242 : 143-150.

Feldman, H.B., Szczepanik, P.A., Havre, P., Corrall, R.J.M., Yu, L.C., Rodman, H.M., Rosner, B.A., Klein, P.D. & Landau, B.R. (1977) Stereospecificity of the hydrogen transfer catalysed by human placental aldose reductase. *Biochim. Biophys. Acta* 480 : 14-20.

Felsted, R.L. & Bachur, N. R. (1980) Mammalian carbonyl reductases. *Drug Metab. Rev.* 11 : 1-60.

Flynn, T.G. (1982) Aldehyde reductases: Monomeric NADPH-dependent oxidoreductases with multifunctional potential. *Biochem. Pharmacol.* 31 : 2705-2712.

Fujii, Y., Watanabe, K., Hayashi, H., Urade, Y., Kuramitsu, S., Kagamiyama, H. & Hayaishi, O. (1990) Purification and charactezisation of ρ -crystallin from japanese common bullfrog lens. *J. Biol. Chem.* 265 : 9914-9923.

Hearl, W.G. & Churchich, J.E. (1985) A mitochondrial $NADP^+$-dependent reductase related to the 4-aminobutryate shunt. Purification, characterization and mechanism. *J. Biol. Chem.* 260 : 16361-16366.

Hoffman, P.L.,Wermuth, B. & von Wartburg, J.-P. (1980) Human brain aldehyde reductases: Relationship to succinic semialdehyde reductase and aldose reductase. *J. Neurochem.* 35 : 354-366.

Ichinose, H., Katoh, S., Sueoka, T., Titani, K., Fujita, K. & Nagatsu, T. (1991) Cloning and sequencing of cDNA encoding human sepiapterin reductase - an enzyme involved in tetrahydrobiopterin biosynthesis. *Biochem. Biophys. Res. Commum.* 179 : 183-189.

Jarabak, J. & Talalay, P. (1960) Stereospecificity of hydrogen transfer by pyridine nucleotide-linked hydroxysteroid dehydrogenases. *J. Biol. Chem.* 235 : 2147-2151.

Jarabak, J., Luncsford, A. & Berkowitz, D. (1983) Substrate specificity of three prostaglandin dehydrogenases. *Prostaglandins* 26 : 849-868.

Katoh, S. (1971) Sepiapterin reductase from horse liver: Purification and properties of the enzyme. *Arch. Biochem. Biophys.* 146 : 202-214.

Katoh, S. & Sueoka, T. (1984) Sepiapterin reductase exhibits a NADPH-dependent dicarbonyl reductase activity. *Biochem. Biophys. Res. Commum.* 118 : 859-866.

Katoh, S. & Sueoka, T. (1989) Properties of carbonyl reductase activity of sepiapterin reductase, an enzyme involved in the biosynthesis of tetrahydrobiopterin. *Prog. Clin. Biol. Res.* 290 : 381-395.

Katoh, S., Sueoka, T. & Yamada, S. (1982) Direct inhibition of brain sepiapterin reductase by a catecholamine and an indoleamine. *Biochem. Biophys. Res. Commum.* 105 : 75-81.

Katoh, S., Sueoka, T. & Yamada, S. (1983) Inhibition of brain sepiapterin reductase by a catecholamine and an indoleamine, *in*: "Chemistry and Biology of Pteridines", Blair, J.A., ed., pp 789-793, de Gruyter, Berlin.

Kaufman, S. (1963) The structure of the phenylalanine-hydroxylation cofactor. *Proc. Natl. Acad. Sci. U.S.A.* 50 : 1085-1093.

Levine, R.A., Kapatos, G., Kaufman, S. & Milstien, S. (1990) Immunological evidence for the requirement of sepiapterin reductase for tetrahydrobiopterin biosynthesis in brain. *J. Neurochem.* 54 : 1218-1224.

Markwell, M.A.K., Haas, S.M., Bieber, L.L. & Tolbert, N.E. (1978) A modification of the lowry procedure to simplify protein determination in membrane and lipoprotein samples. *Anal. Biochem.* 87 : 206-210.

Matsubara, M. & Akino, M. (1964) On the presence of sepiapterin reductase different from folate and dihydrofolate reductase in chicken liver. *Experientia* 20 : 574-575.

Matsubara, M., Tsusue, M. & Akino, M. (1963) Occurrence of two different enzymes in the silkworm, *Bombyx mori*, to reduce folate and sepiapterin. *Nature (London)* 199 : 908-909.

Matsubara, M., Katoh, S., Akino, M. & Kaufman, S. (1966) Sepiapterin reductase. *Biochim. Biophys. Acta* 122 : 202-212.

Mayer, B., John, M., Heinzel, B. Werner, E.R., Wachter, H., Schultz, G. & Böhme, E. (1991) Brain nitric oxide synthase is a biopterin- and flavin-containing multi-functional oxido-reductase. *FEBS Lett.* 288 : 187-191.

Milstien, S. & Kaufman, S. (1989a) The biosynthesis of tetrahydrobiopterin in rat brain. Purification and characterization of 6-pyruvoyl tetrahydropterin (2'-oxo) reductase *J. Biol. Chem.* 264 : 8066-8073.

Milstien, S. & Kaufman, S. (1989b) Immunological studies on the participation of 6-pyruvoyl tetrahydropterin (2'-oxo) reductase, an aldose reductase, in tetrahydrobiopterin biosynthesis. *Biochem. Biophys. Res. Commum.* 165 : 845-850.

Molowa, D.T., Shayne, A.G. & Guzelian, P.S. (1986) Purification and characterization of chlordecone reductase from human liver *J. Biol. Chem.* 261 : 12624-12627.

Nagai (Matsubara), M. (1968) Studies on sepiapterin reductase: Further characterization of the reaction product. *Arch. Biochem. Biophys.* 126 : 426-435.

Nakayama, T., Hara, A. & Sawada, H. (1982) Purification and characterization of a novel pyrazole-sensitive carbonyl reductase in guinea pig lung. *Arch. Biochem. Biophys.* 217 : 564-573.

Nakayama, T., Yashiro, K., Inoue, Y., Matsuura, K., Ichikawa, H., Hara, A. & Sawada, H. (1986) Characterization of pulmonary carbonyl reductase of mouse and guinea pig. *Biochim. Biophys. Acta* 882 : 220-227.

Nichol, C.A., Smith, G.K. & Duch, D.S. (1985) Biosynthesis and metabolism of tetrahydrobiopterin and molybdopterin. *Annu. Rev. Biochem.* 54 : 729-764.

Niederwieser, A. (1987) Atypical phenylketonuria with tetrahydrobiopterin deficiency, *in:* "Unconjugated Pterins In Neurobiology. Basic and Clinical Aspects", Lovenberg, W. & Levine, R.A., eds., pp 107-118 Taylor & Francis, London.

Oyama, R., Katoh, S., Sueoka, T., Suzuki, M., Ichinose, H., Nagatsu, T. & Titani, K. (1990) The complete amino acid sequence of the mature form of rat sepiapterin reductase. *Biochem. Biophys. Res. Commum.* 173 : 627-631.

Pailhoux, E.A., Martinez, A., Veyssiere, G.M. & Jean, C.G. (1990) Androgen-dependent protein from mouse vas deferens. cDNA cloning and protein homology with the aldo-keto reductase superfamily *J. Biol. Chem.* 265 : 19932-19936.

Park, Y.S., Heizmann, C.W., Wermuth, B., Levine, R.A., Steinerstauch, P., Guzman, J. & Blau, N. (1991) Human carbonyl and aldose reductases: New catalytic functions in tetrahydrobiopterin biosynthesis. *Biochem. Biophys. Res. Commum.* 175 : 738-744.

Pawlowski, J.E., Huizinga, M. & Penning, T.M. (1991) Cloning and sequencing of the cDNA for rat liver 3α-hydroxysteroid/dihydrodiol dehydrogenase. *J. Biol. Chem.* 266 : 8820-8825

Penning, T.M., Mukharji, I., Barrows, S. & Talalay, P. (1984) Purification and properties of a 3α-hydroxysteroid dehydrogenase of rat liver cytosol and its inhibition by anti-inflammatory drugs. *Biochem. J.* 222 : 601-611.

Poulsom, R. (1986) Inhibition of hexonate dehydrogenase and aldose reductase from bovine retina by sorbinil, statil, MD 79175 and valproate. *Biochem. Pharmacol.* 35 : 2955-2959.

Rondeau, J.M., Tête-Favier, F., Podjarny, A., Reymann, J.M., Barth, P., Biellmann, J.F. & Moras, D. (1992) Novel NADPH-binding domain revealed by the crystal structure of aldose reductase. *Nature (London)* 355 : 469-472.

Rumigny, J.F., Maitre, M., Cash, C. & Mandel P. (1980) Specific and non-specific succinic semialdehyde reductases from rat brain: Isolation and properties. *FEBS Lett.* 117 : 111-116

Ryle, C.M. & Tipton, K.F. (1985) Kinetic studies with the low-K_m aldehyde reductase from ox brain. *Biochem. J.* 227 : 621-627.

Ryle, C.M., Dowling, T.G. & Tipton, K.F. (1984) Purification and properties of low-K_m aldehyde reductase from ox brain. *Biochim. Biophys. Acta* 791 : 155-163.

Schieber, A. Frank, R.W. & Ghisla, S. (1992) Purification and properties of prostaglandin 9-ketoreductase from pig and human kidney. Identity with human carbonyl reductase. *Eur. J. Biochem.* 206 : 491-502

Smith, G.K. (1987) On the role of sepiapterin reductase in the biosynthesis of tetrahydrobiopterin. *Arch. Biochem. Biophys.* 255 : 254-266.

Smith, G.K. & Nichol, C.A. (1986) Synthesis, utilization, and structure of the tetrahydropterin intermediates in the bovine adrenal medullary *de novo* biosynthesis of tetrahydrobiopterin *J. Biol. Chem.* 261 : 2725-2737.

Smith, G.K., Duch, D.S., Edelstein, M.P. & Bigham, E.C. (1992) New inhibitors of sepiapterin reductase. Lack of an effect of intracellular tetrahydrobiopterin depletion upon *in vitro* proliferation of two human cell lines. *J. Biol. Chem.* 267 : 5599-5607.

Steinerstauch, P., Wermuth, B., Leimbacher, W. & Curtius, H.-Ch. (1989) Human liver 6-pyruvoyl tetrahydropterin reductase is biochemically and immunologically indistinguishable from aldose reductase. *Biochem. Biophys. Res. Commum.* 164 : 1130-1136.

Sueoka, T. & Katoh, S. (1982) Purification and characterization of sepiapterin reductase from rat erythrocytes. *Biochim. Biophys. Acta* 717 : 265-271.

Sueoka, T. & Katoh, S. (1985) Carbonyl reductase activity of sepiapterin reductase from rat erythrocytes. *Biochim. Biophys. Acta* 843 : 193-198.

Walton, D.J. (1973) Stereochemistry of reduction of D-glyceraldehyde catalysed by a nicotinamide adenine dinucleotide phosphate dependent dehydrogenase from skeletal muscle. *Biochemistry* 12 : 3472-3478.

Watanabe, K., Yoshida, R., Shimizu, T. & Hayaishi, O. (1985) Enzymatic formation of prostaglandin $F_{2\alpha}$ from prostaglandin H_2 and D_2. Purification and properties of prostaglandin F synthetase from bovine lung. *J.Biol. Chem.* 260 : 7035-7041.

Watanabe, K., Fujii, Y., Nakayama, K., Ohkubo, H., Kuramitsu, S., Kagamiyama, H., Nakanishi, S. & Hayaishi, O. (1988) Structural similarity of bovine lung prostaglandin F synthase to lens ε-crystallin of the european common frog. *Proc. Natl. Acad. Sci. U.S.A.* 85 : 11-15.

Wermuth, B. (1981) Purification and properties of an NADPH-dependent carbonyl reductase from human brain. Relationship to prostaglandin 9-keto reductase and xenobiotic ketone reductase. *J. Biol. Chem.* 256 : 1206-1213.

Wermuth, B., Münch, J.D.B. & von Wartburg, J.-P. (1979) Stereospecificity of hydrogen transfer of aldehyde reductase. *Experientia* 35 : 1288-1289.

Wermuth, B., Bohren, K.M., Heinemann, G., von-Wartburg, J.-P. & Gabbay, K.H. (1988) Human carbonyl reductase. Nucleotide sequence analysis of a cDNA and amino acid sequence of the encoded protein. *J. Biol. Chem.* 263 : 16185-16188.

Wilson, D.T., Bohren, K.M., Gabbay, K.H. & Quiocho, F.A. (1992) An unlikely sugar substrate site in the 1.65 Å structure of the human aldose reductase holoenzyme implicated in diabetic complications. *Science* 257 : 81-84.

Winters, C.J., Molowa, D.T. & Guzelian, P.S. (1990) Isolation and characterization of cloned cDNAs encoding human liver chlordecone reductase. *Biochemistry* 29 : 1080-1087.

POLYMORPHISMS OF THE ALDOSE REDUCTASE LOCUS (ALR2) AND SUSCEPTIBILITY TO DIABETIC MICROVASCULAR COMPLICATIONS

Ashok Patel, Suvina Ratanachaiyavong, B. Ann Millward and Andrew G. Demaine

Department of Medicine, King's College School of Medicine and Dentistry
Denmark Hill, London SE5 8RX

INTRODUCTION

The reason why only some patients with type 1 diabetes develop microvascular complications, such as retinopathy and nephropathy, is unclear. Prolonged exposure to hyperglycemia is the primary factor associated with the development of most of these complications (Brownlee et al., 1990). Good metabolic control can help to slow their relentless progression in some patients but does not prevent them entirely (Larkins et al., 1992). Therefore genetic determinants of tissue susceptibility as well as environmental factors are probably important in the predisposition to diabetic microvascular complications.

A rise in tissue sorbitol secondary to concentration-dependent activation of polyol pathway activity by glucose and an accompanying fall in tissue myo-inositol and Na^{+}-K^{+} ATPase activity are thought to be important in the pathogenesis of long term diabetic complications (Greene et al., 1987).

Aldose reductase (ALR2;EC 1.1.1.21) is the first enzyme in the polyol pathway and it catalyses the NADPH-dependent reduction of hexose sugars to their corresponding alcohols. It is part of a family of monomeric aldo-keto reductases, all between 35-40 kilodaltons, which show broad substrate specificities for aldehydes and ketones, including xenobiotics as well as endogenous compounds. Its physiological role has not yet been elucidated, but it is expressed in the cell cytoplasm of a number of human tissues including lens, kidney, liver, placenta and brain. Under normal physiological conditions this pathway is thought to play a minor role in glucose metabolism, but in hyperglycemic conditions cells undergoing insulin-independent uptake of glucose produce significant quantities of sorbitol. Since sorbitol does not readily diffuse out of the cell and its conversion to fructose is slow, the resulting intracellular accumulation of sorbitol causes hyperosmotic stress leading to loss of cellular integrity. This excessive flux through the polyol pathway is postulated, by some, to be the primary cause of diabetic complications (Dvornik, 1987).

A number of aldose reductase inhibitors have been shown to reverse decreased axonal transport and improve nerve conduction velocity (Tomlinson et al., 1984), to delay or prevent the onset of sugar cataracts (Kinoshita et al., 1981) and to ameliorate glomerular hyperfiltration (Goldfarb et al., 1986), in animals with either chemically-induced or genetic diabetes. However, the efficacy of these drugs on the treatment of diabetic complications in patients has still to be fully demonstrated (Masson et al., 1990).

Human ALR2 cDNA has been cloned from a number of tissues (Bohren et al., 1989, Chung et al., 1989, Graham et al., 1989 and Nishimura et al 1990) and strong homology has been found at the nucleotide and protein level to aldehyde reductase, 2,5-diketogluconic acid reductase, and P-crystallin. This suggests that these proteins are evolutionarily related and are collectively known as the aldo-keto reductase super-gene family (Bohren et al., 1989).

The gene for the functional human ALR2 has been localised to chromosome 7q35 and consists of ten exons extending over 18 kilobases (kb) of DNA, giving rise to a 1.3 kb RNA transcript (Graham et al., 1991a and 1991b). A pseudogene of ALR2 has recently been assigned to chromosome 3 (Brown et al., 1992). We report here studies using amplified regions of the ALR2 as probes to screen for restriction fragment length polymorphisms (RFLPs) and to investigate their frequency in a population of patients with type I diabetes and well defined microvascular complications.

METHODS

A. Subjects

One hundred and thirty British Caucasoid patients with type I diabetes mellitus as defined by the National Diabetes Data Group (1979) were obtained from the Diabetic Clinic at King's College Hospital and from the Renal Unit, Dulwich Hospital. Local ethical committee approval was obtained. The patients were classified according to their microvascular complications as follows:

(i) Uncomplicated patients (N = 41) - patients who have had type I diabetes for at least 20 years but are free of retinopathy (< 5 dots or blots per fundus), nephropathy (urine Albustix negative on 3 consecutive occasions over 12 months).

(ii) Nephropaths (N = 53) - patients with type I diabetes, for at least 10 years, with persistent proteinuria (urine Albustix positive on at least 3 consecutive occasions over 12 months or, 3 successive total urinary protein excretion rates greater than 0.5g/24 hours), in the absence of haematuria or infection on mid-stream urine samples. Diabetic nephropathy was always associated with diabetic retinopathy.

(iii) Retinopaths (N = 36) - patients with type I diabetes and retinopathy and no proteinuria (urine Albustix negative on 3 consecutive occasions over 12 months). Diabetic retinopathy was defined as more than 5 dots or blots per eye, hard or soft exudates, new vessels or fluorescein angiographic evidence of maculopathy or previous laser treatment or vitrectomy for new vessels, maculopathy or vitreous haemorrhage. Fundoscopy was performed by both a diabetologist and opthalmologist.

The clinical features of these groups are shown in table 1.

Fifty-seven British Caucasoid blood donors with no evidence of diabetes or renal disease were used to obtain normal control frequencies of alleles and genotypes.

Table 1. Clinical features of the patient sub-groups

	Uncomplicated	Nephropaths	Retinopaths
Male:Female	25:16	27:26	17:19
Age at onset of disease (years)	20.1 (1-51)	17.3 (3-40)	21.1 (4-40)
Duration of disease (years)	34.8 (20-65)	29.1 (13-49)	34.4 (17-60)

Mean figures and range (in brackets).

B. Experimental Methods

High molecular weight DNA was prepared from 35-50 ml of peripheral blood using standard proteinase-K digestion/phenol extraction. The DNA (5-10 micrograms) was digested with 10-20 units of the appropriate restriction endonuclease following the manufacturer's protocol (Bethesda Research Laboratories, Paisley, Scotland). The DNA fragments were separated by size in 0.6% agarose gels transferred and immobilised in nylon membranes (Hybond-N+) using a modified method of Southern (1975).

The membranes were hybridised with a denatured random-primer ^{32}P-dCTP labelled probe for 16-20 hours in a solution containing 6X SSC, 5X Denhardt's solution, 5% dextran sulphate, 0.2 mg/ml denatured salmon sperm DNA at 65°C. After hybridisation, the filters were washed in 0.2 X SSC, 0.5% SDS at 65°C for 30-45 minutes and subsequently placed between Cronex lightening plus intensifying screens with Kodak XAR5 film at -70°C. Films were developed after 1-4 days.

C. DNA probes

Three DNA probes 5'ALR2, 3'ALR2 and ALR2EX2 were constructed using the polymerase chain reaction. Probes 5'ALR2 and 3'ALR2 were prepared using the following pairs of oligonucleotide primers, AR1 and AR2, and AR3 and AR4 respectively. The corresponding positions of these primers on the published sequence of human placental ALR2 cDNA (Graham et al., 1989) are shown in brackets. ALR2EX2 was prepared using primers AR5 and AR6; the corresponding positions on the genomic sequence of the ALR2 gene are also shown in brackets (Graham et al., 1991b):

AR1 5' GCAAGCCGTATCCTGCTCAAC (position 39 to 60)
AR2 5' CCAGTGAATAAGGTAGAGGTC (position 372 to 351)
AR3 5' AGTATAAGCCTGCAGTTAAC (position 565 to 585)
AR4 5' CACAGACCCTCCAGTTCCT (position 934 to 914)
AR5.5' GAGGTCCCGTCCACTGACTCC (position 7735 to 7756)
AR6 5' TTGCTGACGATGAAGAGCTCC (position 7898 to 7877)

Probe 5'ALR2 spanned a region of 333 bp on the ALR2 cDNA but when primers AR1 and AR2 were compared to the ALR2 genomic sequence, they were found to cover a region of approximately 8.2 kb, including intron 1 (7.1 kb) and exons 1, 2 and 3. Probe 3'ALR2 spanned approximately 3.5 kb of genomic DNA which included exons 6,7,8 and 9, whilst probe ALR2EX2 was 163 bp covering most of exon 2 (figure 1A). The restriction sites of Pst I and BamHI on the ALR2 gene are shown in figure 1B.

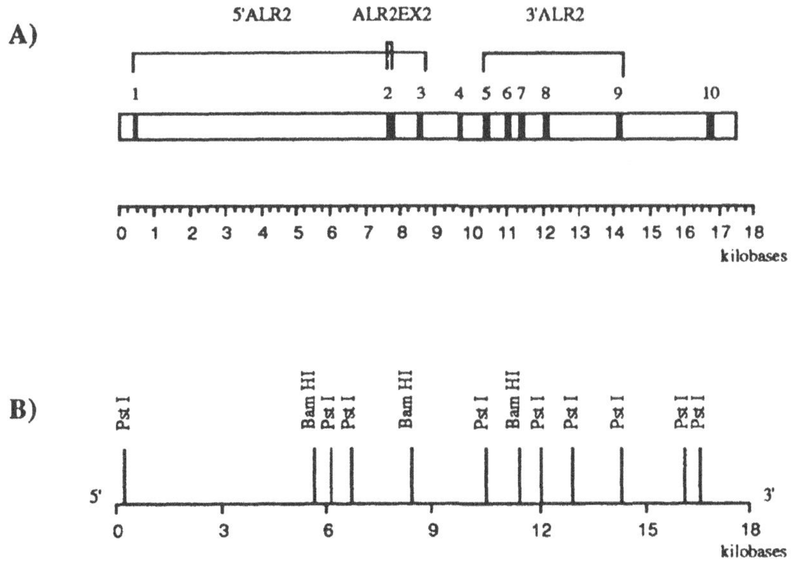

Figure 1 A. Schematic representation of the organization of the ALR2 gene showing regions spanned by probes B. Restriction map of the ALR2 gene.

One microgram of genomic DNA from a normal individual was used to amplify the probes using these primer combinations. The conditions for the amplification were: 94°C for 2 min, 55°C for 2 min and 72°C for 3 min. After amplification 50% of the product was purified by two rounds of electrophoresis through 1% low melting point (LMP) agarose followed by purification through an Elutip column.

D. Statistical Analysis

The Chi-Squared test was used to compare the data obtained for the various groups and where necessary the P values were corrected for the number of comparisons (Pc).

RESULTS

Probes 5'ALR2 and 3'ALR2 were used in conjunction with a number of restriction endonucleases to identify polymorphisms of the ALR2 gene. Screening with 7 different restriction endonucleases (BamHI, SstI, HindIII, TaqI, MspI, EcoRI, and PstI) identified two polymorphisms. A polymorphism was detected using 3'ALR2 and the restriction endonuclease BamHI with allelic fragments of 9.2 or 8.2 kb and genotypes of either 9.2/8.2, 9.2/9.2 or 8.2/8.2 kb. These polymorphic fragments were not detected with either the 3'ALR2 or ALR2EX2 probes. The size of the fragments correlated with those reported in a preliminary study by other workers using the same restriction endonuclease (Gabbay KH., 1990). A complex hybridisation pattern is found because the probe detects both the functional and pseudo- ALR2 genes. The pseudogene is distinct from the functional ALR2 gene because it does not have any introns. Using the 5'ALR2 probe two polymorphic allelic fragments of 5.7 and 5.3 kb with corresponding genotypes of either 5.7/5.3, 5.7/5.7 or 5.3/5.3 kb were detected. This RFLP was not detected using the ALR2EX2 or the 3'ALR2 probes, indicating that the polymorphic site is at the 5' end of the ALR2 gene in the non-coding region. This correlates with the restriction map of the ALR2 locus which shows that there is a PstI restriction site between exon 1 and exon 2. The fragments detected using both PstI and BamHI enzymes correspond to those predicted from the restriction map of the ALR2 locus (Fig 1B).

The allelic and genotype frequency of the 5'ALR2/PstI RFLPs in 113 of the 130 patients and 57 normal controls are shown in Fig 3. No significant differences were found between any of the genotypes or alleles and either type I diabetes or the presence of diabetic microvascular complications.

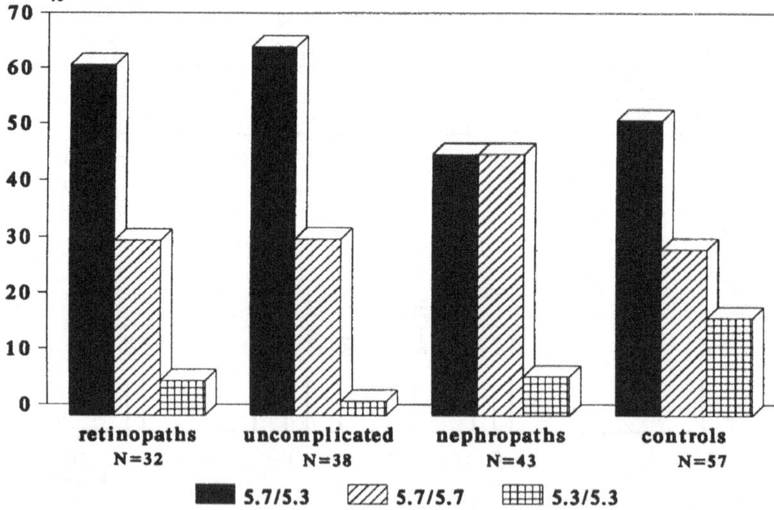

Figure 2. Distribution of 5'ALR2/PstI genotypes.

The distribution of the 3'ALR2/BamHI genotypes and alleles was determined in 89 of the 130 patients and 51 of the 57 normal controls (Fig.3). Insufficient DNA was available to type all the samples with the 3'ALR2 probe. New DNA samples were obtained where possible but in many cases the individuals were unavailable. There was a significant increase in the frequency of the 8.2/8.2 kb genotype in the retinopaths compared to the uncomplicated (66.7% vs. 32.0% $p<0.007$, $Pc=0.037$) and nephropath groups (66.7% vs 34.9% $p<0.006$, $Pc=0.03$) respectively. The 5'ALR2 and 3'ALR2 genotypes in the normal control population were in Hardy-Weinberg equilibrium.

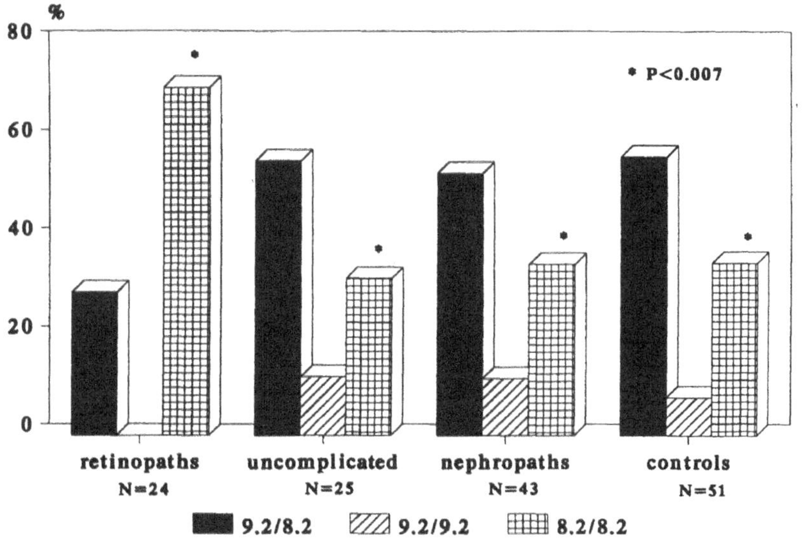

Figure 3. Distributions of 3'ALR2/BamHI genotypes.

Seventy one patients and 25 normal controls were typed for both the 5'ALR2 and 3'ALR2 RFLPs and the frequency of the genotypes is shown in Fig 4. The combination of both the PstI and BamHI alleles allowed four ALR2 haplotypes to be defined of the following combinations: 5.7-9.2, 5.3-9.2, 5.7-8.2 and 5.3-8.2 (PstI-BamHI) which give rise to nine ALR2 genotypes. The 5.7-8.2/5.3-8.2 genotype was found in 52.9% of the retinopaths and in none of the uncomplicated patients ($Chi^{+}=12.0$, $p<0.0006$, $Pc<0.007$).

DISCUSSION

It has been suggested by many workers that ALR2 may contribute towards the pathogenesis of diabetic microvascular complications (Gabbay KH., 1973, Winegrad AI., 1987 and Kinoshita et al., 1988). However, much of these data have been derived from animal models of diabetes and from long-term cell cultures. Now that the gene for ALR2 has been cloned and sequenced, its importance in diabetic microvascular complications in humans can be addressed. We have attempted to answer the question of whether the ALR2 locus confers susceptibility by identifying polymorphisms of the gene and analysing their frequency in a clinically well-defined patient population. These results suggest that the ALR2 or an adjacent locus contributes to the susceptibility to diabetic retinopathy. However, it should be noted that no association was found with those patients with nephropathy, all of whom also had retinopathy. This is somewhat surprising and may reflect differences in the provoking factors (environmental and genetic) leading to nephropathy plus retinopathy compared to retinopathy alone.

Recent reports suggest that the enzyme activity of ALR2 in neutrophils is increased in patients with severe microvascular complications but not in severe neuropathy (Dent et al., 1991a and 1991b). Unfortunately, aldose reductase activity was not measured in this study. The activity of ALR2 may be genetically determined and it is crucial to address the relationship between genotype and enzyme activity in future studies.

The complex pattern of hybridisation seen with the probes indicates that there are a number of genes coding for ALR2. Recently, an ALR2 pseudogene has been characterised and localised to chromosome 3. This pseudogene is 90% homologous at the nucleotide level to ALR2 but it contains no introns. In the rat there appears to be a multi-gene family encoding ALR2 with three pseudogenes containing no introns and one putative functional gene mapped on chromosome 4 (Graham et al., 1991).

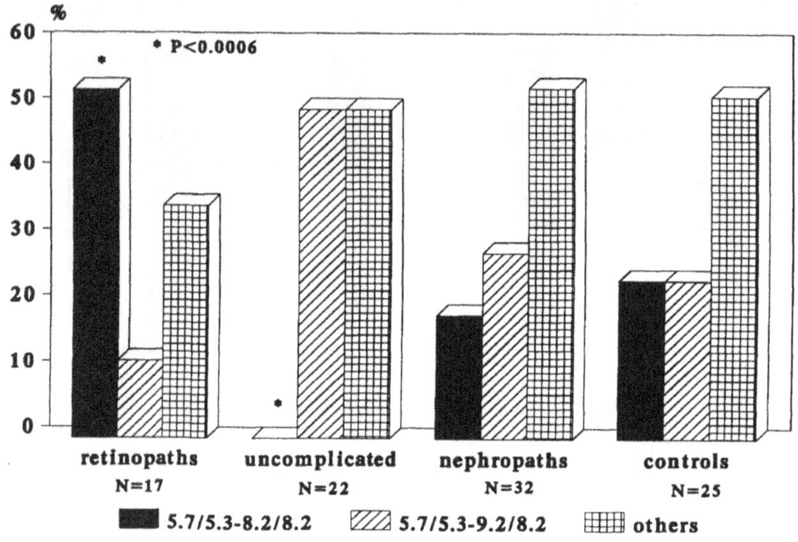

Figure 4. Distribution of 5'ALR2/PstI-3'ALR2/BamHI genotypes.

It is possible that the association between ALR2 and diabetic complications may be due to an adjacent locus. It is interesting that the gene for ALR2 maps to chromosome 7q35, a region that also codes for the T-cell antigen receptor TCR-Cß chain genes (Kronenburg et al., 1986). We have recently shown that a TCR-Cß is associated with diabetic microvascular complications (Hibberd et al., 1992). Work is currently in progress to investigate the relationship between the TCR-Cß locus, ALR2 and microvascular complications.

Finally these results suggest that a polymorphism of the ALR2 gene is associated with a group of patients who developed diabetic retinopathy without nephropathy. If confirmed, it may be possible for the first time to identify diabetic patients who may be at risk of developing retinopathy alone. In the future it might be possible to target this group for prophylactic treatment.

ACKNOWLEDGEMENTS

The patient samples were kindly provided by Dr Watkins, King's College Hospital and Professors Keen and Viberti, Guy's Hospital. This work was partly funded by the British Diabetic Association and the British Council for the Prevention of Blindness.

REFERENCES

Bohren KM, Bullock B, Wermuth B, Gabbay KH. The aldo-keto reductase superfamily. cDNAs and deduced amino acid sequences of human aldehyde and aldose reductases. J Biol Chem 1989; 264: 9547-9551

Brown L, Hedge PJ, Markham AF, Graham A. A human aldehyde dehydrogenase (aldose reductase) pseudogene: nucleotide sequence analysis and assignment to chromosome 3. Genomics 1992; 13: 465-468

Brownlee M. Glycosylation products as toxic mediators of diabetic complications. Ann Rev Med 1990; 42:159-166

Chung S, LaMendola J. Cloning and sequence determination of human placental aldose reductase gene. J Biol Chem 1989;
264: 14775-14777

Dent MT, Tebbs SE, Gonzalez AM, Ward JD, Wilson RM. Neutrophil aldose reductase activity and its association with established microvascular complications. Diabetic Med. 1991a; 8: 439-442.

Dent, MT, Veves, A, Tebbs SE, Gonzalez AM, Malik RA, Boulton AJM, et al. Neutrophil aldose reductase activity as a potential marker for neuropathy and cataract in diabetes. Diabetic Med. 1991b; 8: 911-916.

Dvornik D. Hyperglycemia in the pathogenesis of diabetic complications.In:Porte D.(Ed), Aldose reductase inhibition: An approach to the prevention of diabetic complications. McGraw-Hill,New York. 1987; pp70-151.

Gabbay KH The sorbitol pathway and the complications of diabetes. New Engl J Med 1973; 288: 831-836

Gabbay KH. .in 'US-Japan aldose reductase workshop'. Hotta N (ed), 1990.

Goldfarb S, Simmons DA, Kern E. Amelioration of glomerular hyperfiltration in acute experimental diabetes by dietary myo-inositol and aldose reductase inhibition. Trans Assoc Am Physicians 1986; 99: 67-72

Graham A, Brown L, Hedge PJ, Gammack AJ, Markham AF. Structure of the human aldose reductase gene. J Biol Chem 1991a; 266: 6872-6877

Graham A, Heath P, Morten JEN, Markham AF. The human aldose reductase gene maps to chromosome region 7q35. Human Genet 1991b; 86: 509-514

Graham A, Hedge PJ, Powell SJ, Riley J, Brown L, Gammack AJ, et al. Nucleotide sequence of cDNA for human aldose reductase. Nucleic Acids Res 1989; 17: 8368

Graham C, Szpirer C, Levan G, Carper D. Characterization of the aldose reductase-encoding gene family in rat. Gene 1991; 107: 259-267

Greene DA, Lattimer SA, Sima AAF. Sorbitol, phosphoinositides, and sodium-potassium-ATPase in the pathogenesis of diabetic complications. New Engl J Med 1987; 316: 599-606

Hibberd ML, Millward BA, Wong FS, Demaine AG. T-cell receptor constant ß chain polymorphisms and type I diabetes. Diabetic Medicine (in press)

Kinoshita JH, Kador P, Datiles M. Aldose reductase in diabetic cataracts. JAMA 1981; 246: 257-261

Kinoshita JH, Nishimura C. The involvement of aldose reductase in diabetic complications. Diabetes Metab Rev 1988; 4: 323

Kronenberg M, Siu G, Hood LE, Shastri N. The molecular genetics of the T-cell antigen receptor. Ann Rev Immunol 1986; 5: 529-591

Larkins RG, Dunlop ME. The link between hyperglycaemia and diabetic nephropathy. Diabetologia 1992; 35: 499-504.

Masson EA, Boulton AJM. Aldose reductase inhibitors in the treatment of diabetic neuropathy. A review of the rationale and clinical evidence. Drugs 1990; 39: 190-202

National Diabetes Data Group. Classification and diagnosis of diabetes mellitus and other categories of glucose intolerance. Diabetes 1979; 28: 1039-1044

Nishimura C, Matsuura Y, Kokai Y, Akera T, Carper D, Morjana N, et al. Cloning and expression of human aldose reductase. J Biol Chem 1990; 265: 9788-9792

Southern EM. Detection of specific sequences among DNA fragments separated by gel electrophoresis. J Mol Biol. 1975; 98: 507-513

Tomlinson DR, Moriarty RJ, Mayer JH. Prevention and reversal of defective axonal transport and motor nerve conduction velocity in rats with experimental diabetes by treatment with the aldose reductase inhibitor sorbinil. Diabetes 1984; 33: 470-476.

Winegrad AI. Does a common mechanism induce the diverse complications of diabetes? Diabetes 1987; 36: 396-406

POLYCYCLIC AROMATIC HYDROCARBONS AND PHENOLIC ANTIOXIDANTS DO NOT SIGNIFICANTLY INDUCE CARBONYL REDUCTASE IN HUMAN CELL LINES

Barbara Ruepp and Bendicht Wermuth

Dept. of Cinical Chemistry
University of Berne, Inselspital
CH-3010 Berne

INTRODUCTION

Carbonyl reductase (E.C. 1.1.1.184) is a cytosolic, monomeric oxidoreductase belonging to the short-chain alcohol dehydrogenase superfamily (Wermuth, 1992). It is widely distributed in human and animal tissues (Wirth and Wermuth, 1992). Characteristically, it occurs in three size heterogeneous forms with apparent molecular weights between 30'000 and 33'000. Evidence from sequence analysis suggests that the heterogeneity arises from posttranslational modification of lysine residues (Forrest et al., 1990). The nature of the modification and its physiological significance, however, are not yet known. Carbonyl reductase catalyzes the NADPH-dependent reduction of a variety of aldehydes, ketones and quinones among which quinones are most efficiently reduced. Based on the broad substrate specificity, including quinones derived from polycyclic aromatic hydrocarbons (PAH), we suggested that the enzyme may be involved in the detoxication of quinones and other reactive carbonyl compounds (Wermuth et al., 1986).

Quinones are ubiquitously distributed in the environment and a number of them exert toxic effects in biological systems. These compounds are reduced by single electron transfer to the semiquinones which easily react with oxygen yielding cytotoxic and mutagenic oxygen radicals. Alternatively, quinones can take up two electrons, yielding hydroquinones which are accessible for conjugation with glucuronate or sulfate. Enzymes mediating the two-electron transfer may, therefore, protect the cell against quinone-induced toxicity. In addition to carbonyl reductase, NAD(P)H:quinone reductase (NQR, E.C. 1.6.99.2) catalyzes the two-electron reduction of quinones in human tissues. Various studies have demonstrated the protective effect of the enzyme against quinone-induced oxidative stress (Wefers et al., 1984; Chesis et al., 1984). Characteristic of enzymes involved in the detoxication of foreign compounds, PAHs and phenolic antioxidants induce the biosynthesis of NQR (Talalay and Benson, 1982). The responsible xenobiotic and antioxidant responsive elements were recently identified upstream of the NQR gene (Favreau and Pickett, 1991; Rushmore et al., 1991; Jaiswal, 1991).

Little is known about the effect of PAHs and the phenolic antioxidants on carbonyl reductase. Forrest et al. (1990) reported a 3-4-fold increase in carbonyl reductase mRNA after exposure of MCF-7 breast cancer and HepG2 liver tumor cells to sudan I, ß-naphthoflavone and t-butyl-4-hydroxyanisole. Enzyme protein and activity, however, were not determined. In this study, we investigated the effect of 3-methylcholanthrene (3-MC), ß-naphthoflavone (ß-NF) and t-butyl-4-hydroxyanisole (BHA) on the expression and activity of carbonyl reductase in MCF-7, HepG2 and in keratinocyte- derived HaCaT cells. For comparison, NQR protein and enzyme activity were also determined.

EXPERIMENTAL PROCEDURE

HaCaT, MCF-7 and HepG2 cells were grown in 100 mm diameter culture dishes in Eagle's minimal essential medium supplemented with 10% fetal calf serum, 200 U/ml penicillin and 5 mg/ml chlortetracycline at 37°C, 80% humidity and 5% CO_2. At near confluency the cell cultures were exposed to fresh medium containing 3-MC, ß-NF and BHA, respectively, at concentrations indicated in the figure legends, and incubated for 24 and 48 hours. Controls with no addition and controls treated with dimethylsulfoxide (DMSO), which was used as solvent for 3-MC, ß-NF and BHA, were routinely included. The cells were harvested by scraping, suspended in 10 mM Tris/HCl, pH 7.4 and disrupted by sonification. After centrifugation, the supernatants were subjected to SDS-PAGE and the proteins were blotted onto nylon membranes. The membranes were incubated with antibodies against carbonyl reductase and NQR, and bound antibodies were visualized using horseradish peroxidase-protein A conjugate, H_2O_2 and diaminobenzidine.

Quinone reductase activity was assayed spectrophotometrically at 340 nm. Reaction mixtures consisted of 0,1 M sodium phosphate buffer, pH 7.0, 50 µM NADPH and 200 µM menadione. Carbonyl reductase activity was estimated from the fraction of total enzyme activity that was inhibited by 10 µM rutin. Similiarly, NQR activity was determined by including 10µM dicoumarol in the assay medium.

RESULTS AND DISCUSSION

As shown in Fig.1, both carbonyl reductase and NQR were detectable in the control cultures of all three cell lines. The two closely spaced slower migrating bands correspond to the low and intermediate molecular weight forms of carbonyl reductase (Wermuth, 1981), the lower band represents NQR. In the extracts of all three cell lines, low molecular weight forms (Mr: 12'000-17'000) of both enzymes which most probably represent degradation products were sometimes detectable (Fig. 1, panels B,C). In addition, HepG2 cells contained an unidentified antigen with an apparent molecular weight of 75'000 which cross-reacted with the antibodies against NQR (Fig. 1, panel C). Reflecting the results from Western blot analysis, carbonyl reductase and NQR activities were detectable in all cell lines (Fig. 2). HaCaT and MCF-7 contained similiar total activities (20-30 mU/mg protein) with equal contributions of the two reductases, whereas in HepG2 cells total activity was more than twice the activity of the two other cell lines (ca. 60 mU/mg) and the NQR activity was clearly higher than carbonyl reductase activity. The high level of NQR expression is typical for tumor-derived cells (Cresteil and Jaiswal, 1991) but is in sharp contrast to normal liver tissue where NQR accounts only for a minor part of the total quinone reductase activity (Wermuth et al., 1986).

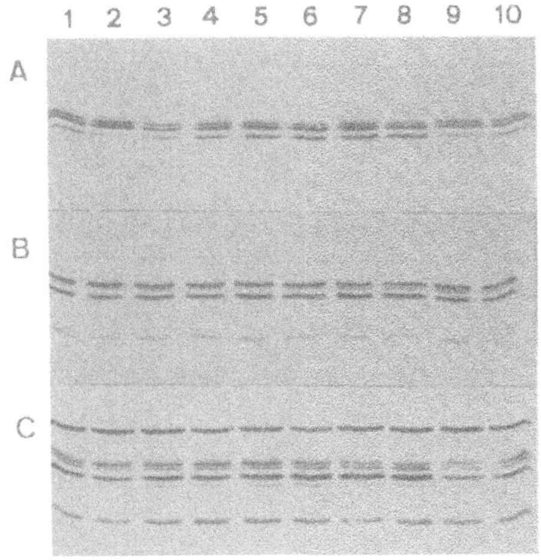

Figure 1. Western blot analysis of extracts from (A) HaCaT, (B) MCF-7 and (C) HepG2 cells after exposure to inducing agents for 48 h. Sixty micrograms of cytoplasmic protein were subjected to SDS-PAGE, transferred to nylon membranes and probed with a mixture of antibodies against carbonyl reductase and NQR. Lanes 1,2, controls; 3,4, 0,1% dimethylsulfoxide; 5,6, 3.7 μM 3-methylcholanthrene; 7,8, 50 μM ß-naphthoflavone; 9,10, 60 μM t-butylhydroxyanisole. For identification of bands see text.

Treatment of the cell cultures with the two PAHs, 3-MC and ß-NF caused an up to two-fold increase in total menadione reductase activity which was more pronounced with ß-NF than 3-MC and was highest in the HepG2 cells (Fig. 2). Inclusion of the specific carbonyl reductase and NQR inhibitors rutin and dicoumarol, respectively, revealed that the higher activity was mainly due to an increase in NQR activity (Fig. 3). With the exception of an approximately 1.3-fold increase in HaCaT and HepG2 cells in the presence of ß-NF, carbonyl reductase activity was not affected by the two PAHs. No increase in either carbonyl reductase or NQR activity was detectable when the phenolic antioxidant BHA was added to the cell cultures.

In agreement with the results of the enzyme activity determinations, Western blot analysis indicated an increase in NQR protein in PAH-treated cells which was most pronounced in HepG2 cells (Fig. 1). A slight increase was also noted for carbonyl reductase. In a time dependence study over a period of five days, the increase in CR and NQR protein was detectable already after 24 h and reached a plateau at 48 h. More pronounced than the quantitative effects, however, ß-NF and, to a lesser extent, 3-MC caused qualitative changes in the expression of the carbonyl reductase isozymes. As shown in Fig. 1 the proportion of the two molecular forms were changed in favor of the low molecular weight species in the presence of the PAHs. The cause of this change is not known. The posttranslational modification of carbonyl reductase, however, seems to be a relatively slow, time-dependent process which could lead to the accumulation of the newly synthesized low molecular weight form. Alternatively, the posttranslational conversion of the low to the intermediate molecular weight form may be impeded by the PAHs.

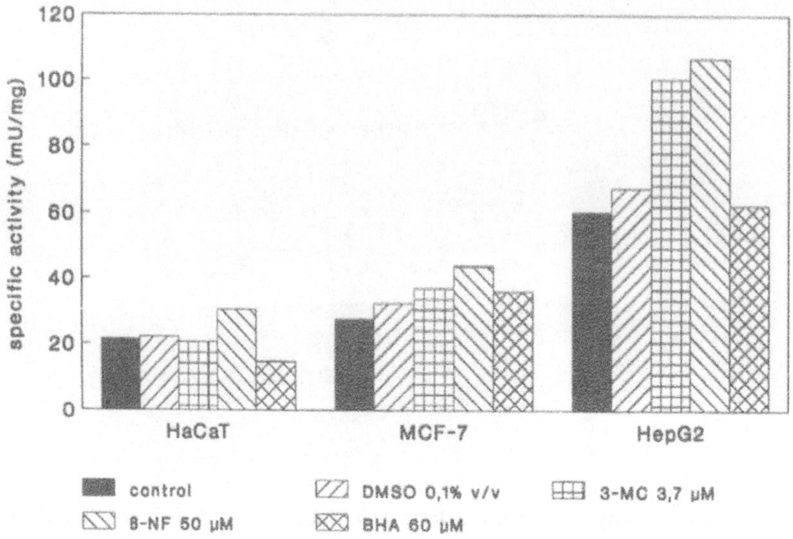

Figure 2. Effect of 3-methylcholanthrene (3-MC), ß-naphthoflavone (ß-NF) and t-butylhydroxyanisole (BHA) on total quinone reductase activity in HaCaT, MCF-7 and HepG2 cells. Cells were incubated with the inducing agents for 48 h, disrupted by sonification, and total quinone reductase activity was determined using 200µM menadione as substrate. D, dimethylsulfoxide.

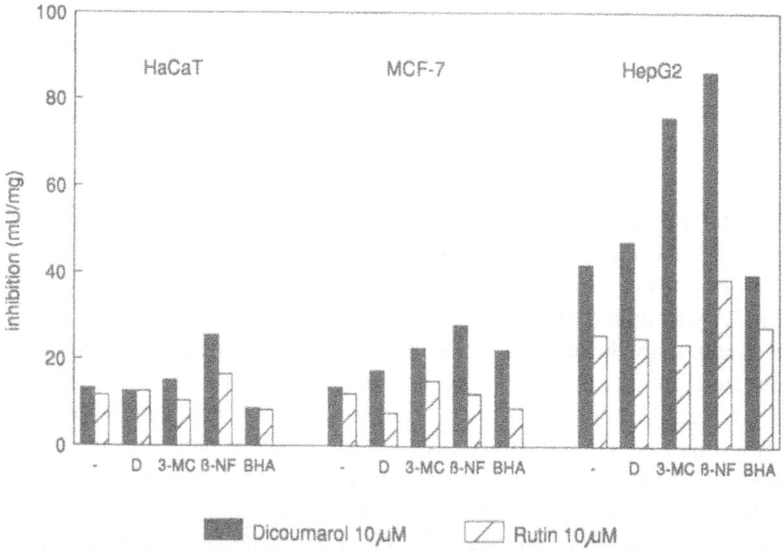

Figure 3. Effect of 3-methylcholanthrene (3-MC), ß-naphthoflavone (ß-NF) and t-butylhydroxyanisole (BHA) on carbonyl reductase and NQR activities in HaCaT, MCF-7 and HepG2 cells. The contribution of carbonyl reductase and NQR to the total quinone reductase activity was estimated from the activity inhibited by rutin and dicoumarol, respectively. The sum of the two activities may be greater than the total quinone reductase activity due to partially overlapping inhibitor sensitivities of the two enzymes. D, dimethylsulfoxide.

In summary, we showed that PAHs and phenolic antioxidants do not significantly induce the synthesis of carbonyl reductase in HaCaT, MCF-7 and HepG2 cells. One reason may be that the synthesis of the protein is already fully stimulated in the three cell lines. However, enzyme activities similiar to those in HepG2 cells are present in normal liver tissue and the enzyme also occurs at high concentrations in normal skin and breast endothelial cells (Wirth and Wermuth, 1992). Moreover, while this work was in progress, Forrest et al. (1991) reported the nucleotide sequence of the carbonyl reductase gene. In contrast to the NQR gene, the 5'-non coding region of the carbonyl reductase gene contains no xenobiotic and antioxidant responsive elements and also lacks the TATA and CAAT box. The constitutive expression of carbonyl reductase in most human tissue and the absence of regulatory elements strongly suggest a housekeeping function of the carbonyl reductase enzyme which is not subject to stimulation by xenobiotics and phenolic antioxidants.

REFERENCES

Chesis, P.L., Levin, M.T., Ernester, L., Ames, B.N., 1984, Mutagenicity of quinones: pathways of metabolic activation and detoxication, Biochemistry, 81:1696.

Cresteil, T., Jaiswal, A. K., 1991, High levels of expression of the NAD(P)H:quinone oxidoreductase (NQO1) gene in tumor cells compared to cells of the same origin, Biochem. Pharmacol., 42:1021.

Favreau, L.V., Pickett, C.B., 1991, Transcriptional regulation of the rat NAD(P)H:quinone reductase gene, J. Biol. Chem., 266:4556.

Forrest, G.L., Akman, S., Krutzik, S., Paxton, R.J., Sparkes R.S., Doroshow, J., Felsted, R.L., Glover, C.J., Mohandas, T., Bachur, N.R., 1990, Induction of a human carbonyl reductase gene located on chromosome 21, Biochim. Biophys. Acta, 1048:149.

Forrest, G.L., Akman, S., Doroshow, J., Rivera, H., Kapplan, W.D., 1991, Genomic sequence and expression of a cloned human carbonyl reductase gene with daunorubicin activity, Mol. Pharmacol., 40:502.

Jaiswal, A.K., 1991, Human NAD(P)H:quinone oxidoreductase (NQO1) gene structure and induction by dioxin, Biochemistry, 30:10647.

Rushmore, T.H., Morton, M.R., Pickett, C.B., 1991, The antioxidant responsive element, J. Biol. Chem., 266:11632.

Talalay, P., Benson, A.M., 1982, Elevation of quinone reductase activity by anticarcinogenic antioxidants, Adv. Enzyme Regul., 20:287.

Wefers, H. Komai, T., Talalay, P., Sies, H., 1984, Protection against reactive oxygen species by NAD(P)H:quinone reductase induced by the dietary antioxidant butylated hydroxyanisole (BHA), FEBS Lett., 163:3.

Wermuth, B., 1981, Purification and properties of an NADPH-dependent carbonyl reductase from human brain, J. Biol. Chem., 256:1206.

Wermuth, B., 1992, NADP-dependent 15-hydroxyprostaglandin dehydrogenase is homologous to NAD-dependent 15-hydroxyprostaglandin dehydrogenase and other short chain alcohol dehydrogenases, Prostaglandins, in press.

Wermuth, B., Platt, K.L., Seidel, A., Oesch, F., 1986, Carbonyl reductase provides the enzymatic basis of quinone detoxication in man, Biochem. Pharmacol., 35:1277.

Wirth, H., Wermuth, B., 1992, Immunohistochemical localization of carbonyl reductase in human tissues, J.Histochem. Cytochem., in press.

THE PURIFICATION AND PROPERTIES OF A NOVEL CARBONYL REDUCING ENZYME FROM MOUSE LIVER MICROSOMES

Edmund Maser

Department of Pharmacology and Toxicology
School of Medicine, University of Marburg
Karl von Frisch-Strasse, 3550 Marburg/Lahn, Germany

INTRODUCTION

Carbonyl reduction is a metabolic pathway being widely distributed in living matter and many endogenous compounds such as prostaglandins, biogenic amines and steroids, as well as xenobiotic aromatic and aliphatic aldehydes and ketones are converted to the corresponding alcohols prior to their further metabolism and/or elimination (Felsted and Bachur, 1980). The enzymes mediating carbonyl reduction belong to the aldo-keto reductase family and comprise carbonyl reductase (EC 1.1.1.184), aldehyde reductase (EC 1.1.1.2) and aldose reductase (EC 1.1.1.21), which share common features such as monomeric structure (30-40 kDa), cytosolic subcellular localization and a cosubstrate specificity for NADPH (Wermuth, 1985). Furthermore, enzymes like dihydrodiol dehydrogenase (EC 1.3.1.20) and hydroxysteroid dehydrogenases (3α-, 3ß-, 17ß-) were also shown to be involved in reductive metabolism of carbonyl compounds or, as has been supposed, isozymes of the latter two groups might even be identical to enzymes described previously as aldo-keto reductases (Pietruszko and Chen, 1976; Penning et al., 1984; Sawada et al., 1988).

In addition to cytosolic activities reduction of carbonyl compounds also takes place in membraneous fractions of cells, i.e. rabbit liver mitochondria (Ahmed et al.,1979) and microsomes of guinea pig (Sawada and Hara, 1978; Usui et al., 1984) and rat liver (Kahl, 1970; Hara et al., 1987). The number of distinct reductases in a tissue and whether or not they are the same as similar enzymes in other tissues is of particular relevance with regard to the search for the physiological role of these enzymes. However, in contrast to cytosolic representatives little is known on membrane-bound carbonyl reducing enzymes, and their physiological role remains to be determined.

Enzymology and Molecular Biology of Carbonyl Metabolism 4
Edited by H. Weiner, Plenum Press, New York, 1993

In previous investigations it was found that the diagnostic cytochrome P450 inhibitor metyrapone is rapidly reduced to the alcohol metabolite metyrapol in mouse and human liver (Maser and Legrum, 1985; Maser and Netter, 1989; Maser et al., 1991). This study describes the purification of the metyrapone reducing enzyme from female mouse liver microsomes to homogeneity together with its characterization and identification as a novel membrane-bound carbonyl reducing enzyme. Enzyme characterization was performed by evaluating the substrate specificity, inhibitor sensitivity, cofactor requirement and subcellular localization. In addition, the physiological function of the liver microsomal metyrapone reductase was investigated. Furthermore, the occurrence of homologous enzymes in rat, guinea pig and human liver microsomes was tested immunologically using an antibody against the purified microsomal mouse liver metyrapone reductase in the immunoblot analysis.

MATERIALS AND METHODS

Materials. Livers of female NMRI mice, Wistar rats and guinea pigs of the Hartley strain were used for the experiments. Human liver microsomes from both sexes were kindly supplied by O. Pelkonen, Oulu (SF) and had been derived from legal medical biopsies or from livers of individuals post mortem. Enzyme purification was carried out with octyl-sepharose CL-4B and CM-sepharose CL-6B from Pharmacia (Freiburg, FRG), DEAE-cellulose from E. Merck (Darmstadt, FRG) and hydroxyapatite SC from Serva (Heidelberg, FRG). Peroxidase conjugated anti rabbit IgG antibodies were obtained from Dakopatts (Hamburg, FRG). All other chemicals used in the experiments were reagent grade and were obtained from commercial suppliers.

Preparation of liver microsomes. After perfusing the livers with an ice cold isotonic solution of KCl they were homogenized in 4 volumes of 20 mM Tris-HCl buffer, pH 7.4, containing 250 mM sucrose, 1mM EDTA, 1 mM phenylmethanesulfonylfluoride (PMSF). After centrifugation (105,000 x g) the microsomal pellet was washed once with 0.15 M KCl to remove glycogen and other proteins and then resuspended in the homogenization buffer without PMSF.

Purification of the enzyme. The microsomal suspension (about 20 mg of protein/ml) was diluted with an equal volume of a 10 mM sodium phosphate buffer, pH 7.2, containing 1 mM EDTA, 1 M NaCl, 40% glycerol (w/v) and 0.4 % (w/v) of the nonionic detergent Emulgen 913. The solution was gently stirred for 30 min and subsequently centrifuged at 218,000 x g for 45 min. The supernatant was adjusted to 0.4 % (w/v) of sodium cholate before being applied to the octyl-sepharose CL-4B column.

In order to separate the metyrapone reductase from cytochrome P-450 and cytochrome P-450 reductase a method according to Kling et al. (1985) was applied. The following buffers were used: Buffer A: 10 mM sodium phosphate, 1mM EDTA, 500 mM NaCl, 20 % (w/v) glycerol, 0.5 % (w/v) sodium cholate, pH 7.4; Buffer B: 10 mM sodium phosphate, 1 mM EDTA, 400 mM NaCl,20% (w/v) glycerol, 0.4 % (w/v) sodium cholate, 0.1 % (w/v) Emulgen 913, pH 7.4; Buffer C: 10 mM sodium phosphate, 1 mM EDTA, 20 % (w/v) glycerol, 2 % (w/v) Emulgen 913, pH 7.4.

Solubilized microsomes were applied to the octyl-sepharose CL-4B column (1.6 x 20 cm) previously equilibrated with 300 ml of buffer A. Elution was achieved with buffer A until the end of peak 2, then with buffer B until the end of peak 3 followed by buffer C eluting peak 4. The elution profile was monitored measuring the absorbance of the fractions at the wavelength of 417 nm. The column flow rate was 50 ml/h and the volume per fraction 5 ml. The metyrapone reductase activity coincides only with peak 3, the fractions of which were collected, concentrated through an Amicon PM-10 membrane and dialysed against 5 mM sodium phosphate buffer, pH 7.4.

The dialysed enzyme solution was applied to a column (1.6 x 20 cm) packed with DEAE-cellulose, equilibrated with 5 mM sodium phosphate buffer, pH 7.4. The column was washed with the equilibration buffer and the adsorbed metyrapone reductase was eluted with a 30 mM phosphate buffer, pH 7.4, at a flow rat of 30 ml/h. Enzymatically active fractions were directly applied on a hydroxyapatite column (1.6 x 5 cm), equilibrated with a 5 mM phosphate buffer, pH 7.4. The column was rinsed with the equilibration buffer, then with a 150 mM phosphate buffer, pH 7.4. The enzyme was eluted with the 150 mM phosphate buffer, pH 7.4, containing 0.1 % Triton X-100. The fractions with high enzyme activity were pooled, concentrated through an Amicon PM-10 membrane and dialysed against 5 mM sodium phosphate buffer, pH 6.6.

The dialysed pool was loaded onto a column (1.6 x 20 cm) of CM-sepharose equilibrated with 5 mM phosphate buffer, pH 6.6. The column was washed with the equilibration buffer, then step by step with 20 mM, 40 mM and 60 mM phosphate buffer, pH 6.6. The enzyme was eluted with a 60 mM phosphate buffer, pH 7.4. Throughout the purification, the temperature was kept at 4° C.

Enzyme assays. 10 μl of metyrapone reductase solution were preincubated in 50 mM sodium phosphate buffer, pH 7.4. For inhibitor studies 10 μl of the respective inhibitor were added to a final concentration of 0.5 or 1 mM. Inhibitors which were not sufficiently soluble in buffer were dissolved in ethanol or 0.04 M NaOH. Control velocities were determined in the presence of appropriate quantities of the solvents. After the preincubation period of 3 min the reaction was started by adding 10 μl of the substrate (final concentrations cf. Tables) and 10 μl of the respective cosubstrate (final concentrations: NADH 3.2 mM; NADPH 3.2 mM; NADPH-regenerating system: $NADP^+$ 0.8 mM, G-6-P 6 mM, G-6-P-DH 0.35 U, $MgCl_2$ 3 mM) to a final volume of 50 μl. The reduction was stopped by mixing the reaction sample with 150 μl of ice cold acetonitrile. The samples were centrifuged for 6 min at 8,000 x g in the cold and 20 μl of the supernatant served for metabolite determination by HPLC-analysis.

Reductase activity using menadione as substrate was assayed photometrically in a Kontron Uvikon 810 spectrophotometer by following the decrease in the absorption of NADPH at 340 nm. Each 1.0 ml cuvette contained equal amounts of 10 mM phosphate buffer, pH 7.4, 0.4 mM NADPH and substrate concentrations between 3 - 400 μM.

Metabolite determination by HPLC. After enzymic conversion oxidized or reduced metabolites were detected on a BioRad (Munich, FRG) reversed phase HPLC system, with an Octadecyl-Si 100 polyol (Serva, Heidelberg, FRG) matrix column (4.5 mm x 25 cm), an UV monitor and an HRLC integration software (BioRad, Munich, FRG). Using an eluent of 30 % acetonitrile (v/v) in 0.1 % acetic acid, pH 7.4, (flow rate: 1 ml/min) the following retention times were achieved, metyrapol: 6.4 min, metyrapone: 10.3 min, p-nitromethylphenylcarbinol: 11.3 min, p-nitroacetophenone: 18.8 min, p-nitrobenzalcohol: 8.2 min, p-nitrobenzaldehyde: 14.8 min. Substances were monitored at 254 nm.

HPLC separation of glucocorticoids was achieved using a methanol/H_2O (58:42) eluent and a flow rate of 0.5 ml/min. Under these conditions glucocorticoids elute as follows, cortisone: 15 min, cortisol: 17 min, corticosterone: 16 min, dehydrocorticosterone: 30 min. Glucocorticoids were monitored at 262 nm.

SDS-polyacrylamide gel electrophoresis. Sodium dodecyl sulphate-polyacrylamide gel electrophoresis was carried out as described by Laemmli (1970) using 10% acrylamide in the separating gel. Protein bands were visualized applying the silver stain technique according to Ansorge (1985).

Preparation of antisera. Immunization of rabbits and preparation of antisera was carried out as described previously (Maser and Netter, 1991).

Immunoblot. Electrophoretically separated proteins were transferred to nitro cellulose sheets (Kyse-Anderson, 1984). Antisera against the microsomal metyrapone reductase were diluted 1 : 1,000 and incubated with protein saturated nitrocellulose sheets. Antigen-antibody complexes were detected by peroxidase conjugated secondary antibodies specific for rabbit IgG (dilution 1: 1,000). These complexes were visualized by the peroxidase reaction (chloronaphthol method).

Protein determination. Protein concentration was determined by the methods of Lowry et al. (1951) or Bradford (1976) using bovine serum albumin as standard.

Table 1. Purification of microsomal mouse liver metyrapone reductase.

Step	Total protein (mg)	Total activity (μmol/30min)	Specific activity (μmol/30min/mg)	Recovery (%)
Microsomes	960.0	15.36	0.016	100.0
Solubilized microsomes	416.5	33.74	0.081	219.7
Octyl-Sepharose CL-4B	78.9	25.41	0.322	165.4
DEAE-Cellulose	6.9	3.89	0.564	25.3
Hydroxyapatite	2.6	3.12	1.202	20.3
CM-Sepharose CL-6B	0.5	2.91	5.824	19.0

The reductase activity was assayed in the standard reaction mixture containing 50 mM sodium phosphate buffer, pH 7.4, an NADPH-regenerating system and 2.3 mM metyrapone. The solubilized microsomal suspension used in this reaction mixture did not contain sodium cholate.

RESULTS

Purification. Table 1 summarizes the purification procedure of the microsomal metyrapone reducing enzyme. Solubilization of the microsomal fraction in 0.2% (w/v) Emulgen 913 already resulted in a 5 fold increase of metyrapol formation, indicating that the enzyme is bound to the membranes of the endoplasmic reticulum in a latent state. Also each of the following steps resulted in a severalfold increase in specific activity. With an NADPH-regenerating system as electron supply it is enhanced to about 360 fold compared to that in microsomes.

Table 2. Specific activity of microsomal mouse liver metyrapone reductase using different reducing systems throughout the purification.

	Formation of metyrapol (μmol/30min/mg).		
	NADH	NADPH	NADPH-reg.sys.
Microsomes	0.005	0.009	0.016
Solubilized microsomes	0.023	0.036	0.081
Octyl-Sepharose CL-4B	0.103	0.238	0.322
DEAE-Cellulose	0.259	0.381	0.564
Hydroxyapatite	0.334	1.150	1.202
CM-Sepharose CL-6B	1.577	4.018	5.824

The reductase activity was assayed in the standard reaction mixture containing 50 mM sodium phosphate buffer, pH 7.4, 2.3 mM metyrapone and either 3.2 mM NADH, 3.2 mM NADPH or an NADPH-regenerating system, respectively. The solubilized microsomal suspension used in this reaction mixture did not contain sodium cholate.

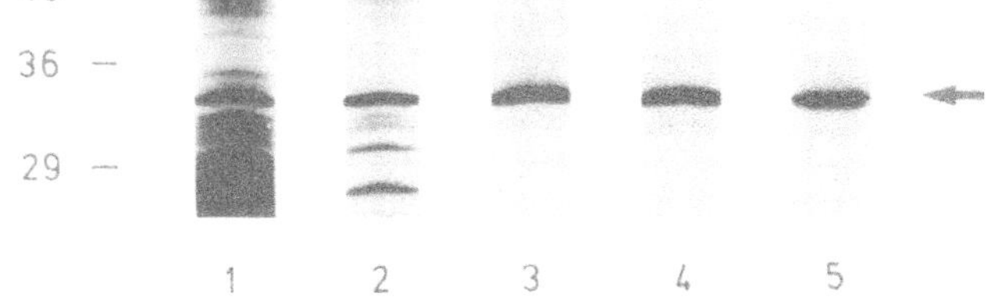

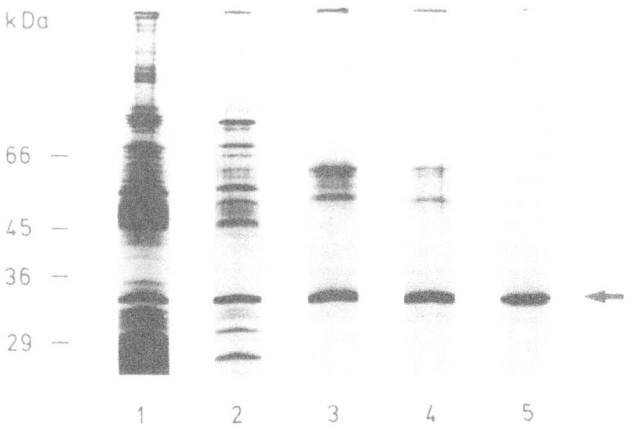

Figure 1. SDS-polyacrylamide gel electrophoresis of the purification steps of the microsomal metyrapone reductase. Lane S = molecular mass standards (values in kDa on the left); lane 1 = protein profile of mouse liver microsomes; lane 2 = after octyl-sepharose; lane 3 = after DEAE-cellulose; lane 4 after hydroxyapatite; lane 5 = after CM-sepharose. Visualizing of protein bands was carried out by the silver stain technique. For details see Materials and Methods.

Table 2 compares the specific activity of the enzyme using either NADH, NADPH or an NADPH-regenerating system as electron donors. The ratio of activity of the different electron donors remains nearly constant throughout the purification procedure. From this it becomes obvious that either NADPH or NADH can deliver the required reducing equivalents, although reduction is weaker with NADH. Highest activity is obtained with the NADPH-regenerating system. Incubations containing both cofactors at the same time reveal that there is no synergistic effect of NADH and NADPH.

A sample taken from the various purification steps was subjected to SDS-polyacrylamide gel electrophoresis and subsequent silver stain, as shown in Figure 1. Lane 1 represents the total quantity of microsomal protein, which was markedly decreased after affinity chromatography on an octyl-sepharose CL-4B column (lane 2). Further purification was achieved with DEAE-cellulose (lane 3), hydroxyapatite (lane 4) and CM-sepharose CL-6B (lane 5).

The last step of the purification procedure yielded a single band in the 34 kDa region as compared to the protein standard, indicating that the microsomal metyrapone reductase was purified to homogeneity.

Immunoblot analysis. For estimating whether or not microsomal metyrapone reductase is also present in rat, guinea pig and human liver the immunoblot technique was applied. Polyclonal antibodies raised in rabbits against the microsomal metyrapone reductase specifically crossreacted with one single protein band in the liver microsomal fractions of all species (Figures 2 and 3) corresponding to the respective enzyme in the 34 kDa molecular weight region (32 kDa in guinea pig liver). The common antigenic determinants point to the existence of a homologous enzyme in microsomes of rat, guinea pig and human liver.

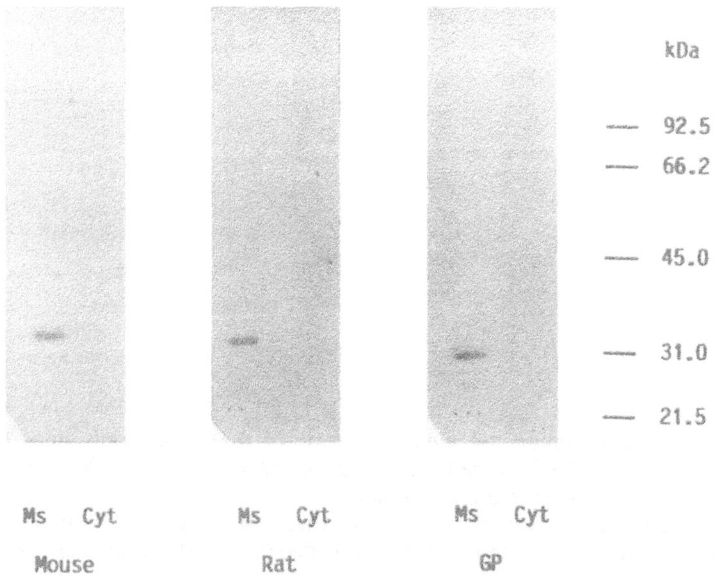

Figure 2. Immunoblot of microsomal and cytosolic fractions of mouse, rat and guinea pig liver using polyclonal antibodies against microsomal metyrapone reductase from mouse liver.

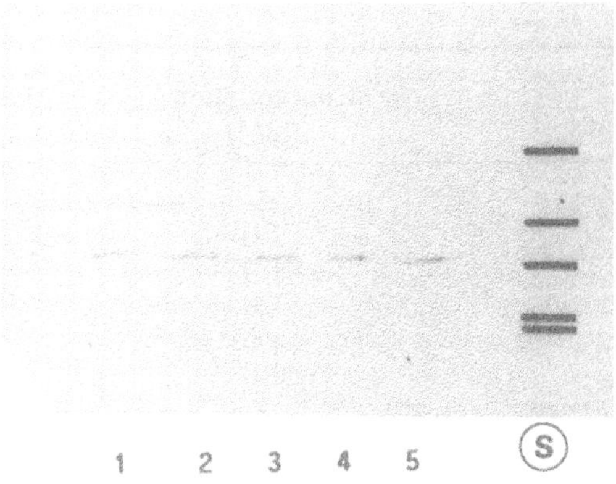

Figure 3. Immunoblot of microsomal fractions of 5 individual human livers. Polyclonal antibodies against the microsomal metyrapone reductase from mouse liver were incubated as primary antibody. (S = molecular mass standard: albumin, bovine = 66 kDa; albumin, egg = 45 kDa; glyceraldehyde-3-phosphate dehydrogenase = 36 kDa; carboanhydrase = 29 kDa; trypsinogen = 24 kDa).

Inhibition. Table 3 lists some compounds which were tested as inhibitors of purified metyrapone reductase. Dicoumarol, which is known to inhibit NADP(H):quinone-oxidoreductase, decreased metyrapone reduction to about 33 %. Phenobarbital, an inhibitor of aldehyde reductase, and indomethacin also affected metyrapol formation and lowered it to about 72 % and 52 %, respectively. Pyrazole, a potent inhibitor of alcohol dehydrogenase had only little effect on metyrapol formation, whereas quercitrin, a flavonoid, showed no inhibition.

As also shown in Table 3, purified metyrapone reductase is strongly inhibited by steroid compounds, such as androsterone, testosterone, 5ß-dihydrotestosterone, 5α-dihydrotestosterone, progesterone and androstandione. The competitive nature of this inhibition (determined for the latter three compounds) gave rise to the suggestion that steroids are substrates of this enzyme and that the physiological role of this enzyme is endogenous steroid metabolism. But when tested, neither of these substances was metabolized by metyrapone reductase, as indicated by TLC, HPLC or indirect spectrophotometric recording of pyridine nucleotide oxidoreduction.

Substrate specificity. Finally it could be demonstrated that microsomal metyrapone reductase is able to catalyze the reversible 11ß-oxidoreduction of glucocorticoids. As shown in Table 4 the 11-oxo forms cortisone and dehydrocorticosterone are reduced to the respective 11ß-alcohols in the presence of the reduced pyridine nucleotides NADH and NADPH and, reversely, cortisol and corticosterone are oxidized to the 11-oxo forms by this enzyme in the presence of NAD or NADP. Enzyme kinetic studies revealed that xenobiotic carbonyl substances such as p-nitroacetophenone, p-nitrobenzaldehyde and menadione are also substrates of this enzyme (Table 4). However, whereas this enzyme shows reversibility of glucocorticoid oxidoreduction, xenobiotic substrates are only reduced. No reverse oxidation was detected with respective alcohol products as substrates.

Table 3. Effects of inhibitors on purified microsomal metyrapone reductase.

Inhibitor	Concentration (mM)	Relative enzyme activity (%)
None	---	100
Dicoumarol	1.0	33
Indomethacin	1.0	52
Phenobarbital	1.0	72
Pyrazole	1.0	86
Quercitrin	1.0	100
Testosterone	0.5	23
Androstandione	0.5	30
Progesterone	0.5	35
5ß-Dihydrotestosterone	0.5	39
5α-Dihydrotestosterone	0.5	47
Androsterone	0.5	57

Enzyme activities were assayed in the standard reaction mixture containing 50 mM sodium phosphate buffer, pH 7.4, 1 mM metyrapone (0.5 mM in the case of steroid compounds), a NADPH-regenerating system and 0.5 or 1 mM of the respective inhibitor. The percentages were calculated from uninhibited control experiments. Control velocities were determined in the presence of appropriate quantities of the solvents.

Table 4. Substrate specificity of purified microsomal metyrapone reductase.

Substrate	Cosubstrate	Vmax (μmol/30min/mg)	Km (mM)
Metyrapone	NADPH	0.33	0.50
	NADH	0.24	1.02
p-Nitrobenzaldehyde	NADPH	0.79	1.19
	NADH	0.06	7.40
p-Nitroacetophenone	NADPH	n.d.	1.67
	NADH	--	--
Menadione	NADPH	0.65	0.012
	NADH	--	--
Corticosterone	NADP	5.40	0.25
	NAD	6.02	0.34
Cortisol	NADP	2.28	0.18
	NAD	2.08	0.18
Dehydrocorticosterone	NADPH	0.051	0.041
	NADH	0.030	0.081
Cortisone	NADPH	0.048	0.065
	NADH	0.013	0.070

Enzyme activities were assayed as described in "Methods". Substrate concentrations varied between 25 μM - 10 mM (with menadione 3 μM - 400 μM). Reduced or oxidized metabolite detection was performed by HPLC analysis (except for menadione).
n.d. = not determined
-- = no activity detectable

DISCUSSION

To date only few membrane-associated carbonyl reducing enzymes have been isolated, i.e. from guinea pig and rat liver microsomes (Sawada et al., 1979; 1981; Usui et al., 1984). The objective of these investigations was the purification and characterization of a novel microsomal carbonyl reducing enzyme from mouse liver. The characterization was performed by investigating the substrate specificity, inhibitor sensitivity, cofactor requirement and subcellular localization in order to elucidate the physiological function of this enzyme. The ketone metyrapone proved a useful substrate, since it was shown to be mainly reduced in vivo to the respective alcohol, rather than being oxidized at the nitrogen of the two pyridine rings (Maser and Legrum, 1985; Usansky and Damani, 1985). Moreover, the method of HPLC analysis for direct determination of the alcohol product metyrapol is clearly advantageous in comparison to the indirect spectrophotometric recording of pyridine nucleotide oxidation at 340 nm usually employed.

As shown in the results, the specific activity of microsomal metyrapone reductase was enhanced about 360 fold after the purification. SDS-polyacrylamide gel electrophoresis of the respective enzyme preparations during the purification procedure gave, after the last step, a single band in the 34 kDa region, indicating that the reductase was purified to homogeneity.

Carbonyl reducing enzymes have been shown to be principally dependent on NADPH as an electron donating system (Wermuth, 1985). In this study metyrapol formation occurred with either NADPH or NADH as cofactor, although activity with NADH was weaker. Highest activity was obtained with an NADPH-regenerating system. The ratio of the different cosubstrates remains nearly constant throughout the chromatographic steps of purification, providing evidence that the purified enzyme can use both NADPH or NADH. The potency of dual cofactor exploitation is also true for p-nitrobenzaldehyde as substrate, but not for p-nitroacetophenone and menadione, which are reduced exclusively in the presence of NADPH.

Immunoblot analysis, using polyclonal antibodies against the mouse liver microsomal metyrapone reductase, revealed the existence of a homologous enzyme in rat, guinea pig and human liver microsomes.

Inhibitor studies showed that the flavonoid quercitrin, which is designated as an inhibitor of carbonyl reductase (EC 1.1.1.184) does not inhibit metyrapol formation. Phenobarbital, which was defined as an inhibitor of aldehyde reductase (EC 1.1.1.2), decreased the enzyme activity to about 72 %. Other inhibitors of metyrapone reduction were indomethacin and dicoumarol, the specific inhibitors of dihydrodiol dehydrogenase (EC 1.3.1.20) and NAD(P)H:quinone-oxidoreductase (EC 1.6.99.2), respectively. Pyrazole, the specific inhibitor of alcohol dehydrogenase did not decrease metyrapol formation.

Steroid hormones, such as androsterone, testosterone, 5α-dihydrotestosterone, 5ß-dihydrotestosterone, progesterone and androstandione proved to be the strongest inhibitors of purified metyrapone reductase. The competitive nature of this inhibition (determined for the latter three compounds) and the fact that microsomal metyrapone reductase of mouse liver exhibits functional as well as immunological relationships to a procaryontic 3α-hydroxysteroid dehydrogenase from *Pseudomonas testosteroni* (Maser et al., 1992; Oppermann et al., this volume) gave rise to the suggestion that steroids are physiological substrates of the liver enzyme. But when tested, neither of these substances was metabolized by microsomal metyrapone reductase, as indicated by TLC, HPLC or indirect spectrophotometric recording of pyridine nucleotide oxidoreduction.

Concentrating on other steroids as possible substrates it could finally be demonstrated that this microsomal enzyme catalyzes the reversible 11ß-oxidoreduction of glucocorticoids. In detail, the 11-oxo forms cortisone and dehydrocorticosterone are reduced to the respective 11ß-alcohols in the presence of the reduced pyridine nucleotides NADH and NADPH and, reversely, cortisol and corticosterone are oxidized to the 11-oxo forms by this enzyme in the presence of NAD or NADP. Enzyme kinetic studies revealed that xenobiotic carbonyl substances such as p-nitroacetophenone, p-nitrobenzaldehyde and menadione are also substrates of this enzyme.

Until now, several steroid dehydrogenases, which catalyze the hydroxy-oxidation or oxo-reduction at C-3 and/or C-17 of the steroid nucleus are described to be also involved in reductive xenobiotic carbonyl metabolism (Sawada et al., 1979; 1988; Pietruszko and Chen, 1976; Penning et al., 1984; Nakagawa et al., 1989). This report provides evidence that, in addition to these 3α-, 3ß- and 17ß-hydroxysteroid dehydrogenases, another group of hydroxysteroid dehydrogenases is capable of xenobiotic carbonyl reduction, the physiological function of which is the oxidoreduction at C-11 of the steroid nucleus. However, in contrast to the reversibility of oxidoreduction with glucocorticoids, no reverse reaction of respective xenobiotic alcohol metabolites by this enzyme could be detected.

The subcellular localization in the endoplasmic reticulum, the molecular mass of 34 kDa and the ability of this enzyme to catalyze the reversible 11ß-oxidoreduction of glucocorticoids suggest this enzyme to be 11ß-hydroxysteroid dehydrogenase, which was described recently in rat liver (Lakshmi and Monder, 1988). This would support the hypothesis that microsomal metyrapone reductase, together with 3α-hydroxysteroid dehydrogenase from *Pseudomonas testosteroni*, belongs to the short chain alcohol dehydrogenase family (Maser et al., 1992; Oppermann et al., this volume). However, comparison of the N-terminal amino acid sequence of the mouse liver enzyme with that of other known proteins by search of the Heidelberg EMBL data bank did not reveal any significant homology.

Taken together, a membrane-bound carbonyl reducing enzyme has been purified from mouse liver endoplasmic reticulum and is demonstrated to catalyze the reversible 11ß-oxidoreduction of glucocorticoids. Therefore, its pysiological role might be the regulation of glucocorticoid activation and inactivation. On the other hand, this enzyme mediates the reductive metabolism of xenobiotic carbonyl compounds, and probably plays a role in detoxication processes. Thus, it shows properties similar to the previously described 3α-, 3ß- and 17ß-hydroxysteroid dehydrogenases and contributes to an expanding list of pluripotent enzymes involved in reductive xenobiotic carbonyl metabolism as well as being specific towards their physiological steroid substrates.

REFERENCES

Ahmed, N.K., Felsted, R.L., and Bachur, N.R., 1979, Comparison and characterization of mammalian xenobiotic ketone reductases, *J. Pharmacol. Exp. Ther.* **209**:12.

Ansorge, W., 1985, Fast and sensitive detection of protein and DNA bands by treatment with potassium permanganate *J. Biochem. Biophys. Meth.* **11**:13.

Bradford, M.M., 1976, A rapid and sensitive method for the quantitation of microgram quantities of protein utilizing the principle of protein-dye binding, *Anal. Biochem.* **72**:248.

Felsted, R.L. and Bachur, N.R., 1980, Mammalian carbonyl reductases, *Drug Metab. Rev.* **11**:1.

Hara, A., Usui, S., Hayashibara, M., Horiuchi, T., Nakayama, T., and Sawada, H., 1987, Microsomal carbonyl reductase in rat liver. Sex difference, hormonal regulation and characterization, in: "Enzymology and Molecular Biology of Carbonyl Metabolism," H. Weiner and T.G. Flynn, eds., Alan R. Liss, New York, p.401.

Kahl, G.F., 1970, Experiments on the metyrapone reducing microsomal enzyme system, *Naunyn-Schmiedebergs Arch. Pharmacol.* **266**:61.

Kling, L., Legrum, W., and Netter, K.J., 1985, Induction of liver cytochrome P-450 in mice by warfarin, *Biochem. Pharmacol.* **34**:85.

Laemmli, U.K., 1970, Cleavage of structural proteins during the assembly of the head of bacterophage T4, *Nature* **227**:680.

Lakshmi, V. and Monder, C., 1988, Purification and characterization of the corticosteroid 11ß-dehydrogenase component of the rat liver 11ß-hydroxysteroid dehydrogenase complex, *Endocrinology* **123**:2390.

Lowry, O.H., Rosebrough, N.J., Farr, A.L., and Randall, R.J., 1951, Protein measurements with the Folin phenol reagent, *J. Biol. Chem.* **193**:265.

Maser, E. and Legrum, W., 1985, Alteration of the inhibitory effect of metyrapone by reduction to metyrapol during the metabolism of methacetin in vivo in mice, *Naunyn-Schmiedeberg's Arch. Pharmacol.* **331**:283.

Maser, E. and Netter, K.J., 1989, Purification and properties of a metyrapone reducing enzyme from mouse liver microsomes - this ketone is reduced by an aldehyde reductase, *Biochem. Pharmacol.* **38**:3049.

Maser, E. and Netter, K.J., 1991, Reductive metabolism of metyrapone by a quercitrin-sensitive ketone reductase in mouse liver cytosol, *Biochem. Pharmacol.* **41**:1595.

Maser, E., Oppermann, U., Bannenberg, G., and Netter, K.J., 1992, Functional and immunological relationships between metyrapone reductase from mouse liver microsomes and 3α-hydroxysteroid dehydrogenase from Pseudomonas testosteroni, *FEBS Lett.* **297**:196.

Maser, E., Gebel, T., and Netter, K.J., 1991, Carbonyl reduction of metyrapone in human liver, *Biochem. Pharmacol.* **42**:S93.

Nakagawa, M., Tsukada, F., Nakayama, T., Matsuura, K., Hara, A., and Sawada, H., 1989, Identification of two dihydrodiol dehydrogenases associated with 3(17)α-hydroxysteroid dehydrogenase activity in mouse kidney, *J. Biochem.* **106**:633.

Penning, T.M., Mukharji, I., Barrows, S., and Talalay, P., 1984, Purification and properties of a 3α-hydroxysteroid dehydrogenase of rat liver cytosol and its inhibition by anti-inflammatory drugs, *Biochem. J.* **222**:601.

Pietruszko, R. and Chen, F.F., 1976, Aldehyde reductase from rat liver is a 3α-hydroxysteroid dehydrogenase, *Biochem. Pharmacol.* **25**:2721.

Sawada, H. and Hara, A., 1978, Studies on metabolism of bromazepam, *Drug Metab. Dispos.* **6**:205.

Sawada, H., Hara, A., Hayashibara, M., and Nakayama, T., 1979, Guinea pig liver aromatic aldehyde-ketone reductases identical with 17ß-hydroxysteroid dehydrogenase isozymes, *J. Biochem. Tokyo* **86**:883.

Sawada, H., Hara, A., Hayashibara, M., Nakayama, T., Usui, S., and Saeki, T., 1981, Microsomal reductase for aromatic aldehydes and ketones in guinea pig liver. Purification, characterization, and functional relationship to hexose-6-phosphate dehydrogenase, *J. Biochem.* **90**:1077.

Sawada, H., Hara, A., Nakayama, T., Nakagawa, M., Inoue, Y., Hasebe, K., and Zhang, Y., 1988, Mouse liver dihydrodiol dehydrogenases: identity of the predominant and a minor form with 17ß-hydroxysteroid dehydrogenase and aldehyde reductase, *Biochem. Pharmacol.* **37**:453.

Usansky, J.I. and Damani, L.A., 1985, The in vivo metabolism of metyrapone in the rat, in: "Biological Oxidation of Nitrogen in Organic Molecules," J.W. Gorrod and L.A. Damani, eds., Ellis Horwood, Chichester, p.231.

Usui, S., Hara, A., Nakayama, T., and Sawada, H., 1984, Purification and characterization of two forms of microsomal carbonyl reductase in guinea pig liver, *Biochem. J.* **223**:697.

Wermuth, B., 1985, Aldo-keto reductases, in: "Enzymology of Carbonyl Metabolism 2", T.G. Flynn and H. Weiner, eds., Alan R. Liss, New York, p.209.

PROPERTIES AND STEREOSELECTIVITY OF CARBONYL REDUCTASES INVOLVED IN THE KETONE REDUCTION OF WARFARIN AND ANALOGUES

J.J.R. Hermans, and H.H.W. Thijssen

Dept. Pharmacology, University of Limburg
P.O. Box 616, 6200 MD, Maastricht, the Netherlands

INTRODUCTION

Reduction of carbonyl groups plays an important role in the detoxification of xenobiotic compounds (Wermuth et al, 1986; Ziegler, 1988). The enzymes responsible for this metabolic route are widely distributed amongst species and tissues (Ahmed et al, 1979; Felsted and Bachur, 1980). Next to their role in detoxification, they are also involved in the processing of endogenous compounds, as e.g. steroids, prostaglandins, bile acids, glucuronic acid etc. (Pietruszko and Chen, 1976; Felstedt and Bachur, 1980; Wermuth, 1981; Hara et al, 1982; Kudo et al, 1990).

Warfarin (figure 1) is a chiral compound which is extensively used as anticoagulant and as rodenticid. In man, warfarin is eliminated by biotransformation, which involves an oxidative route, mediated by cytochromes P450, and a reductive route, i.e. reduction of the acetonyl side chain to the diastereomeric alcohols (Lewis et al, 1974). When applied orally, about 15-20% of an oral dose is recovered in urine as the alcohols. In man, ketone reduction is more extensive for R-warfarin than for S-warfarin, indicating substrate stereoselectivity of the enzymes involved. In addition, in the ketone reduction of warfarin, product stereoselectivity is observed, since the alcohols found are mainly in the S-configuration (Lewis et al, 1974; Banfield et al, 1983). The 4'-nitro analogue of warfarin, acenocoumarol is also reported to be converted to its alcohols in vivo in man (Dieterle et al, 1977), but information on stereochemical aspects is lacking. In a previous study we described the in vitro acetonyl side chain reduction of warfarin and some of its 4'-substituted analogues in subcellular fractions from various species (Hermans and Thijssen, 1989). The reaction proved to be NADPH dependent and was conducted by microsomal and cytosolic reductases. In microsomal fractions, large differences in alcohol formation rates as well as in substrate- and product stereoselectivity were observed between species and between the warfarin analogues. In cytosolic fractions however, a general pattern of substrate and product stereoselectivity was seen in that the R-enantiomers were preferred as a substrate and the product formed was mainly in the S-configuration. Another difference between microsomal and cytosolic warfarin reducing enzymes is that in the rat, the cytosolic warfarin reductase activity is enhanced by pretreatment with phenobarbitone and methylcholanthrene (table 1), which is not the case in rat liver microsomes.

Enzymology and Molecular Biology of Carbonyl Metabolism 4
Edited by H. Weiner, Plenum Press, New York, 1993

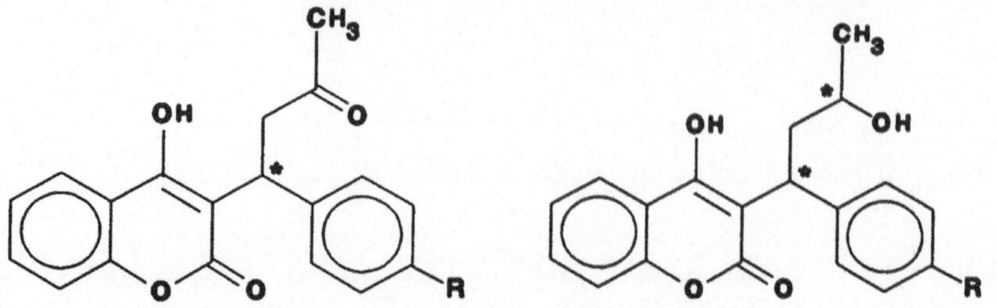

Figure 1. Structural formulas of warfarin, its 4'-hydrogen substituted analogues and the alcohols. R=H: warfarin; R=NO_2: 4'-nitrowarfarin; R=Cl: 4'-chlorowarfarin; R=COH_3: 4'-methoxywarfarin; * indicates a stereocentre.

Table 1. The effect of the enzyme inducers phenobarbitone (PB) and methylcholanthrene (MC) on rat liver cytosolic ketone reduction of the R-enantiomers of warfarin and analogues.

Substrate	Control	PB	MC
R-warfarin	17±4	173±11	125±24
R-4'-nitrowarfarin	77±1	370±5	222±5
R-4'-chlorowarfarin	26±2	214±11	86±4

Data are expressed as the mean ± SEM formation rates (pmol/mg protein x minute) of the RS alcohols. Taken from Hermans and Thijssen, 1989, with permission of the publisher.

In order to investigate this metabolic route in more detail, an attempt has been made to purify the cytosolic reductase from rabbit liver (Hermans and Thijssen, 1992). Some features of the warfarin reducing enzymes will be discussed here, and data of the stereochemical aspects of the ketone reduction of 4'-nitrowarfarin in human liver cytosol and microsomes will be presented.

MATERIALS AND METHODS

Synthesis of the warfarin analogues and their alcohols as well as resolution of the enantiomers was performed as described previously (Hermans and Thijssen, 1989; Baars et al, 1990). Enantiomeric purity was higher than 98.5 %. Human liver samples (table 1) were obtained post mortem via the Department of Pathology, University Hospital Maastricht, the Netherlands, with the approval of the Ethics Committee. Microsomal and cytosolic fractions were prepared according to standard methods. The partial purification of the rabbit liver cytosolic warfarin reductase was conducted by ammonium sulfate precipitation (75% saturation), gelfiltration, DEAE-Sephacel chromatography and hydroxyl appatite chromatography as previously described (Hermans and Thijssen, 1992). For the determination of the ketone reduction of the warfarin analogues an HPLC method was used, as described previously (Hermans and Thijssen, 1992). Briefly, 25 μl of the partially purified rabbit liver cytosolic enzymes, human liver microsomes

or cytosol, were incubated for 30 minutes at 37°C in the presence of substrate (1 mM unless stated otherwise) and NADPH generating system (final NADPH concentration 1 mM) in 0.1 M sodium-phosphate buffer (final volume 100 μl). The reaction was stopped by adding 250 μl 0.5 M acetic acid and one of the warfarin analogues was added as internal standard. Then, an extraction with 2 ml petroleumether/dichloromethane was performed. After evaporation of the organic phase, the residue was taken up in 50 μl of a 1/1 v/v mixture of acetonitrile and water. 10 μl was used for analysis on HPLC. The HPLC system consisted of a Chromspher C18 column (200x3.0 mm) as stationary phase and a 3/7 v/v mixture of acetonitrile and an aequous solution of 0.1 % acetic acid, brought to pH=4.80 with ammonia as the mobile phase. The alcohols were quantified by determination of the peak area ratio (versus internal standard) of their UV absorption at 303 nm, related to a calibration curve.
In the inhibition experiments, the substrate, NADPH, and inhibitor concentrations were 1 mM. Complete incubation mixtures, without substrate were preincubated for 5 minutes at 37 °C, before the reaction was started by the addition of the substrate. Inhibitors were either dissolved in 0.1 M sodium-phosphate buffer, pH=7.4, or in 10% v/v ethanol in sodium-phosphate buffer. The final ethanol concentration was 2.5%, which was without influence on the reductase activities.

RESULTS AND DISCUSSION

Rabbit liver cytosol was used as the enzyme source to purify the cytosolic warfarin reductase, because of its high specific activity. By the DEAE chromatographic step, two fractions were resolved (figure 2) that were able to reduce the acetonyl side chain of warfarin and its analogues. At the final stage of purification, the specific activities of the obtained reductase fractions were enhanced 90 fold and 240 fold respectively but the preparations were still not homogeneous. Attempts for further purification of the enzymes by the use of affinity chromatography with Blue or Red Sepharose failed since the enzymes could not be eluted from these media in their active form. In addition, rapid loss of activity during processing at 0 °C was observed.

The obtained reductase fractions displayed a marked substrate and product stereoselectivity in reducing the ketone group of warfarin and its 4'-substituted analogues (figure 3). At a substrate concentration of 2.5 mM, the R-enantiomers were reduced at rates that were 15 to >250 (fraction

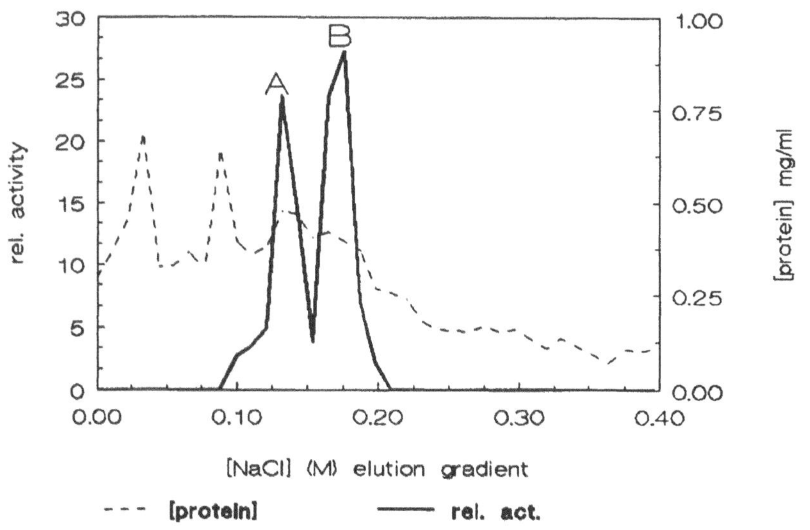

Figure 2. DEAE-Sephacel fractionation pattern of the rabbit liver cytosolic warfarin reductase activity.

A) and 10 to 160 (fraction B)times higher than those of the S-enantiomers. From the S-enantiomers, only the S-alcohols were produced, with the exception of S-4'-chlorowarfarin, of which some R-alcohol was produced by fraction B. Using the R-enantiomers as a substrate the formation rates of the S-alcohols were 20 to 86 (fraction A) and 10 to 160 (fraction B) times higher than the formation rates of the R-alcohols.

Both reductases were NADPH dependent, which was absolute for reductase B. At a cofactor concentration of 5 mM, fraction A showed 15 times lower reduction rates if NADH in stead of NADPH was added. Using various R-4'-nitrowarfarin concentrations the 'real' Km values for the cofactor NADPH, determined by replots, were 4 and 9 μM for fraction A and B respectively. The pH optimum for the reduction of the acetonyl side chain of the warfarin analogues was between 7.0 and 7.5. In the pH range between 6 and 8.5, the reverse reaction, oxidation of the alcohols to the ketones was not observed, even at alcohol and $NADP^+$ concentrations of 5 mM.

Although the stereoselectivity of the ketone reduction of warfarin analogues by both reductases fractions appeared to be grossly similar, substitution of the 4'-hydrogen of warfarin differently affected the enzyme kinetic parameters of the reduction of the R-enantiomers by the two reductase fractions (table 2). 4'-Substitution did not appear to have a great impact on the Vmax values of fraction A but the Km values of 4'-nitro and 4'-chlorowarfarin were about 6 times lower than those of 4'-methoxywarfarin or warfarin itself. In contrast, in the case of fraction B, the Km values of the warfarin analogues did not differ grossly but the Vmax values ranged about 9 fold among the warfarin analogues tested. For both enzyme fractions however, the intrinsic clearances of the R-enantiomers of 4'-nitro and 4'-chlorowarfarin were higher than those of the R-enantiomers of warfarin or its 4'-methoxy analogue. The observed stereoselectivity as well as the higher intrinsic clearancesof 4'-nitrowarfarin and 4'-chlorowarfarin are in agreement with the general pattern that we observed previously in the cytosolic fractions of a variety of species (Hermans and Thijssen, 1989).

The effect of some selected inhibitors on the ketone reduction of R-4'-nitrowarfarin by both rabbit liver cytosolic reductase fractions is listed in table 3. The alcohol dehydrogenase inhibitor pyrazole had no effect on both reductase fractions. Also, the DT-diaphorase inhibitor dicoumarol, which is structurally related to warfarin, was without effect. Therefore, it does not seem likely that the reductases can be classified as alcohol dehydrogenases or as DT-diaphorases. The flavonoid quercetin was a potent inhibitor of both reductases, whereas sodium-barbitone was without effect. As such, these reductases should be classified as ketone reductases (Ahmed et al, 1979). Indomethacin, furosemide and prostaglandin E2 also inhibited both reductases, whereas menadione inhibited fraction A only.

Table 2. Enzyme kinetic parameters of the rabbit liver cytosolic carbonyl reductase fractions using the R-enantiomers of warfarin and analogues as a substrate.

Substrate	Fraction A			Fraction B		
	Vmax	Km	Cli	Vmax	Km	Cli
R-Warfarin	1.33	1.34	1.0	1.75	0.41	4.3
R-4'-Nitrowarfarin	1.92	0.22	8.7	11.6	0.35	33
R-4'-Chlorowarfarin	1.02	0.23	4.4	6.9	0.28	25
R-4'-Methoxywarfarin	2.04	1.46	1.4	1.3	0.34	3.8

Kinetic parameters were obtained by direct fitting of the Michaelis-Menten equation to the data. Vmax is given in nmol/mg protein x min. Km in mM and intrinsic clearances (Cli) in μl/mg protein x min. From Hermans and Thijssen, 1992, with permission of the publisher.

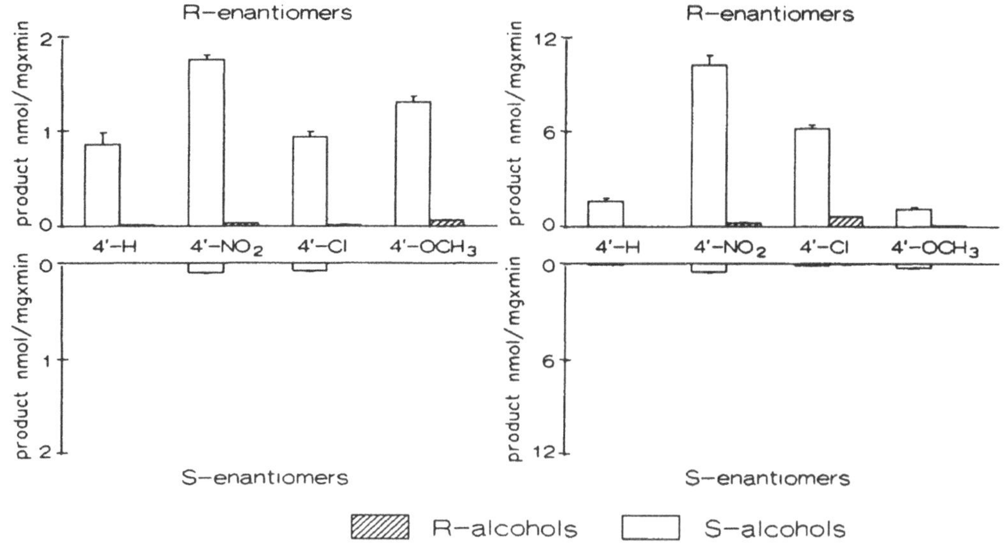

Figure 3. Substrate and product stereoselectivity of the reduction of warfarin and some of its 4'-hydrogen substituted analogues by the two rabbit liver cytosolic reductase fractions resolved by DEAE-Sephacel chromatography. From Hermans and Thijssen, 1992, with permission of the publisher.

Table 3. Inhibitor profile of two cytosolic reductases resolved from rabbit liver using R-4'-nitrowarfarin as a substrate.

	% inhibition (SD)	
Inhibitor	fraction A	fraction B
Pyrazole	14 (10)	4 (2)
Dicoumarol	4 (1)	5 (1)
Na-Barbitone	2 (3)	0 (1)
Quercetin	88 (2)	88 (1)
Indomethacin	75 (1)	87 (1)
Prostaglandin E2	62 (5)	78 (4)
Furosemide	98 (1)	45 (5)
Menadione	74 (4)	3 (5)

Data are expressed as mean (SD) of three incubations. The inhibitor and NADPH concentration was 1 mM; the substrate concentration was 0.5 mM. Taken from Hermans and Thijssen, 1992, with permission of the publisher.

The rabbit liver cytosolic enzyme fractions were able to reduce carbonyl groups of a variety of compounds (table 4), including aldehydes and ketones. In general, the substrate specifity of fraction A appeared to be broader than that of fraction B. Fraction A but not fraction B was able to reduce the quinones menadione and vitamin K1. As determined by HPLC, the product of the reduction of vitamin K1 by fraction A is its hydroquinone, which is the biologically active form of the vitamin. It does however not seem likely that this enzyme fraction plays a significant role in the generation of the hydroquinone of vitamin K1 under physiological conditions since vitamin K1 is a lipophilic compound (which is expected to reside in mainly in membranes) which is normally reduced to its hydroquinone by microsomal dithiol dependent reductases (Fasco and Principe, 1980). Estrone was a substrate for reductase fraction A, whereas androsterone was a substrate for fraction B. Based on their low reduction rates as well as their moderate potency as inhibitors of the reduction of R-4'-nitrowarfarin (data not shown), it does not seem likely that these compounds are physiological substrates for the warfarin reductases, although a role in the processing of (other) steroids of these enzymes can not be excluded. Both enzyme fractions reduced prostaglandin E2, which was also shown to be an inhibitor of the reduction of 4'-nitrowarfarin (table 3). The product formed was prostaglandin F2α. Kinetic analysis revealed the inhibition by prostaglandin E2 to be of the competitive type with Ki values, similar to the Km values that prostaglandin E2 had as a substrate for the enzyme fractions (about 150 μM). Vmax values were 3.2 nmol/mgxmin for fraction A and 1.5 nmol/mgx min for fraction B. However, since the Km values of these enzymes towards prostaglandin E2 are relatively high, and because prostaglandins are processed by other, more efficient routes (Lands, 1979), the significance of the role of these enzymes in prostaglandin metabolism is questionable. The function of NADPH dependent carbonyl reductases that are (also) able to reduce prostaglandin E2 has been questioned previously by Wermuth et al (1981) and Chang and Tai (1981). Based on the relatively broad substrate specifity, it seems more likely that the warfarin reducing enzymes are involved in detoxification processes (of e.g. quinones in the case of enzyme fraction A). The fact that the rat liver cytosolic warfarin reducing activity is enhanced by the enzyme inducers phenobarbitone and methylcholanthrene would also suggest this.

Table 4. Reduction rates of the rabbit liver cytosolic reductase fractions using various substrates.

	Fraction A	Fraction B
R-4'-Nitrowarfarin	1.5 ± 0.7	8.5 ± 0.5
p-Nitroacetophenone	22 ± 1.5	6.1 ± 0.5
p-Nitrobenzaldehyde	14 ± 2.1	22.5 ± 4.3
Prostaglandin E2	3.2 ± 0.5	1.2 ± 0.1
Menadione	33 ± 0.4	ND
Vitamin K1	1.2 ± 0.2	ND
Estrone	0.6 ± 0.1	ND
Androsterone	ND	0.8 ± 0.1
DL-Glyceraldehyde	3.0 ± 0.1	ND
Haloperidol	3.5 ± 0.5	ND

Data are expressed as the mean amount of oxidized NADPH (nmol/mg protein x min.) ± SEM. The substrate and NADPH concentrations were 1 mM. ND: not detectable.
From Hermans and Thijssen, 1992, with permission of the publisher.

To obtain information on the stereochemical aspects of the acetonyl side chain reduction of warfarin analogues in humans, this reaction has been studied in subcellular fractions from 9 subjects (table 5) using 4'-nitrowarfarin as a substrate. The ketone reduction of 4'-nitrowarfarin in human liver cytosolic fractions is shown in figure 4. Using R- or S-4'-nitrowarfarin as a substrate, the mean formation rate of the S-alcohols is about 15 times higher than the formation rates of the R-alcohols. Considering substrate stereoselectivity however, there is a considerable variation between the subjects. In samples C and H, no substrate stereoselectivity is observed, whereas in sample G the alcohol formation rate was about 11-fold higher for R-4'-nitrowarfarin than for S-4'-nitrowarfarin. The overall alcohol formation rate of the R-enantiomer of 4'-nitrowarfarin was about twice as high as that of the S-enantiomer.

Table 5. Information on the sources of human liver samples.

Code	Age	Sex	Death cause
A	36	Female	Subarachnoidal bleeding
B	90	Female	Brain damage
C	0.5	Male	Sepsis (Salmonella Typhii.)
D	52	Male	Traffic accident
E	61	Female	Contusio cerebri
F	70	Male	Lung cancer
G	86	Male	Thyroid gland cancer
H	91	Female	Sepsis
I	fetus (39 wks)	Male	Subarachnoidal bleeding

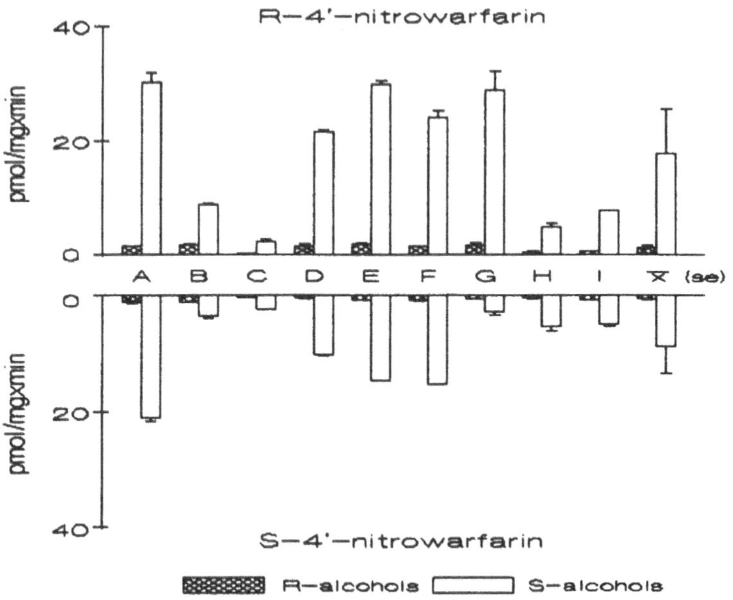

Figure 4. Reduction of the enantiomers of 4'-nitrowarfarin in human liver cytosolic fractions. Data are expressed as the mean ± SEM of two experiments. The final bar indicates the overall mean ± SEM.

Reductase activity was present in all cytosolic preparations, including those obtained from the fetal and neonatal liver, which may point to an important but still unknown function of the reductase(s) involved in the ketone reduction of warfarin analogues. The observed interindividual differences in substrate stereoselectivity may point to the presence of multiple human liver cytosolic enzymes, with different stereochemical requirements that are able to reduce the acetonyl side chain of 4'-nitrowarfarin. This hypothesis is supported by the observations of Moreland and Hewick (1975) that the substrate stereoselectivity of warfarin ketone reduction in kidney cytosol was about 10 times higher than that of liver cytosol obtained from one and the same person. From figure 4 it is clear that the stereoselectivity of 4'-nitrowarfarin reduction is qualitatively similar but quantitatively less pronounced as that observed for the rabbit liver cytosolic reductases. This may point to the fact that the enzyme fractions reducing the ketonic group of warfarin analogues obtained from rabbit liver cytosol possess stricter stereochemical requirements than those present in human liver cytosol.

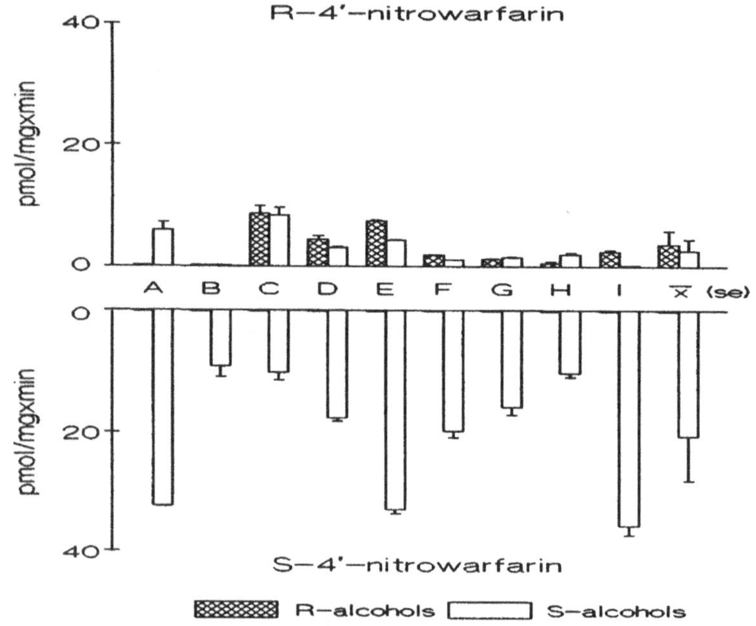

Figure 5. Reduction of the enantiomers of 4'-nitrowarfarin in human liver microsomal fractions. Data are expressed as the mean ± SEM. The final bar indicates the overall mean ± SEM.

The ketone reduction of 4'-nitrowarfarin by human liver microsomes is shown in figure 5. The observed pattern clearly differs from that in the liver cytosolic fractions. In all samples, except for sample C, selectivity for the reduction of the S-enantiomer is evident: the reduction rate of S-4'-nitrowarfarin is about 3 times higher than that of R-4'-nitrowarfarin. The S-enantiomer is almost exclusively reduced to the S-alcohol. In contrast, the R-enantiomer is reduced both to the R- and S-alcohols. Using R-4'-nitrowarfarin as a substrate, microsomes from subject B did not show any alcohol formation, whereas microsomes from subject A and I formed only the S- and R-alcohol respectively. Formation rates of the alcohols from R-4'-nitrowarfarin did not correlate (correlation coefficients of about 0.3) with the alcohol formation rate of the S-enantiomer, which may indicate that different enzymes are responsible for the microsomal reduction of R- and S-4'-nitrowarfarin. In microsomes, as well as in cytosol no differences between the sexes in alcohol formation rates or in stereoselectivity were apparent.

Figure 6 shows the effect of some selected inhibitors on the S-alcohol formation of R- and S-4'-nitrowarfarin in human liver cytosolic and microsomal fractions. The formation of none of the alcohols was inhibited by sodium-barbitone in both cytosol and in microsomes. In contrast, indomethacin and to a lesser extent prostaglandin E2 inhibited the formation of the RS and SS alcohols in cytosol as well as in microsomes. Quercetin inhibited the formation of the RS and the SS alcohol in cytosol, but in microsomes this compound only inhibited the formation of the RS alcohol. The observed inhibitor pattern in human liver cytosol is similar to that observed for the two partially purified rabbit liver cytosolic reductases (table 4). This and the (qualitative) similarities in the stereoselectivity may point to a possible relationship between the human and rabbit liver cytosolic reductases. Whether this is indeed the case and what the real function of the warfarin reducing enzymes might be is at present still unknown but will be the goal for future research.

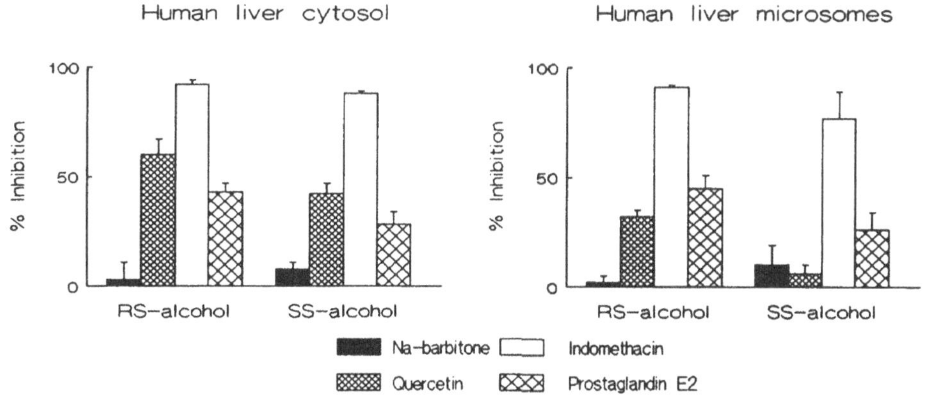

Figure 6. The effect of some selected inhibitors on human liver microsomal and cytosolic 4'-nitrowarfarin reduction.

REFERENCES

Ahmed, N.K., Felsted, R.L., and Bachur, N.R., 1979, Comparison and characterization of mammalian xenobiotic ketone reductases, *J. Pharmacol. Exp. Ther.* 209: 12.

Baars, L.G.M., Schepers, M.T., Hermans, J.J.R., Dahlmans, H.H.J., and Thijssen, H.H.W., 1990, Enantioselective structure-pharmacokinetic relationship of ring substituted warfarin analogues in the rat, *J. Pharm. Pharmacol.* 42: 861.

Banfield, C., O'Reilly, R., Chan, E., and Rowland, M., 1983, Phenylbutazone-warfarin interaction in man: further stereochemical and metabolic considerations, *Br. J. Clin. Pharmacol.* 16: 669.

Chang, D.G.-B., and Tai, H.-H., 1981: Prostaglandin 9-ketoreductase/type II 15-hydroxy-prostaglandin dehydrogenase is not a prostaglandin specific enzyme, *Biochem. Biophys. Res. Comm.* 101: 898.

Dieterle, W., Faigle, J.W., Sulc, M., and Theobald, W., 1977, Biotransformation and pharmacokinetics of acenocoumarol (Sintrom), *Eur. J. Clin. Pharmacol.* 11: 367.

Fasco, M.J., and Principe, L.M., 1980, Vitamin K1 hydroquinone formation catalyzed by a microsomal reductase system. *Biochem. Biophys. Res. Comm.* 97: 1487.

Felsted, R.L., and Bachur, N.R., 1980, Mammalian carbonyl reductases, *Drug Metab. Rev.* 11: 1.

Hara, A., Deyashiki, Y., Nakagawa, M., Nakayama, T., and Sawada, H., 1982, Isolation of proteins with carbonyl reductase activity and prostaglandin 9-keto-reductase activity from chicken kidney. *J. Biochem.* 92: 1753.

Hermans, J.J.R., and Thijssen, H.H.W., 1989, The in vitro ketone reduction of warfarin and analogues: Substrate stereoselectivity, product stereoselectivity and species differences, *Biochem. Pharmacol.* 38: 3365.

Hermans, J.J.R., and Thijssen, H.H.W., 1992, Stereoselective acetonyl side chain reduction of warfarin and analogues: partial characterization of two cytosolic carbonyl reductases, *Drug Metab. Dispos.* 20: 268.

Kudo, K., Amuro, Y., Hada, T., and Higashino, K., 1990, Purification and properties of 3α-hydroxysteroid dehydrogenase as a 3-keto bile acid reductase from human liver cytosol. *Biochim. Biophys. Acta* 1046: 12.

Lands, W.E.M., 1979, The biosynthesis and metabolism of prostaglandins, *Ann. Rev. Physiol.* 41: 633.

Lewis, R.J., Trager, W.F., Chan, K.K., Breckenridge, A., Orme, M., Rowland, M., and Schary, W., 1974, Warfarin: stereochemical aspects of its metabolism and the interaction with phenylbutazone, *J. Clin. Invest.* 53: 1607.

Pietruszko, R., and Chen, F.-F., 1976, Aldehyde reductase from rat liver is a 3α-hydroxysteroid dehydrogenase. *Biochem. Pharmacol.* 25: 2721.

Wermuth, B., 1981, Purification and properties of an NADPH-dependent carbonyl reductase from human brain. Relationship to prostaglandin 9-ketoreductase, *J. Biol. Chem.* 256:1206.

Wermuth, B., 1986, Platts, K.L., Seidel, A., and Oesch, F., 1986, Carbonyl reductase provides the enzymatic basis of quinone detoxification in man. *Biochem. Pharmacol.* 35: 1277.

Ziegler, D.M., Detoxication: oxidation and reduction, 1988, in: The Liver: Biology and Pathobiology, Arias, I.M., Jakoby, W.B., Popper, H., Schachter, D., and Shafritz, D.A., eds, 2nd edn, New York.

ACTIVATION OF PULMONARY CARBONYL REDUCTASE BY AROMATIC AMINES AND PYRIDINE RING-CONTAINING COMPOUNDS

Akira Hara,[1] Masaki Sakai,[1] Toshihiro Nakayama,[1]
Yoshihiro Deyashiki,[1] and Hideo Sawada[2]

[1]Laboratory of Biochemistry, Department of Manufacturing Pharmacy, Gifu Pharmaceutical University, Mitahora-higashi, Gifu 502, Japan

[2]Gifu College of Medical Technology, Seki, Gifu 501-32, Japan

INTRODUCTION

Lung of mice, guinea-pigs and pigs contains a tetrameric carbonyl reductase composed of Mr 24,000 subunits which is specifically localized in some epithelial cells of bronchus and alveolus (Nakayama et al., 1982; 1986; Matsuura et al., 1990; Oritani et al., 1992). The pulmonary enzyme has the following outstanding features which are different from those of the monomeric and dimeric enzymes in the other tissues (Sawada et al., 1979; Wermuth 1981; Hara et al., 1982; 1986; Nakayama et al., 1985; Iwata et al., 1990). The lung enzyme shows broad specificity and relatively high affinity for aliphatic and aromatic carbonyl compounds with either NADPH or NADH as a cofactor and is inhibited by pyrazole. Furthermore, it oxidizes secondary alcohols (Nakayama et al., 1986; Matsuura et al., 1988; 1989), and mediates the dismutation of aldehydes to the corresponding acids (Matsuura et al., 1989; Hara et al., 1991). Therefore, this enzyme has been suggested to play important roles in pulmonary metabolism of xenobiotic and endogenous carbonyl compounds.

In recent comparative studies of the kinetic properties with pig lung carbonyl reductase, we observed that the enzyme is an allosteric enzyme exhibiting negative cooperativity with respect to carbonyl substrates (Oritani et al., 1992) and that fatty acids such as linolenic acid and arachidonic acid act as allosteric activators abolishing the negative homotoropic interaction (Hara et al., 1992). Therefore, we interested in the regulation of pulmonary carbonyl reductase, and re-examined the effects of fatty acids on the kinetic properties of the enzymes of guinea-pig and mouse lung, but the guinea-pig and mouse enzymes did not show the cooperativity and were not significantly activated by the fatty acids. However, we found that N-containing aromatic compounds, particularly phenanthrolines, were potent activators for the enzymes from the three animal lungs. In this study, we characterized the activation mechanism and show that the tetrameric carbonyl reductases possess binding site(s) for basic N-containing aromatic compounds which are probably distinct from the fatty acid binding site and the active site.

Enzymology and Molecular Biology of Carbonyl Metabolism 4
Edited by H. Weiner, Plenum Press, New York, 1993

EXPERIMENTAL PROCEDURES

Arachidonic acid was purchased from Kurita Water Industries (Tokyo Japan) and other fatty acids were from Sigma Chemicals. Aromatic amines and heterocyclic compounds were obtained from Nacalai Tesque (Kyoto, Japan) and Wako Pure Chemical Industries (Osaka, Japan). Paraquat, monoquat and diquat were gifts from Dr. M. Iwasaki (Panapharm Laboratories, Kumamoto, Japan). Pulmonary carbonyl reductases from pigs (Oritani et al., 1992), mice and guinea-pigs (Nakayama et al., 1986) were purified to homogeneity as described.

The enzyme activity was assayed spectrophotometrically or fluorometrically by recording the NADPH oxidation as described (Oritani et al., 1992). The standard assay was performed in 0.2 M potassium phosphate buffer, pH 7.0, containing 0.1 mM NADPH, 1 mM 4-nitroacetophenone and enzyme. Kinetic data were analyzed as described (Oritani et al., 1992; Hara et al., 1992). One unit of the enzyme activity was defined as the amount catalyzing the oxidation of 1 µmol of NADPH per min at 25°C.

Analytical gel filtration was carried out on a Sephadex G-100 column (1.2 x 90 cm) in 0.2 M potassium phosphate, pH 7.0 with or without 1.0 mM 1,10-phenanthroline (OP). Polyacrylamide gel electrophoresis with sodium dodecyl sulfate was performed according to the method of Weber and Osborn (1969). The relative molecular masses of the non-activated and activated enzymes were estimated by the gel filtration and sodium dodecyl sulfate-polyacrylamide gel electrophoresis using the molecular mass standards as described (Oritani et al., 1992).

RESULTS AND DISCUSSION

Specificity of Activation by N-Containing Aromatic Compounds

When the effects of various amines on the NADPH-linked 4-nitroacetophenone reductase activity of the pig lung carbonyl reductase was examined at pH 7.0, the enzyme activity was stimulated by aromatic amines and pyridine ring-containing aromatic compounds listed in Table 1. In the aromatic amines, tertiary amines such as aminopyrine, chlorpromazine and promethazine were more stimulatory to the enzyme than primary and secondary amines.

Table 1. The effects of aromatic amines and pyridine ring-containing aromatic compounds on 4-nitroacetophenone reductase activity of pig lung carbonyl reductase.

Compound	Concentration (mM)	Stimulation (%)	Compound	Concentration (mM)	Stimulation (%)
Aniline	5	78	OP	1	536
2-Phenylenediamine	5	100	1,7-Phenanthroline	1	408
2-Methylaniline	5	111	4,7-Phenanthroline	1	395
4-Methoxyaniline	5	179	2,2'-Dipyridyl	5	267
Diphenylamine	1	170	4,4'-Dipyridyl	5	129
Indole	5	100	2,6-Lutidine	5	162
Aminopyrine	5	177	Paraquat	2	4
Promethazine	5	347	Monoquat	6	10
Chlorpromazine	1	276	Diquat	5	0

On the other hand, the enzyme activity was not activated by 5 mM concentrations of aliphatic amines such as amino acids, catechol amines and polyamines. Thus, the hydrophobic benzene ring of the amines appears to be critical for the carbonyl reductase-stimulating effects of the amines. The enzyme activity was also stimulated to a significant extent by several pyridine ring-containing aromatic compounds. Although 1,10-phenanthroline (OP) and 2,2'-dipyridyl are chelating agents, the activation may be independent of the chelating ability, because the isomers of the compounds similarly activated the pig lung enzyme, in which no metal has been detected (Oritani et al., 1992). In contrast, no activation was observed by the addition of quarternary amines, such as paraquat, monoquat and diquat, which have structures similar to the dipyridyls and phenanthrolines. Taken together, these data suggest that a relatively specific interaction occurs between these N-containing aromatic compounds and the enzyme. Furthermore, they show that, as well as the hydrophobic aromatic ring, the presence of the weak basicity due to the nitrogen atom seems to be a critical determinant of this interaction.

Since OP is a chelating agent, we examined whether the OP-metal complex is effective as an activator of the pig lung enzyme. OP quickly activated the enzyme activity when it was added to the reaction mixture without the activator. When zinc ions were added to the reaction mixture which had been initiated in the presence of OP, the activity rapidly decreased to the level of the non-activated enzyme, and further addition of EDTA led to re-activation of the enzyme (Figure 1). The result indicates that OP behaves as an instantaneous and reversible activator,

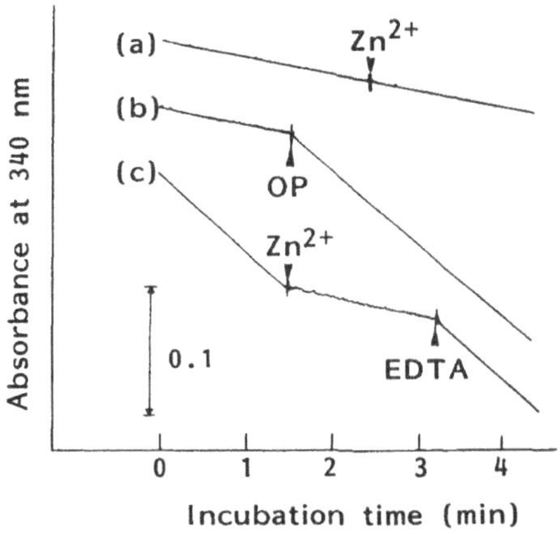

Figure 1. Effects of zinc ions and EDTA on the activation of carbonyl reductase by OP. The enzyme activity was monitored at 340 nm. Curve (a), the addition of 1 mM zinc sulfate did not affect the rate of the reaction mixture without OP. Curve (b), the reaction was started in the absence of OP, and then 1.25 mM OP was added to the mixture. Curve (c), 1 mM zinc sulfate was added to the reaction mixture at 1.5-min after the reaction was initiated in the presence of 1.25 mM OP, and then 1 mM EDTA was further added to the reaction mixture.

and supports the importance of the weak basicity in the activator molecule for the binding to the enzyme.

The phenanthrolines and 2,2'-dipyridyl also stimulated the activity of carbonyl reductases from guinea-pig and mouse lung in a dose-dependent manner. When the data were analyzed by the double-reciprocal plots of stimulation percentage versus activator concentration, the plots were linear and the Smax (maximum stimulation percentage) and Ka(concentration giving 1/2 Smax) values for the activator can be calculated from the intercepts of y and x axes, respectively, of the plots. These values for the compounds are summarized in Table 2. Although there is a difference in Smax and Ka values among the three enzymes, the Smax/Ka values of the respective enzymes for the three phenanthrolines were similar, but were higher than that for 2,2'-dipyridyl. On the other hand, 2,2'-dipyridyl and OP have been reported not to activate the activity of monomeric and dimeric carbonyl reductases from human and animal tissues (Sawada et al., 1979; Hara et al., 1982; Nakayama et al., 1985; Hara et al., 1986; Iwata et al., 1990). Therefore, the pyridine ring-containing compounds may be specific activators for the pulmonary enzymes.

Table 2. The potency of phenanthrolines and 2,2'-dipyridyl on activation of carbonyl reductases from pig, guinea-pig and mouse lung.

Compound	Pig enzyme			Guinea-pig enzyme			Mouse enzyme		
	Smax (%)	Ka (mM)	Smax/Ka	Smax (%)	Ka (mM)	Smax/Ka	Smax (%)	Ka (mM)	Smax/Ka
OP	1790	2.0	895	806	2.7	299	2110	15	140
1,7-Phenanthroline	996	1.2	830	388	1.3	298	389	3.1	125
4,7-Phenanthroline	866	1.2	722	132	1.1	120	220	1.7	129
2,2'-Dipyridyl	820	12	66	523	11	48	822	25	33

Effect of OP on the Conformation of Carbonyl Reductase

Analytical Sephadex G-100 filtration and sodium dodecyl sulfate-polyacrylamide gel electrophoresis showed that the molecular weight (Mr 90,000) and subunit size (Mr 24,000) of the pig lung enzyme activated by 2.5 mM OP were the same as those of the native enzyme, which indicates that the activation by OP is not resulted from changes in molecular weight and subunit composition of the tetrameric enzyme.

A difference between the native enzyme and activated enzyme became, however, apparent in their thermal stability (Figure 2). While the native pig lung enzyme was gradually inactivated at 47 °C and the inactivation was protected by the addition of NADPH, the addition of OP accelerated the thermal-inactivation of the enzyme in both the absence and presence of NADPH. In contrast, OP protected against the thermal-inactivation of the guinea-pig enzyme, and the protective effect increased with elevating OP concentration. The complete protection was achieved by the addition of 2 mM OP. The results, although OP exhibited opposite effects on the thermal stability of the two enzymes, suggest that OP binds to both the free enzyme and the enzyme-NADPH binary complex, and that the binding of OP causes some conformational changes of the enzymes.

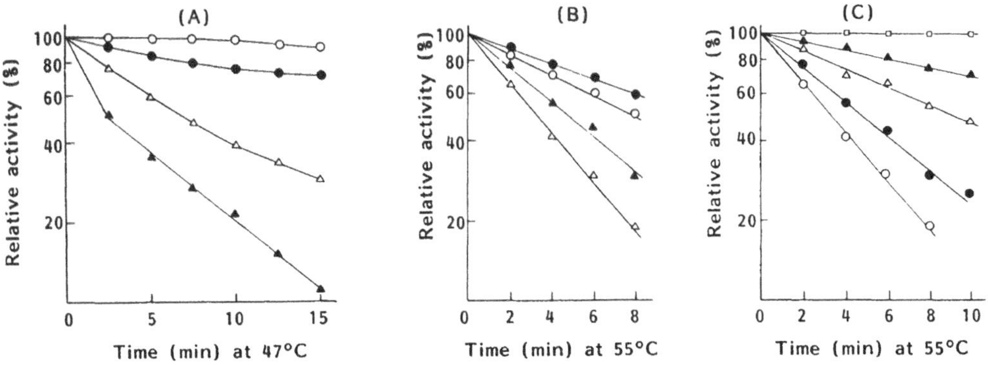

Figure 2. Effect of OP on the thermal inactivation of carbonyl reductases. The enzyme was diluted to its concentration of 0.1 mg/ml with 50 mM potassium phosphate buffer, pH 7.0 (△) or the buffer containing the following ligands, and incubated at 47°C or 55°C for indicated times, then the residual activity was determined. (A) Inactivation of the pig enzyme. Ligand: 2.5 mM OP (▲), 20 μM NADPH (○), and 2.5 mM OP plus 20 μM NADPH (●). (B) Inactivation of the guinea-pig lung enzyme. Ligands: 0.25 mM OP (▲), 0.5 μM NADPH (○), and 0.25 mM OP plus 0.5 μM NADPH (●). (C) Protection by OP against the inactivation of the guinea-pig enzyme. OP concentrations: 0 mM (○), 0.25 mM (●), 0.5 mM (△), 1.0 mM (▲), and 2.0 mM (□).

Effect of OP on the Kinetic Properties of Carbonyl Reductase

Pig lung carbonyl reductase has been shown to exhibit negative co-operativity with respect to carbonyl substrate but not with respect to NADPH (Oritani et al., 1992). The interaction of OP with the enzyme at different concentrations of several carbonyl substrates was analyzed by Hill plot and the kinetic constants are summarized in Table 3. In each

Table 3. Effect of OP on the kinetic constants in the reduction of carbonyl compounds by pig lung carbonyl reductase.

Substrate	$[S]_{0.5}$ (μM)	Vmax (units/mg)	Hill coefficient	Vmax/$[S]_{0.5}$ (units/mg/mM)
Acetone	200	6.12	1.01	31
	(19)	(1.53)	(0.70)	(81)
4-Nitroacetophenone	55	18.8	0.99	342
	(2.4)	(1.67)	(0.75)	(696)
Cyclohexanone	32	13.6	1.07	425
	(1.6)	(1.99)	(0.67)	(1243)
1-Propanal	68	17.3	0.98	254
	(4.3)	(2.16)	(0.68)	(502)
Pyridine-4-aldehyde	150	22.6	1.02	151
	(19)	(3.93)	(0.65)	(217)

The constants were determined in the presence of 1.25 mM OP, and those in parentheses were in the absence of OP.

case the slopes of Hill plots were almost 1.0, which suggests that OP is an allosteric activator abolishing the negative interaction of the enzyme for carbonyl substrates. It is apparent that the Vmax and $[S]_{0.5}$ values were increased on the OP-induced activation of the enzyme. Thus, the catalytic efficiency (Vmax/$[S]_{0.5}$) of the enzyme decreased on the activation. On the other hand, the double-reciprocal plots of the activity versus NADPH concentration at increasing concentrations of OP resulted in a family of linear plots intersecting above the x axis (Figure 3A); OP decreased the Km for NADPH. Secondary plots of the slopes and intercepts of the double-reciprocal plots against OP concentration were hyperbolic (Figure 3B), but replots of 1/Δ slope and 1/Δ intercept versus 1/[OP] were linear (Figure 3C). The results are consistent with a case of nonessential activation system (Figure 4), in which the kinetic constants can be determined (Segel, 1975). The factors, α and β were determined to be 0.3 and 2.3, which suggests that the OP-enzyme interaction increases the affinity for NADPH and the catalytic rate by factors of 3.3 and 2.3 respectively.

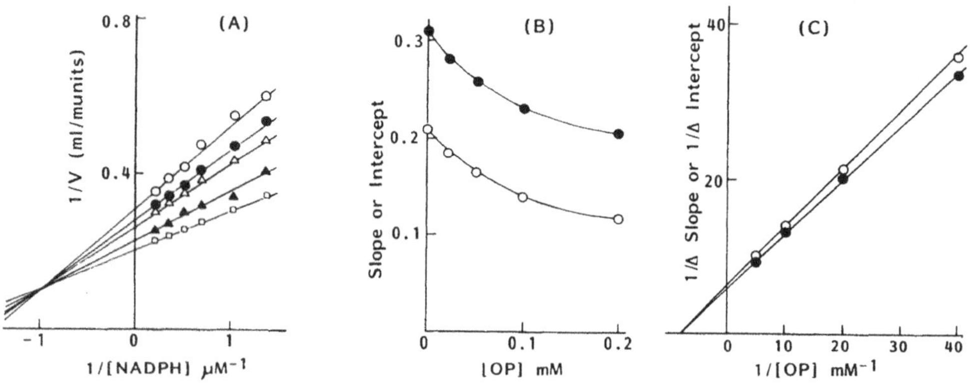

Figure 3. Effect of OP on the reductase activity of pig lung carbonyl reductase as a function of NADPH concentration. (A) Double-reciprocal plots of velocity versus NADPH concentration. OP concentrations: 0 mM (○), 0.025 mM (●), 0.05 mM (△), 0.1 mM (▲), and 0.2 mM (□). (B) Plots of the slopes (○) and the intercepts (●) of the reciprocal plots versus OP concentration. (C) Replot of the 1/change in slope (○) or intercept (●) against 1/[OP].

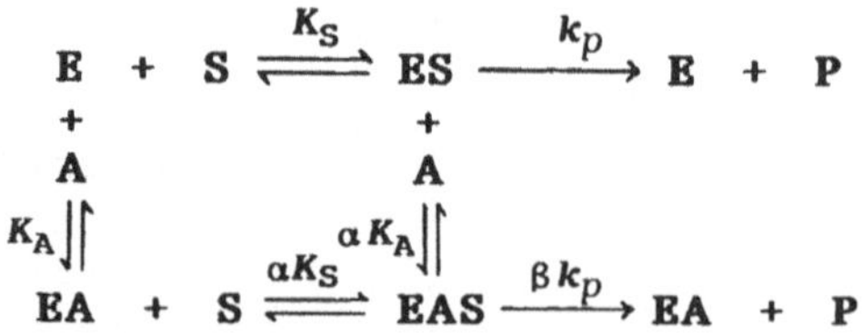

Figure 4. Scheme for nonessential activation. E, enzyme; A, activator (OP); S, substrate (NADPH); P, product.

Similar kinetic analyses were carried out with the guinea-pig lung carbonyl reductase. Since the enzyme does not exhibit cooperativity, the effect of OP can be analyzed by the double-reciprocal plots of the activity versus concentration of carbonyl substrate or cofactor. The activation patterns by OP were of uncompetitive types when either the carbonyl substrate or NADPH was used as the varied substrate (Figure 5). Such an uncompetitive activation can exist under a special case of the mixed type of activation, in which $\alpha = \beta$ in Figure 4 (Dixon et al., 1979). Therefore, the data suggest that OP binds to all the kinetic enzyme forms in the ordered mechanism (Matsuura et al., 1988) and that the OP binding site of the enzyme is different from the active site. The values of α and β, determined in the first step of the enzyme mechanism were 25, which implies that the OP-enzyme interaction decreases the affinity for NADPH by a factor of 0.04 and increases the catalytic rate by a factor of 25.

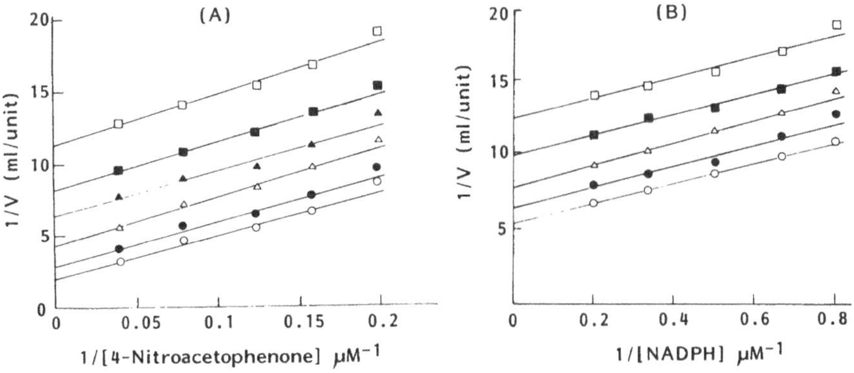

Figure 5. Double reciprocal plots of velocity versus concentration of 4-nitroacetophenone (A) or NADPH (B) in the presence of OP. The reductase activity of the guinea-pig enzyme was determined with a fixed concentration of NADPH (0.1 mM) or 4-nitroacetophenone (1 mM). OP concentrations: 0 mM (□), 0.125 mM (■), 0.25 mM (▲), 0.5 mM (△), 1.0 mM (●), and 1.6 mM (○) in (A); 0 mM (□), 0.1 mM (■), 0.2 mM (△), 0.35 mM (●), and 0.5 mM (○) in (B).

Comparison of the Binding Sites for Fatty Acids and OP

Fatty acids, particulary cis-unsaturated fatty acids, have been shown to be potent activators of pig lung carbonyl reductase (Hara et al., 1992). The Smax/Ka values for fatty acids are much higher than that for OP. For examples, the value for arachidonic acid is 41,000 $\%mM^{-1}$, whereas that for OP was 581 $\%mM^{-1}$. In this regard, fatty acids are more potent activators of the pig lung enzyme. However, arachidonic acid and linolenic acid, potent activators for the pig lung enzyme, gave nominal (1.2-2 fold) activation of the activity of the enzymes from guinea-pig and mouse lung. In constrast, basic N-containing aromatic compounds are common activators for carbonyl reductases from pig, guinea-pig and mouse lung, and particularly OP gave high stimulation of 800-2100%. This suggests that all the enzymes have specific binding site(s) for the basic N-containing aromatic compounds.

Although the effect of OP on the thermal stability and the kinetic properties is similar to that of arachidonic acid in the case of the pig lung enzyme (Hara et al., 1992), there is a marked difference in structural requirements for the interaction with the enzyme between N-containing aromatic compounds and fatty acids; the critical part of N-containing aromatic compounds may be the nitrogen atom with weak basicity, whereas that of fatty acids has been thought to be the acidic carboxyl group (Hara et al., 1992). The difference in structural requisite for activation between the activators suggests that the tetrameric carbonyl reductase has two distinct activator-binding sites. To confirm the presence of the two distinct binding sites, we examined the stimulatory effects by mixture of two different activators. Oleic acid and arachidonic acid are expected to compete for a site on the enzyme. Thus, oleic acid was slightly inhibitory to the stimulation by arachidonic acid (Figure 6A). When OP was added to the reaction mixture containing arachidonic acid, the stimulatory effects by the two activators were additive. Similarly, the effects of 2,2'-dipyridyl and arachidonic acid on the activation of the enzyme by OP were examined (Figure 6B). The additive stimulatory effects were observed by the addition of arachidonic acid but not by that of 2,2'-dipyridyl.

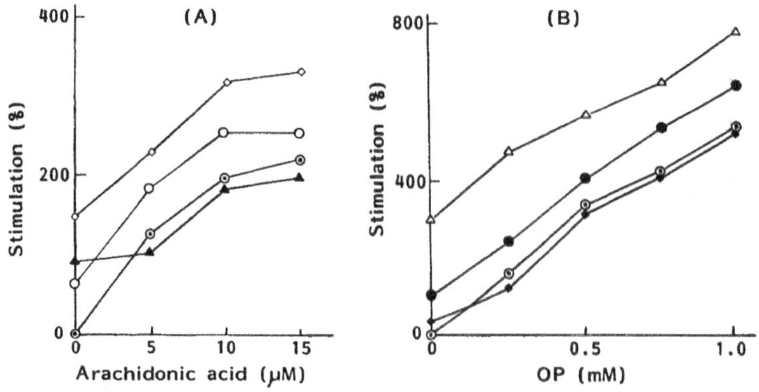

Figure 6. Effects of the combined activators on the stimulatory effects by arachidonic acid and OP. The reductase activity of pig lung carbonyl reductase was determined in the presence of different concentrations of arachidonic acid (A) or OP (B), as well as one of the combined activators. The control activity (◉) was assayed only with arachidonic acid or OP. The combined activators are 2.5 μM oleic acid (▲), 0.1 mM OP (○), 0.2 mM OP (◇), 0.2 mM 2,2'-dipyridyl (◇), 3 μM arachidonic acid (●), and 10 μM arachidonic acid (△).

Thus, tetrameric carbonyl reductase probably has two separate and specific classes of binding sites or regions for fatty acids and the N-containing aromatic compounds. Fatty acids are physiologically important modulators, especially of the pig lung enzyme, whereas the physiological significance of the activation by N-containing aromatic compounds is unclear at present, because such activators tested in this study were all xenobiotics. However, the activation by N-containing aromatic compounds is a common feature of the tetrameric carbonyl reductases. Further studies on natural activators, which structurally mimic the xenobiotic N-containing aromatic compounds, may lead to elucidation of the regulation and function of the pulmonary enzyme in a metabolic pathway.

ACKNOWLEDGMENTS

This work was supported by grants from the Ministry of Eduction, Science and Culture of Japan, and from the Suzuken Memorial Foundation.

REFERENCES

Dixon, M., Webb, E. C., Thorne, C. J. R., and Tipton, K. F., 1979, in "Enzymes," Longman Group, London, p. 381.
Hara, A., Deyashiki, Y., Nakagawa, M., Nakayama, T., and Sawada, H., 1982, Isolation of proteins with carbonyl reductase activity and prostaglandin-9-ketoreductase activity from chicken kidney, J. Biochem., **92**:1753.
Hara, A., Oritani, H., Deyashiki, Y., Nakayama, T., and Sawada, H., 1992, Activation of carbonyl reductase from pig lung by fatty acids, Arch Biochem. Biophys., **292**:548.
Hara, A., Nakayama, T., Deyashiki, Y., Kariya, K., and Sawada, H., 1986, Carbonyl reductase of dog liver: Purification, properties, and kinetic mechanism, Arch. Biochem. Biophys., **244**:238.
Hara, A., Yamamoto, H., Deyashiki, Y., Nakayama, T., Oritani, H., and Sawada, H., 1991, Aldehyde dismutation catalyzed by pulmonary carbonyl reductase: Kinetic studies of chloral hydrate metabolism to trichloroacetic acid and trichloroethanol Biochim. Biophys. Acta, **1075**:61.
Iwata, N., Inazu, N., Takeo, S., and Satoh, T., 1990, Carbonyl reductases from rat testis and vas deferens: Purification and characterization, Eur. J. Biochem., **193**:75.
Matsuura, K., Hara, A., Sawada, H., Bunai, Y., and Ohya, I., 1990, Localization of pulmonary carbonyl reductase in guinea pig and mouse: Enzyme histochemical and immunochemical studies, J. Histochem. Cytochem., **38**:217.
Matsuura, K., Naganeo, F., Hara, A., Nakayama, T., and Sawada, H., 1989, Pulmonary carbonyl reductase: Metabolism of carbonyl products in lipid peroxidation, in "Enzymology and Molecular Biology of Carbonyl Metabolism 2," H. Weiner and T. G. Flynn, eds., Alan R. Liss, New York, p. 335.
Matsuura, K., Nakayama, T., Nakagawa, M., Hara, A., and Sawada, H., 1988, Kinetic mechanism of pulmonary carbonyl reductase, Biochem. J., **252**:17.
Nakayama, T., Hara, A., and Sawada, H., 1982, Purification and characterization of a novel pyrazole-sensitive carbonyl reductase in guinea pig lung, Arch. Biochem. Biophys., **217**:564.
Nakayama, T., Hara, A., Yashiro, K., and Sawada, H., 1985, Reductases for carbonyl compounds in human liver, Biochem. Pharmacol., **34**:107.
Nakayama, T., Yashiro, K., Inoue, Y., Matsuura, K., Ichikawa, H., Hara, A., and Sawada, H., 1986, Characterization of pulmonary carbonyl reductases of mouse and guinea pig, Biochim. Biophys. Acta, **882**:220.
Oritani, H., Deyashiki, Y., Nakayama, T., Hara, A., Sawada, H., Matsuura, K., Bunai, Y., and Ohya, I., 1992, Purification and characterization of pig lung carbonyl reductase, Arch. Biochem. Biophys., **292**:539.
Sawada, H., Hara, A., Kato, F., and Nakayama, T., 1979, Purification and properties of reductases for aromatic aldehydes and ketones from guinea pig liver, J. Biochem., **86**:871.

Segel, I. H., 1975, in "Enzyme Kinetics," John Willy & Sons, New York, p.227.
Weber, K., and Osborn, M., 1969, The reliability of molecular weight determinations by dodecy sulfate-polyacrylamide gel electrophoresis, J. Biol. Chem., **244**:4406.
Wermuth, B., 1981, Purification and properties of an NADPH-dependent carbonyl reductase from human brain, J. Biol. Chem., **256**:1206.

UNIQUE DIHYDRODIOL SPECIFIC DEHYDROGENASE OF BOVINE LIVER : INHIBITION STUDIES AND COMPARISON WITH ALDO/KETO REDUCTASE

Hirofumi Nanjo, Tohru Nishinaka,
Makoto Nagai, Tomoyuki Terada,
Tadashi Mizoguchi, and Tsutomu Nishihara

Laboratory of Biochemistry,
Faculty of Pharmaceutical Sciences,
Osaka University,
1-6 Yamada-oka, Suita, Osaka 565, Japan

INTRODUCTION

Dihydrodiol dehydrogenase (EC 1.3.1.20: DD) catalyzes the dehydrogenation of the dihydrodiols of benzo(a)pyrene and benzo(a) anthracene in the presence of $NADP^+$ and forms *ortho*-quinone (Vogel *et al.*, 1980; Smithgall *et al.*, 1988a). We previously reported on the purification of three multiple forms of bovine liver DD, DD1, DD2 and DD3, using benzenedihydrodiol (*trans*-1,2-dihydrobenzene-1,2-diol) which is a model substrate for DD with $NADP^+$ as a coenzyme (Nishinaka *et al.*, 1991). DD1 and DD2 have been identified with 3α-hydroxysteroid dehydrogenase (Nanjo *et al.*, 1992) and high-*K*m aldehyde reductase (Terada *et al.*, 1985), respectively. On the other hand, DD3 is a unique enzyme which can specifically catalyze the dehydrogenation of benzenedihydrodiol and naphthalenedihydrodiol and has no activity toward other alcohols, aldehydes, ketones and quinones which are well known substrates for aldo/keto reductases. The *K*m-value of DD3 for benzenedihydrodiol was the lowest among the three enzymes. Thus, DD3, judging from its substrate specificity and inhibitor sensitivity, could not be classified into any known DD including carbonyl and aldehyde reductases and steroid dehydrogenases (Balcsak *et al.*, 1983; Hara *et al.*, 1986; Sawada *et al.*, 1989). The most interesting property of DD3 is an immunological crossreactivity with DD1. Properties of bovine liver cytosolic DD were summarized in Table 1.

In this report, we demonstrated the further characterization of DD3 and DD1 in the view points of inhibition study and partial primary structure.

MATERIALS AND METHODS

Materials

$NADP^+$ were purchased from Oriental Yeast, Co. Steroids and oxidized glutathione (GSSG) were obtained from Sigma Chemical Co. 5,5'-dithiobis(2-nitrobenzoic acid) (DTNB), cyanogen bromide and *Achromobacter lyticus* lysyl endopeptidase were bought from Wako Pure Chemical Industries. Dithiothreitol (DTT), N-ethylmaleimide (NEM) and iodoacetate were the products of Nacalai Tesque. Nick-column was from Pharmacia-LKB. Benzenedihydrodiol (*trans*-1,2-dihydrobenzene-1,2-diol) was synthesized according to the method of Platt and Oesch (1977). The other reagents were the highest grade commercially available.

Purification of Dihydrodiol Dehydrogenase (DD)

Bovine liver DDs were purified from the cytosolic fraction according to the method as previously reported (Nishinaka *et al*, 1991).

Enzyme Assay

The DD3 and DD1 activities were measured by monitoring the increase in absorbance at 340 nm under the following reaction mixture; DD3: 0.067 mM $NADP^+$ and 1.0 mM benzenedihydrodiol in 100 mM glycine/NaOH buffer, pH 9.5, DD1: 0.067 mM $NADP^+$ and 0.1 mM androsterone in the same buffer. The assay was initiated by addition of the enzyme solution at 25°C.

Treatment of DD3 and DD1 with SH-Reagents and GSSG

DD3 and DD1 were reduced by the pretreatment with 10 mM DTT for 30 min at 25°C and were gelfiltrated with Nick-column to remove excess DTT. Then, the reduced enzymes were incubated with SH-reagents or GSSG at 25°C. Aliquots of the reaction mixtures were taken out at appropriate time for measurement of their remaining activity.

Digestion of the Proteins

Lyophilized DD3 and DD1 (1 nmol) were hydrolyzed by 1% cyanogen bromide in 70% formic acid at 25°C for 24 h. Peptides were separated with a VYDAC-Protein C4 column (4.6 x 250 mm) linked to a gradient high performance liquid chromatography system (LKB). Peptides were eluted with

Table 1. Properties of bovine liver DD (DD1, DD2 and DD3).

	DD1	DD2	DD3
Separation by Q-Sepharose	unadsorbed	50mM NaCl	75mM NaCl
Molecular weight	35,000	36,500	35,500
Typical substrate	Androsterone	D-Glucuronate	Benzenedihydrodiol
Km value for benzene-dihydrodiol (mM)	2.3	20	0.18
Vmax for benzene-dihydrodiol (units/mg)	0.25	0.55	1.3
Immunoreactivity			
anti DD1	+	-	+
anti DD3	+	-	+

a gradient of 0-60% acetonitrile in the presence of 0.1% trifluoroacetic acid (1 ml/min, 2 h) and were detected by monitoring the absorbance at 220 nm.

DD3 was digested with *Achromobacter lyticus* lysyl endopeptidase ([DD]/ [protease]=200/1) in 100 mM Tris-HCl buffer, pH 9.0, containing 0.1% SDS. The digestion was allowed to proceed at 35°C for 2 h. Peptide separation was performed by using a Wakosil 5C18 column (4.6 x 250 mm, Wako). Peptides were eluted with a gradient of 0-60% acetonitrile in the presence of 0.1 % trifluoroacetic acid (0-30%, 1 ml/min, 1 h, and 30-60%, 1 ml/min, 2 h) and were detected by monitoring the absorbance at 220 nm.

Peptide Sequencing

The peptide sequencing was carried out by using Simadzu PSQ-1 automatic gas phase protein sequencer.

RESULTS

Inactivation of DD3 and DD1 by SH-reagents and GSSG

Rat liver DD, which was identified with 3α-hydroxysteroid dehydrogenase and has no multiple forms (Smithgall, *et al.*, 1988b), was inactivated by biological disulfides, GSSG and L-cystamine (Terada *et al.*, 1991). In order to clarify the role of cysteine residue(s) of bovine liver DD, the purified three enzymes were incubated with SH-reagents, 0.1 mM DTNB, 0.1 mM NEM and 10 mM iodoacetate. DD3 and DD1 were inactivated by these reagents. Especially, DTNB was a strong inactivator for DD3 (84 % inactivation) and DD1 (86 % inactivation). DTNB-inactivation of DD3 and DD1 was dependent on incubation-time following pseudo first order reaction (Fig. 1). In addition, these two enzymes were also inactivated by 5 mM GSSG and reactivated with the following the 10 mM DTT-treatment (data not shown). These results suggested that DD3 and DD1 lost their activities due to the modification of their essential cysteine residue(s) by these reagents.

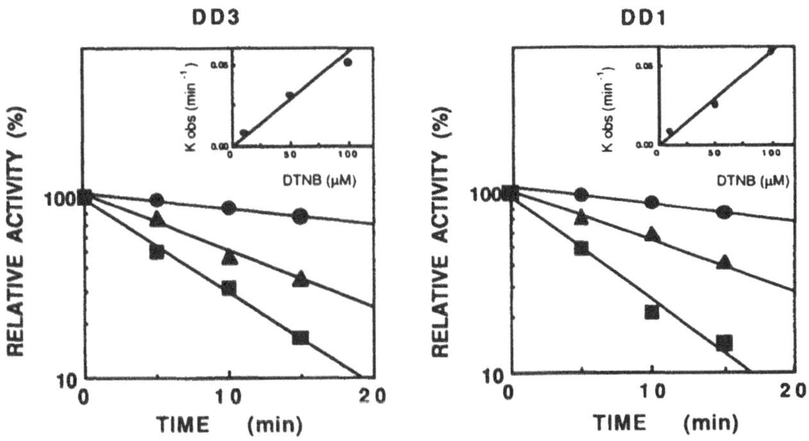

Figure 1. Time-dependent inactivations of DD3 and DD1 by DTNB. The reduced enzymes were incubated with 10 (●), 50 (▲) and 100 μM (■) of DTNB at 25°C. Inserted graphs at upper right of each figures were plotted *K*obs-values against the concentration of DTNB.

Table 2. Protection of DD3 and DD1 against inactivation by DTNB.

Addition		Inhibition (%)	
		DD3	DD1
0.1 mM DTNB		84	86
+ 0.33 mM	$NADP^+$	2	15
+ 0.50 mM	Benzenedihydrodiol	87	-
+ 0.50 mM	Androsterone	-	85

The DTT-reduced enzyme was incubated at 25°C for 15 min with DTNB in the presence of protector.
-, Not determined.

Additionally, DD3 and DD1 were incubated with DTNB in the presence of coenzyme or substrate. Coenzyme ($NADP^+$) protected the enzymes against inactivation effectively. However, typical substrates (benzenedihydrodiol for DD3 and androsterone for DD1) had no effect on this inactivation (Table 2). This result showed that the cysteine residue(s) located at or nearby the coenzyme binding site of these enzymes. On the other hand, DD2 which has been identified with high-*K*m aldehyde reductase was not inactivated by these reagents at all.

Inhibition of DD3 by Steroids

DD3 has no activity for the steroids including androsterone and 5α-androstane-3,17-dione though DD3 is immunologically crossreacted with anti-DD1 serum (Nishinaka *et al.*, 1991). Interestingly, DD3 was inhibited competitively by the steroids such as androsterone and 5α-androstane-3,17-dione which were good substrates for DD1. DD3 had low *K*i-values for androsterone, 5β-androstan-3α-ol-17-one, 5α-androstane-3,17-dione and 5β-androstane-3,17-dione (6, 31, 11 and 41 μM, respectively), which were in the same level as the *K*m-values of DD1 for these steroids (Table 3). This result showed that the substrate binding sites of DD3 and DD1 might be similar in their structures.

Partial Primary Structure of DD3 and DD1

As both DD3 and DD1 were expected to have great similarity in their structures on the basis of inhibition studies and immunological studies, we attempted to compare both these enzymes in the primary structure. These enzymes were subjected to cyanogen bromide hydrolysis and lysyl endopeptidase digestion because the N-terminals of these enzymes were blocked. And then, peptide fragments were separated by reversed-phase HPLC as described in MATERIALS AND METHODS. These peptides were applied to the automatic protein sequencer. We succeeded in the determination of partial amino acid sequences of both DD3 (49 residues) and DD1 (20 residues) .

Partial amino acid sequences of DD3 and DD1 were compared with those of rat liver 3α-hydroxysteroid/dihydrodiol dehydrogenase (Pawlowski *et al.*, 1991) and the enzymes of aldo/keto reductase family. As shown in Fig. 2,

Table 3. Apparent *K*i Values of DD3 for Various Steroids

Inhibitor	*K*i(μM)	
Androsterone	6	(DD1; *K*m=4.8 μM)
5β-Androstan-3α-ol-17-one	31	(DD1; *K*m=7.4 μM)
5α-Androstane-3,17-dione	11	(DD1; *K*m=3.1 μM)
5β-Androstane-3,17-dione	41	(DD1; *K*m=2.9 μM)

*K*i values were estimated from Lineweaver-Burk plot using 0.067 mM $NADP^+$ as a coenzyme and benzenedihydrodiol as a substrate.

the sequences of the peptides obtained by cyanogen bromide hydrolysis and lysyl endopeptidase digestion had large homology with the C-terminal sequences of rat liver 3α-hydroxysteroid/dihydrodiol dehydrogenase and bovine lung prostaglandin F synthase (Watanabe *et al.*, 1988) which are well known members of aldo/keto reductase family.

DISCUSSION

DD have been purified from many tissues and animals (Balcsak *et al.*, 1983; Matsuura *et al.*, 1987; Hara *et al.*, 1990). Most of DDs were identified with 3α-hydroxysteroid dehydrogenase, 17β-hydroxysteroid dehydrogenase and aldehyde reductase judging from their substrate specificities and inhibitor sensitivities (Penning *et al.*, 1984; Hara *et al.*, 1986; Sawada *et al.*, 1989). Bovine liver DD3, however, is a unique enzyme which can not be assigned to any known DD. In this paper, we demonstrated the further characterization of this unique enzyme; inhibition study, partial amino acid sequence, comparing with the property of DD1, namely, 3α-hydroxysteroid dehydrogenase (Nanjo *et al.*, 1992).

DD3 has a cysteine residue(s) for its activity which located at or nearby the coenzyme binding site. The same feature was found in DD1. The activities of DD3 and DD1 might be regulated by thiol/disulfide exchange reaction as same as the case of rat liver 3α-hydroxysteroid/dihydrodiol dehydrogenase (Terada *et al.*, 1991).

Additionally, DD3 was competitively inhibited by the steroids which were good substrates for DD1, i.e. androsterone and 5α-androstane-3,17-dione, and also inhibited by quinones, i.e. camphorquinone and 9,10-phenanthrenequinone (data not shown). Taking this result into consideration, the substrate binding sites of DD3 and DD1 may have high similarity to each other. It's expected that small differences in their active sites made different substrate specificity.

We determined the partial primary structures of DD3 and DD1 by cyanogen bromide hydrolysis and lysyl endopeptidase digestion.These partial amino acid sequences showed great homology with C-terminal sequences of rat liver 3α-hydroxysteroid/dihydrodiol dehydrogenase (54% in DD3 and 55% in DD1) and bovine lung prostaglandin F synthase (81% in DD3 and 75% in DD1) which are members of aldo/keto reductase family. Similarities in the amino acid sequences are also seen to those of human placenta (Chung *et al.*, 1989), bovine lens (Petrash *et al.*, 1989) and rat lens aldose reductases (Carper *et al.*, 1987) more than 50%. In addition, Penning and coworkers (1991) estimated the active site of rat 3α-hydroxysteroid/dihydrodiol dehydrogenase

```
DD3             QLEK
                ----
                ||||
PGFS 170       KQLEK 174
                ||||
R.DD 171       RQLEK 175


DD1                            KAIDGLNKNKRYYEFLV-V-HPEYPF
                               ||||||| |:|| | |: | ||||||
DD3             ENIQVFDFELTPEDMKAIDGLN-NMRYNELLLGVGHPEYPFVEEY
                ---------------------------------
                || ||||||||||||||||||| | || :  :|:||||||| |||
PGFS 277       KENMQVFDFELTPEDMKAIDGLNRNIRYYDFQKGIGHPEYPFSEEY 322
                ||  |||:||:||||||:||||||||| :          || ||  |
R.DD 278       KELTQVFEFQLASEDMKALDGLNRNFRYNNAKYFDDHPNHPFTDE  322
```

Figure 2. Comparison of partial amino acid sequences between DD3, DD1, prostaglandin F synthase and rat liver 3α-hydroxysteroid/dihydrodiol dehydrogenase. PGFS, bovine lung prostaglandin F synthase; R.DD, rat liver 3α-hydroxysteroid/dihydrodiol dehydrogenase. Solid and broken underlines indicate cyanogen bromide-hydrolyzed and lysyl endopeptidase-digested peptides of DD3 and DD1, respectively. Exact matches are signified with a |, and conserved substitutions are shown with a :. -, not identified.

with affinity labeling techniques by using the substrate analogues and demonstrated that its active site located neraby the C-terminal region. In order to characterize the differences in substrate specificity between DD3 and other aldo/keto reductases, the further study on the of active site of DD3 should be needed.

In conclusion, DD3 which is a unique dihydrodiol specific enzyme, as well as DD1, is a member of aldo/keto reductase family.

REFERENCES

Bolcsak, K. L., and Nerland, D. E., 1983, Purification of mouse liver benzene dihydrodiol dehydrogenase, *J. Biol. Chem.*, **258**:7252.

Carper, D., Nishimura, C., Shinohara, T., Dietzchold, B., Wistow, G., Craft, C., Kador, P., and Kinoshita, J.H., 1987, Aldose reductase and ρ-crystallin belong to the same protein superfamily as aldehyde reductase, *FEBS Lett.*, **171**:209.

Chung, S., and LaMendola, J., 1989, Cloning and Sequence determination of human placental aldose reductase gene, *J. Biol. Chem.*, **264**:14775.

Hara, A., Hasebe, K., Hayashibara, M., Matsuura, K., Nakayama, T., and Sawada, H., 1986, Dihydrodiol dehydrogenase in guinea pig liver, *Biochem. Pharmacol.* **35**:4005.

Hara, A., Taniguchi, H., Nakayama, T., and Sawada, H., 1990, Purification and properties of multiple forms of dihydrodiol dehydrogenase from human liver, *J. Biochem.*, **108**:250.

Matsuura, K., Hara, A., Nakayama, T., Nakagawa, M., and Sawada, H., 1987, Purification and properties of two multiple forms of dihydrodiol dehydrogenase from guinea-pig testis, *Biochim. Biophys. Acta*, **912**:270.

Nanjo, H., Terada, T., Umemura, T., Nishinaka, T., Mizoguchi, T., and Nishihara T., 1992, Characterization of bovine liver cytosolic 3α-hydroxysteroid dehydrogenase and its aldo-keto reductase activity, *Int. J. Biochem.*, **24**:815.

Nishinaka, T., Terada, T., Umemura, T., Nanjo, H., Mizoguchi, T., and Nishihara T., 1991, Study on dihydrodiol dehydrogenase (I) Molecular forms of the enzyme and the presence of a dihydrodiol specific enzyme in bovine liver cytosol, in "Enzymology and Molecular Biology of Carbonyl Metabolism 3," Weiner, H.,Wermuth, B., and Crabb,D.W.,eds., Plenum Press, New York, p.165.

Pawlowski, J. E., Huizinga, M., and Penning, T.M., 1991, Cloning and sequencing of cDNA for rat liver 3α-hydroxysteroid/dihydrodiol dehydrogenase, *J. Biol. Chem.*, **266**:8820.

Penning, T.M., Mukharji, I., Barrows, S., and Talalay, P., 1984, Purification and properties of a 3α-hydroxysteroid dehydrogenase of rat liver cytosol and its inhibition by anti-inflammatory drugs, *Biochem. J.*, **222**:601

Penning, T. M., Abrams, W. R., and Pawlowski, J. E., 1991, Affinity labeling of 3α-hydroxysteroid dehydrogenase with 3α-bromoacetoxyandrosterone and 11α-bromoacetoxyprogesterone, *J. Biol. Chem.*, **266**:8826.

Petrash, J. M., and Favello, A. D., 1989, Isolation and Characterization of cDNA clones encoding aldose reductase, *Curr.Eye Res.*, **8**:1021.

Platt, K. L., and Oesch, F., 1977, An improved synthesis of *trans*-5,6-dihydroxy-1,3-cyclohexadiene (*trans*-1,2-dihydroxy-1,2-dihydrobenzene), *Synthesis*, **7**:449.

Sawada, H., Hara, A., Nakagawa, M., Tsukada, F., Ohmura, M., and Matsuura, K., 1989, Separation and properties of multiple forms of dihydrodiol dehydrogenase from hamster liver, *Int. J. Biochem.*, **37**:2852.

Smithgall, T. E., Harvey, R. G. and Penning, T. M., 1988a, Stereoscopic identification of *ortho*-quinones as the products of polycyclic aromatic *trans*-dihydrodiol oxidation catalyzed by dihydrodiol dehydrogenase, *J.Biol.Chem.*, **263**:1814.

Smithgall, T. E., and Penning, T. M., 1988b, Electrophoretic and immunochemical characterization of 3α-hydroxysteroid/dehydrogenase of rat tissues, *Biochem. J.*, **254**:715.

Terada, T., Kohno, T., Samejima, T., Hosomi, S., Mizoguchi, T., and Uehara, K., 1985, Purification and properties of beef liver aldehyde reductase catalyzing the reduction of D-erythrose 4-phosphate, *J. Biochem.*, **97**:79.

Terada, T., Shinagawa, K., Umemura, T., Nishinaka, T., Hosomi, S., Nanjo H., Mizoguchi, T., and Nishihara T., 1991, Study on dihydrodiol dehydrogenase (II) modulation of dihydrodiol dehydrogenase activity by biological disulfides, in "Enzymology and Molecular Biology of Carbonyl Metabolism 3," Weiner, H., Wermuth, B., and Crabb,D.W., eds., Plenum Press, New York, p.177.

Vogel, K., Bentley, P., Platt, K-L., Oesch, F., 1980, Rat liver cytoplasmic dihydrodiol dehydrogenase, *J. Biol. Chem.*, **255**:9621.

Watanabe, K., Fujii, Y., Nakayama, K., Ohkubo, H., Kuramitsu, S., Kagamiyama, H., Nakanishi, S., and Hayaishi, O., 1988, Structual similarity of bovine lung prostaglandin F synthase to lens ε-crystallin of european common frog, *Proc. Natl. Acad. Sci. USA*, **85**:11.

CARBONYL REDUCTION BY 3α-HSD FROM *COMAMONAS TESTOSTERONI* - NEW PROPERTIES AND ITS RELATIONSHIP TO THE SCAD FAMILY

Udo C.T. Oppermann, Karl J. Netter and Edmund Maser

Department of Pharmacology and Toxicology, School of Medicine
Philipps University Marburg, D 355 Marburg, FRG

INTRODUCTION

The efforts to characterize the enzymes involved in carbonyl metabolism have led in the past decade to the establishment of several classes of proteins, e.g. the aldo-keto reductase superfamily or to several classes of alcohol dehydrogenases (Jörnvall et al., 1981; Persson et al., 1991; Krozowsky, 1992; Bohren et al., 1989, Wales and Fewson, 1990). The relationship or ancestry to procaryotic proteins was revealed (Baker, 1991; Bohren et al., 1989) and led to the conclusion that - in the case of the short chain alcohol dehydrogenase superfamily (SCAD) - some vertebrate-type cellular communication and signal transduction pathways already have procaryotic and lower eucaryotic analogs (Baker, 1992; Lenard, 1992) which might have evolved from the confrontation of procaryonts with hydrophobic molecules like flavonoids or perhydrocyclopentanophenanthrenes i.e. steroids (Baker, 1992; Karlson, 1983). Vertebrate-type steroid hormone action is a consequence of intracellular hormone receptor specificity and its availability in the cell, hormone-receptor interaction with its cis acting HRE's (hormone responsive elements) and for example in the case of mineralocorticoid action the existence of 11ß-hydroxysteroid dehydrogenases which regulate the intracellular active hormone level (Beato, 1989; Truss et al., 1992; Whorwood et al., 1992; Roy, 1992). This resembles in many aspects the situation in Rhizobia/plant signal transduction, where plant flavonoids and bacterial compounds act as signal molecules (Kondorosi, 1991; Baker, 1991). The complex process of signalling is achieved by different gene products, some of them e.g. NodG and fixR, belonging to the short chain alcohol dehydrogenase family. The relationship of these and other steroid dehydrogenases to other procaryotic and eucaryotic proteins and their membership to the SCAD family is established and underlines the importance of this protein family in uni- and multicellular physiology.

The present paper deals with new functional and structural properties of a 3α-hydroxysteroid dehydrogenase, which most likely belongs to the SCAD family, isolated from the gram negative bacterium Comamonas testosteroni*.

MATERIALS AND METHODS

Materials: 3α-(E.C.1.1.1.50) and 3ß-(1.1.1.51) hydroxysteroid dehydrogenase containing enzyme preparations (Pseudomonas testosteroni ATCC 11996) and steroid substrates were from Sigma, Deisenhofen, FRG or Boehringer Mannheim, FRG. 3α-hydroxy desogestrel and 3-oxo desogestrel were from Organon, Oss, The Netherlands, metyrapone was from Fluka, Buchs, Switzerland. α- and ß-ecdysone were kind gifts of J. Koolman, Inst. Physiol. Chem., Marburg. The respective 3-dehydro ecdysteroids were obtained by complete 3ß-hydroxy oxidation with 3ß-hydroxysteroid dehydrogenase from Comamonas testosteroni (E.C. 1.1.1.51), and isolated on preparative TLC with silica coated plates (60 F, Merck, Darmstadt, FRG). The 3-dehydroecdysteroid bands were visualized at 254 nm, scraped off and eluted with methanol. Trans-benzene dihydrodiol was a kind gift from K.L. Platt, Inst. Toxicol., Mainz, FRG, cis-benzene dihydrodiol was from Sigma, Deisenhofen, FRG. All other chemicals were of p.a. grade and obtained from Merck, Darmstadt, FRG.

Enzyme assays: Carbonyl reduction of metyrapone, nitroacetophenone and nitrobenzaldehyde as well as alcohol oxidation of metyrapol and nitrobenzalcohol were carried out in 50 μl volumes containing different concentrations of substrates, 90 μM cofactor (NADH or NAD+) and 10 mM sodium phosphate buffer, pH 7.4, incubated for 10-30 minutes at 25°C and diluted with the 3 fold volume of icecold acetonitrile and centrifuged. The supernatant was analyzed on a RP HPLC system with a C18 column (Serva, Heidelberg, FRG, 5x25 mm) and 30% (v/v) acetonitrile in 0.1% ammonium acetate, pH 7.0 as mobile phase. Metabolites were detected at 254 nm and the produced amount of metabolite was determined by comparison with known standards. Steroid substrate reactions were performed in 500 μl volumes and the absorbance change of cofactor at 340 nm was determined in a Kontron UV/VIS spectrophotometer. Steroid metabolite detection and identification with known standards on TLC (mobile phase: chloroform/ethylacetate 4:1 (v/v) or in the case of ecdysteroids chloroform/methanol 4:1 (v/v)) on silica plates with subsequent detection under UV light or spraying with concentrated H_2SO_4- were performed in order to exclude false results obtained from unspecific cofactor oxidoreduction. Kinetic data were determined by use of the Graph Pad PC software.

Growth of *Comamonas testosteroni* ATCC 11996 and induction: Comamonas testosteroni strain ATCC 11996 was obtained from the DSM (Deutsche Sammlung für Mikroorganismen, Braunschweig, FRG) and grown under the conditions described by Marcus and Talalay (Marcus and Talalay, 1956). Induction experiments were carried out with 300 mg/l steroid or with 100 mg/l of different polycyclic aromatic inducers. Cells were grown for 18 hours after inoculation and induction, centrifuged for 10 minutes at 10.000 x g and resuspended in 10 mM sodium phosphate buffer, pH 7.4. Disruption of cells was achieved with a Braun homogenisator (Braun, Melsungen, FRG) by use of glass pearls and cooling at 4°C for 1 minute. Cell debris was sedimented at 10.000 x g for 10 minutes and the resulting cell free extract was passed over a Sephadex G 25 column before further use in enzyme assays, SDS Page and Western blot.

SDS PAGE and Western blotting: SDS PAGE was carried out as described by Laemmli (Laemmli, 1970), transfer of separated proteins onto nitrocellulose (Schleicher&Schuell, Dassel, FRG) was achieved by use of the semi-dry technique (Kyhse-Anderson, 1984).

Native gel electrophoresis (NGE) and activity staining: Native gel electrophoresis was performed with the same buffer system as in SDS PAGE, except that SDS was omitted. Total acrylamide (T) was 8% in the separating gel. Activity staining and western blot were as earlier described (Maser et al. 1992). Antibodies against HSD 28 and MLMR were obtained as described earlier (Oppermann et al., 1992).

RESULTS AND DISCUSSION

Purification and identification of 3α-hydroxysteroid dehydrogenase as carbonyl reductase: Crude extracts of Comamonas testosteroni cells containing 3α-hydroxysteroid dehydrogenase were subjected to native gel electrophoresis. After completion of electrophoresis the border segments were cut off and stained for 3α-hydroxysteroid dehydrogenase and metyrapone reductase activity. Two main bands of 3α-hydroxysteroid dehydrogenase activity can be detected, carbonyl reduction of metyrapone takes place in the same region as 3α-hydroxysteroid dehydrogenase activity. Staining with Coomassie Blue shows a complex protein pattern. Western blotting and subsequent incubation with an antibody raised against the microsomal mouse liver metyrapone reductase (MLMR) (Maser and Netter, 1989) reveals that the immunoreactive band (band A) corresponds to one of the two bands detected by 3α-hydroxysteroid dehydrogenase and metyrapone reductase activity staining. The results of this experiment are shown in Figure 1. For further studies both bands in NGE were cut out from the gel, homogenized in a microcup and suspended in a small volume of 50 mM sodium phosphate buffer, pH 7.4. After 2 days at 4°C the slurry was transferred to a new tube, perforated at the bottom with a cannula and filled with glasswool. The enzyme containing suspension was centrifuged through the glasswool into a new tube. The resulting enzyme preparation was checked for homogeneity by SDS PAGE. Figure 2 shows the result and efficiency of this purification procedure: band A in NGE corresponds to a protein with a molecular mass of 28 kDa and is termed in the following HSD 28 and used in further experiments, band B corresponds to a protein with a molecular mass of 24 kDa (HSD 24). Furthermore, band A is the immunoreactive 3α-hydroxysteroid dehydrogenase discovered in NGE as revealed by subsequent immunoblotting with anti MLMR antibodies (Figure 2).

Functional properties of HSD 28: Enzyme assays performed with the HSD 28 fraction revealed that a wide variety of steroids with different A/B ring fusions, B ring conformations and several substituents could be substrates for reversible 3α-OH/3oxo conversions. 3oxo-4ene substrates like testosterone, progesterone and 3oxo-desogestrel could generally not be reduced, but surprisingly 3α-alcohol oxidation of 3α-OH-desogestrel could be doubtlessly carried out. Besides the vertebrate-type steroids several other naturally occurring steroids are substrates for HSD 28: The fungal antibiotic fusidic acid is efficiently oxidized, the insect type steroids 3-dehydro-ecdysone (3 dehydro-α-ecdysone) and 3-dehydro-20-OH-ecdysone (3-dehydro-ß-ecdysone) are reduced to the respective 3α-OH-ecdysteroids. No reactions are observed when HSD 28 was tested for oxidoreduction at other sites than C3 of the steroid nucleus. The positions tested include OH/oxo groups at C2, C7, C11, C17, C20, C21. Furthermore, no dihydrodiol dehydrogenase activity could be detected, neither with trans-benzene-dihydrodiol nor with cis- benzene-dihydrodiol. In this respect HSD 28 can be clearly distinguished from the indomethacin sensitive mammalian 3α-hydroxysteroid dehydrogenase (Pawlowski et al., 1991). Several xenobiotic carbonyl compounds like metyrapone, nitrobenzaldehyde and nitroacetophenone could be reduced, but the respective alcohol oxidation could only hardly be carried out or not at all. Table 1 summarizes the obtained data.

Inducibility of HSD 28: The broad substrate specificity of HSD 28 suggests a general role in degradation processes of different aromatic and alicyclic compounds. Comamonas testosteroni is able to grow on several organic carbon sources and can degrade them (Sondossi et al., 1992; Busse et al., 1992; Garcia-Valdes et al., 1988). In order to reveal any inducing effect of steroids or PAHs on the expression of HSD 28 induction experiments were carried out. Figure 3 summarizes the results.

The immunoblot shows that only testosterone, progesterone and lanosterol are able to induce an immunoreactive band as revealed with an antibody against HSD 28. The existence of a functionally active enzyme in the fractions containing immunoreactive material is revealed by performing enzyme assays of 3α-hydroxysteroid dehydrogenase activity and carbonyl reduction of metyrapone. This clearly indicates that the regulatory elements controlling the expression of HSD 28 have strict requirements for the inducer. A further interesting aspect concerning induction of ATCC 11996 cells is the fact that testosterone changes the ability of the bacterium to grow in the presence of the fungal steroid antibiotic fusidic acid, which is also metabolized by HSD 28 as mentioned above. Figure 4 gives the result.

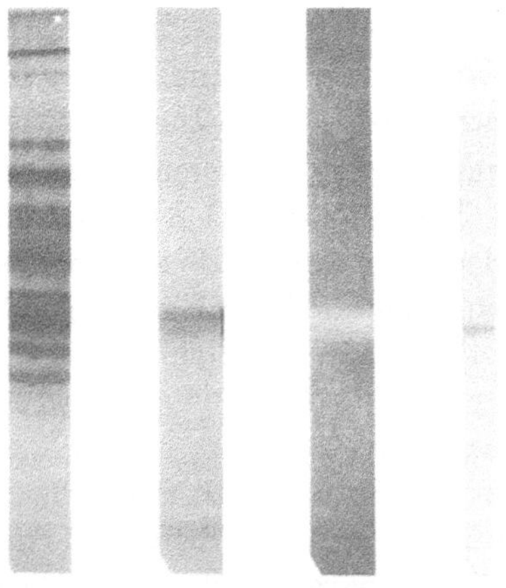

Figure 1: Native gel electrophoresis of extract from Comamonas testosteroni. Lane 1 gives the Coomassie stained protein pattern, lanes 2 and 3 show 3α-OH steroid dehydrogenase activity, lane 4 immunoblot with anti mouse liver metyrapone reductase antibodies.

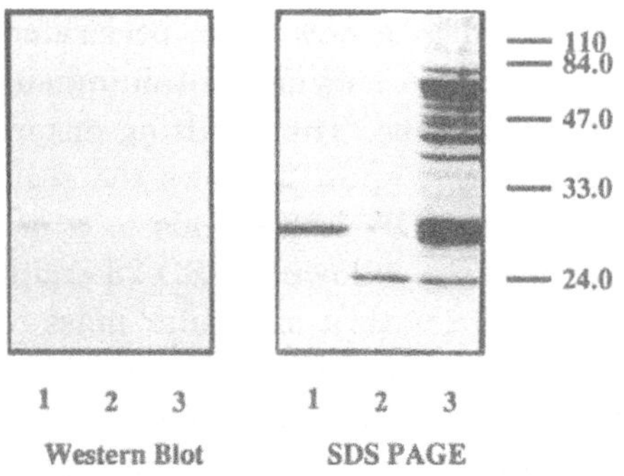

Figure 2: Purification and identification of 3α-OH steroid dehydrogenases from Comamonas testosteroni extract: Lane 1 is Band A from NGE (cf figure 1), lane 2 is Band B from NGE. Lane 3 is extract before NGE. A western blot with the same fractions was performed with an anti mouse liver metyrapone reductase antibody and shows crossreactions only at the molecular weight of 28 kDa.

Table 1

3-Oxosteroid reduction			3α-Hydroxysteroid oxidation		
Substrate	**Km**	**Vmax**	**Substrate**	**Km**	**Vmax**
androstandione	35.7	52.3	androsterone	24.4	10.3
-	-	-	epiandosterone	20.7	10.1
5α–DHT	16.1	48.4	androstandiol	32.8	41.7
5β-DHT	11.1	67.6	-	-	-
testosterone	n.a.	-	-	-	-
5α-pregnanolone	28.7	5.0	-	-	-
5β-pregnanolone	22.0	12.1	-	-	-
3-oxo desogestrel	n.a.	-	3α-OH desogestrel	15.2	8.6
progesterone	n.a.	-	-	-	-
-	-	-	cholic acid	30.1	10.2
-	-	-	fusidic acid	3.0	18.1
dehydro-α–ecdysone	25.0	1.0	-	-	-
dehydro-β-ecdysone	35.0	0.5	-	-	-
Xenobiotic carbonyl reduction			**Xenobiotic alcohol oxidation**		
Substrate	**Km**	**Vmax**	**Substrate**	**Km**	**Vmax**
metyrapone	610	0.63	metyrapol	2100	0.005
nitrobenzaldehyde	1090	0.05	nitrobenzalcohol	n.a.	
nitroacetophenone	800	0.01	-	-	-

n.a.: no activity detectable, Vmax: µmol*min-1*mg-1,-:not determined, Km: µmol/l, DHT: dihydrotestosterone

Control cells (uninduced, column 1) have an Ic 50 value of about 3 µM, i.e. at this concentration of fusidic acid 50 % of the cells survive compared to cells grown without antibiotic. Steroid induction during growth and all plating procedures changes the Ic 50 value more than 5 fold, i.e. at a concentration of around 16 µM fusidic acid 50% of cells can survive, a value at which uninduced cells cannot survive (column 4). The induction process must occur very fast in the bacterium, because presence or absence of steroid during dilution and plating procedures rapidly alters the values obtained for uninduced and induced cells (column 2 and 3). These results clearly show that steroid induction gives the bacterium the ability to survive in an environment which would be lethal for the uninduced phenotype.

Comparison of HSD 28 to procaryotic members of the SCAD superfamily: Cofactor requirement, molecular weight and N-terminal sequence analysis (data not shown) suggest that HSD 28 belongs to the short chain alcohol dehydrogenase family, like 3ß-hydroxysteroid dehydrogenase from Comamonas testosteroni (Yin et al., 1991). In order to reveal any similarities to other procaryotic members of this family,

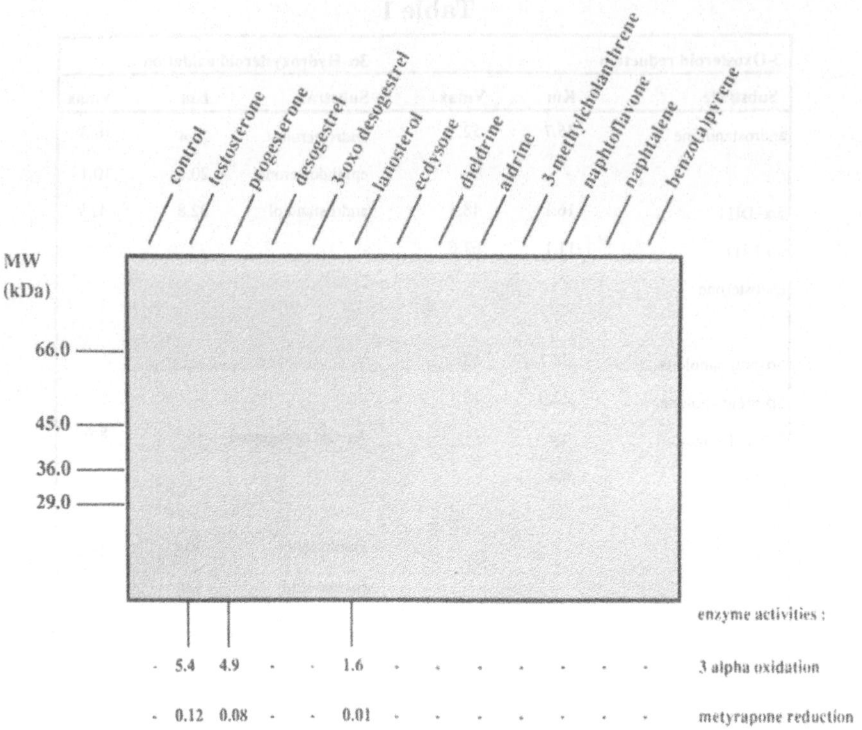

Figure 3: **Comamonas testosteroni cells ATCC 11996 were grown and prepared either with steroids (300 mg/l) or aromatic hydrocarbons (100 mg/l), dissolved in a small volume of methanol under the conditions described in Methods. Western blot with anti HSD 28 antibodies was performed, enzyme activities of the respective fractions are shown below the blot and given as specific activities in nmol/min/mg protein.**

structural and functional comparisons where performed as can be seen in figure 5.

The immunoblot analysis of HSD 28, 3ß-OH steroid dehydrogenase from Comamonas testosteroni ATCC 11996 and extracts of control and genistein induced extracts of Bradyrhizobium japonicum (which contain fixR, a member of the SCAD family) with antibodies against HSD 28 shows that no structural similarities exist in these fractions. Moreover they can be distinguished by their ability for carbonyl reduction of metyrapone. Metyrapone is reduced to its alcohol metabolite by HSD 28 and the extracts of Bradyrhizobia, but not by the 3ß-hydroxysteroid dehydrogenase from Comamonas testosteroni. HSD 28 mediated metyrapone reduction can be inhibited by steroids, in contrast to metyrapone reduction in Bradyrhizobia which is insensitive to steroid inhibition. These results confirm the diversity among the procaryotic members of the SCAD family. This becomes very distinct when HSD 28 is compared to 3ß-hydroxysteroid dehydrogenase from Comamonas testosteroni. A detailed comparison is given in table 2.

Both enzymes are expressed in ATCC 11996 cells when grown with steroids as inducers, both act on the 3-hydroxy group of the steroid nucleus but on the respective epimer. Interestingly the 3α- and 3ß-hydroxysteroid dehydrogenases from Comamonas testosteroni act sequentially in the conversion of ecdysteroids. Ecdysteroids are first converted by 3ß-hydroxysteroid dehydrogenase to 3-dehydroecdysteroids and these

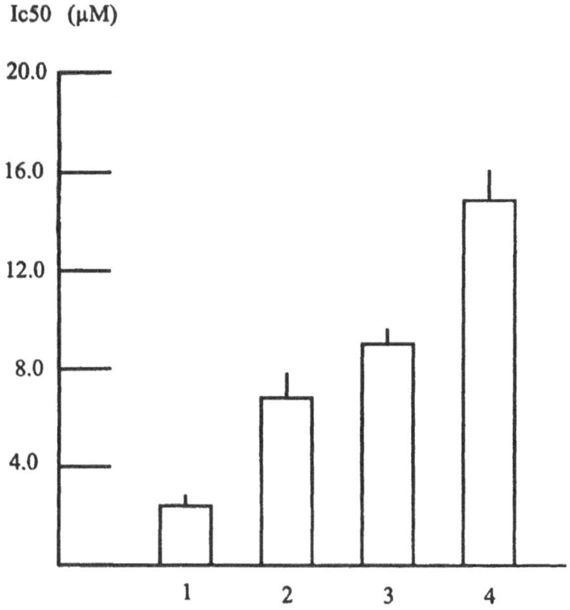

Figure 4: **Influence of steroid induction on the antibiotic resistance against fusidic acid in ATCC 11996 cells. Cells were grown in the absence of steroid (columns 1,2) or with testosterone (300 mg/l) (columns 3 and 4). Cells were plated in duplicate on agar plates with different concentrations of fusidic acid and the number of colonies was determined after growth for 36 hours at 30°C. Steroid was added (2,4) or omitted (1,3) through plating procedures and indicates that induction occurs very fast. (n=4)**

compounds are reduced by 3α-hydroxysteroid dehydrogenase to the 3-epi-ecdysteroid. The production of dehydroecdysteroids and 3-epi-ecdysteroids is considered to be the inactivation step of moulting hormone action in insects, although the first step, the production of dehydro-ecdysone is carried out in insects by ecdysone oxidase (Koolman, 1985; Thompson et al., 1985). Comparison of carbonyl reduction by HSD 28 and insect larvae homogenate reveals that this reaction is inhibited in both fractions by ecdysteroids in a mixed-type manner. The carbonyl reduction of dehydroecdysteroids by HSD 28 as well as structural homologies, revealed by Western blot, suggest that this important reaction is carried out in insects by a functionally and structurally related carbonyl reducing enzyme (Oppermann et al., 1992). This suggests an evolutionary relationship between HSD 28 and the carbonyl (dehydroecdysteroid reductase) from insects. Corresponding experiments with vertebrate liver microsomes also clearly show, that there are functional and structural homologies between HSD 28 and proteins in these fractions, i.e. anti HSD 28 antibodies reveal the same crossreaction pattern as anti MLMR antibodies, moreover metyrapone reduction present in these fractions is susceptible to steroid inhibition (Oppermann et al., 1992).

The results presented above lead to several conclusions: The substrate specificity of 3α-hydroxysteroid dehydrogenase from Comamonas testosteroni is not restricted to androstane or pregnane derivatives (Roe and Kaplan, 1969), but includes insect-type and fungal steroids. Moreover, carbonyl reduction of several xenobiotics is also carried out. This wide range of substrates suggests an important role for HSD 28 in the degradation of organic compounds which occur in the biosphere of the bacterium. This becomes clear when the reaction described above for ecdysteroids is considered. The

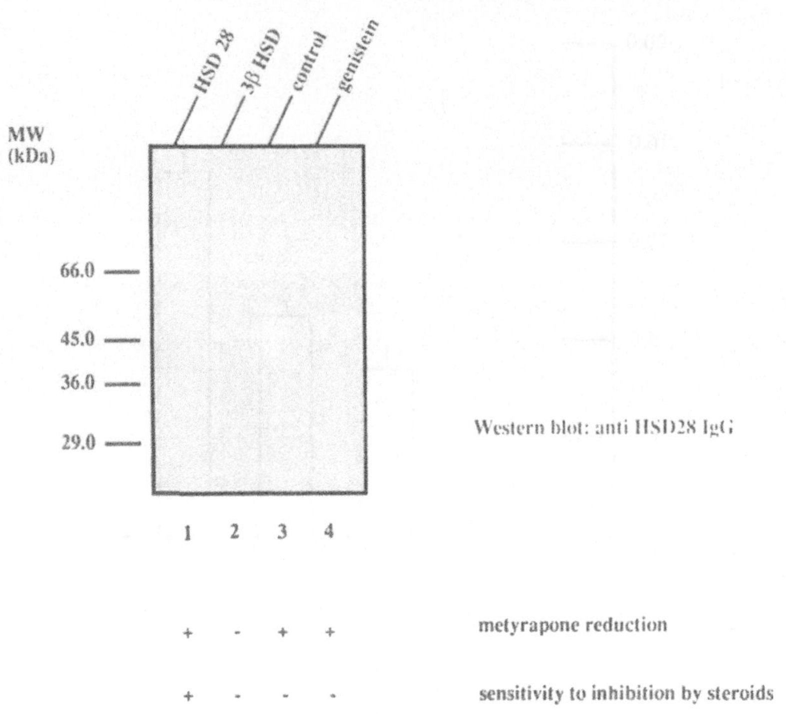

Figure 5: **Immunological and functional comparison of procaryotic members of SCAD family: HSD 28 (Lane 1), 3β HSD (Lane 2) and extracts of induced (4) and control (3) Bradyrhizobia jap. bacteria were separated, blotted and probed with an anti HSD 28 antibody. The blot clearly shows that no structural homologies to HSD 28 can be discovered. In the lower two rows the ability for metyrapone reduction is indicated.**

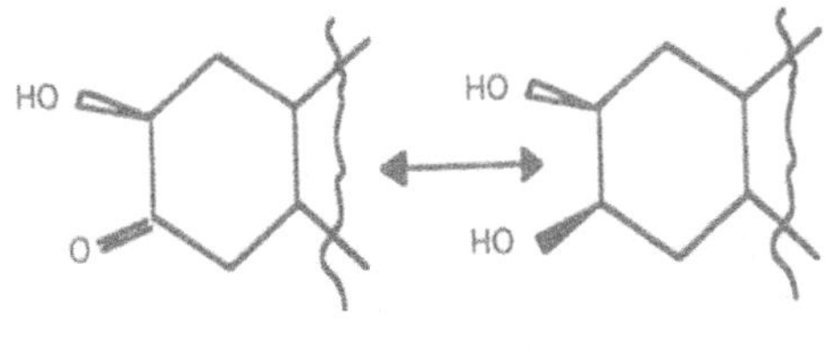

Figure 6: **Reaction carried out by HSD 28 at the steroid A ring of ecdysteroids. The same reaction occurs in the degradation of alicyclic hydrocarbons before ring cleavage.**

oxidoreduction of hydroxy/oxo groups adjacent to another hydroxy group is one of the important steps in the microbial degradation of alicyclic hydrocarbons and occurs before the cleavage of the alicyclic ring (Trudgill, 1990). The reaction is illustrated in Figure 6.

The ubiquitous occurrence of Comamonas testosteroni in heavily polluted marine sediments as well as in the gut of vertebrates and man (Garcia-Valdes et al., 1988, Barbaro et al., 1987) and its ability for degradation of a wide variety of organic matter proposes an important function for this bacterium and its enzymes involved in degradation processes for the respective environment.

The role of HSD 28 in the intracellular bacterial steroid signalling is not understood and needs further investigation. The structural and functional homologies

Table 2

	3α-HSD	3β-HSD
Molecular mass	28 kDa	26 kDa
Substrates	steroids	steroids
Enzyme activity	3α-OH/3-oxo conversion	3β-OH/3-oxo conversion
Xenobiotic carbonyl reduction:		
metyrapone	+	-
nitrobenzaldehyde	+	+
Immunoreactivity to:		
anti HSD 28 IgG	+	-
anti MLMR IgG	+	-
Steroid inducible	+	+
Member of SCAD family	probably	yes
Conversion of ecdysteroids	3-dehydro-ecdysteroids to 3-epi-ecdysteroids	ecdysteroids to 3-dehydro-ecdysteroids

with eucaryotic proteins as well as its membership to the SCAD family suggest a common ancestor for these proteins and might allow a more detailed classification of the short chain alcohol dehydrogenase family based upon structural and functional criteria.

Abbreviations

SCAD: short chain alcohol dehydrogenase; MLMR: microsomal mouse liver metyrapone reductase; HSD 28: 3α-hydroxysteroid dehydrogenase with molecular mass of 28 kDa; HSD 24: 3α-hydroxysteroid dehydrogenase with molecular mass of 24 kDa; α-ecdysone: 2ß, 3ß, 14,α, 22α, 25ß-pentahydroxycholest- 7-ene- 6-one; ß-ecdysone: 2ß, 3ß, 14,α, 20, 22α, 25ß-hexahydroxycholest- 7- ene-6- one; 3-epi-ecdysteroids: 2ß, 3α, 14, (20), 22, 25- hydroxycholest-7-ene-6-one; dehydroecdysone: 2ß, 3oxo, 14, 22, 15-hydroxycholest-7-ene-6-one; PAH: polycyclic aromatic hydrocarbon; androstandione: 5α, 3, 17, androstandione; androsterone: 5α, 3α, 17ß, androstanolone; epiandrosterone: 5ß, 3α, 17ß, androstanolone; androstandiol: 5α, 3α, 17ß androstandiol; pregnanolone: pregnan-20α-ol-3-one; desogestrel: 13-ethyl-11-methylene-18, 19-dinor-pregn-4-ene-20-yn-17-ol; cholic acid: 3α, 7α, 12α-trihydroxy-5ß-cholan-24-oic acid; fusidic acid: 16-(acetyloxy)-3α, 11α, dihydroxy-29-dammara-17(20), 24-diene-21-oic acid;

* Comamonas testosteroni (Tamaoka et al, 1987) instead of Pseudomonas testosteroni as introduced by Marcus and Talalay (1956) is used throughout the text.

Acknowledgements

The authors thank J. Koolman for fruitful discussions, M. Parniske and R. Kape for kind supply with Bradyrhizobia japonicum, and A. Bamberger and I. Bohl for excellent assistance in photographical work.

REFERENCES

Baker, M.E., 1991, Genealogy of regulation of human sex and adrenal function, prostaglandin action, snapdragon and petunia flower colors, antibiotics, and nitrogen fixation: functional diversity from two ancestral dehydrogenases, Steroids,56:354.

Baker, M.E, 1992, Evolution of regulation of steroid-mediated intercellular communication in vertebrates: Insights from flavonoids, signals that mediate plant-Rhizobia symbiosis, J.Steroid Biochem.Mol.Biol.,41(3-8):301.

Barbaro, D.J., Mackowiak, P.A., Barth, S.S., Southern, P.M., 1987, *Pseudomonas testosteroni* infections, Rev.Infect.Dis.,9, 124.

Beato, M, Gene regulation by steroid hormones, Cell,56:335.

Bohren, KM, Bullock, B, Wermuth, B, Gabbay, KH, 1988, The aldo-keto reductase superfamily, J. Biol. Chem.,264:9547.

Bradford, MM, 1976, A rapid and sensitive method for the quantitation of microgram quantities of protein utilizing the principle of protein-dye binding, Anal. Biochem.,72:248.

Busse, HJ, El-Banna, T, Oyaizu, H, Auling, G, 1992, Identification of xenobiotic degrading isolates from the beta subclass of the proteobacteria by a polyphasic approach including 16S rRNA partial sequencing, J.Syst.Bacteriol., 42(1):19.

Garcia-Valdes, E, Cozar, E, Rotger, R, Lalucat, J, Ursing, J, 1988, New naphtalene degrading Pseudomonas strains, Appl.Environm.Microbiol.,5410:2478.

Jörnvall, H, Persson, M, Jeffery, J, 1981, Alcohol and polyol dehydrogenases are both divided into two protein types, and structural properties cross-relate the different enzyme activities within each type, Proc.Natl.Acad.Sci.USA,78(7):4226.

Karlson, P, 1983, Why are so many hormones steroids?, Hoppe Seyler`s Z.Physiol.Chem.,364:1067.

Kondorosi, A, Overview on genetics of nodule induction: Factors controlling nodule induction by *Rhizobium meliloti*, in "Advances in molecular genetics of plant-microbe interactions 1", H. Hennecke and D.P.S. Verma eds., p. 111.

Koolman, J, 1985, Ecydsone oxidase, Meth. Enzymol., 111:419.

Krozowski, Z, 1992, 11-beta hydroxysteroid dehydrogenase and the short chain alcohol dehydrogenase (SCAD) superfamily, Mol.Cell.Endocrinol., 84:c25.

Kyhse-Anderson, J, 1984, Electroblotting of multiple gels. A simple apparatus without buffer tank for rapid transfer of proteins from polyacrylamide to nitrocellulose, J.Biochem.Biophys.Meth.,10:203.

Laemmli, UK, 1970, Cleavage of structural proteins during the assembly of the head of bacteriophage T4, Nature,227:680.

Lenard, J, 1992, Mammalian hormones in microbial cells, TIBS, 17(4):147.

Marcus, PI, Talalay, P, 1955, Induction and purification of α and ß hydroxysteroid dehydrogenases, J. Biol. Chem., 661.

Maser, E, Netter, KJ, 1989, Purification and properties of a metyrapone reducing enzyme from mouse liver microsomes-this ketone is reduced by an aldehyde reductase, Biochem.Pharmacol., 38(18):3049.

Maser, E, Oppermann, UCT, Bannenberg, G, Netter, KJ, 1992, Functional and immunological relationships between metyrapone reductase from mouse liver microsomes and 3α-hydroxysteroid dehydrogenase from Pseudomonas testosteroni, FEBS Lett., 297(12):196.

Oppermann, UCT, Maser, E, Hermans, JJR, Koolman, J, Netter, KJ, 1992, Homologies between enzymes involved in steroid and xenobiotic carbonyl reduction in vertebrates, invertebrates and procaryonts, J.Steroid.Biochem.Mol.Biol. (in press).

Pawlowski, J.E., Huizinga, M., Penning, T.M., 1991, Cloning and sequencing of the cDNA for rat liver 3α-hydroxysteroid dehydrogenase/dihydrodiol dehydrogenase, J.Biol.Chem., 266(14):8820.

Persson, B, Krook, M, Jörnvall, H, 1991, Characteristics of short-chain alcohol dehydrogenases and related enzymes, Eur. J. Biochem.,200:537.

Raimondi-Genti, S, Tolmasky, ME, Patrito, LC, Flury, A, Actis, LA, 1991, Molecular cloning and expression of the ß-hydroxysteroid dehydrogenase gene from Pseudomonas testosteroni, Gene, 105:43.

Ringold, HJ, Bellas, T, Clark, A, 1967, Adamantone as a probe for the dimensions and characteristics of the substrate binding pocket of certain alcohol dehydrogenases, Biochem. Biophys. Res. Commun., 27(3):361.

Roe, CR, Kaplan, NO, 1969, Purification and substrate specificities of bacterial hydroxysteroid dehydrogenases, Biochemistry, 8(12):5093.

Roy, AR, 1992, Regulation of steroid hormone action in target cells by specific hormone-inactivating enzymes, Proceedings of the Society for Experimental Biology, 199(3):265.

Sondossi, M, Sylvestre, M, Ahmad, D, 1992, Effects of chlorobenzoate transformation on the Pseudomonas testosteroni biphenyl and chlorobiphenyl degradation pathway, Appl.Environm.Microbiol., 58(2):485.

Tamaoka, J, Duk, H, Komagata, K, 1987, Reclassification of Pseudomonas acidovorans den Dooren de Jong 1926 and Pseudomonas testosteroni Marcus and Talalay 1956 as Comamonas acidovorans comb.nov. and Comamonas testosteroni comb.nov., with an amended description of the genus Comamonas, Int.J.Syst.Bacteriol., 37(1):52.

Thompson, MJ, Weirich, GF, Svoboda, JA, 1985, Ecdysone epimerase, Meth. Enzymol., 111:437.

Trudgill, PW, 1984, Microbial degradation of the alicyclic ring, in:" Microbial degradation of organic compounds", D.T. Gibson ed., Microbiology series, Vol. 13, M. Dekker, Inc., New York/Basel, p.131.

Truss, M, Chalepakis, G, Pina, B, Barettino, D, Brueggemeier, U, Kalff, M, Slater, EP, Beato, M, 1992, Transcriptional control by steroid hormones, J.Steroid Biochem.Mol.Biol., 41(3-8):241.

Wales, MR, Fewson, CA, 1990, Comparison of the primary structure of NAD(P)-dependent bacterial alcohol dehydrogenases, in "Enzymology and Molecular biology of Carbonyl metabolism 3",H. Weiner and T.G. Flynn, eds. Alan R. Liss, New York, p.193.

Whorwood, CB, Franklyn, JA, Sheppard, MC, Stewart, PM, 1992, Tissue localization of 11ß-hydroxysteroid dehydrogenase and its relationship to the glucocorticoid receptor, J.Steroid Biochem.Mol.Biol., 41(1):21

Yin, SJ, Vagelopoulos, N, Lundquist, G, Jörnvall, H, 1991, Pseudomonas 3ß-hydroxysteroid dehydrogenase, Eur. J. Biochem., 197:359.

SUBSTRATE SPECIFICITY OF ALCOHOL DEHYDROGENASES

Bryce V. Plapp, David W. Green, Hong-Wei Sun,
Doo-Hong Park and Keehyuk Kim

Department of Biochemistry
The University of Iowa
Iowa City, IA 52242

INTRODUCTION

Alcohol dehydrogenases have important physiological functions. Thus, it is important to determine the substrate specificities of these enzymes. Usually, a variety of substrates are chosen for a survey, and the most reactive substrates are identified by steady-state kinetic methods. Ethanol is a good substrate for the liver and yeast enzymes and is certainly one of the most important physiological substrates. However, these enzymes have broad, indeed promiscuous, specificities, acting on primary, secondary, and cyclic substrates, even steroids and hemiacetals. Thus, we should ask which compounds are the "normal" or "natural" substrates, and what good, important substrates have not been discovered? Some of these substrates may be involved in the pathogenesis of alcoholism.

The zinc-containing alcohol dehydrogenases are widely distributed, being found in animals, plants, fungi, and bacteria, and are part of a diverse family. As more alcohol dehydrogenases are isolated and sequenced, determining their specificities becomes more important, as there are multiple enzymes in a species (six so far identified in humans), and the physiological roles should be identified. Sequence alignments and evolutionary trees show that there are common structural features that have been conserved for over 1000 million years. Figure 1 presents an evolutionary tree, updated from Sun and Plapp (1992) to include 58 alcohol dehydrogenases. These enzymes have been subjected to selective pressure, and they must have important metabolic functions. Their substrate specificities differ since the amino acid residues in the active sites differ.

Since the three-dimensional structures of the horse EE and human $\beta\beta$ liver enzymes have been determined (Eklund and Brändén, 1987; Hurley *et al.*, 1991), it is possible to use molecular modeling to predict potential substrates. For instance, cyclohexanol derivatives have been docked into the active site of the horse enzyme, and the presumed mode of productive binding roughly correlates with the observed activities (Horjales and Brändén, 1985). Such studies are based on the current understanding of the determinants of specificity, which can be stated simply. The size and shape of the substrate binding pocket determine which substrates can fit, and the hydrophobic and polar interactions determine the affinity and specificity.

Enzymology and Molecular Biology of Carbonyl Metabolism 4
Edited by H. Weiner, Plenum Press, New York, 1993

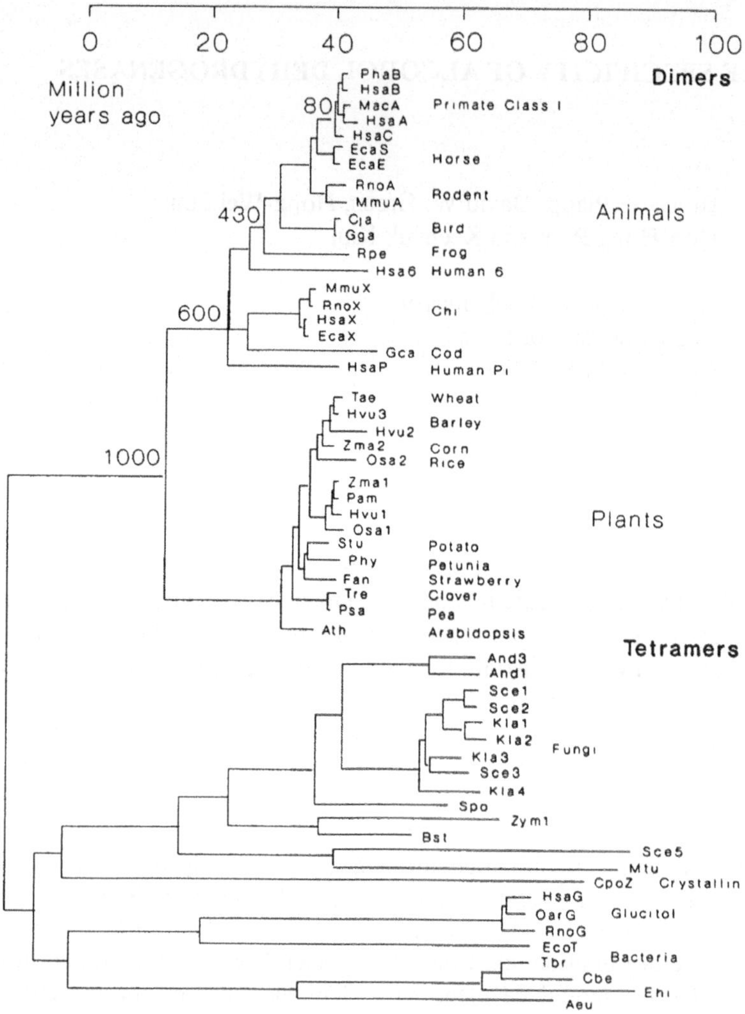

Figure 1. Phylogenetic tree for 57 members of the Zn-containing alcohol dehydrogenase family. The three letter abbreviations represent the first letter of the genus and the first two letters of the species. The fourth character represents the isoenzyme. The abbreviations for 47 sequences are given in Sun and Plapp (1992). The identities and sources for the new sequences are as follows: *Bst*, *Bacillus stearothermophilus* (Sakoda and Imanaka, 1992); *Cbe, Clostridium beijerinckii* (Rifaat and Chen, 1992); *Gca, Gadus callarius*, cod (Danielsson *et al.*, 1992); *Ehi, Entamoeba histolytica* (Kumar *et al.*, 1992); *Kla2, Kluyveromyces lactis* (Shain *et al.*, 1992); *Kla3, Kla4, K. lactis* (Saliola *et al.*, 1991); *Mtu, Mycobacterium tuberculosis bovis BCG* (Stelandre *et al.*, 1992); *Phy, Petunia hybrida* (Gregerson *et al.*, 1991); *Sce5, Saccharomyces cerevisiae* (Van der Linden *et al.* 1992).

Today, these concepts can be tested by using protein engineering to make directed substitutions of specific amino acid residues in the structures. In general, the results of such studies support the theories that direct interactions between the enzyme and substrate determine specificity. In addition, these studies show that reactivity can be affected indirectly by amino acid residues that do not directly contact the substrate, due to structural perturbations or effects on conformational changes that occur at various stages of the catalytic reaction. The purpose of this work is to illustrate how changes both at the active site and distant from it can affect substrate specificity.

MATERIALS AND METHODS

A set of C programs modified to run on a VAX 6410 were used to generate the progressive sequence alignment and the evolutionary tree (Feng and Doolittle, 1990).

The gene for yeast alcohol dehydrogenase I and the cDNA's for the horse liver EE and SS isoenzymes and the rhesus monkey liver α enzyme have been cloned and expressed from plasmids in yeast or *E. coli* (Bennetzen and Hall, 1982; Park and Plapp, 1991; Light *et al.*, 1992). Mutations were made by the oligonucleotide-directed, two primer method of Zoller and Smith (1984) using single-stranded DNA in M13 phage. The mutations were confirmed by sequencing the DNA and in most cases by protein sequence analysis or amino acid compositions. The enzymes were purified to homogeneity by precipitation with polyethylene glycol and chromatography on ion exchange and hydrophobic interaction media (Gould and Plapp, 1990; Park and Plapp, 1991; Light *et al.*, 1992). The concentration of active sites was determined by titration with NAD^+ in the presence of pyrazole. More than 50% of the subunits in the purified enzymes were capable of binding coenzyme.

Substrate specificities were studied by steady-state kinetics in a physiological buffer of 83 mM potassium phosphate and 40 mM KCl at pH 7.3 and 30 °C (Cornell, 1983). For such work, the concentration of coenzyme was fixed at a high, almost saturating level. In principle, the concentrations of both substrate and coenzyme should be varied so that true kinetic constants can be determined. V and K_m values were calculated with the programs of Cleland (1979).

Transient kinetics were studied with a BioLogic SFM3 stopped-flow instrument, which allows concentrations of substrates to be varied in a systematic way. The KINSIM and FITSIM programs (Barshop *et al.*, 1983; Zimmerle and Frieden, 1989) were used to estimate rate constants for binding and hydrogen transfer steps from the transient data.

The FRODO program (Jones, 1985) on an Evans and Sutherland PS350 was used for molecular modeling.

RESULTS AND DISCUSSION

Specificity of the α-Isoenzyme

Comparison of the amino acid sequences of the α-isoenzymes from human and rhesus monkey liver showed the interesting feature that Phe-93 (as found in other class I enzymes) in the substrate pocket was substituted with Ala (Ikuta *et al.*, 1986; von Bahr-Lindström *et al.*, 1986; Light *et al.*, 1992). The enzymes also have Gly-47 instead of Arg-47, which interacts with the pyrophosphate of the coenzyme. The enlarged substrate pocket should accommodate larger substrates, and kinetic studies showed that secondary alcohols were good substrates, with cyclohexanol being the best (Stone *et al.*, 1989; Light *et al.*, 1992). These observations agree with model building studies, which show that Ala-93 would provide a binding pocket into which cyclohexanol would fit well (Figure 2). Nevertheless, the α-isoenzyme was not reactive with 3-keto or 3β-hydroxy 5β-steroids. The steroids may be sterically hindered by Thr-48, which is replaced by Ser in the liver enzymes active on steroids. All of the liver alcohol dehydrogenases active on steroids have Ser at position 48.

Inverting the Specificity of Yeast Alcohol Dehydrogenase

Yeast alcohol dehydrogenase is most specific for small, primary alcohols. Ethanol is the best substrate, and longer chain alcohols have reduced activity. The *S. cerevisiae* isoenzyme I has no detectable activity on cyclohexanol or benzyl alcohol. Changing Met-294 to Leu in the cytoplasmic isoenzyme I altered the pattern of specificity on primary alcohols so that it resembled the pattern found for isoenzyme II (Ganzhorn *et al.*, 1987). The results can be explained by the molecular modeling of the active site, and thus they

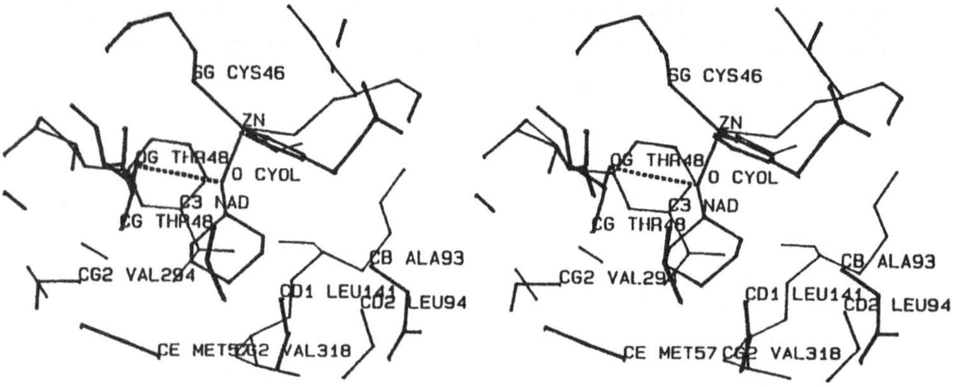

Figure 2. Stereoview of cyclohexanol bound in a productive orientation in a model of the active site of rhesus monkey α liver alcohol dehydrogenase (Light *et al.*, 1992). The hydroxyl group of cyclohexanol (*CYOL*) is ligated to the catalytic zinc (*ZN*) and hydrogen bonded (*dashed line*) to the hydroxyl group of Thr-48 (*OG*). The methyl group of Ala-93 is *CB ALA93*.

support the modeling, at least for the region near the catalytic zinc (Figure 3). Further inspection of the model suggests that it might be possible to convert the yeast enzyme into one that has a specificity similar to the monkey α enzyme. With the substitutions Thr48Ser, Trp93Ala, and Trp57Met, the pocket should be increased enough to accommodate longer chain alcohols and cyclohexanol.

The Trp93Ala mutation dramatically inverted the specificity of yeast alcohol dehydrogenase I on primary alcohols (Figure 4). Relative to the wild-type enzyme, the Trp93Ala enzyme has 350-fold lower activity on ethanol and 3-fold more activity on hexanol. Its specificity pattern resembles that of the monkey α enzyme. The altered specificity probably reflects the increased volume of the pocket. Ethanol could be bound in a variety of modes, reducing the probability of binding in a productive mode. Hexanol may fill the pocket better and make good contacts with the amino acid residue side chains. These interpretations are consistent with the general theory that the best substrate for an enzyme fits well into the binding pocket.

Although the Trp93Ala yeast enzyme has a substrate binding pocket enlarged enough to accommodate hexanol, it does not react well with cyclohexanol. The mutant yeast enzyme has a rate enhancement specificity (V/K_m) of 0.04 $M^{-1}s^{-1}$, whereas the values for the horse and monkey enzymes are 5,500 and 130,000 $M^{-1}s^{-1}$, respectively. Perhaps the pocket in the yeast enzyme is shaped like a bottle, with limited access through a bottleneck. The model of the yeast enzyme is somewhat uncertain because this enzyme is a tetramer and the arrangement of the subunits is not known. Furthermore, the yeast enzyme has a deletion of 21 amino acid residues corresponding to residues 119-139 in the liver enzyme. This loop forms part of the bottom of the substrate binding pocket.

Determinants of Specificity on Steroids

Although ethanol is a good substrate for yeast and liver alcohol dehydrogenases, much larger alcohols are also oxidized by various liver isoenzymes. The horse liver SS isoenzyme

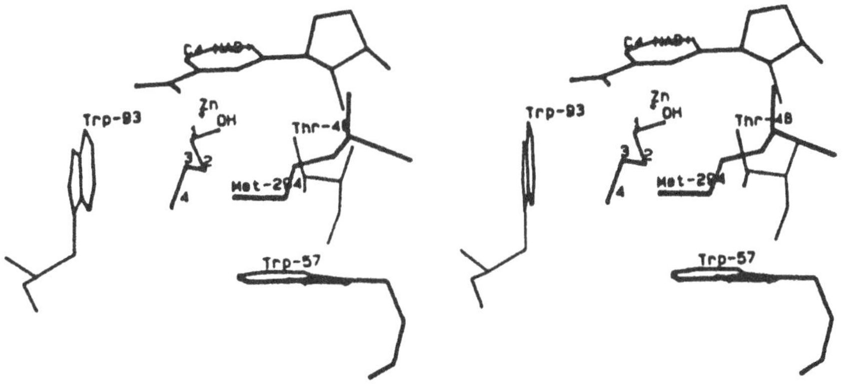

Figure 3. Model for the active site of yeast alcohol dehydrogenase. Butanol is bound in mode presumed to be productive for transfer of the pro-R hydrogen to C4 of the nicotinamide ring. The hydroxyl group (*OH*) is ligated to the zinc and hydrogen bonded to Thr-48. The side chains of Thr-48, Trp-57, Trp-93, and Met-294 appear to make very close interactions with carbons 3 and 4 of the butanol.

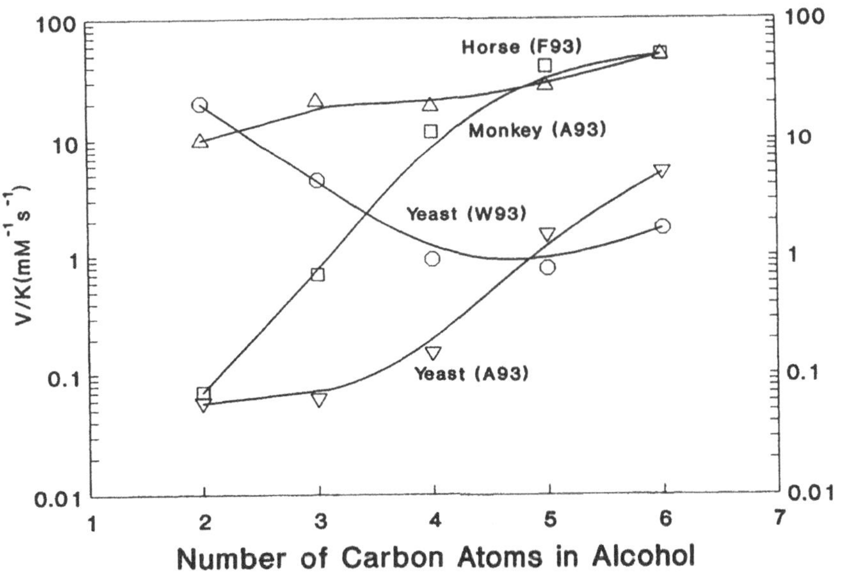

Figure 4. Specificity of alcohol dehydrogenases with aliphatic, primary alcohols from ethanol to hexanol. Data for the liver enzymes are taken from Light *et al.* (1992). Data for the wild-type yeast enzyme were obtained by Ganzhorn *et al.* (1987) and Green (1988), and those for the Trp93Ala mutant of the yeast enzyme by Sun (1992).

oxidizes 3β-hydroxysteroids, whereas the EE isoenzyme does not. Both isoenzymes are active on ethanol. Understanding the basis for the difference in specificity was facilitated by determining the complete amino acid sequence of the SS isoenzyme from the cloning and sequencing of the cDNA (Park and Plapp, 1991). The two isoenzymes differ in only 10 amino acid residues, and computer graphics modeling of the active site leads to an explanation for the differences in activity. A model of the EE isoenzyme with a steroid built into the active site is shown in Figure 5. The substrate binding pocket of the horse liver enzyme is shaped like a barrel or a cylinder and can bind large alcohols (Horjales and Brändén, 1985). The steroid is positioned so that hydrogen transfer could occur. However, the side-chain of Leu-116 appears to be too close to the steroid to permit binding in such a mode. Moving the side chain to a more favorable location requires changes in the peptide backbone of the protein. Although proteins are flexible, and side chains can easily move upon binding different substrates, the binding energy may not overcome the conformational energy, and productive binding may not occur.

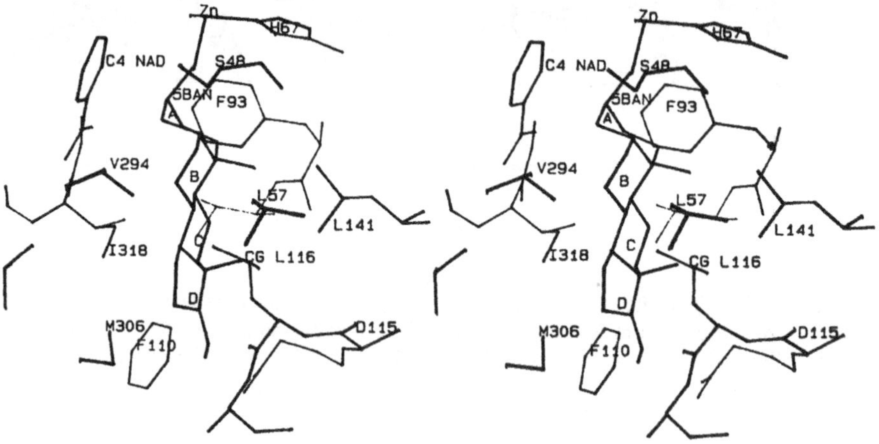

Figure 5. Model of the active site of the horse liver EE isoenzyme with the steroid 5β-androstane-3β,17β-diol (*5BAN*, rings labeled *A-D*) positioned for hydrogen transfer (Park and Plapp, 1991). The hydroxyl group of the steroid is linked to the zinc. Amino acid residues are identified by the one letter codes. The side chain of Leu-116 is too close to the steroid, but the other residues make favorable contacts.

Of the ten differences in the EE and SS isoenzymes, the most critical difference appears to be the deletion of Asp-115 in the SS isoenzyme. The deletion must change the peptide backbone, and the neighboring Leu-116 must be in a different position than it is in EE isoenzyme. This explanation has been supported by a directed mutation of the EE isoenzyme where Asp-115 is deleted (Park and Plapp, 1992). The mutated enzyme has about one-third of the catalytic efficiency of the SS isoenzyme on 5β-androstane-3β,17β-diol. Since some of the other substitutions could also contribute to the specificity, a series of chimeric enzymes were made which could reflect the evolution of catalytic specificity.

Conversion of EE isoenzyme to SS also includes the Thr94Ile, Arg101Ser, and Phe110Leu substitutions, which are all near the substrate binding pocket (see Figure 5). None of these residues directly contacts the steroid in the model, but they could affect activity indirectly through local structural changes. A chimeric enzyme (ESE) was produced that had the amino- and carboxyl-terminal sequences of the EE isoenzyme and the internal sequence of the SS isoenzyme, including the four substitutions close to the substrate pocket (94, 101, 110, and 115). Surprisingly, the ESE enzyme had 5-fold higher activity than the SS isoenzyme on the steroid. Furthermore, changing three residues in the amino terminal sequence of the SS isoenzyme (S → E: Gln17Glu, Ala43Thr, and Ala59Thr) or one residue in the carboxyl terminal region (Lys366Glu) increased the activity by 2-fold. These differences must be due to indirect effects of amino acid residues. Although changing amino acid residues that are in direct contact with the substrate may have the largest effects on activity, distant residues can indirectly modulate activity. Such effects may explain why yeast alcohol dehydrogenase II has a K_m for ethanol that is 1/10 of the K_m for isoenzyme I (Ganzhorn *et al.*, 1987).

The horse SS, human $\gamma\gamma$, and rat liver enzymes are active on steroids. However, comparison of the amino acid residues in their active sites suggests that the basis for this activity is different in each case. A common feature is Ser-48, and changing Thr-48 to Ser in the human $\beta\beta$ enzyme produced an enzyme with activity on steroids (Höög *et al.*, 1992). For these three natural and two mutant enzymes active on steroids, then, the common residues near the substrate are Ser-48, Leu-57, Phe-93, Leu-116, Val-294, and Met-306. Since the human enzymes have Asp-115, the explanation for the positioning of Leu-116 must be different. Perhaps Gly-117 in the human enzymes (as compared to Ser in the horse enzymes) allows more flexibility in the peptide backbone so that Leu-116 can move to a position that does not sterically hinder binding of steroid. Relative to the human enzymes, the rat liver enzyme has an insertion of one residue, four substitutions, and no residue corresponding to Asp-115 in the loop containing residues 112 to 119. These changes could also affect the positioning of Leu-116.

On Determining Rate Constants for Substrate Binding and Reaction

Substrate specificity is usually expressed in terms of V/K_m, which represents the overall bimolecular rate constant for binding of substrate, hydrogen transfer, and release of product. For the simplest mechanism:

$$\text{E-NAD}^+ \underset{k_{-2}}{\overset{k_2}{\rightleftharpoons}} \text{E-NAD}^+\text{-RCH}_2\text{OH} \underset{k_{-3}}{\overset{k_3}{\rightleftharpoons}} \text{E-NADH-RCHO} \underset{k_{-4}}{\overset{k_4}{\rightleftharpoons}} \text{E-NADH}$$

V/K_m is given by $k_2k_3k_4/(k_{-2}k_4 + k_{-2}k_{-3} + k_3k_4)$. This parameter is a measure of catalytic efficiency, and it is not affected by non-productive binding, which would decrease V and K_m proportionately. However, we should like to measure the affinity for the substrate and the rates of hydrogen transfer in the forward and reverse reactions.

Steady-state kinetics cannot yield rate constants for the isomerization of the central complex, except in special cases, but transient kinetics can provide the necessary data. Using high concentrations of enzyme, transient reactions in a stopped-flow instrument can be used to determine rate constants for binding and dissociation of coenzymes and substrates, and for the chemical interconversion steps.

In order to estimate all of the rate constants for the reaction shown above, we have studied the forward and reverse reactions (oxidation and reduction) with varied concentrations of substrates and fixed levels of coenzyme. Data for the reactions of benzyl alcohol and benzaldehyde are shown in Figure 6. Each curve can be analyzed to obtain a first order rate constant for the exponential phase, and, for benzyl alcohol oxidation, the zero order phase for the steady-state reaction. KINSIM and FITSIM were used to simulate and fit all of the progress curves simultaneously with the mechanism given above and a set of rate constants that best describe the data. (The programs originally were developed for a VAX computer, and we have modified them to run on IBM personal computers.) The fits are usually very good, giving small standard errors and similar values from different experiments. Different alcohols give different rate constants (Lee *et al.*, 1988; Sekhar and

Plapp, 1990). It should be noted, however, that additional steps could be kinetically significant, for instance proton transfer and conformational changes around the zinc when water is exchanged for substrate. A mechanism can always be more complicated than we initially propose.

Such studies (at pH 8 and 30 °C) show that the rate constants for hydride transfer for aliphatic primary alcohols are similar, 400 to 600 s^{-1}, and that the rates of dissociation of alcohol or aldehyde decrease as the length of the carbon chain increases. Thus, the dissociation constant for ethanol from the ternary complex with enzyme and NAD^+ is 2.3 mM, and K_d with 1-butanol is 8.6 μM. Cyclohexanol has a hydride transfer rate of 190 s^{-1}

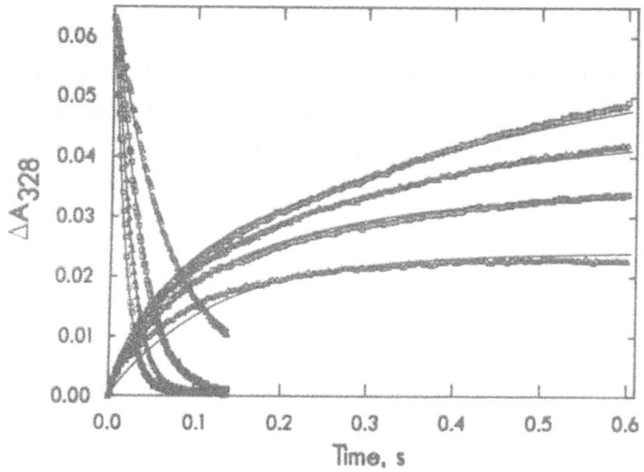

Figure 6. Transient kinetic studies of the oxidation of benzyl alcohol and reduction of benzaldehyde catalyzed by horse liver alcohol dehydrogenase, at 25 °C. The final concentrations in the 1 cm reaction cuvette were as follows: 11 μN enzyme in 46 mM sodium phosphate buffer, pH 7.0, 11 μM NADH and 0.013, 0.03, 0.06, and 0.12 mM benzaldehyde or 1 mM $NAD^+$and 0.05, 0.1, 0.2, and 0.4 mM benzyl alcohol. As the concentrations of substrates increase, the reactions went faster. Rate constants estimated with KINSIM/FITSIM were $k_2 = 3.1 \times 10^6\ M^{-1}s^{-1}$, $k_{-2} = 120\ s^{-1}$, $k_3 = 32\ s^{-1}$, $k_{-3} = 280\ s^{-1}$, $k_4 = 74\ s^{-1}$, and $k_{-4} = 2.9 \times 10^6\ M^{-1}s^{-1}$, with standard errors less than 10% of the values. The points are data and the lines are simulated.

and a K_d of 1.7 mM. Benzyl alcohol binds tightly, with a K_d of 16 μM, but has a relatively slow hydride transfer rate of 32 s^{-1}.

CONCLUSIONS AND PROSPECTS

The principles governing substrate specificity are being elucidated by x-ray crystallography, kinetics, and structure-function studies. Thus the prospects are good for predicting substrate specificities for enzymes when a three-dimensional structure is known or can be modeled by homology. Further progress will use calculations of the energetics of binding, the molecular dynamics of the enzyme and its complexes, and the quantum chemistry of the reaction (Tapia, 1988).

ACKNOWLEDGMENTS

This work was supported by Grants AA00279 and AA06223 from the National Institute on Alcohol Abuse and Alcoholism, United States Public Health Service. We thank Dr. Hans Eklund for assistance with the molecular modeling.

REFERENCES

Barshop, B. A., Wrenn, R. F., and Frieden, C., 1983, Analysis of numerical methods for computer simulation of kinetic processes: Development of KINSIM--A flexible, portable system, *Anal. Biochem.* 130:134-145.

Bennetzen, J. L., and Hall, B. D., 1982, The primary structure of the *Saccharomyces cerevisiae* gene for alcohol dehydrogenase, *J. Biol. Chem.* 157:3018-3025.

Cleland, W. W., 1979, Statistical analysis of enzyme kinetic data, *Methods Enzymol.* 63:103-138.

Cornell, N. W., 1983, Properties of alcohol dehydrogenase and ethanol oxidation *in vivo* and in hepatocytes, *Pharmacol. Biochem. Behav.* 18 Suppl. 1:215-221.

Danielsson, O., Eklund, H., and Jörnvall, H., 1992, The major piscine liver alcohol dehydrogenase has class-mixed properties in relation to mammalian alcohol dehydrogenases of classes I and III, *Biochemistry* 31: 3751-3759.

Eklund, H, and Brändén, C.-I., 1987, Alcohol dehydrogenase, *in* "Biological Macromolecules and Assemblies: Volume 3--Active Sites of Enzymes," F. A. Jurnak and A. McPherson, eds., John Wiley, New York.

Feng, D.-F., and Doolittle, R. F., 1990, Progressive alignment and phylogenetic tree construction of protein sequences, *Methods Enzymol.* 183:375-387.

Ganzhorn, A. J., Green, D. W., Hershey, A. D., Gould, R. M., and Plapp, B. V., 1987, Kinetic characterization of yeast alcohol dehydrogenases: Amino acid residue 294 and substrate specificity, *J. Biol. Chem.* 262:3754-3761.

Gould, R. M., and Plapp, B. V., 1990, Substitution of arginine for histidine-47 in the coenzyme binding site of yeast alcohol dehydrogenase I, *Biochemistry* 29:5463-5468.

Green, D. W., 1988, Substrate specificity and structure of alcohol dehydrogenase, Ph. D. Thesis, The University of Iowa, Iowa City IA.

Gregerson, R., McLean, M., Beld, M., Gerats, A. G., and Strommer, J., 1991, Structure, expression, chromosomal location and product of the gene encoding ADH1 in petunia, *Plant Mol. Biol.* 17:37-48.

Horjales, E., and Brändén, C.-I., 1985, Docking of cyclohexanol-derivatives into the active site of liver alcohol dehydrogenase using computer graphics and energy minimization, *J. Biol. Chem.* 260:15445-15451.

Hurley, T. D., Bosron, W. F., Hamilton, J. A., and Amzel, L. M., 1991, Structure of human $\beta_1\beta_1$ alcohol dehydrogenase: Catalytic effects of non-active-site substitutions, *Proc. Natl. Acad. Sci. USA* 88:8149-8153.

Höög, J.-O., Eklund, H., and Jörnvall, H., 1992, A single-residue exchange gives human recombinant ββ alcohol dehydrogenase γγ isoenzyme properties, *Eur. J. Biochem.* 205:519-526.

Ikuta, T., Szeto, S., and Yoshida, A., 1986, Three human alcohol dehydrogenase subunits: cDNA structure and molecular evolutionary divergence, *Proc. Natl. Acad. Sci. USA* 83:634-638.

Jones, T. A., 1985, Interactive computer graphics: FRODO, *Methods Enzymol.* 115:157-171.

Kumar, A., Shen, P.-S., Descoteaux, S., Pohl, J., Bailey, G., and Samuelson, J. C., 1992, Cloning and expression of the NADP-dependent alcohol dehydrogenase of *Entamoeba histolytica*, GenBank M88600.

Lee, K. M., Dahlhauser, K. F., and Plapp, B. V., 1988, Reactivity of horse liver dehydrogenase with 3-methylcyclohexanols, *Biochemistry* 27:3528-3532.

Light, D. R., Dennis, M. S., Forsythe, I. J., Liu, C.-C., Green, D. W., Kratzer, D. A., and Plapp, B. V., 1992, α-Isoenzyme of alcohol dehydrogenase from monkey liver: Cloning, expression, mechanism, coenzyme and substrate specificity, *J. Biol. Chem.* 267:12592-12599.

Park, D.-H., and Plapp, B. V., 1991, Isoenzymes of horse liver alcohol dehydrogenase active on ethanol and steroids: cDNA cloning, expression, and comparison of active sites, *J. Biol. Chem.* 266:13296-13302.

Park, D.-H., and Plapp, B. V., 1992, Interconversion of E and S isoenzymes of horse liver alcohol dehydrogenase. Several residues contribute indirectly to catalysis, *J. Biol. Chem.* 267:5527-5533.

Rifaat, M. M., and Chen, J.-S., 1992, Cloning and sequence analysis of a gene encoding a medium-chain zinc-containing alcohol dehydrogenase from the Gram-positive anaerobe *Clostridium beijerinckii* NRRL B593, GenBank M84723.

Sakoda, H., and Imanaka, T., 1992, Cloning and sequencing of the gene coding for alcohol dehydrogenase of *Bacillus stearothermophilus* and rational shift of the optimum pH, *J. Bacteriol.* 174: 1397-1402.

Saliola, M., Gonnella, R., Mazzoni, C., and Falcone, C., 1991, Two genes encoding putative mitochondrial alcohol dehydrogenases are present in the yeast *Kluyveromyces lactis, Yeast* 7:391-400.

Sekhar, V. C., and Plapp, B. V., 1990, Rate constants for a mechanism including intermediates in the interconversion of ternary complexes of horse liver alcohol dehydrogenase, *Biochemistry* 29:4289-4295.

Shain, D. H., Salvadore, C., and Denis, C. L., 1992, Evolution of the alcohol dehydrogenase (ADH) genes in yeast: Characterization of a fourth ADH in *Kluyveromyces lactis, Mol. Gen. Genet.* 232:479-488.

Stelandre, M., Bosseloir, Y., Maes, P., De Bruyn, J., and Content, J., 1992, Alcohol dehydrogenase gene of *Mycobacterium bovis* BCG: Cloning, sequence analysis and expression in *Escherichia coli*, GenBank X63450.

Stone, C. L., Li, T.-K., and Bosron, W. F., 1989, Stereospecific oxidation of secondary alcohols by human alcohol dehydrogenases, *J. Biol. Chem.* 264:11112-11116.

Sun, H.-W., 1992, Evolution and substrate specificities of alcohol dehydrogenases, Ph. D. Thesis, The University of Iowa, Iowa City IA.

Sun, H.-W., and Plapp, B. V., 1992, Progressive sequence alignment and molecular evolution of the Zn-containing alcohol dehydrogenase family, *J. Mol. Evol.* 34:522-535.

Tapia, O., 1988, Theoretical chemistry studies on the catalytic mechanism of liver alcohol dehydrogenase, *J. Mol. Catal.* 47:199-210.

Van Der Linden, C. G., Maurer, C. T. C., Planta, R. J., and Van Vliet-Reedijk, J. C., 1992, Hypothetical zinc-type alcohol dehydrogenase-like protein in HMR 3'-region, GenBank P25377.

von Bahr-Lindström, H., Höög, J.-O., Hedén, L.-O., Kaiser, R., Fleetwood, L., Larsson, K., Lake, M., Holmquist, B., Holmgren, A., Hempel, J., Vallee, B. L., and Jörnvall, H., 1986, cDNA and protein structure for the α subunit of human liver alcohol dehydrogenase, *Biochemistry*, 25-2465-2470.

Zimmerle, C. T., and Frieden, C., 1989, Analysis of progress curves by simulations generated by numerical integration, *Biochem. J.* 258:381-387.

Zoller, M. J., and Smith, M., 1984, Oligonucleotide-directed mutagenesis: A simple method using two oligonucleotide primers and a single-stranded DNA template, *DNA* 3:479-488.

THE INFLUENCE OF PH ON THE SUBSTRATE SPECIFICITY AND STEREOSELECTIVITY OF ALCOHOL DEHYDROGENASE FROM HORSE LIVER

Hans W. Adolph, Martin Kiefer
and Michael Zeppezauer

FR 12.4 Biochemie
Universität des Saarlandes
W-6600 Saarbrücken
Federal Republic of Germany

INTRODUCTION

Alcohol dehydrogenase from horse liver (HLADH) catalyzes the reversible oxidation of primary and secondary alcohols. The substrate specificity of the EE isozyme is very broad and has been characterized in depth (for review see Jones and Beck, 1976).

HLADH has been frequently used as an enantioselective catalyst for organic syntheses. For most substrates its stereoselectivity is very high and exclusively *pro-R* for primary alcohols with the exception of 1-octanol for which Shapiro et al. (1983) could show that the (*pro-S*) hydrogen atom can be exchanged also in a HLADH-NAD+/NADH-diaphorase recycling system in D_2O at a rate of 2 - 4% that found for (*pro-R*) exchange. For some prochiral cyclic ketones the stereoselectivity of HLADH is low resulting in both enantiomeric alcohols as products (Graves et al., 1965; Willaert et al., 1988; see also Jones and Beck, 1976).

The smallest substrates showing a discrimination of enantiomeric forms are represented by (*S*)-2-butanol and (*R*)-2-butanol. Both are substrates of the horse liver enzyme (EE isozyme) but only the (*S*)-enantiomer is a substrate for ADH from yeast as first shown by Dickinson and Dalziel (1967). We could show that the reduction of 2-butanone results in both alcohols with (*S*)-2-butanol as the major product (Adolph et al., 1991). A graphic representation of the configuration of the productive ternary complex HLADH-NADH-ketone is shown in Figure 1.

The enzymatic determination of product composition has taken advantage of the fact that yeast ADH exclusively oxidizes (*S*)-2-butanol while HLADH oxidizes both enantiomeric alcohols.

Since the stereospecificity for the system (*S*)-2-butanol, (*R*)-2-butanol/2-butanone is low with an approximately ten-fold excess of the (*S*)-alcohol over (*R*)-2-butanol as the products of 2-butanone reduction it can be regarded as a very sensitive system to study the influences of external parameters on the stereoselectivity of the enzyme.

Enzymology and Molecular Biology of Carbonyl Metabolism 4
Edited by H. Weiner, Plenum Press, New York, 1993

Using a combination of steady-state and stopped-flow kinetic experiments and comparisons of free energies of activation and binding for a systematically varied set of substrates, we were able to determine the complete set of rate constants of the kinetic mechanism shown in Figure 2. It is an ordered bi-bi mechanism for a pair of enantiomeric alcohols and the corresponding ketone which can bind in two different ternary complex configurations resulting in two different alcohols as the products.

The corresponding data for pH 8.5 (Adolph et al., 1991) were used to show which single steps of the catalytic mechanism govern the substrate specificity and stereoselectivity of HLADH. In contradiction to Dickinson and Dalziel (1967) it became evident that the hydride transfer step is not generally rate-limiting for the oxidation of secondary

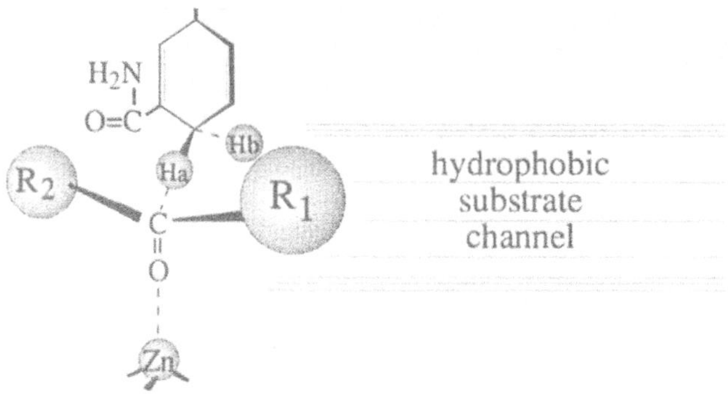

Figure 1. Graphic representation of the configuration of productive ternary complexes HLADH-NADH-ketone. R_1 and R_2 are alkyl residues. In the case of an aldehyde, R_2 is hydrogen. For 2-butanone in a (*pro-S*)-configuration R_1 corresponds to $-C_2H_5$ and R_2 to CH_3. For the (*pro-R*)-configuration, R_1 = $-CH_3$ and R_2 = $-C_2H_5$.

alcohols. For the reverse reaction the hydride transfer step is rate-limiting. The stereoselectivity of 2-butanone reduction is mainly determined by a difference of the free energies of binding for the two different configurations of the HLADH-NADH-ketone ternary complex. The fact that 2-butanone, bound weaker in a (*pro-R*)-configuration, shows a lower free energy of activation as compared to the (*pro-S*)-configuration, reduces slightly the selectivity effect due to the different binding constants. This result finds its counterpart on the alcohol side of the reaction scheme. The weaker bound (*R*)-2-butanol is oxidized at a 6-fold increased rate as compared to (*S*)-2-butanol.

In the past, several studies concerning external influences on substrate specificity and stereoselectivity of alcohol dehydrogenases like pH, temperature, dielectricity of the solvent have been published. For the case of HLADH Willaert et al. (1988) stated that a change of pH would change the stereoselectivity of the reduction of 3-cyano-4,4-dimethylcyclohexanone.

It was our aim to find out whether the stereoselectivity of 2-butanone reduction were influenced by a pH variation and how the discrimination of substrates were affected. Thus the complete set of experiments with native HLADH at pH 8.5 (Adolph et al., 1991) was reinvestigated at pH 7.0 also.

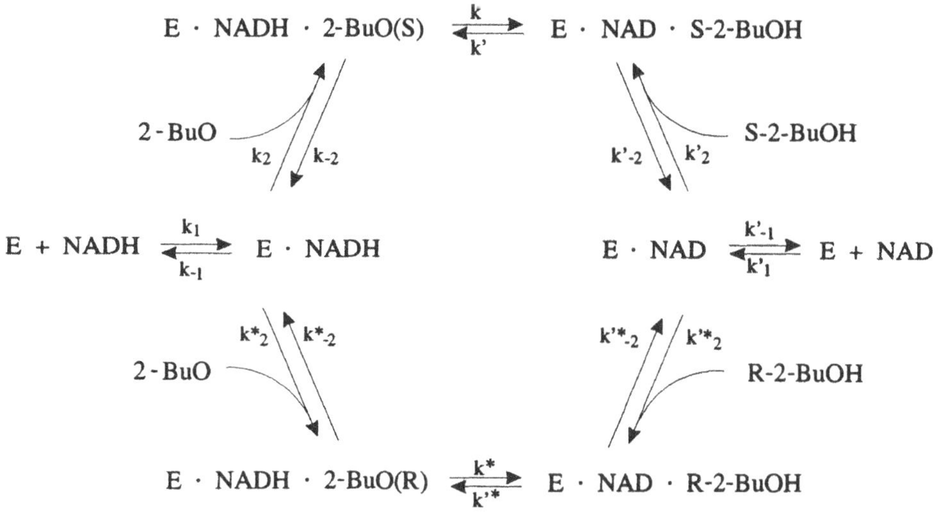

Figure 2. Ordered bi-bi mechanism for the system (*S*)-2-butanol, (*R*)-2-butanol/2-butanone. 2-BuO, 2-butanone; 2-BuO(S), (*pro-S*)-configuration of 2-butanone; 2-BuO(R), (*pro-R*)-configuration of 2-butanone; S-2-BuOH, (*S*)-2-butanol; R-2-BuOH, (*R*)-2-butanol.

Some results obtained at pH 7.0 were confusing in comparison to the data at pH 8.5 and a partial reinvestigation of data at pH 8.5 was necessary. Especially acetone and isopropanol have to be fractionally distilled several times to reach a purity which allows determinations of reliable data. As a consequence of the reinvestigation differing values for k_{cat} and the hydride transfer rate constant of acetone were obtained and some dependent rate constants, obtained from calculations were also changed. Most changes are minor and the overall effect of the concerting rate constants is not influenced significantly. Thus all conclusions drawn remain valid and can be extended to the new data found for pH 7.0. It is now possible to show which single steps of the catalytic mechanism are affected by a pH variation and to what extent substrate specificity and stereoselectivity are influenced.

MATERIALS AND METHODS

Rapid kinetic measurements were carried out on a stopped-flow photometer system (model SF-51 from HiTech Scientific). Data processing was performed on an IBM-compatible personal computer using the program IS.1 from HiTech. In all kinetic experiments the reactions were detected via absorbance measurements at 340 nm in steady-state experiments and at the isosbestic point of free and enzyme-bound NADH at 328 nm in single-turnover experiments, respectively. Experimental procedures and the theoretical background for the calculation of the individual rate constants from stopped-flow and steady-state kinetic data has been described in detail (Adolph et al., 1991).

Alcohol dehydrogenase from horse liver was purified as follows: Fresh horse liver was homogenized and extracted with 50 mM K-phosphate pH 7.0. After centrifugation a heat denaturation at 50°C for 15 min was performed with the supernatant followed by a $(NH_4)_2SO_4$ precipitation (40% saturation at 4°C). After centrifugation further $(NH_4)_2SO_4$ was added to give 70% saturation. After centrifugation the precipitate was dissolved in 10 mM TRIS/HCl pH 8.2 and dialyzed against the same buffer. HLADH was applied to DEAE cellulose equilibrated and immediately eluted with the same buffer. After dialysis against 5 mM K-phosphate pH 7.0, the enzyme was loaded on a CM sepharose column equilibrated and eluted with the same buffer to remove unbound proteins. Subsequent elution with 15 mM K-phosphate pH 7.0 gave the major EE-isoenzyme which was concentrated by ultrafiltration and crystallized twice as described earlier (Adolph et al., 1991).*

Concentrations of the enzyme have been determined by active site titrations according to Einarsson et al. (1976). The coenzymes NAD+ and NADH were purchased from Boehringer/Mannheim in the highest available quality. The substrates were also obtained in the highest available quality but had to be further purified by fractionated distillation to reduce unspecific side reactions. These side reactions are due to impurities (possibly primary alcohols in the secondary alcohols and aldehydes in the case of ketones) which are also substrates of the enzyme and metabolized at a much higher rate as compared to the substrates under study. In the case of acetone these impurities could not be totally removed resulting in a biphasic steady-state and transient state behaviour. A real steady-state due to pure acetone reduction was reached after 15 min of reaction time under the experimental conditions used. The single-turnover experiments had to be evaluated using fits of a biexponential function. These difficulties made it necessary to reinvestigate the corresponding data for this substrate at pH 8.5 also.

RESULTS AND DISCUSSION

In Table 1 the experimentally obtained kinetic data for the different substrates investigated are listed and data for pH 7.0 and 8.5 are compared.

All rate constants of an ordered bi-bi mechanism can be determined, provided the following data are known: k_{cat} and K_m for forward and reverse reaction, the dissociation rate constants of NAD+ (k_{-1}') and NADH (k_{-1}) from the enzyme-coenzyme binary complex and the rate constants of the hydride transfer step ($k^{(\prime)}$) for both directions of the reaction. From the definitions of k_{cat} and K_m in terms of rate constants of an ordered bi-bi mechanism obtained by Plapp (1973) the following equations for the association and dissociation rate constants of substrates can be obtained:

$$k_{-2} = -(k'_{cat}k'k_{-1} + k'_{cat}kk_{-1})/(k'_{cat}k' + k'_{cat}k_{-1} - k'k_{-1}). \quad (1)$$

$$k'_{-2} = -(k_{cat}kk'_{-1} + k_{cat}k'k'_{-1})/(k_{cat}k + k_{cat}k'_{-1} - kk'_{-1}). \quad (2)$$

$$k_2 = k'_{-1}(k'_{-2}k + k_{-2}k'_{-2} + k'k_{-2})/K_m(kk'_{-2} + kk'_{-1} + k'k'_{-1} + k'_{-1}k'_{-2}). \quad (3)$$

$$k'_2 = k_{-1}(k'_{-2}k + k_{-2}k'_{-2} + k'k_{-2})/K'_m(k'k_{-2} + k'k_{-1} + kk_{-1} + k_{-1}k_{-2}). \quad (4)$$

The dissociation rate constant of NADH in 0.1 M TES pH 7.0 has been determined in stopped-flow experiments detecting the quenching of protein fluorescence due to coenzyme binding. The plot of apparent rate constants versus NADH concentration is linear as indicative for a second order reaction resulting in a dissociation rate constant of

*From this column two additional isoenzymes can be eluted: the heterodimeric ES isoenzyme with 20 mM K-phosphate pH 7.0, which can be further purified by crystallization, and the SS isoenzyme with 25 mM K-phosphate, which has to be further purified by affinity chromatography on AMP-sepharose as described by Andersson et al. (1974). The SS-isoenzyme obtained in this way has been used by Hubatsch et al. to determine the amino acid sequence of the protein (this volume).

NADH of 3.8 s^{-1} as ordinate intercept. The dissociation rate constant of NAD^+ was assumed to be 90 s^{-1} as adopted from Sekhar and Plapp (1988). It was verified by model calculations that changes of coenzyme dissociation rate constants in the range of 20% do not affect the calculated data significantly.

In single turnover experiments it is possible to detect the ternary complex interconvertion step selectively. The rate constants obtained from substrate concentration-dependent measurements were ascribed to the hydride transfer step (Adolph et al., 1991).

With the thus obtained kinetic constants it is possible to calculate the association and dissociation rate constants of substrates according to equations (1-4) for normal substrate pairs (e. g. isopropanol-acetone, 3-pentanol-3-pentanone). This simple calculations are not sufficient to determine all rate constants involved in Figure 2 for the system (*S*)-2-butanol, (*R*)-2-butanol/2-butanone. We could show that the experimentally observed rate constants of 2-butanone reduction (steady-state and single-turnover kinetics) are mainly due to the formation of (*S*)-2-butanol proceeding from a HLADH-NADH-2-butanone(*pro-S*) ternary complex (compare Figure 1). (*R*)-2-butanol is produced in a minor amount and the contribution of its formation to the observed rate constants can be ignored in a first approximation.

By comparison of structurally related substrates (acetone, 2-butanone, 3-pentanone) it could be shown that a systematic variation of the substrate structure results in a systematic variation of free energies of activation for the hydride transfer step and the free energies of binding for the formation of a HLADH-NADH-ketone ternary complex. This finding allows us to calculate the free energies of activation and binding for the reduction of 2-butanone from a (*pro-R*)-configuration, which are experimentally not accessible, according to equation 5.

$$\Delta G^0[\text{2-butanone}(pro\text{-}R)] = \Delta G^0[\text{acetone}] + (\Delta G^0[\text{3-pentanone}] - \Delta G^0[\text{2-butanone}(proS)]) \quad (5)$$

With the free energies obtained in this way the complete set of rate constants shown in Figure 2 can be determined. The reliability of this method for the studied substrates has been investigated in depth (Adolph et al., 1991).

It could be additionally shown that the ratio of product concentrations (c) obtained from 2-butanone reduction can be calculated directly from the single rate constants using a calculation of apparent k_{cat} and K_m values (with the corresponding equations from Plapp (1973)) and application of equation (6).

$$c[(S)\text{-2-butanol}]/c[(R)\text{-2-butanol}] = (k_{cat}/K_m)_{[(pro\text{-}S)\text{-2-butanone}]}/(k_{cat}/K_m)_{[(pro\text{-}R)\text{-2-butanone}]} \quad (6)$$

The product compositions obtained are identical to those obtained from numerical simulations using the scheme in Figure 2 and in good agreement with experimentally obtained data.

The product compositions are 93.0% (*S*)-2-butanol and 7.0% (*R*)-2-butanol at pH 7.0 and 92.6% (*S*)-2-butanol and 7.4% (*R*)-2-butanol at pH 8.5, respectively. These results indicate that there is virtually no effect of a pH variation on the stereoselectivity of 2-butanone reduction catalyzed by HLADH.

In Table 2 the substrate specific constants of the ordered bi-bi mechanism for the substrates investigated are shown and the data for pH 7.0 and 8.5 are directly compared. It is well-known that an ordered bi-bi mechanism is only a simplified model for the kinetic mechanism of HLADH. At fixed conditions (pH, temperature) it describes sufficiently well the kinetic behaviour, the substrate specificity and the stereoselectivity of the enzyme. A change of external parameters, however, can reveal that a particular reaction step of the ordered bi-bi mechanism can be composed of several distinguishable single steps. A detailed analysis of the presented data allows some far-reaching conclusions concerning the kinetic mechanism of the enzyme.

Table 1. Experimental data obtained with different secondary alcohols and the corresponding ketones for LADH. k_{cat} and K_m are the results of steady-state experiments; k(H) is the rate constant for the ternary complex interconversion, obtained from single-turnover experiments.

substrate	k_{cat}		K_m		k_{cat}/K_m		k(H)	
	pH 7.0	pH 8.5	pH 7.0	pH 8.5	pH 7.0	pH 8.5	pH 7.0	pH 8.5
	s^{-1}	s^{-1}	mM	mM	$M^{-1}s^{-1}$	$M^{-1}s^{-1}$	s^{-1}	s^{-1}
isopropanol	0.091	0.274	9.4	7.10	9.68	38.6	0.13	0.33
acetone	0.056	0.067	53.3	66.0	1.05	1.02	0.057	0.070
(*S*)-2-butanol	0.42	1.0*	0.51	1.35*	823	740*	1.90	2.50*
(*R*)-2-butanol	0.56	2.0*	6.9	7.5*	81.2	267*	8.5	15.2*
2-butanone	0.10	0.12*	11.4	15.0*	8.77	8.00*	0.131	0.122*
3-pentanol	0.827	2.26*	2.1	1.6*	394	1410*	6.6	12.5*
3-pentanone	0.29	0.36*	52.8	75.0*	5.49	4.80*	0.39	0.37*

*(Adolph et al., 1991)

Table 2. Substrate-specific constants of LADH for the ordered bi-bi mechanism. k, rate constant of ternary complex interconversion; k_2, association rate constant of substrate; k_{-2}, dissociation rate constant of substrate; K_D, dissociation constant of substrate.

substrate	k		k_2		k_{-2}		K_D	
	pH 7.0	pH 8.5	pH 7.0	pH 8.5	pH 7.0	pH 8.5	pH 7.0	pH 8.5
	s^{-1}	s^{-1}	$M^{-1}s^{-1}$	$M^{-1}s^{-1}$	s^{-1}	s^{-1}	mM	mM
acetone	0.057	0.070	9.86	42.5	0.474	2.79	48.1	65.0
isopropanol	0.13	0.33	915	1130	10.9	9.09	11.9	8.1
2-butanone(*pro-S*)	0.131	0.122*	66.8	176*	0.672	2.53*	10.1	14.4*
(*S*)-2-butanol	1.9	2.5*	4230	53900*	6.58	171*	1.55	3.2*
2-butanone(*pro-R*)	0.170	0.217	3.63	12.69	0.762	4.07	210	321
(*R*)-2-butanol	8.5	15.2*	25400	129000	2150	6950	84.6	54.0
3-pentanone	0.39	0.37*	30.3	81.9*	1.33	5.81*	44.0	70.9*
3-pentanol	6.6	12.5*	1980	66800*	20.5	544*	10.4	8.14*

*(Adolph et al., 1991)

Is the Hydride Transfer Step Influenced by a pH Variation?

A literal interpretation of the term "hydride transfer step" means the transfer of a hydride ion from an alcohol to NAD+ or from NADH to an aldehyde or ketone. In such a reaction no protonation/deprotonation step is involved. A pH dependency would thus imply that the geometry or the electronic properties of the reactive ternary complex had to be changed upon a pH variation. In this case, both the forward and the reverse reaction should be affected.

From the data in Table 2 it can be seen that the rate constants for the ternary complex interconversion of HLADH-NADH-ketone to HLADH-NAD+-alcohol are pH-independent in the range of experimental error. This does not hold for the reverse reaction. For alcohol oxidations the rate constants are higher by factors of 1.3 to 2.5 at pH 8.5 as compared to pH 7.0, depending on the substrate structure.

Such an asymmetric pH dependency of the ternary complex isomerization step has been described earlier by Kvassman and Pettersson (1978) for the system benzylalcohol/benzaldehyde and was interpreted in terms of an additional step in the kinetic mechanism, i. e. the deprotonation of the alcohol prior to the hydride transfer step.

From the findings shown here, it is obvious that the hydride transfer step itself cannot be influenced by pH at the alcohol side if it is not also influenced at the ketone side, supposing the transition state theory is applicable to describe the hydride transfer as a one-step reaction. A change of the free energy of activation for the hydride transfer step should change the rate constants for both directions of the reaction. Thus an additional protonation/deprotonation step seems conceivable.

Kvassman and Pettersson (1978) integrated a protonation equilibrium of the alcohol into the ordered bi-bi mechanism. It has to be emphasized that the findings shown here reflect a kinetic and not a thermodynamic problem. There is no experimental evidence for a biphasic transient phase in the case of secondary alcohols which should be observed if the fraction of protonated alcohol would be higher at pH 7.0 as compared to pH 8.5 and the oxidation would proceed at two different rates for protonated and deprotonated alcohol. The non-existence of such a biphasic behaviour indicates that the deprotonation is rate-determining for every single molecule of alcohol which has to be oxidized.

This implies that the rate limitation is a secondary process involving groups of the enzyme which are possibly part of the proton relay system responsible for the transfer of the abstracted proton from the catalytic center of the enzyme to the surrounding solvent. There are several candidates which could be involved: Ser-48 and His-51 as suggested by Kvassman and Pettersson (1978) as well as Glu-68 which could act as a fifth ligand of the catalytic metal ion as suggested by E. Cedergren-Zeppezauer for the active site-specifically Cd(II)-substituted enzyme (as cited in Bauer et al., 1991).

The pH influence is different for different substrates which means that the substrates themselves influence the protonation/deprotonation properties of enzymic groups involved in these processes. A rough estimation of the magnitude of rate constants additionally involved in the ternary complex interconversion is in the range of 5 to 30 s^{-1} for the substrates investigated. Better estimations presuppose more data at different pH values.

What are the consequences for secondary alcohol oxidations? At least in the cases of 2-butanol and 3-pentanol the hydride transfer step is not alone rate-limiting for the ternary complex interconversion. It has to be separated into deprotonation and hydride transfer with the deprotonation partly rate-limiting for this step of the reaction sequence.

For the ketones it seems obvious that the rate constants observed in the single turnover experiments directly reflect the hydride transfer step which is rate-limiting for k_{cat} at pH 8.5. At pH 7.0 however k_{cat} is reduced by up to 30% as compared to the hydride transfer rate constant. This reduction is due to a slow step following the hydride transfer step and coupled to alcohol release from the HLADH-NAD+-alcohol ternary complex.

How is Substrate Binding Influenced by a pH Variation?

A comparison of the association and dissociation rate constants of substrates (k_2 and k_{-2}, respectively, in Table 2) shows significant influences of a pH variation. For all substrates investigated (with the exception of isopropanol) both rate constants are lower at pH 7.0 as compared to pH 8.5.

For the ketones k_2 is lowered by factors of approximately 3 and k_{-2} by factors of approximately 5 resulting in K_D values ($K_D = k_{-2}/k_2$) which are lowered at pH 7.0 to give 60-70% the values found for pH 8.5. This strengthening of binding is found to be independent of the substrate structure. This looks different for the alcohols. Although the rate constants k_2 and k_{-2} are both reduced at pH 7.0, the relative effects are different for the different alcohols. The association rate constants are higher at pH 8.5 by factors of 1.2 up to 34. The dissociation rate constants are nearly identical for isopropanol and higher by factors of up to 27. The calculated K_D values are higher at pH 7.0 with the exception of (*S*)-2-butanol which shows only half the value found for pH 8.5.

The interpretation of these findings seems to be simple for the ketones. Unidirectional effects of a comparable magnitude are found for all substrates and the K_D values are almost identical to the K_m values found in steady-state experiments. This is due to the rate-limiting hydride transfer step. The reduced rates of substrate binding and release at low pH which are also found for the alcohols, can be ascribed to changed properties of the enzyme with pH. The most surprising observation are the very low rate constants found for association and dissociation of the ketones (the dissociation is partly rate-limiting for k_{cat} in secondary alcohol oxidations). This result is most likely due to the hydrophobic nature of these substrates. The slow association is possibly caused by non-productive binding in the hydrophobic substrate channel of the enzyme. The event observed experimentally is the oxidation of NADH and thus the association rate constants obtained from the calculations reflect only the fraction of ketones bound in a productive ternary complex configuration. The very slow dissociation may also be linked to the hydrophobic nature of ketones and secondary alcohols, for which the solubilization could be rate limiting for the dissociation step.

In the case of the alcohols a second complication has to be taken into account in data interpretation. Production and decomposition of a ketone are directly coupled to the production or decomposition of NADH, which is observed experimentally. In case of the alcohols a protonation/deprotonation step has to be integrated and thus the species obtained directly after ketone reduction (disappearance of NADH) is an alkoxide complex. The dissociation of the alcohol deserves a protonation step to be performed first. Therefore it is likely that the protonation is involved in the calculated dissociation rate constant in a similar way as shown for the rate constants of ternary complex interconversion. Such a protonation step can be influenced significantly by the structure of the HLADH-NAD+-alcohol complex, hence this could explain the strong dependency of the pH effect on the substrate structure.

For a more detailed interpretation additional data at different pH values are necessary to allow calculations of pK_a values and the separation of hydride transfer step and deprotonation as well as dissociation of the alcohol and protonation of the alkoxide.

Influence of pH on Substrate Specificity and Stereoselectivity

As shown above, the stereoselectivity of 2-butanone reduction is not affected by a pH variation within experimental error. The determining factors for discrimination of the two pathways of 2-butanone reduction are combinations of free energies of binding and activation with the free energies of binding as the dominating factors. The difference of free energies of binding for both ternary complex configurations is 7.5 $kJmol^{-1}$ at pH 7.0 and 7.7 $kJmol^{-1}$ at pH 8.5. The identical free energy changes at both pH values result from the fact that pH-dependent changes of association and dissociation rate constants are unidirectional and of the same order of magnitude for all the substrates investigated. The kinetic effects observed do not contain substrate-specific protonation/deprotonation processes.

For the free energies of activation the differences between (*pro-R*) and (*pro-S*) reduction are very small with 0.6 $kJmol^{-1}$ at pH 7.0 and 1.4 $kJmol^{-1}$ at pH 8.5 and thus of minor significance for the observed stereoselectivity of 2-butanone reduction. Also in the hydride transfer step no substrate-specific protonation/deprotonation step is involved as can be seen by comparison of the other ketones investigated.

In case of the alcohols pH effects can be observed. Comparison of the k_{cat}/K_m values at both pH values shows that there is no unidirectional influence. On the contrary these values show a strong dependency on the substrate structure indicating that substrate-specific protonation/deprotonation processes are involved. As shown above, it is most likely that some step related to the alcohol-alkoxide equilibrium is partly rate-limiting for the ternary complex interconversion and the alcohol release.

The absence of unidirectional effects, as found for the ketones, results in a changed selectivity concerning the enantiomeric alcohols (*S*)-2-butanol and (*R*)-2-butanol. According to equation (6) the oxidation of (*S*)-2-butanol is favoured by a factor of 10 as compared to (*R*)-2-butanol at pH 7.0; the same factor is 2.8 at pH 8.5. This pH effect is mainly due to the extremely lowered dissociation constant of (*S*)-2-butanol and the slightly increased value for (*R*)-2-butanol at pH 7.0 as compared to pH 8.5. The strong reduction of K_D for (*S*)-2-butanol however is caused mainly by a reduced dissociation rate constant.

These considerations show that a pH dependency of substrate specificity (in terms of a discrimination of different substrates) and stereoselectivity presupposes the existence of a protonation/deprotonation step for which the rate constants change differently with pH for different substrates or enzyme-substrate complex configurations, respectively. For the substrates investigated here this can be seen for the secondary alcohols, but there is no effect of a pH variation on the stereoselectivity of ketone reductions.

ACKNOWLEDGEMENTS

This work was supported by Deutsche Forschungsgemeinschaft, Universität des Saarlandes (Zentrale Forschungskommission), and Fonds der Chemischen Industrie.

REFERENCES

Adolph H. W., Maurer P., Schneider-Bernlöhr H., Sartorius C., and Zeppezauer M., 1991, Substrate specificity and stereoselectivity of horse liver alcohol dehydrogenase. Kinetic evaluation of binding and activation parameters controlling the catalytic cycles of unbranched, acyclic secondary alcohols and ketones as substrates of the native and active-site-specific Co(II)-substituted enzyme, *Eur. J. Biochem.* 201:615.

Andersson I., Jörnvall H., Åkeson Å., and Mosbach K., 1974, Separation of isozymes of horse liver alcohol dehydrogenase and purification of the enzyme by affinity chromatography on an immobilized AMP-analogue, *Biochem. Biophys. Acta* 364:1.

Bauer R., Adolph H. W., Andersson I., Danielsen E., Formicka G., and Zeppezauer M., 1991, Coordination geometry for cadmium in the catalytic zinc site of horse liver alcohol dehydrogenase: studies by PAC spectroscopy, *Eur. Biophys. J.* 20:215.

Dickinson F. M. and Dalziel K., 1967, Substrate specificity and stereoselectivity of alcohol dehydrogenases, *Nature* 214:31.

Einarsson R., Widell L., and Zeppezauer M., 1976, Active-site titration of horse liver alcohol dehydrogenase by dual wavelength spectrophotometry, *Anal.Lett.* 9:815.

Graves J. M. A., Clark A., and Ringold H. J., 1965, Stereochemical aspects of the substrate specificity of horse liver alcohol dehydrogenase, *Biochemistry* 4:2655.

Jones J. B. and Beck J. F., 1976, Asymmetric syntheses and resolutions using enzymes, *in:* "Applications of Biochemical Systems in Organic Chemistry", J. B. Jones, C. J. Sih, D. Perlman, Eds., Wiley-Interscience, New York.

Kvassman J. and Pettersson G., 1978, Effect of pH on the process of ternary-complex interconversion in the liver-ADH reaction, *Eur. J. Biochem.* 87:417.

Plapp B. V., 1973, On calculation of rate and dissociation constants from kinetic constants for the ordered bi-bi mechanism of liver alcohol dehydrogenase, *Arch. Biochem. Biophys.* 156:192.

Sekhar V. C. and Plapp B. V., 1988, Mechanism of binding of horse liver alcohol dehydrogenase and nicotinamide adenine dinucleotide, *Biochemistry* 27:5082.

Shapiro S., Arunachalan T., and Caspi E., 1983, Equilibration of 1-octanol with alcohol dehydrogenase. Evidence for horse liver alcohol dehydrogenase responsibility for exchange of the 1-pro-S hydrogen atom, *J. Am. Chem. Soc.* 105:1642.

Willaert J. J., Lemière G. L., Joris L. A., Lepoivre J. A., and Alderweireldt F. C., 1988, Enzymatic in vitro reduction of ketones. 15. The influence of reaction conditions on the stereochemical course of HLADH-catalyzed reductions: 3-cyano-4,4-dimethylcyclohexanol as a sensitive probe, *Bioorg. Chem.* 16:223.

THE CATALYTIC SPECIFICITY OF LIVER ALCOHOL DEHYDROGENASE: VITAMIN A ALCOHOL AND VITAMIN A ALDEHYDE ACTIVITIES

Y. Pocker and Hong Li

Department of Chemistry, BG-10
University of Washington
Seattle, WA 98195 U.S.A.

INTRODUCTION

Liver alcohol dehydrogenase (LADH, E.C.1.1.1.1) is an NAD^+/NADH dependent enzyme with a broad substrate specificity. It catalyzes the reversible oxidation of a wide variety of alcohols to the corresponding aldehydes and ketones, as well as the oxidation of certain aldehydes to their carboxylic acids (Jörnvall, 1970; Brändén, *et al.*, 1975; Klinman, 1981; Eklund and Brändén, 1983; Zeppezauer, 1986; Pocker, *et al.*, 1986; Pettersson, 1987; Pocker, 1989). The enzyme's primary and tertiary structures have been determined, as have its catalytic mechanisms, isozyme differences and evolutionary divergences (Eklund and Brändén, 1987; Eklund, *et al.*, 1987; Pocker, 1989, Light, *et al.*, 1992). Although the physiological substrates of alcohol dehydrogenase are unknown, this enzyme catalyzes the interconversion of all-*trans* retinol, Vitamin A alcohol and all-*trans* retinal, Vitamin A aldehyde, Figure 1 (Pocker *et al.*, 1987). In addition, liver alcohol dehydrogenase catalyzes the oxidation of all-*trans* retinal to the corresponding retinoic acid (Pocker *et al.*, 1987).

Vitamin A and its derivatives play fundamental roles in many physiological processes such as growth, anti-infection, gluconeogenesis, cell maintenance, and most notably in the vision process (Bridges, 1984; Packer, 1990a, 1990b). Furthermore, extensive research has shown that various forms of Vitamin A can be used in the prevention and treatment of different types of cancers (Roberts and Sporn, 1984). Moon and Itri (1984) have discussed the use of various retinoids as inhibitors of chemical carcinogenesis in the mammary gland and urinary bladder of experimental animals. The applied biology of the retinoids, including their use in toxicology, immunology, cancer and dermatology, has been recently reviewed (Sporn, *et al.*, 1984a, 1984b; Packer, 1990a, 1990b).

The isomerization between the all-*trans* and the 11-*cis* forms of retinol or retinal, and the interconversion between alcohol and aldehyde are known to play fundamental roles in the vision process (Bridges, 1984; Packer, 1990a, 1990b). These processes are catalyzed by retinol isomerase and retinol oxidoreductase, respectively, with the latter being similar to LADH in terms of structure and function. Thus, an investigation of the retinol-retinal activity of liver alcohol dehydrogenase should further enhance our understanding of the physiological importance of this enzyme. Previous work has focused on the LADH catalyzed interconversion of all-*trans* retinol and all-*trans* retinal (Mezey and Holt, 1971; Pocker and Raymond, 1980). Our present studies are an attempt to delineate the capacity of LADH to catalyze the interconversion of all-*trans* as well as 13-*cis* and 9-*cis* retinols and retinals. Our results indicate that all the retinoids investigated bind at the catalytic zinc ion of LADH. However, due to the spatial restrictions within the ternary complex, 13-*cis* retinol, 13-*cis* retinal, and 9-*cis* retinal display a very low reactivity compared to the corresponding all-*trans* isomers. Furthermore, our kinetic studies shed light on the LADH catalyzed oxidation of all-*trans* retinal to all-*trans* retinoic acid.

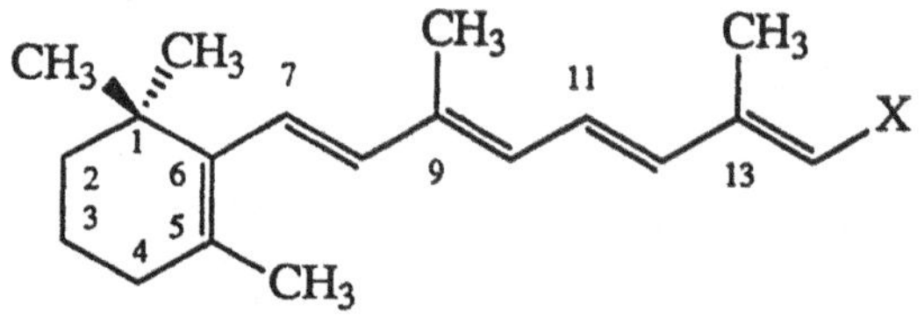

Figure 1. Chemical structures of all-*trans* retinol X=CH_2OH, all-*trans* retinal X=CHO and all-*trans* retinoic acid X=CO_2H.

MATERIALS AND EXPERIMENTAL PROCEDURES

Enzyme and Coenzymes. Horse liver alcohol dehydrogenase was obtained as crystallized and lyophilized preparations from Sigma. The oxidized and reduced forms of the coenzyme β-nicotinamide adenine dinucleotide, NAD^+ and NADH, were purchased from Sigma.

Substrates and other Reagents. All-*trans* retinal, all-*trans* retinol, all-*trans* retinoic acid, 13-*cis* retinal, 13-*cis* retinol and t-octylphenoxypolyethoxyethanol, TX-100, were purchased from Sigma and used as received. The aldehyde 9-*cis* retinal was a generous gift from Hoffmann-La Roche Inc. All other chemicals were either reagent or spectral grade.

Buffers. Unless otherwise specified, aqueous phosphate buffers were used for the preparation of all solutions of enzyme, coenzymes, substrates and inhibitors. Phosphate buffers were prepared by dissolving weighed amounts of K_2HPO_4 and KH_2PO_4 in distilled, deionized water. The phosphate concentration was 30 mM and the ionic strength was maintained at 0.1 by addition of Na_2SO_4. The solution was finally adjusted to the required pH by addition of either dilute NaOH or H_2SO_4.

Enzyme and Coenzyme Solutions. Solutions of enzyme were prepared by dissolving known amounts of LADH in buffer before each experiment. The concentration of LADH was checked spectrophotometrically using a value for the extinction coefficient, ϵ, at 280 nm of 3.6×10^4 $M^{-1}cm^{-1}$. The percent activity of the enzyme was determined by titration of the active sites with NAD^+ and pyrazole (Theorell and Yonetani, 1963). Enzyme solutions prepared in this manner and stored at 0 °C lost no activity within a 6 to 8 hour time period. Coenzyme solutions were made by dissolving weighed amounts of NAD^+ and NADH in buffer and determining the concentration spectrophotometrically using extinction coefficients of 1.78×10^4 $M^{-1}cm^{-1}$ at 260 nm and 6.22×10^3 $M^{-1}cm^{-1}$ at 340 nm for NAD^+ and NADH, respectively. Fresh solutions were prepared for each day's experiments.

Retinol and Retinal Solutions. Due to their low solubility in water, all-*trans* retinol, all-*trans* retinal, all-*trans* retinoic acid, 13-*cis* retinol, 13-*cis* retinal and 9-*cis* retinal stock solutions were prepared using the method described by Mezey and Holt (1971). The substrates and TX-100 were dissolved together in a small amount of acetone, which was subsequently evaporated under a stream of dry nitrogen or argon. The resulting mixture was then dissolved in buffer and diluted to the mark, with the TX-100 concentration in the stock solution being 1.0% (v/v). The final concentration of TX-100 in the kinetic runs was 0.1% (v/v). The concentration of substrate was then ascertained spectrophotometrically using the experimentally determined extinction coefficients listed in Table 1. Stock solutions of the various retinoids were stored under nitrogen in the dark below 5 °C.

Instrumentation. The pH measurements were conducted on a PHM84 Research pH Meter (Radiometer Copenhagen) equipped with a Cole-Palmer Ag/AgCl glass electrode. Spectroscopic and kinetic measurements were performed on a Varian Cary 210 UV-visible spectrophotometer interfaced to an Apple II/e microcomputer and thermostated at 25.00±0.05 °C with a circulating water bath controlled by a Forma-Temp Jr. Model 2095 thermonitor. Reactions involving the retinoids were monitored at 400 nm where the absorbance of NADH and NAD^+ is zero. The extinction coefficients of the various retinoids investigated at 400 nm are listed in Table 1. Since both substrate and product absorb at this wavelength, the reaction rate was calculated using the expression: velocity = $dA_{400} / (\epsilon_{400,product} - \epsilon_{400,substrate})dt$.

Stopped-Flow Apparatus. Fast reactions, those whose half-lives are less than a few seconds, were monitored on an extensively modified Durrum-Gibson stopped-flow spectrophotometer with the temperature kept at 25.00+0.05 °C by a Forma-Temp Model 2095-2 thermonitor (Pocker and Bjorkquist, 1977a, 1977b; Pocker and Fong, 1980; Pocker and Janjić, 1988). The output from the photomultiplier was filtered and amplified by a specially constructed amplification board. The signals were collected by a NEC Powermate 386SX microcomputer through a DAS-20 A/D interface board (MetraByte) with a precision of 12 bits and a conversion rate of up to 50 MHz.

Molecular Modeling Studies. Calculations were performed on SiliconGraphic Personal IRIS computers using InsightII and Discover software packages (Biosym Techonologies, Inc.). The structure of LADH as determined by Eklund and Jones (1984) and given in the Protein DataBase was used. The structures of all the retinoids investigated possessed an *s-cis* conformation about the C6-C7 bond, as depicted in Figure 1 (Dawson and Hobbs, 1990).

RESULTS

Our kinetic studies indicate that the enzyme utilizes an ordered bi-bi mechanism for retinol-retinal interconversion, similar to the one it uses for ethanol oxidation and acetaldehyde reduction, with all product inhibition patterns matching those expected for an ordered bi-bi mechanism (Wratten and Cleland, 1963).

The kinetic parameters were determined from the initial rate equation for an ordered bi-bi mechanism using double reciprocal plots as previously described (Pocker and Raymond, 1980, 1985). Table 2 summarizes the k_{cat} and K_m values for reactions involving the interconversion of all-*trans* and 13-*cis* retinols and retinals. It also contains data on the reduction of 9-*cis* retinal as well as the oxidation of all-*trans* retinal. The K_m values indicate that the interactions between the long carbon skeleton of the retinoid and the hydrophobic pocket of the enzyme provide a major driving force for the binding process. However, the k_{cat} values for 13-*cis* retinol oxidation and reduction, as well as for 9-*cis* retinal reduction, are over 100 fold slower than those of the corresponding all-*trans* isomers.

Table 1. Extinction coefficients of various all-*trans*, 13-*cis* and 9-*cis* retinoids.[a]

	λ_{max} nm	ϵ_{max} $mM^{-1}cm^{-1}$	$\epsilon_{400\ nm}$ $mM^{-1}cm^{-1}$
all-*trans* retinol	325	43.8	0.355
all-*trans* retinal	381	40.5	28.9
all-*trans* retinoate	345	31.2	4.00
13-*cis* retinol	331	39.0	0.297
13-*cis* retinal	378	28.4	22.5
9-*cis* retinal	376	30.4	22.7

[a]Determined in 30 mM phosphate buffer containing 0.1% TX-100, at pH 7.00.

In previous work we have demonstrated that LADH can also behave as an aldehyde dehydrogenase, particularly with respect to formaldehyde oxidation (Pocker and Li, 1990). In this report we show that the enzyme can also oxidize all-*trans* retinal, allbeit at a significantly lower rate than the corresponding oxidation of all-*trans* retinol. Furthermore, as can be seen from the data, the k_{cat} value for the reduction of all-*trans* retinal is *ca.* 100 fold higher than for its oxidation to all-*trans* retinoic acid.

For an ordered bi-bi mechanism, the equilibrium constant, K_{eq}, can be calculated using the Haldane relationship given in Equation 1 (Haldane, 1930; Segel, 1975). The constant K_{eq} was obtained by allowing the reaction to reach equilibrium and then determining substrate and product concentrations spectrophotometrically. Table 3 lists $K_{Haldane}$ and K_{eq} values for the interconversions of all-*trans* retinol and all-*trans* retinal as well as for 13-*cis* retinol and 13-*cis* retinal.

Table 2. Kinetic Parameters for the reactions LADH with retinols and retinals.[a]

reaction	k_{cat} (s^{-1})	$K_m^{retinol}$ (μM)
all-*trans* retinol + NAD^+	1.55 ±0.10	59 ±4
13-*cis* retinol + NAD^+	0.018 ±0.005	257 ±20
reaction	k_{cat} (s^{-1})	$K_m^{retinal}$ (μM)
all-*trans* retinal + NADH	2.50 ±0.10	340 ±40
13-*cis* retinal + NADH	0.018 ±0.005	137 ±10
9-*cis* retinal + NADH	0.023 ±0.005	242 ±20
reaction	k_{cat} (s^{-1})	$K_m^{retinal}$ (μM)
all-*trans* retinal + NAD^+	0.022 ±0.012	260 ±120

[a]Reactions were carried out in 30 mM phosphate buffer containing 0.1% TX-100 at pH 7.00.

Table 3. Equilibrium constants for the LADH catalyzed interconversion of retinol and retinal.

Reaction	$K_{Haldane}$ (M)[a]	K_{eq} (M)[b]
all-*trans* retinol ⇌ all-*trans* retinal	1.5×10^{-9} $\pm 0.2 \times 10^{-9}$	2.0×10^{-9} $\pm 0.4 \times 10^{-9}$
	2.6×10^{-9}[c]	3.3×10^{-9}[d]
13-*cis* retinol ⇌ 13-*cis* retinal	1.4×10^{-9} $\pm 0.2 \times 10^{-9}$	2.1×10^{-9} $\pm 0.3 \times 10^{-9}$

[a]Values were calculated with the constants listed in Table 2. [b]Experiments were performed in 30 mM phosphate buffer containing 0.1% TX-100, at pH 7.50. [c]Determined by Raymond at pH 7.0 (1981). [d]Determined by Bliss (1951).

$$K_{eq} = \frac{[\text{Aldehyde}][\text{NADH}][H^+]}{[\text{Alcohol}][\text{NAD}^+]}$$

$$K_{Haldane} = \frac{k_1 k_2 k_h k_3 k_4 [H^+]}{k_{-1} k_{-2} k_{-h} k_{-3} k_{-4}}$$

$$= \frac{V_f\ K_{mp}\ K_{iq}\ [H^+]}{V_r\ K_{mb}\ K_{ia}} \quad (1)$$

Our $K_{Haldane}$ of 1.5×10^{-9} M at pH 7.50 is close to the value of 2.6×10^{-9} M calculated for the LADH catalyzed oxidation of all-*trans* retinol at pH 7.0 in phosphate buffers (Raymond 1981). For all-*trans* and 13-*cis* retinol-retinal interconversion, there is very good agreement between $K_{Haldane}$ and K_{eq}. Furthermore, our K_{eq} value for the all-*trans* retinol-retinal system of 2.0×10^{-9} M (Table 3), is close to the previously reported value of 3.3×10^{-9} M (Bliss, 1951).

DISCUSSION

The k_{cat} values for all-*trans* retinol oxidation and all-*trans* retinal reduction have the same order of magnitude, with the reduction proceeding about 1.5-2 times faster. Similarly, the reduction of 13-*cis* retinal has the same order of magnitude as the oxidation of 13-*cis* retinol. This is in contrast to the interconversion of methanol-formaldehyde, ethanol-acetaldehyde and other short-chain alcohol-aldehyde pairs, where the reduction proceeds 10-100 times faster than the oxidation.

Contrary to methanol-formaldehyde, ethanol-acetaldehyde and other short-chain alcohol-aldehyde systems, the equilibrium constant for retinol-retinal interconversion is about 200 times larger. Indeed, the equilibrium constant, $K_{eq} = ([\text{NADH}][\text{aldehyde}][H^+])/([\text{NAD}^+][\text{alcohol}])$, for ethanol-acetaldehyde is 9.7×10^{-12} M (Bäcklin, 1958) while for retinol-retinal it is 2.0×10^{-9} M (Table 3). Thus at equilibrium, for a given $[\text{NADH}]/[\text{NAD}^+]$ ratio at the same pH, the [retinal]/[retinol] ratio is significantly higher than the corresponding [acetaldehyde]/[ethanol] value. As the all-*trans* isomers comprise the major portion of the naturally occurring retinoids (Frolik, 1984), this finding may have important physiological consequences. Furthermore, with respect to the various retinoids, the all-*trans* alcohol and aldehyde display the highest reactivity towards LADH. Apparently, the all-*trans* forms can achieve a conformation in the ternary complex that facilitates hydride transfer.

Comparing the kinetic constants of all-*trans* retinal reduction with those of 13-*cis* and 9-*cis* retinal reduction, it is noted that there exist large differences in k_{cat} values, while the K_m values are more nearly similar. It has been shown that the binding of the coenzyme induces a conformational change in the enzyme which increases the lipophilicity of the active site and stabilizes the binding of substrates with hydrophobic chains (Pettersson, 1987). Indeed, previous studies have shown that K_m decreases as the carbon chain length increases (Sund and Theorell, 1963). Thus, interactions between the hydrophobic region of the enzyme and the carbon skeleton of the substrate provide the main driving force in the binding process. Consequently, it is not surprising that all-*trans*, 13-*cis* and 9-*cis* retinals possess reasonably similar binding constants.

Crystallographic studies have revealed that in the enzyme-NAD^+-alcohol ternary complex, the alkoxide binds to the Zn ion through an inner sphere coordination

(Eklund, *et al.*, 1982). The side chains of Ser48 and Phe93 effectively lock the alcohol into position for hydride transfer to occur. For ethanol, the favorable mode of binding enables its *pro*-R hydrogen to face the C4 of NAD^+; pointing the *pro*-S hydrogen towards C4 of NAD^+ results in steric hindrance between ethanol's methyl group and the benzyl group of Phe93.

Our molecular modeling calculations indicate that when 13-*cis* and 9-*cis* retinols are in the correct juxtaposition for hydride transfer, unfavorable steric interactions between enzyme and substrate occur in the active site. When 13-*cis* retinol is bound, the carbon chain is too close to Ile318, Phe93 and Thr94, while for 9-*cis* retinol, unfavorable interactions exist with Phe110, Phe93 and Ile 318. These steric interactions prevent 13-*cis* and 9-*cis* retinol from achieving a favorable conformation for hydride transfer in the ternary complex. This is in contrast to all-*trans* retinol, whose carbon skeleton possesses an elongated linear conformation, Figure 1, an arrangement that produces favorable interactions within the hydrophobic barrel of the enzyme. The active-site of LADH appears to be somewhat flexible, as it can accomondate the 13-*cis* and 9-*cis* retinoids even though these two isomers must overcome unfavorable energetics to yield productive ternary complexes. The net result being that the rate of the hydride transfer is greatly diminished in comparison to that of the all-*trans* isomers, compounds that can easily assume the proper molecular orientation for hydride transfer in the active site. Consequently, of the retinoids investigated, the all-*trans* alcohol and aldehyde exhibit the highest reactivity.

ACKNOWLEDGEMENTS

Acknowledgement is made to the donors of the Petroleum Research Fund, administered by the American Chemical Society, for partial support of this research. Additional support of this research by grants from the National Science Foundation and the Muscular Dystrophy Association are gratefully acknowledged. We also wish to thank Dr. Peter Sorter of Hoffmann-La Roche Inc., Nutley NJ, for his timely gift of 9-*cis* retinal. Finally, we thank Greg Spyridis for his helpful comments and expert assistance in the preparation of this manuscript.

REFERENCES

Bäcklin, K. -I. 1958, The equilibrium constant of the system ethanol, aldehyde, DPN^+, DPNH and H^+, *Acta Chem. Scand.*, 12: 1279.

Bliss, A. F., 1951, The equilibrium between vitamin A alcohol and aldehyde in the presence of alcohol dehydrogenase, *Arch. Biochem. Biophys.*, 31: 197.

Brändén, C. -I., Jörnvall, H., Eklund, H. and Furugren, B., 1975, Alcohol dehydrogenases *in*: "The Enzymes", (Boyer, P. D. ed.) vol. 11, Academic Press, New York, pp 103-190.

Bridges, C. D. B., 1984, Retinoids in photosensitive systems, *in*: "The Retinoids", (Sporn, M. B., Roberts, A. B. and Goodman, D. S. eds.) vol. 2, Academic Press, New York, pp 90-124.

Dawson, M. I. and Hobbs, P. D., 1990, Synthetic retinoic acid analogs: handling and characterization, *in*: "Methods in Enzymology", (Packer, L. ed.) vol. 189 Academic Press, San Deigo. pp 13-43.

Eklund, H. and Brändén, C. -I., 1983, The role of zinc in alcohol dehydrogenases *in*: "Zinc Enzymes", (Spiro, T. G. ed.), John Wiley and Sons, New York, pp 123-152.

Eklund, H. and Brändén, C. -I., 1987, Alcohol dehydrogenase, *in*: "Biological Macromolecules and Assemblies", (Jurnak, F. and McPherson, A. eds.), vol. 3, John Wiley and Sons, New York, pp 73-142.

Eklund, H. Plapp, B. V., Samama, J. -P. and Brändén, C. -I., 1982, Binding of substrate in a ternary complex of horse liver alcohol dehydrogenase, *J. Biol. Chem.*, 257: 14349

Frolik, C. A., 1984, Metabolism of retiniods, *in*: "The Retinoids", (Sporn, M. B., Roberts, A. B. and Goodman, D. S. eds.) vol. 2, Academic Press, New York. pp 177-208.

Haldane, J. B. S., 1930, "Enzymes", Longmans Green, London.

Hinson, J. A. and Neal, R. A., 1972, An examination of the oxidation of aldehydes by horse liver alcohol deydrogenase, *J. Biol. Chem.*, 247: 7106.
Jörnvall, H., 1970, Horse liver alcohol deydrogenase, *Eur. J. Biochem.*, 16: 25.
Klinman, J. P., 1981, Probes and mechanism and transition state structure in the alcohol deydrogenase reaction, *CRC Crit. Rev. Biochem.*, 10: 39.
Light, D. R., Dennis, M. S., Forsythe, I. J., Liu, C. C., Green, D. W., Kratzer, D. A. and Plapp, B. V., 1992, α-Isoenzyme of alcohol deydrogenase from monkey liver: cloning, expression, mechanism, coenzyme, and substrate specificity, *J. Biol. Chem.*, 267: 12592.
Mezey, E. and Holt, P. R., 1971, The inhibitory effect of ethanol on retinol oxidation by human liver and cattle retina, *Experimental and Molecular Pathology.*, 15: 148.
Moon, R. C. and Itri, 1984, Retinoids and cancer, *in*: "The Retinoids", (Sporn, M. B., Roberts, A. B. and Goodman, D. S. eds.) vol. 2, Academic Press, New York. pp 327-372.
Packer, L. ed., 1990a, "Methods in Enzymology" vol. 189 Academic Press, San Diego.
Packer, L. ed., 1990b, "Methods in Enzymology" vol. 190 Academic Press, San Diego.
Pettersson, G., 1987, Liver alcohol deyhdrogenase, *CRC Crit. Rev. Biochem.*, 21: 349.
Pocker, Y., 1989, Alcohol dehydrogenase: structure catalysis and site directed mutagenesis, *in*: "Metal Ions in Biological Systems", (Sigel, H., ed.) Marcel Dekker, New York, pp 335-358.
Pocker, Y. and Bjorkquist, D. W., 1977a, Stopped-flow studies of carbon dioxide hydration and bicarbonate dehydration in H_2O and D_2O. Acid-base and metal ion catalysis, *J. Am. Chem. Soc.* 99: 6537.
Pocker, Y. and Bjorkquist, D. W., 1977b, Comparative studies of bovine carbonic anhydrase in H_2O and D_2O. Stopped-flow studies of the kinetics of interconversion of carbon dioxide and bicarbonate, *Biochemistry* 16: 5698.
Pocker, Y. and Fong, C. T. O, 1980, Kinetics of inactivation of erythrocyte carbonic anhydrase by sodium 2,6-pyridinedicarboxylate, *Biochemistry* 19: 2045.
Pocker, Y. and Janjič, N., 1988, Differential modification of specificity in carbonic anhydrase, *J. Biol. Chem.* 263: 6169.
Pocker, Y. and Li, H., 1990, Kinetics and mechanism of methanol and formaldehyde interconversion and formaldehyde oxidation catalyzed by liver alcohol deydrogenase, *in*: "Enzymology and Molecular Biology of Carbonyl Metabolism 3", (Weiner, H., Wermuth, B. and Crabb, D. W. eds.) Plenum Press, New York, pp 315-325.
Pocker, Y., Li, H. and Page, J. D., 1987, Liver alcohol deydrogenase: metabolic and energetic aspects, *in*: "Advances in Biomedical Alcohol Research", (Lindros, K. O., Ylikahri, R. and Kiianmaa, K. eds.) Pergamon Press, London, pp 181-185.
Pocker, Y. and Raymond, K. W., 1980, Kinetics and mechanistic studies of vitamin A alcohol to vitamin A aldehdye by liver alcohol deydrogenase. The inhibition by ethanol and pyrazole, *in*: "Alcohol and Aldehyde Metabolizing Systems-IV", (Thurman, R. G. ed.) Plenum Press, New York, pp 137-150.
Pocker, Y. and Raymond, K. W., 1985, Liver alcohol deydrogenase: substrate inihibition and competition between substrates, *Alcohol* 2:3.
Pocker, Y., Raymond, K. W. and Thompson III, W. H., 1986, Liver alcohol deydrogenase: kinetic and mechanistic studies, *in*: "Zinc Enzymes", (Bertini, I., Luchinat, C. Maret, W. and Zeppezauer, M. eds.) Birkhäuser, Boston, pp 435-449.
Raymond, K.W.,1981, Horse liver alcohol dehydrogenase: kinetic and mechanistic studies, Ph.D. Dissertation, University of Washington, Seattle, WA.
Roberts, A. B. and Sporn, M. B., 1984, Cellular biology and biochemistry of the retinoids, *in*: "The Retinoids", vol. 2, (Sporn, M. B., Roberts, A. B. and Goodman, D. S. eds.) Academic Press, New York, pp 210-287.
Segel, I. H., 1975, "Enzyme Kinetics", John Wiley and Sons, New York.
Sporn, M. B., Roberts, A. B. and Goodman, D. S. eds.,1984a, "The Retinoids", vol. 1, Academic Press, New York.
Sporn, M. B., Roberts, A. B. and Goodman, D. S.,1984b, "The Retinoids", vol. 2, Academic Press, New York.
Sund, H. and Thorell, H., 1963, Alcohol dehydrogenases, *in*: "The Enzymes", (Boyer, P. D., Lardy, H. and Myrback, K. eds.) vol. 7, Academic Press, New York, pp 25-83.
Theorell H. and Yonetani, T., 1963, Liver alcohol dehydrogenase-DPN-pyrazole complex: a model of a ternary intermediate in the enzyme reaction. *Biochemische Zeitschrift*, 338: 537.
Wratten, C. C. and Cleland, W. W., 1963, Product inhibition studies on yeast and liver alcohol dehydrogenases, *Biochemistry*, 2: 935.
Zeppezauer, M., 1986, The metal environment of alcohol dehydrogenase: aspects of chemical speciation and catalytic effiency in a biological catalyst, *in*: "Zinc Enzymes", (Bertini, I., Luchinat, C. Maret, W. and Zeppezauer, M. eds.) Birkhäuser, Boston, pp 417-434.

A SYNTHETIC APPROACH TO ANALYSIS OF THE STRUCTURAL ZINC SITE OF ALCOHOL DEHYDROGENASE

Tomas Bergman[1], Hans Jörnvall[1], Torleif Härd[2], Barton Holmquist[3], and Bert E. Vallee[3]

[1]Department of Chemistry I, Karolinska Institutet, S-104 01 Stockholm, Sweden
[2]Center for Structural Biochemistry, Karolinska Institutet NOVUM, S-141 57, Huddinge, Sweden
[3]Center for Biochemical and Biophysical Sciences and Medicine Harvard Medical School, Boston, MA 02115, USA

INTRODUCTION

The mammalian alcohol dehydrogenases are medium-sized dimeric zinc-containing enzymes with a subunit molecular mass of about 40 kDa. Each subunit contains two zinc atoms, one of which is part of the catalytic site, whereas the other appears to fulfil a structural role (Åkeson, 1964; Drum et al., 1967; Drum and Vallee, 1970) and is bound to a separate loop of the protein that contributes to subunit interactions (Eklund et al., 1976). However, the precise role of this non-catalytic zinc is not known. It is tetrahedrally coordinated by four cysteine residues, which are fairly adjacent in a short segment of the polypeptide backbone (Cys97, Cys100, Cys103 and Cys111; cf. Eklund et al., 1976). This structure follows the general rules for ligands to non-catalytic zinc atoms, as deduced from similarities with the relationships in other proteins (Vallee and Auld, 1990; Vallee and Auld, 1991). The cysteines are also conserved in medium-chain (Persson et al., 1991) alcohol dehydrogenases proper (Jörnvall et al., 1987a), but other residues in this region are highly variable and species-specific (Jörnvall, 1985; Jörnvall et al., 1987b). Consequently, the cysteine spacing, i.e. Cys-Xaa$_2$-Cys-Xaa$_2$-Cys-Xaa$_7$-Cys, appears to be of major importance for the protein/zinc complex to form. This structural arrangement differs distinctly from that of other sulfur-containing, zinc-binding motifs, e.g. zinc fingers, zinc twists, and zinc clusters.

The possibility to construct a model system for the structural zinc site of horse liver alcohol dehydrogenase was investigated by means of a synthetic peptide with an amino acid sequence identical to that of the segment in the enzyme subunit, encompassing the four zinc-liganding cysteine residues. The minimal region, spanning from Cys97 to Cys111, was extended by four residues at either end to yield a 23-residue peptide, comprising residues 93 - 115. To determine whether this synthetic protein segment by itself can bind zinc in a stable manner resembling native conditions, or if the balance of the intact enzyme is required, the metal-binding characteristics of the synthetic peptide were probed spectroscopically with zinc and cobalt. As recently reported in part (Bergman et al., 1992) and now extended, the data reveal that the 23-residue peptide binds both these metals in stable structures that mimic the corresponding metal-binding properties of the intact enzyme. The synthetic peptide thus provides a metal-binding model for liver alcohol dehydrogenase.

EXPERIMENTAL PROCEDURES

Synthetic peptides were prepared with an Applied Biosystems 430A instrument using side-chain-protected tertiary butyloxycarbonyl amino acid derivatives (cf. Kent, 1988). Cleavage from the resin and deprotection of the material was effected with hydrogen fluoride for 1 h at 0°C in the presence of scavengers (HF/dimethylsulfide/anisole/p-thiocresol, 10:1:1:0.2). Extraction of the peptide was performed with 30% acetic acid and the extract was washed with diethylether before lyophilization. The crude preparations were purified by reverse-phase HPLC (Vydac C_{18}, 250 x 22 mm, 10 ml/min) using a linear acetonitrile gradient (0 - 60%, 45 min) in aqueous 0.1% trifluoroacetic acid. N-terminal acetylation was carried out before cleavage and deprotection by treatment for 10 min with a mixture of acetic anhydride/triethylamine/dichloromethane (9:4:87, by vol.).

Metal-binding experiments were performed at pH 7.5 in 20 mM Hepes buffer containing 50 mM NaCl. Buffer solutions were treated with diphenylthiocarbazone before use as described (Holmquist, 1988) to remove traces of metal ions. Peptide reduction was accomplished with a 250-fold molar excess of dithiothreitol (DTT) for 6 h at 45°C, after which the preparations were stored at -70°C until used. The metal-binding studies employed peptide concentrations in the range 5 to 250 μM and were performed under nitrogen at room temperature. Hepta-hydrates of zinc and cobalt sulfates were from Johnson Matthey Chemicals. Exclusion chromatography on PD-10 (Pharmacia) or Bio-Gel P4 (BioRad) was used to remove DTT and excess metal. Concentrations of peptide and metal were determined by amino acid analysis as described (Bergman et al., 1986) and by atomic absorption spectroscopy (Perkin-Elmer 2380), respectively. Absorption readings and spectra were recorded with a Hewlett Packard 8451A or a Varian Cary 219 instrument. Magnetic circular dichroic

(MCD) spectra were monitored at 4 T in a 1-cm-pathlength cuvette with a Cary 61 spectropolarimeter equipped with a Varian V4145 superconducting magnet. Nuclear magnetic resonance (NMR) spectra were recorded with a Varian Unity 500 instrument at 26°C and 500 MHz. The NMR studies employed 40 mM phosphate buffer, pH 5.8 - 6.2, containing 7 - 8% D_2O. Peptide concentrations were in the range 0.8 to 1.6 mM.

RESULTS AND DISCUSSION

A 23-residue synthetic replica of the segment of horse liver alcohol dehydrogenase responsible for the interaction with the structural zinc atom was prepared by solid-phase peptide synthesis (peptide I). It corresponds to residues 93 - 115 of the parent molecule and represents a four-residue extension of the protein zinc-containing loop (residues 97 - 111) at both ends. A control peptide which also contains cysteine (peptide II), corresponding to residues 1 - 14 of the enzyme subunit, was also synthesized. Since the native protein has an acetyl-blocked N-terminus, peptide II was N-terminally acetylated. To avoid simultaneous ϵ-amino modification

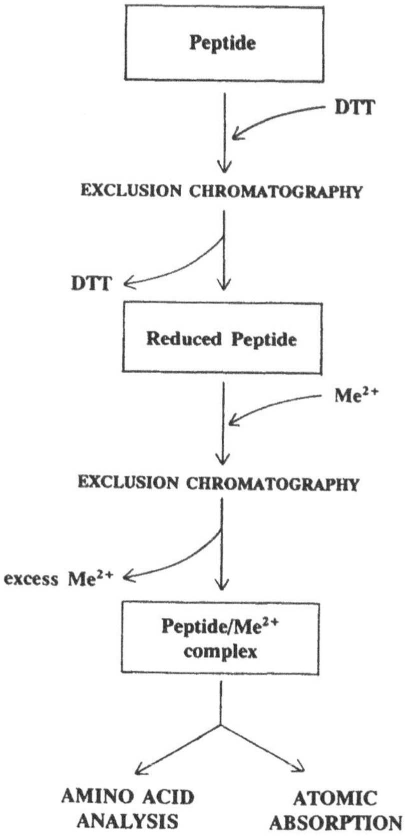

Figure 1. Experimental scheme for peptide reduction, metal incubation and quantification of the peptide/metal complex.

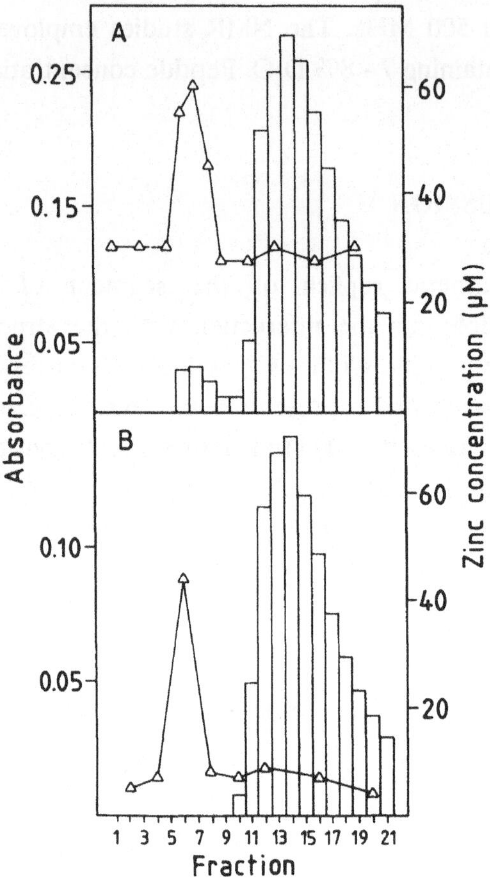

Figure 2. Exclusion chromatography after incubation of the reduced peptides with zinc.
(A) peptide I, (B) peptide II. Fractions were monitored for peptide at 214 nm (△) and for zinc by atomic absorption (bars).

Table 1. Zinc-binding stoichiometry. Molar ratios between bound zinc and peptide I in two fractions collected after zinc incubation and two consecutive exclusion chromatographies (cf. Fig. 2A).

	Zn^{2+}/Peptide I
Initial chromatography	1.24
	1.17
Rechromatography	1.14
	1.19

of its three lysine residues, the reaction was performed with the peptide still resin-bound and side-chain-protected.

The experimental set-up for reduction, metal incubation and quantification is shown in Fig. 1. Peptide stock solutions (1 mM) were submitted to treatment with 250 mM DTT which, before experiments, was removed by exclusion chromatography under a nitrogen atmosphere. Starting with that step, all further manipulations were carried out anaerobically to ensure complete reduction. Fractions containing peptides were mixed with a 2- to 10-fold molar excess of Zn^{2+} after which unbound metal was separated by a second exclusion chromatography (Fig. 1). The content of zinc in the fractions collected was analyzed by atomic absorption spectroscopy, while that of the peptide contents were determined by amino acid analysis. As shown in Fig. 2A, peptide I coelutes with zinc in a molar ratio of 1:1 between bound zinc and peptide I (Table 1). In contrast, the chromatogram of the control peptide (peptide II, Fig. 2B) reveals only the large lagging peak corresponding to unbound zinc. To assess the stability of the peptide I/Zn^{2+} complex, it was subjected to exclusion rechromatography and the data (Table 1) show that the zinc binding and stoichiometry are maintained. We can thus conlude that the interaction between peptide I and Zn^{2+} is specific and that a 1:1 complex is formed.

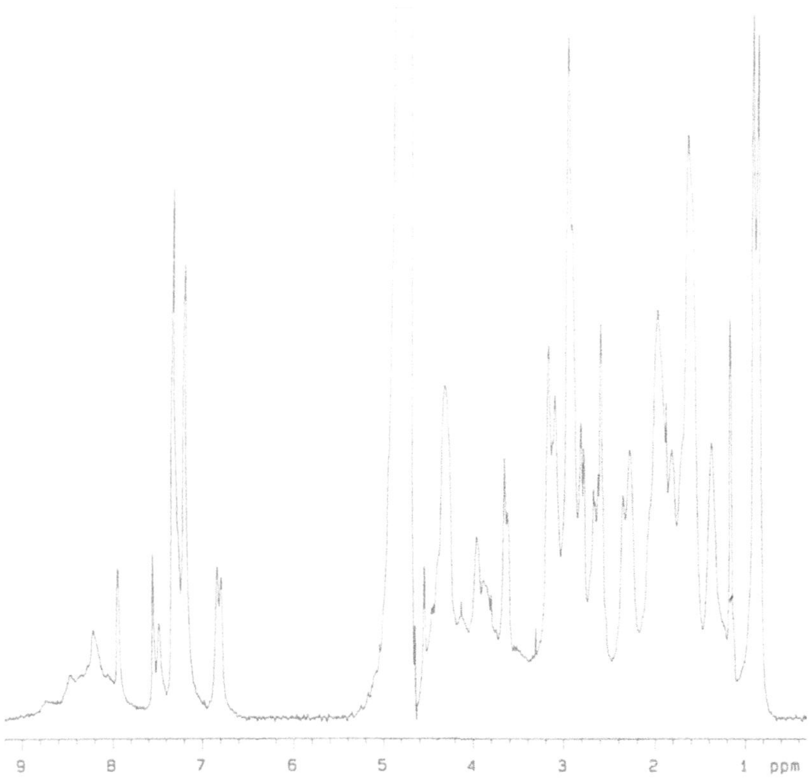

Figure 3. NMR spectrum of the peptide I/Zn^{2+} complex at pH 6.2 and 0.8 mM.

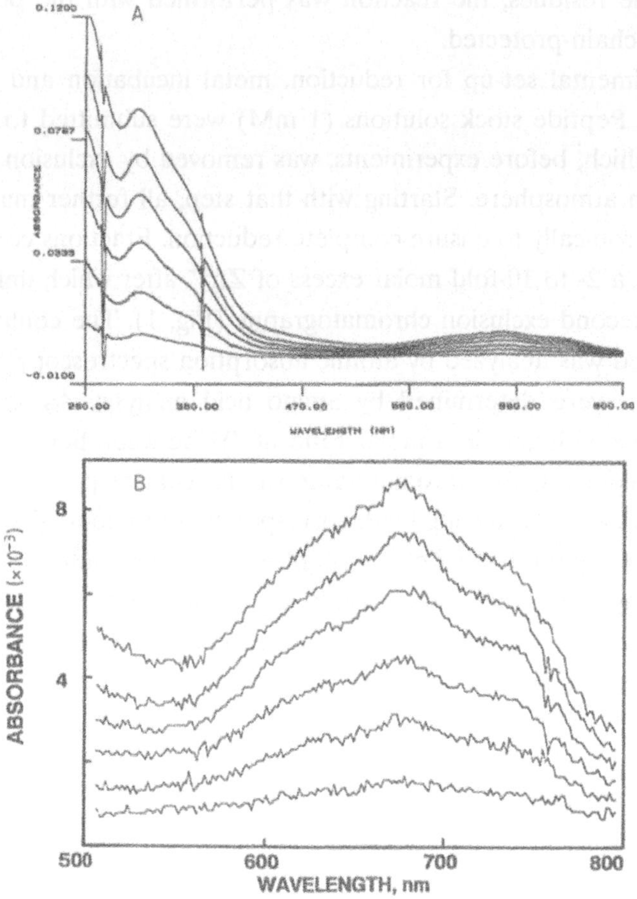

Figure 4. Titration of reduced peptide I (16 μM) with aliquots of Co^{2+} (1 - 12 μM) at pH 7.5. (A) The absorption spectrum (250 - 800 nm) recorded after each addition, corrected for the absorbance of free peptide and for dilution. (B) Spectrum expanded in the long-wavelength region.

Evidence in support for the formation of a structurally distinct peptide/zinc complex was also obtained from NMR spectra. In phosphate buffer, both with and without the addition of hexafluoropropanol (30%, by vol.), and at different pH-values (6.2 and 5.8), the spectra reveal a constant pattern (Fig. 3), suggesting the presence of an unrandomized three-dimensional structure.

Cobalt was used as a spectral probe to closer examine the interaction between peptide I and metal. A solution of the reduced peptide was titrated by stepwise increments of Co^{2+} from concentrations about 10-fold below to 100-fold above that of the peptide. The formation of a peptide I/Co^{2+} complex can readily be monitored by plotting the absorbance spectrum (250 - 800 nm) after each addition. A characteristic chromophore develops, with maximal absorption at 310 and 675 nm, and shoulders at 350, 640 and 720 nm (Fig. 4). The overall absorption pattern of the Co^{2+} complex in the region 550 - 800 nm (Fig. 4B) is characteristic of tetrahedral ligand geometry (Vallee and Galdes, 1984) and the charge-transfer bands below 400 nm (Fig. 4A) indicate sulfur coordination (Vallee and Galdes, 1984; Green and Berg, 1989). Data on the intensity of the charge-transfer bands for various cobalt-substituted proteins reveal an absorptivity contribution per S-Co bond of about 1000 $M^{-1}cm^{-1}$ (Holmquist and Vallee, 1979). Moreover, in different structural-site cobalt-substituted alcohol dehydrogenase isozymes, the absorptivities per S-Co bond at 340 - 346 nm are within the range 730 - 1320 $M^{-1}cm^{-1}$ (Formicka-Kozlowska et al., 1988). The absorptivity of the peptide I/Co^{2+} complex, 4200 $M^{-1}cm^{-1}$, therefore indicates that all four cysteine-sulfurs of the peptide are coordinated to the metal. The 675 nm absorptivity of the peptide I/Co^{2+} complex, about 500 $M^{-1}cm^{-1}$, is also within the range reported for structural-site cobalt-substituted alcohol dehydrogenases, i.e. 425 - 525 $M^{-1}cm^{-1}$ per cobalt atom at 655 - 660 nm (Formicka-Kozlowska et al., 1988). The visible absorption envelope, with a maximum at 675 nm and two shoulders at 640 and 720 nm (Fig. 4B), reveals a striking similarity to that of human liver $\beta_1\beta_1$ (Formicka-Kozlowska et al., 1988) and horse liver (Sytkowski and Vallee, 1976; Formicka-

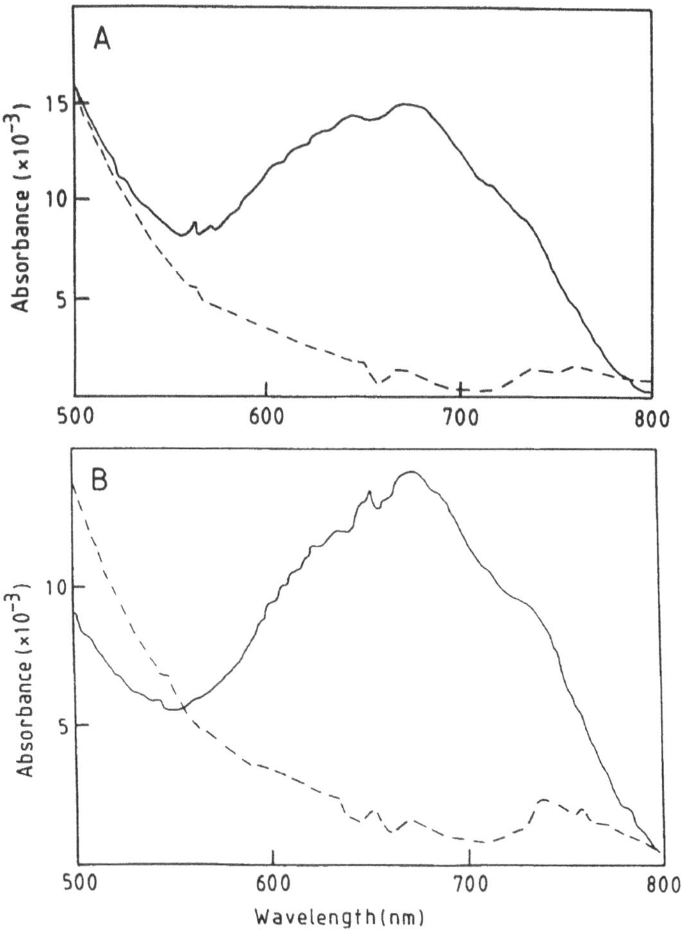

Figure 5. The spectral effects of Zn^{2+} and EDTA additions to the peptide I/Co^{2+} complex.

(A) Addition of an equimolar amount of Zn^{2+} to the cobalt complex. Solid line represents the absorption spectrum before and dashed line after the addition of Zn^{2+}. (B) Addition of a ten-fold excess (1 mM) of EDTA to the cobalt complex. Solid line represents the absorption spectrum before and dashed line after the EDTA addition.

Kozlowska and Zeppezauer, 1988) alcohol dehydrogenases, both cobalt-substituted at the structural site only.

To get an estimate of the affinity of peptide I for cobalt, the complex dissociation constant was calculated to be 2.1 μM at pH 7.5, based on the change in absorbance at 675 nm during titration. Addition of an equimolar amount of zinc to the peptide I/Co^{2+} complex results in abolished absorption (Fig. 5A), indicating a competition between cobalt and zinc for the peptide metal-binding site, in favor of Zn^{2+} complex formation. This higher affinity of peptide I for zinc was evaluated by titration of the cobalt-saturated peptide with Zn^{2+}. An observed decrease in absorbance at 675 nm is consistent with a dissociation constant in the low-nanomolar range. The addition of EDTA to the peptide I/Co^{2+} complex results in metal removal which is evident from a significant decrease in the visible absorbance between 550 and 800 nm (Fig. 5B). This decrease is similar to that for zinc addition (Fig. 5A) and represents the formation of an octahedral cobalt-EDTA complex. Also the charge-

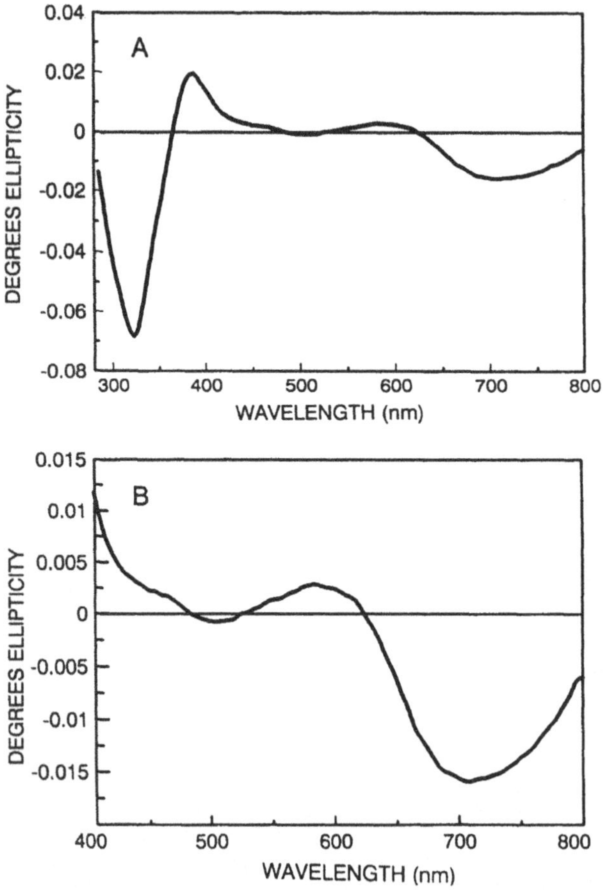

Figure 6. MCD spectrum of the peptide I/Co^{2+} complex obtained with a solution containing 1.7 equivalents of cobalt per peptide.

(A) Full length spectrum (280 - 800 nm) showing the significant sulfur-to-cobalt charge-transfer bands below 400 nm. (B) An expanded view towards the long-wavelength region reveals a pattern characteristic of tetrahedral coordination geometry.

transfer bands at 310 nm are abolished, confirming the disappearance of sulfur coordination and cobalt removal.

MCD spectra are efficient and reliable measures of coordination geometry of cobalt complex ions and cobalt-substituted metalloenzymes (Holmquist et al., 1975; Holmquist and Vallee, 1979). Also the presence of sulfur-ligands is readily detected by the occurrence of ligand-to-metal charge-transfer bands in the MCD spectrum at 230 - 400 nm (Vallee and Holmquist, 1980). The MCD spectrum of the peptide I/Co^{2+} complex (Fig. 6) reveals a characteristic negative ellipticity band near 700 nm and a smaller positive band centered at 580 nm which is typical for MCD spectra of tetrahedral Co^{2+} complexes. This pattern has been observed for many other tetrahedrally coordinated cobalt-substituted metalloproteins (for reviews, cf. Vallee and Holmquist, 1980; Holmquist, 1986). Furthermore, the large negative and less intense positive ellipticity bands below 400 nm with extremes at 320 and 390 nm respectively (Fig. 6), are significant for ligand-to-metal charge-transfer between sulfur and cobalt and consistent with MCD spectra of cobalt-substituted horse liver alcohol dehydrogenase (Vallee and Holmquist, 1980). The results thus indicate a tetrahedral sulfur coordination geometry for the peptide I/Co^{2+} complex.

In conclusion, the metal binding characteristics of a 23-residue synthetic peptide with an amino acid sequence identical to the segment of horse liver alcohol dehydrogenase encompassing the binding site of the structural zinc atom has been examined with spectroscopic methods. It mimics the properties of the intact enzyme and binds Zn^{2+} and Co^{2+} with high affinity. The NMR spectrum reveals a significant and stable structure and both the absorption and MCD spectra of the cobalt complex are consistent with tetrahedral metal coordination by the four conserved cysteine residues in this otherwise variable protein segment.

ACKNOWLEDGMENTS

This work was supported by grants from the Swedish Medical Research Council (project 03X-3532), the Magn. Bergvall Foundation, and the Endowment for Research in Human Biology.

REFERENCES

Bergman, T., Carlquist, M., and Jörnvall, H., 1986, Amino acid analysis by high performance liquid chromatography of phenylthiocarbamyl derivatives, *in*: Advanced Methods in Protein Microsequence Analysis, Wittmann-Liebold, B., Salnikow, J., and Erdmann, V.A., eds., Springer, Berlin/Heidelberg, p. 45.

Bergman, T., Jörnvall, H., Holmquist, B., and Vallee, B.L., 1992, A synthetic peptide encompassing the binding site of the second zinc atom (the "structural" zinc) of alcohol dehydrogenase, *Eur. J. Biochem.*, 205:467.

Drum, D.E., and Vallee, B.L., 1970, Differential chemical reactivities of zinc in horse liver alcohol dehydrogenase, *Biochemistry*, 9:4078.

Drum, D.E., Harrison, J.H., IV, Li, T.-K., Bethune, J.L., and Vallee, B.L., 1967, Structural and functional zinc in horse liver alcohol dehydrogenase, *Proc. Natl. Acad. Sci. USA*, 57:1434.

Eklund, H., Nordström, B., Zeppezauer, E., Söderlund, G., Ohlsson, I., Boiwe, T., Söderberg, B.-O., Tapia, O., Brändén, C.-I., and Åkeson, Å., 1976, Three-dimensional structure of horse liver alcohol dehydrogenase at 2.4 Å resolution, *J. Mol. Biol.*, 102:27.

Formicka-Kozlowska, G., and Zeppezauer, M., 1988, Horse liver alcohol dehydrogenase derivatives containing nickel(II) and cobalt(II) in the noncatalytic metal binding site, *Inorg. Chim. Acta*, 151:183.

Formicka-Kozlowska, G., Schneider-Bernlöhr, H., von Wartburg, J.-P., and Zeppezauer, M.,1988, $H_8Zn(c)_2$ and $Zn(c)_2Co(n)_2$ human liver alcohol dehydrogenase, *Eur. J. Biochem.*, 173:281.

Green, L.M., and Berg, J.M., 1989, A retroviral Cys-Xaa_2-Cys-Xaa_4-His-Xaa_4-Cys peptide binds metal ions: spectroscopic studies and a proposed three-dimensional structure, *Proc. Natl. Acad. Sci. USA*, 86:4047.

Holmquist, B., 1986, Magnetic circular dichroism, *Methods Enzymol.*, 130:270.

Holmquist, B., 1988, Elimination of adventitious metals, *Methods Enzymol.*, 158:6.

Holmquist, B., and Vallee, B.L., 1979, Metal-coordinating substrate analogs as inhibitors of metalloenzymes, *Proc. Natl. Acad. Sci. USA*, 76:6216.

Holmquist, B., Kaden, T.A., and Vallee, B.L., 1975, Magnetic circular dichroic spectra of cobalt(II) substituted metalloenzymes, *Biochemistry*, 14:1454.

Jörnvall, H., 1985, Use of peptides in studies of protein structures and functions, *in*: Synthetic Peptides in Biology and Medicine, Alitalo, K., Partanen, P., and Vaheri, A., eds., Elsevier, Amsterdam, p. 13.

Jörnvall, H., Persson, B., and Jeffery, J., 1987a, Characteristics of alcohol/polyol dehydrogenases: the zinc-containing long-chain alcohol dehydrogenases, *Eur. J. Biochem.*, 167:195.

Jörnvall, H., Höög, J.-O., von Bahr-Lindström, H., and Vallee, B.L., 1987b, Mammalian alcohol dehydrogenases of separate classes: intermediates between different enzymes and intraclass isozymes, *Proc. Natl. Acad. Sci. USA*, 84:2580.

Kent, S.B.H., 1988, Chemical synthesis of peptides and proteins, *Annu. Rev. Biochem.*, 57:957.

Persson, B., Krook, M., and Jörnvall, H., 1991, Characteristics of short-chain alcohol dehydrogenases and related enzymes, *Eur. J. Biochem.*, 200:537.

Sytkowski, A.J., and Vallee, B.L., 1976, Chemical reactivities of catalytic and noncatalytic zinc or cobalt atoms of horse liver alcohol dehydrogenase: differentiation by their thermodynamic and kinetic properties, *Proc. Natl. Acad. Sci. USA*, 73:344.

Vallee, B.L., and Auld, D.S., 1990, Active-site zinc ligands and activated H_2O of zinc enzymes, *Proc. Natl. Acad. Sci. USA*, 87:220.

Vallee, B.L., and Auld, D.S., 1991, Zinc chemistry in function and structure of zinc proteins, *in*: Methods in Protein Sequence Analysis, Jörnvall, H., Höög, J.-O., and Gustavsson, A.-M., eds., Birkhäuser, Basel., p. 363.

Vallee, B.L., and Galdes, A., 1984, The metallobiochemistry of zinc enzymes, *Adv. Enzymol.*, 56:283.

Vallee, B.L., and Holmquist, B., 1980, Circular dichroism and magnetic circular dichroism, *Adv. Inorg. Biochem.*, 2:27.

Åkeson, Å., 1964, On the zinc content of horse liver alcohol dehydrogenase, *Biochem. Biophys. Res. Commun.*, 17:211.

KINETICS OF A GLYCINE FOR ARG-47 HUMAN ALCOHOL DEHYDROGENASE MUTANT CAN BE EXPLAINED BY LYS-228 RECRUITMENT INTO THE PYROPHOSPHATE BINDING SITE

Carol L. Stone[1], Thomas D. Hurley[1], L. Mario Amzel[2], Michael F. Dunn[3], and William F. Bosron[1]

[1]Department of Biochemistry and Molecular Biology, and Medicine
Indiana University School of Medicine
Indianapolis, IN 46202

[2]Department of Biophysics and Biophysical Chemistry
The Johns Hopkins University School of Medicine
Baltimore, MD 21205

[3]Department of Biochemistry
University of California at Riverside
Riverside, CA 92521

INTRODUCTION

Ethanol and other alcohols are metabolized in liver by alcohol dehydrogenase isoenzymes. At least five electrophoretically distinct isoenzymes are found in human liver that can be divided into three classes (Burnell & Bosron, 1989). Class I comprises the α, β, and γ subunits, while class II contains the π subunit and class III contains the χ subunit. In addition, heterogeneity exists at the β and γ loci, producing the β_1, β_2, and β_3 subunits, and the γ_1 and γ_2 subunits. Although the class II and III subunits associate only with themselves to form active homodimers, the class I isoenzymes associate to form both homodimers and heterodimers. Thus, the liver can contain as many as 17 active alcohol dehydrogenase isoenzyme forms.

The human alcohol dehydrogenase isoenzymes vary dramatically in kinetic properties (**Table 1**). Even among the class I isoenzymes, which differ only 7% in amino acid sequence, kinetic parameters vary substantially. For instance, the maximal

Table 1. Kinetic Constants of Human Liver Alcohol Dehydrogenase Isoenzymes (pH 7.5)

	αα	$\beta_1\beta_1$	$\beta_2\beta_2$
Amino Acid 47	Gly	Arg	His
V^f_{max} (min^{-1})	27	9.2	400
K_m^{NAD+} (μM)	13	7.4	180

Kinetic constants are from Burnell et al, 1989.

rate for ethanol oxidation (V^f_{max}) of $\beta_1\beta_1$ is 9.2 min^{-1}, a value about 3-times slower than that of αα, but over 100-times slower than that of $\beta_2\beta_2$. The K_m values for NAD^+ and ethanol (K_m^{NAD+} and K_m^{EtOH}, respectively) also vary considerably among these isoenzymes. The β_1 and β_2 subunits vary in amino acid sequence by only one amino acid--an Arg-47 in β_1 is substituted by a histidine in the β_2 subunit (Hempel et al., 1984; Jörnvall, et al., 1984). The α subunit, 7% different in amino acid sequence from the β subunits, contains a glycine at position 47 (von Bahr-Lindström et al., 1986). The x-ray crystal structure of the β_1(coenzyme) binary complex reveals that, like horse alcohol dehydrogenase (Eklund et al., 1984), the positively charged side chain at Arg-47 interacts with the negatively charged pyrophosphate moiety of the NAD(H) molecule (Hurley, et al., 1991b). Accordingly, changes in the charge character of the amino acid at position 47 are expected to affect steady-state kinetic values for coenzymes.

Table 2. Kinetic Constants of Expressed Alcohol Dehydrogenase Enzymes (pH 7.5)

	β47R	β47K	β47H	β47Q	β47G
Amino Acid 47	Arg	Lys	His	Gln	Gly
V^f_{max} (min^{-1})	3.8	10	370	840	2.1
K_m^{NAD+} (μM)	23	140	200	560	53

Kinetic constants are from Hurley et al. (1990, 1991a).

The effect of amino acid position 47 on the steady-state kinetic parameters of the ββ isoenzymes was investigated by site directed mutagenesis (Hurley, et al., 1990, 1991a). Mutants containing amino acids arginine (β47R or $\beta_1\beta_1$), histidine (β47H or $\beta_2\beta_2$), lysine (β47K), glutamine (β47Q), and glycine (β47G) were expressed in *E. coli*, and the steady-state kinetics were evaluated (**Table 2**). The V^f_{max} value generally correlates with the base strength of the side chain at position 47, in which a decrease in base strength is associated with an increase in the V^f_{max}. As base strength decreases, the K_m^{NAD+} value also increases; the β47Q enzyme exhibits the highest K_m^{NAD+} and V^f_{max} values yet found for a class I human isoenzyme. The β47G mutant, however, does not fit this general relationship. The absence of a basic side chain at position 47 does not substantially increase the K_m^{NAD+} of β47G. Rather, its steady-state kinetic parameters resemble more closely those of mutants with the strongest base strength at the position, β47R and β47K.

To examine why the β47G mutant and αα enzymes exhibit steady-state kinetics similar to that of $\beta_1\beta_1$ (β47R), stopped-flow kinetics were used to evaluate the coenzyme apparent association and dissociation rate constants of the β47R, β47H, and β47G mutants. X-ray crystallography of the β47G(NAD^+) complex was used to visualize the coenzyme binding site structure.

MATERIALS AND METHODS

Stopped-flow Kinetics

The apparent NADH association rate constant (k_{on}^{NADH}, $\mu M^{-1}sec^{-1}$) was determined by mixing enzyme with varying concentrations of NADH and excess 4-*trans*-(N,N-dimethylamino)cinnamaldehyde (DACA; Dunn, et al., 1975; **equation 1**). The increase of absorbance at 464 nm was measured with time. Estimates of k_{on}^{NADH} were determined by linear regression of 1/τ versus NADH concentration (DeTraglia, et al., 1977), using 10 to 14 data determinations at 5 to 7 NADH concentrations. Final concentrations were 3 μM enzyme active sites, 75 μM DACA, and 9 to 50 μM NADH.

$$E + NADH \underset{}{\overset{k_{on}^{NADH}}{\rightleftharpoons}} E{\cdot}NADH + DACA \rightarrow \underset{A464nm}{E{\cdot}NADH{\cdot}DACA} \qquad (1)$$

The apparent NADH dissociation rate constant (k_{off}^{NADH}, sec^{-1}) was determined by monitoring the decrease in enzyme-bound NADH fluorescence (Excitation = 330 nm, Emission = 410 ± 10 nm interference filter; **equation 2**; DeTraglia, et al., 1977). Enzyme, pre-incubated with enough NADH to bind approximately 95% of the active sites, was mixed with varying concentrations of NAD^+ and excess 4-methylpyrazole. Eight to 10 data points at 4 to 5 NAD^+ concentrations were used to determine k_{off}^{NADH} by linear regression of 1/τ versus 1/NAD^+ concentration (Kvassman and Pettersson, 1979). Final concentrations were 3 μM enzyme active sites, 6 μM NADH with β47R and β47G or 25 μM NADH with β47H, 15 mM 4-methylpyrazole, and 0.5 to 17 mM NAD^+.

$$\underset{\substack{Ex330nm \\ Em410nm}}{E{\cdot}NADH} + NAD^+ \overset{k_{off}^{NADH}}{\rightleftharpoons} E{\cdot}NAD^+ + 4\text{-methylpyrazole} \rightarrow E{\cdot}NAD^+{\cdot}4\text{-methylpyrazole} \qquad (2)$$

The apparent NAD^+ association rate constant (k_{on}^{NAD+}, $\mu M^{-1}sec^{-1}$) was determined by mixing enzyme with varying NAD^+ concentrations, and excess *n*-butanol and DACA (**equation 3**; Kvassman and Pettersson, 1979). After NAD^+ binding, the excess *n*-

butanol was rapidly oxidized, forming butanal and NADH. Excess DACA was used to drive butanal dissociation and form the ternary enzyme(NADH,DACA) complex. Complex formation was monitored at 464 nm (Dunn, et al., 1975). Estimates of k_{on}^{NAD+} were determined by linear regression of $1/\tau$ versus NAD^+ concentration, using 14 to 15 values of $1/\tau$ at 7 to 8 NAD^+ concentrations. Final concentrations were 3 μM enzyme active sites, 75 μM DACA, 5 mM *n*-butanol, and 20 to 60 μM NAD^+.

$$E + NAD^+ \underset{}{\overset{k_{on}^{NAD+}}{\rightleftharpoons}} E{\cdot}NAD^+ + n\text{-butanol} \rightarrow E{\cdot}NAD^+{\cdot}n\text{-butanol} \rightarrow E{\cdot}NADH{\cdot}n\text{-butanal} + DACA \quad (3)$$

$$\downarrow\uparrow$$

$$E{\cdot}NADH{\cdot}DACA$$
$$A_{464nm}$$

The apparent NAD^+ dissociation rate constant (k_{off}^{NAD+}, sec^{-1}) was determined by mixing enzyme, pre-equilibrated with enough NAD^+ to saturate 95% active sites, with varying NADH concentrations and excess DACA (**equation 4**). The increase in absorbance at 464 nm was monitored. Estimates of k_{off}^{NAD+} were determined by linear regression of $1/\tau$ versus 1/NADH concentration, using 8 to 14 data points at 5 to 7 NADH concentrations. Final concentrations were 3 μM enzyme active sites, 75 μM DACA, 10 to 250 μM NADH, and 25 μM NAD^+ with β47R and β47G or 250 μM NAD^+ with β47H.

$$E{\cdot}NAD^+ + NADH \overset{k_{off}^{NAD+}}{\rightleftharpoons} E{\cdot}NADH + DACA \rightarrow E{\cdot}NADH{\cdot}DACA \quad (4)$$
$$A_{464nm}$$

Experiments were performed on either a Durrum or a HI-TECH SF-51 stopped-flow instrument. Exponential traces were analyzed with either KINFIT (OLIS; Jefferson, GA) or HI-TECH software, and regression analysis was performed with SAS (Cary, NC). All experiments were performed at 25 °C in 15 mM PIPES, 15 mM BICINE buffer, pH 7.5. Final pH varied by no more than ±0.2 pH unit. Goodness of fit was evaluated by standard error (all estimates exhibited standard errors within 10%), and correlation coefficient. Except where noted, all traces exhibited single exponential behavior. Apparent rate constant values at pH 7.5 were obtained from regression fits of pH-profiles.

Steady-state Kinetics

Steady-state kinetics were performed on a Gilford Response spectrophotometer. Values of V^f_{max} were obtained by varying concentrations of NAD^+ in a mixture containing enzyme and 33 mM ethanol. Values of V^r_{max} were obtained by varying concentrations of NADH in a mixture containing enzyme, and 1 mM acetaldehyde with β47R and β47G, or 10 mM acetaldehyde with β47H. Experiments were performed in either 15 mM PIPES, 15 mM BICINE, pH 7.5, or in 15 mM ACES, 15 mM glycylglycine, 60 mM CAPS, pH 7.5, at 25°C. Initial rates were obtained by monitoring either the formation or depletion of NADH at 340 nm over one to four minutes. Product formation during this time interval was less than 10% the initial substrate concentration. Estimates of V_{max} were obtained from computer programs as described (Stone, et al., 1989); standard errors were within 10% of the constant's value.

Crystallization and Data Collection on β47G

The recombinant enzyme was purified as described (Hurley, et al., 1990). Crystals of the binary β47G(NAD^+) complex were grown using the vapor diffusion method and the sitting drop configuration. Typically, 3 μl of a 10 mg/ml solution of the enzyme was mixed with 3 μl of precipitant solution which contained 50 mM sodium phosphate, pH 7.5, 6 mM NAD^+, and 15% (wt/vol) PEG 8000. The triclinic crystals formed as flat plates and grew to maximal size within one week. Two crystals (approximate dimensions 1.0 x 0.3 x 0.2 mm) were used to collect the native data set at 23 °C with a Nicolet multiwire area detector equipped with a Rigaku Rotaflex RU-200B rotating-anode generator. The data sets were scaled using the software package XENGEN (Howard, et al., 1987). The final merged data set contained 23,820 reflections to 2.38 Å (70% complete; 80% complete to 2.5 Å).

Molecular Replacement and Crystallographic Refinement

The structure of the mutant β47G enzyme was solved by molecular replacement using as the search model the human binary complex dimer with NAD^+ (Hurley, et al., 1991b). The Crowther rotation function (Crowther, 1972), as implemented in the program package, MERLOT (Fitzgerald, 1988), with the data between 10 and 4 Å, indicated the rotation solution was within 3° of the origin. Rigid-body refinement in X-PLOR (Brunger, 1988) was used to properly align the model structure in the new cell.

Refinement of the model structure was accomplished using the program X-PLOR and conjugant gradient refinement protocols. Following each round of refinement $2F_o$-F_c, omit $2F_o$-F_c, and F_o-F_c maps were calculated and the current model was inspected for errors using the program CHAIN (Sack, et al., 1991). The corrected model was then resubmitted for crystallographic refinement until convergence.

RESULTS

The apparent NADH association rate constant (k_{on}^{NADH}) of the mutant enzyme β47G was not appreciably different from either β47R or β47H (**Table 3**); the apparent rate constant varied in magnitude from 2.8 to 5.8 μM^{-1} sec^{-1}. The k_{off}^{NADH} value of β47G (0.08 sec^{-1}) was nearly identical with that of β47R, but was nearly 100-times slower than that of β47H. The ratio of the apparent NADH dissociation and association rate constants (the equilibrium NADH dissociation constant) reflected this dramatic difference in k_{off}^{NADH}. The affinity for NADH of β47G was similar to that of β47R but 100-times higher than that of β47H. The experiments to measure k_{off}^{NADH} of β47G exhibited a second faster exponential, the value of which did not change with varying NAD^+ concentrations. This second exponential was identified as $k_{on}^{4\text{-methylpyrazole}}$, 0.8 sec^{-1} **(equation 2).**

Table 3. Apparent Dissociation and Association Rate Constants for NADH (pH 7.5)

Apparent rate constant	β47G	β47R	β47H
k_{off}^{NADH} (s^{-1})	0.08	0.07	7.5
k_{on}^{NADH} ($\mu M^{-1}s^{-1}$)	3.0	5.8	2.8
$k_{off}^{NADH}/k_{on}^{NADH}$ (μM)	0.03	0.01	2.7

The apparent NAD^+ association rate constant (k_{on}^{NAD+}) of the three mutant enzymes varied from 0.14 to 0.90 $\mu M^{-1}sec^{-1}$ (**Table 4**). A much larger difference occurred among the mutants with the apparent NAD^+ dissociation rate constant (k_{off}^{NAD+}). The k_{off}^{NAD+} value of β47G was 4.9 sec^{-1}, a value about 3-times faster than that of β47R, but was 20-times slower than that of β47H. The difference in individual rate constants was reflected in the ratio of apparent NAD^+ dissociation and association rate constants, in which the affinity for NAD^+ of β47G was 6-times lower than that of β47R, but almost 70-times higher than that of β47H. A comparison of apparent coenzyme dissociation rate constants with steady-state V_{max} values is shown in **Table 5**. With all three mutants, the k_{off}^{NADH} and k_{off}^{NAD+} values coincided with the V^f_{max} and V^r_{max} values, respectively.

Table 4. Apparent Dissociation and Association Rate Constants for NAD^+ (pH 7.5)

Apparent rate constant	β47G	β47R	β47H
k_{off}^{NAD+} (s^{-1})	4.9	1.7	98
k_{on}^{NAD+} ($\mu M^{-1}s^{-1}$)	0.44	0.90	0.14
$k_{off}^{NAD+}/k_{on}^{NAD+}$ (μM)	11	1.9	700

The structure of the β47G enzyme in a binary complex with NAD^+ was solved by molecular replacement using the human $\beta_1\beta_1(NAD^+)$ binary complex as the model structure. The current model, including 34 solvent molecules, possesses an R-factor of 20.1% for the data between 7.0 and 2.5 Å. Although the refinement of the structure has not been completed, it is clear that a rearrangement of the protein structure has occurred in the vicinity of the coenzyme binding site.

As a result of the substitution of glycine for Arg-47, the residues from 226 to 230 have shifted in position relative to their locations in the $\beta_1\beta_1(NAD^+)$ binary complex. The result of this change in structure is that the terminal side chain nitrogen of Lys-228 in β47G interacts through an ordered water molecule with the phosphate oxygens of NAD^+. In $\beta_1\beta_1$, Lys-228 interacted with the 2'-hydroxyl oxygen of the adenine ribose.

Table 5. Rate-limiting Step for Ethanol Oxidation and Acetaldehyde Reduction (pH 7.5)

Kinetic constant (s^{-1})	β47G	β47R	β47H
k_{off}^{NADH}	0.08	0.07	7.5
V^f_{max}	0.04	0.045	7
k_{off}^{NAD+}	4.9	1.7	98
V^r_{max}	3.3	1.5	60

These differences in interactions reflect a shift in C_α position of Lys-228 by about 1.0 Å, and a repositioning of the Lys-228 ε-nitrogen by 1.5 Å in the structure of the mutant enzyme relative to the wild-type enzyme (the overall root mean square difference in the positions of backbone atoms is less than 0.45 Å).

DISCUSSION

While the substitution of Arg-47 in the human $\beta_1\beta_1$ alcohol dehydrogenase by histidine ($\beta_2\beta_2$) or glutamine results in large changes in steady-state parameters, it was surprising that substitution by glycine at position 47 had a relatively small effect on steady-state kinetics (**Table 2**). The similar substitution of Arg-47 by glycine occurs with the human αα isoenzyme, and the coenzyme K_m and V_{max} values of αα are more like human $\beta_1\beta_1$ than $\beta_2\beta_2$ (**Table 1**). Also, the kinetics of an α-like isoenzyme from Rhesus monkey are more like the mutant with Arg-47 than mutants with histidine at position 47 (Light, et al., 1992). The present stopped-flow kinetic analysis of human $\beta_1\beta_1$ clarifies the coenzyme association and dissociation steps that are affected by the substitutions. The glycine substitution at position 47 produces an enzyme with apparent NADH and NAD^+ dissociation rate constants much more similar to $\beta_1\beta_1$ (Arg-47) than to $\beta_2\beta_2$ (His-47) (**Tables 3 and 4**). Also, apparent k_{off}^{NADH} and k_{off}^{NAD+} values of β47G coincide with steady-state V^f_{max} and V^r_{max} values, respectively (**Table 5**). Therefore, the rate-limiting step for ethanol oxidation and acetaldehyde reduction is coenzyme dissociation.

X-ray crystallography of the triclinic β47G(NAD^+) crystals reveals that the structure resembles the triclinic "closed" $\beta_1\beta_1$(NAD^+) or horse EE(NAD^+) complex more closely than the orthorhombic horse EE "open" apoenzyme complex (Hurley et al., 1991b; Eklund et al., 1984; Eklund et al., 1976). In $\beta_1\beta_1$ and horse EE alcohol dehydrogenase, the ε-amino group of Lys-228 interacts with either the 2'- or 3'-ribose hydroxyl. The new structure of β47G indicates that the ε-amino group of Lys-228 in the coenzyme binding site is recruited into the pyrophosphate binding site. Hence, the high negative charge of the pyrophosphate group of NAD^+ is neutralized, in part, by Lys-228 in β47G in contrast to the similar role by Arg-47 in $\beta_1\beta_1$. This explains the higher than expected coenzyme affinity of β47G relative to that of β47H or β47Q. The interaction of the positively charged side chain of Lys-228 with NAD(H) pyrophosphate gives rise to low k_{off}^{NADH} and k_{off}^{NAD+} values of the β47G mutant, so that the V_{max} values for ethanol oxidation and acetaldehyde reduction resemble those of β47R rather than those with the weak base β47H (**Table 5**).

The kinetic mechanism of β47R, β47H and other mutants appears to be Ordered Bi-Bi, while that of β47G, based on initial product inhibition studies, is consistent with a Rapid Equilibrium Random mechanism (Hurley, et al., 1990). The results here indicate that the rate-limiting step is not some internal step, but rather coenzyme dissociation (**Table 5**). If the mechanism is fully random, then acetaldehyde must be co-rate-limiting. Alternatively, the enzyme may follow the hybrid Random/Ordered mechanism proposed to direct the kinetics of the Rhesus monkey α-isoenzyme (Light et al., 1992).

ACKNOWLEDGEMENTS

This work was supported by R37-AA02342, R01-AA07117, DMB-8703697, DMB-9107808, and post doctoral training fellowships to CLS and TDH on T32-AA07642.

REFERENCES

Brünger, A., 1988, Crystallographic refinement by simulated annealing: application to a 2.8 Å resolution structure of aspartate aminotransferase, *J. Mol. Biol.*, 203:803.

Burnell, J. C., and Bosron, W. F., 1989, Genetic polymorphism of human liver alcohol dehydrogenase and kinetic properties of the isoenzymes, in "Human metabolism of alcohol," Volume II, Crow K. E. and Batt R. D., eds., CRC Press, Florida, p. 65.

Crowther, R. A.,1972, The fast rotation function, in "The molecular replacement method," Rossmann M. G., ed., Gordon and Breach, New York, p. 173.

DeTraglia, M. C., Schmidt, J., Dunn, M. F., and McFarland, J. T., 1977, Liver alcohol dehydrogenase-coenzyme reaction rates, *J. Biol. Chem.*, 252:3493.

Dunn, M. F., Biellmann, J. -F., Branlant, G., 1975, Roles of zinc ion and reduced coenzyme in horse liver alcohol dehydrogenase catalysis. The mechanism of aldehyde activation, *Biochemistry*, 14:3176.

Eklund, H., Nordström, B., Zeppezauer, E., Söderlund, G., Ohlsson, I., Boiwe, T., Soderberg, B. O., Tapia, O., Bränden, C.-I., and Åkeson, Å., 1976, Three-dimensional structure of horse liver alcohol dehydrogenase at 2-4 Å resolution, *J. Mol. Biol.*, 102:27.

Eklund, H., Samama, J.-P., and Jones, T. A., 1984, Crystallographic investigations of nicotinamide adenine dinucleotide binding to horse liver alcohol dehydrogenase, *Biochemistry*, 23:5982.

Fitzgerald, P. M. D., 1988, MERLOT, an integrated package of computer programs for the determination of crystal structures by molecular replacement, *J. Appl. Crystallogr.*, 21:273.

Hempel, J., Bühler, R., Kaiser, R., Holmquist, B., de Zalenski, C., von Wartburg, J.-P., Vallee, B., and Jörnvall, H., 1984, Human liver alcohol dehydrogenase. 1. The primary structure of the $\beta_1\beta_1$ isoenzyme, Eur. *J. Biochem.*, 145:437.

Howard, A. J., Gilliland, G. L., Finzel, B. C., Poulos, T. L., Ohlendorf, D. H., and Salemme, F. R., 1987, The use of a imaging proportional counter in macromolecular crystallography, *J. Appl. Crystallogr.*, 20:383.

Hurley, T. D., Edenberg, H. J., and Bosron, W. F., 1990, Expression and kinetic characterization of variants of human $\beta_1\beta_1$ alcohol dehydrogenase containing substitutions at amino acid 47, *J. Biol. Chem.*, 265:16366.

Hurley, T. D., Ehrig, T., Edenberg, H. J., and Bosron, W. F., 1991a, Characterization of human alcohol dehydrogenases containing substitutions at amino acids 47 and 51, *Enzym. Mol. Biol. Carb. Metab.*, 3:271.

Hurley, T. D., Bosron, W. F., Hamilton, J. A., and Amzel, L. M., 1991b, Structure of human $\beta_1\beta_1$ alcohol dehydrogenase: catalytic effects of non-active-site substitutions, *Proc. Natl. Acad. Sci. USA.*, 88:8149.

Jörnvall, H., Hempel, J., Vallee, B. L., Bosron, W. F., and Li, T.-K., 1984, Human liver alcohol dehydrogenase: amino acid substitution in the $\beta_2\beta_2$ Oriental isozyme explains functional properties, establishes an active site structure, and parallels mutational exchanges in the yeast enzyme, *Proc. Natl. Acad. Sci. USA.*, 81:3024.

Kvassman, J., and Pettersson, G., 1979, Effect of pH on coenzyme binding to liver alcohol dehydrogenase, *Eur. J. Biochem.*, 100:115.

Light, D. R., Dennis, M. S., Forsythe, I. J., Liu, C.-C., Green, D. W., Kratzer, D. A., Plapp, B. V., 1992, α-isoenzyme of alcohol dehydrogenase from monkey liver; cloning, expression, mechanism, coenzyme and substrate specificity, *J. Biol. Chem.*, 267:12592.

Sack, J. S., 1988, CHAIN - A crystallographic modelling program, *J. Mol. Graphics*, 6:224.

Stone, C. L., Li, T.-K., and Bosron, W. F., 1989, Stereospecific oxidation of secondary alcohols by human alcohol dehydrogenases, *J. Biol. Chem.*, 264:11112.

von Bahr-Lindström, H., Höög, J.-O., Hedén, L.-O., Kaiser, R., Fleetwood, L., Larsson, K., Lake, M., Holmquist, B., Holmgren, A., Hempel, J., Vallee, B. L., and Jörnvall, H., 1986, cDNA and protein structure for the α subunit of human liver alcohol dehydrogenase, *Biochemistry*, 25:2465.

SITE-DIRECTED MUTAGENESIS OF MAMMALIAN ALCOHOL AND SORBITOL DEHYDROGENASES MAP FUNCTIONAL DIFFERENCES WITHIN THE ENZYME FAMILY

Jan-Olov Höög[1], Christina Karlsson[1], Hans Eklund[2], Robert Shapiro[3] and Hans Jörnvall[1]

[1]Department of Chemistry I, Karolinska Institutet, S-104 01 Stockholm, Sweden, [2]Department of Molecular Biology, Swedish University of Agricultural Sciences, Biomedical Center, S-751 24 Uppsala, Sweden and [3]Center for Biochemical and Biophysical Sciences and Medicine, Harvard Medical School, Boston, MA, USA

INTRODUCTION

Mammalian alcohol dehydrogenases (ADH) and sorbitol dehydrogenase (SDH) belong to the same protein super-family of zinc-containing ADHs (Jörnvall et al., 1989). ADHs have two zinc atoms per subunit, one catalytic and one structural, while SDH only harbors the catalytic zinc atom (Jeffery et al., 1983). The human class I ADH consists of several isozymes that are formed from three different types of subunit, α, β, and γ (Smith et al., 1971; cf. Jörnvall et al., 1989). They combine to form active homo- and heterodimers, which constitute the enzymes responsible for the main ethanol metabolism in liver but exhibit differences in substrate specificity and catalytic efficiency. Steroid dehydrogenase activity has only been shown for isozymes containing the γ subunit (McEvily et al., 1988), and these isozymes are the only ones that can be inhibited by testosterone (Mårdh et al., 1986). Moreover, $\beta\beta$ but not $\gamma\gamma$ ADH has a large pH dependence when ethanol is the substrate (Bosron et al., 1983). In addition to these class I ADHs, at least four more classes of mammalian ADH exist, differing enzymatically, structurally and in organ distribution (Jörnvall et al., 1989; Parés et al., 1992).

Active site models of the human class I ADH subunits have been constructed using computer graphics (Eklund et al., 1987), based on the crystallographically determined structure of the horse liver EE enzyme (Eklund et al., 1976; 1984). Recently, a three-dimensional structure has been determined also directly for the human $\beta\beta$ ADH (Hurley et al., 1991). Residue differences between the isozymes are quite extensive and all enzymatic differences have not been explained (Table 1), e.g. the 3β-hydroxy-5β-steroid activity and the inhibitory effects of testosterone on the human γ isozymes (McEvily et al., 1988; Mårdh et al., 1986). Within the human class I subunits, differences occur at no less than 35 positions, several of which affect residues with a functional role, and 21 of which affect the isozyme pair β/γ (Jörnvall et al., 1989; Eklund et al., 1987).

SDH and ADH have few substrates in common, which is noteworthy in view of the broad substrate specificity of ADH (Maret and Auld, 1988). Based on the fitting of the initially characterized primary structure of an SDH, that of the sheep enzyme, to the three-dimensional model of horse liver ADH, it has been proposed that the coordination of the catalytic zinc is different in the two enzymes (Eklund et al., 1985). Of the two cysteine zinc ligands in ADH, one has been proposed to be glutamate in SDH, while the third ligand to the catalytic zinc is a histidine in both enzymes. The glutamate is conserved in all SDHs examined (Karlsson et al., 1991) and extended X-ray absorption fine structure studies indicate that a glutamate is one of the ligands to the zinc atom (Feiters and Jeffery, 1989).

Table 1. Established differences between the human $\beta\beta$ and $\gamma\gamma$ alcohol dehydrogenases compared to the mutant enzyme with β48S subunits.

	$\beta\beta$	$\gamma\gamma$	β48S
Km for methanol (mM)	6	74	26
Km for cyclohexanol (mM)	14.5	0.042	0.280
pH effect	+	-	(-)
Steroid dehydrogenase activity	-	+	+
Testosterone inhibition	-	+	+
Amino acid residue at position 48	Thr	Ser	Ser
Total number of residue differences β/γ	21		

To map functional differences within the protein family, site-directed mutageneses were performed on the human β ADH subunit (Höög et al., 1992) and of rat SDH. For the class I ADH the pronounced different characteristics for the γ subunit was studied and focused on position 48 by production of recombinant proteins with different residues at this position, chosing both minor differences [Thr→Ser mutations, keeping the hydrogen bonding capacity (Eklund et al., 1984; 1990)] and major differences (Thr→Ala mutations, abolishing the hydrogen bonding pattern). The importance of the zinc-ligands in general and the uncertainty of the ligands to the catalytic zinc in SDH started the site-directed mutagenesis of zinc ligands in SDH, once a cDNA-clone coding for mammalian SDH had been isolated (Karlsson et al., 1991). The question early raised about the zinc ligands in SDH, also involved a Cys as the putative ligand.

The results obtained are conclusive and form direct support for the importance of position 48 in ADH in substrate binding and catalytic efficiency. Furthermore, the mutation of Glu155 in SDH shows that this residue is a true ligand to the catalytic zinc atom.

MATERIALS AND METHODS

Plasmid Constructions

An expression plasmid pADHex, giving non-acetylated $\beta\beta$ ADH in *E. coli*, has been described (Höög et al., 1987) and used for expression of variants of human class I ADH (Höög et al., 1992a). The same expression system was now also used for expression of rat SDH. The cDNA coding for SDH was from two overlapping clones, λSDH1 and λSDH2

(Karlsson et al., 1991). The cDNA fragments were mutated at both the 5' and the 3' ends to give optimal distances when inserted into the expression plasmid. Two different constructs in M13mp8, harboring the 5'-region and the 3'-region, respectively, were used for site-directed mutagenesis. An *Eco*RI site was mutated just ahead of the ATG start codon and a *Pst*I site was mutated just after the TGA stop codon. The two mutated cDNA-fragments were liberated from M13mp8 and ligated into the expression vector, yielding pSDHex used for expression of native recombinant SDH. Glu155 was mutated to Cys and Ala, respectively. A third mutation was created, changing Cys164 to Ala. The mutations were carried through on single-stranded DNA, and followed the protocol of Taylor et al. (1985). Mutated DNA-fragments were liberated from the replicative form of M13mp8 before ligation into pSDHex. For ADH, Thr48 was changed to Ser and Ala, respectively. The methods used followed the same principles as those for SDH and have been described (Höög et al., 1992a). One further mutation was created, changing Cys174 in class I ADH to Glu, for comparison with possible zinc ligands in SDH. The plasmids were isolated according to the Qiagen ion-exchange method (Diagen) and digested DNA was purified with glassmilk (Geneclean; Bio101). All constructions were verified with dideoxy chain termination sequence analysis (Sanger et al., 1977) and restriction mapping.

Protein Purification

Recombinant variants of human ADH were expressed in *E. coli* strain HB101 after a burst of 0.2 mM isopropyl-β-thiogalactopyranoside. An *E. coli* lacI^q strain, TG1, was also tested but strain HB101 gave a slightly higher yield. After harvesting, the cells were disrupted in 1 mM dithiothreitol, 10 mM Tris-Cl, pH 8, and sonicated before centrifugation for 60 min at 48,000x*g*. The supernatant was applied to a DEAE column (DE-52, Whatman). The void volume, containing recombinant ADH, was applied to a 5'AMP-Sepharose column (Pharmacia-LKB Biotechnology), washed with 100 mM Tris-Cl, pH 8, and the recombinant enzyme was eluted with 2.4 mM NAD^+ in 1 mM dithiothreitol, 100 mM Tris-Cl, pH 8. The pooled fractions were concentrated with microsep concentrators (Filtron Scandinavia), before a final purification step on a Superdex-200 Hi-load column (Pharmacia-LKB Biotechnology).

The variants of rat SDH, tetramers with E155C, E155A and C164A as well as native SDH, were expressed in *E. coli* strain TG1 after addition of 0.3 mM isopropyl-β-thiogalacto-pyranoside. After harvesting, the cells were disrupted in 1 mM dithiothreitol, 10 mM Tris-Cl, pH 8, and sonicated before centrifugation. The supernatant was applied to a DEAE column (DE-52, Whatman). The void volume, containing recombinant SDH, was dialyzed against 20 mM phosphate buffer, pH 6.8, and applied to a CM-cellulose column (CM-52, Whatman) equilibrated with the same buffer. The recombinant proteins were eluted with a pH-gradient and emerged at slightly different pH-values (ca. 7.1-7.4). All recombinant proteins were submitted to Western blot analysis and were analyzed for purity by SDS/polyacrylamide gel electrophoresis. Protein concentrations were measured according to the method of Bradford (1976), standardized with bovine serum albumin.

For molecular modelling the program FRODO (Jones, 1985) was used on an Evans and Sutherland PS390 graphics terminal, and was based on the crystallographically determined structure of the horse liver EE ADH (Eklund et al., 1976; 1984).

Kinetic Measurments

Enzymatic activity was determined spectrophotometrically at 340 nm using a Beckman DU 68 or a DU 8 spectrophotometer to follow the NADH formation in 0.1 M glycine/NaOH, pH 10, or in 0.1 M Na-phosphate, pH 7.5 or 7.1, 25°C. The NAD^+ concentration was 2.4 mM in the ADH assays and 1 mM in the SDH assays. The absorption coefficient utilized for NADH was $6.22x10^3$ $M^{-1}cm^{-1}$, and one unit of activity was defined as the amount of enzyme required to produce 1 μmol product per min. Substrates were of analysis grade and were used without further purification. NAD^+ grade III, testosterone, and 3β-hydroxy-5β-androstan-17-one, sorbitol, 2,3-butanediol, fructose, ribitol, L-threitol and xylitol were all from Sigma. 3β-Hydroxy-5β-androstan-17-one and testosterone were dissolved in 50% aqueous acetonitrile, yielding 1.8% acetonitrile in the final reaction mixture.

Table 2. Kinetic constants for native and mutated recombinant human ADH. For experimental conditions see Materials and Methods. N = no activity.

	recombinant dimer with β48T subunits			recombinant dimer with β48S subunits		
Alcohol	K_m (μM)	k_{cat} (min^{-1})	k_{cat}/K_m (min^{-1} mM^{-1})	K_m (μM)	k_{cat} (min^{-1})	k_{cat}/K_m (min^{-1} mM^{-1})
Methanol	6500	8.2	1.3	26000	68	2.6
Ethanol	1100	11	1	310	64	206
Octanol	13	10	769	11	27	2450
Benzyl alcohol	65	12	246	40	37	925
Cyclohexanol	10900	12	1.1	280	25	89
3β-Hydroxy-5β-androstan-17-one	N	N	N	1000	13	15
Ethanol (pH 7.5)	100	8.4	84	160	8.4	52

The program ENZYME (Lutz et al., 1986), a weighted non-linear regression analysis program, was used to fit all lines to data points and to calculate kinetic parameters. All kinetic values given are the mean from at least three different measurements.

Zinc analysis

The zinc content of the homogeneous SDHs were determined by atomic absorption spectroscopy on a Perkin-Elmer 5000 equipped with an HGA 500 graphite furnace and AS 40 autosampler, while the protein content was determined by amino acid analysis using the pico-tag method. Samples were prepared in Centricon-30 microconcentrators (Amicon) that had been rendered metal-free by rinsing once with 2 ml 1 mM HCl and then several times with metal-free 5 mM Hepes, pH 7.6. Various volumes containing about 50 μg of recombinant SDH were added and the volume brought to 500 μl with Hepes buffer. In duplicate samples Zincon (5 μl of a 11.5 mM solution) was added to remove loosely bound zinc. All samples were concentrated to a minimal volume (about 50 μl) before analysis.

RESULTS

Recombinant ADH was expressed from four different plasmids, pADHex, pADHexβ48S, pADHexβ48A and pADHexβ174E, yielding dimers with the non-acetylated forms β48T, β48S, β48A and β174E, respectively. Distances between the ribosome binding site and the ATG start codon were confirmed for all constructions by the dideoxy sequencing method, giving 13 bp in pADHex and pADHexβ174E, and 10 bp in both pADHexβ48S and pADHexβ48A. The latter two constructs were obtained through a mutation of an *Eco*RI restriction site just ahead of the ATG start codon. These two expression plasmids gave an approximately 30% higher yield of protein after induction with isopropyl-β-thiogalactopyranoside in *E. coli* HB101 cells than the two plasmids with 3 bp more between the ribosome binding site and the start codon. A three-step purification scheme, including a final gel-permeation step, gave homogeneous protein with a subunit of approximately 40 kDa, as judged from SDS/polyacrylamide gel electrophoresis. The protein expressed from pADHexβ174E was not stable during purification and was therefore not possible to isolate as a homogeneous protein, but gave a positive Western-blot signal from the crude extract. The mutant with β48A subunits was inactive and the protein was purified according to the scheme for the two active forms. Notably, the β48A protein, although inactive, could be monitored by Western-blot analysis and was also possible to bind to AMP-Sepharose for subsequent elution with NAD^+.

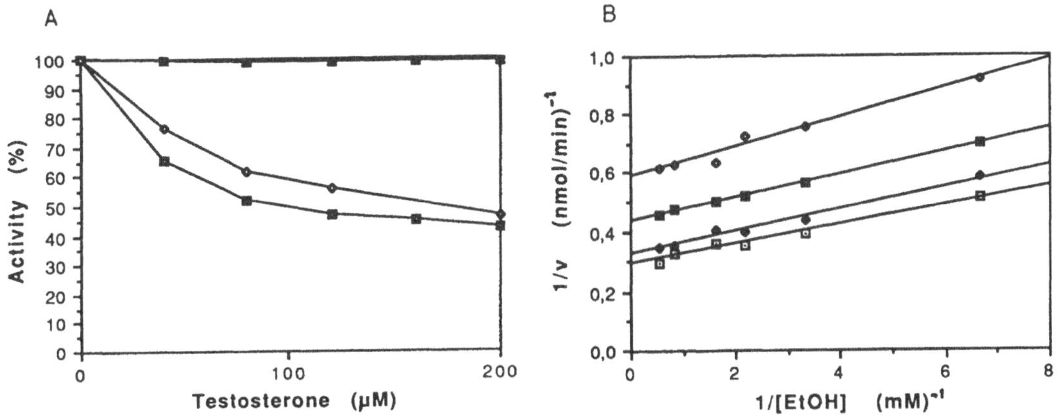

Figure 1. Testosterone inhibition of mutated ADH with β48S subunits.
A: Activity in percent with 3 mM ethanol at different concentrations of testosterone at pH 7.5 (◇), 0.1 M NaP_i-buffer, and pH 10 (□), 0.1 M glycine/NaOH-buffer. Recombinant ADH with β48T subunits, for comparison (◆,■) at both pHs. B. Lineweaver-Burk plot with different concentrations of ethanol at testosterone concentrations of 0 μM (⊡), 20 μM (◆), 40 μM (□) and 80 μM (◇), 25°C, in glycine/NaOH, pH 10.

For the enzyme with β48T subunits, the K_m differences for ethanol at pH 10 and pH 7.5 were about 10-fold but for the mutant with β48S subunits the pH difference was less than two-fold. With methanol and cyclohexanol, large differences were noted between dimers containing β48T and those containing β48S. For the mutated enzyme, with β48S subunits, the K_m-value was lowered from 11 to 0.28 mM with cyclohexanol, and was raised from 6 to 26 mM with methanol. Generally, an increase of between 2 to 8-fold was observed in k_{cat}-values for the mutated variant, which in most cases resulted in high k_{cat}/K_m ratios. Ethanol and cyclohexanol gave rise to the largest increases in catalytic efficiency, about 200 and 80-fold, respectively (Table 2). However, the k_{cat}/K_m ratio for the mutated variant with cyclohexanol is still much lower than that for the liver isolated $\gamma\gamma$ isozyme. Strikingly, the mutated enzyme with β48S subunits, could oxidize 3β-hydroxy-5β-androstan-17-one, although with a high K_m (1 mM) resulting in a low k_{cat}/K_m ratio (3% of that for the $\gamma\gamma$ isozyme). Furthermore, the mutant was inhibited to 50% by testosterone at a concentration of 100 μM (Fig. 1). A double reciprocal plot of 1/v *versus* 1/[EtOH] approaches an uncompetitive inhibition pattern (Fig. 1), with a K_i-value of 69 μM. However, a noncompetitive pattern may also appear possible, with an apparent K_i-value of 86 μM, if another calculation model is used. Dimers with β48T subunits did not show any activity with the 3β-hydroxy-5β-androstan-17-one and were not affected by the testosterone treatment.

Model building shows that the Thr→Ser exchange gives the substrate pocket a larger inner sphere. In the γγ isozyme the active site pocket is even wider, amplified by Val at position 141 instead of Leu in ββ. To some extent this is compensated in ββ by Val at position 318 (at the middle part of the active site pocket) instead of Ile in γγ. The residue at position 48 makes hydrogen bonding contact both with the NAD^+ and the alcohol substrate as outlined in Figure 2. With Ala at this position, it is impossible to obtain the hydrogen bonding bridge from the alcohol substrate over the hydroxyl group at position 48 to the ribose of the NMN part of NAD^+. Furthermore, substrate-docking experiments show that the larger Thr residue can give one extra van der Waals contact with small substrates, thus lowering the K_m-value for methanol. The smaller Ser residue gives a larger space for bulky substrates, explaining the lower K_m-value with cyclohexanol for the isozymes with γ subunits.

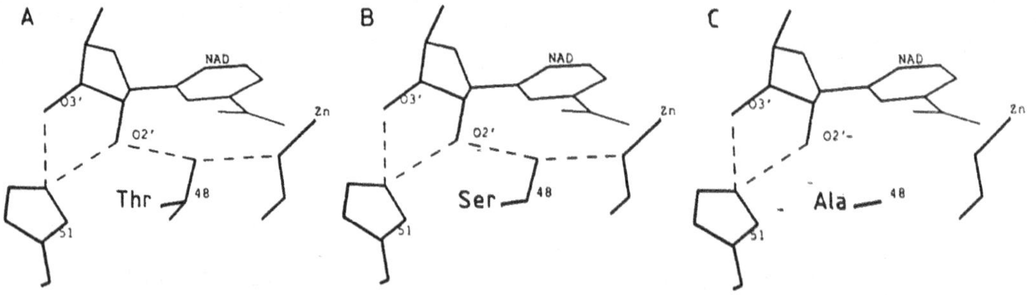

Figure 2. The active center of human ADH from computer graphic modelling. Different amino acid residues at position 48 showing hydrogen bonding between the amino acid residue and the coenzyme/substrate are outlined with broken lines. Ethanol as substrate, and part of the NAD^+ are shown. A: Thr at position 48 (as in the human β subunit and the recombinant protein β48T). B: Ser at position 48 (as in the human γ subunit and in the mutant β48S). C: Ala at position 48 (as in the mutant β48A).

Recombinant SDH was expressed from four different plasmids, pSDHex, pSDH155C, pSDH155A and pSDH164A, yielding tetramers with the non-acetylated forms SDH155E (native), SDH155C, SDH155A and SDH164A, respectively. The distance between the ribosome binding site and the ATG start codon was 10 bp in all expression plasmids due to a mutated *Eco*RI restriction site ahead of the ATG start codon. All proteins were expressed in fairly high yield, about 1.3 mg recombinant SDH per liter medium, and the recombinant proteins were purified to homogeneity in a two-step purification scheme, involving DEAE- and CM-cellulose ion-exchange chromatography. The purified proteins were homogeneous with a subunit size of 40 kDa.

Native, recombinant rat SDH, expressed from pSDHex, with a specific activity of 4.7 U/mg protein, gave kinetic values similar to those determined for the protein isolated from human liver, but all k_{cat}-values were decreased about ten times (Table 3). K_m-values were determined for sorbitol, fructose, 2,3-butanediol, ribitol, L-threitol and xylitol. The mutated variant SDH155A was inactive and the protein was purified according to the protocol for the two active forms. The SDH155C variant showed a decrease in catalytic efficiency (k_{cat}/K_m) of 460-fold due to very high K_m-values (Table 4). The activity of this mutant dropped during the assay from a high value to a relatively low, stable value. The

Table 3. Kinetic constants for recombinant rat SDH. For experimental conditions see Material and Methods. Values for human liver SDH from (Maret and Auld, 1991).

	recombinant rat SDH			human liver SDH		
Substrate	K_m (mM)	k_{cat} (min^{-1})	k_{cat}/K_m (min^{-1} mM^{-1})	K_m (mM)	k_{cat} (min^{-1})	k_{cat}/K_m (min^{-1} mM^{-1})
Sorbitol (pH 10)	0.34	42.6	125	0.67	312	466
Sorbitol (pH 7.1)	0.68	25.2	37	0.62	102	165
Fructose (pH 7.1)	32	312	10	140	2700	19
2,3-butanediol (pH 10)	62	15	0.24	9.5	252	27
Ribitol (pH 10)	0.7	27.6	39	3.4	280	82
L-threitol (pH 10)	1.7	30	18	2.7	318	118
Xylitol (pH 10)	0.13	36.6	279	0.22	354	1609

addition of 0.2 mM $ZnSO_4$ to the assay mixture stabilized the enzyme and no significant drop in activity could be observed. The mutant SDH164A showed almost the same K_m-values for sorbitol and xylitol as the native form, but the k_{cat}-values were increased almost four times (Table 4).

The zinc content of recombinant SDH was determined by atomic absorption spectrometry. The SDH155E, SDH155C, SDH155A and SDH164A proteins contained 0.83, 0.96, 0.84 and 0.88 mol zinc per mol protein, respectively (Table 5). Treatment with a weak zinc chelator, Zincon, did not affect these values.

DISCUSSION

Adult human liver isozymes of class I ADH have two major subunit types, β and γ. They both give active enzymes with ethanol and have related properties, thus constituting typical isozyme forms differing in a few of their properties. Some of these have been the subject of many studies, i.e. the activities towards methanol, ethanol, and cyclohexanol for which the enzymatic parameters differ widely, affecting both K_m-values and turnover numbers. In addition, the γ subunit shows steroid dehydrogenase activity and is also inhibited by testosterone, while the β subunit is not affected by steroids (Mårdh et al., 1986; McEvily et al., 1988). Furthermore, the isozymes differ in pH-dependence, which is large for $\beta\beta$ with ethanol but not for $\gamma\gamma$ (Bosron et al., 1983). All these differences between isozymes with 21 residue exchanges are now being directly demonstrated in the single-residue mutant, β48S, which essentially converts the β type subunit into a γ type, accounting for much of the total isozyme differences.

Regarding the cyclohexanol differences between β and γ type isozymes, the mutant with β48S subunits has acquired most of the properties of $\gamma\gamma$. Thus, the K_m-value with $\gamma\gamma$ ADH is lower than that with $\beta\beta$ by a factor of about 350, of which the mutant accounts for a major part (a factor of about 40). The k_{cat}-value is increased as for $\gamma\gamma$, resulting in a more efficient enzyme, but the k_{cat}/K_m ratio is considerably lower than that for the $\gamma\gamma$ isozyme. Regarding the second substrate difference, that towards methanol, the differences in K_m-values between the native isozymes are in a direction opposite to that towards cyclohexanol, again with the mutant in between, with functional properties more of the γ type than of the β type (Tables 1 and 2). The k_{cat}-value is also increased for methanol, resulting in a k_{cat}/K_m ratio closer to that for γγ ADH than that for the ββ enzyme. With ethanol as substrate the k_{cat}/K_m ratio is increased about 200-fold and is very similar to that for the γγ isozyme. The same mutation, Thr→Ser, at a position corresponding to position 48 in the human enzyme has been created in the highly different yeast ADH (Creaser et al., 1991), but the results then obtained differ from those in this report. However, a reasonable explanation is that the characteristics of native yeast ADH deviate in many respects from the human enzyme, e.g. at position 57 the human enzyme has Leu which affects the substrate binding pocket (Table 6), while the corresponding position in the yeast enzyme has Trp. This may be compatible with the fact that the two different ADHs do not behave similarly.

γ Subunit-containing isozymes can oxidize steroid alcohols such as 3β-hydroxy- 5β-androstan-17-one. The mutated enzyme with β48S subunits oxidizes this steroid with a K_m of 1 mM. This is 6 times higher than for the γγ isozyme, resulting in a lower k_{cat}/K_m ratio. However, the presence of this activity as such classifies the mutant enzyme as of the γγ type. Finally, testosterone was tested as an inhibitor and appeared to be uncompetitive for the mutated protein (Fig. 1), while the liver-isolated protein with γ subunits shows a clearly

Table 4. Kinetic constants for recombinant SDH. All measurements were performed at pH 10 in glycine/NaOH.

	recombinant SDH			SDH155C			SDH164A		
Substrate	K_m (mM)	k_{cat} (min^{-1})	k_{cat}/K_m (min^{-1} mM)	K_m (mM)	k_{cat} (min^{-1})	k_{cat}/K_m (min^{-1} mM)	K_m (mM)	k_{cat} (min^{-1})	k_{cat}/K_m (min^{-1} mM)
Sorbitol	0.34	42.6	125	220	60	0.27	0.25	156	624
Xylitol	0.1	36.6	366	100	120	1.2	0.11	132	1200

noncompetitive pattern with a low K_i-value (Mårdh et al., 1986). Like liver-isolated ββ ADH, the recombinant ADH with β48T subunits, was not inhibited by testosterone. In all aspects discussed above, the single amino acid exchange turns the dimer with β48S subunits into an enzyme with characteristics of the γγ isozyme type.

All known zinc-containing ADHs, from any class or species, have Thr or Ser at position 48 (Jörnvall et al., 1987; Cederlund et al., 1991). Ser is, however, not the only requirement for steroid dehydrogenase activity. The common class I ADH from horse (E type subunit) has Ser at position 48 without showing steroid dehydrogenase activity. The horse S type enzyme was the first ADH that was shown to have this type of activity towards steroids (Pietruszko et al., 1966) and corresponds in this respect to the human γ form. The rat class I ADH has also been shown to oxidize different 3β-hydroxy steroids (Cronholm et al., 1975), and has Ser at position 48 (Jörnvall, 1974; Crabb & Edenberg, 1986), like the horse E and S subunits, the human γ subunit, and the present mutant β48S. From comparisons of these primary structures, position 115 has been postulated as one of those restricting the space in the active site pocket for steroid alcohols and shows a gap within the horse S subunit (Jörnvall, 1970; Park and Plapp, 1991). The conclusion is that Ser at position 48 is required for steroid activity but not the only requirement (Table 6), and that space in the active site pocket is crucial (Höög et al., 1992b).

Table 5. Zinc stoichiometries for recombinant SDH.

Sample	Zn (g-at/mol)	k_{cat}/K_m (sorbitol, pH 10)
Recombinant SDH	0.83	125
SDH155C	0.96	0.27
SDH155A	0.84	inactive
SDH164A	0.88	624

On the basis of the horse EE three-dimensional structure it has been suggested that Ser48 should play a role in proton transfer between the active center and the environment (Eklund et al., 1976; 1982). The Ser side-chain hydrogen-bonds to the oxygen of the substrate and to the O_2' of the ribose, which is further hydrogen-bonded to His51 [(Eklund et al., 1982), Fig. 2]. Substitutions of the residue corresponding to His51 with Gln or Glu in yeast ADH (Plapp et al., 1991) and with Gln in the human $\beta\beta$ isozyme (Ehrig et al., 1991) drastically reduce the activity. It has further been suggested that the Ser side-chain will polarize the substrate during the reaction (Eklund et al., 1982). Calculations have shown that the effect of Ser48 on the hydride transfer was to decrease the energy gap between the reactant and the product of this reaction step. The present mutant β48A and a similar yeast ADH variant (Plapp et al., 1991) demonstrate that removal of the side-chain oxygen results in an inactive enzyme. The lower activity observed with Thr rather than Ser at position 48 is probably an effect of small electrostatic differences between the two side-chains.

In terms of mechanisms, it is possible to interpret both the minor effects and the opposite directions of the $\beta\beta$ and $\gamma\gamma$ differences towards methanol, cyclohexanol and hydroxysteroids by the exchange at position 48. Thus, $\gamma\gamma$ has a larger substrate pocket than $\beta\beta$ also from the difference at position 141, where Val in γ instead of Leu in β makes the substrate pocket of γ larger than that of β48S. With a small substrate like methanol, this size difference contributes less van der Waals interactions in the case of γ and gives space for extra water. A large substrate pocket has previously been concluded also to explain the differences with ethanol between classes I and III ADH enzymes (Eklund et al., 1990), with water reducing the efficiency of hydride transfer in the oxidation of substrate (Dunn et al., 1982). In conclusion, the present results constitute a direct illustration of the differential roles of the substrate binding residues, proving the crucial importance of position 48 and correlating the isozyme differences with the class differences.

Model-building of SDH early raised the question of zinc ligands (Eklund et al., 1985). Two of the ligands, a Cys and a His, could be established by homology to class I ADH. The third ligand, corresponding to Cys174 in ADH, was more difficult to ascertain. Glu155 was a strong candidate, but the neighboring Cys164 was also stated as an alternative. All results have indicated that Glu should be the ligand, i.e. sequence determinations of SDH from different species (Karlsson et al., 1991), lack of labelling by carboxymethylation (Johansson et al., 1992) and extended X-ray absorption fine structure (Feiters and Jeffery, 1989). The results obtained with the two mutants, SDH155C and SDH155A, show that Cys can replace Glu as a ligand, but the efficiency of the enzyme decreases markedly. Cys at this position must adjust the polypeptide backbone due to a smaller side-chain and thereby give rise to changes in the active site pocket. With an Ala at the same position, a residue

Table 6. Amino acid residues lining the substrate-binding cleft of class I ADH. Species and residues in italics indicate those that have been shown to have steroid dehydrogenase activity. Δ indicates a deletion.

	inner part of binding cleft				middle and outer part of binding cleft						
Position	*48*	*93*	*140*	*141*	*57*	*115*	*116*	*117*	*294*	*318*	*110*
Horse E	Ser	Phe	Phe	Leu	Leu	Asp	Leu	Ser	Val	Ile	Phe
Horse S	*Ser*	*Phe*	*Phe*	*Leu*	*Leu*	Δ	*Leu*	*Ser*	*Val*	*Ile*	*Leu*
Human α	Thr	Ala	Phe	Leu	Met	Asp	Val	Ser	Val	Ile	Tyr
Human β	Thr	Phe	Phe	Leu	Leu	Asp	Leu	Gly	Val	Val	Tyr
Human γ	*Ser*	*Phe*	*Phe*	*Val*	*Leu*	*Asp*	*Leu*	*Gly*	*Val*	*Ile*	*Ile*
Baboon β	Thr	Phe	Phe	Val	Leu	Asp	Leu	Ser	Val	Val	Tyr
Rabbit	*Ser*	*Phe*	*Phe*	*Ile*	*Ile*	*Asp*	*Leu*	*Gly*	*Val*	*Ile*	*Phe*
Mouse	Ser	Phe	Phe	Ile	Leu	Asp	Leu	Leu	Val	Ile	Tyr
Rat	*Ser*	*Phe*	*Phe*	*Leu*	*Leu*	*Asn*	*Leu*	*Thr*	*Val*	*Ile*	*Tyr*
β48Ser	*Ser*	*Phe*	*Phe*	*Leu*	*Leu*	*Asp*	*Leu*	*Gly*	*Val*	*Val*	*Tyr*

that cannot ligand zinc, an inactive enzyme is obtained. However, it is not likely that Cys164 should ligand the zinc atom according to the kinetic data determined for the mutant SDH164A. This variant is highly active towards sorbitol and xylitol, and the efficiency is even increased (Table 4). This could probably be explained by the fact that oxidation of a free Cys at position 164 cannot occur. The results from the zinc analysis did not give a clear pattern to the kinetic results. All variants harbor one zinc atom per subunit of SDH (Table 5). This may indicate that a zinc atom can be liganded to the two remaining residues, Cys44 and His69, that are liganding the zinc atom in SDH. However, just two amino acid residues as ligands to a zinc atom have not been shown in any protein (Vallee and Auld, 1990). Anyhow, it can be concluded from these results that Glu155, corresponding to Cys174 in ADH, is liganding the catalytic zinc in SDH. The corresponding mutation in class I ADH, Cys174 to Glu, did not resulted in a stable protein and no further information was obtained from recombinant β174E enzyme. A possible explanation is that the large Glu causes too many changes in the polypeptide backbone and that thereby the protein is degraded by proteases.

In summary, it is possible to map functional differences within the zinc-containing alcohol dehydrogenase protein family with site-directed mutageneses of single amino acid residues. The change, Thr to Ser, at position 48 in class I ADH explains the steroid effects and Glu155 in SDH can be verified as a zinc ligand.

ACKNOWLEDGEMENTS

This work was supported by grants from the Swedish Medical Research Council, the Swedish Alcohol Research Fund and the Magn. Bergvall Foundation.

REFERENCES

Bosron, W.F., Magnes, L.J. & Li T.-K. (1983) Kinetic and electrophoretic properties of native and recombined isoenzymes of human liver alcohol dehydrogenase. *Biochemistry,* 22:1852.

Bradford, M.M. (1976) A rapid and sensitive method for the quantitation of microgram quantities of protein utilizing the principle of dye binding. *Anal. Biochem.*, 72:248.

Cederlund, E., Peralba, J.M., Xavier, P. & Jörnvall, H. (1991) Amphibian alcohol dehydrogenase, the major frog liver enzyme. Relationships to other forms and assessment of an early gene duplication separating vertebrate class I and class III alcohol dehydrogenases. *Biochemistry,* 30:2811.

Crabb, D.W. & Edenberg, H.J. (1986) Complete amino acid sequence of rat liver alcohol dehydrogenase deduced from the cDNA sequence. *Gene,* 48:287.

Creaser E.H., Murali, C & Britt, K.A. (1990) Protein engineering of alcohol dehydrogenases; effects of amino acid changes at position 93 and 48 of yeast ADH1. *Prot. Engng.,* 3:523.

Cronholm, T., Larsén, C., Sjövall, J., Theorell, H. & Åkeson, Å. (1975) Steroid oxidoreductase activity of alcohol dehydrogenase from horse, rat, and human liver. *Acta Chem. Scand.,* B29:571.

Dunn, M.F., Dietrich, H., MacGibbon, A.K.H., Koerber, S.C. & Zeppezauer. M. (1982) Investigation of intermediates and transition states in catalytic mechanisms of active site substituted cobalt (II), nickel (II), zinc (II), and cadmium (II) horse liver alcohol dehydrogenase. *Biochemistry,* 21:354.

Eklund, H., Nordström, B., Zeppezauer, E., Söderlund, G., Ohlsson, I., Boiwe, T., Söderberg, B.-O., Tapia, O., Brändén, C.-I. & Åkeson, Å. (1976) Three-dimensional structure of horse liver alcohol dehydrogenase at 2.4 Å resolution. *J. Mol. Biol.,* 102:27.

Eklund, H., Plapp, B., Samama, J.P. & Brändén, C.-I. (1982) Binding of substrate in a ternary complex of horse liver alcohol dehydrogenase. *J. Biol. Chem.,* 257:14349.

Eklund, H., Samama, J.-P. & Jones, T.A. (1984) Crystallographic investigations of nicotinamide adenine dinucleotide binding to horse liver alcohol dehydrogenase. *Biochemistry,* 23:5982.

Eklund, H., Horjales, E., Jörnvall, H. , Brändén, C.-I. & Jeffery, J. (1985) Molecular aspects of functional differences between alcohol and sorbitol dehydrogenases. *Biochemistry,* 24:8005.

Eklund, H., Horjales, E., Vallee, B.L. & Jörnvall, H. (1987) Computer-graphics interpretations of residue exchange between α, β and γ subunits of human-liver alcohol dehydrogenase class I isozymes. *Eur. J. Biochem.,* 167:185.

Eklund, H., Müller-Wille, P., Horjales, E., Futer, O., Holmquist B., Vallee, B.L., Höög, J.-O., Kaiser, R. & Jörnvall, H. (1990) Comparison of three classes of human liver alcohol dehydrogenase. Emphasis on different substrate binding pockets. *Eur. J. Biochem.,* 193:303.

Ehrig, T., Hurley, T.D., Edenberg, H.J. & Bosron, W.F. (1991) General base catalysis in a glutamine for histidine mutant at position 51 of human liver alcohol dehydrogenase. *Biochemistry,* 30:1062.

Feiters, M.C. & Jeffery. J. (1989) Zinc environment in sheep liver sorbitol dehydrogenase. *Biochemistry,* 28:7257.

Hurley, T.D., Bosron, W.F., Hamilton, J.A. & Amzel, L.M. (1991) Structure of human $\beta_1\beta_1$ alcohol dehydrogenase: catalytic effects of non-active site substitutions. *Proc. Natl. Acad. Sci. USA,* 88:8149.

Höög, J.-O., Weis, M., Zeppezauer, M., Jörnvall, H. & von Bahr-Lindström, H. (1987) Expression in *Escherichia coli* of active human alcohol dehydrogenase lacking N-terminal acetylation. *Biosci. Rep.,* 7:969.

Höög, J.-O., Eklund, H. & Jörnvall, H. (1992a) A single-residue exchange gives human recombinant $\beta\beta$ alcohol dehydrogense $\gamma\gamma$ isozyme properties. *Eur. J. Biochem.,* 205:519.

Höög, J.-O., Vagelopoulos, N., Yip, P.-K., Keung, W.M. & Jörnvall, H. (1992b) Isozyme development in mammalian class I alcohol dehydrogenase. cDNA cloning, functional correlations, and lack of evidence for genetic isozymes in rabbit. *Eur. J. Biochem.,* submitted.

Jeffery, J., Chester, J., Mills, C., Sadler, P.J. & Jörnvall, H. (1984) Sorbitol dehydrogenase is a zinc enzyme. *EMBO J.,* 3:357.

Johansson, J., Fleetwood, L. & Jörnvall, H. (1992) Cysteine reactivity in sorbitol dehydrogenase and aldehyde dehydrogenase. Differences towards the pattern in alcohol dehydrogenase. *FEBS Lett.,* 303:1.

Jones, T.A. (1985) Interactive computer graphics: FRODO. *Methods Enzymol.,* 115:157.

Jörnvall, H. (1970) Horse liver alcohol dehydrogenase. On the primary structures of the isoenzymes. *Eur. J. Biochem.,* 16:41.

Jörnvall, H. (1974) Functional aspects of structural studies of alcohol dehydrogenases. In *Alcohol and Aldehyde Metabolizing Systems* (Thurman, R.G., Yonetani, T., Williams, J.R. & Chance, B., eds.) pp. 23-32, Academic Press, New York.

Jörnvall, H., von Bahr-Lindström, H. & Jeffery, J. (1984) Extensive variations and basic features in the alcohol dehydrogenase - sorbitol dehydrogenase family. *Eur. J. Biochem.,* 140:17.

Jörnvall, H., Persson, B. & Jeffery, J. (1987) Characteristics of alcohol/polyol dehydrogenases. The zinc-containing long-chain alcohol dehydrogenases. *Eur. J. Biochem.,* 167:195.

Jörnvall, H., von Bahr-Lindström, H. & Höög, J.-O. (1989) Alcohol dehydrogenases - structure. In *Human Metabolism of Alcohol* (Batt, R.D. & Crow, K.E., eds.) vol. 2, pp. 43-64, CRC press, Boca Raton FL.

Karlsson, C., Jörnvall, H, & Höög, J.-O. (1991) Sorbitol dehydrogenase: cDNA coding for the rat enzyme; variations within the alcohol dehydrogenase family independent of quaternary struture and metal content. *Eur. J. Biochem.,* 198:761.

Lutz, R.A., Bull, C. & Rodbard, D. (1986) Computer analysis of enzyme-substrate-inhibitor kinetic data with automatic model selection using IBM-PC compatible computers. *Enzyme,* 36:197.

Maret, W. & Auld, D.S. (1988) Purification and characterization of human liver sorbitol dehydrogenase. *Biochemistry,* 27:1622.

McEvily, A.J., Holmquist, B., Auld, D.S. & Vallee, B.L. (1988) 3β-Hydroxy-5β-steroid dehydrogenase activity of human liver alcohol dehydrogenase is specific to γ-subunits. *Biochemistry,* 27:4284.

Mårdh, G., Falchuk, K.H., Auld, D.S. & Vallee, B.L. (1986) Human class II (π) alcohol dehydrogenase has a redox-specific function in norepinephrine metabolism. *Proc. Natl. Acad. Sci. USA, 83:*2836.

Parés, X., Cederlund, E., Moreno, A., Saubi, N., Höög, J.-O. & Jörnvall, H. (1992) Class IV alcohol dehydrogenase (the gastric enzyme). Structural analysis of human $\sigma\sigma$-ADH reveals class IV to be variable and confirms the prescence of a fifth mammalian alcohol dehydrogenase class. *FEBS Lett.,* 277:115.

Park, D.-H. & Plapp, B.V. (1991) Isoenzymes of horse liver alcohol dehydrogenase actives on ethanol and steroids. *J. Biol. Chem.,* 266:13296.

Pietruszko, R., Clark, A., Graves, J.M.H. & Ringold, H. (1966) The steroid activity and multiplicity of crystalline horse liver alcohol dehydrogenase. *Biochem. Biophys. Res. Commun.,* 23:526.

Plapp, B.V., Ganzhorn, A.J., Gould, R.M., Green, D.W., Jacobi, T., Warth, E. & Kratzer, D.A. (1991) Catalysis by yeast alcohol dehydrogenase. In *Enzymology and Molecular Biology of Carbonyl Metabolism 3* (Weiner, H., Wermuth, B. & Crabb, D.W. eds.) pp. 241-251, Plenum Press, New York.

Sanger, F., Nicklen, S. & Coulson, A.R. (1977) DNA sequencing with chain termination inhibitors. *Proc. Natl. Acad. Sci. USA,* 74:5463.

Smith, M., Hopkinson, D.A & Harris, H. (1971) Developmental changes and polymorphism in human alcohol dehydrogenase. *Ann. Hum. Genet.,* 34:251.

Taylor, J.W., Ott, J. & Eckstein, F. (1985) The rapid generation of oligonucleotide-directed mutations at high frequency using phosporothioate-modified DNA. *Nucleic Acids Res.,* 13:8765.

HORSE LIVER ALCOHOL DEHYDROGENASE - S - ISOZYME: CONFIRMATION OF THE PRIMARY STRUCTURE BY PROTEIN SEQUENCING AND ION SPRAY MASS SPECTROMETRY

Ina Hubatsch[1], Michael Zeppezauer[1],
Dietmar Waidelich[2] and Ernst Bayer[2]

[1]Fachrichtung Biochemie
Universität des Saarlandes
D-W-6600 Saarbrücken
[2]Institut für Organische Chemie
Universität Tübingen
D-W-7400 Tübingen

INTRODUCTION

Horse liver alcohol dehydrogenase (HLADH) occurs in multiple molecular forms; up to twelve fractions have been separated on starch gels (Lutztorf et al., 1970). The three main isozymes are formed by dimeric combination of the subunits E and S (nomenclature from Pietruszko and Theorell, 1969). Only the alcohol dehydrogenase isozymes containing the S-subunit are active towards 3ß-hydroxysteroids (Pietruszko et al., 1966; Theorell et al., 1966). Therefore a larger substrate binding channel has been suggested (Björkhem et al., 1974).

The biological function of mammalian alcohol dehydrogenases is largely unknown; their elucidation will require a thorough understanding of structure-function relationships in all classes of alcohol dehydrogenases. While the primary structure of the EE-isozyme has been known for a long time (Jörnvall, 1970a), only parts of the S-chain have been sequenced (Jörnvall, 1970b). So far only the tertiary structure of the horse EE isozyme of the class I alcohol dehydrogenases has been determined by X-ray diffraction (Eklund et al., 1976). For its nearest neighbour, the SS isozyme of HLADH, the differences in substrate specifity and kinetic behaviour have been explained on the basis of six reported point mutations of the primary structure (Jörnvall, 1970b), assuming that the tertiary structures of both isozymes are highly similar. However, position 115 was not safely established, leaving open the choice between Ser 115 or a deletion as compared to Asp 115 in the E-chain.

Recently, the cDNA-sequences of both the E and S chain genes of HLADH have been reported (Park and Plapp, 1991). The cDNA- derived sequences reveal a total of ten amino acid exchanges; as will be shown in this communication these findings agree with our results of peptide sequencing.

Enzymology and Molecular Biology of Carbonyl Metabolism 4
Edited by H. Weiner, Plenum Press, New York, 1993

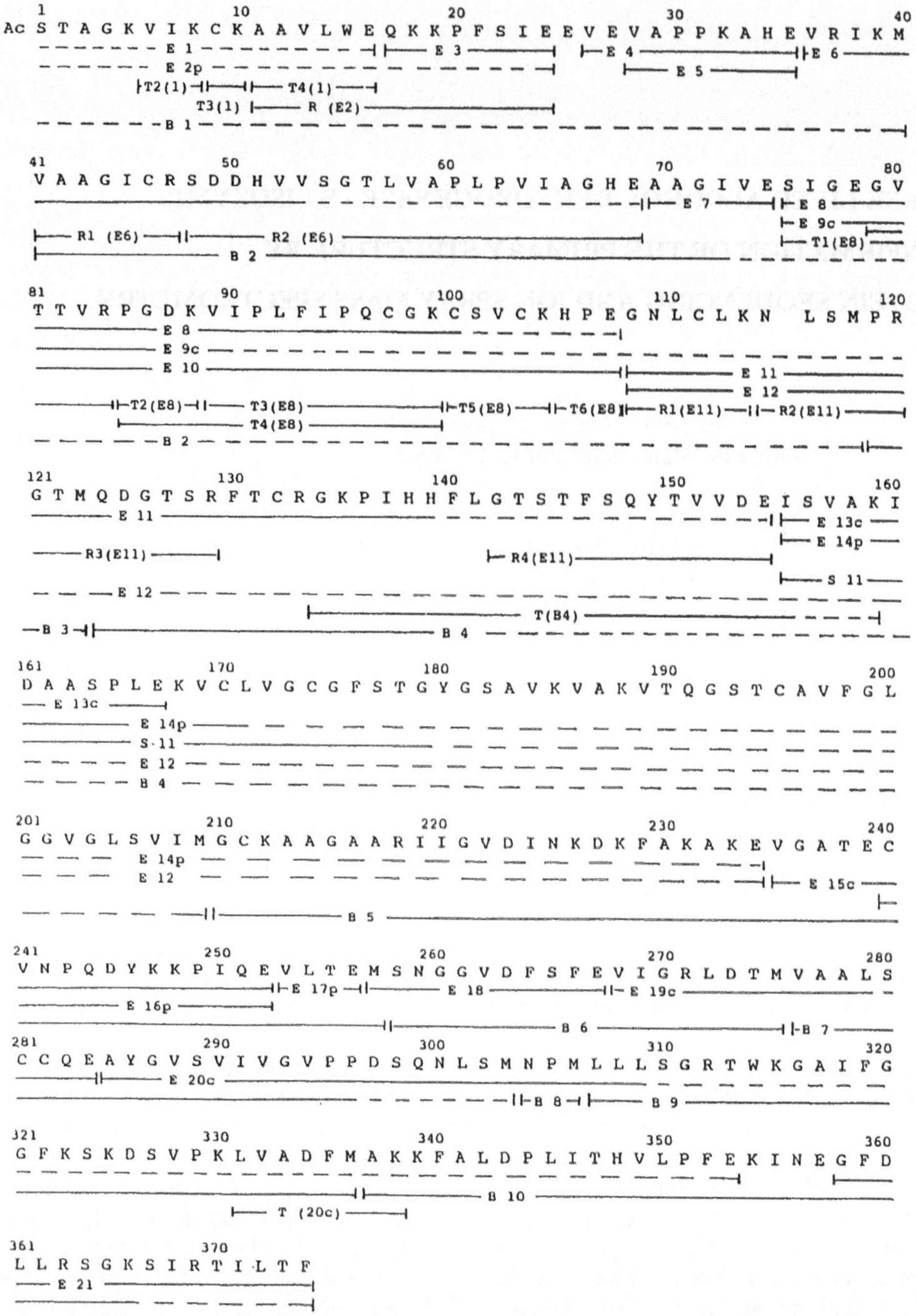

Figure 1. Amino acid sequence of the S-subunit of horse liver alcohol dehydrogenase; peptides are marked by bars as far as sequenced. Peptides derived from glu-proteolytic digestions are indicated as E (Ec only from the carboxymethylated, Ep only from the pyridylethylated protein), peptides from tryptic or arg-proteolytic digestions as T or R, respectively, peptides derived from CNBr-cleavage as B. The glu-proteolytic peptide sequenced after blotting is indicated as S 11.

MATERIAL AND METHODS

The SS-isozyme was prepared from fresh horse liver as described by Adolph et al. (this issue). Cysteine-SH groups were blocked by carboxymethylation according to Jörnvall (1970c) or by pyridylethylation. For the latter procedure the protein was reduced with 2-mercaptoethanol (110-fold molar excess over total cysteine) in 6 M guanidine-HCl, 0.25 M Tris + 0.1 mM EDTA pH 8.5 for 4 h at 37 °C under argon; then 4-vinylpyridine (Aldrich; 20% molar excess over the total of SH-groups) was added, flushed with argon and incubated for 4 h at 37 °C. After reaction the protein was dialyzed against 2 mM HCl and lyophilized.

CNBr-fragmentation was carried out in 70 % formic acid with a 60-fold molar excess of CNBr (Sigma) over methionine for 24 h under argon. During enzymatic fragmentations the enzyme : substrate ratio was 1 : 50. Glu-proteolytic digestion by Staph. aureus protease V8 (Sigma) was carried out in 1 M guanidine-HCl, 0.1 M NH_4HCO_3 pH 7.8 at room temperature for 9 h; tryptic digestion (TPCK-treated trypsin from Serva) in 0.1 M NH_4HCO_3 + 0.1 mM $CaCl_2$ pH 8.1 for 5 h at 37 °C and arg-proteolytic digestion with Endoproteinase Arg C (Boehringer) in 0.1 M NH_4HCO_3 pH 8 at 37 °C over night.

Peptides were purified by RP-chromatography on HPLC (Waters 600) over a C_{18}-column (μ-bondapak, 3.9 mm x 30 cm, Waters); for elution 0.1 % TFA/water to 80 % CH_3CN + 0.1 % TFA/ water gradients were used. Precipitated peptides were dissolved in 1 % SDS and separated by SDS-PAGE (6 M urea, 16.5% C, 6% T) and electroblotted on PVDF (10 % methanol/10 mM CAPS pH 11, 50 V const., 90 min). Mixtures of peptides, which remained unresolved on RP-HPLC were subsequently rechromatographed over a gel permeation column (TSK 3000 SW 7.5 mm x 600 mm) with 0.1% TFA as eluent.

Amino acid analysis by RP-HPLC after pre-column derivatization with PITC was performed as described by Bidlingmeyer et al. (1984). For sequencing an Applied Biosystems sequencer 473A was used; the chemicals (sequential grade) for Edman degradation were obtained from Applied Biosystems.

MS was done with an API III TAGA 6000 E instrument with ion spray ionization source (SCIEX, Thornhill, Toronto).

RESULTS AND DISCUSSION

Compared with the primary structure of the E-subunit and the primary structure of the S-subunit proposed by Jörnvall (1970b), we have found four additional exchanges by protein sequencing: T 43 → A 43, T 59 → A 59, I 172 → V 172, T 277 → A 277 (amino acid numbering as in the E-chain). The deletion of D 115 of the E-chain was confirmed. The results of the primary structure determination are shown in Fig. 1.

Neither the carboxymethylated nor the pyridylethylated protein could be dissolved completely under the conditions used for glu-specific proteolysis. From the supernatant 16 fragments of the pyridylethylated protein and 17 fragments of the carboxymethylated derivative were isolated by HPLC. Not all fragments of both digestions were identical as indicated in fig. 1. The precipitate was dissolved in 1 % SDS, submitted to SDS-PAGE and electroblotted onto a PVDF membrane. This procedure yielded 14 fragments between 4 and 35.5 kDalton. From the blotted 8 kDalton-fragment S 11 an initial sequence of 25 amino-acids was determined.

Peptides E 6 and E 11, which proved too long for complete sequencing were therefore cleaved by Endoproteinase Arg-C to obtain subfragments. Single non-specific cleavages occured during this digestion, but the C-terminal sequences of peptides E 11 and E 6 could be established by this way. Peptide E 8 was redigested with trypsin and yielded the six subfragments T 1 (E8) to T 6 (E8).

Peptide E 21 was concluded to be the C-terminal peptide because its C-terminal residue is F, which should not be a cleavage site for the Staph. aureus V8-protease. Peptides E 1 and E 2 were not affected by Edman degradation; so it was concluded, that they were derived from the N-terminus of the S-subunit, because it is known that the enzyme is N-terminally acetylated. They were cleaved by trypsin (peptide E 1) resp. Arg-protease (peptide E 2) and the resulting peptides were submitted to Edman-degradation; their amino acid sequences correlate with sequence datas of the N-terminal part of the S-subunit given by Jörnvall (1970b). For peptide E 1 a molecular mass of 1850 ± 0.5 Dalton was obtained by ion-spray-MS. This agrees with N-acetylation and an amino acid sequence identical to the E-chain.

After CNBr-treatment of the S-subunit 10 peptides were isolated. Peptide B 4 was isolated in such a little amount, that a complete primary structure determination including further fragmentation of this peptide, was not achieved. Ion-spray-MS shows a molecular mass of 9218.95 ± 3 Dalton for this peptide, which is in accordance with the assumed molecular weight of 9216.17 Dalton calculated from the amino acid composition.

MS-data of the N-terminal CNBr-peptide have also been used to check the first 40 amino acids because of missing overlaps under the sequenced glu-proteolytic peptides. There is no indication of a deletion or insertion compared to the E-chain.

Generally, the substituted amino acids in the S-chain are less charged and more hydrophobic than in the E-chain.

A cumulation of exchanges (amino acids 94, 101, 110, 115) occurs in the loop, which is located at one side of the substrate binding pocket. As Park and Plapp (1991) have concluded from computer modelling studies based on the EE crystal structure and the cDNA derived sequence of the SS isozyme, the deletion of D 115, which is located in this loop, may facilitate the binding of the steroid substrates in the S subunit by moving V 116 away, which is facilitated by the exchange of F 110 by L.

Such regions with structural differences are also interesting with regard to a differentiation of the isozymes by immunological methods. Immunological differences between the two subunits have been early observed by Pietruszko and Ringold (1968) using polyclonal antisera. Over the last years, there have been attempts to generate monoclonal antibodies, partially by using synthetic peptides as antigens (Zeppezauer et al., 1990), in the hope to get antibodies with more restricted specifities. But also with monoclonal antibodies, cross reactivities were observed (Adinolfi et al.,1991) reflecting farreaching similarities between the primary structures within this class of dehydrogenases (Jörnvall et al, 1984).

Table 1. Molecular mass of MS-analyzed peptides

peptide	sequence position (amino acid number)	mass estimation by MS	theoretical mass from amino acid composition
E 1	1 - 16	1850.00 ±0.50	1851.02
B 1	1 - 40	4574.71 ±0.41	4574.40
B 2	41 - 118	8352.97 ±1.05	8353.41
B 4	124 - 209	9218.95 ±3.01	9216.75
B 5	210 - 257	5311.29 ±1.41	5310.75

REFERENCES

Adinolfi, A., Leone, N., Swallow, D.,Hopkinson, D.A., Zapponi, M.C., Ferri, G., Camardella, L., and Ronchi, S., 1991, Identification of a conserved epitope in class I alcohol dehydrogenase isoenzymes using monoclonal antibodies, *Exp. Clin. Immunogenet.*, 8:96.

Bidlingmeyer, B.A., Cohen, S.A., and Tarvin, T.L., 1984, Rapid analysis of amino acids using pre-column derivatization methods, *J. Chrom.*, 336:93.

Björkhem, I., Jörnvall, H., and Åkeson, Å., 1974, Oxidation of -hydroxylated fatty acids and steroids by SS-isozyme of liver alcohol dehydrogenase, *Biochem. Biophys. Res. Commun.* 57:870.

Eklund, H., Nordström, B., Zeppezauer, E., Söderlund,.G., Ohlsson, J., Boiwe, T., Tapia, O., Brändén, C.-I., and Åkeson, Å., 1976, Three dimensional structure of horse liver alcohol dehydrogenase at 2.4 Å resolution, *J. Mol. Biol.*, 102:27.

Jörnvall, H., 1970a, Horse liver alcohol dehydrogenase: the primary structure of the protein chain of the ethanol-active isozyme, *Eur. J. Biochem.*, 16:25.

Jörnvall, H., 1970b, Horse liver alcohol dehydrogenase: on the primary structure of the isozymes, *Eur. J. Biochem.*, 16:41.

Jörnvall, H., 1970c, Horse liver alcohol dehydrogenase: the primary structure of the N-terminal part of the protein chain of the ethanol active isozyme, *Eur. J. Biochem.*, 14:521.

Jörnvall, H., von Bahr-Lindström, H., and Jeffrey, J., 1984, Extensive variations and basic features in the alcohol dehydrogenase - sorbitol dehydrogenase family, *Eur. J. Biochem.*, 140:17.

Lutztorf, U.M., Schürch, P.M., and von Wartburg, J.-P., 1970, Heterogenity of horse liver alcohol dehydrogenase, *Eur. J. Biochem.*, 17:497.

Park, D.H. and Plapp, B.V., 1991, Isoenzymes of horse liver alcohol dehydrogenase active on ethanol and steroids, *J. Biol. Chem.*, 266:13296.

Pietruszko, R.,Clark, A., Graves, J.M.H., and Ringold, H.J., 1966, The steroid activity and multiplicity of crystalline horse liver alcohol dehydrogenase, *Biochem. Biophys. Res. Commun.*, 23:526.

Pietruszko, R. and Ringold, H.J., 1968, Antibody studies with the multiple enzymes of horse liver alcohol dehydrogenase, *Biochem. Biophys. Res. Commun.*, 33:497.

Pietruszko, R. and Theorell, H., 1969, Subunit composition of horse liver alcohol dehydrogenase, *Arch. Biochem. Biophys.*, 131:288.

Theorell, H., Taniguchi, S., Åkeson, Å., and Skursky, L., 1966, Crystallisation of a separate steroid-active liver alcohol dehydrogenase, *Biochem. Biophys. Res. Commun.*, 24:603.

Zeppezauer, M.,Rawer, S., Schönberger, A., Rapp, W., and Bayer, E., 1990, Hydrophilic polystyrene-polyoxyethylene graft polymer beads as carrier of antigenic structures for in vivo and in vitro immunization techniques, *in*: Peptides 1990, Proceedings of the 21st European Peptide Symposium, E. Girald and D. Andreu, eds., Escom Science Publishers B.V., Leiden, The Netherlands, 1991.

MIXED SUBSTRATE EXPERIMENTS WITH CLASS III (χ) ALCOHOL DEHYDROGENASES FROM HUMAN AND PIG LIVER AND STOMACH

Edwin P. Carr,[1] P.W. Napoleon Keeling,[2] and Keith F. Tipton[1]

[1]Department of Biochemistry
[2]Department of Clinical Medicine
Trinity College
University of Dublin
Dublin 2
Ireland

INTRODUCTION

Class III (χ)–alcohol dehydrogenase (ADH; EC 1.1.1.1) was first identified in human liver homogenates by Parés and Vallee (1981). Its properties differ considerably from other ADH classes. It can only be saturated by long chain alcohols such as octanol, is not inhibited by 4-methyl pyrazole and migrates anodically on starch gels electrophoresed at both pH 8.6 and 7.6 (Vallee and Bazzone, 1983). Class III (χ)-ADH is ubiquitous in its tissue distribution and has been detected in human liver (Parés and Vallee, 1981; Vallee and Bazzone, 1983), stomach (Moreno and Parés, 1991), brain (Beisswenger *et al.*, 1985) and placenta (Parés and Vallee, 1984). Immunological studies by Adinolfi *et al.* (1984) have demonstrated its presence in all human tissues. In the rat class III (χ)-ADH has been designated ADH-2, based on electrophoretic properties, and its presence has been detected in all tissues (Julià *et al.*, 1987; Boleda *et al.*, 1989).

Glutathione dependent formaldehyde dehydrogenase (FDH; EC 1.2.1.1) was first described from bovine and chicken livers by Strittmatter and Ball (1955). It has since been purified from human liver (Uotila and Koivusalo, 1974) and its presence has been detected in a variety of different organs, animals, plants and micro-organisms (Uotila and Koivusalo, 1989). As with class III (χ)-ADH, FDH is a cytosolic enzyme. The enzyme catalyses the oxidation of formaldehyde in the presence of reduced glutathione (GSH) and NAD^+ according to the following reversible reaction (Uotila and Koivusalo, 1974; Uotila and Koivusalo, 1989).

$$\text{Formaldehyde} + \text{GSH} + NAD^+ \rightleftharpoons \text{S-formylglutathione} + \text{NADH} + H^+ \qquad (1)$$

Enzymology and Molecular Biology of Carbonyl Metabolism 4
Edited by H. Weiner, Plenum Press, New York, 1993

The true substrate for the enzyme is S-hydroxymethylglutathione, an adduct formed from formaldehyde and GSH in a nonenzymic reaction as outlined below (Uotila and Mannervik, 1979).

$$\text{Formaldehyde} + \text{GSH} \rightleftharpoons \text{Adduct}$$

$$\frac{[\text{Formaldehyde}]\,[\text{GSH}]}{[\text{Adduct}]} = K_d \tag{2}$$

The reaction has a reported dissociation constant (K_d) of 1.5 m$\underline{M}$ at pH 8.0 (Uotila and Koivusalo, 1974).

Recent studies (Koivusalo *et al.*, 1989; Koivusalo and Uotila, 1991; Holmquist and Vallee, 1991; Gutheil *et al.*, 1992) have provided evidence indicating that class III (χ)-ADH and FDH are identical enzymes. The main evidence provided has been by means of sequencing work with some kinetic studies and the enzyme source has been limited to rat and human liver as well as *E. coli* . Since, as stated earlier, both class III (χ)-ADH and FDH are ubiquitous in their tissue distribution it seems likely that class III (χ)-ADH and FDH will prove to be identical irrespective of the tissue source studied. In this work the enzyme from two different tissues in two different animals was used in order to investigate this hypothesis.

In this report we utilised the method of mixed substrate experiments (see Houslay *et al.*, 1974; Dixon and Webb, 1979) to provide futher evidence that class III (χ)-ADH and FDH are indeed the same enzyme. In this method it is possible to determine whether one or more enzymes are acting on two different substrates. The two substrates used were octanol and S-hydroxymethyl glutathione. Partially purified class III (χ)-ADH (FDH) preparations were obtained from human and pig liver and stomach.

EXPERIMENTAL

Tissue and Reagents

Post mortem human liver and stomach (<24 h.) were obtained from Beaumount Hospital, Dublin. Pig liver and stomach were obtained from Coyle Bros. Ltd., Dublin. Ion exchange media (DEAE-sepharose CL6B), octanol, GSH and glutathione-independent formaldehyde dehydrogenase (EC 1.2.1.46) were obtained from Sigma Chemical Co. Formaldehyde (ANALAR grade) was purchased from BDH Ltd. and NAD^+ (grade II, 98%) was purchased from Boehringer Mannheim.

Assay Methods

The assay mixture for class III (χ)-ADH contained 2 m$\underline{M}$ NAD^+, enzyme and octanol. The assay mixture for FDH contained 2 m$\underline{M}$ NAD^+, 1 m$\underline{M}$ GSH, enzyme and formaldehyde. The reduction of NAD^+ was monitored at 340nm at 37°C, in 0.1 $\underline{M}$ phosphate buffer, pH 8.0, in a Hewlett Packard diode array spectrophotometer. The measured absorbance changes were related to NADH concentration using a molar extinction co-efficient of 6220 $\underline{M}^{-1}.cm^{-1}$ (Dawson *et al.*, 1983). One unit of enzyme activity is defined as that amount of enzyme which catalyses the formation of 1 μmol of NADH per minute under the conditions described above. Protein concentration was determined by the method of Markwell *et al.* (1978). Formaldehyde concentration was determined by a spectrophotometric method using glutathione-independent formaldehyde dehydrogenase from *Pseudomonas putida* The concentrations of S-hydroxymethylglutathione were determined by using the expression in equation 2.

Partial Purification of Class III (χ)-ADH from Human Liver and Stomach

Both tissues were homogenised in 10 m<u>M</u> Tris/HCl, pH 9.0, containing 0.5 m<u>M</u> DTT. It was found that during the homogenisation procedure the pH of the stomach fraction dropped considerably due to the presence of acid in parietal cells. Both homogenates were centrifuged at 10,000**g** for 30min to remove cellular debris and particulate material and the supernatants were further spun at 85,000**g** for 90min to pellet the microsomes. The resultant supernatants, cytosolic fractions, were dialysed overnight against homogenisation buffer before application to a DEAE sepharose CL-6B column (45x2 cm) which had been previously equilibrated with homogenisation buffer. The columns were washed with buffer to elute class I & II activity. Class III (χ)-ADH (FDH) was eluted off the column by application of a 0-200mM salt gradient. Fractions containing activity were pooled. All procedures were performed at 4°C.

Partial Purification of Class III (χ)-ADH from Pig Liver and Stomach

The purification procedures for the pig enzymes were identical to those used for the human enzymes except that the buffer used was at a pH of 8.0.

RESULTS AND DISCUSSION

Electrophoretic Analysis

Starch gel electrphoresis of human and pig liver and stomach fractions eluted from DEAE-sepharose was performed (Figure 1). An intense anodic band was present in all fractions. This staining pattern is typical for class III (χ)-ADH. A more anodic faint second band was also present in all fractions. The second band was in close proximity to the first band. A second band has been reported in previous studies (Parés and Vallee, 1981; Vallee and Bazzone, 1983). Isoelectric focussing of FDH followed by activity staining has revealed the presence of one major and two minor bands (Uotila and Koivusalo, 1987). These authors state that the major band accounts for 70% of total activity and the two minor bands account for 25% and 5% of total activity respectivly. It would appear that the faint band seen on the starch gels in the present study may correspond to the minor band which represents 25% of total FDH activiy. The substrate used in the staining procedure was pentanol and this could account for the failure to observe a third band on the activity-stained starch gels.

Kinetic Studies

Octanol and S-hydroxymethylglutathione were assayed over a range of substrate concentrations to determine their kinetic constants (K_m and V_{max} values) with respect to human and pig class III (χ)–ADH (FDH) from liver and stomach. Data were analysed by non-linear regression of the initial rates using the computer program *Enzfitter* . All the fractions exhibited typical saturation kinetics for both substrates (Table 1).

The previously reported K_m values for octanol oxidation by class III (χ)-ADH are of the order of 0.4 m<u>M</u> (Vallee and Bazzone, 1983). The K_m values recorded here are in the range of 1.4-1.7 m<u>M</u>. However, it is important to note that the assay conditions used here differed

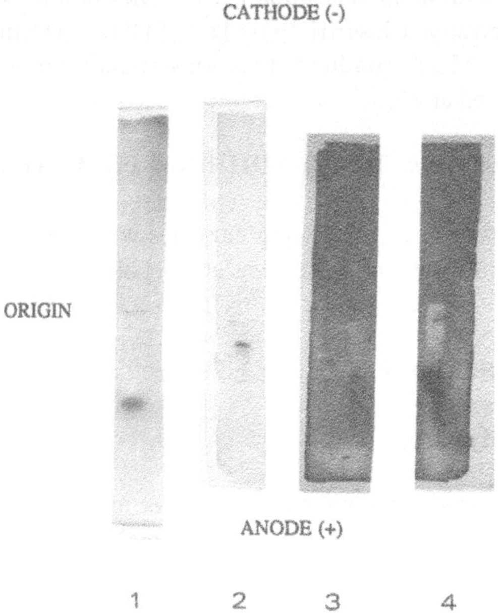

Figure 1. Starch Gel Electrophoresis of class III (χ)-ADH (FDH) from pig stomach (1), pig liver (2), human stomach (3) and human liver (4). Samples were loaded on the origin and gels were electrophoresed for 5h at 720V. The gels containing samples from pig stomach, human stomach and human liver were electrophoresed at pH 8.6. The gel containing the pig liver sample was electrophoresed at pH 8.0. Following electrophoresis gels were stained in a mixture containing 0.6m<u>M</u> NAD^+, 0.24m<u>M</u> nitro blue tetrazolium, 0.065m<u>M</u> phenazine methosulphate, 11m<u>M</u> pyruvate and 0.1<u>M</u> pentanol in 50m<u>M</u> Tris HCl, pH 8.6.

from those used by Parés and Vallee (1981) for octanol oxidation by class III (χ)-ADH which were 0.1 M glycine NaOH, pH 10.0 and 25°C. The previously reported K_m value for S-hydroxymethylglutathione oxidation by FDH was 8 μM (Uotila and Koivusalo, 1974), a value which is approximately one third of the values obtained in the present work. Similar assay conditions were employed in the present work except that the assay temperature used was 37°C not 25°C. The possibility also exists that the formaldehyde solution contained impurities.

Table 1. Kinetic constants for octanol and S-hydroxymethylglutathione oxidation by class III (χ)-ADH (FDH).

	Enzyme Source	K_m	V_{max}
Octanol		(mM)	
	Human Stomach	1.44 ± 0.15	0.0240 ± 0.0021
	Human Liver	1.50 ± 0.17	0.0254 ± 0.0033
	Pig Stomach	1.41 ± 0.22	0.0353 ± 0.0012
	Pig Liver	1.70 ± 0.22	0.0294 ± 0.005
S-hydroxymethylglutathione		(μM)	
	Human Stomach	17.94 ± 1.1	0.103 ± 0.0013
	Human Liver	26.40 ± 1.8	0.110 ± 0.0014
	Pig Stomach	26.70 ± 2.9	0.125 ± 0.0075
	Pig Liver	24.40 ± 2.1	0.145 ± 0.0027

Values shown are the mean ± range of two experiments each of which were performed in triplicate
Vmax values are expressed as μmol NADH formed. $min^{-1}.mgprotein^{-1}$.

Mixed Substrate Experiments

When an enzyme is not absolutely specific for a single substrate the possibility arises that it may act on two different substrates present at the same time. This cannot be considered simply as a case of two independent parallel reactions, for the same active centres act on both substrates and therefore there will be competition between them. If we consider an enzyme species, E, catalysing the conversion of two substrates A and B, as far as A is concerned B may be considered a competitive inhibitor and vice versa. The rate of conversion of A (v_a) will be given by the following equation.

$$v_a = \frac{V_a}{1 + \frac{K_a}{a}\left(1 + \frac{b}{K_b}\right)} \tag{3}$$

and that of B (v_b):

$$v_b = \frac{V_b}{1 + \frac{K_b}{b}\left(1 + \frac{a}{K_a}\right)} \tag{4}$$

where K_a and V_a are the K_m and V_{max} values for substrate A respectively, and K_b and V_b are the K_m and V_{max} values for substrate B respectively. The sum of reactions (3) and (4) is given in equation (5).

$$v_t = v_a + v_b = \frac{V_a \frac{a}{K_a} + V_b \frac{b}{K_b}}{1 + \frac{a}{K_a} + \frac{b}{K_b}} \qquad (5)$$

In any mixture containing concentrations a and b of A and B, respectively, the total velocity, v_t, must lie between the two velocities which would be obtained with these concentrations of A and B separately. The method can be more quantitative if the relative concentrations of the two substrates are known (i.e if the K_m values have been determined). If both substrates are present at a concentration equal to their respective K_m values then from equation (5)

$$v_t = \frac{V_a + V_b}{3} \qquad (6)$$

When the activities towards the two substrates are assayed separately:

$$v_a = \frac{V_a}{2} \qquad v_b = \frac{V_b}{2} \qquad (7)\ \&\ (8)$$

Substituting equation 7 and 8 into equation 6

$$v_t = \frac{2v_a + 2v_b}{3}$$

$$= \frac{2\,(v_a + v_b)}{3} \qquad (9)$$

Therefore when both substrates are present at a concentration equal to their respective K_m values, the initial rate observed will be 2/3 (66.7%) the sum of the initial rates when each substrate is present separately. In contrast if distinct enzymes were responsible for the oxidation of the two substrates, the rate determined with the substrate mixture would be equal to $v_a + v_b$.

The reaction rates of human and pig stomach and liver class III (χ)-ADH when octanol and S-hydroxymethylglutathione were present in the assay at their K_m concentrations were determined. The initial rates determined are expressed as percentages of the sum of the reaction rates when the activity towards each substrate, at the same concentration, was determined separately (see Table 2).

CONCLUDING REMARKS

The similar electrophoretic profiles and kinetic properties of class III (χ)-ADH between the different tissue sources demonstrates the ubiqutous nature of this enzyme. The data

Table 2. Ratio of the rate obtained by a mixture of octanol and S-hydroxymethyl glutathione, compared to the sum of the rates when each substrate was assayed separately.[1]

Enzyme Source	Ratio x 100%
Human Stomach	69.8 ± 1.4
Human Liver	64.5 ± 1.8
Pig Stomach	69.9 ± 1.9
Pig Liver	65.9 ± 0.8

Values shown are the mean ± range of two experiments each of which was performed in triplicate.
[1]The concentration of substrate used was equivalent to the previously determined Km values shown in Table 1.

presented in the mixed substrate experiments are consistent with a single enzyme being responsible for the oxidation of both substrates. In conclusion the present results provide further evidence supporting the view that class III (χ)-ADH and FDH are the same enzyme.

REFERENCES

Adinolfi, A., Adinolfi, M. and Hopkinson, D.A. (1984) Immunological and biochemical characterization of the human alcohol dehydrogenase χ-ADH isoenzyme. *Ann. Hum. Genet.* 48, 1-10.

Beisswenger, T.B., Holmquist, B. and Vallee, B.L. (1985) χ-ADH is the alcohol dehydrogenase of mammalian brains: implications and inferences. *Proc. Natl. Acad. Sci. USA* 82, 8369-8373.

Boleda, M.D., Julià, P. Moreno, A. and Parés, X. (1989) Role of extrahepatic alcohol dehydrogenase in rat ethanol metabolism. *Arch. Biochem. Biophys.* 274, 74-81.

Dixon, H. and Webb, E.C. (1979) Enzymes 3rd edit., Longman, London, pp72-74.

Gutheil, W.G., Holmquist, B. and Vallee, B.L. (1992) Purification, characterization, and partial sequence of the glutathione-dependent formaldehyde dehydrogenase: A class III alcohol dehydrogenase. *Biochemistry* 31, 475-481.

Holmquist, B. and Vallee, B.L. (1991) Human class III alcohol and glutathione-dependent formaldehyde dehydrogenase are the same enzyme. *Biochem. Biophys. Res. Comm.* 178, 1371-1377.

Houslay, M.D., Garrett, N.J. and Tipton, K.F. (1974) Mixed substrate experiments with human brain monoamine oxidase. *Biochem Pharmacol.* 23, 1937-1944.

Julià, P., Farrés, J. and Parés, X. (1987) Characterization of the three isoenzymes of rat alcohol dehydrogenase. *Eur. J. Biochem.* 162, 179-189.

Koivusalo, M. and Uotila, L. (1991) Glutathione-dependent formaldehyde dehydrogenase (EC 1.2.1.1): Evidence for identity with class III alcohol dehydrogenase. *Adv. Exp. Med. Biol.* 284, 305-313.

Koivusalo, M., Baumann, M. and Uotila, L. (1989) Evidence for the identidy of glutathione-dependent formaldehyde dehydrogenase and class III alcohol dehydrogenase. *FEBS Lett.* 257, 105-109.

Markwell, M.A.K., Hass, S.M., Bieber, L.L. and Tolbert, N.E. (1978) A modification of the Lowry procedure to simplify protein determination in membrane and lipoprotein samples. *Anal. Biochem.* 87, 206-210.

Moreno, A. and Parés, X. (1991) Purification and characterization of a new alcohol dehydrogenase from human stomach. *J. Biol. Chem.* 266, 1128-1133.

Parés, X. and Vallee, B.L. (1981) New human liver alcohol dehydrogenase forms with unique kinetic characteristics. *Biochem. Biophys. Res. Commun.* 98, 122-130.

Parés, X. and Vallee, B.L. (1984) Organ specific alcohol metabolism: placental χ-ADH. *Biochem. Biophys. Res. Commun.* 119, 1047-1055.

Strittmatter, P. and Ball, E.C.J. (1955) Formaldehyde dehydrogenase: a glutathione-dependent enzyme system. *J. Biol. Chem.* 213, 445-461.

Uotila, L. and Koivusalo, M. (1974) Formaldehyde dehydrogenase from human liver. *J. Biol. Chem.* 249, 7653-7663.

Uotila, L. and Koivusalo, M. (1987) Formaldehyde dehydrogenase from human erythrocytes: purification, some properties and evidence for multiple forms. *in* : Enzymology and Molecular Biology of Carbonyl Metabolism: Progress in Clinical and Biological Research vol. 232 (Weiner, H. and Flynn, T.G. eds) Alan R. Liss, Inc. New York, pp 165-178.

Uotila, L. and Koivusalo, M. (1989) Glutathione-dependent oxidoreductases: formaldehyde dehydrogenase. *in*: Glutathione, Chemical, Biochemical and Medical Aspects. Coenzymes and Cofactors, vol. III, part A (Dolphin, D., Poulson, R. and Avramovic, O. eds) John Wiley & Sons, New York, pp 517-551.

Uotila, L. and Mannervik, B. (1979) A steady-state model for formaldehyde dehydrogenase from human liver. *Biochem. J.* 177, 869-878.

Vallee, B.L. and Bazzone, T.S. (1983) Isozymes of human liver alcohol dehydrogenase. *Curr. Top. Bio. Med. Res.* 8, 219-244.

GLUTATHIONE-DEPENDENT FORMALDEHYDE DEHYDROGENASE/CLASS III ALCOHOL DEHYDROGENASE : FURTHER CHARACTERIZATION OF THE RAT LIVER ENZYME

Martti Koivusalo and Lasse Uotila

Department of Medical Chemistry
University of Helsinki
Siltavuorenpenger 10 A
00170 Helsinki, Finland

INTRODUCTION

Glutathione-dependent formaldehyde dehydrogenase (EC 1.2.1.1) was first described from beef and chicken livers (Strittmatter and Ball, 1955). The enzyme was first purified to homogeneity and characterized from human liver by Uotila and Koivusalo (1974a) who showed that the enzyme catalyzes a reversible reaction in which formaldehyde, GSH and NAD are converted into *S*-formylglutathione and NADH. The actual substrate of the enzyme is *S*-hydroxymethylglutathione, a hemimercaptal adduct nonenzymically formed from formaldehyde and GSH (Uotila and Mannervik, 1979). *S*-Formylglutathione is irreversibly hydrolyzed into formate and GSH in the reaction catalyzed by a specific *S*-formylglutathione hydrolase (EC 3.1.2.12) which has also been purified and characterized from human liver (Uotila and Koivusalo, 1974b). Both formaldehyde dehydrogenase and *S*-formylglutathione hydrolase are widely occurring enzymes and have later been isolated and characterized from a number of other sources in addition to human liver (Uotila and Koivusalo, 1989; Uotila, 1989).

Three classes of mammalian alcohol dehydrogenases are distinguished on the basis of their structural and functional properties (Vallee and Bazzone, 1983). Class III alcohol dehydrogenases are characterized by their very low affinity for ethanol but good use of long-chain alcohols as substrates, insensitivity to inhibition by pyrazole and anodic electrophoretic mobility. Class III alcohol dehydrogenases have a broader tissue distribution, with occurrence in most mammalian tissues, than the other alcohol dehydrogenases (Juliá *et al.*, 1987). The physiological functions of the class III alcohol dehydrogenases have remained obscure but the oxidation-reduction of long-chain alcohols/aldehydes

Enzymology and Molecular Biology of Carbonyl Metabolism 4
Edited by H. Weiner, Plenum Press, New York, 1993

(Juliá *et al.*, 1987) and the oxidation of ω-hydroxy fatty acids (Giri *et al.*, 1989) have been suggested.

We have reported evidence (Koivusalo *et al.*, 1989; Koivusalo and Uotila, 1991) which indicates that formaldehyde dehydrogenase and class III alcohol dehydrogenase are actually identical enzymes. In this report we describe some further characteristics of the formaldehyde dehydrogenase/class III alcohol dehydrogenase. Formaldehyde dehydrogenase has been regarded as highly specific for glutathione as the thiol cofactor (Koivusalo and Uotila, 1974; Uotila and Koivusalo, 1989). It has been reported recently that the enzymes from human liver (Holmquist and Vallee, 1991) and *Escherichia coli* (Gutheil *et al.*, 1992) could also use effectively some other thiols. We have reinvestigated the thiol specificity with the rat liver formaldehyde dehydrogenase. We have also investigated whether *S*-hydroxymethylglutathione and long-chain alcohols are competitive substrates for the enzyme and carried out alternate product inhibition experiments. Further we have characterized the anion activation of the forward and reverse reactions of the enzyme. We have reported earlier that chloride and some other anions activate formaldehyde dehydrogenase from human erythrocytes (Uotila and Koivusalo, 1987) and human liver (Uotila and Koivusalo, 1989).

MATERIALS AND METHODS

Formaldehyde dehydrogenase was purified from the cytosolic fraction of rat liver by a combination of 5'AMP-Sepharose affinity chromatography and chromatofocusing in a pH gradient 8-5 as described previously (Koivusalo *et al.*, 1982, Uotila and Koivusalo, 1983). The purified enzyme was homogeneous in SDS-gel electrophoresis giving only one protein band in silver staining (Merrill and Pratt, 1986) of the gel.

Female rats of Wistar strain weighing around 200 g were used as experimental animals. 5'AMP-Sepharose 4B and the materials for chromatofocusing were obtained from Pharmacia, Uppsala, Sweden. NAD(P)(H), nucleotide analogs, GSH, pyrazole, 12-hydroxydodecanoic acid, 1-octanol and 1-octanal were purchased from Sigma Chemical Co., St.Louis, MO, U.S.A. *S*-Formylglutathione was synthesized according to Uotila (1981), 6-mercaptohexanoate and 8-mercaptooctanoate according to Noll *et al.* (1987), and glutathione monoethyl ester according to Anderson and Meister (1989). The final product of the latter contained no unesterified glutathione detectable by thin layer chromatography.

The assay mixture (forward reaction) for formaldehyde dehydrogenase activity contained 1 mM formaldehyde, 1 mM GSH, 1.2 mM NAD and enzyme in 0.1 M sodium pyrophosphate buffer pH 8.0 (Uotila and Koivusalo, 1989) and that for class III alcohol dehydrogenase activity 1 mM 1-octanol, 1.2 mM NAD and enzyme in 0.1 M NaOH-glycine buffer pH 9.6. For the reverse reaction the assay mixtures contained 0.1 M sodium phosphate buffer pH 6.0, 1 mM *S*-formylglutathione or 1 mM 1-octanal, and 0.2 mM NADH or 0.2 mM NADPH. The reduction of NAD(P) or the oxidation of NAD(P)H was monitored at 340 nm and 25 °C either on a Shimadzu UV-240 spectrophotometer or a nine-channel FP-9 Analyzing System (Labsystems Oy, Helsinki, Finland). One unit of activity equals to 1 µmol of NADH produced or oxidized per min. The absorption coefficient of 6220 $M^{-1}\,cm^{-1}$ was used for NAD(P)H at 340 nm. In measurements requiring high sensitivity NAD(P)H was recorded fluorometrically on a Hitachi F-2000 Fluorescence Spectrophotometer. The protein concentrations

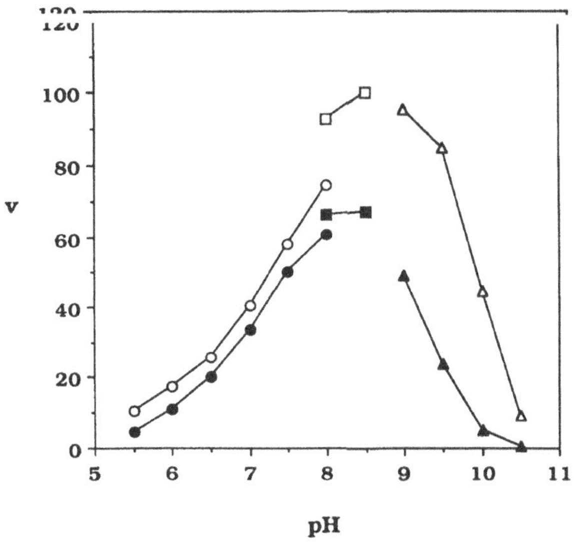

Figure 1. pH-activity profile for the oxidation of formaldehyde by rat liver formaldehyde dehydrogenase using either glutathione (open symbols) or glutathione monoethyl ester (closed symbols) as alternative cofactors. The buffers used (0.1 M) were sodium phosphate (○, ●), sodium pyrophosphate (□, ■) and NaOH-glycine (△, ▲). The assay mixtures contained 1 mM thiol, 1 mM formaldehyde and 1.2 mM NAD.

were determined by the method of Lowry *et al.* (1951) or by a modification of the method of Bradford (1976) with bovine serum albumin as the standard. Formaldehyde was prepared from hexamethylenetetramine and assayed by the chromotropic acid method (Koivusalo, 1956).

RESULTS AND DISCUSSION

Formaldehyde dehydrogenase is highly specific for formaldehyde as the substrate in the glutathione-dependent reaction. Several investigators have reported methylglyoxal as an active substrate (Rose and Racker, 1962, Schütte *et al.*, 1976, Uotila and Koivusalo, 1983) but these results were apparently due to heavy contamination of commercial methylglyoxal preparations with formaldehyde as pointed out by Pourmotabbed and Creighton (1986) for bovine liver formaldehyde dehydrogenase. With our present rat liver enzyme preparations we were also unable to find any activity with methylglyoxal prepared from methylglyoxal dimethylacetal which was free from formaldehyde.

The results of a reinvestigation of the thiol specificity of formaldehyde dehydrogenase from rat liver are presented in Table 1. Glutathione monoethyl ester gave almost as high a maximal velocity at pH 8 as glutathione, but the K_m value for the ester was 20-fold higher and the V_{max}/K_m ratio 30-fold lower than for glutathione (Table 2). The absence of the free carboxyl group of glycine in the ester resulted in several differences in the reactions with the ester or with glutathione as the cofactor. The ester was used much less efficiently than GSH at pH-values higher than 8 (Fig. 1), NADP was poorly used with the ester as nucleotide cofactor at pH 6, and NaCl which activated the enzyme with glutathione, was inhibitory with the ester (Table 3). We found no activity with cysteinylglycine as the thiol cofactor, in contrast to a recent study in which an efficient use of cysteinylglycine by the human enzyme was reported (70 % of the rate with glutathione; Holmquist and Vallee, 1991). On the other hand 6-

Table 1. Thiol specificity of rat liver formaldehyde dehydrogenase[1]

Thiol	Relative activity
Glutathione	100
Glutathione monoethyl ester	70
L-Cysteinyl-L-glycine	0
L-Cysteine	0
2-Mercaptoethanol	0
Dithiothreitol	0
Dithioerythritol	0
6-Mercaptohexanoate	30
8-Mercaptooctanoate	35
Captopril*	8
DL-6,8-Thioctic acid**	3
Coenzyme A	0

[1]The assay mixture contained 1 mM formaldehyde, 1 mM thiol and 1.2 mM NAD in 0.1 M sodium pyrophosphate buffer pH 8.0. The relative activity with glutathione has been set to 100.
*[2S]-1-[3-Mercapto-2-methylpropionyl]-L-proline **DL-α-Lipoic acid

Table 2. Kinetic constants for glutathione and glutathione monoethyl ester as cofactors of rat liver formaldehyde dehydrogenase[1]

Thiol	K_m (μM)	V_{max} (relative)	V_{max}/K_m (relative)
Glutathione	6.0	100	100
Glutathione monoethyl ester	125	78	3.7

[1]The K_m values refer to the total thiol concentration in the assays which were performed in 0.1 M Na-pyrophosphate buffer pH 8.0 with the constant concentrations of 1 mM formaldehyde and 1.2 mM NAD.

Table 3. Pyridine nucleotide specificity and effect of NaCl on rat liver formaldehyde dehydrogenase with glutathione and glutathione monoethyl ester as alternative cofactors[1]

Thiol	Relative activity			
	with NAD	with NAD plus NaCl	with NADP	with NADP plus NaCl
Glutathione	100	237	150	67
Glutathione monoethyl ester	67	52	55	0

[1]The assays were conducted in 0.1 M Na-phosphate pH 6.0, 1 mM thiol, 1 mM formaldehyde and 1.2 mM NAD or 1.2 mM NADP. The concentration of NaCl, when used, was 0.25 M. The relative activity with glutathione and NAD has been set to 100.

mercaptohexanoate and 8-mercaptooctanoate were used as relatively good cofactors by the rat liver enzyme, in agreement with recent reports for the human liver (Holmquist and Vallee, 1991) and *E. coli* (Gutheil *et al.*, 1992) enzymes. Dithiothreitol, mercaptoethanol, cysteine and coenzyme A had no activity as cofactors. Captopril and 2,6-thioctic acid gave only a weak activity, much less than observed for the human enzyme (Holmquist and Vallee, 1991). Glutathione can clearly still be held as the only physiological thiol cofactor for formaldehyde dehydrogenase in mammals as well as in most other sources, especially when the high affinity of glutathione for formaldehyde dehydrogenase, as well as the wide distribution and the high physiological concentration of glutathione are taken into account.

In addition to NAD and NADH formaldehyde dehydrogenase can also use NADP and NADPH as the nucleotide cofactor but only at pH values lower than 7 (Uotila and Koivusalo 1974a, 1989).Thus NADP could not be used in the oxidation of alcohols which requires a high pH even with NAD. Besides NAD(P)(H) the rat liver enzyme could also efficiently use several NAD analogs as the coenzyme. 3-Acetylpyridine adenine dinucleotide gave over 20-fold higher maximal velocity and 3-fold higher V_{max}/K_m ratio, when compared with those obtained with NAD. 3-Acetylpyridine hypoxanthine dinucleotide, thio-NAD and nicotinamide hypoxanthine dinucleotide also gave higher maximal velocities but due to their higher K_m-values, lower V_{max}/K_m ratios than NAD.

We performed several experiments with mixed substrates to see whether *S*-hydroxymethylglutathione, formed from formaldehyde and GSH, and the long-chain alcohol substrates react with the same active center on the enzyme competing with each other. When the effect of 12-hydroxydodecanoate on the formation of *S*-formylglutathione from glutathione and formaldehyde was investigated at 240 nm where the enzymic thioester formation can be recorded in a modified assay (Uotila and Koivusalo, 1974a) a linear competitive inhibition was observed (data not shown). When several constant concentrations of *S*-hydroxymethylglutathione were used with varying concentrations of 12-hydroxydodecanoate, a number of curves converging at the same point on the 1/v-axis were obtained in double reciprocal plots, showing competitive behavior; however, the curves were nonlinear in the presence of the both substrates (Fig. 2). When *S*-formylglutathione was used in the reverse reaction at constant concentrations with varying concentrations of 1-octanal, similar results to the former experiment were obtained (Fig. 3). The primary double reciprocal plots were under the same conditions linear when only one of the competing substrates (*S*-hydroxymethylglutathione or 12-hydroxydodecanoate; *S*-formylglutathione or 1-octanal) was present. These experiments indicate that the alternative substrates compete for the same active center on the enzyme but that the kinetic mechanism is more complex than a simple ordered mechanism. Further experiments are in progress to clarify the situation with mixed substrates.

Uotila and Mannervik (1979, 1980) have proposed for the human formaldehyde dehydrogenase a kinetic model in which the binding of the substrates *S*-hydroxymethylglutathione and NAD to the enzyme is random and glutathione is also needed as an allosteric activator. Juliá *et al.* (1987) have studied the kinetics of human class III alcohol dehydrogenase with 1-octanol, 1-octanal and NAD(H) as substrates and suggested an ordered Bi Bi mechanism with NAD binding first to the enzyme. It was not feasible in the present work to use *S*-formylglutathione as a product inhibitor with *S*-hydroxymethylgluta-

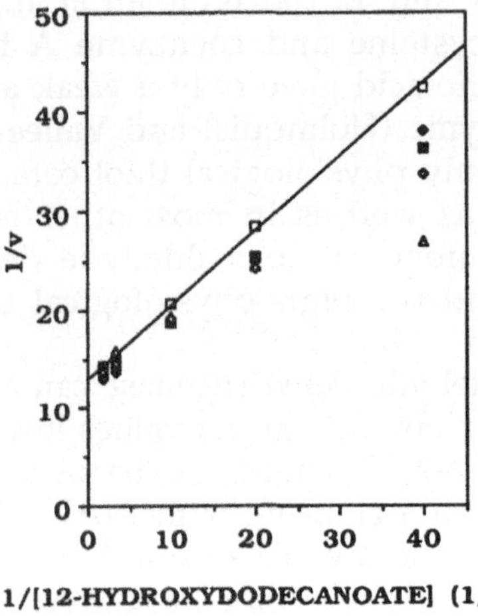

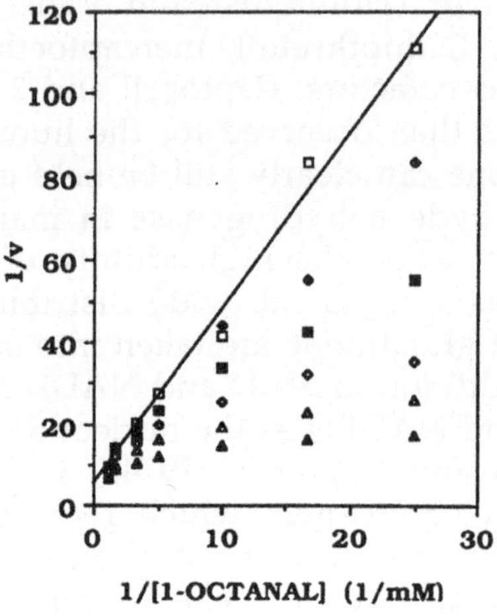

Figure 2. (Left). Effect of *S*-hydroxymethylglutathione on the oxidation of 12-hydroxydodecanoate by formaldehyde dehydrogenase. The concentrations of *S*-hydroxymethylglutathione used were 0 (□), 5 μM (◆), 12.5 μM (■), 25 μM (◇) and 50 μM (△). The buffer used was 0.1 M NaOH-glycine pH 9.6. NAD was constantly used at 1.2 mM.

Figure 3. (Right). Effect of *S*-formylglutathione on the reduction of 1-octanal by formaldehyde dehydrogenase. The concentrations of *S*-formylglutathione used were 0 (□), 75 μM (◆), 100μM (■), 150 μM (◇), 200 μM (△) and 250 μM (▲). The buffer used was 0.1 M sodium phosphate, pH 6.0. NADH was constantly used at 0.2 mM.

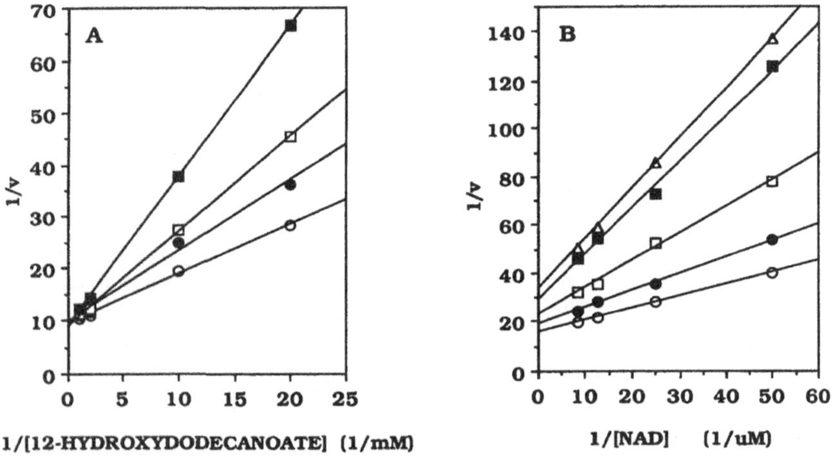

Figure 4. Effect of *S*-formylglutathione as alternate product on the oxidation of 12-hydroxydodecanoate by formaldehyde dehydrogenase. **A**: The concentration of 12-hydroxydodecanoate varied, NAD constantly used at 1.2 mM concentration. The concentrations of *S*-formylglutathione used were 0 (○), 0.4 mM (●), 0.6 mM (□) and 1.0 mM (■). **B**: The concentration of NAD varied, 12-hydroxydodecanoate constantly used at 0.2 mM concentration. The concentrations of *S*-formylglutathione used were 0 (○), 4 μM (●), 8 μM (□), 16 μM (■) and 20 μM (△). NaOH-glycine buffer (0.1 M) pH 9.6 was used in both experiments.

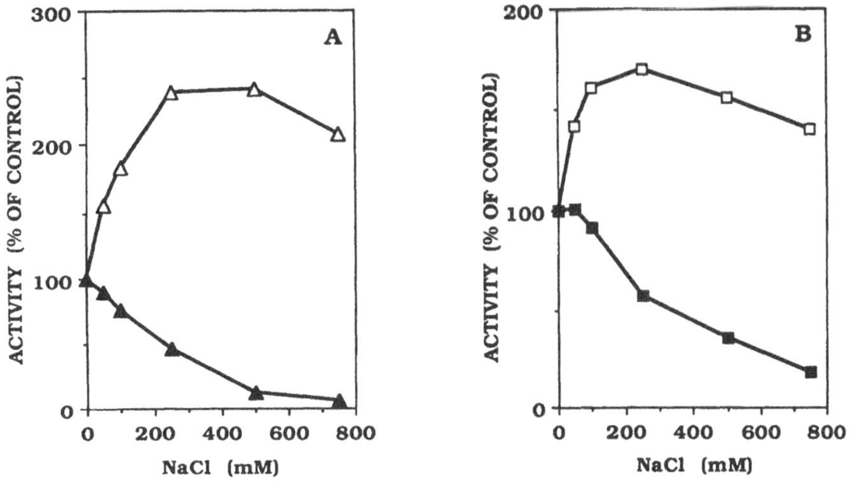

Figure 5. Effect of NaCl on the oxidation of *S*-hydroxymethylglutathione (**A**) and on the reduction of *S*-formylglutathione (**B**) by formaldehyde dehydrogenase. The assay mixtures contained varying amounts of NaCl as well as 1 mM formaldehyde, 1 mM GSH and 1.2 mM NAD (△) or 1.2 mM NADP (▲), 1 mM *S*-formylglutathione and 0.2 mM NADH (□) or 0.2 mM NADPH (■). The buffer used was 0.1 M sodium phosphate pH 6.0.

thione as the substrate, because all thiol ester preparations contain free glutathione which is difficult to remove. The oxidation of 12-hydroxydodecanoate was, however, not affected by glutathione. Thus we studied the effects of *S*-formylglutathione as alternative product inhibitor on the oxidation of 12-hydroxydodecanoate by rat liver formaldehyde dehydrogenase. When the results obtained at a constant saturating NAD concentration were plotted against varying 12-hydroxydodecanoate concentrations, the kinetic pattern was that of competitive inhibition with linear primary double reciprocal plots (Fig. 4A). The slope replot was parabolic (not shown). When NAD concentrations were varied in a similar experiment with a constant nearly saturating 12-hydroxydodecanoate concentration, the kinetic pattern was that of mixed inhibition (Fig. 4B). The replots of both slope and intercept were linear (not shown). These results do not fit an ordered Bi Bi mechanism but could be compatible with a Theorell-Chance type or random mechanism with possible dead-end complexes.

Chloride and several other anions affect the function of alcohol dehydrogenases due to the presence of anion-binding sites at the active centers of the enzymes (Brändén *et al.*, 1975). The effects on the different alcohol dehydrogenase forms differ due to structural differences. We have earlier reported activation of human liver and erythrocyte formaldehyde dehydrogenase by chloride (Uotila and Koivusalo, 1987; 1989). In the present experiments with rat liver enzyme, chloride activated the oxidation of *S*-hydroxymethylglutathione and the reduction of *S*-formylglutathione at pH values lower than 9. At higher pH values chloride was inhibitory. The activation was seen only when NAD(H) was used as coenzyme. The reactions were effectively inhibited when NADP(H) was used to replace NAD(H) (Fig. 5). These effects of chloride differ from those for human liver class I and horse liver alcohol dehydrogenases. The oxidation of alcohols (1-pentanol, 1-hexanol, 1-octanol and 12-hydroxydodecanoate) was also activated by chloride, even at pH values higher than 9 (Fig. 6).

Bromide and iodide salts also activated *S*-hydroxymethylglutathione oxi-

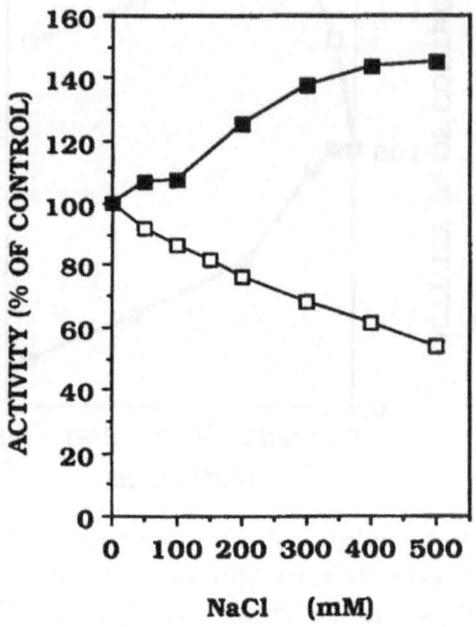

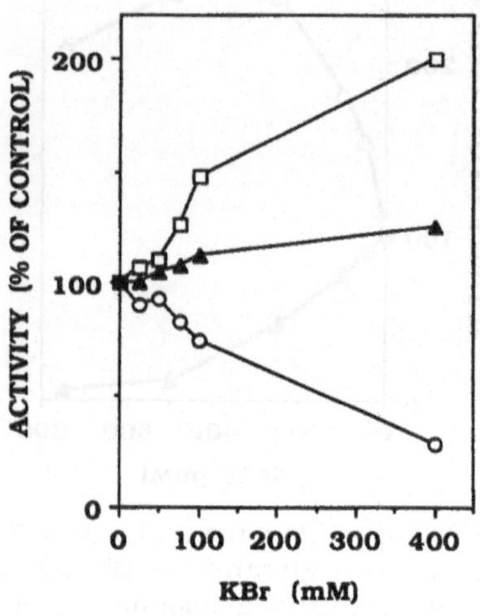

Figure 6. (Left). Effect of NaCl on the oxidation of *S*-hydroxymethylglutathione (□) and of 1-octanol (■) by formaldehyde dehydrogenase. The assay mixtures contained 0.1 M NaOH-glycine pH 10.0, 1.2 mM NAD and either 1 mM formaldehyde and 1 mM GSH or 1 mM 1-octanol.

Figure 7. (Right). Effect of KBr on the oxidation of *S*-hydroxymethylglutathione at pH 8 (□) and at pH 10 (○), and on the oxidation of 12-hydroxydodecanoate at pH 10 (▲) by formaldehyde dehydrogenase. The assays contained 1mM formaldehyde and 1mM GSH or 1 mM 12-hydroxydodecanoate, 1.2 mM NAD and either 0.1 M sodium pyrophosphate pH 8.0 or 0.1 M NaOH-glycine pH 10.0.

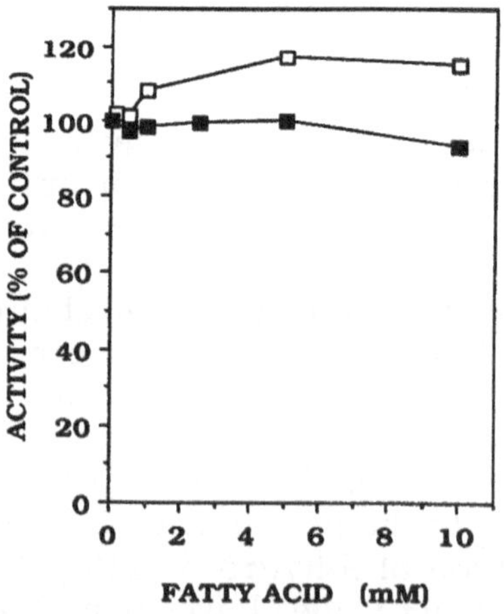

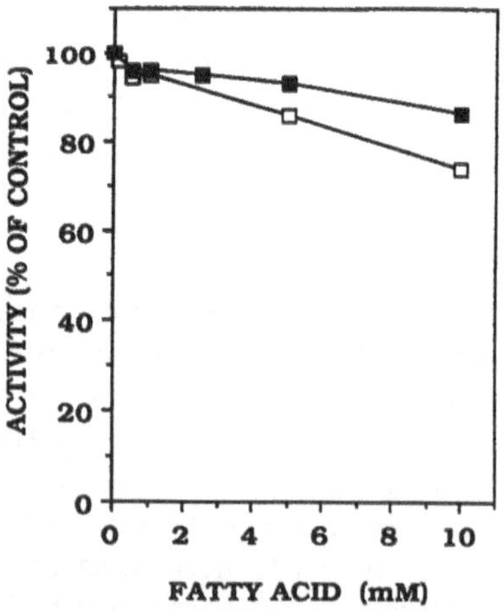

Figure 8. (Left). Effect of hexanoate (□) and octanoate (■) on the oxidation of *S*-hydroxymethylglutathione by formaldehyde dehydrogenase. The assays contained 1 mM formaldehyde, 1 mM GSH, 1.2 mM NAD and 0.1 M sodium pyrophosphate pH 8.0.

Figure 9. (Right). Effect of hexanoate (□) and octanoate (■) on the oxidation of 1-octanol by formaldehyde dehydrogenase. The assays contained 1 mM 1-octanol, 1.2 mM NAD and 0.1 M NaOH-glycine pH 10.0.

dation at pH 8 but were at pH 10 more inhibitory than chloride. Bromide and iodide salts had no effect on the oxidation of alcohols at pH 10 (Fig. 7). Formate, acetate and propionate activated the oxidation of *S*-hydroxymethylglutathione but of longer-chain fatty acids hexanoate activated only slightly and octanoate had no effect (Fig. 8). Hexanoate and octanoate were slightly inhibitory for the oxidation of 1-octanol by rat liver formaldehyde dehydrogenase (Fig. 9). Moulis *et al.* (1991) reported effective activation of human liver class III alcohol dehydrogenase by hexanoate and octanoate in the oxidation of some short-chain alcohols but, in agreement with our results, found octanoate to inhibit the oxidation of 1-octanol.

ACKNOWLEDGEMENTS

This research has been supported by a grant from the University of Helsinki. Mrs. Eija Haasanen has provided skillful technical assistance.

REFERENCES

Anderson, M.E., and Meister, A., 1989, Glutathione monoesters, *Anal.Biochem.* 183:16-20.

Bradford, M.M, 1976, A rapid and sensitive method for the quantitation of microgram quantities of protein utilizing the principle of protein-dye binding, *Anal.Biochem.* 72:248-254.

Brändén, C.-I., Jörnvall, H., Eklund, H., and Furugren, B., 1975, Alcohol dehydrogenases, *in*: "The Enzymes, Third edition, vol.XI," P. Boyer, ed., chapter 3, pp.104-190, Academic Press, New York.

Giri, P.R., Linnoila, M., O'Neill, J.B., and Goldman, D., 1989, Distribution and possible metabolic role of class III alcohol dehydrogenase in the human brain, *Brain Res.* 481: 131-141.

Gutheil, W.G., Holmquist, B., and Vallee, B.L., 1992, Purification, characterization, and partial sequence of the glutathione-dependent formaldehyde dehydrogenase from *Escherichia coli*: A class III alcohol dehydrogenase, *Biochemistry* 31:475-481.

Holmquist, B., and Vallee, B., 1991, Human liver class III alcohol and glutathione dependent formaldehyde dehydrogenase are the same enzyme, *Biochem.Biophys.Res.Commun.* 178:1371-1377.

Juliá, P., Boleda, M.D., Farrés, J., and Parés, X., 1987, Mammalian alcohol dehydrogenase: Characteristics of class III isoenzymes, *Alcohol & Alcoholism*, Suppl. 1:169-173.

Koivusalo, M., 1956, Studies on the metabolism of methanol and formaldehyde in the animal organism, *Acta physiol.Scand.* 39:Suppl.131:31-34.

Koivusalo, M., Baumann, M., and Uotila, L., 1989, Evidence for the identity of glutathione-dependent formaldehyde dehydrogenase and class III alcohol dehydrogenase, *FEBS Lett.* 257:105-109.

Koivusalo, M., Koivula, T., and Uotila, L., 1982, Oxidation of formaldehyde by nicotinamide nucleotide dependent enzymes, *in*:"Enzymology of Carbonyl Metabolism: Aldehyde Dehydrogenase and Aldo/Keto Reductase", H. Weiner and B. Wermuth, eds., pp.155-168, Alan R. Liss, Inc., New York.

Koivusalo, M., and Uotila, L., 1974, Enzymic method for the quantitative determination of reduced glutathione, *Anal.Biochem.* 59:34-45.

Koivusalo, M., and Uotila, L., 1991, Glutathione-dependent formaldehyde dehydrogenase (EC 1.2.1.1): Evidence for the identity with class III alcohol dehydrogenase, *in*: "Enzymology and Molecular Biology of Carbonyl Metabolism 3", H. Weiner, B. Wermuth, and D.W. Crabb, eds., pp. 305-313, Plenum Press, New York.

Lowry, O.H., Rosebrough, N.J., Farr, A.L., and Randall, R.J., 1951, Protein measurement with the Folin phenol reagent, *J.Biol.Chem.* 193:265-275.

Merrill, C.R., and Pratt, M.E., 1986, A silver stain for the rapid quantitative detection of proteins or nucleic acids on membranes or thin layer plates, *Anal.Biochem.* 156:96-110.

Moulis, J.-M., Holmquist, B., and Vallee, B.L., 1991, Hydrophobic anion activation of human liver χχ alcohol dehydrogenase, *Biochemistry* 30:5743-5749.

Noll, K.M., Donnelly, M.I., and Wolfe, R.S., 1987, Synthesis of 7-mercaptoheptanoylthreonine phosphate and its activity in the methylcoenzyme M methylreductase system, *J.Biol.Chem.* 262:513-515.

Pourmotabbed, T., and Creighton, D.J., 1986, Substrate specificity of bovine liver formaldehyde dehydrogenase, *J.Biol.Chem.* 261:14240-14244.

Rose, Z.B., and Racker, E., 1962, Formaldehyde dehydrogenase from baker's yeast, *J.Biol.Chem.* 237:3279-3281.

Schütte, H., Flossdorf, J., Sahm, H., and Kula M.-R, 1976, Purification and properties of formaldehyde dehydrogenase and formate dehydrogenase from *Candida boidinii*, *Eur.J.Biochem.* 62:151-160.

Strittmatter, P., and Ball, E.G., 1955, Formaldehyde dehydrogenase, a glutathione-dependent enzyme system, *J.Biol.Chem.* 213:445-461.

Uotila, L., 1981, Thioesters of glutathione, *Meth.Enzymol.* 77:424-430.

Uotila, L., 1989, Glutathione thiol esterases, *in*: "Coenzymes and Cofactors, vol. III, Glutathione. Chemical, Biochemical and Medical Aspects, part A", D. Dolphin, R. Poulson, and O. Avramovic, eds., pp. 767-804, John Wiley & Sons, Inc., New York.

Uotila, L., and Koivusalo, M., 1974a, Formaldehyde dehydrogenase from human liver. Purification, properties, and evidence for the formation of glutathione thiol esters by the enzyme, *J.Biol.Chem.* 249:7653-7663.

Uotila, L., and Koivusalo, M., 1974b, Purification and properties of *S*-formylglutathione hydrolase from human liver, *J.Biol.Chem.* 249:7664-7672.

Uotila, L., and Koivusalo, M., 1983, Formaldehyde dehydrogenase, *in*:"Functions of Glutathione. Biochemical, Physiological, Toxicological and Clinical Aspects", A.Larsson, S.Orrenius, A.Holmgren, B.Mannervik, eds., pp.175-186, Raven Press, New York.

Uotila, L., and Koivusalo, M., 1987, Multiple forms of formaldehyde dehydrogenase from human red blood cells, *Human Heredity* 37, 102-106.

Uotila, L., and Koivusalo, M., 1989, Glutathione-dependent oxidoreductases: Formaldehyde dehydrogenase, *in*: "Coenzymes and Cofactors, vol.III, Glutathione. Chemical, Biochemical and Medical Aspects, part A", D. Dolphin, R. Poulson and O. Avramovic, eds., pp.517-551, John Wiley & Sons, Inc., New York.

Uotila, L., and Mannervik, B., 1979, A steady-state kinetic model for formaldehyde dehydrogenase from human liver. A mechanism involving NAD^+ and the hemimercaptal adduct of glutathione and formaldehyde as substrates and free glutathione as an allosteric activator of the enzyme, *Biochem.J.* 177:869-878.

Uotila, L., and Mannervik, B., 1980, Product inhibition studies of human liver formaldehyde dehydrogenase, *Biochim.Biophys.Acta* 616:153-157.

Vallee, B.L., and Bazzone, T.J., 1983, Isozymes of human liver alcohol dehydrogenase, *in*: "Isozymes: Current Topics in Biological and Medical Research, vol. 8", M.C. Rattazi, J.C. Scandalios, and G.S. Whitt, eds., pp. 219-244, Alan R. Liss, New York.

CLASS IV ALCOHOL DEHYDROGENASE: STRUCTURE AND FUNCTION

Xavier Parés, Jaume Farrés, Alberto Moreno, Narcís Saubi, M. Dolors Boleda, Ella Cederlund, Jan-Olov Höög, and Hans Jörnvall

Department of Biochemistry and Molecular Biology, Universitat Autònoma de Barcelona, 08193 Bellaterra, Spain, and Department of Chemistry I, Karolinska Institutet, S-104 01 Stockholm, Sweden

INTRODUCTION

Mammalian alcohol dehydrogenases (ADH, EC 1.1.1.1) have been extensively characterized mainly from human hepatic tissue. Initially they were grouped into three classes (I, II, and III), based on their electrophoretic and kinetic properties (Vallee and Bazzone, 1983). Homologous ADH classes were found in the rat and other mammals (Julià et al., 1987). In contrast to the human liver enzyme system, a class II-like activity could never be detected in rat liver. Rather, such an activity was found first in rat stomach (Julià et al., 1987) and later in human stomach (Moreno and Parés, 1991).

Further characterization of the stomach ADH from both rat and human demonstrated that it was not a class II enzyme, leading to the identification of a novel ADH class, class IV (Parés et al., 1990, Parés et al., 1992). Here we compare the kinetic and structural properties of the class IV ADH enzymes isolated from rat and human stomach mucosa. We also discuss the physiological significance of class IV ADH, and its role in the non-hepatic metabolism of ethanol.

EXPERIMENTAL

Purification of Human and Rat Stomach Alcohol Dehydrogenase

Human gastric tissue was obtained from a surgical sample (72 g) and stored at -80 °C until use. Purification of the human gastric enzyme ($\sigma\sigma$-ADH) was carried out by chromatography on DEAE-Sepharose and AMP-Sepharose, as described (Moreno and Parés, 1991). A modification in the second chromatography step was the use of 100 mM Tris-HCl, pH 8.0, 0.5 mM dithiothreitol. Under these conditions, the enzyme binds to the column and is specifically eluted by a linear gradient of 0-0.4 mM NADH in the same buffer.

Rat stomach ADH was purified from whole stomach of Sprague-Dawley rats by

Enzymology and Molecular Biology of Carbonyl Metabolism 4
Edited by H. Weiner, Plenum Press, New York, 1993

chromatography on DEAE-Sepharose and AMP-Sepharose, as described (Julià et al., 1987).

Protein concentration was determined by the Coomassie blue method (Bradford, 1976), using bovine serum albumin as a standard.

Enzyme Assays

Alcohol dehydrogenase activity was determined by monitoring the formation or utilization of NADH at 340 nm and 25°C in a Cary 219 spectrophotometer. Alcohol oxidation was measured in a 1 ml/1 cm light-path cell with 0.1 M sodium phosphate, pH 7.5. Aldehyde reducing activity was measured in a 0.7 ml/0.2 cm light-path cell with 0.1 M sodium phosphate, pH 7.5. The ADH activity in partially purified fractions was measured using 0.1 M glycine-NaOH buffer, pH 10.0, 2.4 mM NAD^+ and 100 mM ethanol as standard conditions. Kinetic results were analysed by using the non-linear regression data analysis program Enzfitter (Leatherbarrow, 1987).

Electrophoresis

Starch gel electrophoresis was performed according to a modification of a previous method (Parés et al., 1985). The gels (24 x 12 x 0.3 cm) contained 11 % starch, 0.74 mM NAD^+ and 20 mM Tris-HCl buffer pH 8.6 for human samples and 20 mM Tris-HCl buffer pH 7.6 for rat samples. Samples were applied to paper wicks (6 x 3 x 0.25 mm) and inserted into the gel at the center of the slabs. Gels were run at 720 V for 5 h at 4°C. Immediately after electrophoresis gels were sliced longitudinally into two slabs, that were stained for ADH activity. The staining solution (250 ml) contained 50 mM Tris-HCl, pH 8.6, either 100 mM pentanol or 100 mM 2-buten-1-ol, 11 mM pyruvic acid, 0.55 mM NAD^+, 0.24 mM NBT and 0.065 mM PMS. Gel slabs were incubated in these solutions at 40°C for about 40 min in the dark. After being stained, the gels were washed with water, photographed and stored at 4°C.

Primary Structure Determination

Pure proteins were [^{14}C]carboxymethylated by solubilization in 8 M urea, 0.4 M Tris, 2 mM EDTA, pH 8.15, reduction with dithiothreitol, and treatment with ^{14}C-labelled iodoacetate (Julià et al., 1988). Samples of the carboxymethylated proteins were cleaved with *Lysobacter* Lys-specific protease and *Pseudomonas* Asp-specific protease. In both cases, digests obtained from the rat protein were directly fractionated by reverse-phase HPLC on Vydac C4 and Ultropac C18 columns as described (Cederlund et al., 1991). Peptides obtained from the human protein were fractionated by reverse-phase HPLC on a C_8 column (Nucleosil 3 C8 100A, 50 x 2 mm). Pure peptides obtained were analyzed for amino acid sequence by degradations in an ABI 477A sequencer with an on-line 120A analyzer, and a MilliGen/Biosearch Prosequencer 6600, also with on-line detection.

RESULTS AND DISCUSSION

Electrophoretic properties

The ADH characteristic of rat and human stomach was first identified by starch gel electrophoresis, using activity staining with pentanol, 2-buten-1-ol or with a high ethanol concentration (250 mM) (Fig. 1). Under these conditions, the most prominent band in rat stomach homogenates is class IV ADH (Fig. 1A, lane 3). This band is absent in liver homogenates (e.g., Fig. 1A, lane 1) and is different from the more cathodic liver enzyme (class I) or the ubiquitous class III (glutathione-dependent formaldehyde dehydrogenase).

The class IV band of the human stomach analysis exhibits, in general, similar intensity to the class I bands (γγ-ADH forms, Fig. 1B, lane 1), and also shows different mobility than the human liver class I and class II (ππ-ADH) enzymes (not shown).

In rat stomach, class IV ADH is very anodic (pI 5.1), while class IV ADH from human stomach (also known as σσ-ADH) is a cathodic enzyme (pI 8.5-8.7). By this technique we do not detect a class II enzyme in rat liver (Fig. 1A, lane 1), although a cDNA coding for it has been found by Jan-Olov Höög and Hans Jörnvall.

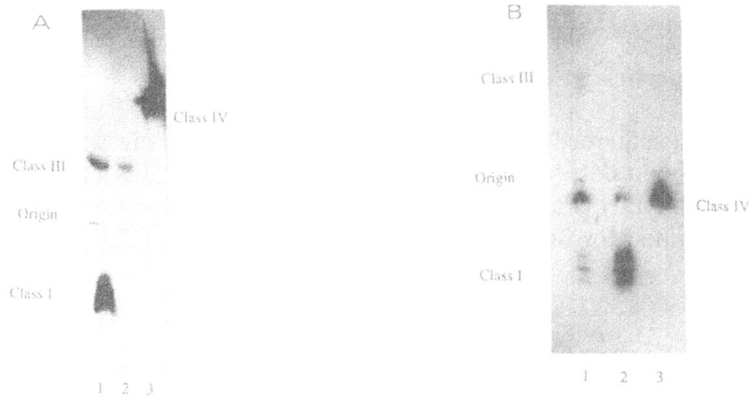

Figure 1. Starch gel electrophoresis of rat (A) and human (B) samples. A, Lane 1: liver homogenate. Lane 2: brain homogenate. Lane 3: stomach homogenate. B, Lane 1: gastric mucosa homogenate. Lane 2: class I-enriched chromatographic fraction. Lane 3: purified class IV ADH.

Kinetic properties

Rat and human class IV alcohol dehydrogenases have been purified by column chromatography, and have been kinetically and structurally characterized. Both enzymes have physicochemical properties similar to those of other zinc-containing, mammalian alcohol dehydrogenases, i.e., they are dimers of 40 kDa subunits with two zinc atoms per monomer (Julià et al., 1987; Moreno and Parés, 1991). Their general kinetic features are also typical of mammalian alcohol dehydrogenases. For instance, coenzyme specificity for NAD(H), pH optimum about 10.0 and more activity with aldehydes than with the corresponding alcohols at neutral pH. Some of the kinetic constants of class IV enzymes at pH 7.5 are indicated in Table 1.

Class IV ADHs from human and rat stomach exhibit a high Km for ethanol, a preference for medium chain and aromatic compounds, and very high kcat values as compared to other ADH classes. m-Nitrobenzaldehyde is one of the best substrates. In general, Km and kcat values are much higher for the rat enzyme than for the human. This is particularly true for ethanol (5,000 mM vs. 40 mM), suggesting very different kinetics for ethanol elimination in rat and human gastric tissues.

Table 1. Kinetic constants of the human and rat stomach enzymes, in 0.1 M sodium phosphate, pH 7.5, at 25°C.

	Human		Rat	
Substrate	Km (mM)	kcat (min^{-1})	Km (mM)	kcat (min^{-1})
Ethanol	40	280	5,000	1,000
Octanol	0.03	170	0.5	300
Acetaldehyde	13	300	does not saturate	
Octanal	0.08	1,400	0.3	23,000
m-Nitrobenzaldehyde	0.04	5,500	1.4	59,000

Structural properties

Amino acid sequence analysis has been performed on class IV ADH. In total, 165 residues have been determined for rat stomach ADH (Parés et al., 1990) and 72 for human stomach ADH (Parés et al., 1992). Structural comparisons corresponding to the regions determined for both enzymes are summarized in Table 2.

Table 2. Structural comparisons between the five known classes of mammalian ADH. Data were taken from Jörnvall et al. (1989), Parés et al. (1990) and Parés et al. (1992). Class V designates the enzyme encoded by the recently reported *ADH6* gene (Yasunami et al. (1991). Values correspond to the regions already sequenced for the human class IV enzyme. Species differences cannot be established for class V since only the human form is known.

A.	Class differences			
	Human class IV versus human class			
	I	II	III	V
	(residue differences, %)			
	32	35	39	38

B.	Species differences			
	Human/rat species variations within class			
	I	II	III	IV
	(residue differences, %)			
	18	25	3	15

The extent of residue differences between class IV and other classes is above 30% in all cases (Table 2, A), even for class II (35 % differences), which initially was thought to be related to the stomach enzyme because of kinetic similarities (Algar et al., 1983, Julià et al., 1987, Moreno and Parés, 1991). The difference values are typical for class distinction and imply that stomach ADH represents a separate ADH class (class IV). When a given class is compared between different species (Table 2, B), the stomach enzymes show 15 % differences, similar to what is found for class I, indicating that the human and rat gastric enzymes belong to the same class, despite of their differences in kinetic properties, and that class IV is widely distributed in mammals. From these comparisons it can also be concluded that classes I, II and IV are variable enzymes, while class III is a more conserved enzyme.

Table 3 summarizes the main features of the four human ADH classes that have been characterized so far. Although their tissue distribution and primary structures are quite different, the human class IV enzyme is very similar to class II ADH in terms of isoelectric point and Km for ethanol. However, class IV has a much lower Ki for 4-methylpyrazole than class II. Therefore, inhibition with pyrazole derivatives could be used as a criterium to differentiate the four ADH classes.

Table 3. Properties of human ADH classes.

Class	Enzymes	Isoelectric point	Km ethanol (mM)	Ki 4-methylpyrazole (μM)
I	Dimers of α,β,γ	> 8.7	0.05-5	0.1-2
II	ππ-ADH	8.6	34	2,000
III	χχ-ADH	6.4	>5,000	No inhibition
IV	σσ-ADH	8.5-8.7	40	10

Physiological role

In the search for the physiological significance of class IV ADH, we studied its tissue distribution in the rat (Boleda et al., 1989). Class IV ADH is mostly distributed in all external epithelia, cornea, skin, and digestive, respiratory and sexual tracts. This localization suggests a protective role for class IV against external toxic alcohols or aldehydes, or against those generated during photochemical or oxidative reactions through lipid peroxidation. Class IV also oxidizes retinol and reduces retinal, and thus a role in the metabolism of retinoids cannot be excluded. Human class IV shows a similar tissue distribution.

An estimation of the relative contribution of each ADH class in rat total ethanol metabolism, using 33 mM ethanol, results in a value of only 4 % for class IV ADH (Boleda et al., 1989). The remaining activity is accounted by class I ADH, mostly the hepatic enzyme. However, some gastrointestinal tissues, such as esophagus and stomach, show similar activity than liver per gram of tissue with 1 M ethanol because of their high content in class IV ADH. A high activity is also found in human digestive organs, being esophagus mucosa a tissue with higher activity per gram of tissue than liver, with 100 mM ethanol. Since these concentrations are reached in the gastrointestinal tube during ethanol consumption, class IV ADH could, at least in part, account for the first-pass metabolism of ethanol (Julkunen et al., 1985). The contribution of gastric ADH to total ethanol metabolism is however small, because of the small amount of mucosa as compared to that of liver tissue.

ACKNOWLEDGEMENTS

This work was supported by grants from the Spanish Dirección General de Investigación Científica y Técnica (Project PB89-0285), the Swedish Medical Research Council (Project 03X-3532, 8639), and the Swedish Alcohol Research Fund.

REFERENCES

Algar, E.M., Seeley, T.-L., and Holmes, R.S., 1983, Purification and molecular properties of mouse alcohol dehydrogenase isozymes, *Eur. J. Biochem.* 137:139-147.

Boleda, M.D., Julià, P., Moreno, A., and Parés, X., 1989, Role of extrahepatic alcohol dehydrogenase in rat ethanol metabolism, *Arch. Biochem. Biophys.* 274:74-81.

Bradford, M., 1976, A rapid and sensitive method for the quantitation of microgram quantities of protein utilizing the principle of protein-dye binding, *Anal. Biochem.* 72:248-254.

Cederlund, E., Peralba, J.M., Parés, X., and Jörnvall, H., 1991, Amphibian alcohol dehydrogenase, the major frog liver enzyme. Relationships to other forms and assessment of an early gene duplication separating vertebrate class I and class III alcohol dehydrogenases, *Biochemistry* 30:2811-2816.

Jörnvall, H., von Bahr-Lindström, H., and Höög, J.-O., 1989, Alcohol dehydrogenase -structure, *in*: "Human Metabolism of Alcohol", Vol. II, K.E. Crow and R.D. Batt, eds., CRC Press, Boca Raton, pp. 43-64.

Julià, P., Farrés, J., and Parés, X., 1987, Characterization of three isoenzymes of rat alcohol dehydrogenase. Tissue distribution and physical and enzymatic properties, *Eur. J. Biochem.* 162:179-189.

Julià, P., Parés, X., and Jörnvall, H., 1988, Rat liver alcohol dehydrogenase of class III. Primary structure, functional consequences and relationships to other alcohol dehydrogenases, *Eur. J. Biochem.* 172:73-83.

Julkunen, R.J., Tannenbaum, L., Baraona, E., Lieber, C.S., 1985, First pass metabolism of ethanol: an important determinant of blood levels after alcohol consumption, *Alcohol* 2:437-441.

Leatherbarrow, R.J., 1987, Enzfitter: A nonlinear regression data analysis program for the IBM-PC, Elsevier Biosoft, Cambridge, United Kingdom.

Moreno, A., and Parés, X., 1991, Purification and characterization of a new alcohol dehydrogenase from human stomach, *J. Biol. Chem.* 266:1128-1133.

Parés, X., Julià, P., and Farrés, J., 1985, Properties of rat retina alcohol dehydrogenase, *Alcohol* 2:43-46.

Parés, X., Moreno, A., Cederlund, E., Höög, J.-O., and Jörnvall, H., 1990, Class IV mammalian alcohol dehydrogenase. Structural data of the rat stomach enzyme reveal a new class well separated from those already characterized, *FEBS Lett.* 277:115-118.

Parés, X., Cederlund, E., Moreno, A., Saubi, N., Höög, J.-O., and Jörnvall, H., 1992, Class IV alcohol dehydrogenase (the gastric enzyme). Structural analysis of human σσ-ADH reveals class IV to be variable and confirms the presence of a fifth mammalian alcohol dehydrogenase class, *FEBS Lett.* 303:69-72.

Vallee, B.L., and Bazzone, T.J., 1983, Isoenzymes of human liver alcohol dehydrogenase, *Isozymes: Curr. Top. Biol. Med. Res.* 8:219-244.

Yasunami, M., Chen, C.-S., and Yoshida, A., 1991, A human alcohol dehydrogenase gene (*ADH6*) encoding an additional class of isozyme, *Proc. Natl. Acad. Sci. USA* 88:7610-7614.

THE OXIDATION OF ALDEHYDES BY HORSE LIVER ALCOHOL DEHYDROGENASE

Gary T.M. Henehan, George L. Kenyon and Norman J. Oppenheimer

Department of Pharmaceutical Chemistry, University of California,
San Francisco, California 94143-0446.

INTRODUCTION

Horse liver alcohol dehydrogenase (HL-ADH) exhibits a broad substrate specificity (Pietruszko, 1979). The physiological role of HL-ADH has proven difficult to define. Indeed, it is not clear whether the enzyme will function as an oxidase or a reductase under physiological conditions (see Weiner, 1991). HL-ADH has been shown to catalyze the following reactions:

Alcohol oxidation: $RCH_2OH + NAD^+ \rightleftharpoons RCHO + NADH + H^+$ (1)

Aldehyde reduction: $RCHO + NADH + H^+ \rightleftharpoons RCH_2OH + NAD^+$ (2)

Aldehyde oxidation: $RCHO + NAD^+ \rightarrow RCOO^- + NADH + 2H^+$ (3)

Aldehyde dismutation: $2RCHO \rightarrow RCH_2OH + RCOO^- + H^+$ (4)

Of these four reactions, (1) and (2) (Dalziel and Dickinson, 1966) have been much more extensively studied than (3) (Hinson and Neal, 1975) and (4) (Dalziel and Dickinson, 1965). This report focusses on the aldehyde oxidation (3) and aldehyde dismutation (4) reactions.

Aldehyde dismutation neither produces nor consumes NADH so the dihydronicotinamide chromophore that absorbs at 340 nm cannot be used to follow the reaction. Rather, the reaction must be monitored either by direct assay of the reaction products, for example by gas liquid chromatography, or the protons produced may be followed continuously by a titrimetric procedure (Dalziel and Dickinson, 1965). Scheme I shows the mechanism proposed to account for aldehyde dismutation by HL-ADH. Most unbranched aliphatic aldehydes exist in aqueous solution as roughly 50:50 mixtures of free aldehyde and aldehyde hydrate with the exception of formaldehyde which is fully hydrated. The hydrated form of the aldehyde substrate (gem-diol), which structurally resembles a secondary alcohol, combines with the E·NAD complex and is oxidized to the corresponding carboxylic acid (Abeles and Lee, 1960). Before the E·NADH complex formed dissociates NADH, it immediately combines with the free form of the aldehyde, reducing it to the corresponding alcohol.

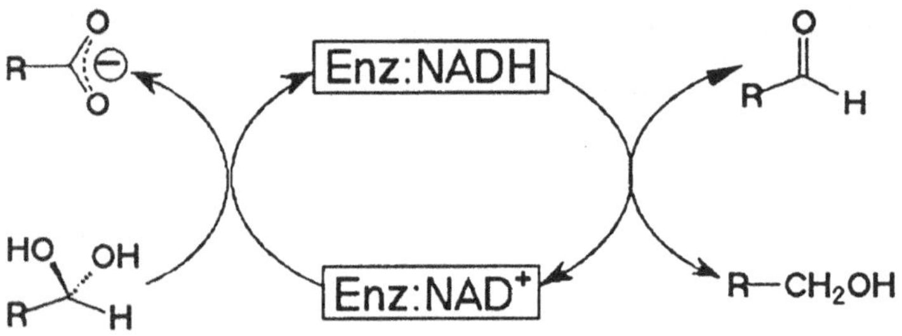

Scheme 1. Proposed mechanism of aldehyde dismutation by HL-ADH (Dalziel and Dickinson, 1965).

Each turn of the cycle produces one molecule of carboxylic acid and one molecule of the corresponding alcohol from two molecules of aldehyde. The acid that accumulates as the reaction proceeds does not appear to inhibit the enzyme (Anderson and Dahlquist, 1982). Clearly, the alcohol generated during dismutation will compete with the hydrated aldehyde for the E·NAD complex thereby slowing the progress of the reaction, especially as the concentration of aldehyde decreases. Nonetheless, Dalziel and Dickinson (1965) have shown that the dismutation reaction does go to completion. Dismutation will occur even with very low concentrations of NAD^+, i.e., comparable with the enzyme concentration. By contrast very high K_m values for the aldehyde substrate (e.g., 100 mM for acetaldehyde) have been reported for dismutation. An interesting feature of the dismutation reaction is that it bypasses the rate-limiting step in the forward direction, i.e., NADH release, which has been shown to increase greatly the rate of the coupled oxidation-reduction activity of HL-ADH (Gupta and Robinson, 1966).

In 1975, Hinson and Neal reported that HL-ADH was capable of catalysing the oxidation of aldehydes *and* net NADH formation with a K_m for NAD^+ of about 2.0 mM and low K_m values for aliphatic aldehyde substrates (e.g., see Table 1). The V_{max} for aldehyde oxidation was found to increase with increasing chain length, with octanal being the best substrate having a V_{max} equal to 7% that of ethanol under the same conditions. All evidence pointed to the same catalytic site conducting both the oxidation of alcohols and aldehydes (see Pettersson, 1987). Other alcohol dehydrogenases including nonmetallo enzymes have been found to carry out aldehyde oxidation, and in the case of the *Drosophila melanogaster* enzyme this activity was even suggested to be its true physiological function (Moxon *et al.*, 1985). More recently, some carbonyl reductases have been reported to conduct aldehyde oxidation (Hara *et al.*, 1991).

Table 1 compares previously reported kinetic constants for the dismutation and aldehyde oxidation reactions when butanal is used as substrate. For the dismutation reaction the K_m value for NAD^+ is low but that for butanal is high while the oxidation reaction shows the opposite pattern. This contrast is also seen with other aldehyde substrates.

Table 1. Comparison of kinetic constants for the aldehyde dismutation and aldehyde oxidation reactions of HL-ADH with butanal.

Activity	Km, NAD (μM)	Km, Butanal (μM)
Aldehyde dismutation*	24	16000
Aldehyde oxidation**	1880	29
Aldehyde oxidation***	2100	40

* Dalziel and Dickinson (1965); ** Hinson and Neal (1975); *** Kinetic constants determined in the present work from the data of Figure 2

This report reexamines the aldehyde oxidation reaction reported by Hinson and Neal (1975) and its relationship to the aldehyde dismutation reaction.

MATERIALS AND METHODS

HL-ADH and NAD^+ were obtained from the Sigma Chemical Company and were used without further purification. Butanal obtained from Aldrich Chemical Company was distilled under a stream of oxygen-free nitrogen and stored at 4°C until use. NAD^+ was standardized by absorbance measurement at 260 nm using an extinction coefficient of 1.8×10^4 l mol^{-1} cm^{-1}in distilled water. All enzyme assays were performed in 0.1M sodium phosphate buffer, pH 7.5 at 37°C, in an assay volume of 1 ml. Spectrophotometric assays were monitored by measuring the increase in absorbance at 340 nm with time due to NADH formation.

Proton nuclear magnetic resonance spectra were acquired at 500 MHz on a General Electric GN-500 instrument. Buffer salts and coenzyme were lyophilized twice from 99.8% D_2O and finally dissolved in 100% D_2O. Freshly distilled aldehyde was dissolved directly in 100% D_2O. Enzyme samples were dissolved in 99.8% D_2O and exchanged into

100% D_2O by diafiltration on Centricon centrifugal microconcentrators (30,000 molecular weight cutoff) with two exchanges. The probe temperature was maintained at 37°C and 5 mm NMR tubes were used. Each time point consisted of 256 scans acquired with a spectral width of ±2700 Hz, using 16K data points. The pulse width was set to correspond to a 45° tip angle, and a 2.0 s post acquisition delay was used to allow for full relaxation of the resonances. For direct comparison with ^1H NMR data spectrophotometric assays were conducted using the same deuterated buffers.

Titrimetric measurements were made on a Radiometer Titralab 11 automatic titration system. The reaction chamber was thermostatted at 37°C, and a stream of nitrogen gas was blown across the reaction mixture surface to exclude atmospheric carbon dioxide. Protons produced during aldehyde oxidation were neutralised by addition of 2.0 mM sodium hydroxide.

RESULTS AND DISCUSSION

A series of assay progress curves for the oxidation of octanal using assay conditions similar to those described by Hinson and Neal, (1975) are shown in Figure 1. As the concentration of enzyme is decreased, a pronounced lag in the assay progress curve is observed.

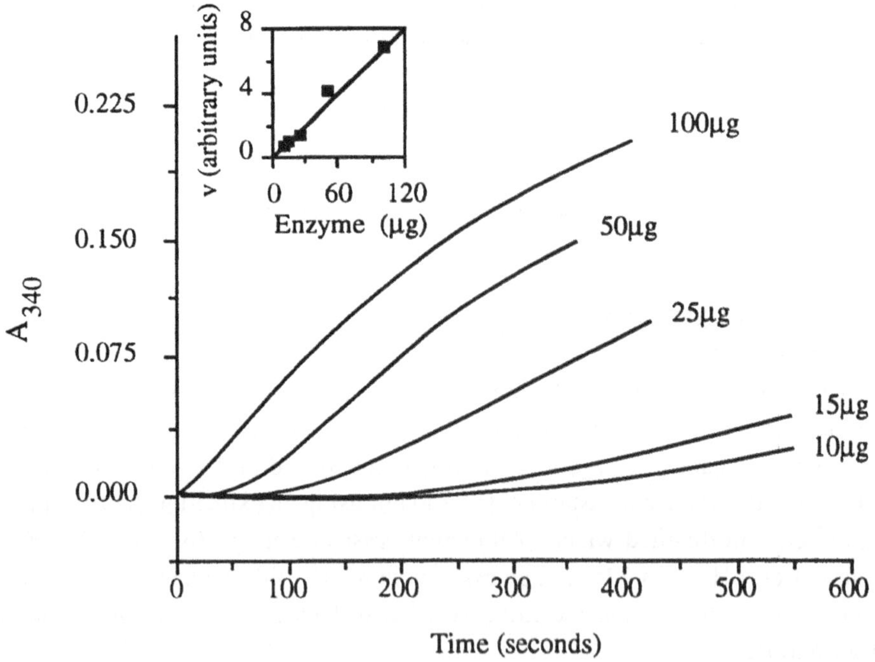

Figure 1. Series of assay progress curves for the oxidation of octanal by different amounts of HL-ADH. Assay conditions: NAD^+, 10 mM; octanal, 0.15 mM; amounts of HL-ADH are indicated in the figure. Reactions were initiated by addition of octanal. The inset shows a plot of the steady-state rate versus enzyme concentration

The lag phase is followed by a linear steady-state portion of the progress curve which ultimately curves off at high absorbance values. Importantly, the linear steady-state rate achieved following the lag is proportional to the enzyme concentration (Fig. 1, inset). The steady-state rate exhibits hyperbolic kinetic behaviour (Fig 2 (a) and (b)), i.e., it fully follows Michaelis-Menten kinetics and yields kinetic parameters (see Table 1) consistent with those observed at higher enzyme concentrations where there is no lag.

The duration of the lag phase is independent of NAD^+ concentration. It is directly proportional to aldehyde substrate concentration and inversely proportional to enzyme concentration (Fig. 3, (a) and (b)).

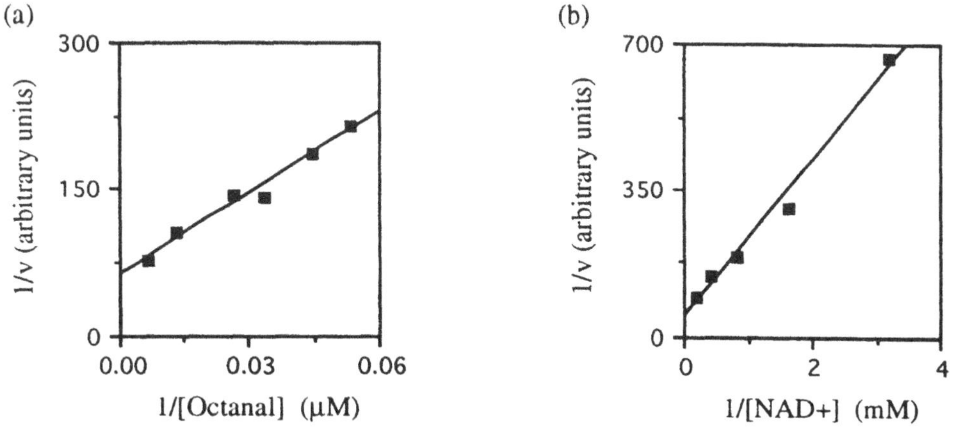

Figure 2. Hyperbolic behaviour of steady-state rate of aldehyde oxidation: (a) NAD^+ was held constant at 10 mM while octanal was varied between 18.75 μM - 150 μM; (b) octanal was held constant while NAD^+ was varied between 0.31 - 5.0 mM. Reactions were initiated by aldehyde addition.

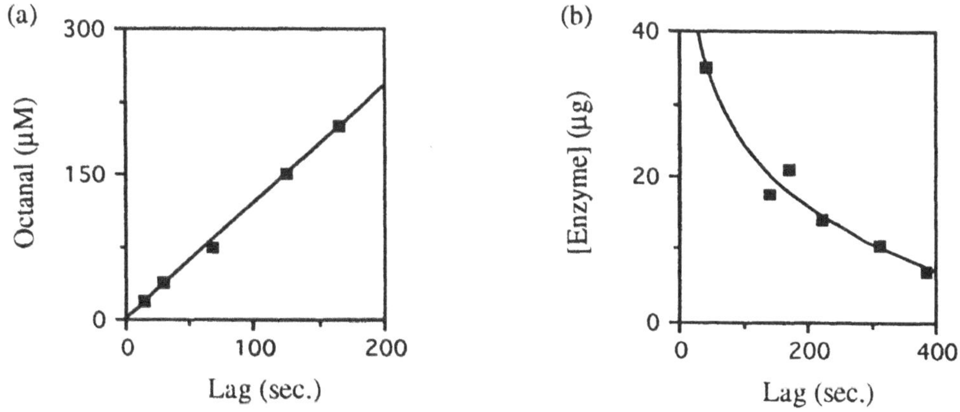

Figure 3. Variation of assay progress curve lag with substrate (a) and enzyme (b) concentrations. Assay conditions were: (a) NAD^+, 10 mM; HL-ADH, 20 μg; (b) NAD^+, 10 mM; Octanal, 0.15 mM. Reactions were initiated by aldehyde addition.

Similar results were obtained with propanal and butanal except that lag phases with these aldehydes tended to be longer. Lags of this nature in enzyme assay progress curves may be due to a variety of reasons such as: (a) cooperative substrate binding, (b) rate-limiting isomerisation of enzyme complexes, (c) slow dissociation of enzyme into a more active form in the assay medium, (d) artefact, due, for instance, to contamination with an inhibitor that is slowly turned over before the assay substrates can bind to the enzyme (see Dixon and Webb, 1979, for more detailed review). The possibility that contamination with an inhibitor might explain the lag phenomenon was excluded when repeated distillations of the aldehyde substrates, the use of higher purity cofactor preparations, exhaustive dialysis of the enzyme solution and changes of buffer salts all failed to affect the lag phase duration. The other possibilities are more difficult to exclude but lags of this nature have not previously been reported for alcohol dehydrogenase with other substrates.

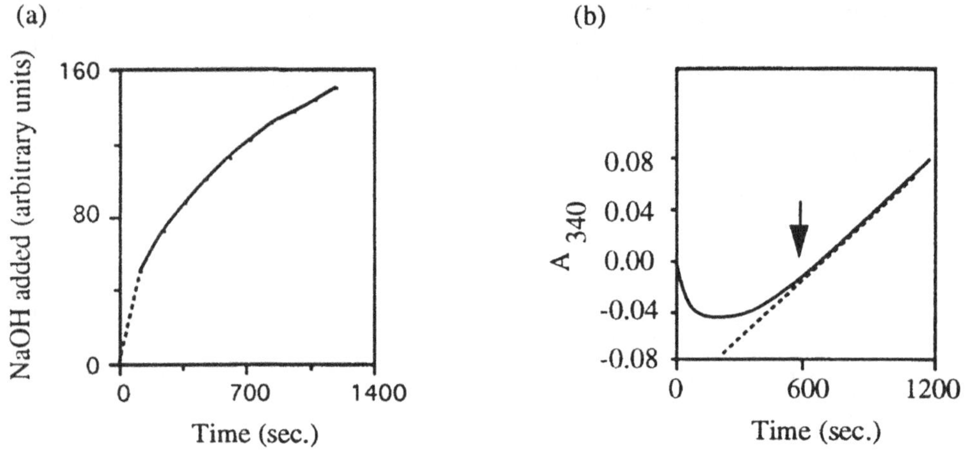

Figure 4. Comparison of titrimetric and spectrophotometric assays of aldehyde oxidation: (a) Aldehyde oxidation monitored titrimetrically. The dashed line indicates the initial portion of the assay progress curve where measurements were uncertain (see text for details). (b) The same assay monitored spectrophotometrically. The dashed line is a tangent to the curve where the steady-state rate begins. The arrow indicates roughly where the lag ends. Assay conditions were identical for both assay methods: NAD^+, 10 mM; butanal, 1.0 mM; [HL-ADH], 50 μg/ml.

The protons produced allow these reactions to be followed titrimetrically. Fig. 4(a) shows an assay progress curve followed by monitoring acid production under conditions where a lag phase of several minutes was observed for the spectrophotometric assay (Fig 4(b)). The initial rate of acid production could not be measured titrimetrically because of fluctuations in the electrode response making this assay unsuitable for determination of kinetic parameters by initial rates methods. It is shown here only as a qualitative measure of acid production. An initial decrease in A_{340nm} was observed when assays were started by addition of aldehyde substrate (Fig, 4(b)), implying the presence of NADH in the assay medium prior to initiation of the reaction by aldehyde addition. Indeed, when NADH was added to the assay medium prior to initiation the initial decrease was larger, and proportional to the amount of NADH added. The initial decrease was not observed when reactions were started by enzyme addition. NADH produced in the latter stages of the reaction are not reoxidised. These results suggest that NADH is being produced in the assay medium possibly by a blank reaction. The occurrence of a blank reaction is known for HL-ADH but it has not been fully characterised (see Hadorn *et al.*, 1974). Another possibility is that a pre-steady-state burst of NADH formation occurs that is slowly reoxidised in the steady-state. These possibilities are currently under investigation. In any case, these NADH concentrations, while significant, are unlikely to be responsible for the large amount of substrate turnover observed during the lag phase (see below, Fig 5).

Clearly, substantial amounts of acid are being produced during the lag phase. The production of acid without concomitant NADH formation strongly indicates that HL-ADH is catalysing dismutation of the aldehyde substrate during the lag period.

To establish unequivocally the time course of changes occurring in the assay components during the lag phase we investigated the oxidation of butanal by ^1H NMR. The methylene protons of butanal (free aldehyde form), butanol and butyric acid give rise to readily identifiable, and clearly separated peaks in the ^{1}NMR spectrum at 2.49 ppm, 3.60 ppm and 2.13 ppm, respectively. Moreover, the aldehyde concentrations used fall well within the sensitivity limitations of ^1H NMR and the lag can be adjusted, by judicious choice of enzyme and substrate concentrations, to allow optimum data acquisition. As shown in Figure 5(b), the aldehyde peak decreases until by roughly 0.7 hours the resonance is almost undetectable with the acquisition parameters used. A concomitant appearance of the resonances of the acid and alcohol in a 1:1 ratio are observed. These spectra reveal that the majority of the aldehyde has dismutated to the corresponding alcohol and acid during the lag in the spectrophotometric assay when there is no net NADH production. In other words, *before the steady-state production of NADH commences, the aldehyde has almost completely dismutated and the aldehyde concentration has fallen to less than 5-10 % of its initial value.*

Substrate turnover is much more rapid in the lag phase than when steady state NADH production commences. In the example cited in Figure 5 (b), in the first 30 min. during the lag roughly 5 μmol of butanal are oxidized to butyric acid, whereas in the next 30 min. after steady-state production of NADH commences, only about 20 nmol of NADH is generated.

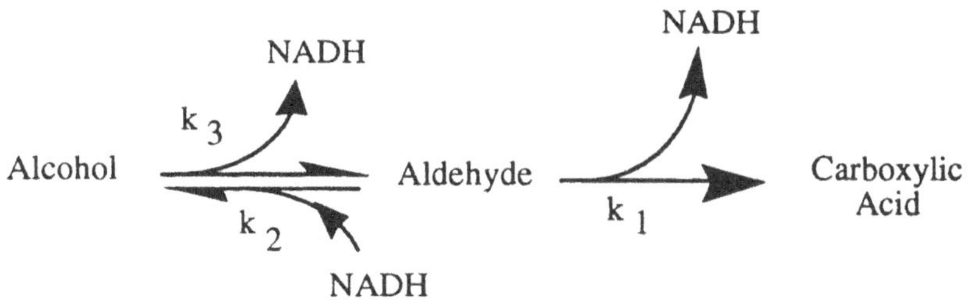

Scheme 2. Aldehyde dismutation mechanism; k_1, k_2 and k_3 represent rate constants. Curved arrows show the direction in which NADH is formed or consumed. Of course, in the initial phases of the reaction when pure dismutation is occurring cofactor will remain enzyme bound and will not dissociate into the assay medium.

The properties of HL-ADH catalyzed oxidation of aldehydes may be explained by the mechanism of Scheme 2. In the early stages of the reaction the enzyme is catalysing pure dismutation as indicated by k_1 and k_2. As the level of alcohol increases, its reoxidation to aldehyde, denoted by k_3, becomes more significant. At some point due to rising alcohol levels and falling aldehyde levels the rate of aldehyde reduction becomes equal to the rate of alcohol oxidation, i.e., $k_3\,[NAD^+]\,[alcohol] = k_2\,[NADH]\,[aldehyde]$. A steady-state equilibrium is established between alcohol formed and aldehyde substrate.

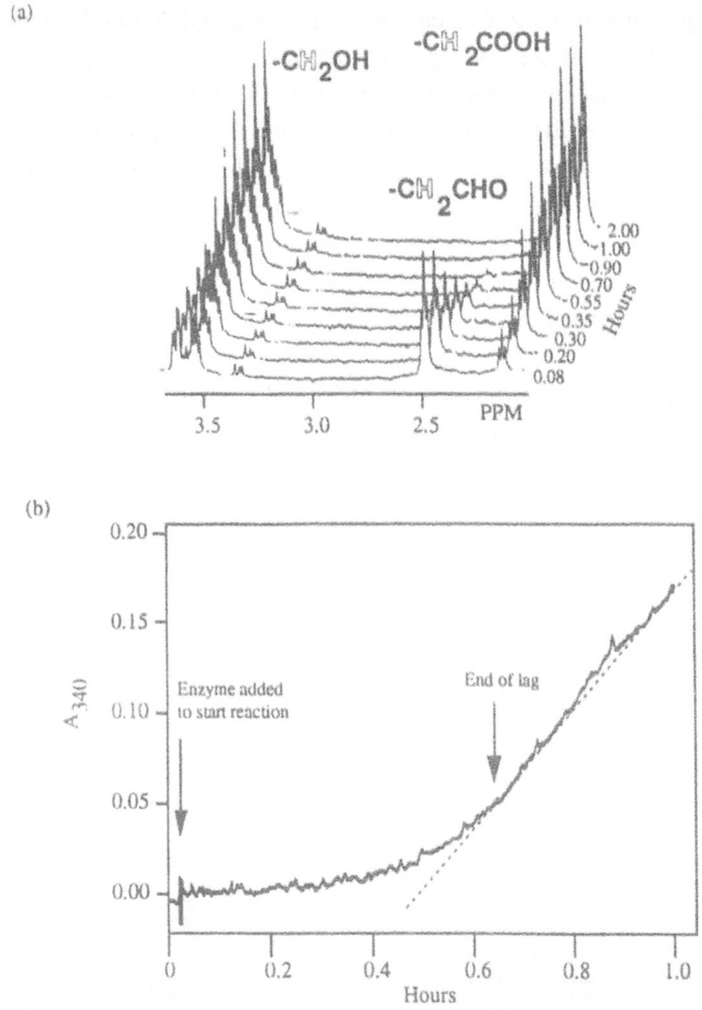

Figure 5. Oxidation of butanal by HL-ADH monitored both spectrophotometrically and by ^1H NMR: (a) butanal oxidation monitored by ^1H NMR. Note: the resonance due to the aldehyde hydrate which represents about half the aldehyde in solution is not shown. (b) butanal oxidation monitored spectrophotometrically. Assay conditions were identical for both assays: NAD^+, 10 mM; butanal, 10 mM; [HL-ADH], 100 μg/ml.

Aldehyde continues to be oxidized to acid and NADH produced cannot be reoxidized without disturbing the equilibrium. Equilibrium will lie towards NAD^+/alcohol at neutral pH values. Thus aldehyde dismutation and aldehyde oxidation are seen to be the same reaction viewed at different points along the assay progress curve.

The model explains the variation of the lag phase with substrate and enzyme concentration (Fig. 3). Increasing the aldehyde concentration lengthens the lag by increasing the time it takes the enzyme to process the extra aldehyde before reaching equilibrium. Increasing the enzyme concentration in the assay mixture decreases the time needed to attain equilibrium, hence shortening the lag. Indeed, it is likely that the lag went unnoticed by earlier workers (Hinson and Neal, 1975) who used high enzyme concentrations in their assays which gave lags of such short duration that they were completed during the assay mixing time.

Obviously, failure to account for substrate depletion in the lag phase will lead to inaccurate estimations of kinetic constants. However, it is somewhat surprising that the steady-state rate following the lag exhibits hyperbolic kinetics. For NAD^+ this observation is consistent with a mass action effect. The net rate of NADH production depends on the equilibrium aldehyde concentration and the rate at which it is oxidized. While the NAD^+ concentration may be saturating, the dismutation reaction that occurs during the lag may leave the steady-state aldehyde concentration well below its K_m. Increasing NAD^+ will shift the alcohol/aldehyde equilibrium to the right, favoring a higher steady-state concentration of aldehyde which increases the net production of NADH. The dependence of the rate of NADH production on increasing NAD^+ will continue until increases in NAD^+ concentration shift the steady-state concentration of aldehyde to above its K_m, whereupon the reaction reaches saturation. Therefore the reported high K_m for NAD^+ is an indirect measurement of a value that is a function of the K_m of *aldehyde*; it has no direct relation to the binding constant or affinity of NAD^+. The hyperbolic kinetics with the aldehyde substrate may be similarly explained. A difficulty with this explanation is that it suggests much lower K_m values for aldehyde substrates than those in the literature (see Table 1). Whether this is the case or whether the system exhibits further, as yet unrevealed, complexities are the subjects of current studies.

The proposed mechanism explains the experimental observations and is consistent with existing knowledge of HL-ADH kinetics (see Pettersson, 1987). Verification of the mechanism awaits the development of a suitable assay for the dismutation phase of the reaction as well as a detailed kinetic analysis of this complex system.

The question as to the true kinetic parameters for HL-ADH towards NAD^+ and aldehyde substrates arises. For NAD^+ the true K_m is likely to be very low and identical with that measured for the dismutation reaction (Dalziel and Dickinson, 1965), consistent with the observation that dismutation will proceed at NAD^+ concentrations comparable with the enzyme concentration. For aldehyde substrates the true K_m is that pertaining in the early stages of the reaction when dismutation is occurring and should be identical with the K_m measured at low NAD^+ concentrations for the dismutation reaction. The K_m value for the dismutation phase was not measured in these studies because of assay difficulties mentioned previously (see Fig. 4). Dalziel and Dickinson (1965) have presented the only study in the literature on the kinetics of aldehyde dismutation. These workers reported K_m values for butanal (K_m = 16 mM) and acetaldehyde (K_m = 100 mM). The differences in the K_m values of theses aldehydes cannot be attributed to their degree of hydration, which is the same, but may reflect a sensitivity to aldehyde chain length. The role of aldehyde chain length and degree of hydration in aldehyde dismutation merits further investigation. The variation in K_m for aldehyde makes it difficult to exclude, on the basis of available data, a role for HL-ADH in aldehyde oxidation.

The physiological relevance of aldehyde transformation by HL-ADH will best be assessed when more detailed substrate specificity data are available for a wide range of aldehyde substrates. It is clear from Figure 5 that aldehyde oxidation with net NADH production takes place in the presence of high concentrations of the corresponding alcohol which will compete with the aldehyde for oxidation by the E·NAD^+ complex, presumably reducing V_{max}. That net oxidation is observed at all is a tribute to the efficacy of the reaction.

These studies, although preliminary, raise many important questions about aldehyde transformation by alcohol dehydrogenases. There is increasing evidence that aldehyde oxidation by enzymes thought of as alcohol dehydrogenases is widespread (Moxon *et al.*, 1985; Hara *et al.*, 1991). The findings reported here are, therefore, not only relevant to HL-ADH but to other alcohol dehydrogenases and carbonyl reductases.

SUMMARY

A lag phase in the spectrophotometric assay progress curve of aldehyde oxidation by HL-ADH was observed and characterised.

The aldehyde oxidation and aldehyde dismutation reactions were shown to be related, and a mechanism to explain net aldehyde oxidation was proposed.

The spectrophotometric assay was shown to be unsuitable for measurement of kinetic parameters for aldehyde oxidation by HL-ADH, and kinetic constants previously determined were shown to be in error.

Existing data on the aldehyde dismutation reaction are insufficient to discount a role for HL-ADH in aldehyde transformation *in vivo*.

ACKNOWLEDGEMENTS

This work supported in parts by NIH grants: AR-17323 (GLK) and GM-22982 (NJO).

REFERENCES

Abeles, R.H. and Lee, H.A., 1960, The dismutation of formaldehyde by liver alcohol dehydrogenase, *J. Biol. Chem.*, **235**: 1499-1503.

Anderson, D.C. and Dahlquist, F.W., 1982, ^{19}F Nuclear magnetic resonance observations of aldehyde dismutation catalysed by horse liver alcohol dehydrogenase, *Arch. Biochem. Biophys.*, **217**: 226-235.

Dalziel, K. and Dickinson, F.M., 1965, Aldehyde mutase, *Nature.*, **206**: 255-257.

Dalziel, K. and Dickinson, F.M., 1966, The kinetics and mechanism of liver alcohol dehydrogenase with primary and secondary alcohols as substrates, *Biochem. J.*, **100**: 34-46.

Dixon, M. and Webb, E.C., 1979, *Enzymes*, 3rd Edition, Longman, London.: 457-460.

Gupta, N. K. and Robinson, W.G., 1966, Coupled oxidation-reduction activity of liver alcohol dehydrogenase, *Biochim. Biophys. Acta.*, **118**: 431-434.

Hadorn, M., John, V.A., Meier, F.K. and Dutler, H., 1975, Kinetic equivalence of the active sites of alcohol dehydrogenase from horse liver, *Eur. J. Biochem.*, **54**: 65-73.

Hara, A., Yamamoto, H., Deyashiki, Y., Nakayama, T., Oritani, H., and Sawada, H., 1991, Aldehyde dismutation catalysed by pulmonary carbonyl reductase: kinetic studies of chloral hydrate metabolism to trichloroacetic acid and trichloroethanol, *Biochim. Biophys. Acta.*, **1075** (1): 61-67.

Hinson, J. A. and Neal, R.A., 1975, An Examination of Octanol and Octanal Methabolism to Octanoic Acid by Horse Liver Alcohol Dehydrogenase, *Biochim. Biophys. Acta.*, **384**: 1-11.

Moxon. L.N., Holmes, R.S., Parsons, P.A., Irving, M.G. and Doddrell, D.M., 1985, Purification and molecular properties of alcohol dehydrogenase from *Drosophila melanogaster*: evidence from NMR and kinetic studies for function as an aldehyde dehydrogenase, *Comp. Biochem. Physiol.*, **80 B**: 525-535.

Pettersson, G. 1987, Liver alcohol dehydrogenase, *CRC Crit. Rev. Biochem.*, **21**: 349-389.

Pietruszko, R., 1979, Nonethanol Substrates of Alcohol Dehydrogenase, In: *Biochemistry and Pharmacology of Ethanol.*, Vol 1, (Majchrowicz, E, and Noble, E.P., eds.), Plenum Press, New York, p. 87-106.

Weiner, H. 1989, Role of alcohol and aldehyde dehydrogenases *in vivo*: speculations on their natural substrates, In: *Human Metabolism of Alcohol 2.*, (Crow, K.E. and Batt, R.D., eds), CRC Press, Boca Raton, p. 147-160.

EFFECT OF GLYCATION UPON ACTIVITY OF LIVER ALCOHOL DEHYDROGENASE

Donald J. Walton and Brian H. Shilton

Department of Biochemistry
Queen's University
Kingston, Ontario, Canada K7L 3N6

INTRODUCTION

We recently showed that the class I alcohol dehydrogenase (ADH) of human and horse liver is glycated *in vivo* (McPherson et al., 1988) mainly at Lys-231 (Shilton and Walton, 1991), and this was ascribed to the proximity of the imidazole group of His-348. Lys-231 is also the main site of glycation *in vitro*, but when phosphate is included in the medium a small degree of glycation occurs at Lys-228 (Walton and Shilton, 1991). In this case the acid-base catalyst is probably a phosphate ion, bound to the anion-binding region of the coenzyme-binding site.

Several years ago it was reported (Tsai and White, 1983) that the catalytic activity of horse ADH was increased on incubation of ADH with glucose or fructose for times of less than 1 h. It seemed possible that activation might also occur when the enzyme is glycated *in vivo*. This might be important in diabetic subjects, and could contribute to the increased rate of ethanol metabolism that results from the administration of fructose (Scholz and Nohl, 1976).

We now present the results of a study of the effects of glycation of lysines 228 and 231 upon the enzymatic activity of ADH.

METHODS

The Enzyme Used

Horse liver ADH, which contained mainly the EE isozyme, was obtained from Sigma.

Phenylboronate Affinity Chromatography of ADH

In vivo glycated ADH was separated from non-glycated enzyme by adsorption on

Glyco-Gel B (Pierce Chemical Co.), followed by elution with sorbitol (McPherson et al., 1988).

Determination of Enzymatic Activity

Enzymatic activity was determined spectrophotometrically, by measuring the rate of NADH formation in a solution of NAD^+ (2 mM) and ethanol (0.5 M) in Tris-phosphate buffer (0.1 M in Tris; pH 9) at 22°. A unit of activity is the quantity that catalyses the formation of 1 μmol of NADH/min. Note that different conditions were used in the kinetics study (see below).

Acetimidylation of ADH

ε-Amino groups of lysyl residues other than 228 were blocked by treating the enzyme with ethyl acetimidate in the presence of NAD^+ and pyrazole, by the method described by Plapp et al.(1973). The properties of the resulting acetimidate (AI-ADH) were similar to those reported by these authors: (a) A trinitrobenzenesulphonic acid based assay showed that only 5 of the 60 amino groups of the dimer were unblocked. (b) Treatment with methyl picolinimidate by their procedure caused a 16-fold increase of enzymatic activity. The ε-amino group of Lys-228 of AI-ADH was therefore free, since activation has been shown to be due to imidylation of this particular group (Dworschack et al., 1975).

Titration of Active Sites

A fluorometric method (Yonetani and Theorell, 1962) was used. A solution of enzyme (*ca.* 0.5 μM in active sites) and isobutyramide (0.1 M) in sodium phosphate buffer (50 mM; pH 8) was titrated with NADH (30 μM). Fluorescence (excitation 330 nm; emission 410 nm) was measured throughout, equivalence being indicated by an inflection in the titration curve.

Steady-State Kinetics

Solutions for reaction rate measurements (all made at 30°) were made in sodium phosphate buffer (50 mM, pH 8) containing EDTA (6.3 mM). Concentrations of active sites were 0.15-0.4 nM, except for that of glyceraldehyde-modified AI-ADH, which was 1-3 nM. Concentrations of coenzyme and substrate ranged from *ca.* 0.5 to 10 times the values of the Michaelis constants. For ethanol oxidation, reaction rates were measured for sixty-four combinations of coenzyme and substrate concentrations, and likewise for the reverse reaction. NADH was assayed fluorometrically (excitation 340 nm; emission 460 nm) or spectrophotometrically for determining rates of substrate oxidation or reduction, respectively. Data were analyzed in terms of an ordered bi-bi mechanism, in which E, A, B, P and Q represent enzyme, NAD^+, ethanol, acetaldehyde and NADH, respectively.

Least squares non-linear regression analysis (SigmaPlot 4.1; Jandel Scientific) was used to fit values of concentration and initial velocity, v_1 or v_2, to equation (1) for ethanol oxidation, or to the corresponding equation for acetaldehyde reduction; values of the maximum velocities, V_1 and V_2, the Michaelis constants, K, and the inhibition constants, K_i, were thus obtained.

$$v_1 = \frac{V_1[\mathrm{A}][\mathrm{B}]}{K_{ia}K_b + K_a[\mathrm{B}] + K_b[\mathrm{A}] + [\mathrm{A}][\mathrm{B}]} \qquad (1)$$

RESULTS

Effect of in Vivo Glycation Upon ADH Activity

Glycation of ADH *in vivo* might be physiologically significant if its catalytic activity were affected. This possibility was explored by comparing the specific activities of samples of native ADH, in which 90% of the molecules are not glycated, with the fraction retained by agarose-phenylboronate, which contains mainly glycated molecules.

The kinetics constants of the *in vivo* glycated fraction were similar to those of the native, unfractionated enzyme (Table 1).

Activation of AI-ADH by Reaction With Glyceraldehyde

Derivatization of Lys-228 by non-carbohydrates often results in activation of the enzyme (Dworschack et al., 1975; Sogin and Plapp, 1975), and glycation at this position occurs *in vitro* (see above). Therefore the effect of glycation of Lys-228 upon enzymatic activity was examined, using the derivative, AI-ADH, in which the only reactive amino group is that of Lys-228, as explained in Methods.

We then set out to prepare a derivative of AI-ADH in which the ε-amino group of Lys-228 was fully glycated, to use in the kinetics study. Although the ultimate goal was to examine the effect of glycation by glucose, its reactivity was too small to achieve complete modification of Lys-228 in a reasonable time. Therefore the more reactive aldose, DL-glyceraldehyde, was used initially, since the mechanisms of glycation by glyceraldehyde and hexoses are similar (Acharya and Manning, 1980). Both involve formation of a Schiff base, followed by an Amadori rearrangement, unless a reductant is included in the medium.

Table 1. Kinetics constants for native and modified ADH.

Constant	Native ADH	Glycated ADH[1]	AI-ADH	Glyceraldehyde/ AI-ADH[2]
		μM ± SD		
K_a	3.5 ± 0.6	3.7 ± 0.4	2.8 ± 0.3	170 ± 20
K_{ia}	37 ± 12	32 ± 3	17 ± 3	480 ± 50
K_b	240 ± 70	210 ± 20	110 ± 14	3100 ± 250
K_p	170 ± 14	180 ± 10	130 ± 7	2400 ± 200
K_q	6.4 ± 0.6	5.2 ± 0.4	4.2 ± 0.3	130 ± 10
K_{iq}	0.4 ± 0.9	0.8 ± 0.6	0.5 ± 0.3	18 ± 4.5
		s^{-1} ± SD		
$V_1/[E]_o$[3]	3.3 ± 0.1	3.9 ± 0.1	2.5 ± 0.1	22 ± 1
k_4[4]	2.6	6.6	3.1	64
$V_2/[E]_o$	41 ± 1	43 ± 2	31 ± 1	460 ± 21
k_{-1}	35	34	15	62

[1]The *in vivo* glycated fraction which was retained by agarose-phenylboronate.
[2]AI-ADH which had been incubated with glyceraldehyde, sodium cyanoborohydride and sodium phosphate for 2 h. For conditions, see legend to Fig. 1.
[3]$[E]_o$ is the normality of the enzyme, determined by active site titration.
[4]Rate constants were calculated thus: $k_4 = V_2K_{iq}/[E]_oK_q$; $k_{-1} = V_1K_{ia}/[E]_oK_a$.

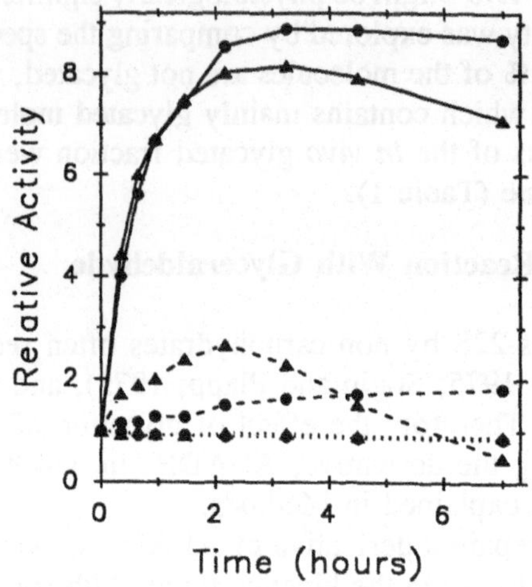

Figure 1. Effect of anions upon activation of AI-ADH by glyceraldehyde. Solutions (pH 7.4; at 22°) containing AI-ADH (4 μM in active sites), 0.1 M Hepes, 3 mM sodium azide, and either 0.1 M sodium phosphate (triangles) or 0.1 M sodium sulphate (circles) were incubated at 22°. In addition, individual solutions contained: 25 mM DL-glyceraldehyde (broken lines); 25 mM DL-glyceraldehyde and 0.1 M $NaBCNH_3$ (solid lines); 0.1 M $NaBCNH_3$ (dotted lines). Activity was assayed as described in Materials and Methods. Relative activity = activity (for ethanol oxidation)/activity at the start.

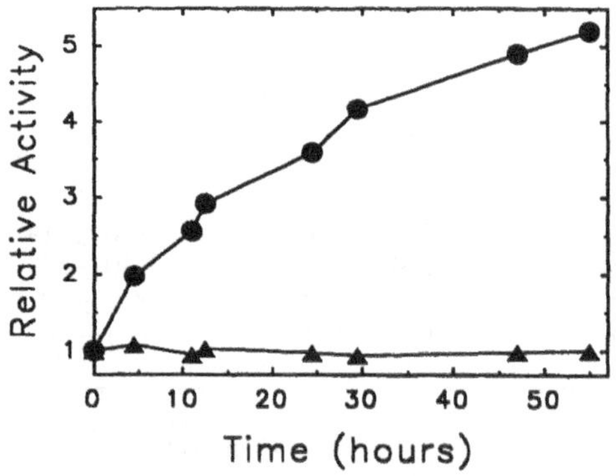

Figure 2. Activation of AI-ADH by galactose. Solutions (pH7.4; at 22°) contained AI-ADH (4 μM in active sites), 0.1 M Hepes, 0.1 M sodium phosphate, 3 mM sodium azide, and either 0.1 galactose (triangles), or 0.1 M galactose and 0.1 M $NaBCNH_3$ (circles). Relative activity = activity (for ethanol oxidation)/activity of enzyme incubated under identical conditions, but in the absence of galactose.

AI-ADH was incubated with glyceraldehyde in a medium containing sodium phosphate. Sodium cyanoborohydride was included to reduce the resulting Schiff base *in situ*, thereby avoiding the slow Amadori rearrangement step. The enzymatic activity increased 8-fold in 2 h, and then remained relatively steady (Fig. 1; second curve from top). It was assumed that complete derivatization of Lys-228 had occurred to give a relatively stable, *N*-(1-deoxyglycerol-1-yl) derivative, which was would be suitable for the ensuing study of reaction kinetics.

The kinetics constants and turnover numbers for AI-ADH (Table 1) were similar to those for the native enzyme, except for k_{-1}, which was smaller for the derivative. The values of all of the constants for glyceraldehyde treated AI–ADH were considerably larger than those for native ADH. It was concluded that the different stages of the catalytic mechanism were not affected significantly by acetimidylation, but that they were strongly influenced by the subsequent reaction of Lys-228 with glyceraldehyde.

Effect of Buffer Anions Upon Activation by Glyceraldehyde

DL-Glyceraldehyde was used again for the next set of experiments, because of its high reactivity. Its effects upon the activity of AI-ADH are shown in Fig. 1. Incubation of AI-ADH with glyceraldehyde in the presence of phosphate led to a 2.7-fold increase of enzymatic activity for the first 2 h, followed by a decline. Replacement of phosphate by sulphate resulted in a slower increase of activity for the first 2 h, and a continued increase for the next 5 h. The inclusion of sodium cyanoborohydride in the medium resulted in an 8-fold activation during the first 2 h, the rate of increase of activity being independent of the type of anion present. Controls conducted in the absence of glyceraldehyde showed that buffer salts and cyanoborohydride did not affect the enzymatic activity.

Effect of Hexoses Upon Activity of AI-ADH

The effects of D-glucose and D-galactose were then studied. The medium contained sodium phosphate, and the concentration (0.1 M) of each hexose was four times greater than that of glyceraldehyde in the previous study. In the presence of sodium cyanoborohydride, galactose increased the activity 5-fold in 55 h (Fig. 2); glucose probably caused a small degree of activation, but this could not be substantiated. The activity of AI-ADH was not altered by incubation with galactose (Fig. 2) or glucose for 55 h in the absence of sodium cyanoborohydride.

Effect of Hexoses Upon Activity of Native ADH

In the presence of cyanoborohydride, D-glucose, D-galactose and 6-deoxy-D-galactose activated AI-ADH at increasingly high rates (Fig. 3). Presumably the relative rates of activation reflect rates of glycation which, in turn, depend upon the fraction of each hexose that is present in the acyclic form (Angyal, 1984; Bunn and Higgins, 1981). In the absence of cyanoborohydride none of the three hexoses affected the enzymatic activity significantly.

DISCUSSION

The data in Table 1 indicate that *in vivo* glycation had not affected the enzymatic activity of the enzyme. It was concluded that addition of hexose to Lys-231, the principal glycation site, did not affect catalysis, presumably because this residue is too far away from the active site to influence binding of the coenzyme or substrate molecule, or to affect the rate of hydranion transfer.

The investigation of the effect of glyceraldehyde upon the activity of AI-ADH supported the conclusions that were made as a result of our site specificity study (Walton and Shilton, 1991). In the absence of cyanoborohydride, the activity of AI-ADH, which presumably reflects the extent of derivatization of Lys-228, increased more rapidly in the presence of phosphate than in the presence of sulphate. It was assumed that immobilized phosphate catalyzed the rearrangement of the Schiff base of Lys-228, but that sulphate did not, since it did not act as a proton acceptor. When AI-ADH was incubated with glyceraldehyde and cyanoborohydride, glycation was interrupted at the Schiff base stage. In this case the increase of activity was independent of the nature of the anion, since the phosphate-dependent Amadori rearrangement was avoided.

The similarity in the values of $V_1/[E]_o$ and k_4 for catalysis of ethanol oxidation by the native enzyme (Table III) has been noted previously (Dalziel and Dickenson, 1966; Plapp et al., 1973) and is considered to be due to the fact that release of NADH from the

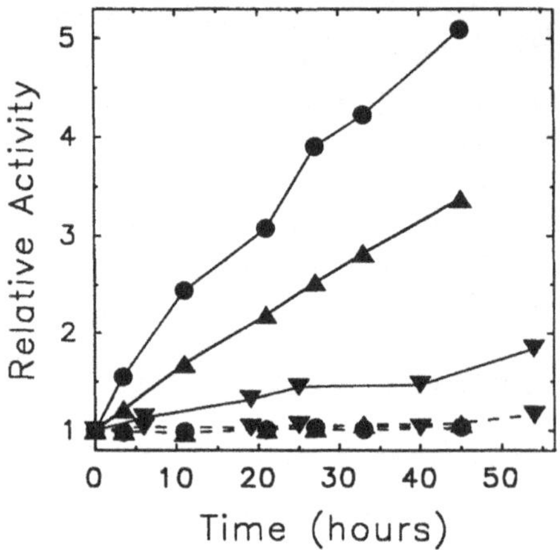

Figure 3. Activation of AI-ADH by galactose. Solutions (pH 7.4; at 22°) contained native ADH (4 μM in active sites), 0.1 M Hepes, 0.1 M sodium phosphate and 3 mM sodium azide, together with 0.2 M D-galactose (circles), 6-deoxy-D-galactose (triangles) or D-glucose (inverted triangles). 0.1 M $NaBCNH_3$ was either present in the medium (solid lines) or absent from it (broken lines). Relative activity is defined as in the legend for Fig. 2.

enzyme is rate controlling. For glyceraldehyde-treated AI-ADH, the value of k_4 was larger than that of $V_1/[E]_o$, and was larger than k_4 for AI-ADH. Therefore, glycation of Lys-228 had increased the rate of release of NADH from the enzyme to the extent that this step was no longer rate controlling. Activation of ADH by blocking Lys-228 with non-carbohydrate groups produces a similar change in the mechanism (Plapp et al., 1973).

For the reduction of acetaldehyde in the presence of the native enzyme, values of $V_2/[E]_o$ and k_{-1} were similar, since release of NAD^+ was rate controlling. For glyceraldehyde-treated AI–ADH, $k_{-1} < V_2/[E]_o$, which implies that the overall rate of reaction was faster than the release of NAD^+, the rate controlling step. For other derivatives of ADH this paradox has been explained in terms of a modified mechanism which includes isomeric or inactive enzyme-NAD^+ complexes (Plapp et al., 1973). Regardless of the precise nature of the mechanism, it is clear that glyceraldehyde activates AI-ADH in a similar manner to pyridoxal (Sogin and Plapp, 1975) and methyl picolinimidate (Plapp et al., 1973). In all cases the attachment of a group to Lys-228 weakens the binding of coenzyme, thereby facilitating its release at the last stage of the oxidative and the reductive reaction sequences.

It was assumed that, in the presence of cyanoborohydride, the hexoses acted similarly to glyceraldehyde, activating AI-ADH by reacting with Lys-228. Native ADH was probably activated by the same mechanism, but this was not confirmed.

The impetus for the kinetics part of this study was the report by Tsai and White (1983) which stated that when native ADH was incubated with glucose or fructose (37 mM) for times of less than 1 h, the activity was increased by *ca.* 25%, and then remained steady. Unfortunately, we could not substantiate this, and were unable to detect any activation using the conditions which they described, or those given in the legend to Fig. 3.

Lys-228 may have been insufficiently glycated to exert a measurable effect upon the activity. Based upon radioactivity measurements, it was shown that a 55-h incubation of ADH with [^{14}C]glucose in the presence of phosphate resulted in incorporation of 0.2 mol of hexose per mol of ADH. Judging from the site specificity study (Walton and Shilton, 1991), 8% of this hexose was attached to Lys-228. Therefore only *ca.* 16 mmol of hexose were attached to Lys-228, per mol of enzyme. It is also possible that even when a molecule of ADH acquires a 1-deoxyhexulosyl group at Lys-228, activation is weak or nonexistent. Further work is required to clarify this point.

In conclusion, glycation of Lys-231, which occurs *in vivo* and *in vitro*, does not affect the activity of liver ADH, and is unlikely to be physiologically significant. Glycation of Lys-228 *in vitro*, by glyceraldehyde, or by hexoses in the presence of sodium cyanoborohydride, weakens coenzyme and substrate binding, thus causing significant activation. It has not been possible to demonstrate activation by glycation of Lys-228 with hexoses in the absence of sodium cyanoborohydride.

Acknowledgment

We thank the Canadian Diabetes Association for financial support of this work.

REFERENCES

Acharya, A.S., and Manning, J.M., 1980, Reactivity of the amino groups of carbonmonoxyhemoglobin S with glyceraldehyde, *J. Biol. Chem.* **255**:1406.

Angyal, S.J., 1984, The composition of reducing sugars in solution, *Adv. Carbohydr. Chem. Biochem.* **42**:15.

Bunn, H.F., and Higgins, P.J., 1981, Reaction of monosaccharides with proteins: possible evolutionary significance, *Science* **213**:222-224.

Dalziel, K., and Dickenson, F.M., 1966, The kinetics and mechanism of liver alcohol dehydrogenase with primary and secondary alcohols as substrates, *Biochem. J.* **100**:34.

Dworschack, R., Tarr, G., and Plapp, V.P., 1975, Identification of the lysine residue modified during the activation by acetimidylation of horse liver alcohol dehydrogenase, *Biochemistry* **14**:200.

McPherson, J.D., Shilton, B.H., and Walton, D.J., 1988, Evidence for glycation of horse liver alcohol dehydrogenase in vivo , *Biochem. Biophys. Res. Comm.* **152**:711.

Plapp, B.V., Brooks, R.L., and Shore, J.D., 1973, Horse liver alcohol dehydrogenase: amino groups and rate-limiting steps in catalysis, *J. Biol. Chem.* **248**:3470.

Scholz, R., and Nohl, H., 1976, Mechanism of the stimulatory effect of fructose on ethanol oxidation in perfused rat liver, *Eur. J. Biochem.* **63**:449.

Shilton, B.H., and Walton, D.J., 1991, Sites of glycation of human and horse liver alcohol dehydrogenase in vivo, *J. Biol. Chem.* **266**:5587.

Sogin, D.C., and Plapp, B.V., 1975, Activation and inactivation of liver alcohol dehydrogenase with pyridoxal compounds, *J. Biol. Chem.* **250**:205.

Tsai,. C.S., and White, J.H., 1983, Activation of liver alcohol dehydrogenase by glycosylation, *Biochem. J.* **209**:309.

Walton, D.J., and Shilton, B.H., 1991, Site specificity of protein glycation, *Amino Acids* 1:199.
Yonetani, T., and Theorell, H., 1962, On the ternary complex of liver alcohol dehydrogenase with reduced coenzyme and isobutyramide. Effect of *p*-chloromercuriphenyl sulfonate and stability of the complex, *Arch. Biochem. Biophys.* 99:433.

KINETICALLY SPECIFIC SPIN-LABEL SUBSTRATES OF LIVER ALCOHOL DEHYDROGENASE AND OF LIVER ALDEHYDE DEHYDROGENASE

William E. Boisvert,[1] Devkumar Mustafi,[1] Seppo Kasa,[1] Marvin W. Makinen[1] Howard J. Halpern,[2] Cheng Yu,[2] Eugene Barth,[2] and Miroslav Peric[2]

[1]Department of Biochemistry and Molecular Biology
[2]Department of Radiation and Cellular Oncology
The University of Chicago
Chicago, Illinois 60637

INTRODUCTION

The metabolism of alcohol in the mammalian liver involves two enzymes: (i) alcohol dehydrogenase, which converts an alcohol to its corresponding aldehyde product through a readily reversible reaction, and (ii) aldehyde dehydrogenase, which catalyzes the oxidation of the aldehyde to a carboxylic acid through an essentially irreversible step. For both enzymes the reaction proceeds at the level of a ternary complex formed by the enzyme, the nicotinamide adenine dinucleotide coenzyme, and the substrate. In the case of alcohol dehydrogenase, the ternary complex is formed with coordination of the alcohol or the aldehyde to the active site Zn^{2+}, and no covalent interaction occurs between the substrate and the protein. In this complex, hydride transfer proceeds directly between the substrate and the nicotinamide ring of the coenzyme that is responsible for chemical conversion of alcohol to aldehyde (Morris et al., 1980; Pettersson, 1987). In aldehyde dehydrogenase, hydride transfer from the aldehyde to the coenzyme results in formation of an acylenzyme intermediate in which a thiol group of a cysteinyl residue serves as a nucleophile (Tu and Weiner, 1988). It is likely that the conformations of the bound substrate for both enzymes are critically important in directing the path of the reaction. However, hitherto there has been relatively little structural data of sufficient resolution and accuracy to define catalytically competent conformations in alcohol dehydrogenase (Cedergren-Zeppezauer et al., 1982; Eklund et al., 1982), and at present X-ray crystallographic studies of aldehyde dehydrogenase are only at a preliminary stage (Rone et al., 1991; Hurley et al., 1993). Incisive progress towards understanding the molecular basis of the catalytic action of both enzymes would be obtained through high resolution studies of the structural relationships of active site residues with bound substrates in catalytically competent conformations.

In recent years we have endeavored to develop a method of three-dimensional

Enzymology and Molecular Biology of Carbonyl Metabolism 4
Edited by H. Weiner, Plenum Press, New York, 1993

structure determination and conformational analysis that can be applied to molecules in solution through the application of electron nuclear double resonance (ENDOR) spectroscopy, and we have shown that ENDOR of nitroxyl spin-labeled molecules can be incisively applied to determine structure and conformation of molecules with an accuracy that is exceeded only by that of single crystal X-ray diffraction methods (Wells and Makinen, 1988; Mustafi et al., 1990a; Wells et al., 1990; Mustafi et al., 1991). We have also developed new organic and enzymatic synthetic procedures to produce and modify a series of nitroxyl spin-label compounds that are highly reactive specific substrates of liver alcohol dehydrogenase (ADH) and of liver aldehyde dehydrogenase (ALDH). The ENDOR method, applied in conjunction with kinetic and cryoenzymologic techniques to identify conditions where enzyme reaction intermediates can be temporally isolated and stabilized in solution (Makinen and Fink, 1977; Fink and Cartwright, 1991), provides a means to characterize the molecular structures of true intermediates of enzyme catalyzed reactions.

In this communication we review results of kinetic and ENDOR spectroscopic studies of the spin-labels 3-(2,2,5-5-tetramethyl-1-oxypyrrolinyl)-2-propen-1-ol (**I**) and 3-(2,2,5,5-tetramethyl-1-oxypyrrolinyl)-2-propenal (**II**). Both compounds are specific substrates of ADH, and the spin-labeled aldehyde is also a substrate of ALDH. These synthetic spectroscopic substrate probes thus provide a means to investigate the catalytically competent active site structures of both enzymes. While spectroscopic substrate probes are employed ordinarily only for *in vitro* studies, we have also found that the spin-labels **I** and **II** can be employed as direct spectroscopic probes of the metabolism of alcohol through application of a novel electron paramagnetic resonance (EPR) imaging technique (Halpern et al., 1989; Halpern and Bowman, 1992) to monitor their time dependent organ distribution in intact, anesthetized mice. The use of these substrates for both *in vitro* and *in vivo* studies, thus, may provide a possibly unique opportunity to construct a pharmacokinetic model of the metabolism of alcohol and how it is influenced by pathological states.

METHODS AND MATERIALS

In Figure 1 we have illustrated the chemical bonding structures of the spin-labels **I** and **II** and the specifically deuterated analogues employed in this investigation. The synthesis of these compounds and their chemical characterization have been described elsewhere (Koch et al., 1979; Mustafi et al., 1992a,b).

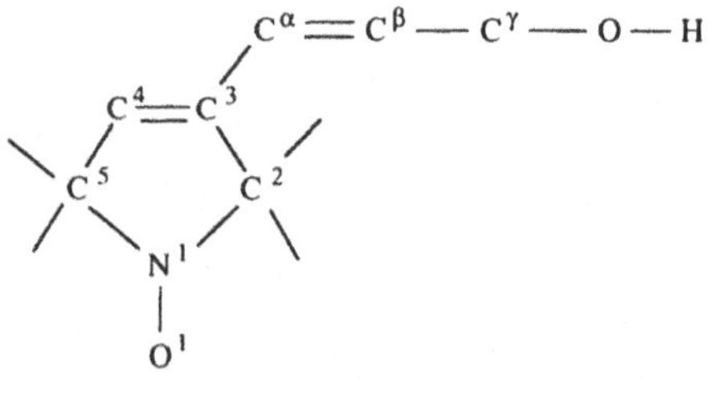

I. $SL-C^{\alpha}H=C^{\beta}H-C^{\gamma}H_2-OH$

Ia. $SL-C^{\alpha}D=C^{\beta}H-C^{\gamma}H_2-OH$

Ib. $SL-C^{\alpha}H=C^{\beta}H-C^{\gamma}HD-OH$

Ic. $SL-C^{\alpha}H=C^{\beta}H-C^{\gamma}D_2-OH$

II. $SL-C^{\alpha}H=C^{\beta}H-C^{\gamma}HO$

IIa. $SL-C^{\alpha}D=C^{\beta}H-C^{\gamma}HO$

IIb. $SL-C^{\alpha}H=C^{\beta}H-C^{\gamma}DO$

Figure 1. (Left) Illustration of the carbon atom skeletal structure of 3-(2,2,5,5-tetramethyl-1-oxypyrrolinyl)-2-propen-1-ol (**I**). (Right) The series of specifically deuterated analogues of **I** and of 3-(2,2,5,5-tetramethyl-1-oxypyrrolinyl)-2-propenal (**II**) used in this investigation. SL refers to the 2,2,5,5-tetramethyl-1-oxypyrrolinyl moiety.

Initial velocity data to estimate the steady state kinetic parameters k_{cat} and k_{cat}/K_M for the oxidation of **I** and the reduction of **II** catalyzed by ADH were collected and analyzed as described in earlier studies (Makinen et al., 1983; Maret and Makinen, 1991). Crystalline horse liver ADH was used as previously described (Maret and Makinen, 1991). The low K_M mitochondrial form of ALDH purified from bovine liver was a gift from Professor H. Weiner.

EPR and ENDOR spectra were collected with use of a Bruker 200D X-band spectrometer equipped with a broad banded (1-100 MHz) ENDOR/Triple accessory and an Oxford Instruments ESR10 liquid helium cryostat. The conditions for collection of proton ENDOR spectra were essentially identical to those described earlier (Wells and Makinen, 1988; Mustafi et al., 1990a,b; Wells et al., 1990; Mustafi et al., 1991). The collection of EPR imaging data with a custom built 250 MHz EPR spectrometer and the use of a resonator for restraint of intact, anesthetized mice have been described earlier (Halpern et al., 1989; Halpern et al., 1992).

Molecular modeling and torsion angle search calculations were carried out with SEARCH incorporated into the program SYBYL (G. R. Marshall, personal communication), as described earlier (Wells and Makinen, 1988; Mustafi et al., 1990a; Wells et al., 1990; Mustafi et al., 1991).

RESULTS AND DISCUSSION

Steady-State Kinetic Studies

It is well recognized that mammalian ADH and ALDH react with a variety of substrates differing widely in molecular structure and that ethanol and, correspondingly, acetaldehyde are by no means the most catalytically reactive substrates. In this respect the spin-labels **I** and **II** employed in this investigation may be expected to be kinetically reactive substrates of these enzymes, especially since they should be structurally similar to the well characterized substrate N,N′-dimethylcinnamaldehyde (Dunn and Hutchison, 1973) through the olefinic side chain containing the terminal alcohol or aldehyde functional group.

Figure 2 illustrates a hyperbolic plot of initial velocity data characterizing the activity of ADH in the oxidation of **I** and the reduction of **II** under conditions of saturating coenzyme concentration. For the ADH catalyzed oxidation of **I**, the data yielded the steady-state kinetic parameters $k_{cat} \sim 14.4\ (\pm 0.5)\ sec^{-1}$ and $K_M \sim 3.79\ (\pm 0.26) \times 10^{-3}$ M at pH 8. This pH region corresponds to a maximum plateau value of k_{cat} for the oxidation of benzyl alcohol (Maret and Makinen, 1991). Similarly, at pH 7, corresponding to a maximum in activity for reduction of aldehydes, the steady-state data yielded values of $k_{cat} \sim 32.1\ (\pm 1.2)\ sec^{-1}$ and $K_M \sim 1.48\ (\pm 0.13) \times 10^{-3}$ M for the reduction of **II**. The upper panel of Figure 2 also shows a plot of initial velocity data for the ADH catalyzed oxidation of the deuterated analogue **Ic**. For **Ic** the steady-state kinetic parameters were $k_{cat} \sim 9.1\ (\pm 0.2)\ sec^{-1}$ and $K_M \sim 3.20\ (\pm 0.19) \times 10^{-3}$ M. The isotope effect, thus, is seen in k_{cat} and corresponds to a ratio of $k_H/k_D \sim 1.6\ (\pm 0.1)$ comparable to that observed for benzyl alcohol (Dworschack and Plapp, 1977).

In Table 1 we list the values of steady-state kinetic parameters of **I** and **II** and compare them to those of other classical substrates. This comparison shows that **I** and **II** are as catalytically reactive as other synthetic substrates employed hitherto with bulky residues.

The oxidation of **II** catalyzed by ALDH proceeds considerably more sluggishly, and we have been able to only estimate a turnover rate of $\sim 0.03\ sec^{-1}$ at pH 9.0 on a subunit basis (in the presence of Mg^{2+}). No estimate of K_M could be obtained. However, we have observed that oxidation of the spin-labeled analogue 5-(2,2,5,5-

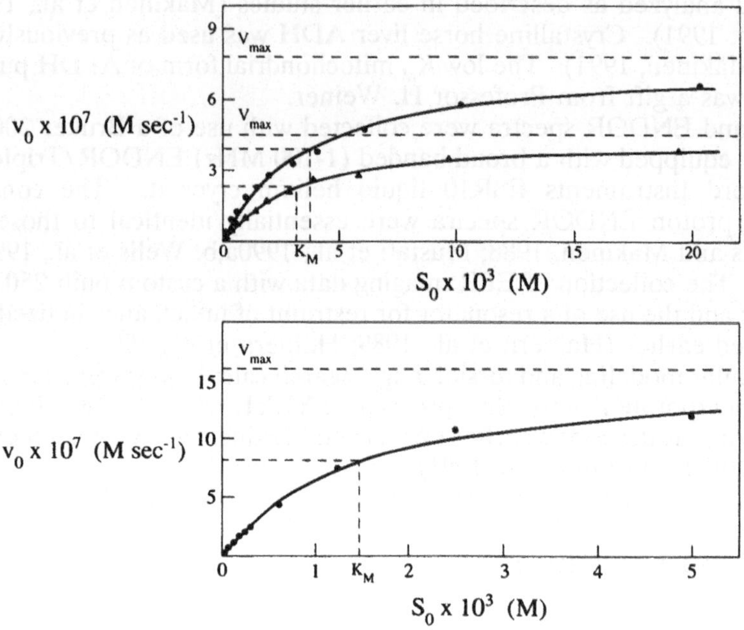

Figure 2. Comparison of initial velocity data (v_0) as a function of substrate concentration (S_0). The data were fitted to a rectangular hyperbola and V_{max} and K_M values, indicated by the dashed lines in the ordinate and abscissa, respectively, are calculated by a non-linear least-squares fitting program using Simplex and Marquardt algorithms, developed by Prof. J. Westley of The University of Chicago. Top: Oxidation of **I** (●) and its deuterated analogue **Ic** (▲) catalyzed by ADH. Bottom: Reduction of **II** catalyzed by ADH.

Table 1. Comparison of steady-state kinetic parameters of substrates of ADH.

	Substrate	pH	K_M (10^{-3} M)	k_{cat} (sec^{-1})
A. Alcohol oxidation	SL-propen-1-ol (**I**)	8.0	3.79 (± 0.26)	14.4 (± 0.5)
	Ethyl alcohol[a]	8.0	0.16	3.2
	Benzyl alcohol	8.0	0.029 (± 0.008)	4.52 (± 0.17)
B. Aldehyde reduction	SL-propenal (**II**)	7.0	1.48 (± 0.13)	32.1 (± 1.2)
	Acetaldehyde[a]	7.1	0.41	125
	Benzaldehyde[b]	8.0	0.043 (±0.007)	18 (± 1)
	p-nitrobenzaldehyde[b]	8.0	0.0038 (± 0.0007)	3.9 (± 0.2)

[a]Dalziel (1963); [b]Dworschack and Plapp (1977).

tetramethyl-1-oxypyrrolinyl)-2,4-pentadienal (Mustafi et al., 1992a) catalyzed by ALDH proceeds readily with steady-state kinetic parameters of $k_{cat} \sim 0.3\ sec^{-1}$ and $K_M \sim 2.9 \times 10^{-6}$ M at pH 9 in the presence of 10^{-3} M Mg^{2+}.

ENDOR and Molecular Modeling Studies

The EPR absorption spectrum of a nitroxyl spin-label is a composite of three sets of spectral components, differing by their projection of the ^{14}N nuclear moment onto the laboratory magnetic field $\mathbf{H}_0$. These components are designated by the values of $m_I = +1, 0, -1$. ENDOR spectra are obtained by partially saturating an EPR transition with the incident microwave power and simultaneously sweeping the sample with a radio frequency (rf) field. The interaction of a magnetic nucleus with the unpaired electron results in a shift of its resonance position from the Larmor frequency that is dependent on (dipolar) anisotropic hyperfine (hf) interactions and (Fermi contact) isotropic hf interactions. This shift is most accurately measured by the ENDOR method.

In the left-hand panel of Figure 3, we have illustrated the EPR absorption spectrum and the proton ENDOR spectra of **II**. Microwave power saturation of the low-field feature of the EPR spectrum (we term this setting A) of a nitroxyl spin-label selects molecules for ENDOR such that $\mathbf{H}_0$ is oriented perpendicularly to the oxypyrrolinyl ring. On the other hand, saturation of the central feature of the EPR spectrum (which we term setting B) selects molecules of all orientations. For a nearby nucleus in or near the plane of the nitroxyl spin-label exhibiting axially symmetric hf interactions, setting A yields the perpendicular hf coupling while setting B yields both parallel and perpendicular hf couplings. On this basis selective saturation of both types of EPR absorption features yields a complete assignment of the ENDOR absorptions of the nucleus. This method of selective saturation of the low-field and central absorption features of the EPR spectrum of a nitroxyl spin-label provides a particularly incisive stratagem for the assignment of structure and conformation. In the lower left-hand panel of Figure 3, proton ENDOR spectra are also illustrated for analogues of **II** in which the proton at the α-position or at the γ-position is selectively substituted with deuterium. The absence of a resonance feature through deuterium substitution for a given proton confirms its assignment in the spectrum.

In Figure 3 it is seen that each proton is associated with only two pairs of resonance features, confirming the presence of an axially symmetric hf field. Under the constraints $|A_{iso}| << |A_{\parallel}|, |A_{\perp}|$ and $A_{\parallel}^D > 0 > A_{\perp}^D$ which we have shown to obtain under these conditions, the dipolar hyperfine coupling (hfc) components $A_{\parallel}^D$ and $A_{\perp}^D$ are then calculated from the relationships $(A_{\parallel} + 2A_{\perp}) = 3A_{iso}$, $A_{\parallel}^D = A_{\parallel} - A_{iso}$, and $A_{\perp}^D = A_{\perp}^D - A_{iso}$. The hfc components are then related to the electron-nucleus separation by the relationship where the first right-hand term designates the

$$A = \frac{g_e\, g_n\, \beta_e\, \beta_n}{hr^3}\,(3\cos^2\theta - 1) + A_{iso}$$

dipolar hf coupling; $\theta = 0°$, for the parallel hfc component; and $\theta = 90°$ for the perpendicular hfc component. The other constants have their classically defined values.

In the right panel of Figure 3, we have illustrated the proton ENDOR spectra of **I**. The resonance features of the methylene protons are not easily identifiable and severely overlap. However, the resonance features of the γ-protons could be identified by use of 3-(2,2,5,5-tetramethyl-1-oxypyrrolinyl)-2-(1-^{2}H)propenol obtained through ADH catalyzed reduction of **IIb** with NADH. This reaction is known to yield an enantiomorphically pure RCHDOH analog (Levy et al., 1957). On this basis two well-resolved parallel and perpendicular hfc components for $H^{\gamma 1}$ were resolved. Finally, when the proton ENDOR spectrum of 3-(2,2,5,5-tetramethyl-1-oxypyrrolinyl)-2-(1-$^{2}H_2$)propenol was collected in an aprotic solvent, the resonance features of the

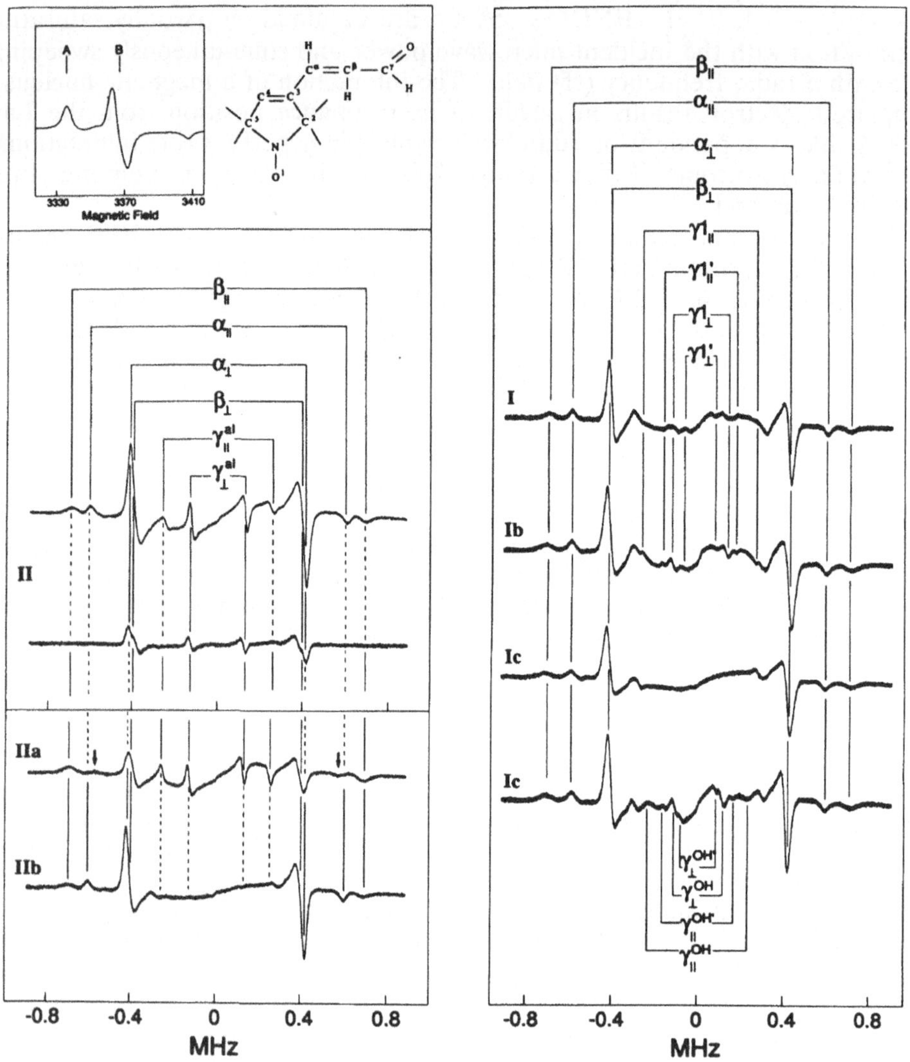

Figure 3. (Left). Proton ENDOR spectra of **II** and its deuterated analogues in ($^{2}H_4$)methanol. In the top left-hand corner, the first derivative EPR spectrum of **II** is shown. In this spectrum, A and B mark the H_0 settings that were saturated for ENDOR. In the upper left-hand panel, the chemical bonding structure of **II** is also shown. In the left-hand panel, the top and bottom spectra of **II** were recorded with H_0 at B and A, respectively. In the lower panel, ENDOR spectra of **IIa** and **IIb** were recorded with H_0 at setting B. For each of the three protons in the propenal side chain, the maximum and minimum ENDOR splittings that correspond to the two principal hfc components $A_{\parallel}$ and $A_{\perp}$ are identified in the stick diagram. The weak features indicated by arrows are assigned to the vinylic proton of the oxypyrrolinyl ring (Mustafi et al., 1990b).

(Right). Proton ENDOR spectra of **I** and its deuterated analogues with H_0 at setting B. The top three spectra were recorded with the samples dissolved in ($^{2}H_4$)methanol while the bottom most spectrum was taken with **Ic** in ($^{2}H_6$)DMSO:($^{2}H_8$)toluene:(^{2}H)chloroform (50:25:25 v/v). The ENDOR line splittings for protons in the propen-1-ol side chain are identified in the stick diagram.

hydroxyl proton could be resolved, as is seen in the lower most spectrum. The spectrum shows at least two sets of hfc components for the hydroxyl proton which are labeled γ^{OH} and $\gamma^{OH'}$. The results for the enantiomorphically pure RCHDOH analogue and of the RCD_2OH analogue together show that there are at least two distinct conformational rotamers of the $-CH_2OH$ group. In contrast, the ENDOR spectrum of spin-labeled propenal showed only one conformational species.

In Table 2 we have summarized the observed hfc components for each side-chain proton of **I** and **II** and have listed their isotropic and dipolar hfc components and corresponding electron-nucleus separations. Since the hfc components for $H^{\gamma 2}$ could not be unambiguously identified, the hf couplings for this proton are not included in the listing.

Table 2. Summary of hfc components and estimated electron-proton distances in 3-(2,2,5,5-tetramethyl-1-oxypyrrolinyl)-2-propen-1-ol (**I**) and 3-(2,2,5,5-tetramethyl-1-oxypyrrolinyl)-2-propenal (**II**).

Proton[a]	$A_\parallel$	$A_\perp$	A_{iso} (MHz)	$A_\parallel^D$	$A_\perp^D$	r[b] (Å)
Compound **I**						
α	1.199	0.836	-0.158	1.357	-0.678	4.90 ± 0.02
ß	1.431	0.836	-0.080	1.511	-0.756	4.71 ± 0.02
γ1	0.512	0.256	0.000	0.512	-0.256	6.76 ± 0.06[c]
γ1′	0.340	0.156	0.009	0.331	-0.165	7.82 ± 0.10[c]
OH	0.469	0.236	-0.001	0.470	-0.235	6.96 ± 0.06[d]
OH′	0.331	0.167	-0.001	0.332	-0.166	7.82 ± 0.08[d]
Compound **II**						
α	1.214	0.809	-0.135	1.349	-0.674	4.91 ± 0.02
ß	1.394	0.794	-0.065	1.459	-0.729	4.76 ± 0.02
γ^{al}	0.515	0.262	-0.003	0.518	-0.259	6.75 ± 0.04

[a]ENDOR line pairs for each proton are assigned from ENDOR spectra of compounds **I** and **II** as shown in Figure 3.

[b]An uncertainty of 0.010-0.016 MHz due to the line width of each absorption is included in the calculation of electron-proton distances.

[c]These two sets of resonance features, labeled γ1 and γ1′, come from one γ proton and are assigned unambiguously to two conformations.

[d]The two sets of resonance features for the hydroxyl proton, labeled γ^{OH} and $\gamma^{OH'}$, are assigned unambiguously to two conformations.

To assign the conformations of **I** and **II**, we have carried out a computational analysis on the basis of torsion angle search calculations, using the ENDOR determined electron-proton distances as constraints. The method entails finding all of the conformations within hard-sphere, van der Waals limits that are compatible with the set of electron-proton distances measured for a given molecule. The beginning set of atomic coordinates is constructed on the basis of X-ray defined molecular fragments and calculated idealized positions of ENDOR determined hydrogens. The results of the ENDOR distance constrained torsion angle calculations to determine the conformation of **II** are illustrated in Figure 4 in the form of a three-dimensional angle map. Each axis represents 0-360° of rotation for the designated torsion angle. It is seen from the angle map that the olefinic double bond of the side chain lies almost exactly in the plane of the oxypyrrolinyl ring ([C(4)=C(3) – $C^{\alpha}=C^{\beta}$] = 168° ± 5°) and the aldehyde group lies in an *s-trans* all-planar orientation with respect to the olefinic bond ([$C^{\alpha}=C^{\beta}-C^{\gamma}=O$] = 180° ± 15°). The distance constraints also are compatible with a conformer in which [$C^{\alpha}=C^{\beta}-C^{\gamma}=O$] = 60° ± 15°. While

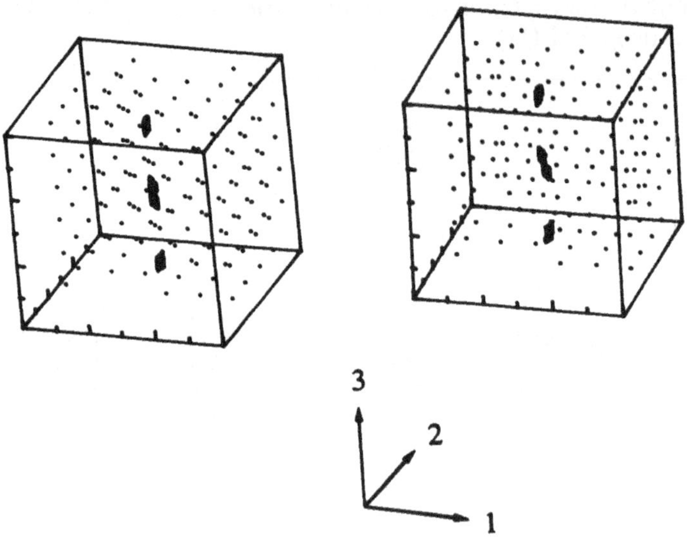

Figure 4. Stereodiagram of the angle map of **II** defining the conformation of the propenal side chain. The axes 1-3 represent 0-360° rotation around the C(3)–C^{α}, $C^{\alpha}=C^{\beta}$, and $C^{\beta}-C^{\gamma}$ bonds, respectively. Each dot in the cube represents a set of three calculated torsion angles. The low-density dots represent the conformational space allowed by nonbonded van der Waals interactions, while the small regions of high density dots represent the conformational space allowed by nonbonded van der Waals constraints together with the ENDOR determined distance constraints to the α, ß, and γ^{al} protons.

this conformer is allowed by the ENDOR distance constraints, we consider it as energetically unlikely or of low population since there would be significantly increased steric hindrance between the carbonyl oxygen and the C^{α} atom.

The ENDOR determined distance constraints to H^{α} and H^{β} in **I** similarly resulted in an essentially planar *trans* conformation of the olefinic $C^{\alpha}=C^{\beta}$ group with respect to the double bond in the oxypyrrolinyl ring. However, the ENDOR distance constraints for $H^{\gamma 1}$ and $H^{\gamma 1'}$ and for $H^{\gamma OH}$ and $H^{\gamma OH'}$ resulted in three conformers with respect to rotation around the $C^{\beta}-C^{\gamma}$ and C^{γ}–O bonds. These conformers are illustrated in Newman diagram form in Figure 5.

We have applied a novel EPR imaging technique to investigate whether the spin-labels **I** and **II** could be used to monitor the metabolism of alcohol in an intact laboratory animal. These experiments have met with encouraging success, and because of the potential advantage of this approach to studying the influence of pathological states and inhibitors of alcohol metabolism employed in the treatment of alcoholism, we present preliminary results to illustrate the method.

EPR spectroscopic imaging involves measurements of an injectable substrate which can be used to target an organ or physiological compartment. Our measurements have taken advantage of a low frequency (250 MHz) custom built EPR spectrometer designed to measure the concentrations of free radicals in living tissue

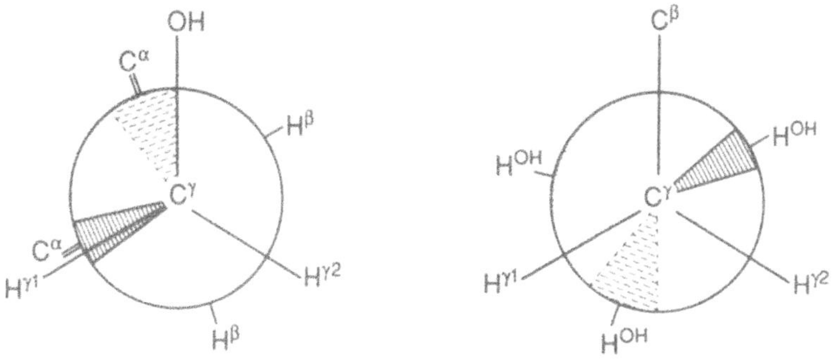

Figure 5. Newman diagrams illustrating the range of the ENDOR-constrained dihedral angles [$C^{\alpha}=C^{\beta}-C^{\gamma}-O$] and [$C^{\beta}-C^{\gamma}-O-H$] in spin-labeled propenol. The diagram on the left corresponds to a projection down the $C^{\gamma}-C^{\beta}$ bond and that on the right illustrates the projection down the $C^{\gamma}-O$ bond. In the left diagram the area indicated by solid lines corresponds to the range of [$C^{\alpha}=C^{\beta}-C^{\gamma}-O$] angle compatible with the ENDOR-determined distance constraint $r^{\gamma 1}$ and the area indicated by dashed lines is obtained from $r^{\gamma 1'}$. The average positions of C^{α} and H^{β} for these two conformers are also indicated in the diagram. In the diagram on the right three regions of conformational space are illustrated. The area with the solid line hatch marks is obtained using $r^{\gamma 1}$ and r^{OH} as constraints (conformer **A**), while the area with the dashed lines is obtained using $r^{\gamma 1}$ and $r^{OH'}$ (conformer **B**). The area indicated by dots is obtained by application of the distance constraints $r^{\gamma 1'}$ and r^{OH} (conformer **C**).

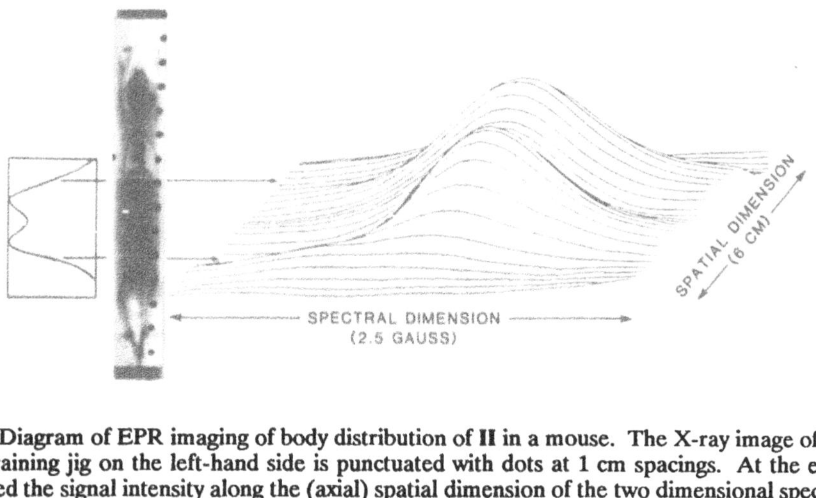

Figure 6. Diagram of EPR imaging of body distribution of **II** in a mouse. The X-ray image of the mouse in its restraining jig on the left-hand side is punctuated with dots at 1 cm spacings. At the extreme left is illustrated the signal intensity along the (axial) spatial dimension of the two dimensional spectral-spatial image. The arrows extending from the left-hand axial dimension spectrum to the two-dimensional image indicate the levels of the liver and bladder in the mouse.

samples and whole organ preparations (Halpern et al., 1989; Halpern and Bowman, 1992). Figure 6 illustrates an X-ray picture of a mouse in the restraining jig that is placed in the resonator of the spectrometer. The two-dimensional spectral-spatial image collected after intraperitoneal injection of **II** (0.7 ml of a 30 mM solution in isotonic saline) into a diazepam (20 mg/kg)/ketamine (45 mg/kg) anesthetized C3H mouse is shown to illustrate the distribution of the spin-label in the abdomen of the mouse. Figure 7 illustrates the time dependent variation of the organ distribution of **I** and **II** injected separately and of 2,2,5,5-tetramethyl-1-oxypyrrolinyl-3-carboxamide. There are pronounced changes in the organ distribution of each spin-label with time which reflect their different metabolic fates. It is seen that the spin-labels are initially accumulated in the liver whereafter the spin-labels and their metabolic products are excreted into the bladder. (In separate experiments we have shown that the spin-label 3-(2,2,5,5-tetramethyl-1-oxypyrrolinyl)-2-propenoic acid as the ALDH catalyzed oxidation product of **II** is rapidly released into the bladder.) The differential retention of the spin-labeled alcohol and aldehyde in the liver is consistent with their different metabolic fates. Since we have determined the turnover rates for oxidation of **I** and **II** separately with purified enzymes, this information together with monitoring of the time dependent changes in organ distribution of each spin-label and its metabolic product should permit construction of a multi-compartment pharmacokinetic model of alcohol metabolism. The use of spin-labels **I** and **II**, thus, provides an unusual opportunity to model the metabolism of a substance in an intact animal without the need for surgical intervention.

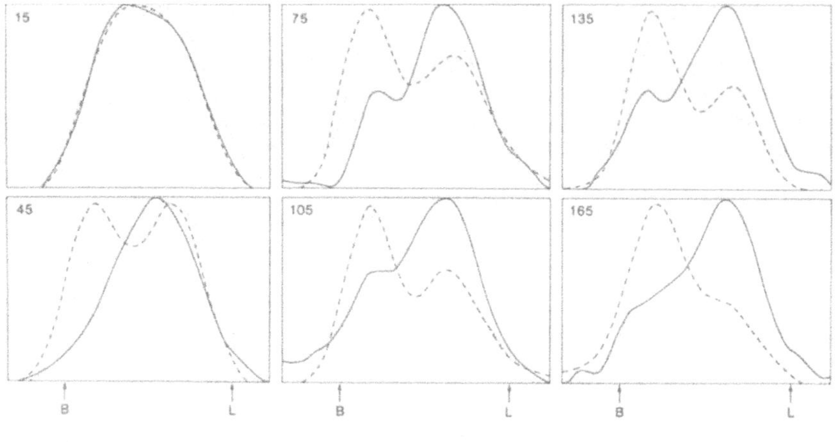

Figure 7. Comparison of axial dimension spectra of the organ distribution of **I** (—), **II** (- -), and of the oxypyrrolinyl spin-label carboxamide (···) in a mouse as a function of time. The number in the upper left-hand corner of each panel indicates the time in minutes elapsed after IP injection of 0.7 ml of 30 mM spin-label solution in saline. The arrows labeled B and L indicate approximate coordinates of the lower and upper boundaries of the bladder and liver, respectively.

ACKNOWLEDGMENTS

We thank Mary Kulberg for expert editorial assistance in manuscript preparation. This work was supported by grants of the National Institutes of Health (AA06374, MH47809, and CA50679).

REFERENCES

Cedergren-Zeppezauer, E., Samama, J. P., and Eklund, H., 1982, Crystal structure determinations of coenzyme analogue and substrate complexes of liver alcohol dehydrogenase: Binding of 1,4,5,6-tetrahydronicotinamide adenine dinucleotide and *trans*-4-(*N,N'*-dimethylamino)cinnamaldehyde to the enzyme, *Biochemistry* 21:4895.

Dalziel, K., 1963, Kinetic studies of liver alcohol dehydrogenase and pH effects with coenzyme preparations of high purity, *J. Biol. Chem.* 238:2850.

Dunn, M. F. and Hutchison, J. S., 1973, Roles of zinc ion and reduced coenzyme in the formation of a transient chemical intermediate during the equine liver alcohol dehydrogenase catalyzed reduction of an aromatic aldehyde, *Biochemistry* 12:4882.

Dworschack, R. T. and Plapp, B. V., 1977, pH, isotope, and substituent effects on the interconversion of aromatic substrates catalyzed by hydroxybutyrimidylated liver alcohol dehydrogenase, *Biochemistry* 16:2716.

Eklund, H., Plapp, B. V., Samama, J. P., and Branden, C. I., 1982, Binding of substrate in a ternary complex of horse liver alcohol dehydrogenase, *J. Biol. Chem.* 257:14349.

Fink, A. L. and Cartwright, S. J., 1991, Cryoenzymology, *CRC Crit. Rev. Biochem.* 11:145.

Halpern, H. J., Spencer, D. P., vanPolen, J., Bowman, M. K., Nelson, A. C., Dancy, E. M., and Teicher, B. A., 1989, Imaging radio frequency electron-spin-resonance spectrometer with high resolution and sensitivity for *in vivo* measurements, *Rev. Sci. Instrum.* 60:1040.

Halpern, H. J. and Bowman, M. K., 1991, Low-frequency electron paramagnetic resonance spectrometers: MHz range, in "EPR Imaging and *in vivo* EPR," G. A. Eaton and S. Eaton, eds., CRC Press, Orlando, p. 45.

Halpern, H. J., Peric, M., Nguyen, T. D., Bowman, M. K., Lin, Y. J., and Teicher, B. A., 1991, *In vivo* O_2 sensitive imaging at lower frequencies, *Physica Medica* 7:39.

Hurley, T, Yang, Z., Bosron, W. F., and Weiner, H., 1993, Crystallization and preliminary X-ray analysis of aldehyde metabolizing enzymes, *Proc. 6th Intern. Workshop Enzymol. and Mol. Biol. of Carbonyl Metabolism*, Plenum, New York.

Koch, T. R., Kuo, L. C., Douglas, E. G., Jaffer, S. and Makinen, M. W., 1979, Synthesis of chromophoric spin-label enzyme substrates useful for cryoenzymology, *J. Biol. Chem.* 254:12310.

Levy, H. R., Loewns, F. A., and Vennesland, B., 1957, The optical rotation and configuration of a pure enantiomorph of ethanol-1-*d*, *J. Am. Chem. Soc.* 79:2949.

Makinen, M. W. and Fink, A. L., 1977, Reactivity and cryoenzymology of enzymes in the crystalline state, *Annu. Rev. Biophys. Bioeng.* 6:301.

Makinen, M. W., Maret, W., and Yim, M. B., 1983, Neutral metal-bound water is the base catalyst in liver alcohol dehydrogenase, *Proc. Natl. Acad. Sci. (USA)* 80:2584.

Maret, W. and Makinen, M. W., 1991, The pH variation of steady-state kinetic parameters of site-specific Co^{2+}-reconstituted horse liver alcohol dehydrogenase: A mechanistic probe for the assignment of metal-linked ionizations, *J. Biol. Chem.* 266:20636.

Marshall, G. R., personal communication. Detailed information on the program package can be obtained from Tripos Associates, Inc., 1600 S. Hanley Road, St. Louis, Missouri 63144 USA.

Morris, R. G., Saliman, G., and Dunn, M. F., 1980, Evidence that hydride transfer precedes proton transfer in the liver alcohol dehydrogenase catalyzed reduction of *trans*-4-(*N,N*-dimethylamino)cinnamaldehyde, *Biochemistry* 19:725.

Mustafi, D., Wells, G. W., Sachleben, J. R., and Makinen, M. W., 1990a, Structure and conformation of spin-labeled amino acids in frozen solutions determined by electron nuclear double resonance. 1. Methyl *N*-(2,2,5,5-tetramethyl-1-oxypyrrolinyl-3-carbonyl)-L-alanate, a molecule with a single preferred conformation, *J. Am. Chem. Soc.* 112:2558.

Mustafi, D., Wells, G. B., Joela, H., and Makinen, M. W., 1990b, Assignment of proton ENDOR resonances of nitroxyl spin-labels in frozen solution, *Free Radical Res. Commun.* 10:95.

Mustafi, D., Joela, H., and Makinen, M. W., 1991, The effective position of the electronic point dipole of the nitroxyl group of spin-labels determined by ENDOR spectroscopy, *J. Magn. Reson.* 91:497.

Mustafi, D., Boisvert, W. E., and Makinen, M. W., 1992a, Synthesis of conjugated polyene carbonyl derivatives of nitroxyl spin-labels and determination of their molecular structure and conformation by electron nuclear double resonance, *J. Am. Chem. Soc.*, submitted for publication.

Mustafi, D., Boisvert, W. E., and Makinen, M. W., 1992b, Multiple rotamers of the allyl alcohol in nitroxyl spin-labeled propenol: Determination of molecular structure and conformation by electron nuclear double resonance, *J. Am. Chem. Soc.*, submitted for publication.

Pettersson, G., 1987, Liver alcohol dehydrogenase, *CRC Crit. Rev. Biochem.* 21:349.

Rone, J. P., Hempel, J., Kuo, I., Lindahl, R., and Wang, B. C., 1991, Preliminary crystallographic analysis of Class 3 rat liver aldehyde dehydrogenase, *Proteins: Struct. Func. Genet.* 8:305.

Tu, G. C. and Weiner, H., 1988, Identification of the cysteine residue in the active site of horse liver mitochondrial aldehyde dehydrogenase, *J. Biol. Chem.* 263:1212.

Wells, G. B. and Makinen, M. W., 1988, ENDOR determined molecular geometries of spin-labeled fluoroanilides in frozen solution, *J. Am. Chem. Soc.* 110:6343.

Wells, G. B., Mustafi, D., and Makinen, M. W., 1990, Structure and conformation of spin-labeled amino acids in frozen solutions determined by electron nuclear double resonance. 2. Methyl *N*-(2,2,5,5-tetramethyl-1-oxypyrrolinyl-3-carbonyl)-L-tryptophanate, a molecule with multiple conformations, *J. Am. Chem. Soc.* 112:2566.

FLUORESCENCE STUDIES OF TERNARY COMPLEXES OF LIVER ALCOHOL DEHYDROGENASE

Y. Pocker and Joe D. Page

Department of Chemistry
University of Washington
Seattle, WA 98195

INTRODUCTION

Investigations concerning the binding and coordination of alcohols in the active site of liver alcohol dehydrogenase (LADH) have long been of interest (Klinman, 1981; Petterson, 1986; Pocker, 1989). These investigations have yielded valuable information about the nature of catalysis effected by zinc metalloenzymes. Such studies, however, have been limited by the fact that neither NAD^+ nor most simple alcohols possess a convenient chromophore for monitoring complex formation.

Attempts at circumventing this difficulty have included the use of chromophoric substrate alcohols such as 3-hydroxy-4-nitrobenzyl alcohol and vitamin A which have been used to probe the active site during binding (MacGibbon et al., 1987; Pocker et al., 1986). Studies with cobalt (II) replacing the zinc ion in the active site of LADH have shed light on the ionization of the alcohol at the metal center (Koerber et al., 1983; Sartorius et al., 1987). ESR studies employing a catalytically inactive spin labeled NAD^+ analog have elucidated many aspects of alcohol positioning within the active site (Mildvan and Weiner, 1969a, 1969b). In the present account, we present a method using an NAD^+ analog that possesses full activity as a coenzyme and introduces only minor perturbations, thus mimicking the action of LADH and NAD^+ with simple alcohols.

In the course of studying the kinetic effects of chlorinated alcohols: 2-chloroethanol, 2,2,2-trichloroethanol; and chlorinated aldehydes: chloroacetaldehyde and trichloroacetaldehyde (chloral hydrate) on the mechanistic behavior of LADH, we observed some unexpected coenzyme reactivity. In particular, the rate limiting step, the release of NADH for the alcohol oxidation reactions and the release of NAD^+ for the aldehyde reduction reactions was much slower for the halogenated substrates than for ethanol and acetaldehyde. A review of the literature revealed that NAD^+ reacts with chloroacetaldehyde to form an etheno adduct. Because chloroacetaldehyde was used as a substrate, and also appeared as a product of 2-chloroethanol oxidation, it is possible that some of our enzyme kinetic data may have been affected by this adduct formation reaction. Given the fluorescent nature of this coenzyme, it struck us as a useful spectroscopic probe, because NAD^+ lacks a convenient chromophore for monitoring complex formation.

The fluorescent NAD^+ analog 1,N^6-ethenoadenine dinucleotide, εNAD^+ (Figure 1), was first synthesized by N. J. Leonard in 1972 and has been tested for its coenzyme activity with several dehydrogenases (Barrio et al., 1972; Secrist et al., 1972a, 1972b; Greenfield et al., 1975; Suhadolnik et al., 1977; Luisi et al., 1975). Techniques utilizing the fluorescence enhancement of εNAD^+ upon binding to a series of NAD^+ requiring dehydrogenases have been reported (Luisi et al., 1973; Baici et al., 1974). The etheno analogs, εNAD^+ and

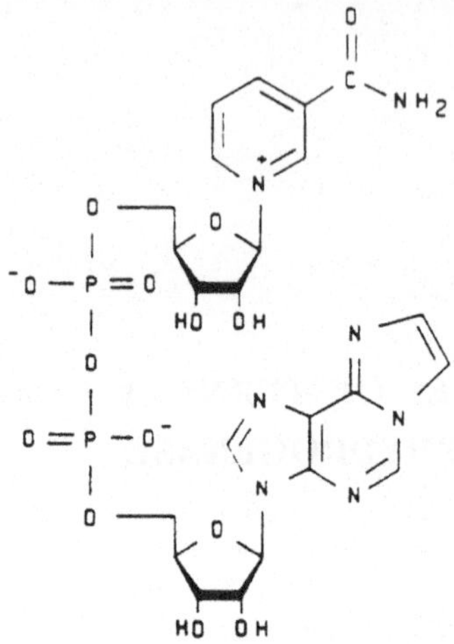

Figure 1. Chemical structure of εNAD+.

εNADH are effective coenzymes with LADH; in fact εNAD+ exhibits an activity comparable to that of NAD+ for the oxidation of ethanol with LADH (data not shown). The fluorescent nature of εNAD+, specifically its enhanced fluorescence emission upon binding to LADH, has been exploited in binding studies with LADH inhibitors. We report a direct spectroscopic method for determining the dissociation constant of εNAD+ from a ternary complex of LADH, a strong inhibitor, and εNAD+ with εNAD+ acting as the reporter ligand. From this method, the pH dependent binding affinity of εNAD+ in the presence pyrazole, and 2,2,2-trichloroethanol in the LADH-εNAD+-inhibitor ternary complex have been determined. The results provide new and valuable information concerning the pK_a of εNAD+ release from ternary complexes of LADH.

MATERIALS AND METHODS

Buffers. All solutions were prepared in deionized water that was 30 mM in buffer, adjusted to an ionic strength of 0.1 with Na_2SO_4. MES buffer was used for pH values below 6.0; sodium phosphate was used for values 6.0 to 8.5; and CHES was used for pH values 8.6 to 10.0.

Enzyme Solutions. LADH (E.C. 1.1.1.1) from horse livers was obtained from Boehringer Mannheim. Prior to binding studies, the percent activity of the enzyme was determined by titration of its active sites with NAD+ and pyrazole (Theorell and Yonetani, 1963). LADH from Boehringer Mannheim was exhaustively dialyzed for 48 hours against pH 7 buffer to remove all traces of ethanol. This enzyme typically showed a 90-95% binding capacity of its active sites. When stored at 4 °C, it retained full activity for at least two weeks. LADH concentrations were determined using ε_{280} = 36 $mM^{-1}cm^{-1}$ and then corrected for its percent activity. LADH contains two active sites per molecule, its concentration is reported in terms of its active site normality, *e*.

Kinetics. Preliminary kinetic results of the oxidation of chloro-ethanols and the reduction of chloro-acetaldehydes by NAD+/NADH and LADH were obtained by performing initial rate kinetics. For each alcohol and aldehyde, kinetics for five substrate concentrations, and five coenzyme concentrations were determined. The solutions were prepared by first premixing the alcohol/NAD+ and aldehyde/NADH in a pH 7.0 phosphate buffer. Reactions were initiated by injecting a small aliquot (10-20 µl) of LADH into 3.00 ml of the reaction mixture. The reaction course was monitored by the appearance or disappearance of NADH at 340 nm.

Coenzyme Solutions. NAD+ (grade IIIc, 99%) and εNAD+ (98%, C-18 rplc purified) were obtained from Sigma. Concentrations of NAD+ and εNAD+ were determined using ε_{260} = 18.0 $mM^{-1}cm^{-1}$ and ε_{265} = 10 $mM^{-1}cm^{-1}$, respectively (Luisi et al., 1975). All coenzyme solutions were prepared immediately prior to use and discarded at the end of the day. εNADH was prepared by first synthesizing εNAD^+ according to the procedures of Barrio *et al.* (1972). The εNAD^+ was purified on a Dowex 1X-200 anionic exchange column by eluting with a formic acid gradient. The UV and fluorescence spectra of this compound matched those of the literature (Barrio et al.,1972; Luisi et al., 1975). εNADH was then produced by enzymatic reduction of εNAD^+ using yeast alcohol dehydrogenase and ethanol (Theorell and Yonetani, 1963; Lee and Everse, 1973). The product was collected and stored as the barium salt. The barium ion was removed prior to use by precipitation with Na_2SO_4. The UV spectrum of εNADH was similar to that previously reported and its fluorescence spectrum resembled that of εNAD^+ with a fluorescence maximum at 420 nm (Greenfield et al., 1975; Luisi et al., 1975) . The molecular weight of the barium salt of this compound was determined to be 825 g/mol by FAB-MS. Its extinction coefficient was determined by monitoring the reduction of trans-4-(N,N-dimethylamino)cinnamaldehyde with εNADH and LADH (Rafter and Colowick, 1957). By reacting a six fold excess of trans-4-(N,N-dimethylamino)cinnamaldehyde (1.6×10^{-5} M) relative to εNADH in the presence of 4×10^{-8} N LADH and allowing to reaction to proceed for 10 half lives, a value of $\varepsilon_{340} = 7.6 \pm 0.2$ $mM^{-1}cm^{-1}$ was determined.

Binding Studies. εNAD^+ is ideally suited for binding studies as its fluorescence is enhanced 9-13 fold upon binding LADH (Luisi et al. 1975). The dissociation (or association) constant of εNAD^+ from the LADH-εNAD^+-inhibitor complex can be determined by monitoring the fluorescence change for each addition of εNAD^+ to LADH saturated with an inhibitor. Because fluorescence emission is diminished in solutions of high optical density due to absorption of the exciting light, these titrations were performed in NMR tubes with diameters less than 4 mm. The tubes were tightly secured in the fluorimeter and mixing was accomplished with a long plastic plunger.

Fluorescence Titrations. It has long been known that when NADH forms complexes with LADH, its UV and fluorescence spectra are perturbed (Theorell and Bonnischsen, 1951; Boyer and Theorell, 1956). In contrast, NAD^+ is not fluorescent which has hindered studies of its complex formation with LADH. However, by utilizing the fluorescence enhancement of εNAD^+ when bound to complexes of LADH, we have been able to determine binding constants of εNAD^+ in the presence of various LADH inhibitors. εNAD^+ dissociation constants were calculated using the Scatchard binding model :

$$\upsilon = \frac{NK_a(\varepsilon NAD_{tot} - e\upsilon)}{1 + K_a(\varepsilon NAD_{tot} - e\upsilon)} \tag{1}$$

where: $\upsilon = (\Delta F/e)/(\Delta F_{max}/e)$ = Molar ratio of bound ligand per total active site concentration (Luisi et al., 1975).
N = Number of binding sites per active site.
K_a = εNAD^+ association constant. In the absence of inhibitor, εNAD^+ binding is negligible. This value corresponds to one half saturation of the active sites. The reciprocal of this value is the εNAD^+ dissociation constant from the ternary complex.
e = Active site concentration at a given fluorescence measurement.
$\Delta F/e$ = Observed fluorescent enhancement per active site concentration.
$\Delta F_{max}/e$ = Maximum fluorescence enhancement per active site concentration.

To ensure that εNAD^+ quenching effects did not contribute to the binding curves, it was necessary to determine the concentrations of εNAD^+ that cause significant quenching effects. It was determined that 135 μM εNAD^+ exhibits a 5-6% deviation from linearity, and 160 μM εNAD^+ shows less than a 10% deviation. To avoid these effects, the highest concentration of εNAD^+ used in titrations with pyrazole was about 100 μM (no quenching effects). Titrations with 2,2,2-trichloroethanol required more εNAD^+ to reach saturation. The upper limit of εNAD^+ in these titrations was typically 135 μM. Additionally, it was determined (at pH 6.5) that εNAD^+ fluorescence enhancement was not observed in the presence of only LADH, but that the presence of an inhibitor also was necessary to observe fluorescence enhancement. This is due to the weak or negligible binding of εNAD^+ to LADH in the absence of an inhibitor.

A typical titration with the inhibitor pyrazole was performed by first establishing a baseline with 1.80 ml of 6.0 μN LADH and 120 μM pyrazole in the NMR tube at room temperature.

A 5 to 50 μl injection of (approx 1.0 mM) εNAD+ was then injected into the tube followed by mixing with a plastic plunger and the fluorescence measured. This process was repeated 15 times until the enzyme was saturated with about 20 fold excess of εNAD+. A typical titration with 2,2,2-trichloroethanol as the inhibitor was performed by first establishing a baseline with 1.80 ml of 4.5 μN LADH and 20 mM 2,2,2-trichloroethanol in the NMR tube at room temperature. With this inhibitor a 30-40 molar excess of εNAD+ was necessary to saturate the enzyme. While it is desirable to go to higher εNAD+ concentrations, quenching limits the amount of excess εNAD+ that may be used. Once a baseline is established, a titration can be completed in 15-20 minutes. Titrations of this type were performed for pyrazole, 2,2,2-trifluoroethanol and 2,2,2-trichloroethanol. The pH dependence of the εNAD+ dissociation constants with pyrazole and 2,2,2-trichloroethanol were determined from pH values 5.6 to 10 in half pH unit increments.

Instrumentation. Inhibitor binding experiments and fluorescence spectra of εNAD+ and εNADH were measured on the Perkin Elmer 650-10s Fluorescence Spectrophotometer. All other kinetic and equilibrium measurements were performed on a Cary 210 UV-visible double beam spectrophotometer interfaced to an Apple II/e microcomputer and thermostated with a Forma Temp Jr. constant temperature bath. ^{1}H and ^{13}C NMR were taken on a Varian VXR 300 MHz NMR spectrophotometer. Fast atom bombardment mass spectra (FAB-MS)of barium εNADH were measured on a VG-70 SEQ. A Radiometer Model PHM84 research pH meter equipped with a Cole-Palmer Ag/AgCl glass electrode was used to determine buffer pH values. Linear regression analysis of kinetic data was performed on an IBM Personal Computer AT. Modeling of kinetic data was done on a VAX/VMS mainframe using BMDPAR (1988) of BMDP Statistical Software, Inc. (Los Angeles, CA).

RESULTS

The fluorescence titration for the binding of pyrazole to LADH at pH 7.0 as reported by εNAD+ is shown in Figure 2. Quantitation of this data is achieved by subtracting the intrinsic fluorescence of the coenzyme from the observed fluorescence upon pyrazole binding. Solutions of εNAD+ in the pH range from 6-10 have an intrinsic fluorescence. Upon binding to LADH, the stacking interactions of the ethenoadenine and nicotinamide rings are removed causing the fluorescence to be enhanced. The intrinsic fluorescense of εNAD+ can be determined from the slope in the saturated linear region of the observed fluorescence curve, extrapolating to zero concentration (Figure 2, inset). εNAD+ dissociation constants were determined by non-linear regression of the fluorescent enhancement curve as a function of the εNAD+ and active site concentrations using equation 1. The $\Delta F_{max}/e$ ratio can be estimated by this analysis and used to normalize the binding titrations to a maximum relative fluorescence of 1.0. Because εNAD+ has an intrinsic fluorescence, this technique is not amenable to the determination of weak dissociation constants because the total observed fluorescence must be distinguishable from the intrinsic fluorescence of the cofactor.

LADH is known to form a tight ternary complex with NAD+ and pyrazole in which pyrazole occupies the substrate binding site and bonds to the catalytic zinc ion. At pH 7.00, pyrazole has a dissociation constant from the ternary complex of about 0.1 to 0.2 μM (Theorell and Yonetani, 1963; Theorell et al., 1969). εNAD+ also forms a tight ternary complex with LADH and pyrazole. Analysis of the data in Figure 1 using the binding model described by equation 1 yields a value of 4.2 ± 0.4 μM was determined in fair agreement with Luisi *et al.*, who report a value less than 1 μM for this complex (Luisi et al., 1975).

Using this fluorescence titration technique, it is possible to gain information about the enzymatic groups which control binding in the LADH-εNAD+ binary complex. Since εNAD+ and NAD+ are equally charged in the pH region under investigation (pH 5.5-10), it may be possible to extrapolate these results to the LADH-NAD+ binary complex. By measuring the dissociation constant of εNAD+ from the LADH-εNAD+-pyrazole ternary complex as a function of pH (Table I), it was possible to determine a pK_a of 7.6 ± 0.2 for a group within this complex that controls εNAD+ dissociation. This result is in accord with other studies that have shown a group with a pK_a 7.6 in the LADH-NAD+ complex controls binding (Evans and Shore, 1980; Andersson et al., 1981a, 1981b). At pH values greater than 8.0, it was noted that pre-saturating concentrations of εNAD+ in the presence of LADH and pyrazole required 10 to 15 minutes to exhibit maximal fluorescence emission. However, at pH values 7.0 and lower, this fluorescence emission reached its maximum immediately. This result suggests a slow time dependent binding of the inhibitor to form the ternary complex at high pH values. This may be due to the ionization of the zinc bound water molecule in the LADH-εNAD+ complex which would slow inhibitor binding.

UV techniques are available for measuring the dissociation constant of εNAD+ from the LADH-NAD+-pyrazole ternary complex because a shoulder in the 290-300 nm region of the

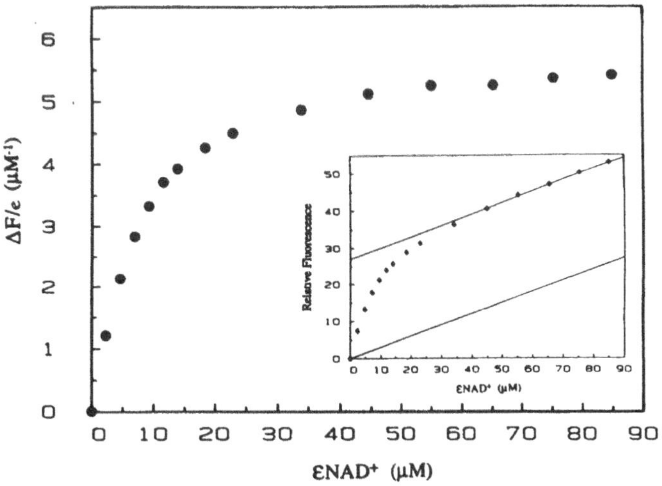

Figure 2. Fluorescence titration of LADH using εNAD+ and pyrazole at pH 7.0. The titration was initiated with 1.8 ml of 5.56 μM LADH dimer and 0.20 mM pyrazole. The titration proceeded by making 5-25 μl injections of 0.859 mM εNAD+ and monitoring the fluorescence emission at 410 nm (2 nm slit). Titrations were performed in an NMR tube with an excitation wavelength of 300 nm (5 nm slit). The εNAD+ dissociation constant was calculated from the data using equation 1. Inset, the baseline fluorescence of εNAD+ was determined from the line drawn from the saturating region of the curve and intersecting the origin. This baseline fluoresecence was subtracted to generate the observed binding curve.

UV spectrum develops when this complex is formed. However, most inhibitors that form ternary complexes with LADH-NAD+ do not possess such a chromophoric advantage. Dissociation constants for these inhibitors must be determined by kinetic inhibition experiments which are labor intensive and time consuming. For example, 2,2,2-trifluoroethanol (TFE) is a potent substrate analog inhibitor of LADH whose dissociation constant must be measured by kinetic techniques because it has no apparent chromophore. The spectroscopic properties of the etheno modification has allowed workers to characterize difference spectra of ternary complexes of LADH containing TFE (Subramanian et al., 1981). Using εNAD+ ,we have measured the dissociation constant of εNAD+ from the LADH-εNAD+-TFE ternary complex at pH 7.0 to be 3.2 ± 0.2 μM.

TABLE I. εNAD+ Dissociation Constants from the LADH-εNAD+-Pyrazole Ternary Complex[a].

pH	K_d (μM)	pK_d
5.62	57 ± 6	4.24
6.00	28 ± 7	4.55
6.48	11.6 ± 0.9	4.94
7.00	4.2 ± 0.4	5.38
7.44	2.9 ± 0.2	5.54
8.00	1.5 ± 0.3	5.82
8.57	0.7 ± 0.1	6.15
9.00	2.0 ± 0.4	5.70
9.67	1.6 ± 0.3	5.80

[a] Dissociation constants were measured using the fluorescence titration technique described in the text using εNAD+ obtained from Sigma. Values were measured at room temperature. Dissociation constants are the average of three determinations. All measurements were made in 30 mM buffer: pH value 5.62 was carried out in MES buffer, pH values 6-8 were carried out in phosphate buffer, pH values 8.6-10 were carried out in CHES buffer. The ionic strength of each buffer was adjusted to 0.1 with Na_2SO_4.

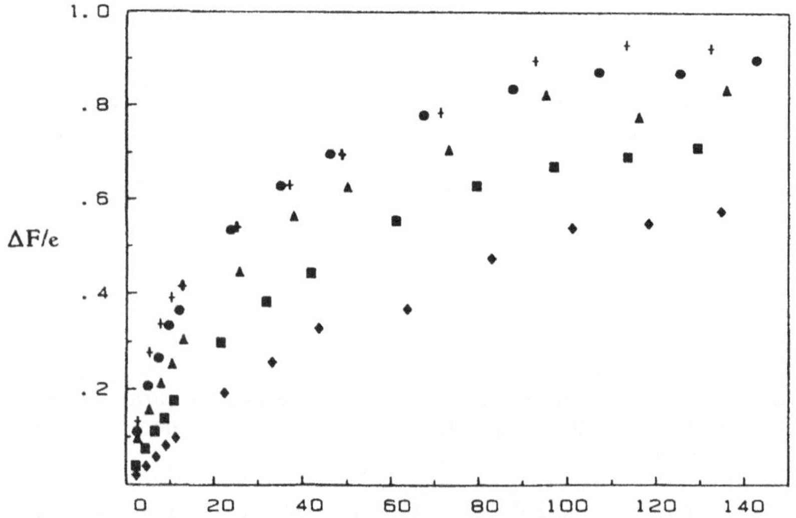

Figure 3. Titration of LADH with εNAD+ in the presence of saturating (20 mM) 2,2,2-trichloroethanol as a function of pH. The baseline fluorescence of εNAD+ has been subtracted from each curve. The curves shown are for pH values: 6.0 (♦), 6.6 (▬), 7.0 (▲), 7.44 (•), 8.0 (+). The change in fluorescence per active site, normalized to 1.0, is plotted along the y-axis. Curves for higher pH values overlap the pH 8.0 curve. The observed data was analyzed by Scatchard analysis to determine the dissociation constant of εNAD+

A representative alcohol, 2,2,2-trichloroethanol (TCE), was investigated to determine if other alcohols which are known to be inhibitors of LADH can be monitored for their capacity to form ternary complexes with LADH-εNAD+ (Pocker and Page, 1990). In a control experiment it was determined that 4 μN LADH incubated with 20 mM TCE at 0 °C, pH 7.00 for 30 minutes showed an approximate 5% loss in coenzyme binding capacity. After 2.3 hours, 20% of the original binding was lost, and after 14 hours 35% of the original binding capacity is lost. Because the fluorescence titrations are completed in about 20 minutes, a negligible binding capacity loss will be incurred by using TCE as an inhibitor. The binding potential of LADH toward this inhibitor was studied as a function of pH in an effort to identify the enzymatic groups that control εNAD+, and inhibitor binding. Fluorescence titrations were performed from pH 6 to 10 in half pH increments and the data is shown in Figure 3.

DISCUSSION

The fluorescence titration results show that the formation of a ternary complex can be measured with a strong inhibitor using εNAD+ as the reporter ligand. This method takes advantage of the fact that upon binding to LADH, in the presence of a strong inhibitor, εNAD+ becomes bound in an elongated conformation causing an enhancement of its fluorescence emission. The enhanced fluorescence is used to monitor the progress of the titration until no further enhancement is observed when the enzyme's binding sites become saturated. It was found that pyrazole and the substrate analog inhibitors 2,2,2-trifluoroethanol and 2,2,2-trichloroethanol could be used in this technique to monitor ternary complex formation. Given the similar characteristics of NAD+ and εNAD+ these results demonstrate that εNAD+ can be used as a reliable probe, and because NAD+ lacks a useful chromophore, the use of εNAD+ in fluorescence studies with LADH should yield valuable mechanistic insights.

Using this fluorescence titration method the dissociation constant of εNAD+ from the LADH- εNAD+-pyrazole ternary complex was determined in half pH increments from pH 5.6 to 9.6. The behavior of the dissociation constant as a function of pH from the ternary complex is controlled by a group with a pK_a of about 7.6. Several studies have shown that a group with a pK_a of 7.6 in the LADH-NAD+ binary complex is also responsible for controlling the binding of certain substrates and inhibitors. It has been shown that 2,2-bipyridine binds to LADH in the presence of saturating NAD+ with a pK_a of 7.6, and that the rate of dissociation of NAD+ from the LADH-NAD+ complex is controlled by a group exhibiting a pK_a of 7.6 (Evans and Shore, 1980; Kvassman and Pettersson, 1979; 1980). The elimination of the pH dependence of NAD+ binding by imidazole coupled with crystallographic evidence that imidazole replaces a water molecule bound to the active site zinc implicates the zinc bound water as part of the proton relay system.

The effect of pH on εNAD+ dissociation from the LADH-εNAD+-trichloroethanol ternary complex was also determined as shown in Figure 3. A plot of pK_a versus pH yields an inflection at pH 6.9. It has been shown that the pK_a of an alcohol is significantly lowered when it is bound in the ternary complex. The lower the pK_a of the free alcohol, the lower will be the pK_a of the alcohol bound in the ternary complex (Pettersson, 1986; Pocker and Page, 1990). We are currently investigating whether the pK_a of 6.9 that controls εNAD+ dissociation from the ternary complex is actually a macroscopic average of the pK_a of the zinc bound water molecule and the pK_a of trichloroethanol bound in the ternary complex.

The mechanism by which a group or network with a pK_a of 7.6 in the LADH-NAD+-inhibitor complex controls NAD+ dissociation has been the topic of much discussion. The exact identity of this group is still in question. It has been hypothesized from crystallographic studies that a proton relay system consisting of the hydroxyl groups of Ser 48 and the nicotinamide ribose of the coenzyme and His 51 form a hydrogen bonded system that shuttles the proton from the alcohol through the hydroxyl groups to His 51 and then to the solvent (Eklund et al., 1982; 1984). It has been suggested that the pK_a values determined from binding studies reflect the pK_a of this proton relay system (Hennecke and Plapp, 1983). This is supported by a study in which His 51 of human LADH was mutated to glutamine (Ehrig et al., 1991). The resulting mutant showed a six fold loss in activity which could be restored by the addition of buffers that could act as general bases. Thus, His 51, which is about 6Å from the alcohol hydroxyl, must be operating through a hydrogen bonded system. This suggests that changing any of the four components of the proton relay system *i.e.* the substrate/inhibitor/water, Ser 48, nicotinamide ribose and His 51 will result in a perturbation of the observed pK_a of LADH in binding studies of its ternary complexes.

Acknowledgements

We would like to thank the donors of the Petroleum Research Fund, administered by the American Chemical Society, for partial support of this research. We also thank the Muscular Dystrophy Association and the National Science Foundation for their partial support.

REFERENCES

Andersson, P., Kvassman, J., Olden, B. and Pettersson, G., 1981a, Synergism between coenzyme and carboxylate binding in liver alcohol dehydrogenase, *Eur. J. Biochem.* 118, 119-123.

Andersson, P., Kvassman, J., Lindstörm, A., Olden, B. and Pettersson, G., 1981b, Effect of pH on pyrazole binding to liver alcohol dehydrogenase, *Eur. J. Biochem.* 114: 549.

Baici, A. Luisi, P.L., Olomucki, A., Doublet, M.O., and Klincak, J., 1974, Influence of ligands on the coenzyme dissociation constants in octopine dehydrogenase, *Eur. J. Biochem.* 46: 59.

Barrio, J.R., Secrist, J.A., and Leonard, N.J., 1972, A fluorescent analog of nicotinamide adenine dinucleotide, *Proc. Nat. Acad. Sci., U.S.A.* 69: 2039.

Boyer, P.D., and Theorell, H., 1956, The change in reduced diphosphopyridine dinucleotide (DPNH) fluorescence upon combination with liver alcohol dehydrogenase, *Acta Chem. Scand.* 10: 447.

Dunn, M.F., and Hutchison, J.S., 1973, Roles of zinc ion and reduced coenzymes in the formation of a transient chemical intermediate during the equine liver alcohol dehydrogenase catalyzed reduction of an aromatic aldehyde, *Biochemistry* 12: 4882.

Ehrig, T., Hurley, T.D., Edenberg, H.J., and Bosron W.F., 1991, General base catalysis in a glutamine for histidine mutant at postion 51 of human liver alcohol dehydrogenase, *Biochemistry* 30: 1062.

Eklund, H., Samama, J.-P., and Jones, T.A., 1984, Crystallographic investigations of nicotinamide adenine dinucleotide binding to liver alcohol dehydrogenase, *Biochemistry* 23: 5982.

Eklund, H., Plapp, B.V., Samama, J.-P., and Brändén, C.-I., 1982, Binding of substrate in a ternary complex of horse liver alcohol dehydrogenase, *J. Biol. Chem.* 257: 14349.

Evans, S.A. and Shore, J.D., 1980, The role of zinc bound water in liver alcohol dehydrogenase catalysis, *J. Biol. Chem.* 255: 1509.

Greenfield, J.C., Leonard, N.J., and Gumport, R.I., 1975, Nicotinamide 3, N^4-ethenocytosine dinucleotide, an analog of nicotinamide adenine dinucleotide synthesis and enzyme studies, *Biochemistry* 14: 698.

Hennecke, M., and Plapp, B.V., 1983, Involvement of histidine residues in the activity of horse liver alcohol dehydrogenase, *Biochemistry* 22: 3721.

Klinman, J.P., 1981, Probes and mechanism and transition state structure in the alcohol dehydrogenase reaction, *CRC Crit. Rev. Biochem.* 10, 39.

Koerber, S.C., MacGibbon, A.K.H., Dietrich, H., Zeppezauer, M., and Dunn, M.F., 1983, Active-site cobalt (II)-substituted liver alcohol dehydrogenase: characterization of intermediates in the reduction of p-nitrobenzaldehyde by rapid scanning ultraviolet visible spectroscopy, *Biochemistry* 22: 3424.

Kvassman, J., and Pettersson, G., 1980, Unified mechanism for proton-transfer reactions affecting the catalytic activity of liver alcohol dehydrogenase, *Eur. J. Biochem.* 103: 565.

Kvassman, J., and Pettersson, G., 1979, Effect of pH on coenzyme binding to liver alcohol dehydrogenase, *Eur. J. Biochem.* 100: 115.

Lee, C. and Everse, J., 1973, Studies of the properties of 1, N^6-ethenoadenine derivatives of various coenzymes, *Arch. Biochem. Biophy.*, 157: 8.

Luisi, P.L., Olomucki, A., Baici, A., and Karlovic, D., 1973, Fluorescence properties of octopine dehydrogenase, *Biochemistry* 12: 4100.

Luisi, P.L., Baici, A., Bonner, F.J., and Aboderin, A.A., 1975, Relationship between fluorescence and conformation of εNAD^+ bound to dehydrogenases, *Biochemistry* 14: 362.

MacGibbon, A.K.H., Koerber, S.C., Pease, K., and Dunn, M.F., 1987, Characterization of a transient intermediate formed in the liver alcohol dehydrogenase catalyzed reduction of 3-hydroxy-4-nitrobenzyaldehyde, *Biochemistry* 26: 3058.

Mildvan, A.S., and Weiner, H., 1969a, Interaction of a spin labeled analogue of nicotinamide adenine dinucleotide with alcohol dehydrogenase, *J. Biol. Chem.* 244: 2465.

Mildvan, A.S., and Weiner, H., 1969b, Interaction of a spin labeled analogue of nicotinamide adenine dinucleotide with alcohol dehydrogenase. II. Proton relaxation rate and electron paramagnetic resonance studies of binary and ternary complexes, *Biochemistry* 8: 552.

Pettersson, G., 1986, Ionization properties of zinc bound ligands in liver alcohol dehydrogenase, *in* :Zinc Enzymes (Bertini, I., Luchinat C., Maret, W., and Zeppezauer, M., eds.), Birkhäuser Boston.

Pocker Y., and Page, J.D., 1990, Zinc-activated alcohols in ternary complexes of liver alcohol dehydrogenase, *J. Biol. Chem.* 265: 22101.

Pocker, Y., Raymond, K.W., and Thompson, W.H.,III., 1986, Liver alcohol dehydrogenase: kinetic and mechanistic studies *in* : Zinc Enzymes (Bertini, I., Luchinat, C., Maret, W., and Zeppezauer, M., eds.) Birkäuser, Boston, p 435.

Pocker, Y., 1989, Liver alcohol dehydrogenase: structure, catalysis, and site-directed mutagenesis *in* :Metal Ions in Biological Systems (Sigel, H., ed.) vol 25, Marcel Dekker, New York.

Rafter, G.W., and Colowick, S.P., 1957, Enzymatic properties of DPNH and TPNH, *in* : Methods in Enzymology 3: 887.

Sartorius, C., Gerber, M., Zeppezauer, M., and Dunn, M.F.,1987, Active-site cobalt (II)-substituted horse liver alcohol dehydrogenase: characterization of intermediates in the oxidation and reduction processes as a function of pH, *Biochemistry* 26: 871.

Secrist, J.A., Barrio, J.R., and Leonard, N.J., 1972a, Fluorescent modification of adenosine 3',5'-monophosphate: spectroscopic properties and activity in enzyme systems, *Science* 177: 279.

Secrist, J.A., Barrio, J.R., Leonard, N.J., and Weber, G., 1972b, Fluorescent modification of adenosine containing coenzymes. Biological activities and spectroscopic properties, *Biochemistry* 11: 3499.

Subramanian, S., Ross, J.B.A., Ross, P.D., and Brand, L., 1981, Investigation of the nature of enzyme-coenzyme interactions in binary and ternary complexes of liver alcohol dehydrogenase with coenzymes, coenzyme analogs, and substrate analogs by UV absorption and phosphorescence spectroscopy, *Biochemistry* 20: 4086.

Suhadolnik, R.J., Lennon, M.B., Uematsu, T., Monahan, J.E., and Baur, R., 1977, Role of adenine ring and adenine ribose of nicotinamide adenine dinucleotide in binding and catalysis with alcohol, lactate, and glyceraldehyde-3-phosphate dehydrogenase, *J. Biol. Chem.* 252: 4125.

Theorell, H., and Bonnichsen, R., 1951, Studies on liver alcohol dehydrogenase I. Equlibria and initial reaction velocities, *Acta Chem. Scand.* 5, 1105.

Theorell, H., and McKinely McKee, J.S., 1961, Liver alcohol dehydrogenase, *Acta Chem. Scand.* 15: 1811.

Theorell, H., and Yonetani, Y., 1963, Liver alcohol dehydogenase-DPN-pyrazole complex : a model of the ternary intermediate in the enzyme reaction, *Biochem. Z* 338: 537.

Theorell, H., Yonetani, T., and Sjöberg B., 1969, On the effects of some heterocyclic compounds on the enzymatic activity of liver alcohol dehydrogenase, *Acta Chem. Scand.* 23: 255.

EVOLUTIONARY RELATIONSHIPS OF BRANCHED CHAIN AND NON-SPECIFIC ALCOHOL AND ALDEHYDE DEHYDROGENASES

David W. Crabb, Natalia Y. Kedishvili, Kirill M. Popov, Paul Rougraff, Yu Zhao, and Robert A. Harris

Departments of Biochemistry and Molecular Biology and of Medicine
Indiana University School of Medicine
Indianapolis, IN. USA 46202-5121

INTRODUCTION

As more protein sequences have become available in recent years, it has become clear that they can be grouped in families and superfamilies. These groupings may represent conservation of structural motifs over time or convergent evolution; nonetheless, analysis of the sequences can point to important residues and regions involved in protein function. This information can direct site-specific modifications of the proteins to study function and can also help to identify the function of unknown proteins discovered by cloning techniques.

Our laboratory is interested in enzymes involved in branched chain amino acid degradation. Valine catabolism proceeds via reactions reminiscent of both fatty acid and alcohol oxidation (Davis, 1986). The hydroxy acid 3-hydroxyisobutyrate (HIBA) is generated in the mitochrondrion from methylacrylyl CoA and further oxidized by HIBA dehydrogenase (HIBDH, E.C.1.1.1.31) to methylmalonate semialdehyde (MMSA). MMSA is the substrate for MMSA dehydrogenase (MMSDH, E.C.1.2.1.27), a coenzyme A-dependent enzyme that decarboxylates the substrate and generates propionyl-CoA. We have cloned these enzymes' cDNAs from rat liver cDNA libraries. Analysis of the sequence of MMSDH showed that it is clearly related to the non-specific aldehyde dehydrogenases (ALDHs). We therefore also examined the relationship of HIBDH to the alcohol dehydrogenase (ADH) gene families and report structural similarities of the valine catabolic enzymes with short chain ADHs and with ALDH.

METHODS

The descriptions of the cloning of HIBDH (Rougraff et al., 1989) and of MMSDH (Kedishvili et al., 1992) have been published. Each cDNA was cloned by determining the N-terminal amino acid sequence of the purified protein, synthesizing degenerate oligonucleotide probes, and screening a rat liver cDNA library. In the case of MMSDH we used degenerate primers to obtain a PCR product corresponding to the 5' end of the cDNA, which was then used in screening the library. The cDNA inserts were sequenced by conventional methods. The sequences were compared with GENBANK translated protein sequences using the BLAST network service within the National Center for Biotechnology Information at the National Library of Medicine. Groups of proteins were aligned and phylogenetic trees were generated using Doolittle's programs (SCORE, PREALIGN, TREE, and BLEN)(Feng and Doolittle, 1990) running on a Silicon Graphics computer. The alignment of HIBDH with *Drosophila* ADH was used as the starting point for manual alignment with two other members of the short chain ADH family as reported by Jörnvall (Persson et al., 1991).

RESULTS

Relationship of HIBA Dehydrogenase to the Alcohol Dehydrogenase Gene Family

HIBDH is a homodimeric enzyme that requires NAD^+ for activity. It is inactive with other alcohols or hydroxy acids (e.g., various short chain and branched chain alcohols, lactate, or malate). Previous studies demonstrated that it is sensitive to inhibition by sulfhydryl reagents such as iodoacetate or mercurials, but is insensitive to metal ion chelators (EDTA, EGTA, α,α' dipyridyl)(Rougraff et al., 1988). The kinetic mechanism is ordered BiBi. The mature enzyme subunit is 300 residues in length, shorter than the long chain zinc-dependent ADHs or the microbial iron-activated ADHs and somewhat larger than the insect and bacterial short chain, non-zinc dependent ADHs (although several members of this family are in the range of 280-320 residues (Persson et al., 1991)). This length is also similar to that of malate and lactate dehydrogenases. HIBDH was aligned with representatives of the long chain enzyme (pi, chi, and gamma ADH, and sorbitol dehydrogenase), short chain enzymes (*Drosophila melanogaster* ADH, *Klebsiella* ribitol dehydrogenase and *Bacillus* glucose dehydrogenase) and the microbial iron-activated ADHs (yeast ADH4, *Zymomonas* ADH, *E. coli* propanediol dehydrogenase (fucose dehydrogenase), and the $NADP^+$-dependent ADH from *Clostridium butyricum*). The characteristics of the three classes of alcohol/polyol dehydrogenases are summarized in Table 1. We also examined relationships with lactate and malate dehydrogenases

Table 1. Classes of Alcohol/Polyol Dehydrogenases

Long chain zinc enzymes		Iron activated enzymes	Short chain enzymes
Dimeric	Tetrameric	Yeast ADH4 *Clostridium* ADH	*Drosophila* ADH
Animal ADH	Yeast ADH Sorbitol DH	*Zymomonas* ADH2 *E. coli* fucose DH	*Bacillus* glucose DH *Klebsiella* ribitol DH numerous hydroxysteroid dehydrogenases
Approximate subunit size:			
370 aa		380 aa	250 aa

and with 3-hydroxyacyl CoA dehydrogenase, which catalyzes a very similar reaction.

Overall, pairwise comparisons did not show very strong similarities between HIBDH and any of the alcohol dehydrogenases (Table 2). Pairwise identities between HIBDH and lactate dehydrogenase (LDH) isozymes or malate dehydrogenase (MDH) isozymes were 12-14% and 15-17%, respectively. Progressive alignments showed only 9-12% identity between HIBDH and LDH or MDH when these five proteins plus *Drosophila* ADH were aligned, suggesting that HIBDH was more closely related to the ADH family than to LDH or MDH. Progressive alignments between HIBDH and the alcohol dehydrogenases revealed at most 10-15% identities. The residues that coordinate zinc in the long chain ADHs (cys46, cys174, and his67; and the four cysteines in the region of residues 97-111) are not conserved, which is not unexpected given that HIBDH does not require metals for activity. The phylogenetic tree that is generated by these programs suggested that HIBDH may be related to either the short chain or iron-activated ADHs, more so than to the long chain ADHs (Figure 1).

Since HIBDH does not appear to be iron-dependent (Rougraff et al., 1988), we examined the relationship of HIBDH with the short chain alcohol dehydrogenases in more detail. Alignment of these four enzymes by the TREE program showed positional identities between HIBDH and at least one of the other enzymes of 24% and positional homology (identities plus conservative substitutions) of 39%. The GxGxxG motifs that probably are part of the NAD^+ binding domains are aligned in the N-termini of these enzymes. In addition, all four enzymes have a gap in the region of the structural zinc binding domain of the long chain ADHs (containing

Table 2. Sequence comparisons of HIBDH with other dehydrogenases

	Alignment with HIBA Dehydrogenase				
	Pairwise	Progressive		Pairwise	Progressive
Chi ADH	17	12	Drosophila ADH	21	13
Gamma ADH	19	11	Glucose DH	14	8
Pi ADH	20	10	Ribitol DH	15	8
Sorbitol DH	16	9	Hydroxyacyl CoA DH	16	8
Zymomonas ADH2	21	15			
Yeast ADH4	17	13			
E coli Fucose DH	21	14			
Clostridium ADH	21	10			
Lactate dehydrogenase (porcine)					
M isozyme	14	9			
H isozyme	12	10			
Malate dehydrogenase (porcine)					
Cytosolic	17	10			
Mitochondrial	15	12			

Pairwise comparisons were performed with the SCORE program. The progressive alignments were performed using the TREE program for the ADHs, polyol dehydrogenases and hydroxyacyl CoA dehydrogenases with HIBDH) and for the lactate and malate dehydrogenases with *Drosophila* ADH and HIBDH. The abbreviation DH stands for dehydrogenase.

cysteines 97, 100, 103, and 111, not shown) when short and long chain enzymes are aligned.

Jörnvall recently described an extended family of short chain alcohol dehydrogenases that also includes microbial and vertebrate hydroxysteroid dehydrogenases, dihydrodiol dehydrogenases, acetoacetyl-CoA dehydrogenases, and hydroxyprostaglandin dehydrogenase, as well as the short chain ADHs (Persson et al., 1991). Over 20 examples are now known. After aligning HIBDH with *Drosophila* ADH, these two sequences were aligned with ribitol dehydrogenase and glucose dehydrogenase (Figure 2). Five of the six highly conserved residues characteristic of this family are present in HIBDH, namely the glycines at 14, 19, and 132, the aspartate at 64 and the tyrosine at 152 (these residues are numbered according to the sequence of *Drosophila* ADH). The lysine that is found at position 156 in all the other members of this family is a valine in HIBDH.

These comparisions suggest that HIBDH may be related to the short chain, non-zinc dependent ADH family. However, some salient differences remain. *Drosophila* ADH is insensitive to thiol reagents, and some of the short chain dehydrogenases completely lack cysteine residues; on the other hand, HIBDH is sensitive to thiol reagents and has a reactive cysteine in the active site (Rougraff and Harris, unpublished results). 3-hydroxyacyl CoA dehydrogenase may also be more distantly related to these enzymes. Although not shown, this enzyme also has characteristically spaced glycines in the N-terminus. It lacks the characteristic tyrosine 152 but has a lysine at position 156; overall its number of identities with *Drosophila* ADH is lower than the identities between the insect ADH and HIBDH. An additional enzyme, the peroxisomal acyl hydratase/hydroxyacyl dehydrogenase that catalyzes two reactions in the peroxisomal fatty acid oxidation pathway, shares

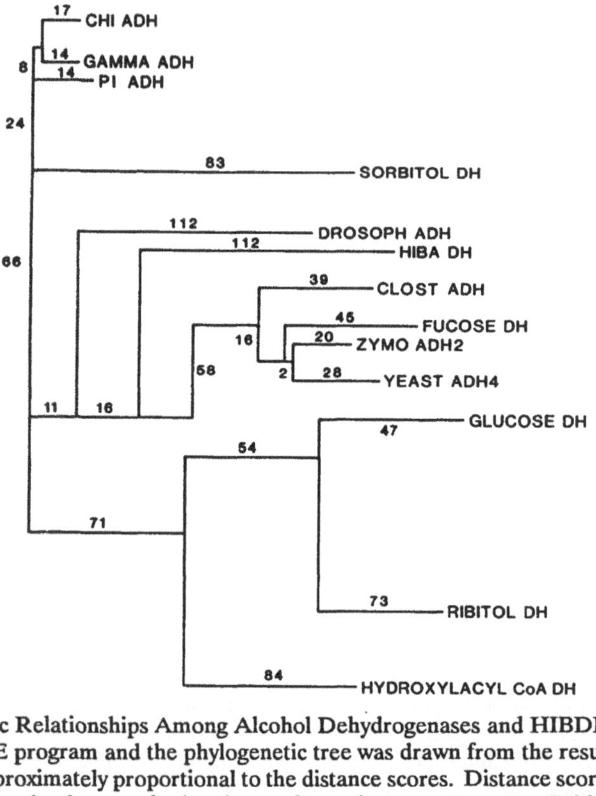

Figure 1. Phylogenetic Relationships Among Alcohol Dehydrogenases and HIBDH. The ADHs were aligned with the TREE program and the phylogenetic tree was drawn from the results of the alignment. Branch lengths are approximately proportional to the distance scores. Distance scores are dimensionless numbers which indicate the degree of relatedness of protein sequences. See Table 1 for full names of enzymes.

```
gldh                    M Y K D L E G K V V V I T G S S T G L G K S M A
ridh      M K H S V S S M N T S L S G K V A A I T G A A S G I G L E C A
hiba                          A S K T P V G F I - G - L G N M G N P M A
dadh                    S F T L T N K N V I F V A G - L G G I G L D T S      23

gldh I R F A T E K A K V V - V N Y R S K E D E A N S V L E E E I K K V
ridh R T L L G A G A K V V - L I D R E G E K L N K L V A E L G E N A F
hiba K N L I K H G Y P L I L Y D V F - - - P D V C K E F K E A G E Q V
dadh K E L L K R D L K N L V I L D R I E N P A A I A E L K A I N P K V      56

gldh G G E A I A V K G D V T V E S D V I N L V Q S A - - - - - - I K -
ridh - - - - - A L Q V D L M Q A D Q V D N L L Q G I - - - - - - L Q -
hiba - - - A S S - P A D V A E K A D R I I T M L P S S M N S I E V Y S
dadh - - - T V T F P Y D V T V P I A E T T K L L K T - - - - - - I F A      81

gldh E F G K L - - D V M I N N A G M E N P V S S H E M S L S D W N K V
ridh L T G R L - - D I F H A N A G A Y I G G P V A E G D P D V W D R V
hiba G A N G I L K K V K K - G S L L I D S S T I D P S - - V S K E L A
dadh Q L K T V - - D V L I N G A G I L D D H Q I E R T - - I A V N Y T     110

gldh I D T N L T G A F L G S R E A I K Y F V E - - - - - N D I K G T V
ridh L H L N I N A A F R C V R S V L P H L I A - - - - - Q K - S G D I
hiba K E V E K M G A - - - V F - M D A P V S G G V G A A R S - - G N L
dadh G L V N T T T A - - - I L D F W D K R K G G - - - - - P - - G G I     133

gldh I N M S S V - H E W K I P - - - - - - - - - W P L F V H Y A A S K
ridh I F T A V I - A G V V P V - - - - - - - - - I W E - P V Y T A S K
hiba T F M V G G V E N E F A A A Q E L L G C M G S N V L - - Y C G A V
dadh I C N I G S V - T G F N A I - - - - - - - - Y Q V - P V Y S G T K     156

gldh G G M K L M T E T L A L E Y A P K G I - R V N N I G P G A - I N T
ridh F A V Q A F V H T T R R Q V A Q Y G V - R V G A V L P G P - V V T
hiba G S G Q S A K I C N N M L L A I S M I G T A E A M N L G I R S G L
dadh A A V V N F T S S L A K L A P I T G V - T A Y T V N P G I - T R T     187

gldh P I N A E K - - - - - - - - F A D P E E R - - - A D V E S M I P M
ridh A L L D D W - - - - - - - - P K A K M D E - - - - - - - - - A L A
hiba D P K L L A K I L N M S S G R C W S S D T Y N P V P G V M D G V P
dadh T L V H K F - - - - - - - - N S W L D V E P Q V A E K L L A H P T     211

gldh G Y - I G E P E E - - - - I A - A V A W L A S S E A S - Y V T G I
ridh N G S L M Q P I E - - - - V A E S V L F M V T R S K N - - V T V R
hiba S S N N Y Q G G F G T T L M A K D L G L A Q D S A T S T K T P I L
dadh Q P S L A C A E N - - - - - F V K A I E L N Q N G A I W - K L D L     239

gldh T L F A D G - - - - - - - - - - - - G M T Q Y P S F E A G R G
ridh D I V I L P - - - - - - - - - - - - N S V D L
hiba L - G S V A H Q I Y R M M C S K G Y S K K D F S S V F Q Y L R E E E T F
dadh G T L E A I Q - - - - - - - - - - - W T K H W D S G I                  255
```

Figure 2. Alignment of HIBDH with short chain ADHs. The alignment was performed using the TREE program to align *Drosphila* ADH with HIBDH, and then fitting these sequences to the alignments published by Jörnvall (Persson et al., 1991). Double underlines indicate identity between all four proteins; underlines indicate the location of three matches among the four proteins. The numbers are for the sequence of the *Drosophila* ADH (dadh); gldh is glucose dehydrogenase, ridh is ribitol dehydrogenase, and hiba is HIBDH.

structural similarities with the mitochondrial 3-hydroxyacyl CoA dehydrogenase, and thus may also be related to the short chain ADHs. It is also possible that these enzymes are related merely by their common usage of pyridine nucleotide coenzymes.

Relationship of Methylmalonate Semialdehyde Dehydrogenase to the Aldehyde Dehydrogenase Gene Superfamily

MMSDH is a tetrameric enzyme found in mitochondria. It is NAD^+- and CoA-dependent. Its substrates are limited to methylmalonate and malonate semialdehydes (Goodwin et al., 1989). Like several other ALDHs, this enyzme exhibits esterase activity, but produces CoA esters rather than a carboxylic acid as

Table 3. Sequence Identities Between MMSDH and Non-Specific ALDHs

	Alignment with Methylmalonate Semialdehyde Dehydrogenase	
	Pairwise	Progressive
Cytosolic ALDH	32	31
Mitochondrial ALDH	31	30
ALDH-x	31	32
Tumor ALDH	23	22
Microsomal ALDH	25	21

The sequences are those of the rat enzymes (except for human ALDH-x) and alignments were performed with the SCORE and TREE programs.

product (Popov et al., 1992). It is sensitive to inhibition by sulfhydryl reagents. The subunit molecular weight is 58 kD and the cloned cDNA shows a subunit size of 503 residues. The relationship between MMSDH and known non-specific ALDHs was stronger than that between HIBDH and ADHs. Pairwise alignments showed identities of 23-32% between MMSDH and class 1, 2, and 3 ALDHs, ALDH-x, and the microsomal ALDH (Table 3), and percent identities in the progressive alignments were similar to the pairwise identities. All six enzymes could be aligned with overall 10% identity of residues in all proteins. MMSDH is clearly part of the ALDH superfamily.

We then sought additional aldehyde and semi-aldehyde dehydrogenases in the database for comparison. The database search revealed similarity with *E. coli* betaine ALDH and with a pea protein called hypothetical protein 3. Betaine ALDH catalyzes the conversion of betaine aldehyde to betaine and is important in the response of *E. coli* to osmotic stress (Falkenberg and Strom, 1990). The hypothetical protein was cloned by subtraction methods from wilted (water-stressed) pea plants and had previously been recognized as being similar to ALDHs (Guerrero et al., 1990). A number of semialdehyde dehydrogenases were aligned with MMSDH. We realized that many of the enzymes named semialdehyde dehydrogenases in fact function in the synthesis of amino acids via NADPH-dependent reduction of

Table 4. Identities Among Synthetic Semialdehyde Dehydrogenases (SADHs)

	PROA	DHAY	DHAS	LYS2
E. coli Glutamate SADH (PROA)	-			
Yeast Aspartate SADH (DHAY)	17	-		
E. coli Aspartate SADH (DHAS)	13	20	-	
Yeast Aminoadipate SADH (LYS2)	17	17	15	
Rat MMSDH	21	18	20	18

Table 5. Identities Among Catabolic Semialdehyde Dehydrogenases (SADHs)

	HYDM	SSDH	PCDH
Pseudomonas Hydroxymuconic SADH (HYDM)	-		
E. coli Succinic SADH (SSDH)	38	-	
Yeast Pyrroline 5-carboxylate DH (PCDH)	26	24	-
Rat Methylmalonate SADH (MMSD)	31	29	25

semialdehydes. This group is shown in Table 4; we have called them synthetic semialdehyde dehydrogenases or reductases. These enzymes are not very related to one another nor to MMSDH. On the other hand, three other "catabolic" semialdehyde dehydrogenases, succinic and hydroxymuconic semialdehyde dehydrogenases and pyrroline 5-carboxylate dehydrogenase, are clearly related to one another and to MMSDH (Table 5). Phylogenetic relationships among the non-specific ALDHs and the semialdehyde dehydrogenases are shown in Figure 3. Pyrroline 5-carboxylate dehydrogenase is not included in this tree because when it was present, negative branch lengths in its vicinity were encountered that could not be eliminated by modifying the tree morphology. *E. coli* succinate semialdehyde dehydrogenase branches from the tree immediately below the cluster of HYDM and DHAB (not shown).

The overall similarities between the ALDHs are much more striking than those among the ADHs (Table 6). Seven glycines are conserved among the ALDHs, as well as the ALDH active site motif ([SAG]xFxxxGQxCx[AGN]. We also noticed a highly conserved motif toward the C-terminus of the proteins (EEIFGP). This appears to be highly characteristic of ALDHs; searching the GENBANK database with this sequence identified a number of prokaryotic ALDHs that were not included in the initial analyses. Of note, the residue aligning with glutamate 268 in ALDH1 is an asparagine in MMSDH; this residue was heretofore considered essential for ALDH function, possibly by acting as a base to deprotonate the active site cysteine.

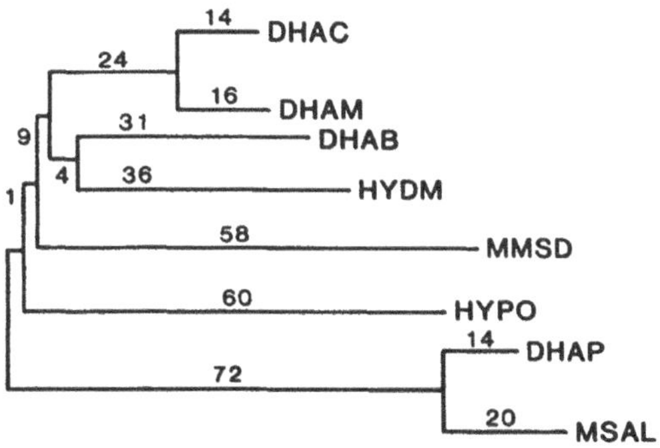

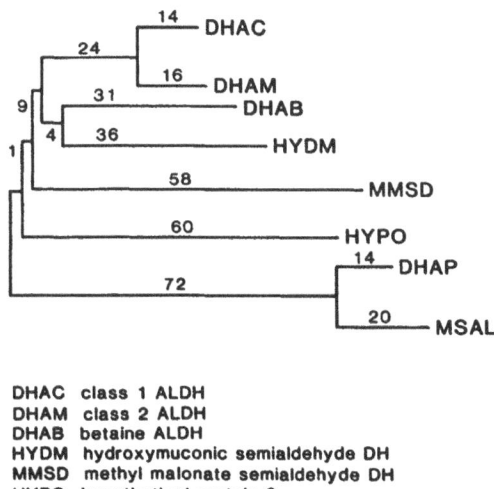

DHAC class 1 ALDH
DHAM class 2 ALDH
DHAB betaine ALDH
HYDM hydroxymuconic semialdehyde DH
MMSD methyl malonate semialdehyde DH
HYPO hypothetical protein 3
DHAP class 3 ALDH
MSAL microsomal ALDH

Figure 3. Phylogenetic Relationships Among ALDHs. The analysis was performed with the TREE program. The lengths of the branches are approximately proportional to the distance score.

Table 6. Summary of Structural Features of MMSDH

Conservation of glycines 160, 186, 245, 250, 449, 467, 478 among classes I-III and microsomal ALDH and MMSDH
Conservation of active site cysteine and ALDH active site motif
Glutamate 268 ---> asparagine
Highly conserved motif in C-terminus: EEIFGP

CONCLUSIONS

ADH and ALDH gene families include enzymes of widely divergent substrate specificities. The sequence comparisons shown in this study suggest that MMSDH is part of the ALDH family and that NAD^+-utilizing semialdehyde dehydrogenases that function predominantly in the direction of oxidation of semialdehydes are related to the non-specific ALDHs. HIBDH is less closely related to the ADH family, but is more similar to these enzymes than to other enzymes that catalyze the dehydrogenation of α-hydroxy-acids, such as lactate and malate dehydrogenases. Among the ADHs, HIBDH shows similarities in enzymatic activity and structure to the short chain, "insect type" ADHs. This relationship can now be tested by specifically mutating conserved residues and examining changes in enzymatic specificity and function.

ACKNOWLEDGEMENTS

This work was carried out with the support of grants from the National Institutes of Health (DK 40441 to RAH), the National Institute on Alcoholism and Alcohol Abuse (AA K02 00081 to DWC), the American Heart Association, Indiana Affiliate (postdoctoral fellowships to NYK and KMP), and the March of Dimes (predoctoral fellowship to YZ).

REFERENCES

Davis EJ, 1986. Pathways of branched chain amino acid metabolism in mammals. In: Odessey R (ed.), Probelms and Potential of Branched Chain Amino Acids in Physiology and Medicine. Amsterdam: Elsevier, pp. 3-29

Falkenberg P. and Strom A.R., 1990, Purification and characterization of osmoregulatory betaine aldehyde dehydrogenase of Escherichia coli, *Biochim.Biophys.Acta* 1034:253-259.

Feng D.-F. and Doolittle R.F., 1990, Progressive alignment and phylogenetic tree construction of protein sequences, *Methods.Enzymol.* 183:375-387.

Goodwin G.W., Rougraff P.M., Davis E.J., and Harris R.A., 1989, Purification and characterization of methylmalonate-semialdehyde dehydrogenase from rat liver, *J.Biol.Chem.* 264:14965-14971.

Guerrero F.D., Jones J.T., and Mullet J.E., 1990, Turgor-responseive gene transcription and RNA levels increase rapidly when pea shoots are wilted. Sequence and expression of three inducible genes, *Plant Mol.Biol.* 15:11-26.

Kedishvili N.Y., Popov K., Rougraff P.M., Zhao Y., Crabb D.W., and Harris R.A., 1992, CoA-dependent methylmalonate semialdehyde dehydrogenase- unique member of the aldehyde dehydrogenase superfamily, *J.Biol.Chem.*, in press.

Persson B., Krook M., and Jörnvall H., 1991, Characteristics of short-chain alcohol dehydrogenases and related enzymes, *Eur.J.Biochem.* 200:537-543.

Popov K.M., Kedishvili N.Y., and Harris R.A., 1992, Coenzyme A- and NADH-dependent esterase activity of methylmalonate semialdehyde dehydrogenase, *Biochim.Biophys.Acta* 1119:69-73.

Rougraff P.M., Paxton R., Kuntz M.J., Crabb D.W., and Harris R.A., 1988, Purification and characterization of 3-hydroxyisobutyrate dehydrogenase from rabbit liver, *J.Biol.Chem.* 263:327-331.

Rougraff P.M., Zhang B., Kuntz M.J., Harris R.A., and Crabb D.W., 1989, Cloning and sequence analysis of a cDNA for 3-hydroxyisobutyrate dehydrogenase, *J.Biol.Chem.* 264:5899-5903.

ENZYME AND ISOZYME DEVELOPMENTS WITHIN THE MEDIUM-CHAIN ALCOHOL DEHYDROGENASE FAMILY

Hans Jörnvall[1], Olle Danielsson[1], Hans Eklund[2], Lars Hjelmqvist[1], Jan-Olov Höög[1], Xavier Parés[3] and Jawed Shafqat[1]

[1]Department of Chemistry I, Karolinska Institutet, S-104 01 Stockholm, Sweden, [2]Department of Molecular Biology, Swedish University of Agricultural Sciences, S-751 24 Uppsala, Sweden, and [3]Department of Biochemistry, Autonomous University of Barcelona, E-08193 Bellaterra (Barcelona), Spain

INTRODUCTION

Alcohol dehydrogenases and related enzymes constitute a complex system of proteins derived from gene duplications at minimally four different levels. The system includes proteins of different type regarding family relationships and overall organization. It also includes different enzymes within each family, as well as different classes of the enzymes, and different isozymes within the classes, apart from allelic variants. We have studied these relationships, starting with the horse liver alcohol dehydrogenase (Jörnvall, 1970, Eklund *et al.*, 1976), distinguishing the parallel evolution of separate enzyme types (Jörnvall *et al.*, 1981) and successively characterizing both the "medium-chain" (Jörnvall *et al.*, 1987) and "short-chain" (Persson *et al.*, 1991) alcohol dehydrogenase families. Recently, we have characterized several novel forms, including both mammalian enzymes and those from other vertebrate lines, as well as from further, distantly related sources. Together, this has made it possible to deduce relationships of the functional and structural organization of the enzyme system, tracing gene duplications, original forms, and functional properties.

Recent advances include the characterization of the amphibian enzyme, which allowed us to evaluate one of the class origins of the mammalian alcohol dehydrogenases (Cederlund *et al.*, 1991). Other recent steps were the characterization of the fish enzyme, revealing a protein with partly hybrid properties (Danielsson *et al.*, 1992), and distinction of novel isozymes in a reptilian line, further illustrating origins of multiplicities (Hjelmqvist *et al.*, 1992). Together with data from similar advances in the knowlegde of the structures of the human classes of the enzyme (Parés *et al.*, 1992, Eklund *et al.*, 1990, Hurley *et al.*, 1991; Yasunami *et al.*, 1991), the "classical", or medium-chain, alcohol dehydrogenases now constitute a well-established protein family with characterized relationships of general interest. Here, we will concentrate on three major aspects, the organizational pattern, the origins, and the unique properties, and

Enzymology and Molecular Biology of Carbonyl Metabolism 4
Edited by H. Weiner, Plenum Press, New York, 1993

correlate them with other data to integrate the results into functional interpretations of the enzyme system.

PATTERN

The organization of the medium-chain alcohol dehydrogenase protein family, as far as presently recognized, is summarized in Fig. 1. The family is organized into several levels, and they are fairly distinctly separated regarding both structural and functional properties.

Level 1: Separate lines within the super-family

The most distant relationships are those that concern the separate lines (top level, Fig. 1). Two lines are fairly well characterized and known in at least one three-dimensional protein structure each and many primary structures. They are the short-chain dehydrogenase family, and the medium-chain alcohol dehydrogenase family. The latter was initially called the long-chain family (Jörnvall *et al.*, 1987), before other, still longer-chain families were recognized.

The short-chain dehydrogenase family includes a large group of different enzymes, among them several sugar, steroid, prostaglandin, and alcohol dehydrogenases. This multi-member family of fairly distantly related enzymes has been recently reviewed (Persson *et al.*, 1991), and the first crystallographic structure of one such enzyme has been reported (Ghosh *et al.*, 1991). The reaction mechanism and active site relationships are different from those of the other alcohol dehydrogenase protein families, and a conserved (Krook *et al.*, 1990), reactive (Krook *et al.*, 1992) tyrosine residue has been functionally highlighted, and confirmed as important by site directed mutagenesis (Ensor and Tai, 1991; Albalat *et al.*, 1992). The domain organization is different from that for the medium-chain, classical, liver alcohol dehydrogenases. The short-chain dehydrogenases have the characteristic β-stranded mononucleotide binding fold (Rossmann *et al.*, 1974) in the N-terminal part (Thatcher and Sawyer, 1980; Persson *et al.*, 1991; Ghosh *et al.*, 1991), rather than in the C-terminal half as the molecules of the medium-chain line. Thus, the molecular architecture is unique to each of the protein families derived from the ancestors outlined at the top level in Fig. 1. Structural similarities between the families involve only part of the molecules and presumably reflect molecular building units that were rearranged, constituting founders for each of the separate lines. The segments of maximum similarity involve the mononucleotide binding units, which are wide-spread among many enzymes, as known since long (Rossmann *et al.*, 1975), and correspond to positional identities at about the 20% level. However, since these segments only involve parts of the entire subunits, the overall residue identity is well below 20% between the separate super-families. Among alcohol dehydrogenases proper, the short-chain line is thus far known in the form of the insect enzyme. For the Drosophila enzyme, several species have been characterized and crystallographic work is in progress (cf. Villarroya *et al.*, 1989). Apart from the short-chain family, not further treated here, and the medium-chain line discussed below, additional alcohol dehydrogenase protein families also exist and include a long-chain family (Inoue *et al.*, 1989), a family of iron-activated enzymes (Scopes, 1983; Williamson and Paquin, 1987), and other lines. Collectively, they are referred to as "other lines" in Fig. 1 but represent further subdivisions and have yet not been studied in great detail.

The medium-chain alcohol dehydrogenase line is the one that includes the classical alcohol dehydrogenases, such as the liver and yeast enzymes, but also the plant enzymes and several bacterial enzymes. In common, these enzymes, typically contain a catalytically active zinc atom at the active site. However, the presence of zinc is not an absolute characteristic of the family, and one of the protein members, ζ-crystallin, exhibits a typical overall protein homology but lacks the zinc ligands and, as far as is known, the zinc atom (Borrás *et al.*, 1989). Even when zinc is present, one of the ligands to the catalytic zinc atom appears to be variable, being definitely Cys in alcohol dehydrogenase, but concluded to be Glu in sorbitol dehydrogenase, and Asp in threonine dehydrogenase (Eklund *et al.*, 1976, 1985; Höög et al., 1992a; Aronson *et al.*, 1989). Furthermore, the alcohol dehydrogenases proper exhibit also a second zinc atom per subunit ("structural" zinc), not present in sorbitol dehydrogenase (Jeffery *et al.*, 1984) or the other family members. Hence, the medium-chain alcohol dehydrogenase family, apart from constituting a fraction of the complex enzyme system, also itself exhibits complex relationships with variations in metalloenzyme properties and active site structures.

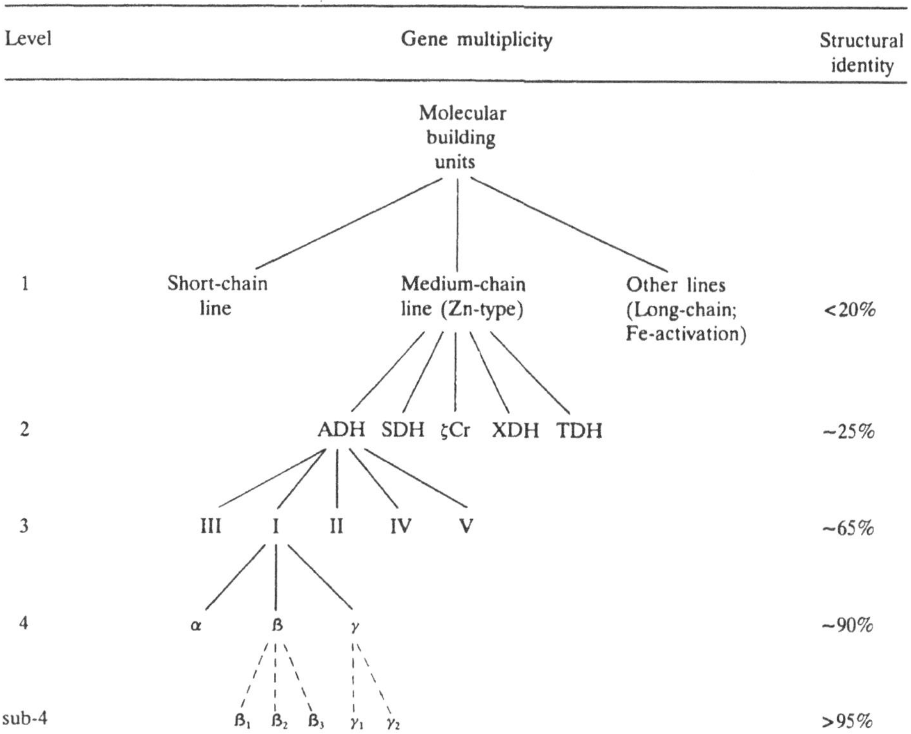

Figure 1. Schematic representation of alcohol dehydrogenase relationships, emphasizing the medium-chain alcohol dehydrogenase family.

Four different levels of relationship are clearly distinguished, the top one giving rise to large families of different organizational patterns, later levels to further divergence within each line, short-chain dehydrogenases, long-chain enzymes and some other lines not further detailed here. ADH, alcohol dehydrogenase; SDH, sorbitol dehydrogenase; ζCr, zeta-crystallin; XDH, xylitol dehydrogenase; TDH, threonine dehydrogenase. I-V, ADH classes; α, β, γ, ADH subunits.

Level 2: Different enzymes

The second level of gene duplications represents the one that gave rise to the different enzymes. Those of the medium-chain line now treated are shown in Fig. 1. Known variants of these enzymes exhibit similar but low overall identity, in the range of 25%, independent of which pairs that are compared. This similarity in values suggests that the corresponding enzyme origins are derived from several similarly-leveled duplications, as shown in Fig. 1. The low absolute values further show that this level is distant, probably beyond the eukaryotic/prokaryotic separation. In line with this, the different enzymes (Fig. 1) are now found in both prokaryots (threonine dehydrogenase) and eukaryots (sorbitol dehydrogenase, ζ-crystallin, xylitol dehydrogenase) or both (alcohol dehydrogenase). Although distantly related and with no good substrates in common, hence constituting truly different enzymes, the separate proteins at this level are clearly related conformationally. Thus, model building is possible (Eklund *et al.*, 1985), and relationships regarding zinc-binding properties, subunit interactions and active site structures can be deduced from comparisons (Eklund *et al.*, 1985; Borrás *et al.*, 1989; Aronson *et al.*, 1989; Kötter *et al.*, 1990). The ζ-crystallin is a particularly intriguing protein. Initially discovered as a lens crystallin (Huang *et al.*, 1987), later deduced to exhibit distant similarities with alcohol dehydrogenases (Borrás *et al.*, 1989), it has recently been shown also to have NADPH:quinone oxidoreductase enzymatic activity (Rao *et al.*, 1992), in spite of the differences, and in spite of probably lacking the catalytic zinc atom. It has further been shown to be derived from a mammalian gene, expressed also in the liver of several species (cf Huang *et al.*, 1990), thus linking it to other large groups of crystallins that derive from enzyme genes recruited (Wistow and Piatigorsky, 1987) from other functions.

Interestingly, the separate enzymes at level 2 exhibit differences in quaternary structure, as well as in occurrence of internal deletions/insertions. Thus, sorbitol dehydrogenase lacks a 20-residue segment corresponding to a separate surface loop (Eklund *et al.*, 1985), ζ-crystallin a 50-residue segment including the same loop, and threonine and xylitol dehydrogenases also a part corresponding to that segment (Aronson *et al.*, 1989; unpublished together with Penttilä and Persson). Not only do these lacking loops correlate with surface elements, but the borders are at largely identical positions in the different cases, and the proteins that lack the extra loop characteristically occur as tetramers. It therefore appears possible that the absence of the loop exposes a surface contributing to dimer-dimer interactions, as also experimentally demonstrated by comparisons of susceptibilities to proteases (Roumi *et al.*, 1992).

In summary, the different enzymes within the family originated at an early stage and exhibit characteristic properties, distinguishable in quaternary structure, substrate specificity, internal deletions/insertions, and presence/number of zinc atoms per subunit.

Level 3: Different classes

This level of complexity derives from a later stage of gene duplications. Classes, I, II, and III were first discerned (Vallee and Bazzone, 1983), while class IV was recently detected as separate, represented by the stomach-expressed form of alcohol dehydrogenase (Holmes, 1988; Yin *et al.*, 1990; Parés *et al.* 1990; Moreno and Parés, 1991), and this also defined a DNA-derived structure (Yasunami *et al.*, 1991) as class V (Parés *et al.*, 1992). By studying the variability between the enzymes from different vertebrate lines, it is possible to evaluate the origins of these duplications. In this manner, the class III/I separation has been traced and was preliminarily given a distant origin at about the time of vertebrate (and liver) evolution (Cederlund *et al.*, 1991), which is know considered a minimal value and should probably be even more distant (below). Recently, the nature of the present fish class I- and III-like enzymes has also traced the distant evolutionary connection between these enzyme forms (Danielsson and Jörnvall, 1992).

The classes represent forms clearly intermediate between the separate enzymes at the level above and the isozymes at the level below (Fig. 1). They are dissimilar enough not to form mixed dimers and therefore not to cause cross-class subunit interactions. Nevertheless, and in contrast to the present-day products of the second level of gene duplications, the different forms at the class level have several substrates in common. Consequently, class I-V were initially defined and named from their ethanol activity in common, although that activity is indeed often highly different. The distinction of classes is therefore relevant, giving a link between anciently derived enzymes with now separate substrates, and recently derived isozymes with substrates even more closely in common. Hence, the class distinction is a typical characteristic, and resembles the situation with other proteins where "classes" have been coined, like for example for major histocompatibility antigens, aldehyde dehydrogenases, and glutathione transferases.

Level 4: Different isozymes

The fourth level of gene duplication in the evolution towards present-day medium-chain alcohol dehydrogenases constitutes the level which gave rise to isozyme formations. This level is fairly recent, as for isozymes of many other proteins, and has thus far been characterized essentially for the mammalian class I enzymes. The extent of positional identity between the isozyme subunits is high, around 90%, and the isozyme pattern is frequently taxon-specific because of separate duplicatory origins after the ancestral separations of corresponding lines. Human alcohol dehydrogenase of class I is represented by three genes, each synthesizing one protein subunit, while at least thus far only two subunit forms have been characterized from horse, and one from several rodents and other lines.

In addition to these four levels, further multiplicity derives from allelic variability. At least three alleles have been characterized for the human class I β form and two for the class I γ form (cf Jörnvall *et al.*, 1989). Further complexity, with both additional levels and additional multiplicity at each level, is not excluded. In particular, additions at each of the four levels have been discovered during the last few years. Therefore, yet further forms will probably soon be encountered. In agreement with this expectation, recent analysis of reptilian and piscine alcohol dehydrogenases have established that further duplications did occur (Hjelmqvist *et al.*, 1992; Danielsson and Jörnvall, 1992), and the cod enzymes establish a gene duplication also in the class III line.

In summary, the alcohol dehydrogenase system represents a number of present-day different proteins, related through a series of duplications and subsequent divergence. The alcohol dehydrogenase super-family soon represents one of the better characterized enzyme types among proteins in general. Species variations have been extensively defined, and just in Fig. 1, no less than ten different human gene products exhibit some ethanol dehydrogenase activity, representing alcohol dehydrogenases, still disregarding the short-chain forms also occurring in nature but not known in human tissues as ethanol dehydrogenases and not detailed in Fig. 1. Hence, the system is complex, with patterns of repeated gene duplications, as also observed in for example the globin and glutathione transferase protein families (Fig. 2).

ORIGINS

The organizational pattern in Fig. 1 does not *per se* correlate the separate levels with specific times. In order to derive at time estimates for gene duplications, the rate of divergence must be evaluated from analysis of species variants, or forms must be investigated in species originating from times at (or before) the duplication and hence having hybrid properties (or lacking one of the products). In the case of alcohol de-

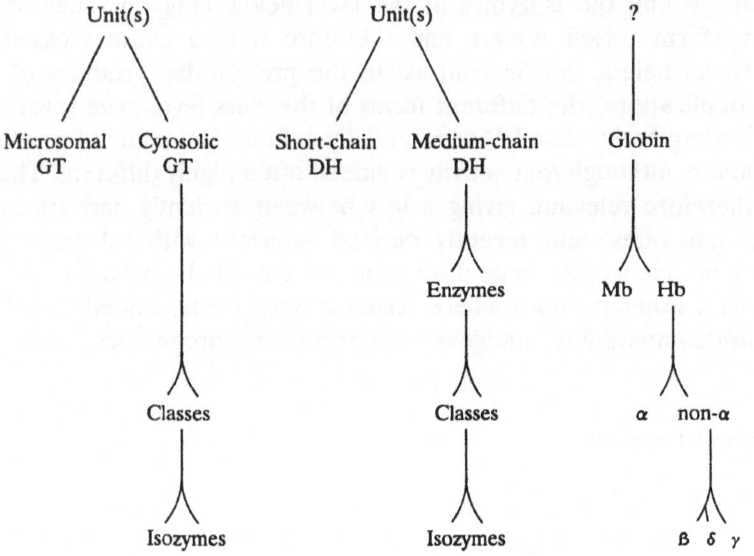

Figure 2. Similarity in patterns between the alcohol dehydrogenase family and other well-characterized protein families.

The medium-chain alcohol dehydrogenase, family (center; DH, dehydrogenase) has an organizational pattern related to that for families like glutathione transferases, GT (left;cf Mannervik *et al.*, 1985) and globins (right; cf Dayhoff *et al.*, 1972; Mb, myoglobin; Hb, hemoglobin). Only levels of branch positions, their numbers and relative lengths differ. Bent arrows indicate gene duplications and subsequent divergence.

hydrogenase, both conditions have been met regarding one of the duplications at level 3, and conclusions largely agree.

Thus, the number of species variants characterized is now quite extensive for class I and class III. For class I, ten different mammalian species have been structurally characterized (Höög *et al.*, 1992b; Sun and Plapp, 1992), all, however, too similar not to be useful for timing of the gene duplications leading to the class distinctions. Similarly, three avian lines have been structurally reported (Kaiser *et al.*, 1990; Estonius *et al.*, 1990; Hiremath *et al.*, 1992), also too similar to the mammalian forms for useful information about the origins. However, the enzyme of amphibian class I alcohol dehydrogenase is sufficiently different to allow a correlation with the class differences and hence a rough time estimate; a class III/I gene duplication at minimally about 450 MYA was then deduced (Cederlund *et al.*, 1991). This value is still highly approximate, since it was derived from comparison of just one amphibian enzyme structure with the divergence within the mammalian classes. The value should be regarded as a minimal estimate and the time could possibly be longer, by a factor of perhaps two or slightly more, which would include also the plant enzymes. We are presently studying the class I and III relatives from these sources in order to arrive at further certainty in the value for the class III/I separation. But in spite of the preliminary nature of the presently available timing for the III/I separation, the available estimate definitely shows that the class separation is a distant event, making it likely that the classes are well represented throughout the mammalian system and much or all of the vertebrate system. The present minimal timing of about 450 MYA (Cederlund *et al.*, 1991) roughly correspond to that of early vertebrate evolution. Reflecting the extensive changes that then happened, many protein families were altered at that time. Traces of hybrid alcohol dehydrogenase forms may perhaps still be expected in the piscine enzyme, since fishes originated early in the vertebrate evolution. We therefore studied the cod enzyme and found the piscine line still to have the two classes activity-wise, with class I- and class III-type functional properties, and corresponding residues at the substrate binding sites, explaining the activities (Danielsson *et al.*, 1992; Danielsson and Jörnvall, 1992). However, **both** enzymes were overall most closely related to class III in residue identity. In other words, of the two enzyme types in cod liver that correspond to alcohol dehydrogenase, one is a typical class III enzyme in both structure and function, like any vertebrate class III form, while the other is a "hybrid" molecule, functionally like the mammalian class I forms, but overall residue-wise still rather of the class III type.

Thus, both the finding of class-hybrid alcohol dehydrogenase forms and the preliminary dating attempts from divergence agree in distant estimates for the class III/I duplication. Even if the time estimate later will have to be extended by a factor 2-fold or so, to include also the plant lines before the duplication, further analysis of the class I and III forms (Yin *et al.*, 1991) has already confirmed the conclusion of the separate nature of the classes and shows that the class III enzyme is more constant, by a factor of about 3. In conclusion, the class III form appears to constitute the more original form, definitely derived from prevertebrate times, whereas the liver ethanol dehydrogenase activity of class I apparently was the form emerging after the class III/I duplication (Danielsson and Jörnvall 1992). The forms have diverged, first in function (enzymatically they are class-separated already in the fish line) and later also in structure (distinct overall residue relationships in the amphibian/reptilian/avian lines). This order of change, more rapidly in substrate specificity than in overall structure, is compatible with the fact that functional properties are governed by a limited number of residues and require fewer changes at the substrate-binding pocket than the overall changes in all residues.

Apart from establishing a distant origin for the class duplications, these results also suggest that ethanol dehydrogenase activity has evolved repeatedly in nature. Thus, the classical liver alcohol dehydrogenase, with considerable ethanol activity, appears to have originated by a gene duplication before that leading to the other well-known alcohol dehydrogenase with ethanol activity, from yeast, which reflects divergence at level 2 of Fig. 1 rather than at level 3 (yeast/class I alcohol dehydrogenases have 20-25% residue identity, cf Jörnvall *et al.*, 1978). Further, the insect alcohol dehydrogenase, also with considerable ethanol activity but representing a short-chain dehydrogenase separated at level 1, has a still more distant original ancestral connection with all the medium-chain enzymes. These relationships are emphasized in Fig. 3. Consequently, the enzymes with similar present-day activity (and identical EC numbers) are not always directly related, but reflect branches from separate levels. In other words, a functional convergence towards similar substrates (or at least lack of functional divergence) is observed in the case of ethanol dehydrogenase activity in nature, at the same time as structural divergence has occurred in the usual manner.

This multiplicity in origin for an enzyme activity is not unique to the alcohol dehydrogenase family, but may present properties of large protein families in general. The relationships are however nicely illustrated in the case of the alcohol dehydrogenase super-family and start to be discernable in greater detail because of the many known structures.

CORRELATION WITH OTHER DATA

Unique properties

The family lines exhibit separate properties in each case, reflecting their origin rather than their present substrate specificity. For example, the short-chain alcohol dehydrogenases, are of course more closely related to other short-chain enzymes than they are to other ethanol dehydrogenases. In several cases, similar organization, including reactive residues and proteolytically sensitive regions have already been proven within the former group (Krook *et al.*, 1992). Similarly, regarding the medium-chain enzymes, yeast alcohol dehydrogenase has a missing segment (Jörnvall *et al.*, 1978) and a quaternary structure, resembling those of sorbitol dehydrogenase with another specificity (cf Eklund *et al.*, 1985; Roumi *et al.*, 1992) rather than those of mammalian alcohol dehydrogenases. In the same manner, the liver class I enzymes are more closely related to the class III enzymes than to other ethanol dehydrogenases.

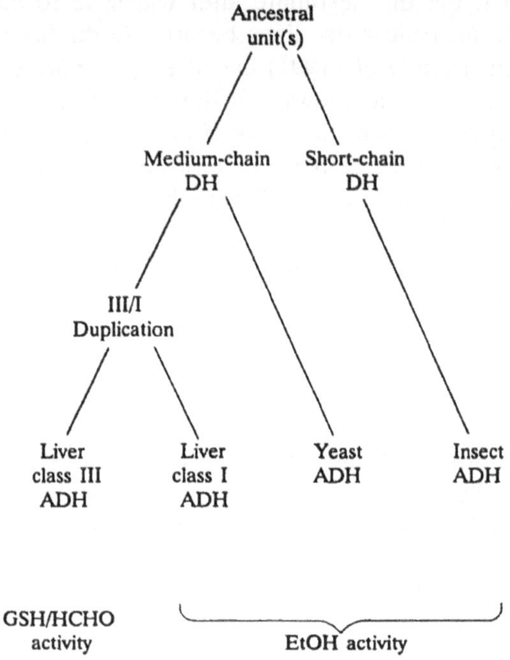

Figure 3. Multiplicity of alcohol dehydrogenases.
Present-day enzymes with ethanol dehydrogenase activity (EtOH) have differently leveled ancestral connections, whereas the glutathione-dependent formaldehyde dehydrogenase activity (GSH/HCHO) of class III appears to be ubiquitous, of common, ancient origin (ADH, alcohol dehydrogenase; DH, dehydrogenase). Hence, the ethanol dehydrogenase activity now observed in several branches (right) may suggest a functional convergence, or at least lack of a similar divergence in function as the one in structure.

Apart from these line-specific features each line also exhibits properties typical of protein families in general. One such property is a tendency to conserve glycine residues, reflecting the fact that space is critical in protein folding. This Gly conservation is clearly noticeable in the medium-chain alcohol dehydrogenase line (Jörnvall *et al.*, 1987), resembling the situation in for example cytochrome c (Smith and Margoliash, 1964), but less clear in short-chain dehydrogenases, although visible also there (Persson *et al.*, 1991). Presumably, the short-chain dehydrogenases are so distantly related (compatible with their low extent of residue identity, ~20%) that divergence has progressed so far that conformational similarities are also fewer and not even the Gly conservation easily observed any more. Another common property visible in many protein families and well demonstrated in the medium-chain line is the fact that deletions/insertions frequently affect superficially positioned segments, often entire loops, as early shown for both the alcohol and sorbitol dehydrogenases (Jörnvall *et al.*, 1978; Eklund *et al.*, 1985).

In summary, the short-chain dehydrogenase line, the medium-chain dehydrogenase line, and the other lines, each constitute typical protein families, together being part of the super-family represented in Fig. 1.

Genetic regulation

The organizational pattern of alcohol dehydrogenases as emphasized in Fig. 1, the similarities with other families as emphasized in Fig. 2, and the multiple divergence are not the only aspects where the alcohol dehydrogenases represent properties typical of large protein families in general. Thus, they also illustrate separate genetic controls, with apparently constituent forms (class III), inducible forms (at least some genes of class I) and genes differently expressed during ontogenesis, with fetal and adult forms (Smith *et al.*, 1971; cf Duester, 1991). In this respect, the medium-chain alcohol dehydrogenase family resembles properties of the globin family or the cytochrome P450 family. In conclusion, the whole organization of the alcohol dehydrogenases suggest that it constitutes a complex system with both structural, functional, and genetic variations similar to those in several other protein families. The similarities in pattern suggest old and common metabolic functions, thus generalizing the role of alcohol dehydrogenases in nature and linking them to many other enzyme families of basic importance in general metabolism.

FUNCTIONAL CONCLUSIONS

What then is the function of liver alcohol dehydrogenase? The presently recognized overall patterns give some new insight into the metabolic roles. Thus:

* Alcohol dehydrogenase activity is clearly not species-specific, not liver-specific, and as far as concerns general occurrence, more or less universal, although structurally derived from separate lines and branches (Fig. 1). Hence, the function of alcohol dehydrogenase is likely to represent fundamental and distantly derived metabolic routes.

* The ethanol activity appears to have separate origins (Fig. 3), suggesting that just this activity is not always critical, since it has arisen or become expressed repeatedly and appears not to be of a continuous, common presence, as the activities of several other enzymes of central importance in basic metabolic pathways. It would therefore not be surprising to find living forms lacking much or all of the ethanol dehydrogenase activity, as has been reported for a strain (Burnett and Felder, 1978) of one mammal (deer mouse) and even more so for invertebrates (unpublished). Apparently, low ethanol dehydrogenase activity *per se* is not incompatible with life.

* Nevertheless, some activities are much more constant. The evolutionary pattern suggests class III to constitute the original form existing before the class III/I duplication now traced. Furthermore, class III is also structurally a more constant enzyme (Yin *et al.*, 1991), one that apparently occurs in all life forms (Uotila and Koivusalo, 1989; Gutheil *et al.*, 1992), and one that is constitutive (Smith, 1986). Hence, the most original function may be linked to the glutathione-dependent formaldehyde dehydrogenase activity of class III, rather than to the "typical" ethanol activity of alcohol dehydrogenases. Metabolically, this in turn relates the whole enzyme system to other enzyme systems, also centering on glutathione and reactive molecules, like glutathione transferases and cytochromes P450. Therefore, not only the overall molecular organization and

the evolutionary scheme resemble that of the glutathione transferases and cytochromes P450, but also the basic concept of several substrates. In conclusion, the basic function of the alcohol dehydrogenase family may involve defense mechanisms common to many life forms. In addition, the repeated duplications, divergence, and the possible, functional convergence has later given additional roles, like for example that in regulation of retinoic acid formation and thereby control of vertebrate differentiation as suggested for class I (Duester *et al.*, 1991). The basic relationships are supported by the general pattern, and the later divergence is compatible with the well-known lack of a single, generally recognized substrate for liver alcohol dehydrogenase. This lack also makes alcohol dehydrogenase resemble common defense enzymes rather than those in glycolytic and other specific pathways.

ACKNOWLEDGMENTS

Support by grants from the Swedish Medical Research Council, the Swedish Natural Science Research Council, the Swedish Alcohol Research Fund, the Spanish Dirección General de Investigación Cientifica y Técnica, Endowment for Research in Human Biology (Boston, USA), Karolinska Institutet, Wenner-Gren Foundation, and Magn. Bergvall Foundation is gratefully acknowledged.

REFERENCES

Albalat, R., Gonzàlez-Duarte, R. and Atrian, S., 1992, Protein engineering of Drosophila alcohol dehydrogenase. *FEBS Lett.*, in press.

Aronson, B.D., Somerville, R.L., Epperly, B.R. and Dekker, E.E., 1989, The primary structure of *Escherichia coli* L-threonine dehydrogenase. *J. Biol. Chem.* 264:5226-5232.

Borrás, T., Persson, B. and Jörnvall, H., 1989, Eye lens ζ-crystallin relationships to the family of "long-chain" alcohol/polyol dehydrogenases. Protein trimming and conservation of stable parts. *Biochemistry* 28:6133-6139.

Burnett, K.G. and Felder, M.R., 1978, Genetic regulation of liver alcohol dehydrogenase in *Peromyscus*. *Biochem. Genet.* 16:443-454.

Cederlund, E., Peralba, J.M., Parés, X. and Jörnvall, H., 1991, Amphibian alcohol dehydrogenase, the major frog liver enzyme. Relationships to other forms and assessment of an early gene duplication separating vertebrate class I and class III alcohol dehydrogenases. *Biochemistry* 30:2811-2816.

Danielsson, O. and Jörnvall, H., 1992, Enzymogenesis: classical liver alcohol dehydrogenase. Origin from the glutathione-dependent formaldehyde dehydrogenase line. *Proc. Natl. Acad. Sci. USA*, in press.

Danielsson, O., Eklund, H. and Jörnvall, H., 1992, The major piscine liver alcohol dehydrogenase has class-mixed properties in relation to mammalian alcohol dehydrogenases of classes I and III. *Biochemistry* 31:3751-3759.

Dayhoff, M.D., Hunt, L.T., McLaughlin, P.J. and Jones, D.D., 1972, Gene duplications in evolution: the globins, *in*: "Atlas of Protein Sequence and Structure", Natl. Biomed. Res. Foundation, Silver Spring, pp. 17-30.

Duester, G., 1991, Human liver alcohol dehydrogenase gene expression, *in*: "Drug and Alcohol Abuse Reviews," Watson R.R., ed., Humana Press, pp. 375-402.

Duester, G., Shean, M.L., Mc Bride, M.S. and Stewart, M.J., 1991, Retinoic acid response element in the human alcohol dehydrogenase gene ADH3: implication for regulation of retinoic acid synthesis. *Mol. Cell. Biol.* 11:1638-1646.

Eklund, H., Nordström, B., Zeppezauer, E., Söderlund, G., Ohlsson, I., Boiwe, T., Söderberg, B.-O., Tapia, O., Brändén, C.-I. and Åkeson, Å., 1976, Three-dimensional structure of horse liver alcohol dehydrogenase at 2.4 Å resolution. *J. Mol. Biol.* 102:27-59.

Eklund, H., Horjales, E., Jörnvall, H., Brändén, C.-I. and Jeffery, J., 1985, Molecular aspects of functional differences between alcohol and sorbitol dehydrogenases. *Biochemistry* 24:8005-8012.

Eklund, H., Müller-Wille, P., Horjales, E., Futer, O., Holmquist, B., Vallee, B.L., Höög, J.-O., Kaiser, R. and Jörnvall, H., 1990, Comparison of three classes of human liver alcohol dehydrogenase. Emphasis on different substrate-binding pockets. *Eur. J. Biochem.* 193:303-310.

Ensor, C.M. and Tai, H.-H., 1991, Site-directed mutagenesis of the conserved tyrosine 151 of human placental NAD^+-dependent 15-hydroxyprostaglandin dehydrogenase yields a catalytically inactive enzyme. *Biochem. Biophys. Res. Commun.* 176:840-845.

Estonius, M., Karlsson, C., Fox, E.A., Höög, J.-O., Holmquist, B., Vallee, B.L., Davidson, W.S. and Jörnvall, H., 1990, Avian alcohol dehydrogenase: the chicken liver enzyme. Primary structure, cDNA cloning, and relationships to other alcohol dehydrogenases. *Eur. J. Biochem.* 194:593-602.

Ghosh, D., Weeks, C.M., Grouchulski, P., Duax, W.L., Erman, M., Rimsay, R.L. and Orr, J.C., 1991, Three-dimensional structure of holo $3\alpha,20\beta$-hydroxysteroid dehydrogenase: a member of a short-chain dehydrogenase family. *Proc. Natl. Acad. Sci. USA* 88:10064-10068.

Gutheil, W.G., Holmquist, B. and Vallee, B.L., 1992, Purification, characterization, and partial sequence of the glutathione-dependent formaldehyde dehydrogenase from *Escherichia coli*: a class III alcohol dehydrogenase. *Biochemistry* 31:475-481.

Hiremath, L.S., Kessler, P.M., Sasaki, G.C. and Kolattukudy, P.E., 1992, Estrogen induction of alcohol dehydrogenase in the uropygial gland of mallard ducks. *Eur. J. Biochem.* 203:449-457.

Hjelmqvist, L., Ericsson, M., Shafqat, J., Carlquist, M., Siddiqi, A.R., Höög, J.-O. and Jörnvall, H., 1992, Reptilian alcohol dehydrogenase. Heterogeneity relevant to class multiplicity of the mammalian enzyme. *FEBS Lett.* 298:297-300.

Holmes, R.S., 1988, Alcohol dehydrogenases and aldehyde dehydrogenases of anterior eye tissues from humans and other mammals. *in*: "Biomedical and Social Aspects of Alcohol and Alcoholism," K. Kuriyama, A. Takada and H. Ishii, eds., Elsevier, Amsterdam, pp. 51-57.

Huang, Q.-L., Russell, P., Stone, S.H. and Zigler, J.S., Jr., 1987, Zeta-crystallin, a novel lens protein from the guinea pig. *Curr. Eye Res.* 6:725-732.

Huang, Q.L., Du, X.Y., Stone, S.H., Amsbaugh, D.F., Datiles, M., Hu, T.S. and Zigler, J.S., Jr., 1990, Association of herediatry cataracts in strain 13/N guinea-pigs with mutation of the gene for ζ-crystallin. *Exp. Eye Res.* 50:317-325.

Hurley, T.D., Bosron, W.F., Hamilton, J.A. and Amzel, L.M., 1991, Structure of human $\beta_1\beta_1$ alcohol dehydrogenase: catalytic effects of non-active site substitutions. *Proc. Natl. Acad. Sci. USA* 88:8149-8153.

Höög, J.-O, Karlsson, C., Eklund, H., Shapiro, R. and Jörnvall, H., 1992a, Site-directed mutagenesis of mammalian alcohol and sorbitol dehydrogenases map functional differences within the enzyme family. *in:* This vol., in press.

Höög, J.-O., Vagelopoulos, N., Yip, P.-K., Keung, W.M. and Jörnvall, H., 1992b, Isozyme developments in mammalian class I alcohol dehydrogenase. cDNA cloning, functional correlations, and lack of evidence for genetic isozymes in rabbit. Submitted.

Inoue, T., Sunagawa, M., Mori, A., Imai, C., Fukuda, M., Takagi, M. and Yano, K., 1989, Cloning and sequencing of the gene encoding the 72-kilodalton dehydrogenase subunit of alcohol dehydrogenase from Acetobacter aceti. *J. Bacteriol.* 171:3115-3122.

Jeffery, J., Chester, J., Mills, C., Sadler, P.J. and Jörnvall, H., 1984, Sorbitol dehydrogenase is a zinc enzyme. *EMBO J.* 3:357-360.

Jörnvall, H., 1970, Horse liver alcohol dehydrogenase. The primary structure of the protein chain of the ethanol-active isoenzyme. *Eur. J. Biochem.* 16:25-40.

Jörnvall, H., Eklund, H. and Brändén, C.-I., 1978, Subunit conformation of yeast alcohol dehydrogenase. *J. Biol. Chem.* 253:8414-8419.

Jörnvall, H., Persson, M. and Jeffery, J., 1981, Alcohol and polyol dehydrogenases are both divided into two protein types, and structural properties cross-relate the different enzyme activities within each type. *Proc. Natl. Acad. Sci. USA* 78:4225-4230.

Jörnvall, H., Persson, B. and Jeffery, J., 1987, Characteristics of alcohol/polyol dehydrogenases. The zinc-containing long-chain alcohol dehydrogenases. *Eur. J. Biochem.* 167:195-201.

Jörnvall, H., von Bahr-Lindström, H. and Höög, J.-O., 1989, Alcohol dehydrogenase - structure. *in*: "Human Metabolism of Alcohol" Vol. II, K.E. Crow and R.D. Batt, eds., CRC Press, Boca Raton, pp. 43-64.

Kaiser, R., Nussrallah, B.A., Dam, R., Wagner, F.W. and Jörnvall, H., 1990, Avian alcohol dehydrogenase. Characterization of the quail enzyme, functional interpretations, and relationships to the different classes of mammalian alcohol dehydrogenase. *Biochemistry* 29:8365-8371.

Kötter, P., Amore, R., Hollenberg, C.P. and Ciriacy, M., 1990, Isolation and characterization of the *Pichia stipitis* xylitol dehydrogenase gene, *XYL2*, and construction of a xylose-utilizing *Saccharomyces cerevisiae* transformant. *Curr. Genet.* 18:493-500.

Krook, M., Marekov, L. and Jörnvall, H., 1990, Purification and structural characterization of placental NAD^+-linked 15-hydroxyprostaglandin dehydrogenase. The primary structure reveals the enzyme to belong to the short-chain alcohol dehydrogenase family. *Biochemistry* 29:738-743.

Krook, M., Prozorovski, V., Atrian, S., Gonzàlez, R. and Jörnvall, H., 1992, Short-chain dehydrogenases: proteolysis and chemical modification of prokaryotic $3\alpha/20\beta$-hydroxysteroid, insect alcohol, and human 15-hydroxyprostaglandin dehydrogenase. *Eur. J. Biochem.*, in press.

Mannervik, B., Ålin, P., Guthenberg, C., Jensson, H., Tahir, M.K., Warholm, M. and Jörnvall, H., 1985, Identification of three classes of cytosolic glutathione transferase common to several mammalian species. Correlation between structural data and enzymatic properties. *Proc. Natl. Acad. Sci. USA* 82:7202-7206.

Moreno, A. and Parés, X., 1991, Purification and caracterization of a new alcohol dehydrogenase form human stomach *J. Biol. Chem.* 266:1128-1133.

Parés, X., Moreno, A., Cederlund E., Höög, J.-O. and Jörnvall, H., 1990, Class IV mammalian alcohol dehydrogenase. Structural data of the rat stomach enzyme reveal a new class well separated from those already characterized. *FEBS Lett.* 277:115-118.

Parés, X., Cederlund, E., Moreno, A., Saubi, N., Höög, J.-O. and Jörnvall, H., 1992, Class IV alcohol dehydrogenase (the gastric enzyme). Structural analysis of human $\sigma\sigma$-ADH reveals class IV to be variable and confirms the presence of a fifth mammalian alcohol dehydrogenase class. *FEBS Lett.* 303:69-72.

Persson, B., Krook, M. and Jörnvall, H., 1991, Characteristics of short-chain alcohol dehydrognases and related enzymes. *Eur. J. Biochem.* 200:537-543.

Rao, P.V., Krishna, C.M. and Zigler, J.S., Jr., 1992, Identification and characterization of the enzymatic activity of ζ-crystallin from guinea pig lens. *J. Biol. Chem.* 267:96-102.

Rossmann, M.G., Moras, D. and Olsen, K.W., 1974, Chemical and biological evolution of a nucleotide-binding protein. *Nature* 250:194-199.

Rossmann, M.G., Liljas, A., Brändén, C.-I. and Baneszak, L.J., 1975, Evolutionary and structural relationships among dehydrogenases. *The Enzymes, 3 edn,* 11:61-102.

Rouimi, P., Loomes, K. and Jörnvall, H., 1992, Comparative proteolysis of sorbitol and alcohol dehydrogenases. Submitted.

Scopes, R.K., 1983, An iron-activated alcohol dehydrogenase. *FEBS Lett.* 156:303-306.

Smith, E.L. and Margoliash, E., 1964, Evolution of cytochrome c. *Fed. Proc.* 23:1243-1247.

Smith, M., 1986, Genetics of human alcohol and aldehyde dehydrogenases. *Adv. Hum. Genet.* 15:249-290.

Smith, M., Hopkinson, D.A. and Harris, H., 1971, Developmental changes and polymorphism in human alcohol dehydrogenase. *Ann. Hum. Genet.* 34:251-271.

Sun, H.-W. and Plapp, B.V., 1992, Progressive sequence alignment and molecular evolution of the Zn-containing alcohol dehydrogenase family. *J. Mol. Evol.* 34:522-535.

Thatcher, D.R. and Sawyer, L., 1980, Secondary-structure prediction from the sequence of *Drosophila melanogaster* (fruitfly) alcohol dehydrogenase. *Biochem. J.* 187:884-886.

Uotila, L. and Koivusalo, M., 1989, Formaldehyde dehydrogenase, *in*: "Coenzymes and Cofactors, Glutathione: Chemical, Biochemical and Medical Aspects", Vol III, part A, D. Dolphin, R. Poulson, and O. Avramovic, eds., Wiley, New York, pp. 517-551.

Vallee, B.L. and Bazzone, T.J., 1983, Isozymes of human liver alcohol dehydrogenases. *Isozymes: Curr. Top. Biol. Med. Res.* 8:219-244.

Villarroya, A., Juan, E., Egestad, B. and Jörnvall, H., 1989, The primary structure of alcohol dehydrogenase from *Drosophila lebanonensis.* Extensive variation within insect "short-chain" alcohol dehydrogenases lacking zinc. *Eur. J. Biochem.* 180:191-197.

Williamson, V.M. and Paquin, C.E., 1987, Homology of *Saccharomyces cerevisiae* ADH4 to an iron-activated alcohol dehydrogenase from *Zymomonas mobilis. Mol. Gen. Genet.* 209:374-381.

Wistow, G. and Piatigorsky, J., 1987, Recruitment of enzymes as lens structural proteins. *Science* 236:1554-1556.

Yasunami, M., Chen, C.-S. and Yoshida, A., 1991, A human alcohol dehydrogenase gene (ADH6) encoding an additional class of isozyme. *Proc. Natl. Acad. Sci. USA* 88:7610-7614.

Yin, S.-J., Wang, M.-F., Liao, C.-S., Chen, C.-M. and Wu, C.-W., 1990, Identification of a human stomach alcohol dehydrogenase with distinctive kinetic properties. *Biochem. Int.* 22:829-835.

Yin, S.-J., Vagelopoulos, N., Wang, S.-L., and Jörnvall, H., 1991, Structural features of stomach aldehyde dehydrogenase distinguish dimeric aldehyde dehydrogenase as a "variable" enzyme. "Variable" and "constant" enzymes within the alcohol and aldehyde dehydrogenase families. *FEBS Lett.* 283:85-88.

TISSUE DISTRIBUTION OF ALCOHOL AND SORBITOL DEHYDROGENASE mRNAs

Mats Estonius[1], Olle Danielsson[1], Jan-Olov Höög[1], Håkan Persson[2] and Hans Jörnvall[1].

Department of Chemistry I[1]
Laboratory of Molecular Neurobiology[2]
Karolinska Institutet,
S-104 01 Stockholm, Sweden

INTRODUCTION

Alcohol dehydrogenase (ADH) of class I is the principal enzyme in liver ethanol oxidation, and has been the subject of much research. It has been studied in many species and is a part of the enzyme system now constituting ADHs at large. The different mammalian ADHs can be divided into at least five classes according to structural properties (Parés et al., 1992). Class I is the classical liver ADH (Vallee and Bazzone, 1983), and class III ADH is the glutathione-dependent formaldehyde dehydrogenase (Koivusalo et al., 1989). ADH of class II shows a higher K_m for ethanol than class I, and exhibits activity toward norepinephrine metabolites (Mårdh et al.,1986), but is less studied than class I and class III ADH. Class IV is a stomach ADH characterized in rat and man (Parés et al., 1990; 1992; Moreno and Parés, 1991), and class V is a DNA-derived human structure recently reported (Yasunami et al., 1991).

All these ADH enzymes belong to a protein family of zinc-containing dehydrogenases, also including sorbitol dehydrogenase (SDH), ζ-crystallin (thus far without known zinc content), and threonine dehydrogenase (Jörnvall et al., 1991). SDH has been linked with pathological conditions such as diabetes and cataract formation (Gabbay, 1973). It has been purified from liver, and in addition, the activity has been found in a variety of tissues such as kidney, brain, cornea and seminal vesicles (cf. Jeffery and Jörnvall, 1988).

A complete insight into the tissue distribution and relative ratios of these zinc-containing enzymes is unknown, but in rat, three different classes of ADH have been mapped by starch gel electrophoresis and activity staining (Julià et al., 1987). Class III has been found in most mammalian tissues, including brain, placenta and testis (Beisswenger et al., 1985; Parés et al., 1984; Dafeldecker and Vallee, 1986). Class II is known to exist in human liver, but has not been proved to exist in other tissues. In rat, however, a cDNA clone for the class II enzyme has been isolated from a liver cDNA library (Höög, 1991). The structure of rat liver SDH has been determined at the cDNA level (Karlsson et al.,

1991). In order to trace physiological roles of these enzymes, and to correlate metabolic functions with specific organs, we have determined the relative occurrence of mRNAs for ADH of classes I, II and III, and for SDH, in 20 different rat tissues.

MATERIALS AND METHODS

Ten male and ten female adult Sprague-Dawley rats (body weight 150 g) were sacrificed by exposure to carbon dioxide. Tissues were dissected and immediately frozen on dry ice/methanol and stored at -70°C until RNA preparation. Thymus, adrenal, spleen, eye, and pancreas from both sexes were pooled before RNA purification. Kidney, stomach, brain, heart, muscle, lung, duodenum, colon, ovary, uterus, small intestine, and liver samples were taken from females. Additional liver samples were taken from males, and pools of testis and epididymis were used. Frozen tissue samples were homogenized in 4 M guanidine isothiocyanate, 0.1 M β-mercaptoethanol, 0.025 M sodium citrate, pH 7.0, and homogenized twice with a Polytron instrument for 15-20 s. Each homogenate was layered over a cushion of 5.7 M CsCl in 0.025 M sodium acetate, pH 5.5, and centrifuged at 100.000x*g* in a Beckman SW 41 rotor overnight at 15°C. Polyadenylated RNA was purified by oligo(dT)-cellulose chromatography, and the material recovered was quantified spectrophotometrically before blot analysis. 20 μg of poly(A)$^+$ RNA from each sample was subjected to electrophoresis in 1% agarose containing 0.7% formaldehyde and 0.1 μg ethidium bromide. After electrophoresis, the gels were examined under UV light to ensure equal amounts of RNA in each sample. The mRNAs were then transferred to nitrocellulose filters, which were hybridized with probes labelled with α[^{32}P]-dCTP to a specific activity of 5×10^8 cpm/μg. A class I ADH probe was amplified from a rat liver cDNA library using PCR, while class II ADH and SDH probes were derived from cDNA clones (Höög, 1991; Karlsson et al., 1991). Class III mRNA was probed with a 382 bp 5' fragment of the corresponding human cDNA (Sharma et al., 1989). Hybridizations were performed at 42°C overnight, using four different sets of filters in 40% formamide, 4 X SSC (1 X SSC is 0.15 M NaCl, 0.015 M sodium citrate, pH 7.0), 1 X Denhardts solution and 10% dextran sulfate. Filters were washed at high stringency (0.1 X SSC, 0.1% SDS, 54°C) before exposure to Kodak X-OMAT films for 20 h.

RESULTS

mRNA from 20 different tissues was prepared, electrophoresed, and blotted onto filters, which were then tested by hybridization with four different probes, complementary to the cDNAs of class I, II, and III ADHs, and SDH. Strong hybridization signals were observed from the filters, supporting the reliability of the methodology. For class I, II and III ADHs several bands appeared on the blots, showing the presence of small or large mRNAs, but here we focus on the major transcripts observed. In general, those of SDH and ADHs of class I and III, were detected in nearly all tissues analyzed, although with great differences in relative amounts. In contrast, class II ADH mRNA was predominantly found in liver and duodenum. mRNAs for class III ADH and for SDH were the ones most similar in occurrence, in common exhibiting a fairly abundant and even distribution.

Class I alcohol dehydrogenase mRNA

For class I (Fig. 1), abundant levels of a 1.9 kb mRNA species were observed in liver, kidney, stomach, duodenum, colon and uterus. Lower, but clearly detectable levels, were present in spleen, testis, lung, epididymis, small intestine, adrenal and eye. Low amounts, corresponding to weak signals, were found in thymus, brain, heart and muscle. No signal was detected from pancreas despite low-stringency washes of the hybridization filters and prolonged exposures of the X-ray films.

Class II alcohol dehydrogenase mRNA

The probe for class II revealed this mRNA to be much more limited in distribution than class I, as judged by the hybridization signals. Class II transcripts were found in only two tissues, liver and duodenum (Fig. 2). Two different-sized mRNA species were detected. The major one had an apparent size of 3.5 kb and the second a size of 1.7 kb. The latter mRNA species corresponds in size to the cDNA coding for rat class II ADH (Höög, 1991). In liver, a clear sex-related difference was now observed, with the smaller RNA hybridizing stronger in male than in female. After prolonged exposure of the filters (for three days), additional bands were seen. Small amounts of the 3.5 kb transcript were detected in kidney and stomach, and trace amounts in spleen. The 1.7 kb transcript was also detected in testis after exposures that had been extended for 3 days.

Class III alcohol dehydrogenase mRNA

Blots with class III ADH mRNA (Fig. 3) revealed strong hybridization signals for a small-sized RNA, in addition to signals for the expected large transcript. The large mRNA species was readily detectable in liver, kidney, stomach, thymus, testis, brain, heart, muscle, lung, duodenum, ovary, adrenal, and, especially in epididymis, colon, uterus and eye. Weak hybridization signals were observed in spleen and small intestine, but no signal was detected from pancreatic samples.

Sorbitol dehydrogenase mRNA

A 2.0 kb sorbitol dehydrogenase mRNA was detected from stomach, thymus, brain, heart, lung, epididymis, colon, ovary, uterus, adrenal and eye (Fig. 4). The highest levels were found in the liver, kidney and testis, while lower levels were found in spleen, muscle and duodenum. No sorbitol dehydrogenase mRNA was detected in the small intestine and pancreas.

DISCUSSION

The zinc-containing dehydrogenases analyzed in this study show quite extensive variations in their mRNA tissue distributions. The class II form was found in liver and duodenum only, while virtually every tissue contained mRNA for class III alcohol dehydrogenase. These inter-dehydrogenase differences in tissue distribution, and the presence of different but precise amounts of the mRNA species in each tissue, may provide clues to the physiological roles of these enzymes in the metabolism of polyols and alcohols.

The class I alcohol dehydrogenase mRNA is widely distributed, but the highest level was observed in liver. Also, in agreement with earlier protein data (Julià et al., 1987; Boleda et al., 1989), colon, duodenum, kidney, and uterus contained high amounts of class I mRNA. In contrast to earlier results, however, one major difference was observed in stomach, where we detected relatively large amounts of class I mRNA, while earlier studies did not show class I ADH in this tissue. In addition, we also found transcripts in spleen. More interestingly, a low level of class I mRNA was detected in brain. Earlier reports have suggested that the brain does not contain class I alcohol dehydrogenase and therefore cannot metabolize alcohol (Beisswenger et al. 1985), but the opposite has also been reported

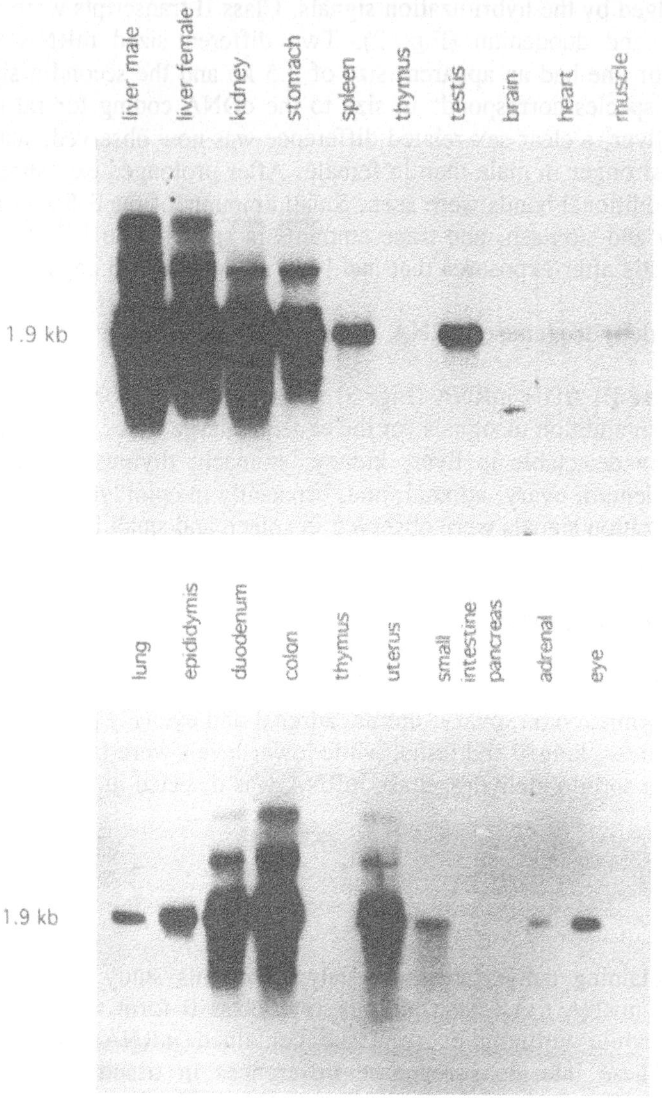

Figure 1. Northern blot analysis of class I ADH in rat tissues. Poly(A)$^+$RNA samples (20 μg/lane) from different tissues were electrophoresed in a formaldehyde-containing agarose gel, blotted onto nitrocellulose filters and hybridized overnight with a PCR-amplified 400 bp rat class I ADH probe. The filters were washed at high stringency before exposure to the X-ray films for 20 h.

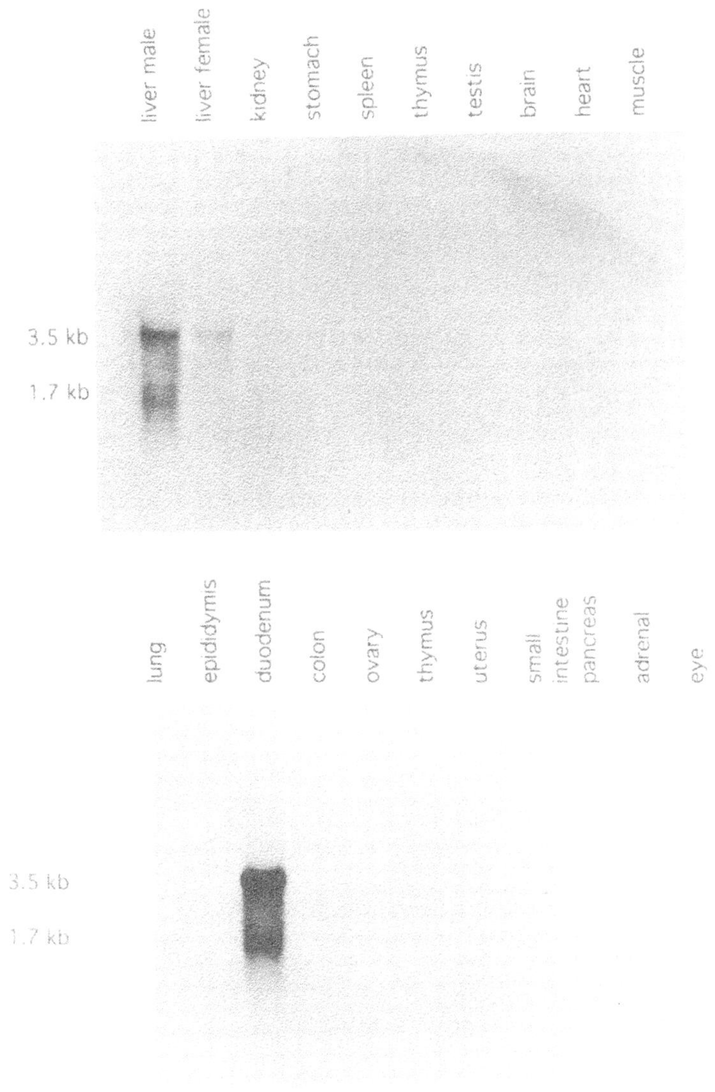

Figure 2. Northern blot analysis of rat class II ADH using a 780 bp *Bst*E I-fragment derived from the rat class II cDNA. For blot conditions, cf. legend to Fig. 1.

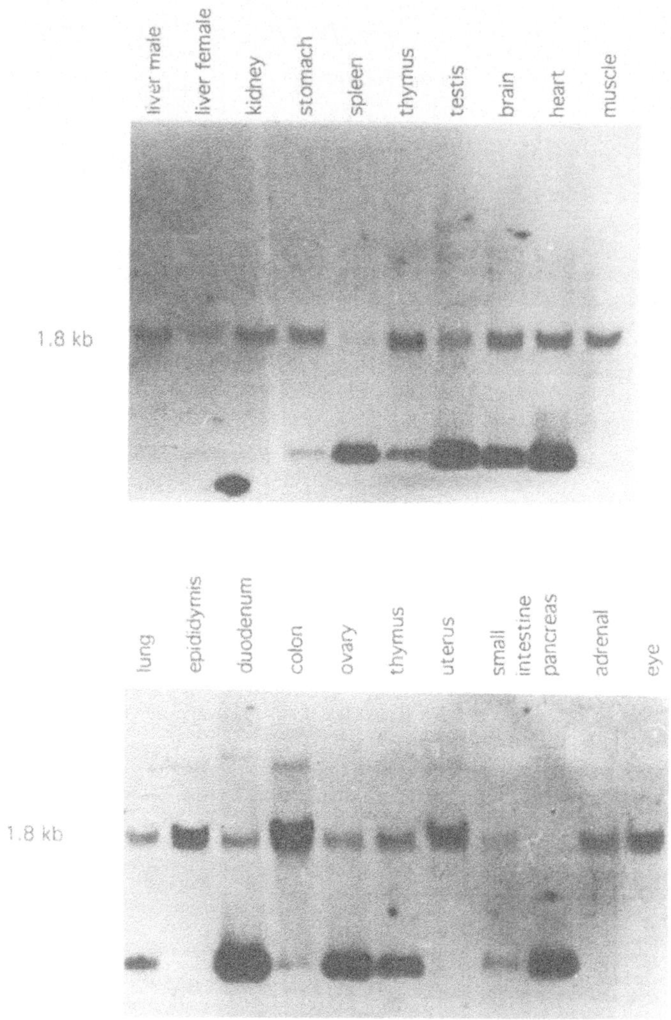

Figure 3. Northern blot analysis of rat class III ADH with a *Eco*R I/*Kpn* I 382 bp probe derived from the 5' region of the human class III cDNA. For blot conditions, cf. legend to Fig. 1.

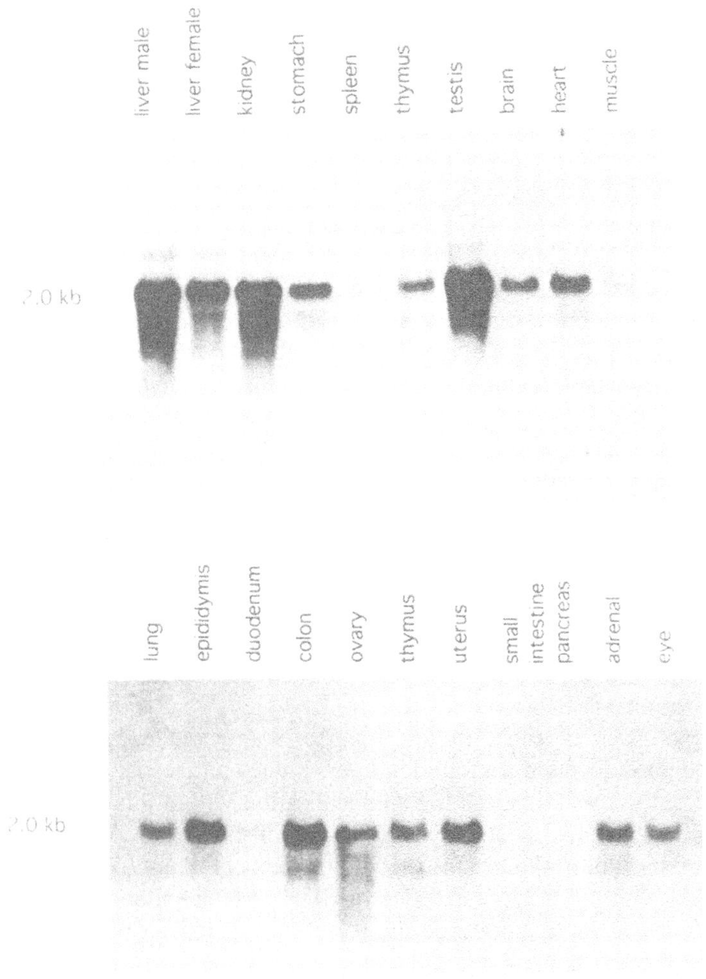

Figure 4. Northern blot analysis of rat SDH utilizing a rat 720 bp *Eco*R I/*Bgl* II cDNA fragment as probe. For blot conditions cf. legend to Fig. 1.

(Bühler et al., 1983; Rout, 1992). The results obtained here strongly support the idea that class I alcohol dehydrogenase is present in the brain. Human liver class I ADHs catalyze the interconversion of intermediary alcohols and aldehydes of dopamine metabolism *in vitro*, with ethanol and acetaldehyde as competing substrates (Mårdh and Vallee, 1986). The physiological significance of these observations is unclear, but considering the now detected class I mRNA in rat brain, the presence of a corresponding brain enzyme may prove important in the interpretation of the interactions between ethanol and neurotransmitters, and the resultant metabolic events.

The class II mRNA expression is virtually limited to duodenum and liver. Activity staining of rat liver homogenates with ethanol and other primary alcohols visualized only class I and class III ADH (Julià et al., 1987). This raises the question whether class II is really ethanol-active in all species. If not, short primary alcohols are also not likely to be physiological substrates for this enzyme type. Since the rat stomach enzyme was found to be ethanol-active, it was initially assumed to be the class II counterpart in rat. However, this activity has later been shown to represent a separate enzyme, class IV (Parés et al., 1990; 1992; Moreno and Parés, 1991). Regarding class II in rat, only the cDNA from liver has been thus far defined (Höög, 1991), but from human liver, class II has been isolated both at the protein and cDNA levels (Li et al., 1977; Höög et al., 1987).

The wide distribution of class III ADH mRNAs is in agreement with earlier reports (Julià et al., 1987; Uotila and Koivusalo, 1989; Rathnagiri et al., 1989). This universal tissue distribution of class III (glutathione-dependent formaldehyde dehydrogenase) parallels its presence in all species investigated, from prokaryotes, yeasts and plants to animals, and reflects its vital role in formaldehyde elimination throughout evolution. mRNAs from the 20 different tissues all show positive hybridization signals to the class III alcohol dehydrogenase probe, but the corresponding 1.8 kb transcript is lacking in pancreas. In this tissue, however, we detected an additional RNA species (Fig. 3). This small RNA is also present in other tissues, but its significance is unclear and it is the subject of investigation. It should be noted that no mRNA for any of the four dehydrogenases studied could be detected in pancreas. This tissue is rich in nuclease activity, which could degrade RNA, but examination of the ethidium bromide stained gels prior to blotting revealed no degradation of pancreatic RNA.

SDH mRNA was detected in all organs investigated except small intestine and pancreas. The highest levels were found in liver, kidney and testis, and the SDH transcripts were also detected in eye, although the sublocalisation within this tissue is unknown. The presence of SDH mRNA in the eye could be of interest, considering the fact that accumulation of sorbitol in the lens of diabetic patients is likely to be a cause of cataracts, and is associated with the relative absence of SDH in lens epithelia (Gabbay, 1973).

In conclusion, we have monitored the relative occurrence of ADH and SDH mRNAs in numerous rat tissues, demonstrated the existence of class I ADH mRNA in mammalian brain, and for the first time described the mRNA tissue distribution of class II ADH.

ACKNOWLEDGEMENTS

This project was supported by grants from the Swedish Medical Research Council, the Swedish Natural Sciences Research Council, the Swedish Alcohol Research Fund and the Magn. Bergvall Foundation.

REFERENCES

Beisswenger, T. B., Holmquist, B., and Vallee, B.L., 1985, χ-ADH is the sole alcohol dehydrogenase isozyme of mammalian brains: Implications and inferences, *Proc. Natl. Acad. Sci. USA*, 82:8369.

Boleda, D. M., Julià, P., and Parés, X., 1989, Role of extrahepatic alcohol dehydrogenase in rat ethanol metabolism, *Arch. Biochem. Biophys.*, 274:74.

Bühler, R., Pestalozzi, D., Hess, M., and von Wartburg, J.-P., 1983, Immunohistochemical localization of alcohol dehydrogenase in human kidney, endocrine organs and brain, *Pharmacol. Biochem. Behav.*, 18:55.

Dafeldecker, W.P., and Vallee, B.L., 1986, Organ-specific human alcohol dehydrogenase: Isolation and characterization of isozymes from testis, *Biochem. Biophys. Res. Commun.*, 134:1056.

Gabbay, K., 1973, The sorbitol pathway and the complications of diabetes, *New Engl. J. Med.*, 288:831.

Höög, J.-O., von Bahr-Lindström, H., Hedén, L.-O., Holmquist, B., Larsson, K., Hempel, J., Vallee, B.L., and Jörnvall, H., 1987, Structure of the class II enzyme of human liver alcohol dehydrogenase: combined cDNA and protein sequence determination of the π subunit, *Biochemistry*, 26:1926.

Höög, J.-O., 1991, Mammalian class II alcohol dehydrogenase: Species and class comparisons at genomic and protein levels, In: Enzymology and Molecular Biology of Carbonyl Metabolism (Weiner, H. et al. eds., Plenum Press, New York), 3:285.

Jeffery, J., and Jörnvall, H., 1988, Sorbitol dehydrogenase, *Adv. Enzymol.*, 61:47.

Julià, P., Farrés, J., and Parés, X., 1987, Characterization of three isoenzymes of rat alcohol dehydrogenase, *Eur. J. Biochem.*, 162:179.

Jörnvall, H., Persson, B., Krook, M., and Hempel, J., 1991, Alcohol and aldehyde dehydrogenases, In: The Molecular Pathology of Alcoholism (Palmer, N. ed., Oxford Univ. Press), pp 130-156.

Karlsson, C., Jörnvall, H., and Höög, J.-O., 1991, Sorbitol dehydrogenase: cDNA coding for the rat enzyme; variations within the alcohol dehydrogenase family independent of quaternary struture and metal content, *Eur. J. Biochem.*, 198:761.

Koivusalo, M., Baumann, M., and Uotila, L., 1989, Evidence for the identity of glutathione-dependent formaldehyde dehydrogenase and class III alcohol dehydrogenase, *FEBS Lett.*, 257:105.

Li, T.-K., Bosron, W.F., Dafeldecker, W.P., Lange L.G., and Vallee, B.L., 1977, Isolation of π-alcohol dehydrogenase of human liver: is it a determinant of alcoholism?, *Proc. Natl. Acad. Sci. USA*, 74:4378.

Moreno, A., and Parés, X., 1991, Purification and characterization of a new alcohol dehydrogenase from human stomach, *J. Biol. Chem.*, 266:1128.

Mårdh, G., and Vallee, B.L., 1986, Human class I alcohol dehydrogenases catalyze the interconversion of alcohols and aldehydes in the metabolism of dopamine, *Biochemistry*, 25:7279.

Mårdh, G., Dingley, A.L., Auld, D.S., and Vallee, B.L., 1986, Human class II (π) alcohol dehydrogenase has a redox-specific function in norepinephrine metabolism, *Proc. Natl. Acad. Sci. USA*, 83:8908.

Parés, X., Farrès, J., and Vallee, B.L., 1984, Organ-specific alcohol metabolism: Placental χ-ADH, *Biochem. Biophys. Res. Commun.*, 119:1047.

Parés, X., Moreno, A., Cederlund, E., Höög, J.-O., and Jörnvall, H., 1990, Class IV mammalian alcohol dehydrogenase. Structural data of the rat stomach enzyme reveal a new class well separated from those already characterized, *FEBS. Lett.*, 277:115.

Parés, X., Cederlund, E., Moreno, A., Saubi, N., Höög, J.-O., and Jörnvall, H., 1992, Class IV alcohol dehydrogenase (the gastric enzyme). Structural analysis of human $\sigma\sigma$-ADH reveals class IV to be variable and confirms the prescence of a fifth mammalian alcohol dehydrogenase class, *FEBS Lett.*, 303:69.

Rathnagiri, P., Linnoila, M., Blanche O'Neill, J., and Goldman, D., 1989, Distribution and possible metabolic role of class III alcohol dehydrogenase in the human brain, *Brain Res.*, 481:131.

Rout, U.K., 1992, Alcohol dehydrogenases in the brains of mice, *Alcoholism: Clin. Exp. Res.*, 16:286.

Sharma, C.P., Fox, E.A., Holmquist, B., Jörnvall, H., and Vallee, B.L., 1989, cDNA sequence of human class III alcohol dehydrogenase, *Biochem. Biophys. Res. Commun.*, 164:631.

Uotila, L., and Koivusalo, M., 1989, Formaldehyde dehydrogenase, In: Coenzymes and Cofactors. Glutathione: Chemical, Biochemical, and Medical Aspects - Part A (Dolphin, D. et al. eds., John Wiley and Sons, New York), 3: 517.

Vallee, B.L., and Bazzone, T.J., 1983, Isozymes of human liver alcohol dehydrogenase, *Isozymes*, 8:219.

Yasunami, M., Chen, C.-S., Yoshida, A., 1991, A human alcohol dehydrogenase gene (*ADH 6*) encoding an additional class of isozyme, *Proc. Natl. Acad. Sci. USA*, 88:7610.

Boleda, D. M., Julià, P., and Parés, X., 1989, Role of extrahepatic alcohol dehydrogenase in rat ethanol metabolism, *Arch. Biochem. Biophys.* 274:74.
Bühler, R., Pestalozzi, D., Hess, M., and von Wartburg, J.-P., 1983, Immunohistochemical localization of alcohol dehydrogenase in human kidney, endocrine organs and brain, *Pharmacol. Biochem. Behav.* 18:55.
Dafeldecker, W. P., and Vallee, B. L., 1986, Organ-specific human alcohol dehydrogenase: Isolation and characterization of isozymes from testis, *Biochem. Biophys. Res. Commun.* 134:1056.
Gabbay, K., 1973, The sorbitol pathway and the complications of diabetes, *New Engl. J. Med.* 288:831.
Höög, J.-O., von Bahr-Lindström, H., Hedén, L.-O., Holmquist, B., Larsson, K., Hempel, J., Vallee, B. L., and Jörnvall, H., 1987, Structure of the class II enzyme of human liver alcohol dehydrogenase: combined cDNA and protein sequence determination of the π subunit, *Biochemistry* 26:1926.
Höög, J.-O., 1991, Mammalian class II alcohol dehydrogenase: Species and class comparisons at genomic and protein levels, In: Enzymology and Molecular Biology of Carbonyl Metabolism (Weiner, H., et al., eds.), Plenum Press, New York, 3:285.
Jeffery, J., and Jörnvall, H., 1988, Sorbitol dehydrogenase, *Adv. Enzymol.* 61:47.
Julià, P., Farrés, J., and Parés, X., 1987, Characterization of three isoenzymes of rat alcohol dehydrogenase, *Eur. J. Biochem.* 162:179.
Jörnvall, H., Persson, B., Krook, M., and Hempel, J., 1991, Alcohol and aldehyde dehydrogenases, In: The Molecular Pathology of Alcoholism (Palmer, T. N., ed.), Oxford Univ. Press, [illegible]
[illegible]
[illegible]

THE ROLE OF ALCOHOL AND ALDEHYDE DEHYDROGENASES IN ALCOHOL-RELATED DISEASES: CLINICAL STUDIES OF MOLECULAR MARKERS

David I.N. Sherman, Roberta J. Ward, and Timothy J. Peters

Department of Clinical Biochemistry
Kings College School of Medicine and Dentistry
London SE5 8RX, U.K.

INTRODUCTION

The rapid advances achieved in the field of molecular genetics over the past few years have had a profound effect on the study of complex conditions as well as single gene disorders. In the field of alcoholism and alcohol - related diseases, the application of these new techniques is now gathering pace. There is general agreement that a significant genetic component exists in alcoholism, as illustrated by the twin and adoption studies performed over the past 20 years, but this clearly does not conform to a simple Mendelian model. The central question that is being addressed concerns the inherited basis for the observed variation in individual responses to ethanol.

Alcoholism and its many associated clinical syndromes comprise an extremely heterogenous set of disorders. In genetic terms, there are many possible phenotypes, which are produced by a combination of inherited and environmental influences. Phenocopies (phenotypes due to environmental effects but indistinguishable from genetically determined traits) may further confound the issue. Inherited predisposition is likely to be determined by several 'major' and 'minor' polymorphic gene loci, with a number of alleles which exhibit varying degrees of penetrance. Racial and gender differences also play a part .Thus it can be seen that any prospective investigator may face greater difficulties in the clinical definition of subjects than in the application of molecular biological techniques, many of which are now routine.

Acetaldehyde may be involved not only in alcohol addiction, for example, by its ability to form condensation products (THIQs) with biogenic amines, but also in the pathogenesis of liver and other organ damage, and in aversion to ethanol via the Oriental flushing reaction. The two enzymes involved in the principal pathway of ethanol metabolism, alcohol dehydrogenase

(ADH) and aldehyde dehydrogenase (ALDH), fulfill the criteria for candidate genes likely to be implicated in alcohol related disorders. Although there is some information available in the literature from phenotypic studies of isoenzymes, the need to obtain liver biopsy specimens in general has precluded the large-scale studies that are necessary for informative results. As the cDNAs for the genes encoding the major isoenzymes of ADH and ALDH have been cloned, they are particularly suitable for clinical genetic studies.

Two possible approaches can be used to investigate the role of candidate genes in a disorder. The simplest is to attempt to demonstrate an association between a gene marker and a particular phenotype in a selected population. In the case of ADH and ALDH, the relevant polymorphic loci are ADH_2, ADH_3, $ALDH_2$ and possibly $ALDH_1$. Specific alleles can be genotyped by designing oligonucleotide primers from the known cDNA sequence to amplify genomic DNA extracted from leucocytes by the Polymerase Chain Reaction. Alleles are then distinguished by slot blot hybridisation with oligonucleotide probes, or by the use of restriction endonucleases. Alternatively, Restriction Fragment Length Polymorphisms (RFLPs), or other markers such as minisatellite polymorphisms or tandem repeats, can be used. These can yield information about coding or non-coding regions of the gene, or even sequences some distance from the gene in question. If linkage between such a marker and alcoholism can be demonstrated in family pedigrees informative for the disorder, this provides powerful evidence for the involvement of a gene. The degree of genetic linkage is normally expressed logarithmically as a LOD score. As yet, there has been no positive study demonstrating significant linkage between a genetic marker and alcoholism in families. This may be partly due to difficulties in phenotype definition and distinguishing the effects of multiple genes, as well as finding suitable families for study.

In association studies comparing alcoholics and controls, there are several potential drawbacks: a) the sample should be large enough that chance findings are minimised, population stratification is reduced and small differences can be detected; b) the population should be racially homogenous, as allele frequencies are likely to differ between races and pathogenesis of alcoholism may differ between races; c) control subjects should match alcoholics in gender composition and age distribution, and preferably be screened to exclude alcoholism. Some authors have argued that in a large sample, with an estimated population frequency of the disorder of 10 - 15%, this is unnecessary.

In this paper we will briefly review recent association studies of ADH and ALDH polymorphism in both alcoholism and alcoholic liver disease, and preliminary data for the involvement of $ALDH_1$ in aversion to alcohol in caucasians.

STUDIES IN ORIENTAL POPULATIONS

Studies of Oriental subjects should be considered separately as the effects of the mutant $ALDH_2^2$ allele must be separated from those of ADH_2 and ADH_3 polymorphism. The $ALDH_2^2$ allele is inherited in an autosomal co-dominant manner, so that homozygotes exhibit marked reduction in $ALDH_2$ enzyme activity (Enomoto *et al.*, 1991). The Japanese population has been studied the most extensively. The 35% of the population possessing the mutant allele, which encodes the inactive form of mitochondrial ALDH, will show elevated blood acetaldehyde levels after alcohol consumption, resulting in numerous unpleasant symptoms including flushing and tachycardia which are aversive to further intake. The majority of those affected, however, are heterozygous for this allele, and show intermediate levels of blood acetaldehyde. In the original studies by Yoshida's group (Shibuya & Yoshida, 1988), $ALDH_2$ genotypes were studied in a sample of 49 non-alcoholic controls. Twenty one were homozygous for wild type $ALDH_2$ ($ALDH_2^1$ / $ALDH_2^1$), 22 were heterozygous ($ALDH_2^1$ / $ALDH_2^2$) and 6 homozygous for the mutant gene ($ALDH_2^2$ / $ALDH_2^2$). In a further study, symptoms of alcohol aversion were sought in 15 unrelated Japanese males (Shibuya, Yasunami and Yoshida, 1989). All of the five homozygotes for the wild type $ALDH_2^1$ allele were non-flushers, and both homozygotes for the mutant $ALDH_2^2$ allele were flushers. Seven of the eight heterozygotes were flushers, indicating that the majority had significantly reduced $ALDH_2$ activity and therefore increased acetaldehyde levels. Out of 23 patients with histologically proven alcoholic liver disease, 20 were homozygous for $ALDH_2^1$, 3 heterozygotes and none homozygotes for $ALDH_2^2$. There was, therefore, a strong association between the mutant allele of the $ALDH_2$ gene and both aversion to ethanol and protection against alcoholic liver disease. Genotyping of the same patient groups for the ADH_2 locus showed allele frequencies of 0.71 for ADH_2^2, and 0.29 for ADH_2^1. The 23 patients with alcoholic liver disease did not show significant differences from the controls. Although the flushers showed an increased frequency of the ADH_2^2 allele, this was not significant in a small sample size.

In a further study of ADH_2, ADH_3 and $ALDH_2$ genotypes in a different population, Thomasson *et al.* (1991) studied 49 male Taiwanese alcoholics (as defined by DSM-III criteria) in comparison with 47 male non-alcoholic controls. The allele frequencies for $ALDH_2$ were similar to those found in Japanese (0.70 for $ALDH_2^1$ and 0.30 for $ALDH_2^2$), but the frequency of the mutant $ALDH_2^2$ allele was markedly reduced in the alcoholics (0.06). ADH_2 frequencies in the controls were similar to those found in Japanese. However, frequencies of the ADH_2^2 and ADH_3^1 alleles were significantly lower in the alcoholics, even when analysed independently from their $ALDH_2$ genotype. As both alleles encode isoenzyme subunits with higher K_m for ethanol, the results suggested that increased rates of acetaldehyde formation were associated with "protection" against the development of alcoholism. No information on clinical parameters of severity, which would have strengthened the case for such an association, were given.

Thus, in Orientals it can be inferred from genetic studies that the rate of acetaldehyde formation plays a crucial part in the development of alcoholism. From the limited information available, the influence of $ALDH_2$ genotype appears to be more important than ADH_2 or ADH_3 genotype. Only those subjects who do not possess the mutant $ALDH_2^2$ allele or symptoms of aversion are able to sustain the levels of alcohol intake required to cause alcoholic liver disease. The role of this allele in predisposition to alcoholic liver disease was examined further by Enomoto *et al.* (1991) in 47 patients, all of whom were interviewed on alcohol intake and had liver biopsies. Not surprisingly, no $ALDH2^2$ homozygotes were found and there were only seven heterozygotes. Interestingly, the severity of histological liver damage appeared to be greater in the small number of heterozygotes, despite a significantly lower mean daily alcohol intake. An inability to metabolise acetaldehyde, perhaps leading to accumulation of immunogenic acetaldehyde-protein adducts, would also therefore appear to predispose Orientals to hepatic damage.

STUDIES IN CAUCASIAN POPULATIONS

Even though the importance of acetaldehyde in addiction and alcohol-related organ damage is accepted, no direct evidence exists as yet to implicate polymorphism of an ALDH gene in alcoholism in Caucasians. More recently we have investigated the incidence and aetiology of the alcohol-related flush in caucasian subjects. Although early studies by Wolff (1972) indicated a low incidence (5-10%) in such subjects, our recent survey of over 200 caucasians revealed that approximately 50% of females and 8% of males show alcohol-induced flushing. Furthermore, there was a high incidence of alcohol-related flushing in other family members. The flushing phenomenon was associated in all cases with a low activity of erythrocyte $ALDH_1$ activity. A test dose of alcohol to affected individuals confirmed the presence of a flushing response, which was of shorter duration (approximately 20 minutes) and was not associated with high blood concentrations of acetaldehyde. The role of $ALDH_1$ within the cell is unknown, although its high K_m for acetaldehyde would preclude a major role in its metabolism. Further studies are in progress to investigate its substrate specificity, its association with the short-lived flush after ethanol consumption and the molecular basis for the decreased $ALDH_1$ activity.

It is likely that the role of ADH, the rate limiting enzyme in ethanol metabolism and acetaldehyde formation, is of greater importance in Caucasians than in Orientals. For this reason attention has focused on polymorphism of the ADH_2 and ADH_3 genes, and in particular the ADH_2^2 and ADH_3^1 alleles. These encode the β_2 and γ_1 subunits, which in the homodimeric form are more active in the metabolism of ethanol *in vitro* than β_1 or γ_2. One might expect that, in contrast to Orientals, these alleles would occur in higher frequencies in alcoholics or patients with alcoholic liver disease as the aversive effects of acetaldehyde accumulation are not operating. Phenotypic studies have suggested that the β_2 subunit is present in only 5 to 20% of Caucasians, and the γ_1 subunit in 50 to 60%, (Bosron & Li, 1986).

Couzigou *et al.* (1991) pursued this question by genotyping 46 French alcoholic patients with cirrhosis as well as 39 controls for the ADH_2 and ADH_3 loci. The frequency of the ADH_2^2 allele was lower than expected (2.5% in the controls and 4% in the cirrhotics), with no homozygotes being detected. No differences in ADH_3 genotype were seen between the two groups. A slightly larger study of a homogenous population from the northeast of England by Day *et al.* (1991) examined 59 patients with alcoholic cirrhosis and 79 controls. The ADH_2^2

genotype was virtually non-existent, occurring in a single control subject and in none of the cirrhotic patients, whereas the $ADH3^1$ allele showed a marginally increased incidence in the cirrhotic group.

In the absence of any information on severity of liver damage or duration of ethanol intake, it can be concluded from both studies that no definite evidence for an association between the ADH3 gene and alcoholic liver disease exists. As far as the ADH2 gene is concerned, there seems even less likelihood that it has a significant role in genetic predisposition in these two populations, although the low $ADH2^2$ allele frequencies appear slightly at odds with the published phenotypic frequencies of the $\beta 2$ subunit. In general, it is difficult to draw firm conclusions from small association studies such as these. The two groups from the Newcastle study were well matched racially, but the gender and age distributions and alcohol intake of the control group were not stated, which may have had an impact on the ADH3 results.

A complementary approach to genotyping is to look at other polymorphic markers such as RFLPs. We have recently studied a two allele PvuII RFLP for the ADH2 gene, demonstrated in leucocyte DNA using a 1.3 Kb genomic ADH36 probe (Smith, 1986), in a group of alcoholics with liver disease and controls. The two alleles, A and B, were denoted by the presence of a 5.1 Kb band and a 3.1/2.9 Kb doublet, respectively, on Southern hybridization. Leucocyte DNA was obtained from 23 non-alcoholic controls and 46 alcoholics, of whom 39 had histological evidence of liver disease (26 cirrhosis with or without alcoholic hepatitis, 5 alcoholic hepatitis alone and 8 fatty liver). Twenty one had clinical and/or questionnaire evidence of dependence, and 19 had a positive family history, with at least one affected first degree relative. The two groups were well matched with respect to gender. The frequency of the B allele was significantly increased in the alcoholic group taken as a whole compared with the control group (0.63 *vs* 0.15; $X^2 = 25.8$, $p < 0.001$). In addition, patients with cirrhosis and/or alcoholic hepatitis showed a greater incidence of the B allele than those with fatty liver alone ($X^2 = 10.8$, $p < 0.05$), demonstrating a correlation with severity of liver damage.

The exact significance of this finding at the molecular level remains to be clarified. The 1.3 Kb ADH36 probe used contains the whole of exon 3 of the $ADH2^2$ allele, with large portions of introns 2 and 3. Genotyping studies for the ADH2 and ADH3 genes are in progress. Given the low frequencies of the $ADH2^2$ allele in the caucasian populations studied to date, it seems unlikely that the B allele corresponds to a mutation in a coding region of the ADH2 gene. However, the strength of the association suggests that this marker merits further study.

CONCLUSION

The role of the mutant allele of the ALDH2 gene in aversion to alcohol and in relation to liver disease in Orientals has been greatly clarified in recent years. Also, evidence is emerging of a relationship between low $ALDH_1$ activity and alcohol-related flushing in Caucasians. In contrast, the relationship between polymorphism of the ADH genes, ethanol metabolism, acetaldehyde accumulation and predisposition to alcohol-related disorders remains unclear, especially in Caucasians, although recent studies have provided encouraging data. It can therefore be seen that further large scale studies of molecular markers for both ADH and ALDH are needed in different racial groups, with an emphasis on meticulous collection of clinical data in terms of alcohol intake, dependency, liver disease etc., if we are to unravel further the role of these genes in genetic predisposition to alcohol related disorders.

REFERENCES

Bosron, W.F., and Li, T-K. (1986). Genetic polymorphism of human liver alcohol and aldehyde dehydrogenases and their relationship to alcohol metabolism and alcoholism. *Hepatology*, 6: 502-510.

Couzigou, P., Fleury, B., Groppi, A., Cassaigne, A., Begueret, J., Iron, A., and the French Group for Research on Alcohol and Liver. (1990). Genotyping study of alcohol dehydrogenase class I polymorphism in french patients with alcoholic cirrhosis. *Alcohol & Alcoholism*, 25: 623-626.

Day, C.P., Bashir, R., James, O.W.F., Bassendine, M.F., Crabb, D.W., Thomasson, H.R., Li, T-K., and Edenberg, H. (1991). Investigation of the role of polymorphisms at the alcohol and aldehyde dehydrogenase loci in genetic predisposition to alcohol-related end-organ damage. *Hepatology*, 14: 798-801.

Enomoto, N., Takase, S., Takada, N., and Takada, A. (1990). Alcoholic liver disease in heterozygotes of mutant and normal aldehyde dehydrogenase-2 genes. *Hepatology*, 13: 1071-1075.

Enomoto, N., Takase, S., Yasuhara, M., and Takada, A. (1991). Acetaldehyde metabolism in different aldehyde dehydrogenase genotypes. *Alcoholism: Clinical and Experimental Research*, 15: 141-144.

Shibuya, A., and Yoshida, A. (1988). Frequency of the atypical aldehyde dehydrogenase gene ($ALDH_2^2$) in japanese and caucasians. *American Journal of Human Genetics*, 3: 741-743.

Shibuya, A., and Yoshida, A. (1988). Genotypes of alcohol-metabolising enzymes in japanese with alcohol liver disease: a strong association of the usual caucasian-type aldehyde dehydrogenase gene ($ALDH_2^1$) with the disease. *American Journal of Human Genetics*, 43: 744-748.

Shibuya, A., Yasunami, M., and Yoshida, A. (1989). Genotypes of alcohol dehydrogenase and aldehyde dehydrogenase loci in japanese alcohol flushers and non-flushers. *Human Genetics*, 82: 14-16.

Smith, M. (1986). Genetics of human alcohol and aldehyde dehydrogenases. In H. Harris & K. Hirchorn (Eds.), *Advances in Human Genetics* (pp. 249-290). New York: Plenum Press.

Thomasson, H.R., Edenberg, H.J., Crabb, D.W., Mai, X-L., Jerome, R.E., Li, T-K., Wang, S-P., Lin, Y-T., Lu, R-B., and Yin, S-J. (1991). Alcohol and aldehyde dehydrogenase genotypes and alcoholism in Chinese men. *American Journal of Human Genetics*, 48: 677-681.

Wolff, P.H. (1973). Vasomotor sensitivity to alcohol in diverse mongoloid populations. *American Journal of Human Genetics*, 25: 193-199.

REGULATION OF THE HUMAN ALCOHOL DEHYDROGENASE GENES *ADH1*, *ADH2* AND *ADH3*: DIFFERENCES IN *CIS*-ACTING SEQUENCES AT CTF/NF-I SITES

Howard J. Edenberg, Celeste J. Brown, and Lu Zhang

Dept. of Biochemistry and Molecular Biology
Indiana University School of Medicine
Indianapolis, IN 46202-5122 U.S.A.

INTRODUCTION

Humans have three closely-related class I alcohol dehydrogenase (ADH) genes, *ADH1, ADH2* and *ADH3*. Each of these encodes an ADH polypeptide of about 40 kDa; these polypeptides form homo- and heterodimeric ADH isozymes that catalyze the reversible oxidation of a wide range of alcohols to their aldehydes (reviewed in Ehrig et al., 1990; Smith, 1986). The individual isozymes differ in kinetic properties and substrate specificities (Ehrig et al., 1990). They catalyze the rate-limiting step in ethanol oxidation, and in so doing they produce acetaldehyde, a toxic metabolite. Thus they are important enzymes in influencing the rate of ethanol oxidation and acetaldehyde production, and through that the physiological and pathological effects of ethanol ingestion.

All three *ADH* genes are expressed at high levels in adult liver (Ehrig et al., 1990; Smith, 1986), where the ADH isozymes carry out the bulk of ethanol oxidation. There are, however, differences in the expression of the three *ADH* genes elsewhere in the body and during liver development. For example, in the liver *ADH1* is expressed early in gestation, followed by *ADH2; ADH3* is detectable in liver only after birth (Smith et al., 1971, 1972). Kidney also shows developmental changes, with *ADH3* predominating prenatally and *ADH2* after birth (Smith et al., 1971). The lung predominantly expresses *ADH2* at all stages. In stomach, *ADH3* predominates in the mucosa and *ADH2* in the muscular layer (Smith et al., 1972; Yin et al., 1988). None of these isozymes are detectable in the brain (Smith, 1986). Thus there must be both common and unique aspects to the regulation of these genes.

We are analyzing the transcriptional control of *ADH1, ADH2* and *ADH3* in order to determine both the *cis*-acting sequences important in their regulation, and the transcription factors that regulate their expression in different tissues. Differences in expression of the *ADH* genes can affect the metabolism of ethanol and other alcohols by various tissues in the body.

We have mapped, by DNase I footprinting, a complex set of *cis*-acting sequences in the proximal 250 to 450 bp upstream of the transcriptional start sites of *ADH1, ADH2* and

Enzymology and Molecular Biology of Carbonyl Metabolism 4
Edited by H. Weiner, Plenum Press, New York, 1993

ADH3. Comparisons among the genes reveal both common footprints and subtle differences, often the result of differences in one or two base pairs. We are determining which transcription factors can bind to each *cis*-acting element by examining the binding of purified factors to these sequences and by analyzing the effects of competition with oligonucleotides containing known binding sites. We analyze binding in several mouse tissues and in cell lines that do and do not express their endogenous *ADH* gene(s). The effects of the *cis*-acting elements on gene expression are analyzed by either transient transfection assays in these cell lines or by in vitro transcription assays in tissue or cell extracts.

Human *ADH* gene regulation has recently been reviewed (Edenberg, 1991; Edenberg and Brown, 1992; Duester, 1991). Proteins contained in liver nuclei bind tightly to many sites in the proximal 450 bp of the promoter (Figure 1). The probable identity of some of these proteins has been deduced by examining either the binding of purified transcription factors to the promoters or the elimination of a footprint by competition with an oligonucleotide containing a consensus sequence for a known protein.

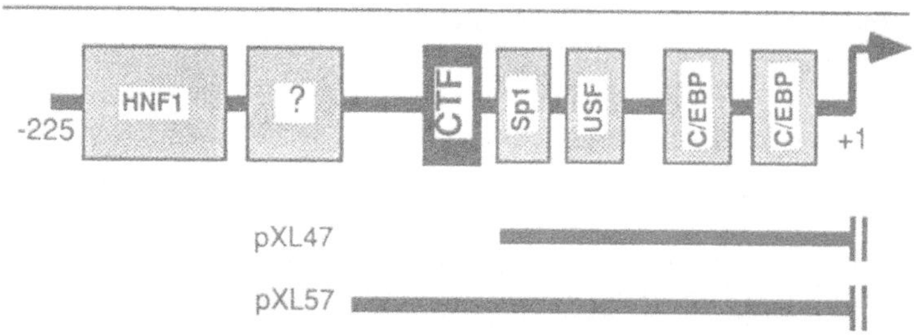

Figure 1. Sites in the *ADH2* promoter at which transcription factors bind. The proximal 225 bp of the *ADH2* promoter is depicted,, with the sites footprinted by proteins in mouse liver nuclear extracts shown. The likely identity of the proteins bound at each site is indicated (Brown et al., 1992a; C. J. B., L. Z. and H. J. E., in preparation). The arrow marks the transcription start site. The portion of the promoter contained in pXL47 and pXL57 are depicted as heavy lines below; both extend to +33 bp.

Proteins of the C/EBP family (C/EBPα, C/EBPβ or LAP and C/EBPδ; Enver et al., 1990; Descombes et al., 1990; Cao et al., 1991; Edenberg and Brown, 1992; C. J. B., L. Z. and H. J. E., in preparation) and DBP (Mueller et al., 1990) can bind to one or both of two sites flanking the TATA box in *ADH1*, *ADH2* and *ADH3* (Stewart et al., 1990, 1991; van Ooij et al., 1992; C. J. B., L. Z. and H. J. E., in preparation). It is likely that NF-Y (Dorn et al., 1987; Raymondjean et al., 1988; Wuarin et al., 1990) can also bind to one or both of these sites. A sequence whose core is GATCACGTG is very important in transcription initiation (Carr and Edenberg, 1990; Brown et al., 1992a). A widespread protein closely related to USF/MLTF (Sawadogo and Roeder, 1985; Miyamoto et al., 1985; Carthew et al., 1985, 1987; Chodosh et al., 1987) or mTFE3 (Roman et al., 1992) is the most likely candidate for binding to this site. Human USF has been shown to bind to the related CACATG site of the rat *ADH-3* promoter and to increase transcription (Potter et al., 1991). The widely expressed transcription factor Sp1 can bind to the G3T sites (GGGTGTGGC; Carr et al., 1989; Carr and Edenberg, 1990) in all three genes (Brown et al., 1992a). A protein from mouse liver nuclei closely resembling Sp1 is the major factor binding to the

G3T sequence (Brown et al., 1992a). The G3T site is very important in transcription initiation (Carr and Edenberg, 1990; Brown et al., 1992a). HNF-1α and HNF-1β (Mendel et al., 1991; Mendel and Crabtree, 1991) can bind to a site in all three genes (Brown et al., 1992b; C. J. B., L. Z. and H. J. E., in preparation). There are several sites for which we do not yet know the identity of the transcription factor.

At several of these sites *ADH1*, *ADH2* and *ADH3* differ slightly in nucleotide sequence. This may result in differences in the affinity of the three genes for the transcription factor, and might also influence which member of a class of transcription factors can bind there. For example, we have found that both purified Sp1 and the mouse liver G3T-binding protein that resembles Sp1 bind to *ADH3* with an affinity severalfold less than for *ADH2* (Brown et al., 1992a). C/EBPα binds to *ADH2* with higher affinity than it binds to either *ADH1* or *ADH3* (C. J. B., L. Z. and H. J. E., in preparation). Since cells and tissues differ not only in which transcription factors they express but also in their level of expression, even small differences in the affinity with which a sequence binds a given factor might have a major influence on the tissues in which each gene is expressed, and on its level of expression.

There is a group of cellular proteins called CTF/NF-I that is involved in both DNA replication and transcription. NF-I was isolated from HeLa cells by affinity chromatography using oligonucleotides containing its binding site in the Adenovirus 2 replication origin as the affinity matrix; it purified as a group of proteins with apparent molecular weights ranging from 52,000 to 66,000 (Rosenfeld and Kelly, 1986). CTF (CCAAT-binding transcription factor) was similarly isolated from HeLa cells with promoter sequences from either the Harvey ras-1 gene or the human α-1 globin gene as the affinity matrix; it purified as a similar group of proteins, which was found to be indistinguishable from NF-I (Jones et al., 1987). CTF/NF-I cDNAs were cloned as a set of alternatively-spliced transcripts, and individual cDNAs were shown to be functional in both DNA replication and transcription activation assays (Santoro et al., 1988). The consensus sequence was defined as 5'-TGG(A/C)N5GCCAA, with the highest affinity sites related to the symmetrical sequence 5'-TTGGCTN3AGCCAA (Gronostajski et al., 1985; Gronostajski, 1986; Jones et al., 1987; de Vries et al., 1987; Rosenfeld et al., 1987; Goyal et al., 1990). Most contacts are on one side of the DNA helix (de Vries et al., 1987). Half-sites can be bound with lower affinity: point mutations in either half-site can reduce binding affinity by four- to ten-fold (Rosenfeld et al., 1987; Jones et al., 1987). A single point mutation converting the TGG to a TCG was reported to reduce binding by ten-fold (Rosenfeld et al., 1987) to 100-fold (Goyal et al., 1990). CTF/NF-I is widely distributed in mouse cell lines, although different cells preferentially contain different isoforms (Goyal et al., 1990).

We have found that an oligonucleotide containing a CTF/NF-I consensus binding site competes with promoter fragments of all three *ADH* genes, eliminating one footprint in each gene. The exact site of the CTF/NF-I-related footprint differs in the three genes, however. This report presents an analysis of the binding of this CTF/NF-I-like protein.

MATERIALS AND METHODS

DNase I Footprinting

DNAse I footprinting (Galas and Schmitz, 1978) was carried out on DNA fragments excised from the *ADH1*, *ADH2* and *ADH3* promoters as previously described (Brown et al., 1992a). The promoter fragments, *ADH1* (-227/+51), *ADH2* (-271/+33), and *ADH3* (-326/+65), were labeled at the 5' end of the coding strand, using [^{32}P]ATP and polynucleotide kinase. Fragments were purified by HPLC.

From .02 to .05 pmol of radiolabeled DNA (10,000 cpm) was incubated at room temperature in a final volume of 20 µl of 10 mM HEPES (pH 7.9), 60 mM KCl, 1 mM dithiothreitol, 1 mM EDTA, 7% glycerol, 2 µg poly(dIdC), with or without 40 µg of protein from mouse liver nuclear extracts (Carr and Edenberg, 1990; Brown et al., 1992a). Where noted, a 100-fold excess of double-stranded competitor oligonucleotide that contains a CTF/NF-I-consensus site (Jones et al., 1987) CCTTTGGCATGCTGCCAATATG (Promega, Madison WI) was added to the incubation before the labeled promoter fragment. After 20 minutes, DNAse I was added in 5 µl of 10 mM HEPES (pH 7.6), 60 mM KCl, 25 mM $MgCl_2$, 5 mM $CaCl_2$, and 50% glycerol. The amount of DNase I added was 0.02 units for incubations in the absence of proteins and 0.2 units when the nuclear protein extract was present. After 2 minutes of DNase I digestion each reaction was stopped with 75 µl of 20 mM Tris pH 7.5, 5 µg tRNA, 20 mM EDTA, 5 mM EGTA, then phenol/chloroform extracted and ethanol precipitated. Samples were resuspended in formamide loading dye and electrophoresed in a 6% polyacrylamide/7 M urea gel.

Transient transfections and CAT assays

Two plasmids in which portions of the *ADH2* promoter were placed in front of the chloramphenicol acetyltransferase gene (*cat*) were compared. pXL47 has previously been described (Carr and Edenberg, 1990); it contains sequences from bp -93 to +33 of *ADH2*. Thus pXL47 contains the two proximal C/EBP sites, the USF site and the G3T site (Figure 1). pXL57 is similar, but extends upstream to bp -131 and thus includes the CTF/NF-I-like site (Figure 1). H4IIE-C3 or HeLa cells in 100 mM dishes were transfected with a total of 25 µg of DNA: 13 µg of test plasmid, 2 µg of pCMV-luciferase (an internal control; de Wet et al., 1987), 5-10 µg of pUC19 and 0-5 µg of pMSV-CEBPwt (a C/EBPα-expression vector; Friedman et al., 1989) by calcium phosphate co-precipitation (Graham and van der Eb, 1983). The DNA was allowed to remain on the cells for 4 h (HeLa cells) or 18 h (H4IIE-C3 cells), then the medium was removed and the cells exposed for 2 minutes to medium containing either 15% (H4IIE-C3) or 20% (HeLa) glycerol. This was replaced with the appropriate growth medium and incubation continued for a total of 48 h. The cells were rinsed, harvested, resuspended in 200 µl of 100 mM K•phosphate (pH 7.8) and 1 mM DTT, and broken by sonication. Thirty µl of each extract was assayed for luciferase activity (de Wet et al., 1987). CAT assays (Gorman et al., 1982) were carried out on aliquots that contained equal amounts of luciferase activity. Incubation was for 4 h, and the acetylated products were separated by thin layer chromatography. Quantitation of the CAT assays was carried out on an AMBIS (San Diego, CA) radioisotope detector. CAT activity is expressed as % chloramphenicol acetylated per 10,000 luciferase units.

RESULTS AND DISCUSSION

DNAse I Footprinting

The sequence just upstream of the G3T box in *ADH2* (-95/-107; see Figure 3) resembles the consensus CTF/NF-I sequence TGG(A/C)N_5GCCAA (Gronostajski et al., 1985; Gronostajski, 1986; Jones et al., 1987; de Vries et al., 1987; Rosenfeld et al., 1987; Goyal et al., 1990). Proteins present in mouse liver nuclear extracts bind to the *ADH2* promoter at a site that overlaps part of this sequence and extends upstream (bp -100/-116; Figure 2, lane 7). Rat liver nuclear extracts also contain a protein that footprints *ADH2* in this region (Stewart et al., 1990, 1991; Winter et al., 1990). The footprint produced by mouse liver extracts can be eliminated by adding an excess of an oligonucleotide that contains a known binding site for the transcription factor CTF/NF-I (see METHODS for sequence) to compete with the *ADH2* promoter for binding (Figure 2). The footprint which can be competed by the CTF/NF-I oligonucleotide is immediately upstream of the G3T site.

The pattern of footprinting of *ADH1* and *ADH3* differs (Figure 2). In both of these genes, the footprint that is eliminated by a 100-fold excess of the competitor CTF/NF-I-oligonucleotide is immediately downstream of another footprint, and not adjacent to the G3T site. [Since the fragments analyzed are of different lengths, homologous sequences do

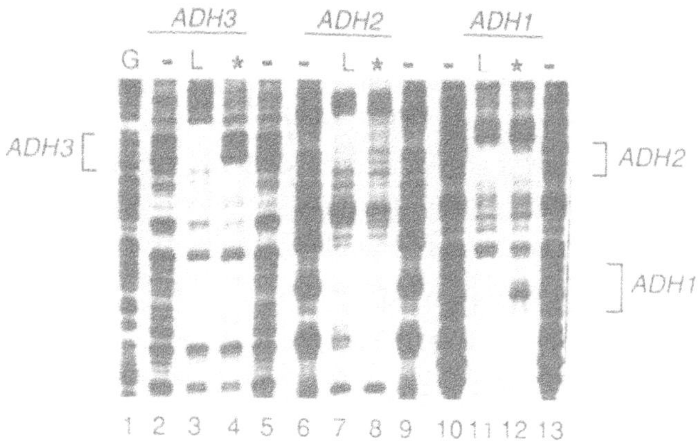

Figure 2. DNase I footprinting of *ADH1, ADH2* and *ADH3*: competition with a CTF/NF-I-consensus oligonucleotide. DNAse I footprinting was carried out as described. The promoter fragments were, from left to right, *ADH3* (lanes 1-5), *ADH2* (lanes 6-9) and *ADH1* (lanes 10-13), all labeled on their coding strand. Lane 1 (G): G+A sequencing reaction. Lanes 2, 5, 6, 9, 10, and 13 (-): DNA fragments in the absence of nuclear extract. Lanes 3, 7, and 11 (L): 40 µg of liver nuclear extract. Lanes 4, 8, and 12 (*): 100-fold excess of double-stranded oligonucleotide containing a CTF/NF-I-consensus site was added to the incubation before the labeled promoter fragment. Brackets indicate the footprint competed by the CTF/NF-I oligonucleotide. Note that the DNA fragments differ in length, therefore homologous sequences do not line up in the gel. Sequences are given in Figure 3.

not align on this gel. The CTF/NF-I site in *ADH3* is actually further away from the transcription start point than the site in *ADH2*, but appears further toward the top of the gel because the large *ADH3* fragment electrophoreses slower. For the same reason, the *ADH3* site looks smaller than the *ADH2* or *ADH1* site, although it covers more bp. See the sequences in Figure 3.] No other footprint in the promoter region extending out to bp -227, -271 and -325 (for *ADH1*, *ADH2* and *ADH3*, respectively) is eliminated by competition with the CTF/NF-I oligonucleotide, although in *ADH3* the competition creates two DNase-I hypersensitive sites further downstream (data not shown). This demonstrates that competition is specifically for a protein that binds to the consensus CTF/NF-I sequence.

Sequence Comparisons

The actual sequences footprinted are underlined in Figure 3. The sequences of both *ADH1* and *ADH3* are very similar to that of *ADH2* in the region between bp -100 and -116 (the *ADH2* sequence is numbered), where the *ADH2* gene is footprinted by the CTF/NF-I-like protein. In this 17 bp sequence *ADH1* differs from *ADH2* by a single T for G transversion. *ADH3* differs from *ADH2* by three substitutions and a single deletion. Yet neither *ADH1* nor *ADH3* are footprinted in that region. It is interesting that the differences make both *ADH1* and *ADH3* more different from the consensus CTF/NF-I sequence. The point mutation at -105 in *ADH1* disrupts the TGG sequence; such a mutation has been shown to reduce the binding affinity of NF-I at least ten-fold (Goyal et al., 1990; Rosenfeld et al., 1987). Differences outside this region at -96, -97, and -98 affect the other half of the consensus.

```
            -145      -135      -125        -115      -105      -95
ADH2    TTTTGAtTGCTGGTTCAGTaCCCTTTT--aTcTGTTTTGACAGTCTGGGAATAATCCA
ADH1    TaTTGAtTGCTGGTTCAtTGCCCTcTTCTTTATGaTTTGACAGTCTGtGAATAATCtA
ADH3    TTTTGAGTGCTGGTTCGGTGCCCATTTCTTTATGATTTGAtAGTC-GaGAAgAATaCg
```

Figure 3. Alignment of CTF/NF-I-related sites. Sequences of *ADH1*, *ADH2* and *ADH3* are aligned between bp -95 and -150 relative to the transcription initiation site; the *ADH2* sequence is numbered. Double underlines mark the regions footprinted in mouse liver extracts that can be eliminated by competition with an oligonucleotide containing a CTF/NF-I-binding site. Singly underlined is an inverted GCAAT sequence (ATTGC) in both *ADH1* and *ADH2* that is mentioned in the text. Nucleotides matching the footprinted sequences (or common to two or three genes in flanking regions) are capitalized.

In the region between bp -121 and -146, all three genes are also quite similar. From -140 to -129 there is another region with some resemblance to a CTF/NF-I site. Partly overlapping is a sequence that resembles a portion of the AP-4 site from SV40 (Mermod et al., 1988). Immediately adjacent to this (-141 to -145) in *ADH2* and *ADH1* is an inverted sequence that matches a C/EBPα-site (GCAAT).

Proteins in the mouse liver extract footprint a large region (28 bp) in *ADH3* that aligns with -121 to -146 of *ADH2* (Figure 2). *ADH1* displays a much smaller footprint, covering only the rightmost 17 bp of the *ADH3* footprint; in this 17 bp sequence, *ADH1* differs from *ADH3* by only two point mutations. *ADH2* differs from both *ADH1* and *ADH3* by a deletion of two bp and by five and three point mutations, respectively, in the rightmost 17 bp, and it is not footprinted between bp -121 and -146 (Figure 2).

It is surprising that the 11 bp at the left side are not footprinted in either *ADH1* or *ADH2*, since both of these genes differ from *ADH3* by only a single point mutation. This point mutation alters the inverted GCAAT sequence (ATTGC) of both *ADH1* and *ADH2* to GCACT (AGTGC in the strand shown; Figure 3). We have found that purified C/EBPα can produce a footprint over the GCAAT sequence in *ADH1*, but not in *ADH2*. The C/EBPα -footprint in *ADH1* does not line up with footprints seen in tissue extracts, rather it overlaps the CTF/NF-I site and the site immediately upstream (C. J. B., L. Z. and H. J. E., in preparation). There may be competition between a transcription factor such as C/EBPα and the CTF/NF-I-like factor that results in CTF/NF-I binding only a shorter sequence in

ADH1, *ADH2* does not footprint in liver extract over this region, even though its sequence is nearly identical to *ADH1*.

Point mutations in one of the two CTF/NF-I-recognition domains can not only reduce the affinity with which the factor binds, but can also reduce the size of the footprint (from 23 bp to 15 bp in the Adenovirus 2 origin sequence) (Jones et al., 1987). This is another possible explanation for the reduced size of the footprint in *ADH1* compared to *ADH3*.

The competition studies depicted in Figure 2 demonstrate that all three genes are binding to a transcription factor that can also bind the consensus CTF/NF-I sequence (de Vries et al., 1987; Jones et al., 1987; Gronostajski, 1986; Santoro et al., 1988). The footprint is large in *ADH3*, and much smaller but overlapping in *ADH1*; a smaller footprint is seen in a different position in *ADH2*. The size of the footprint is consistent with a strong CTF/NF-I-binding site in *ADH3* and weak sites in *ADH1* and *ADH2*. The footprint is seen in liver, kidney, H4IIE-C3 cells, and HeLa cells (C. J. B., L. Z. and H. J. E., in preparation). These data are consistent with the factor being one (or more) of the CTF/NF-I isoforms.

Functional Assays: Transient transfections and CAT assays

The effect of this CTF/NF-I-like site upon expression of *ADH2* was tested by transient transfection assays in two cell lines, H4IIE-C3 cells (a rat hepatoma line that expresses its endogenous ADH at about one-third the level of rat liver; Wolfla et al., 1988) and HeLa cells (a cervical carcinoma line). Cells were cotransfected with a luciferase-encoding vector to allow normalization of the efficiency of transfection, and all data are expressed relative to a constant number of luciferase units.

We compared pXL57, which contains the CTF/NF-I-like site, with pXL47, just 38 bp shorter, which does not (Figure 1). Both plasmids contain the two C/EBP sites flanking the TATA box, and also contains the USF site and the G3T site. pXL47 has previously been shown to direct efficient CAT transcription in transient transfections of H4IIE-C3 cells (Brown et al., 1992a) and in an in vitro assay carried out in nuclear extracts from those cells (Carr and Edenberg, 1990). In a series of seven independent experiments in H4IIE-C3 cells, pXL57 directed only 40% as much CAT expression as pXL47. In three independent experiments in HeLa cells, pXL57 directed only 22% as much CAT expression as pXL47. Thus in two different cell lines, the presence of this CTF/NF-I-like site reduced the level of expression.

Clustering of *cis*-acting elements

The CTF-binding site was originally identified between two Sp1 binding sites in the promoter of the thymidine kinase gene from the herpes simplex virus (Jones et al., 1985). Its presence in *ADH2* just upstream from the G3T site, a variant Sp1-binding site, is interesting. Sp1 sites and CTF or other CCAAT-binding sites are located immediately adjacent to each other in several other genes. The rat γ-fibrinogen gene has a CTF/NF-I site immediately downstream from an Sp1 site (Morgan et al., 1988); it also has a USF binding site immediately upstream of these two sites, a grouping similar to the *ADH* genes. There is a CTF/NF-I site and an inverted G3T site near each other in the human β-globin promoter (Myers et al., 1986). These sites might be clustered to allow the bound transcription factors to interact.

Implications

We have shown that a CTF/NF-I-like protein binds to the promoters of *ADH1*, *ADH2* and *ADH3*, although the exact binding site differs in each gene. Our functional studies using transient transfection into two different cell lines demonstrate that this site reduces transcriptional activity of the *ADH2* promoter. Thus the protein that binds to this site either itself reduces transcription or interacts with other proteins to reduce overall transcription. This is interesting, since in the promoters originally tested, CTF/NF-I acts as a transcriptional activator (Jones et al., 1985, 1987; Santoro et al., 1988).

In our experiments, it is clear that C/EBPα was not the liver protein that footprinted the CTF/NF-I site itself. C/EBPα could footprint a site in *ADH1* that partially overlaps the CTF/NF-I site and a second upstream site found in liver, but the footprint differs from that competed by the CTF/NF-I-consensus oligonucleotide. C/EBPα can produce a much smaller footprint over the rightmost portion of the CTF/NF-I site in *ADH3*, but again this does not align with the footprint seen in liver. In our experiments, C/EBPα did not footprint the *ADH2* promoter anywhere in this region.

C/EBPα has, however, been reported to bind at high concentrations to *ADH2* between bp -111 and -123 (Stewart et al., 1991), a region that partially overlaps the CTF/NF-I site demonstrated here but again does not align with a footprint in mouse liver. A block mutation (from bp -119 to -112) in this site increased CAT activity in HepG2 cells more than two-fold, but surprisingly this mutation did not affect CAT activity when a C/EBPα-expression plasmid was cotransfected (Stewart et al., 1991). Our own functional data, presented above, demonstrate that the CTF/NF-I-like protein acts to reduce expression. We therefore conclude that the transcriptional effect of this sequence is primarily mediated by the CTF/NF-I-like protein, rather than by C/EBPα. We suggest that very high levels of C/EBPα can compete for binding to the CTF/NF-I-site or to an adjacent site and thereby prevent the CTF/NF-I-like protein from repressing transcription.

It is difficult to sort out which transcriptional activators play the important roles under various in vivo situations. Often there are several transcription factors that are capable of binding to a given *cis*-acting element, although with different affinity. The factor bound at a given site may be determined by a competition between different factors, each with a characteristic affinity for the site but expressed at different levels in different cells, in response to different signals or the developmental program, or even at different times of day. This may be further complicated by both positive and negative interactions with the proteins bound at adjacent or overlapping sites. Coexpression of a transcription factor at high levels might allow binding to weak sites that are not occupied in vivo and might displace other factors that normally bind to the sites. There is evidence for this at the CTF/NF-I-site, as discussed above. This suggests caution in interpreting experiments in which there is a very high level of overexpression of a transcription factor.

Comparisons among the *ADH* genes are beginning to identify subtle sequence changes that affect both protein binding and transcriptional activity. This system promises to provide important information about the regulation of tissue- and developmental specificity of gene expression.

ACKNOWLEDGMENTS

This research was supported by NIAAA grant AA06460. C. J. B. was supported by NIAAA training grant T32-AA07463.

REFERENCES

Brown, C.J., Baltz, K.A., and Edenberg, H.J., 1992a, Expression of the human *ADH2* gene: an unusual Sp1-binding site in the promoter of a gene expressed at high levels in liver, *Gene, in press*

Brown, C.J., Zhang, L., and Edenberg, H.J., 1992b, Human alcohol dehydrogenase gene expression: tissue specificity, *J. Cell. Biochem.*, Suppl. 16A:68.

Cao, Z., Umek, R.M., and McKnight, S.L., 1991, Regulated expression of three C/EBP isoforms during adipose conversion of 3T3-L1 cells, *Genes Dev.*, 5:1538.
Carr, L.G., Zhang, K., and Edenberg, H.J., 1989, Protein-DNA interactions in the 5' region of the mouse alcohol dehydrogenase gene *Adh-1*, *Gene*, 78:277.
Carr, L.G. and Edenberg, H.J., 1990, *Cis*-acting sequences involved in protein binding and in vitro transcription of the human alcohol dehydrogenase gene *ADH2*, *J. Biol. Chem.*, 265:1658.
Carthew, R.W., Chodosh, L.A., and Sharp, P.A., 1985, An RNA polymerase II transcription factor binds to an upstream element in the adenovirus major late promoter, *Cell*, 43:439.
Carthew, R.W., Chodosh, L.A., and Sharp, P.A., 1987, The major late transcription factor binds to and activates the mouse metallothionein I promoter, *Genes Dev.*, 1:973.
Chodosh, L.A., Carthew, R.W., Morgan, J.G., Crabtree, G.R., and Sharp, P.A., 1987, The adenovirus major late transcription factor activates the rat gamma-fibrinogen promoter, *Science*, 238:684.
de Vries, E., van Driel, W., van den Heuvel, S.J.L., and van der Vliet, P.C., 1987, Contactpoint analysis of the HeLa nuclear factor I recognition site reveals symmetrical binding at one side of the DNA helix, *EMBO. J.*, 6:161.
de Wet, J.R., Wood, K.V., DeLuca, M., Helinski, D.R., and Subramani, S., 1987, Firefly luciferase gene: Structure and expression in mammalian cells, *Mol. Cell. Biol.*, 7:725.
Descombes, P., Chojkier, M., Lichtsteiner, S., Falvey, E., and Schibler, U., 1990, LAP, a novel member of the C/EBP gene family, encodes a liver-enriched transcriptional activator protein, *Genes Dev.*, 4:1541.
Dorn, A., Bollenkens, J., Staub, A., Benoist, C., and Mathis, D., 1987, A multiplicity of CCAAT box-binding proteins, *Cell*, 50:863.
Duester, G., 1991, Human liver alcohol dehydrogenase gene expression, in: Drug and Alcohol Abuse Reviews, Vol.2: Liver Pathology and Alcohol, R.R. Watson, ed. The Humana Press, p. 375.
Edenberg, H.J., 1991, Molecular biological approaches to studies of alcohol-metabolizing enzymes, in: The Genetic Basis of Alcohol and Drug Actions, J.C. Crabbe and R.A. Harris, eds., Plenum Press, N.Y.. p. 165.
Edenberg, H.J. and Brown, C.J., 1992, Regulation of Human Alcohol Dehydrogenase Genes, *Pharmacogenet.*, *in press.*
Ehrig, T., Bosron, W.F., and Li, T.-K., 1990, Alcohol and aldehyde dehydrogenase, *Alcohol. Alcohol.*, 25:105.
Enver, T., Raich, N., Ebans, A.J., Papayannopoulou, T., Costantini, F., and Stamatoyannopoulos, G., 1990, Developmental regulation of human fetal-to-adult globin gene switching in transgenic mice, *Nature*, 344:309.
Friedman, A.D., Landschulz, W.H., and McKnight, S.L., 1989, CCAAT/enhancer binding protein activates the promoter of the serum albumin gene in cultured hepatoma cells, *Genes Dev.*, 3:1314.
Galas, D.J. and Schmitz, A., 1978, DNAase footprinting: a simple method for the detection of protein-DNA binding specificity, *Nucl. Acids Res.*, 5:3157.
Gorman, C.M., Moffat, L.F., and Howard, B.H., 1982, Recombinant genomes which express chloramphenicol acetytransferase in mammalian cells, *Mol. Cell. Biol.*, 2:1044.
Goyal, N., Knox, J., and Gronostajski, R.M., 1990, Analysis of multiple forms of Nuclear Factor I in human and murine cell lines, *Mol. Cell. Biol.*, 10:1041.
Graham, F.L. and van der Eb, A.J., 1983, A new technique for the assay of infectivity of human adenovirus 5 DNA, *Virology.*, 52:456.
Gronostajski, R.M., Adhya, S., Nagata, K., Guggenheimer, R.A., and Hurwitz, J., 1985, Site-specific DNA binding of Nuclear Factor I: Analyses of cellular binding sites, *Mol. Cell. Biol.*, 5:964.
Gronostajski, R.M., 1986, Analysis of nuclear factor I binding to DNA using degenerate oligonucleotides, *Nucl. Acids Res.*, 14:9117.
Jones, K.A., Yamamoto, K.R., and Tjian, R., 1985, Two distinct transcription factors bind to the HSV thymidine kinase promoter in vitro, *Cell*, 42:559.
Jones, K.A., Kadonaga, J.T., Rosenfeld, P.J., Kelly, T.J., and Tjian, R., 1987, A cellular DNA-binding protein that activates eukaryotic transcription and DNA replication, *Cell*, 48:79.
Mendel, D.B. and Crabtree, G.R., 1991, HNF-1, a member of a novel class of dimerizing homeodomain proteins, *J. Biol. Chem.*, 266:677.
Mendel, D.B., Hansen, L.P., Graves, M.K., Conley, P.B, and Crabtree, G.R., 1991, HNF-1α and HNF-1β (vHNF-1) share dimerization and homeo domains, but not activation domains, and form heterodimers in vitro, *Genes Dev.*, 5:1042.
Mermod, N., Williams, T.J., and Tjian, R., 1988, Enhancer binding factors AP-4 and AP-1 act in concert to activate SV40 late transcription in vitro, *Nature*, 332:557.

Miyamoto, N.G., Moncollin, V., Egly, J.M., and Chambon, P., 1985, Specific interaction between a transcription factor and the upstream element of the adenovirus-2 major late promoter, *EMBO. J.*, 4:3563.

Morgan, J.G., Courtois, G., Fourel, G., Chodosh, L.A., Campbell, L., Evans, E., and Crabtree, G.R., 1988, Sp1, a CAAT-binding factor, and the adenovirus major late promoter transcription factor interact with functional regions of the gamma-fibrinogen promoter, *Mol. Cell. Biol.*, 8:2628.

Mueller, C.R., Maire, P., and Schibler, U., 1990, DBP, a liver-enriched transcriptional activator, is expressed late in ontogeny and its tissue specificity is determined posttranscriptionally, *Cell*, 61:279.

Myers, R.M., Tilly, K., and Maniatis, T., 1986, Fine structure genetic analysis of a β-globin promoter, *Science*, 232:613.

Potter, J.J., Cheneval, D., Dang, C.V., Resar, L.M., Mezey, E., and Yang, V.W., 1991, The upstream stimulatory factor binds to and activates the promoter of the rat class I alcohol dehydrogenase gene, *J. Biol. Chem.*, 266:15457.

Raymondjean, M., Cereghini, S., and Yaniv, M., 1988, Several distinct "CCAAT" box binding proteins coexist in eukaryotic cells, *Proc. Natl. Acad. Sci. USA*, 85:757.

Roman, C., Matera, A.G., Cooper, C., Artandi, S., Blain, S., Ward, D.C., and Calame, K., 1992, mTFE3, an X-linked transcriptional activator containing basic helix-loop-helix and zipper domains, utilizes the zipper to stabilize both DNA binding and multimerization, *Mol. Cell. Biol.*, 12:817.

Rosenfeld, P.J. and Kelly, T.J., 1986, Purification of Nuclear Factor I by DNA recognition site affinity chromatography, *J. Biol. Chem.*, 261:1398.

Rosenfeld, P.J., O'Neill, E.A., Wides, R.J., and Kelly, T.J., 1987, Sequence-specific interactions between cellular DNA-binding proteins and the adenovirus origin of DNA replication, *Mol. Cell. Biol.*, 7:875.

Santoro, C., Mermod, N., Andrews, P.C., and Tjian, R., 1988, A family of human CCAAT-box-binding proteins active in transcription and DNA replication: cloning and expression of multiple cDNAs, *Nature*, 334:218.

Sawadogo, M. and Roeder, R.G., 1985, Interaction of a gene-specific transcription factor with the adenovirus major late promoter upstream of the TATA box region, *Cell*, 43:165.

Smith, M., Hopkinson, D.A., and Harris, H., 1971, Developmental changes and polymorphism in human alcohol dehydrogenase, *Ann. Hum. Genet. Lond.*, 34:251.

Smith, M., Hopkinson, D.A., and Harris, H., 1972, Alcohol dehydrogenase isozymes in adult human stomach and liver: evidence for activity of the *ADH3* locus, *Ann. Hum. Genet. Lond.*, 35:243.

Smith, M., 1986, Genetics of human alcohol and aldehyde dehydrogenases, *Adv. Human Genet.*, 15:249.

Stewart, M.J., McBride, M.S., Winter, L.A., and Duester, G., 1990, Promoters for the human alcohol dehydrogenase genes *ADH1*, *ADH2*, and *ADH3*: interaction of CCAAT/enhancer-binding protein with elements flanking the *ADH2* TATA box, *Gene.*, 90:271.

Stewart, M.J., Shean, M.L., Paeper, B.W., and Duester, G., 1991, The role of CCAAT/enhancer-binding protein in the differential transcriptional regulation of a family of human liver alcohol dehydrogenase genes, *J. Biol. Chem.*, 266:11594.

van Ooij, C., Snyder, R.C., Paeper, B.W., and Duester, G., 1992, Temporal expression of the human alcohol dehydrogenase gene family during liver development correlates with differential promoter activation by hepatocyte nuclear factor 1, CCAAT/Enhancer binding protein α, liver activator protein, and D-element-binding protein, *Mol. Cell. Biol.*, 12:3023.

Winter, L.A., Stewart, M.J., Shean, M.L., Dong, Y., Poellinger, L., Okret, S., Gustafsson, J.A., and Duester, G., 1990, A hormone response element upstream from the human alcohol dehydrogenase gene *ADH2* consists of three tandem glucocorticoid receptor binding sites, *Gene.*, 91:233.

Wolfla, C.E., Ross, R.A., and Crabb, D.W., 1988, Induction of alcohol dehydrogenase activity and mRNA in hepatoma cells by dexamethasone, *Arch. Biochem. Biophys.*, 263:69.

Wuarin, J., Mueller, C., and Schibler, U., 1990, A ubiquitous CCAAT factor is required for efficient in vitro transcription from the mouse albumin promoter, *J. Mol. Biol.*, 214:865.

Yin, S.-J., Cheng, T.C., Chang, C.P., Chen, Y.J., Chao, Y.C., Tang, H.S., Chang, T.M., and Wu, C.W., 1988, Human stomach alcohol and aldehyde dehydrogenases (ALDH): a genetic model proposed for ALDH III isozymes, *Biochem. Genet.*, 26:343.

DNA ELEMENTS MEDIATING RETINOID AND THYROID HORMONE REGULATION OF ALCOHOL DEHYDROGENASE GENE EXPRESSION

Mirna Žgombić and Gregg Duester

La Jolla Cancer Research Foundation
10901 North Torrey Pines Road
La Jolla, CA 92037 USA

ABSTRACT

Vertebrate alcohol dehydrogenase (ADH) plays a role in many alcohol/aldehyde interconversions including the oxidation of retinol to retinaldehyde, the rate-limiting step in the synthesis of retinoic acid. Recent molecular genetic studies on human ADH genes has lent support to a physiological role for ADH in retinoic acid synthesis. A region in the promoter for the human *ADH3* gene was previously shown to function as a retinoic acid response element (RARE), prompting an hypothesis for a positive feedback mechanism for controlling retinoic acid synthesis. The *ADH3* RARE contains three direct AGGTCA repeats which constitute the critical nucleotides of RAREs present in other genes. We compared the *ADH3* RARE to RAREs present in other genes and determined that a region containing two AGGTCA motifs separated by 5 bp was sufficient for regulating gene expression in tissue culture cells. Our experiments also indicate that *ADH3* gene expression is repressed by thyroid hormone receptor in the presence of thyroid hormone. The region of the *ADH3* promoter containing the RARE was found to harbor a negative thyroid hormone response element. Regulation of ADH gene expression by retinoid and thyroid hormones suggests that ADH plays an important role in retinoic acid synthesis.

INTRODUCTION

Vertebrate alcohol dehydrogenase exists as a complex group of isozymes encoded by several distinct genes (Duester, 1991; Yasunami et al., 1991). ADH isolated from various vertebrate sources is involved in the reversible oxidation/reduction of a wide variety of long-chain aliphatic and aromatic alcohols and aldehydes (Eklund et al., 1990; Duester, 1991), but is mostly regarded as catalyzing the rate-limiting step in the metabolism of ethanol, a relatively poor substrate (Li, 1977). Non-vertebrate ADHs such as those present in yeast and plants do not use long-chain aliphatic and aromatic alcohols as substrates (Tsai et al., 1987), and unlike vertebrate ADHs have a preference for ethanol as a substrate. However, since vertebrates in general do not consume ethanol it is likely that other alcohols and aldehydes serve as physiologically important substrates for vertebrate ADH. In particular, the alcohol form of vitamin A (retinol) can be regarded as a physiologically significant substrate for vertebrate ADH. Retinol has a long-chain aliphatic alcohol group and has long been known to be a very good substrate for mammalian ADH (Bliss, 1951; Zachman and Olson, 1961; Mezey and Holt, 1971; Van Thiel et al., 1974). Retinal produced from retinol oxidation is generally required by vertebrates for vision (Futterman, 1963; Wald, 1951), but more importantly vertebrates require retinal for the aldehyde dehydrogenase-catalyzed synthesis of retinoic acid, a known regulator of gene expression (Lee et al., 1991; Glass et al., 1991). Lower animals, plants, and microbes do not appear to use retinoic acid for regulation of gene expression, and do not appear to possess the enzymatic machinery to synthesize retinoic acid. Thus, the evolution of vertebrates utilizing retinoic acid for gene regulation may have been tied to the evolution of forms of ADH and aldehyde dehydrogenase (ALDH) which could participate in the synthesis of this important control molecule.

The effects of retinoic acid are transduced through a nuclear retinoic acid receptor (RAR) which, in the presence of ligand, is transformed into a transcription factor able to bind DNA near the promoters of some genes (Glass et al., 1991). Recently, a RAR gene family (RARα, β, and γ) has been discovered, and differential expression of these receptors is undoubtedly important for correct transduction of the retinoic acid signal in various target tissues (Dollé et al., 1989). Retinoic acid initiates a cascade of events which regulate the transcription of many key genes. Retinoic acid has been shown to induce several genes encoding homeobox-containing proteins which regulate embryonic pattern formation by activating transcription of another set of genes in a morphogenetic cascade (Simeone et al., 1990). Genes encoding extracellular matrix proteins needed for differentiation of many cells, such as laminin B1 and collagen, are also induced by retinoic acid (Vasios et al., 1989). The RARβ gene is induced by retinoic acid in teratocarcinoma and hepatoma cell lines (Hu and Gudas, 1990; De Thé et al., 1990), suggesting that the autoregulation of this retinoic acid receptor species is crucial to differentiation.

Mammalian ADH gene expression is highest in epithelial cells, with highest levels in liver epithelia (Smith, 1986), thus correlating with the fact that retinoic acid has been implicated as a control molecule for differentiation of epithelial tissues (Roberts and Sporn, 1984; Sporn and Roberts, 1991). Thus, it is possible that a major physiological function of ADH in epithelial cells is to catalyze retinoic acid synthesis needed for epithelial cell differentiation. The large amount of ADH in the liver may relate to the fact that this organ is the major site of retinol storage and metabolic turnover (excretion) via conversion to retinoic acid and further oxidized forms (Frolik, 1984; Blomhoff and Wake, 1991).

Human ADH consists of a group of isozymes divided into at least four classes based upon structural and functional distinctions (Duester, 1991; Yoshida et al., 1991). Class I ADH functions as both an ethanol dehydrogenase for detoxification of ingested ethanol (Li, 1977) and as a retinol dehydrogenase for retinoic acid synthesis (Mezey and Holt, 1971). The role, if any, of the other ADH classes in retinol oxidation is unclear, but retinol has been found to be a very poor substrate for class III ADH (Beisswenger et al., 1985). Class I ADH is encoded in humans by three closely related genes *ADH1*, *ADH2*, and *ADH3* (Duester et al., 1986) showing differential tissue-specific and temporal expression during human development (Smith, 1986). Analysis of the regulation of human class I ADH gene expression is now beginning to clarify the importance of ADH in retinoic acid-regulated processes. We recently identified a retinoic acid response element (RARE) in the promoter for the human *ADH3* gene which allows transcriptional activation by retinoic acid (Duester et al., 1991). This connection between ADH and retinoic acid supports the view that ADH plays a significant role in retinol metabolism and suggests the presence of a positive feedback mechanism to regulate retinoic acid synthesis. Interestingly, the other two human class I ADH genes (*ADH1* and *ADH2*) are not regulated by retinoic acid (Duester et al., 1991) and show a different pattern of expression in the developing human fetus, suggesting a special role for *ADH3* in human retinoid homeostasis.

The *ADH3* RARE resembles those previously found in the promoters for the RARβ (De Thé et al., 1990) and laminin B1 genes (Vasios et al., 1991). Both of those RAREs are characterized by the presence of directly repeated copies of the DNA sequence AGGTCA, but differ in the number of repeats or spacing between. We have now further characterized the *ADH3* RARE and determined that it is most closely related to the RARE in RARβ, both consisting of two direct AGGTCA repeats spaced by 5 base pairs (bp). Since thyroid hormone has been shown to exert its effects through a nuclear thyroid hormone receptor (TRα and TRβ isoforms) which recognizes AGGTCA repeats spaced by 4 instead of 5 bp (Umesono et al., 1991; Näär et al., 1991), we tested *ADH3* for thyroid hormone regulation. *ADH3* was not activated by thyroid hormone, but instead this hormone in the presence of its receptor was shown to interfere with ADH gene expression.

EXPERIMENTAL PROCEDURES

Plasmid Constructions

The plasmid constructions *ADH3-cat*(-1102), *ADH3-cat*(-328), *ADH2-cat*(-272), and *ADH2-cat*(-272 RARE) were described previously (Duester et al., 1991), as were the plasmids ΔMTV-cat and ΔMTV-cat(TREp) (Umesono et al., 1988). We used ΔMTV-cat, which contains the promoter for the mouse mammary tumor virus (MTV) fused to the *E. coli cat* reporter gene, to prepare a series of test plasmids to analyze the effect of the human *ADH3* RARE and other response elements on a heterologous promoter. ΔMTV-cat(TREp) contains an inverted repeat (palindrome) of AGGTCA inserted at position -88 bp relative to transcription initiation for ΔMTV-cat; it was previously shown to function as a synthetic thyroid hormone response element (hence TREp), but was also shown to work as a synthetic RARE (Umesono et al., 1988). We synthesized double-stranded oligonucleotides containing regions from several natural and synthetic response elements with a *Bam*HI sticky-end upstream and a *Hin*dIII sticky-end downstream to facilitate directed cloning upstream of the MTV promoter between those two restriction sites. The sequences of the oligonucleotides used are shown in Fig. 1 minus the *Bam*HI and *Hin*dIII sticky-ends. Some of these constructions were previously reported (Harding and Duester, 1992).

Plasmids were purified by alkaline-lysis and Qiagen column chromatography as described by the manufacturer (Qiagen, Inc. Diagen GmbH). All DNA constructions were verified as correct by dideoxynucleotide DNA sequence analysis (Sambrook et al., 1989).

Transfection Conditions

The monkey kidney fibroblast cell line CV-1 was cultured in Dulbecco's modified Eagle's medium (4.5 g glucose per l) containing 10% fetal bovine serum. The differentiated human hepatoma cell line HepG2 (Knowles et al., 1980) was cultured in minimal essential medium containing Earle's salts and non-essential amino acids, and was supplemented with 10% fetal bovine serum. DNA transfections and CAT assays were carried out by the calcium-phosphate co-precipitate method described previously by this laboratory (Duester et al., 1991). For cotransfections with receptor expression plasmids for human RARα (Giguere et al., 1987), RARβ and TRβ (Graupner et al., 1989), 1 μg of the expression plasmid replaced an equal portion of the carrier DNA.

The amount of CAT activity (% acetylation of chloramphenicol) was quantitated by cutting out chloramphenicol substrate spots and acetylated chloramphenicol product spots from the thin layer chromatograms and counting the radioactivity by liquid scintillation. For some experiments the thin-layer chromatograms were analyzed on an AMBIS radioactivity scanner. The % acetylation values of several experiments were averaged to give the reported relative CAT activity values for each test plasmid under the various conditions analyzed.

RESULTS

Retinoid and Thyroid Hormone Response of *ADH3* RARE and Related Elements

The human *ADH3* promoter was previously shown to harbor a RARE near -300 bp relative to the start site of transcription (Duester et al., 1991). This region contains three directly repeated sequences having an imperfect match with the motif 5'-AGGTCA-3' which is responsible for binding the RAR and activating transcription of the *ADH3* gene (Duester et al., 1991). This motif has been shown to be important for RAR binding and transactivation of other genes regulated by retinoic acid. The RAREs in the RARβ (De Thé et al., 1990) and laminin B1 (Vasios et al., 1991) genes show a great deal of sequence identity with the *ADH3* RARE (Fig. 1A). Interestingly, the long version of the *ADH3* RARE and the laminin B1 RARE both have three AGGTCA repeats, whereas the RARβ RARE has only two AGGTCA repeats. For laminin B1 all three repeats are necessary for function, but the RARβ RARE appears to require only two repeats. We have analyzed below which repeats are necessary for *ADH3* RARE function. Also, shown in Fig. 1A is a thyroid hormone response element (TRE) in the α-myosin heavy chain gene which was found to contain direct repeats of the AGGTCA motif, but spaced by 4 instead of 5 bp (Umesono et al., 1991). Thus, the response elements for retinoic acid and thyroid hormone are believed to be closely related, and possibly overlapping.

We have here dissected the *ADH3* RARE and compared it to other response elements to determine which repeats are necessary for function. The function of the *ADH3* RARE (long or short versions) was compared to a few related natural response elements (Fig. 1A) as well as a series of synthetic response elements which all possess two copies of the AGGTCA motif, but differ in the spacing or orientation (Fig. 1B). The various response elements listed (Fig.1A,B) were cloned upstream of the mouse mammary tumor virus (MTV) promoter to enable a direct comparison of their ability to activate transcription of a common promoter (which is normally not responsive to retinoic acid or thyroid hormone) in response to either

A. NATURAL RESPONSE ELEMENTS:

laminin B1 RARE	-433 AGGTGA GCT AGGTTA AGCCCTTAGAAAAA GGGTCA -467
ADH3 long RARE	-284 GGGTCA TTCAG AGTTCA GTTTTCCCTTTG GGGTCA -318
ADH3 short RARE	-284 GGGTCA TTCAG AGTTCA -300
RARβ RARE	-53 GGTTCA CCGAA AGTTCA -37
α-myosin HC TRE	-134 AGGTGA CAGG AGGACA -149

B. SYNTHETIC RESPONSE ELEMENTS:

DR-5	AGGTCA GCATG AGGTCA
DR-4	AGGTCA GCAT AGGTCA
DR-3	AGGTCA GCA AGGTCA
DR-2	AGGTCA GC AGGTCA
DR-1	AGGTCA C AGGTCA
DR-0	AGGTCA AGGTCA
TREp	AGGTCA TGACCT

Figure 1. Sequence of *ADH3* RARE and related response elements. A. Natural response elements from several genes are shown with the arrows indicating regions which are homologous to the motif AGGTCA known to bind both RAR and TR. The numbers indicate the position upstream from the transcription initiation site. B. Synthetic response elements are shown which are based upon the AGGTCA motif. DR-0 to DR-5 are direct repeats of the AGGTCA motif with spacings between the repeats ranging from 0-5 bp. TREp is an inverted repeat of the AGGTCA motif with no spacing between.

of these hormones. This type of heterologous promoter experiment is possible since enhancers (including RAREs and TREs) are separable from their own promoters (i.e. transcription initiation sites) and will retain their function when fused to a different promoter. The MTV promoter was placed upstream of the bacterial *cat* gene which encodes the enzyme chloramphenicol acetyltransferase which can be easily assayed. The responses were analyzed by introduction of these gene fusions via transfection into CV-1 tissue culture cells which efficiently uptake foreign DNA. CV-1 normally has low levels of RAR and TR (Umesono et al., 1988), but the levels can be experimentally increased in transfection reactions by including an expression vector for RARα, RARβ, or TRβ (i.e. cotransfection). After transfection and hormone treatment the cells were harvested and CAT assays were performed to estimate the extent of MTV promoter activity. The MTV promoter alone showed no response to either retinoic acid or thyroid hormone with or without added RAR or TR, but when the TREp element containing an inverted repeat of the motif AGGTCA was fused upstream there was a positive response to both RAR and TR (Fig. 2). TREp is based

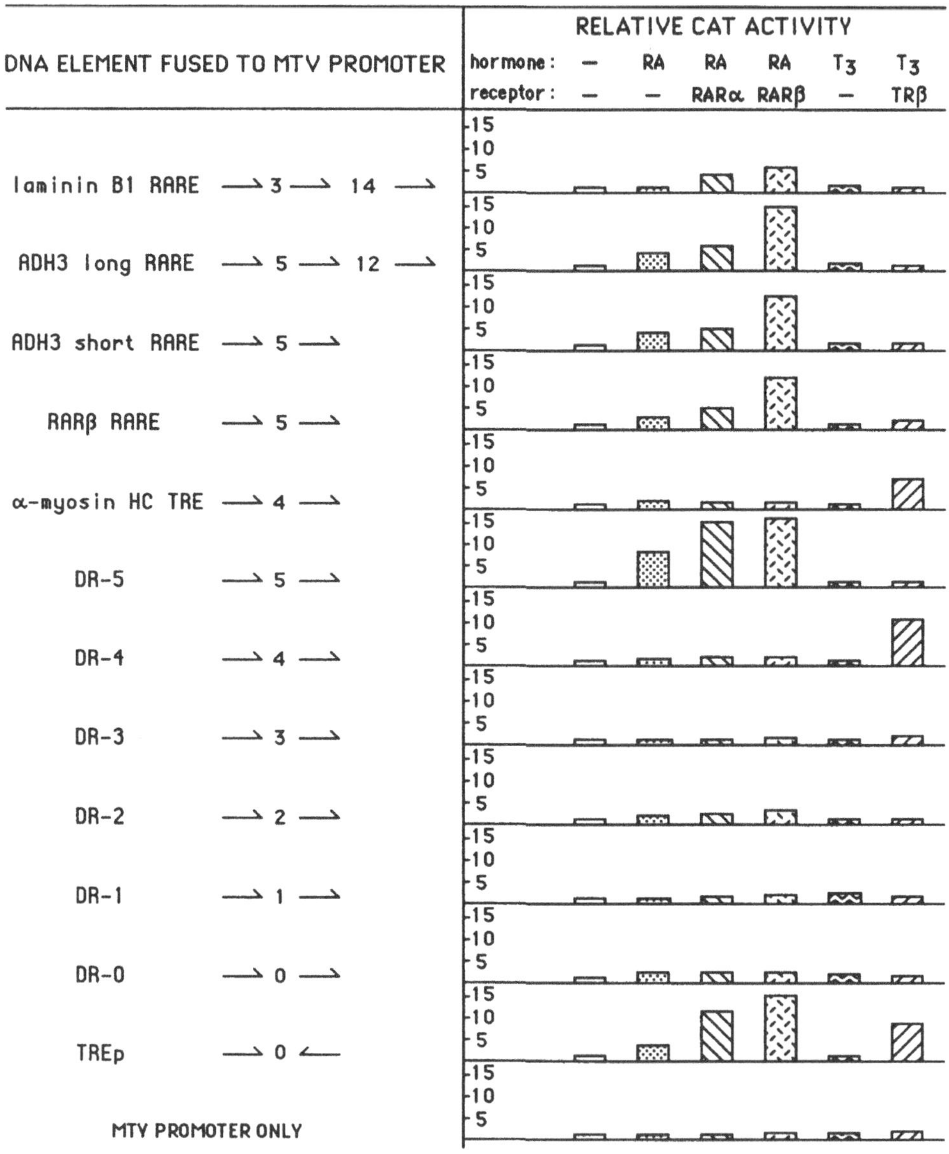

Figure 2. Function of *ADH3* RARE and related response elements. The various MTV-*cat* reporter plasmid constructs listed (containing the specific response elements shown) were transfected into CV-1 cells either in the presence or absence of retinoic acid or thyroid hormone, as well as the presence or absence of RAR or TR. The results of the CAT assays are listed as relative values using the condition with no added hormone nor receptor as the baseline value of 1.0. Arrows indicate the AGGTCA motifs with intervening numbers referring to the number of base pairs between the motifs.

upon a sequence present in the growth hormone gene which has previously been shown to respond positively to both RAR and TR (Umesono et al., 1988). Although the physiological significance of this is unknown, TREp is a useful control since it responds to both hormones.

The two versions of the *ADH3* RARE (long or short) when fused to MTV both exhibited a retinoic acid response (approx. 5-fold) in the absence of cotransfected RAR expression plasmids due to endogenous RAR (Fig. 2). However, addition of RARα or RARβ expression vectors led to inductions of 7- or 15-fold, respectively, for either the long or short *ADH3* RAREs. The RARβ RARE functioned almost identically to the *ADH3*

RARE, but the laminin B1 RARE was consistently less active in activating transcription showing inductions of only about 5-fold even with RAR expression vectors present (Fig. 2). Thus, the *ADH3* RARE behaves much more like the RARβ RARE than the laminin B1 RARE, requiring only the two AGGTCA repeats spaced by 5 bp which appear to activate transcription much better than when spaced as seen in the laminin B1 RARE. None of these RAREs responded to TRβ and thyroid hormone, but the α-myosin heavy chain TRE did respond under those conditions as expected; also this TRE did not respond to RAR and retinoic acid (Fig. 2). This indicates that RAREs and TREs are not always overlapping as was seen for TREp above.

To further analyze the issue of spacing between the AGGTCA repeats, a series of synthetic response elements were analyzed for RAR and TR induction. Basically, the results back up what was seen for the natural response elements in that the direct repeat with a spacing of 5 bp showed a response with RAR and not TR, whereas the direct repeat with a spacing of 4 bp showed the opposite effect (Fig. 2). Spacings of 0-3 bp did not result in significant inductions with either RAR or TR which supports the findings of others (Umesono et al., 1991; Näär et al., 1991). Thus, the *ADH3* RARE clearly falls into the category of RAREs which possess an AGGTCA direct repeat spaced by 5 bp. These RAREs are not able to activate transcription via the TR pathway unlike the TREp element.

Thyroid Hormone Interference of *ADH3* Gene Expression

We showed above that thyroid hormone does not operate positively through the *ADH3* RARE, but we did not rule out the possibility that it could operate negatively. We tested the effects of thyroid hormone and/or retinoic acid on *ADH3-cat*(-1102) expression by cotransfection of human HepG2 cells with RARα and/or TRβ expression vectors. In these experiments the promoter fused to the *cat* gene was the natural *ADH3* promoter with its RARE. We found no transactivation of *ADH3* by TRβ either in the presence or absence of thyroid hormone (Fig. 3) which is consistent with our studies above with the heterologous MTV promoter. This is also consistent with the fact that thyroid hormone does not increase liver ADH activity when given to rats (Mezey and Potter, 1981) or to rat hepatocyte cultures (Crabb et al., 1986). However, thyroid hormone in those studies was shown to reduce ADH activity by as much as 50%, thus prompting us to perform transfection studies in which we noticed that TRβ interferes with the functioning of the *ADH3* RARE when in the presence of thyroid hormone. *ADH3-cat*(-1102) showed a retinoic acid induction of 9.8-fold which was reduced slightly to 8.3-fold by the addition of TRβ, but reduced significantly to 2.4-fold by the simultaneous addition of TRβ and thyroid hormone (Fig. 3).

To further address this issue we tested *ADH3-cat*(-328) which still showed this response, indicating that the negative TRE maps downstream of -328 bp (Fig. 3). To determine if the RARE itself harbors the negative TRE we tested the effect of TRβ and thyroid hormone on the retinoic acid response of the heterologous *ADH2-cat*(-272 RARE) construct in which the *ADH3* RARE was fused to the nonresponsive *ADH2-cat*(-272) construct to create a retinoic acid responsive *ADH2* promoter described previously (Duester et al., 1991). *ADH2-cat*(-272 RARE) had a retinoic acid induction of 5.4-fold which was unaffected by addition of TRβ, but reduced significantly to 2.4-fold by addition of both TRβ and thyroid hormone (Fig. 3). Thus, the negative thyroid hormone response element in *ADH3* maps to the same region as the RARE.

DISCUSSION

Regulation of the human *ADH3* gene by retinoic acid was originally defined by the identification of a RARE in the region between -328 to -272 bp in the *ADH3* promoter which contains three AGGTCA direct repeats. Further characterization of the *ADH3* RARE has now defined a smaller completely functional DNA fragment spanning -300 to -284 bp containing two AGGTCA repeats spaced by 5 bp. Thus, the *ADH3* RARE is nearly identical both structurally and functionally to the RARβ RARE. Like RARβ RARE, *ADH3* RARE did not respond positively in a transfection assay to TR plus thyroid hormone. This was expected since functional TREs have recently been shown to consist of either a direct repeat of AGGTCA spaced by 4 bp, or an inverted repeat (Umesono et al., 1991; Näär et al., 1991). Our work here has confirmed the notion that AGGTCA direct repeats will function as

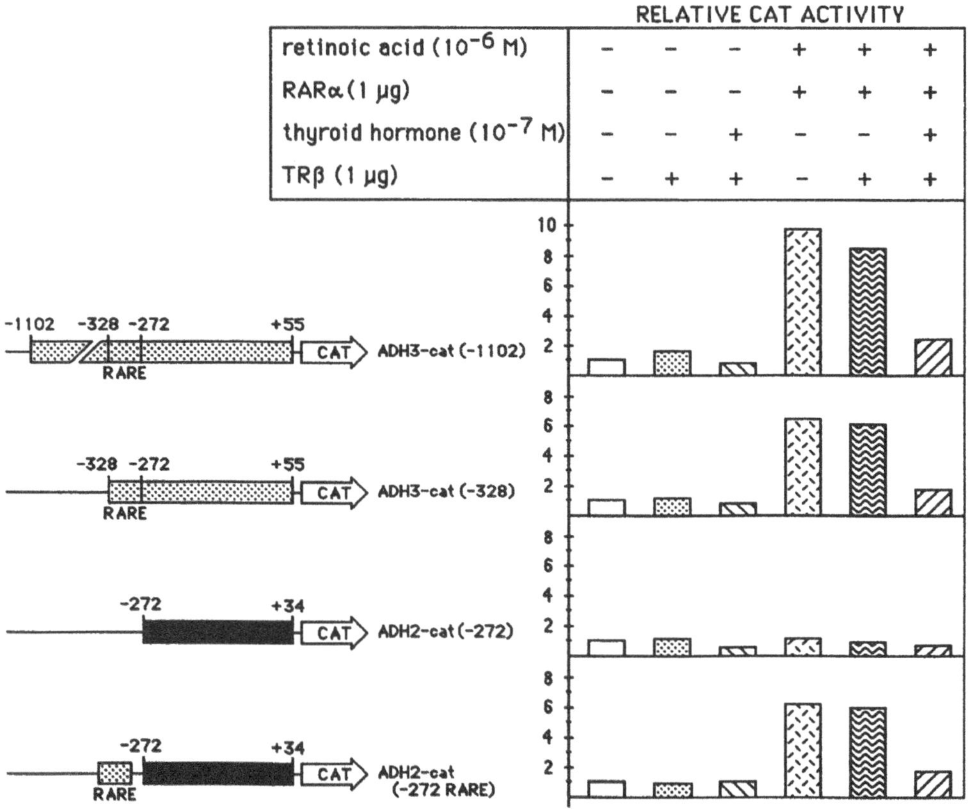

Figure 3. Negative TRE in *ADH3*. Several *ADH-cat* fusions were transfected into HepG2 cells under the different hormone and receptor conditions listed. The *ADH3* region between -328 and -272 bp harbors a RARE which was fused upstream of the *ADH2* promoter as indicated in the construct *ADH2-cat*(-272 RARE). This RARE region also carries with it a negative TRE which counteracts the positive effect of the RARE as indicated by these results. This is a modification of a figure published by Harding and Duester (1992).

a TRE when spaced by 4 bp and as a RARE when spaced by 5 bp. Thus, functional studies indicate that the *ADH3* and RARβ RAREs are designed to work nearly identically, and differently from RAREs composed of an inverted repeat such as TREp (which also functions as a TRE) or RAREs which use a different spacing for the direct repeats such as laminin B1 (which has a less efficient response).

The similarity of the *ADH3* and RARβ RAREs may relate to the physiological roles of ADH and RAR in retinoid homeostasis. The establishment of retinoic acid synthesis is hypothesized to be a highly controlled process involving a positive feedback signal amplification occuring through the retinoic acid feedback induction of the *ADH3* gene as well as the retinoic acid induced autoregulation of the RARβ gene which produces a receptor that transactivates the *ADH3* gene (Fig. 4). These two genes contain a common RARE composed of two AGGTCA repeats spaced by 5 bp and thus appear to be designed to work in synchrony. By utilizing a common RARE, *ADH3* (which generates the retinoic acid signal) and RARβ (which interprets the signal) can be coordinately induced in tissues requiring retinoic acid for their differentiation.

The functioning of the *ADH3* RARE was shown to be reduced by TR in the presence of thyroid hormone. The negative TRE involved was shown to colocalize with the *ADH3* RARE which suggests that TR in the presence of thyroid hormone is able to directly interfere with the actions of RAR and retinoic acid. Interference could be due to binding of TR to the *ADH3* RARE which is very likely considering that the very closely related RARβ RARE has been shown to bind TR strongly despite a lack of transcription activation (Hoffmann et al.,

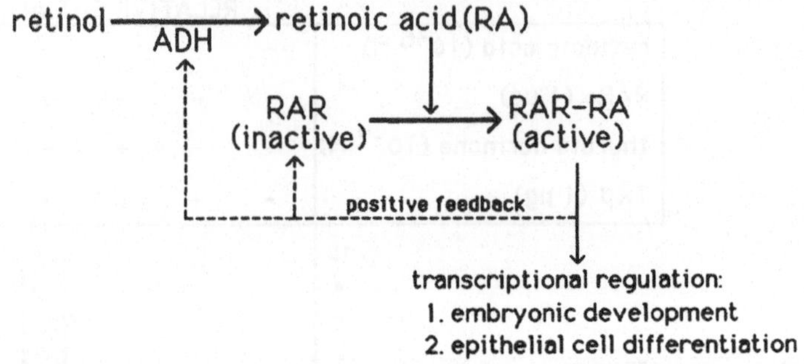

Figure 4. Positive feedback regulation in retinoid homeostasis. The dotted line suggests that ADH and RAR genes are activated by the liganded retinoic acid receptor (RAR-RA) in a positive feedback loop allowing coordinated synthesis of the ligand (RA) and receptor (RAR) once initiated.

1990). Evidently, RARs and TRs both bind as monomers to the core sequence AGGTCA present as direct or inverted repeats with variable spacing, but binding at only certain elements allows a particular receptor to dimerize and achieve a conformation able to impart a hormone-regulated signal to the transcriptional machinery (Leid et al., 1992). The results here on *ADH3* indicate that some elements may bind RAR and TR receptors in a fashion which allows the effects of retinoic acid to be modulated by thyroid hormone.

Thyroid hormone has previously been shown to cause inhibition of ADH activity in rat liver. It has been reported that thyroid hormone treatment of rats leads to reductions in ADH activity of 57% (Hillbom and Pikkarainen, 1970), 40% (Israel et al., 1975), or 50% (Mezey and Potter, 1981). Thyroidectomy leads to 2-fold increases in rat liver ADH activity, but administration of thyroid hormone to thyroidectomized rats once again lowers ADH activity (Mezey and Potter, 1981). Cultured rat hepatocytes treated with 0.1 μM thyroid hormone undergo a 35% reduction of ADH activity indicating that the effect of thyroid hormone is directly on the hepatocyte as opposed to an indirect mechanism (Crabb et al., 1986). Now that it has been determined that thyroid hormone exerts its effects at the transcriptional level, it is tempting to speculate that the rat studies mentioned above are indicative of a thyroid hormone repression of rat liver ADH gene expression. The studies performed here on the human *ADH3* gene support such a model of thyroid hormone repression and even suggest that the mechanism involves an interference with the retinoic acid activation of the gene as shown in Fig. 5.

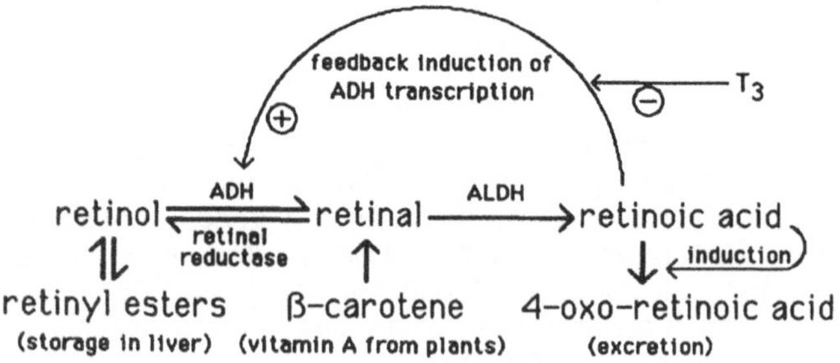

Figure 5. Regulation of retinoic acid synthesis. We have determined that the human *ADH3* gene is induced by retinoic acid as indicated by the 'plus' symbol and have proposed that this ensures a sufficient supply of retinoic acid. We have now shown that thyroid hormone can interfere with retinoic acid induction of ADH as indicated by the 'minus' symbol. Thus, it is possible that thyroid hormone moderates the synthesis of retinoic acid. A modified version of this appeared in Duester et al. (1991).

ACKNOWLEDGEMENTS

We wish to thank G. Graupner and M. Pfahl for the RARβ and TRβ expression plasmids, and R. Evans for the RARα expression plasmid. We thank P. Harding for technical assistance. This work was supported by a National Institute on Alcohol Abuse and Alcoholism research grant AA07261, and a Research Scientist Development Award K02 AA00119 to G.D.

REFERENCES

Beisswenger, T.B., Holmquist, B., and Vallee, B.L., 1985, χ-ADH is the sole alcohol dehydrogenase isozyme of mammalian brains: Implications and inferences, Proc. Natl. Acad. Sci. USA, **82**: 8369-8373.

Bliss, A.F., 1951, The equilibrium between vitamin A alcohol and aldehyde in the presence of alcohol dehydrogenase, Arch. Biochem., **31**: 197-204.

Blomhoff, R. and Wake, K., 1991, Perisinusoidal stellate cells of the liver: Important roles in retinol metabolism and fibrosis, FASEB J., **5**: 271-277.

Crabb, D.W., Bosron, W.F., and Li, T.-K., 1986, Role of the pituitary and neonatal androgenic imprinting in the hormonal regulation of liver alcohol dehydrogenase activity, Biochem. Pharmacol., **9**: 1527-1532.

De Thé, H., Vivanco-Ruiz, M.D.M., Tiollais, P., Stunnenberg, H., and Dejean, A., 1990, Identification of a retinoic acid responsive element in the retinoic acid receptor β gene, Nature, **343**: 177-180.

Dollé, P., Ruberte, E., Kastner, P., Petkovich, M., Stoner, C.M., Gudas, L.J., and Chambon, P., 1989, Differential expression of genes encoding α, β and γ retinoic acid receptors and CRABP in the developing limbs of the mouse, Nature, **343**: 702-705.

Duester, G., Smith, M., Bilanchone, V., and Hatfield, G.W., 1986, Molecular analysis of the human class I alcohol dehydrogenase gene family and nucleotide sequence of the gene encoding the β subunit, J. Biol. Chem., **261**: 2027-2033.

Duester, G., 1991, Human liver alcohol dehydrogenase gene expression: Retinoic acid homeostasis and fetal alcohol syndrome, in " Drug and Alcohol Abuse Reviews, Vol. 2: Liver Pathology," R.R. Watson, ed.,Humana Press, Clifton, New Jersey, p. 375.

Duester, G., Shean, M.L., McBride, M.S., and Stewart, M.J., 1991, Retinoic acid response element in the human alcohol dehydrogenase gene *ADH3*: Implications for regulation of retinoic acid synthesis, Mol. Cell. Biol., **11**: 1638-1646.

Eklund, H., Müller-Wille, P., Horjales, E., Futer, O., Holmquist, B., Vallee, B.L., Höög, J.-O., Kaiser, R., and Jörnvall, H., 1990, Comparison of three classes of human liver alcohol dehydrogenase. Emphasis on different substrate binding pockets, Eur. J. Biochem., **193**: 303-310.

Frolik, C.A., 1984, Metabolism of retinoids, in " The Retinoids, Vol.2," M.B. Sporn, A.B. Roberts, and D.S. Goodman, eds., Academic, Orlando, p. 177.

Futterman, S., 1963, Metabolism of the retina: The role of reduced triphosphopyridine nucleotide in the visual cycle, J. Biol. Chem., **238**: 1145-1150.

Giguere, V., Ong, E.S., Segui, P., and Evans, R.M., 1987, Identification of a receptor for the morphogen retinoic acid, Nature, **330**: 624-629.

Glass, C.K., DiRenzo, J., Kurokawa, R., and Han, Z., 1991, Regulation of gene expression by retinoic acid receptors, DNA Cell Biol., **10**: 623-638.

Graupner, G., Wills, K.N., Tzukerman, M., Zhang, X.-K., and Pfahl, M., 1989, Dual regulatory role for thyroid-hormone receptors allows control of retinoic-acid receptor activity, Nature, **340**: 653-656.

Harding, P.P. and Duester, G., 1992, Retinoic acid activation and thyroid hormone repression of the human alcohol dehydrogenase gene *ADH3*, J. Biol. Chem., **267**: in press.

Hillbom, M.E. and Pikkarainen, P.H., 1970, Liver alcohol and sorbitol dehydrogenase activities in hypo- and hyperthyroid rats, Biochem. Pharmacol., **19**: 2097-2103.

Hoffmann, B., Lehmann, J.M., Zhang, X., Hermann, T., Husmann, M., Graupner, G., and Pfahl, M., 1990, A retinoic acid receptor-specific element controls the retinoic acid receptor-β promoter, Mol. Endocrinol., **4**: 1727-1736.

Hu, L. and Gudas, L.J., 1990, Cyclic AMP analogs and retinoic acid influence the expression of retinoic acid receptor α, β, and γ mRNAs in F9 teratocarcinoma cells, Mol. Cell. Biol., **10**: 391-396.

Israel, Y., Videla, L., Fernandez-Videla, V., and Bernstein, J., 1975, Effects of chronic ethanol treatment and thyroxine administration on ethanol metabolism and liver oxidative capacity, J. Pharmacol. Exp. Ther., **192**: 565-574.

Knowles, B.B., Howe, C.C., and Aden, D.P., 1980, Human hepatocellular carcinoma cell lines secrete the major plasma proteins and hepatitis B surface antigen, Science, **209**: 497-499.

Lee, M.-O., Manthey, C.L., and Sladek, N.E., 1991, Identification of mouse liver aldehyde dehydrogenases that catalyze the oxidation of retinaldehyde to retinoic acid, Biochem. Pharmacol., 42: 1279-1285.

Leid, M., Kastner, P., Lyons, R., Nakshatri, H., Saunders, M., Zacharewski, T., Chen, J.-Y., Staub, A., Garnier, J.-M., Mader, S., and Chambon, P., 1992, Purification, cloning, and RXR identity of the HeLa cell factor with which RAR or TR heterodimerizes to bind target sequences efficiently, Cell, 68: 377-395.

Li, T.-K., 1977, Enzymology of human alcohol metabolism, Adv. Enzymol., 45: 427-483.

Mezey, E. and Holt, P.R., 1971, The inhibitory effect of ethanol on retinol oxidation by human liver and cattle retina, Exp. Mol. Pathol., 15: 148-156.

Mezey, E. and Potter, J.J., 1981, Effects of thyroidectomy and triiodothyronine administration on rat liver alcohol dehydrogenase, Gastroenterology, 80: 566-574.

Näär, A.M., Boutin, J.-M., Lipkin, S.M., Yu, V.C., Holloway, J.M., Glass, C.K., and Rosenfeld, M.G., 1991, The orientation and spacing of core DNA-binding motifs dictate selective transcriptional responses to three nuclear receptors, Cell, 65: 1267-1279.

Roberts, A.B. and Sporn, M.B., 1984, Cellular biology and biochemistry of the retinoids, in " The Retinoids Vol. 2," M.B. Sporn, A.B. Roberts, and D.S. Goodman, eds., Academic, Orlando, p. 209.

Sambrook, J., Fritsch, E.F., and Maniatis, T., 1989, "Molecular cloning: A laboratory manual," Cold Spring Harbor Laboratory, Cold Spring Harbor, New York,

Simeone, A., Acampora, D., Arcioni, L., Andrews, P.W., Boncinelli, E., and Mavilio, F., 1990, Sequential activation of *HOX2* homeobox genes by retinoic acid in human embryonal carcinoma cells, Nature, 346: 763-766.

Smith, M., 1986, Genetics of human alcohol and aldehyde dehydrogenases, Adv. Hum. Genet., 15: 249-290.

Sporn, M.B. and Roberts, A.B., 1991, Interactions of retinoids and transforming growth factor-β in regulation of cell differentiation and proliferation, Mol. Endocrinol., 5: 3-7.

Tsai, C.S., Al-Kassim, L.S., Mitton, K.P., Thompson, L.E., Van Es, C., and White, J.H., 1987, Purification and comparative studies of alcohol dehydrogenases, Comp. Biochem. Physiol. [B], 87B: 79-85.

Umesono, K., Giguere, V., Glass, C.K., Rosenfeld, M.G., and Evans, R.M., 1988, Retinoic acid and thyroid hormone induce gene expression through a common responsive element, Nature, 336: 262-265.

Umesono, K., Murakami, K.K., Thompson, C.C., and Evans, R.M., 1991, Direct repeats as selective response elements for the thyroid hormone, retinoic acid, and vitamin D3 receptors, Cell, 65: 1255-1266.

Van Thiel, D.H., Gavaler, J., and Lester, R., 1974, Ethanol inhibition of vitamin A metabolism in the testes: Possible mechanism for sterility in alcoholics, Science, 186: 941-942.

Vasios, G., Mader, S., Gold, J.D., Leid, M., Lutz, Y., Gaub, M.-P., Chambon, P., and Gudas, L., 1991, The late retinoic acid induction of laminin B1 gene transcription involves RAR binding to the responsive element, EMBO J., 10: 1149-1158.

Vasios, G.W., Gold, J.D., Petkovich, M., Chambon, P., and Gudas, L.J., 1989, A retinoic acid-responsive element is present in the 5' flanking region of the laminin B1 gene, Proc. Natl. Acad. Sci. USA, 86: 9099-9103.

Wald, G., 1951, The chemistry of rod vision, Science, 113: 287-291.

Yasunami, M., Chen, C.-S., and Yoshida, A., 1991, A human alcohol dehydrogenase gene (*ADH6*) encoding an additional class of isozyme, Proc. Natl. Acad. Sci. USA, 88: 7610-7614.

Yoshida, A., Hsu, L.C., and Yasunami, M., 1991, Genetics of human alcohol-metabolizing enzymes, Prog. Nucleic Acid Res. Mol. Biol., 40: 255-287.

Zachman, R.D. and Olson, J.A., 1961, A comparison of retinene reductase and alcohol dehydrogenase of rat liver, J. Biol. Chem., 236: 2309-2313.

MODULATION OF HEPATIC AND RENAL ALCOHOL DEHYDROGENASE ACTIVITY AND mRNA BY STEROID HORMONES IN VIVO

Mona Qulali, Katrina M. Dipple, and David W. Crabb

Departments of Medicine and of Biochemistry and
Molecular Biology
Indiana University School of Medicine
Indianapolis, IN 46202-5121

INTRODUCTION

Alcohol dehydrogenase (ADH) catalyzes the flux-determining step in the oxidation of ethanol and probably also plays a role in the metabolism of steroid alcohols and retinoids. Five classes of ADH have been discovered to date (Ehrig et al., 1990). The class I enzyme is the major liver enzyme, which is also found in other tissues at lower levels. This class of ADH has been studied in greatest detail. The gene for class I ADH has been cloned from humans (Duester et al. 1986), mice (Zhang et al., 1987), and rats (Crabb et al., 1989). In vitro studies have identified promoter elements important for liver specific expression (C/EBP and HNF1 binding sites (Potter et al., 1991; Stewart et al., 1990a; Stewart et al., 1990b; Stewart et al., 1991; van Ooij et al., 1992)) as well as hormone response elements. There are glucocorticoid response elements in the promoter (Duester et al., 1986), and expression of the gene is activiated in hepatoma cells exposed to dexamethasone (Wolfla et al., 1988; Dong et al., 1988; Winters et al., 1990). Retinoic acid response elements are present in the mouse and human ADH3 gene (Duester et al., 1991), and transfection studies demonstrate retinoic acid responsiveness of the promoter. Androgens induced mouse kidney ADH many fold by an effect that is mainly transcriptional (Ceci et al., 1986; Felder et al., 1988). Other hormone responses are less well understood. Estradiol is reported to induce ADH activity in the kidney (Qulali et al., 1991; Dembic and Sabolic, 1982), but not liver, of rats, and thyroid hormone reduces liver ADH activity (Potter and Mezey, 1983; Mezey and Potter, 1981).

We have been interested in the physiological regulation of ADH gene expression. Hormones have multiple effects on tissues, including effects on gene transcription, mRNA stability, and protein synthesis and degradation. We have therefore studied the effects of corticosterone, estradiol, and thyroid hormones in intact animals on ADH activity and protein level, mRNA level, and transcription rate of the ADH gene. Because the expression of the ADH gene is also influenced by the nutritional status of the animal, we paid particular attention to the animals weight gain and food intake to avoid attributing nutritional effects to the hormonal treatment.

METHODS AND MATERIALS

Corticosterone Treatment Protocol

Male Wistar-Kyoto rats were adrenalectomized by the supplier (Harlan-Sprague Dawley Industries, Indianapolis, IN) at 6 weeks of age. They were weight paired at 10.5 weeks of age and treated with a daily subcutaneous injection of either vehicle (olive oil) or corticosterone-21-acetate in oil (10 mg/kg/d) for 10 days. The animals were then sacrificed and liver extracts for enzyme analysis and RNA quantitation were prepared, and nuclei were isolated for nuclear run on assays.

Estradiol Treatment Protocol

Male Wistar-Kyoto rats were given daily intramuscular injections of vehicle (olive oil) or estradiol in oil (1 mg/kg/d) for 10 days begining at age 11.5 weeks. In subsequent experiments the animals received only a single injection of estradiol. After the treatment, the animals were sacrificed and the kidneys were removed, the renal pelvis and ureters were dissected away, and portions of the kidney were used for enzyme assay, Western blotting, RNA preparation, or isolation of nuclei. For the estimation of ADH mRNA half-life, a group of animals received an injection of either oil or 1 mg/kg estradiol in oil. Twenty four hours later, the animals were given a single injection of actinomycin D (1.5 mg/kg) intraperitoneally, and then were sacrificed at 3, 6, and 12 hours to isolate kidney mRNA.

Thyroid Hormone Treatment Protocols

Male Wistar-Kyoto rats were thyroidectomized (n=12) or sham-operated (n=12) at age six weeks by the supplier. Animals were acclimated to our facility with free access to food and water (or 1% $CaCl_2$ for thryoidectomized rats) for two weeks. They were weight-paired then given 13 days of intraperitoneal injections: the sham-operated animals were injected with saline (euthyroid, n=6) or 100 μg thyroxine sodium salt (T_4) kg body weight/day (hyperthroid, n=6). Thyroidectomized animals were injected with saline (hypothyroid, n=6) or 20 μg T4/kg body weight/day as a replacement dose (n=6). Food intake and animal body weight were monitored daily.

Pretreatment plasma thyroxine levels were obtained from tail blood the day before treatment. Posttreatment levels were measured in trunk blood at the time of sacrifice. T_4 assays were kindly performed by Dr. Oei (Indiana University Hospitals, Indianapolis IN) by automated fluorescence polarization immunoassay (TDX, Abbott Laboratories).

ADH Enzyme Assay and Western Blotting

Liver or kidney was homogenized in 50 mM Hepes buffer, pH 8.4 containing 0.5 mM dithiothreitol, and a 100,000g supernatant was prepared. The supernatant was assayed for ADH in 0.5 M Tris, pH 7.2, 2.8 mM NAD^+, and 10 mM ethanol at 25°. One unit catalyzes the reduction of 1 μmol of NAD^+ to NADH per minute. The supernatant was also used for Western blotting. The proteins were fractionated on SDS-polyacrylamide gels, electrophoretically transferred to nitrocellulose, and detected with polyclonal rabbit antiserum raised against affinity purified rat liver ADH. The primary antibody was detected with ^{125}I-labelled protein A (Qulali et al. 1991).

RNA Analysis

Total RNA was isolated by the guanidinium-acid phenol protocol (Chomczynski and Sacchi, 1987). 20 μg of RNA was denatured and electrophoresed on a 1.2% formaldehyde agarose gel, transferred to Nytran membranes by capillary blotting, and hybridized in buffer containing 50% formamide at 42° overnight. The probes used in various experiments were pDC2 (rat ADH cDNA (Crabb and Edenberg, 1986)), CHO-B cDNA (Harpold et al., 1979), and the 28S rRNA cDNA (Chang et al., 1984), which were labelled by random priming. The amount of RNA was estimated by analyzing the washed filters with an AMBIS beta scanner. The cpm in a band of interest was corrected for background radioactivity on the filter. In situ hybridization was performed as described (Qulali et al., 1991).

Nuclear Run-on Assays

Nuclei were isolated by homogenizing the tissues with a Dounce homogenizer and pelleting the nuclei through a sucrose gradient. The run-on RNAs were synthesized in the presence of ^{32}P-UTP and were hybridized with filters bearing various target cDNAs at 65° for 4 days. The filters were then treated with RNAse A, washed, and exposed to X-ray film for 10 days. The target cDNAs were subsequently cut out and counted in a scintillation counter to determine transcription rates as ppm (counts in the target/total radioactivity of the run-on RNA x 10^6).

RESULTS AND DISCUSSION

Effects of Adrenalectomy and Corticosterone Treatment on ADH Activity and Gene Expression

There was no difference in liver ADH activity between control rats, adrenalectomized rats, and adrenalectomized rats treated with corticosterone (Table 1). Western blots of liver extracts developed with antibody specific for class I ADH demonstrated equal amounts of ADH protein in the two adrenalectomized rat groups. Northern blots of liver RNA demonstrated a two-fold increase in ADH mRNA levels in the corticosterone-treated animals when normalized with the level of CHO-B mRNA or 28S rRNA level. Nuclear run-on assays were then performed with nuclei isolated at the time of sacrifice of the animals (i.e., after 10 days of treatment with corticosterone or vehicle). The rate of transcription of the ADH gene, monitored by hybridization to a plasmid containing the 3' 8 kb of the rat ADH gene (Crabb et al., 1989) was 283 ± 115 ppm in control animals vs 428 ± 247 ppm in corticosterone-treated rats. This trend toward increased transcription did not reach statistical significance with a total of 10 animals.

These results suggest that corticosteroids may play a role in maintaining normal levels of ADH mRNA in the adult liver. The difference in mRNA levels may be the result of transcriptional activation of the gene, as suggested by several earlier in vitro studies, and the failure to observe a statistically significant difference in run-on rates may be due to the rather large variation in the results. It is also possible that a larger effect would be seen at earlier time points in the experiment. The fact that ADH activity or protein did not correspond to the mRNA level may reflect differences in rates of mRNA translation or in protein stability between the adrenalectomized rats and those receiving supplemental corticosteroid. It seems unlikely, however, that variation of corticosteroid levels from hypo-adrenal to mildly hyper-adrenal levels would influence ADH activity in liver or contribute to the between individual variation in alcohol elimination rates.

Table 1. Effect of corticosterone on ADH activity in adrenalectomized rats

	Control (10)	Adrenalectomized (6)	Corticosterone-treated (6)
ADH Activity			
U/g liver	1.1 ± 0.1	1.2 ± 0.3	1.2 ± 0.3
U/g protein	7.2 ± 1.0	8.6 ± 1.6	9.0 ± 4.2
U/mg DNA	0.5 ± 0.1	0.5 ± 0.1	0.6 ± 0.1
U/liver	15.8 ± 2.8	12.2 ± 1.8	14.5 ± 3.8

The animals were sacrificed for determination of enzyme activity, preparation of RNA, and isolation of nuclei after 10 days of treatment with vehicle or corticosterone (10 mg/kg/d). Control animals were intact rats of the same age; the other two groups had been adrenalectomized 4 weeks prior to the experiment.

Effects of Estradiol on Kidney ADH Activity and Gene Expression

Estradiol had been previously reported to induce the activity of ADH in the kidney of rats (Dembic and Sabolic, 1982), and we had shown a three-fold induction in both the enzyme activity and mRNA level in female rats treated with pharmacological doses of estradiol (Qulali et al., 1991). In situ hybridization localized the ADH mRNA in female kidney to the medulla and inner cortex (Qulali et al., 1991); in males, the mRNA appeared to be limited to the medulla. Examination of the emulsion autoradiograms showed the ADH mRNA to be present in the tubule cells rather than in endothelium or glomerulus. When kidney tubules were prepared by collagenase digestion of kidney, these preparations also contained ADH mRNA, indicating that the message is mainly expressed in the tubule cells.

Treatment of male rats with estradiol increased the mRNA about 7 fold, while only increasing ADH activity and protein by 3-fold (Table 2). Castration had no effect of kidney ADH activity or mRNA level, demonstrating that the effect of estradiol was not the result of a "pharmacological castration" (not shown). It is known that food restriction or fasting reduces ADH activity in the liver rapidly (Bosron et al., 1984). The estradiol-treated rats did lose weight relative to the controls, which raised the possibility that reduced food intake was responsible for the failure of ADH activity to increase as much as the RNA. When rats were fasted for 48 hours, the kidney ADH activity decreased by 30% while the mRNA was increased by about 2-fold relative to ß-actin or CHO-B mRNAs. Hence, we think it likely that part of the discrepancy between RNA levels and activity is the result of nutritional changes that occur in male rats given high doses of estradiol.

The induction of ADH mRNA was initially studied over a 10 day period. To avoid the complicating changes in nutrition that may occur with this treatment, the animals were given a single dose of estradiol and sacrificed at early time points. We saw that the mRNA began to increase at about 1-2 hours after the injection and was maximal at 24 hours (Figure 1). To examine the mechanism for the induction, we

Table 2. Effects of Estradiol on ADH Activity, Protein Levels, mRNA levels and Rates of ADH Gene Transcription in Male Rat Kidney

	Body Weight	Alcohol Dehydrogenase Activity (U/mg DNA)	mRNA	Transcription Rate (ppm)
Control	416 ± 16	0.10 ± 0.03	2.2 ± 0.2	58 ± 23
Estradiol	309 ± 15*	0.31 ± 0.07*	15.4 ± 1.9*	73 ± 18

The animals were treated for 10 days with estradiol (1 mg/kg/day) or vehicle for the determination of changes in enzyme activity and mRNA level, and for 24 hours (after a single injection of estradiol) for determination of transcription rates by nuclear run-on assay. mRNA level is the ratio of the radioactivity (determined using a two-dimensional ß scanner to analyze the Northern blots) in the ADH mRNA band divided by that in the CHO-B mRNA band. * indicates a significant difference from control animals ($p<0.05$).

isolated kidney nuclei at 1 and 24 hours after an injection of estradiol. There was no difference in the rates of transcription of the ADH gene: transcription rates were 63 ± 25 and 54 ± 18 ppm in control and one hour estradiol-treated rats, and 58 ± 23 and 73 ± 18 ppm in control and 24 hour treated rats. We also attempted to measure the stability of the ADH mRNA after estradiol by the technique of actinomycin D decay curves. As shown in Figure 2, there was a slow decline in the amount ADH mRNA after actinomycin D injection in the estradiol-treated animals, with an approximate half-life of 20 hours. Unfortunately, there was a transient rise in mRNA after actinomycin D treatment in the control group (Figure 2), precluding interpretation of the data.

We conclude that the class I ADH of kidney tubules is rapidly induced by estradiol by a post-transcriptional effect (summarized in Table 2). It will be of interest to study further the mechanism of this induction and the metabolic role of ADH in the kidney.

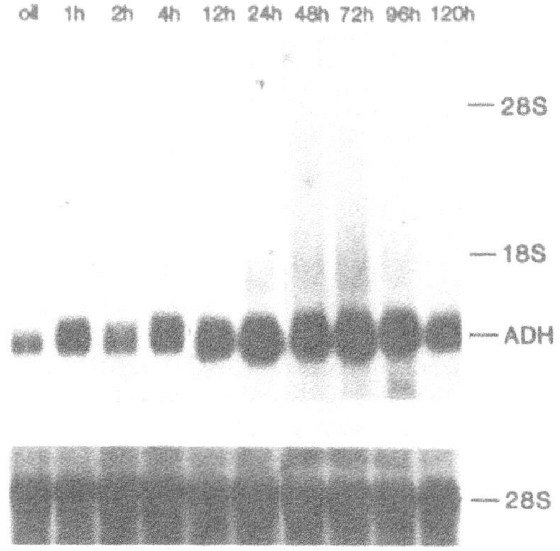

Figure 1. Effect of estradiol on kidney ADH mRNA levels. A single injection of estradiol (1 mg/kg) or vehicle was given at time 0, and an animal from the treated group was sacrificed at the times noted. The kidney RNA was isolated and analyzed by Northern blotting. The top panel shows the ADH mRNA and the lower panel shows the blot reprobed with a 28S rRNA cDNA.

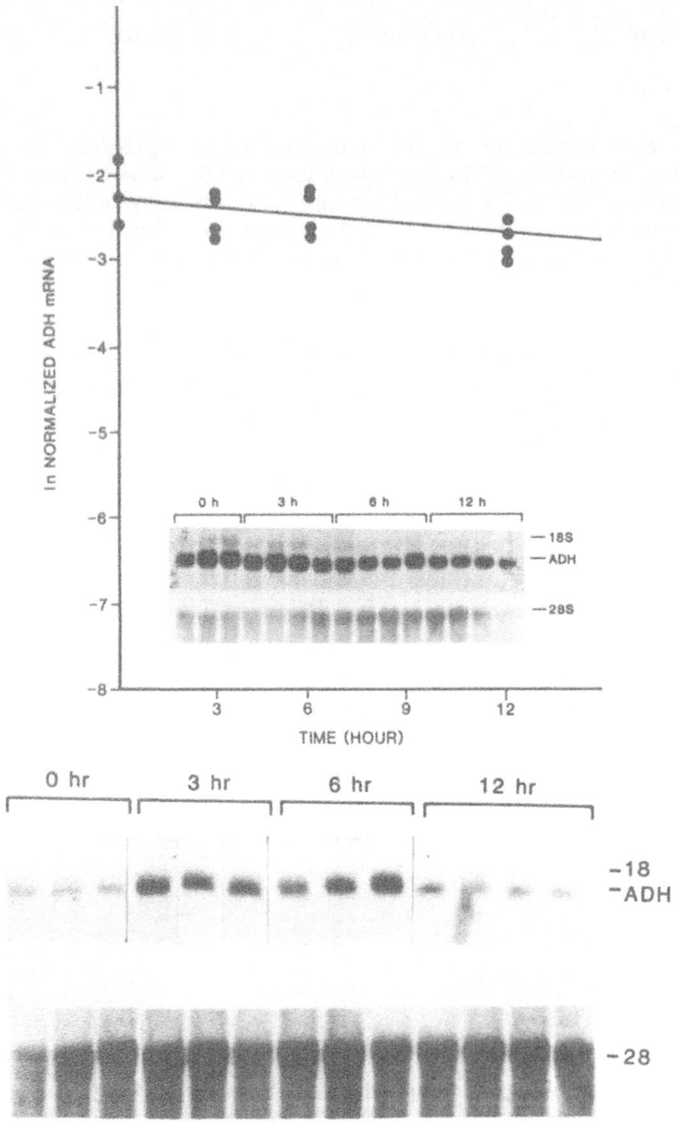

Figure 2. Effect of estradiol on the rate of decay of ADH mRNA in kidney after treatment with actinomycin D. The animals received either 1 mg/kg of estradiol (top panel) or vehicle (lower panel) 24 hours before an injection of actinomycin D. RNA was isolated from the kidney at 3, 6, and 12 hours after the actinomycin D injection, and in the top panel is plotted as the logarithm of the radioactivity of the ADH mRNA normalized with the radioactivity in the 28S rRNA band. Since the mRNA level did not decline in the control group, those data were not replotted.

Effect of thyroid hormone status on ADH activity and expression in liver and kidney

We designed our study of the effect of thyroid hormone on ADH activity to only induce moderate degrees of hyperthyroidism and to control carefully for changes in food intake, in contrast with some earlier reports (Mezey and Potter, 1981). The mean daily food intake of all animals of any experimental group was the same. Thyroidectomy was successful as judged by the low plasma T_4 levels in these animals. The hypothyroid animals did not gain weight in the pre-treatment period, as expected. Treatment of hypothyroid animals with 20 μg of T_4/kg/d restored weight gain and plasma T_4 levels to normal, while treatment of normal rats with five times this dose achieved moderate hyperthyroidism (average T_4 was 20 μg/dl) and reduced weight gain to 1 g/day. ADH activity in both liver and kidney was responsive to changes in thyroxine levels. Hypothyroidism increased ADH activity in liver 70% and in kidney 50%. Hyperthyroidism reduced ADH activithy in liver and kidney by 30 and 40%, respectively. The changes in ADH activity were paralleled by changes in protein level as assayed by Western blotting. Since the changes in activity were less than two-fold, we were not surprised that we were unable to detect differences in the level of ADH mRNA in the liver. However, the molecular weight of the ADH mRNA was unchanged in the thyroid treated animals, and no new species were observed. These experiments indicate that thyroxine indeed modulates ADH activity independent of changes in food intake. It is not clear if the effects are direct effects of the thyroxine or if they are mediated by changes of other hormones (e.g., growth hormone).

In summary, the steroid hormones, corticosterone and estradiol, and thyroxine have effects on the activity and amount of ADH or its mRNA in liver and kidney. In the case of corticosterone, it is likely that both transcriptional and post-transcriptional effects occur. The effect of estradiol is solely post-transcriptional as far as we can tell at present, and the mechanism underlying the effect of thyroxine is unknown. This is in contrast to the effect of androgens on mouse kidney ADH. Although these hormones might modulate differences in ADH activity in the intact animal under pathological endocrine conditions, it seems unlikely that they are responsible for between individual differences in alcohol metabolism or sensitivity to alcohol.

This work was supported by AA 06434 and AA 00081 to DWC, and F30 AA05310 to KMD. The technical assistance of Ruth Ann Ross is gratefully acknowledged.

REFERENCES

Bosron W.F., Crabb D.W., Housinger T.A., and Li T.-K., 1984, Effect of fasting on the activity and turnover of rat liver alcohol dehydrogenase, *Alcoholism:Clin.Exp.Res.* 8:196-200.

Ceci J.D, Lawther R., Duester G., Hatfield W., Smith M., OMalley M.D., and Felder M., 1986, Androgen induction of alcohol dehydrogenase in mouse kidney. Studies with a cDNA probe confirmed by nucleotide sequence analysis, *Gene* 41:217-224.

Chang Y.L., Gutell R., Noller H.F., and Wool I.G., 1984, The nucleotide sequence of a rat 18S ribosomal ribonucleic acid gene and a proposal for the secondary structure of 18S ribosomal ribonucleic acid, *J.Biol.Chem.* 259:224-230.

Chomczynski P. and Sacchi N., 1987, Single step method of RNA isolation by acid guanidinium thiocyanate-phenol-chloroform extraction, *Anal.Biochem.* 162:156-159.

Crabb D.W. and Edenberg H.J., 1986, Complete amino acid sequence of rat liver alcohol dehydrogenase deduced from the cDNA sequence, *Gene* 48:287-291.

Crabb D.W., Stein P.M., Dipple K.M., Hittle J.B., Sidhu R., Qulali M., Zhang K., and Edenberg H.J., 1989, Structure and expression of the rat class I alcohol dehydrogenase gene, *Genomics* 5:906-914.

Dembic Z. and Sabolic I., 1982, Alcohol dehydrogenase activity in rat kidney cortex stimulated by oestradiol, *Biochim.Biophys.Acta* 714:331-336.

Dong Y., Poellinger L., Okert S., Hoog J.O., vonBahr-Lindstrom H., Jornvall H., and Gustafsson J.A., 1988, Regulation of gene expression of class I alcohol dehydrogenase by glucocorticoids, *Proc.Natl.Acad.Sci.USA* 85:767-771.

Duester G., Shean M.L., McBride M.S., and Stewart M.J., 1991, Retinoic acid response element in the human alcohol dehydrogenase gene ADH3: implications for regulation of retinoic acid synthesis, *Mol.Cell.Biol.* 11:1638-1646.

Duester G., Smith M., Bilanchone V., and Hatfield G.W., 1986, Molecular analysis of the human class I alcohol dehydrogenase gene family and nucleotide sequence of the gene encoding the (beta) subunit, *J.Biol.Chem.* 261:2027-2033.

Ehrig T., Bosron W.F., and Li T.-K., 1990, Alcohol and aldehyde dehydrogenase, *Alcohol Alcoholism* 25:105-116.

Felder M.R., Watson G., Huff M.O., and Ceci J.D., 1988, Mechanism of the induction of mouse kidney alcohol dehydrogenase by androgens. Androgen induced stimulation of transcription of the Adh-1 gene, *J.Biol.Chem.* 263:14531-14538.

Harpold M.M., Evans R.M., Salditt-Georgieff M., and Darnell J.E., 1979, Production of mRNA of Chinese hamster cells: relationship of the rate of synthesis to the cytoplasmic concentration of 9 specific mRNA sequences, *Cell* 17:1023-1035.

Mezey E. and Potter J.J., 1981, Effect of thyroidectomy and triiodothyronine administration on rat liver alcohol dehydrogenase, *Gastroenterol.* 80:566-574.

Potter J.J, Mezey E., Christy R.J., Crabb D.W., Stein P.M., and Yang V.W., 1991, CCAAT/Enhancer binding protein binds and activates the promoter of the rat class I alcohol dehydrogenase gene, *Arch.Biochem.Biophys.* 285:246-251.

Potter J.J. and Mezey E., 1983, Effect of thyroidectomy on liver alcohol dehydrogenase in the female rat, *Biochem.Pharmacol.* 32:1132-1134.

Qulali M., Ross R.A., and Crabb D.W., 1991, Estradiol induces class I alcohol dehydrogenase activity and mRNA in kidney of female rats, *Arch.Biochem.Biophys.* 288:406-413.

Stewart M.J, McBride J.S., Winter L.A., and Duester G., 1990a, Promoters for the human alcohol dehydrogenase genes ADH1, ADH2, and ADH3: interactions of CCAAT/enhancer binding protein with elements flanking the TATA box, *Gene* 90:271-279.

Stewart M.J., Shean M.L., and Duester G., 1990b, Trans-activation of human alcohol dehydrogenase gene expression in hepatoma cells by C/EBP molecules bound in a novel arrangement just 5' and 3' to the TATA box, *Mol.Cell.Biol.* 10:5007-5010.

Stewart M.J., Shean M.L., Paeper B.W., and Duester G., 1991, The role of CCAAT/enhancer binding protein in the differential transcriptional regulation of a family of human liver alcohol dehydrogenase genes, *J.Biol.Chem.* 266:11594-11503.

van Ooij C., Snyder R.C., Paeper B.W., and Duester G., 1992, Temporal expression of the human alcohol dehydrogenase gene family during liver development correlates with differential promoter activation by hepatocyte nuclear factor 1, CCAAT/enhancer-binding protein α, liver activator protein, and D-element-binding protein, *Mol.Cell.Biol.* 12:3023-3031.

Winters L.A., Stewart M.J., Shean M.L., Dong Y., and Duester G., 1990, A hormone response element upstream from the human alcohol dehydrogenase gene ADH2 consists of three tandem glucocorticoid receptor binding sites, *Gene* 91:233-240.

Wolfla C.E., Ross R.A., and Crabb D.W., 1988, Induction of alcohol dehydrogenase activity and mRNA in hepatoma cells by dexamethasone, *Arch.Biochem.Biophys.* 263:69-76.

Zhang K., Bosron W.F., and Edenberg H.J., 1987, Structure of the mouse Adh-1 gene and identification of a deletion in a long alternating purine-pyrimidine sequence in the first intron of strains expressing low alcohol dehydrogenase activity, *Gene* 57:27-36.

ALCOHOL- AND ALDEHYDE-DEHYDROGENASE : MODULATION BY BIOGENIC AMINE METABOLITES, NEUROPEPTIDES AND PSYCHOACTIVE AGENTS

F.S. Messiha[1]

[1]Department of Pharmacology and Toxicology
University of North Dakota School of Medicine
Grand Forks, ND 58203 and
Department of Pathology, Division of Toxicology
Texas Tech University Health Sciences Center
School of Medicine
Lubbock, TX 79430

INTRODUCTION

Indirect evidence continues to implicate the biogenic amines (BA) in the pathophysiology of both alcoholism (Abelin et al., 1958; Davis and Walsh, 1970; Cohn, 1973) and affective disorders (Schildkraut, 1965), which prevails to some extent in alcoholism (Mayfield, 1968; 1985). The BA have also been postulated in the mechanism of action of various psychotropic agents (PA) (Carlsson, 1987) used in the management of affective illness and alcoholic patients with underlying affective disorders (Palestine and Alatorre, 1976; O'Brien and Woody, 1980). Some of these PA have potential abuse by alcoholics (Kranzler and Orrok, 1989; Bagnall, 1991) for purposes related to mood stabilization, pain reduction, behavioral reinforcement, self medication and/or for potentiation of alcohol sedation. In addition, various neuropeptides have been suggested in some of ethanol (ET)-evoked responses (Gianoulakis, 1989; Messiha, 1989; Myers, 1989; Summers et al., 1991).

The forgoing indications lend support to the possible interrelationship between BA, ET, PA and neuropeptides. This may be mediated, at least in part, by alcohol-(ADH) and/or aldehyde-dehydrogenase (ALDH) since they, a) primarily metabolize the respective ET and acetaldehyde, b) regulate the corresponding reductive and oxidative pathways of the BA derived aldehyde intermediates (Axelrod et al., 1959; Davis et al., 1967a, 1967b; Ambroziak and Pietruszko, 1991) which possess biological activity (Sabelli et al., 1969), c) are sensitive to the action of some BA and PA (Tipton et al., 1961; Khouw et al., 1963, Skursky et al., 1975; Messiha, 1978a; Messiha, 1985a; 1985b; 1987; Ambroziak and Pietruszko, 1991; Messiha, 1991a; Roig et al., 1991), and d) are modulated by certain analogs of BA-derived condensation products (Messiha et al., 1977; Messiha, 1990a) suggested in the addictive liability of ET (Davis and Walsh, 1970; Cohen, 1973; Myers, 1989).

The present report summarizes previous experimental findings on the effects of some BA, major metabolites, their condensation product analogs, certain PA, enkephalins, and their amino acid residues on hepatic ADH and ALDH. The liver was chosen as the source of the enzymes since it contains high enzymatic activity, is responsible for the major detoxification mechanism of ingested ET and because of certain similarities between the hepatic and the corresponding brain enzymes (Messiha, 1991a). The specific activities of these enzymes were also assessed in relationship to ET preference and ET mediated narcosis in rodents. The results are discussed in reference to the possible role of ADH and ALDH in the behavioral intoxication of ET and of their contribution to the mechanism underlying adverse ET-drug interaction.

Enzymology and Molecular Biology of Carbonyl Metabolism 4
Edited by H. Weiner, Plenum Press, New York, 1993

METHODS

The subjects were adult male Sprague Dawley rats and mice unless otherwise indicated. They were purchased from Holtzman Farm Co., Madison, WI. Animals were housed in a laboratory supplied with alternating 12h of dark and light cycles. They had access to purina pellet food and water *ad libitum* unless otherwise specified. Chemicals were purchased from Sigma Chemical Co., ST. Louis, MO and Aldrich Chemical Co., Milwaukee, WI. The agents studied *in vivo* were dissolved in saline or in an injectable vehicle as indicated and were injected intraperitoneally (i.p.).

Behavioral Studies:

a) Voluntary intake of ET. This behavioral performance test was studied in rats with demonstrable preference to ET drinking over water as the drinking fluid of choice as has been previously described (Messiha, 1978a). The ET drinking fluid was prepared by diluting a 95% ET stock with distilled water to a 5% (w/v). Animals were placed in individual cages. Each cage was supplied with distilled water, 5% ET solution and an empty bottle which were rotated at random once daily to prevent from position preference (Myers et al., 1966).

Fluids and food consumptions were measured once daily (g/24h) and were expressed as percent changes from pre-drug period, designated as base line, as a function of drug administration.

b) Ethanol narcosis. This was studied in mice. The method was standardized to minimize from experimental variables (Messiha et al., 1975). Narcosis was produced by i.p. administration of 25% ET solution diluted with saline, 5 g/kg. The agent tested was either dissolved in the ET solution (simultaneous administration) or was dissolved in saline and given separately before the ET injection at the time intervals indicated. The controls received ET, 5 g/kg, i.p. The duration of narcosis (sleep time) was considered as the time from the animal's loss to regaining of its own righting reflex and were expressed as min ± SEM.

Biochemical Studies:

Animals were sacrificed by decapitation, the liver was quickly excised, blotted dry with filter paper and homogenized in ice cold 0.1 M KCl buffer, pH 6.8 by a Waring blender. The 10% (w/v) homogenates were differentially centrifuged at 4°C to obtain the subcellular mitochondrial (MT) and cytoplasmic (CT) preparations as detailed elsewhere (Messiha and Hughes, 1979). Aliquots of these preparations were used as the source of the enzymes. The protein determination was made by the biuret procedure and NAD or NADP was used as cofactor in the enzymatic assays as indicated. Both L-ADH and L-ALDH were assayed spectrophotometrically (Blair and Vallee, 1966; Blair and Bodley, 1969) and were measured at 25°C. The method of Tottmar and Marchner (1976) was used for the determination of L-MT-ALDH isoenzymes with the apparent low (L) and high (H) K_m. The lineweaver and Burk (1934) method was employed in the kinetic study.

The *in vitro* experiments utilized subcellular rat liver preparations from experimentally naive animals unless otherwise indicated. The enzymatic activity was expressed as means ± SD or SEM of specific activity, nMol/min/mg protein, or as percent changes from corresponding mean control value.

Statistical Analysis:

The two tailed Student's *t* test for independent means was used for the statistical evaluation of the results.

Experimental Design:

In the first set of experiments, adult male Sprague Dawley rats (70 to 80 days old) were used to study the effect of short-term administration of some monoamines and major metabolites on endogenous L-ADH and L-ALDH. These consisted of dopamine (DA), homovanillic acid (HVA), 5-hydroxyindoleacetic acid (5-HIAA), 5-hydroxytryptophol (5-HTOH), normetanephrine (NM), metanephrine (M), 3-methoxy-4-hydroxyphenylethanol (MOPET), 3-methoxy-4-hydroxyphenylglycol (MHPG), 3-methoxytyramine (MT), 3-methoxy-4-hyroxyphenylpyruvic acid (PP), 3-methoxy-4-hydroxyphenylacetic acid (PL), and vanillylmandelic acid (VMA). The agents were dissolved in the vehicle, physiological saline, and were injected in equimolar concentration, 0.5 mMol/kg, i.p. The controls were injected with saline. The agents studied were injected once daily for 7 days and the animals were sacrificed by decapitation 16 hrs after the last dose.

In the second set of experiments, the effects of MOPET on voluntary intake of ET by the rat, the duration of ET-mediated narcosis in mice and the *in vitro* effect of MOPET on mouse L-ADH and L-ALDH were evaluated. Rats with preference to ET drinking were selected and received single injection of MOPET, 0.5 mMol/kg, i.p. The daily fluids consumption (g/24hrs) for the pre-treatment period (designated base line and considered 100%) was compared with measurements made 24 hrs subsequent the MOPET trial.

Ethanol-mediated narcosis was performed in adult male mice. Narcosis was produced by i.p. injection of a narcotic dose of ET, 5 g/kg, containing MOPET, 500 mg/kg, i.p. The controls received ET, 5 g/kg, i.p. Additional group of mice were injected with saline or MOPET, 500 mg/kg, 45 min before receiving the narcotic dose of ET, 5 g/kg, i.p.

In separate experiments the *in vitro* effect of MOPET, 1 mMol, on specific activity of mouse L-ADH and L-CT-ALDH was studied. The capacity of the mouse liver CT preparation to metabolize MOPET as a substrate was also assayed and compared with that of equal concentration of ET, 1 mMol.

In the third set of experiments, the effects of tetrahydropapaverine (THP) on voluntary intake of ET and on subcellular L-ADH and L-ALDH were studied in adult male rats of comparable age. The *in vitro* effect of papaverine (PAP), a THP analog, on specific activities of L-ADH and L-ALDH was also determined in rat liver preparations. Rats with drinking preference to 5% ET were injected initially with THP, 50 mg/kg, i.p. This was then repeated once daily for 5 consecutive days after a 4 day THP-free period. These rats were sacrificed 30 min post the terminal dose along with others of comparable age which were maintained on water or 5% ET solution for a similar period of time and received either saline (control) or identical THP treatment. The livers were excised and kept frozen overnight at -20°C for approximately 16 hrs before they were used for the subcellular fractionation. In the *in vitro* experiments, the effects of THP, 1 mMol, or PAP, 0.1 mMol, on specific activities of rat L-ADH and subcellular L-ALDH were studied.

In the fourth set of experiments, the effects of leucine enkephalin (LNK), methenkephalin (MNK) and the amino acid constituents on ET preference by the rat were studied along with the *in vivo* effects of LNK and MNK on rat L-ADH and L-ALDH. In the ET preference experiment, rats were injected with LNK, MNK, leucine, glycine, phenylalanine or tyrosine, 25 mg/kg, i.p. dissolved in saline. The averages for daily fluid intake for the 5 days preceding treatment were referred to as base line and considered 100%. Changes in fluid consumption was calculated as percent changes from mean base line value.

In the enzymatic study, the *in vivo* effect of LNK in the presence or absence of ET on endogenous L-ADH and L-ALDH was determined. Adult male rats were divided into 4 groups of 5-6 rats each. The first group was injected with saline twice over 3 hrs (control); the second was administered LNK, 25 mg/kg, i.p., twice over same time intervals (LNK-treatment); the third group was injected with saline twice over 3 hrs followed 15 min later by ET, 2.5 mg/kg, i.p., (ET-group) and the fourth group received identical treatment with LNK as in the second group in addition to ET, 2.5 mg/kg, i.p., given 15 min post a second dose of LNK. All animals were sacrificed 2 hrs post the terminal treatment and the livers were processed for the enzymatic assays.

In the fifth set of experiments, the effect of acute dose of pimozide (PMZ), 25 mg/kg, i.p. on voluntary drinking of ET was evaluated in rats with preference to ET. The effects of acute dose regimens of haloperidol (HAL) or PMZ on endogenous female mouse L-ADH and L-ALDH were also determined. The mice were injected with HAL or PMZ, 30 mg/kg, i.p., once daily for two consecutive days and the controls received the injectable vehicle at same time intervals. The drugs were dissolved in chloroform before they were suspended in mineral oil and the chloroform was evaporated over water bath. The mice were sacrificed 30 min after the second treatment and the livers were stored overnight at -20°C before they were used for the subcellular fractionation and the enzymatic assays.

RESULTS

Figure 1 illustrates the effect of short-term administration of equimolar concentration of selected BA and major metabolites on L-ADH, both NAD and NADP-dependent L-ALDH. Mice treated with DA, the major metabolites MT, MOPET or HVA showed little changes in the endogenous enzymatic activity measured compared to saline controls. The L-dopa metabolite PL inhibited L-ADH and NADP-dependent L-ALDH by 28% (p<0.05) and 22% (p<0.05) from controls, respectively. Treatment with MHPG, the major neutral metabolite of norepinephrine and epinephrine, also inhibited L-ADH (15%, p<0.05) and NADP-linked ALDH (17%, p<0.05). The latter was also inhibited by VMA (23%, p<0.01), the major acidic metabolite of norepinephrine and epinephrine. Conversely, treatment with VMA or 5-HTOH resulted in induction of specific activity of L-ADH above corresponding control by 18.4% (p<0.05) and 47% (p<0.001), respectively. The inhibition of NAD-linked L-ALDH by 5-HTOH and MOPET was not statistically significant (p<0.1).

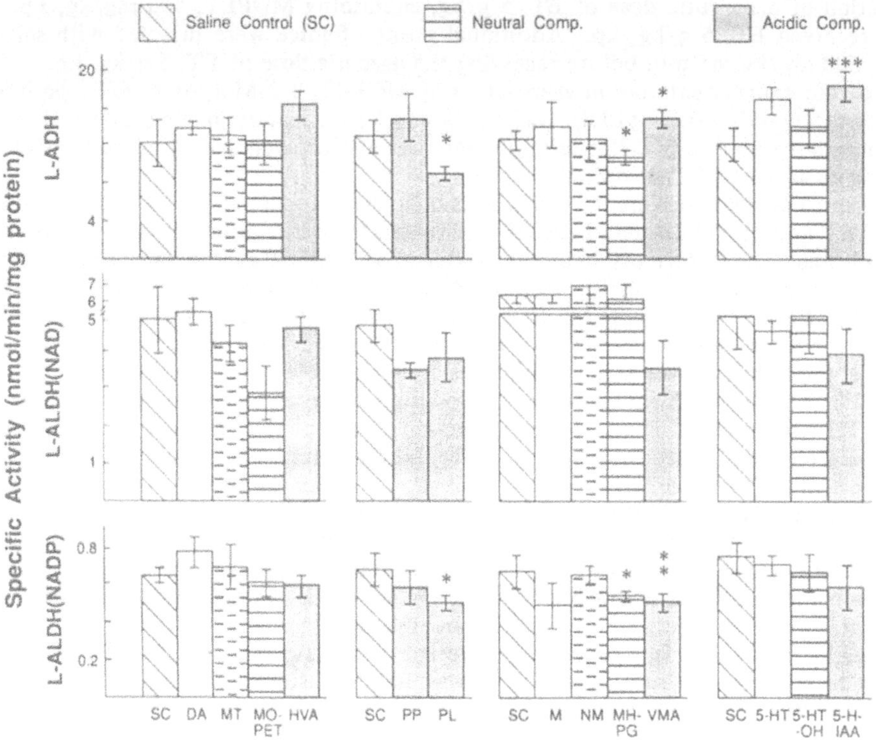

Figure 1. The effect of short-term administration of equimolar doses of dopamine (DA), serotonin (HT) and respective major metabolites (for abbreviation, see text), 0.5 mMol/kg i.p./day for 7 days, on rat liver alcohol-(L-ADH) and aldehyde dehdyrogenase (L-ALDH). Each bar graph represents 4-11 independent assays of the member of animals used. Values are means of specific activity, nMol/min/mg protein. Statistical difference from corresponding saline control. ****p*<0.001, ***p*<0.01, **p*<0.05.

Figure 2 shows the effects of MOPET on some of ET elicited responses (upper panel) and the *in vitro* effect on specific activities of mouse L-ADH and L-ALDH (lower panel). Administration of a single dose of MOPET, 0.5 mMol/kg i.p., decreased voluntary intake of ET by approximately 32% (p<0.05) from pretreatment period. This was accompanied by a 41% (p<0.05) compensatory increases in water consumption.

Figure 2 (upper right panel), shows that mice simultaneously treated with both MOPET, 500 mg/kg, and a narcotic dose of ET showed a prolonged duration of ET-narcosis. This was approximately 1.25 fold greater than the controls (p<0.01). Administration of the same MOPET dose 45 min before an identical narcotic dose of ET had little effect on the duration of ET mediated narcosis compared to controls.

Figure 2 also compares the *in vitro* utilization of ET and MOPET as substrates by an NAD-dependent dehydrogenase (lower left panel) and the effect of MOPET on cytoplasmic L-ADH and L-ALDH (lower right panel). The use of equal substrates concentration of ET or MOPET, 1 mMol, for the dehydrogenase reaction showed specific activity of the same

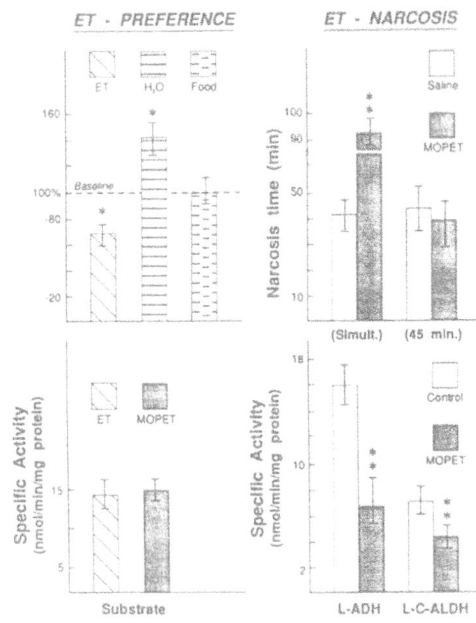

Figure 2. The effect of 3, methoxy-4-hydroxyphenylethanol (MOPET) on voluntary intake of ethanol (ET) by the rat, duration of ET-narcosis in mice, metabolism by an NAD-linked mouse liver cytoplasmic dehydrogenase and *in vitro* effect on liver cytoplasmic alcohol-(L-ADH) and aldehyde dehydrogenase (L-ALDH), see experimental design for details. Values are means ± SEM (n=4-17), **$p<0.01$, *$p<0.05$.

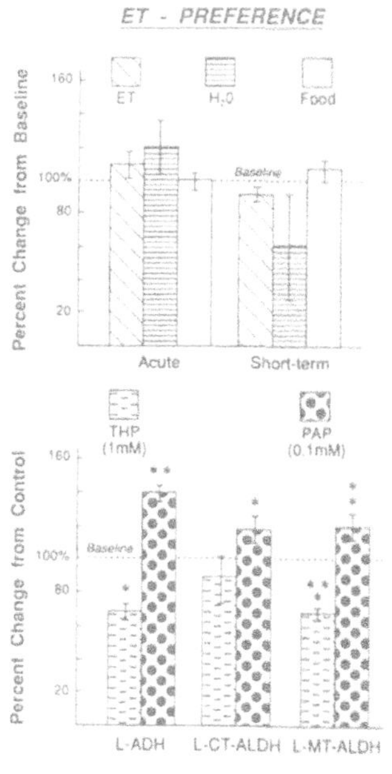

Figure 3. The effect of tetrahydropapaverine (THP) on ethanol (ET) preference by the rat and the *in vitro* effect of THP (1mMol) or papaverine (0.1mMol) (PAP) on subcellular rat liver alcohol-L-(ALDH) and aldehyde dehydrogenase (L-ALDH). See experimental design for details (n=4-17), **$p<0.02$, *$p<0.05$.

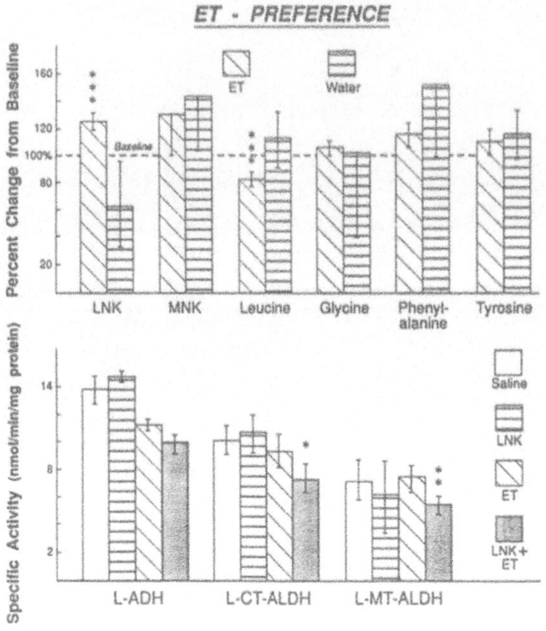

Figure 4. The acute effects of short-term treatment with leucine enkephalin (LNK), metenkephalin (MNK) or the amino acid constituents on ET preference by the rat (upper panel) and on rat liver alcohol-(L-ADH) and aldehyde dehydrogenase (L-ALDH). See experimental design for details, (n=4-6), ***$p<0.025$, **$p<0.02$, *$p<0.05$.

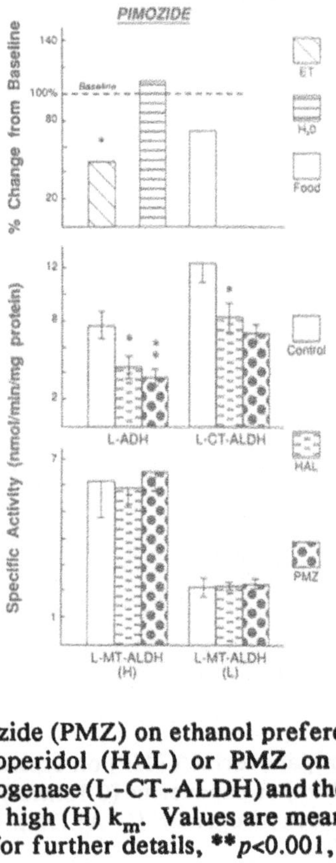

Figure 5. The effect of pimozide (PMZ) on ethanol preference in the rat (upper panel) and of equal acute doses of haloperidol (HAL) or PMZ on mouse liver alcohol-(L-ADH), cytoplasmic aldehyde-dehydrogenase (L-CT-ALDH) and the mitochondrial (MT) isoenzymes with the apparent low (L) and high (H) k_m. Values are means ± SD of specific activity (n=4-5). See experimental design for further details, **$p<0.001$, *$p<0.01$.

Table 1. The effect of chlorpromazine on specific activities of alcohol-(ADH) and aldehyde-(ALDH) dehydrogenase by various enzyme preparations.

Exp.	Enzyme source			ADH	ALDH		Reference
	species	sex	prep.		CT	MT	
In vitro	Rabbit	Male	Liver	↓	-	-	Khouw et al., 1963.
	Horse	-	Liver (crist. prep.)	↓	-	-	Skursky & Kovar, 1975.
	Albino mouse	Female	Liver	↓(NC)	↔	-	Messiha et al., 1983a.
		Female		↓(NC)	-	-	Messiha, 1985b.
		Male		↓(NC)	-	-	
	Black (C_{57}BL/6) mouse	Female		↓(NC)	-	-	
	Albino rat	Male		↓(NC)	-	-	
	Horse	-	Liver (crist. prep.)	↓	-	-	Roig et al., 1991.
In vivo	Black (C_{57}BL/6) mouse (5mg/kg/Dx5D)	Female	Liver	↔	-	↔	Messiha et al., 1983b.
	Albino mouse	Female	Liver	↔	↔	-	Messiha et al., 1983a.
	dose build up (5-30mg/kg/Dx21D)			↔ +UV	↓ (NC)	-	
	Albino rat (5mg/kg/Dx23D)	Male	Brain	↓	↔	-	Messiha, 1985a.

Arrows indicate inhibition (↓), no changes (↔) from controls or not reported (-) results. Noncompetitive (NC) inhibition of alcohol dehydrogenase-(ADH). Subcellular cytoplasmic (CT) and mitochondrial (MT) aldehyde dehydrogenase (ALDH) were indicated along with the commercial horse liver cristaline preparations (crist. prep.) used. Chlorpromazine was injected once daily for the number of days (D) shown in the *in vivo* experiments.

order and magnitude for both agents. MOPET inhibited L-ADH and L-CT-ALDH (lower right panel) by approximately 59% ($p<0.01$) and 40% ($p<0.01$) *in vitro*, respectively.

Figure 3 shows the effect of acute and short-term i.p. administration of THP (upper panel) on the rat ET preference and the *in vitro* effects of THP, 1 mMol, or PAP, 1 mMol, on subcellular rat L-ADH, L-CT-ALDH and L-MT-ALDH (lower panel). The acute and short-term administration of THP did not alter voluntary intake of ET by the rat. Likewise, short-term administration of identical THP dose for same duration of time did not alter endogenous L-ADH or L-CT-ALDH *in vivo* of rats maintained on water or 5% ET drinking fluids.

Figure 3 (lower panel)also shows that THP, 1 mMol, inhibited *in vitro* both L-ADH ($p<0.05$) and L-MT-ALDH ($p<0.001$) but not L-CT-ALDH. Conversely, PAP, a THP analog, enhanced the specific activity of rat L-ADH (42%, $p<0.001$) at 0.1 mM concentration *in vitro*. A similar PAP concentration also increased above control both L-CT-ALDH ($p<0.05$) and L-MT-ALDH ($p<0.02$) *in vitro*.

Figure 4 shows the effects of LNK, MNK or the corresponding amino acid constituents on ET preference in male rats (upper panel) and the *in vivo* acute effects of LNK on subcellular L-ADH and L-ALDH in presence and absence of ET (lower panel). Administration of LNK increased voluntary ET consumption 48 h post treatment ($p<0.02$) which remained statistically significant for a subsequent three days. The MNK dose exerted little effect on ET preference. The administration of the individual amino acid constituents of LNK showed that leucine caused aversion to ET preference ($p<0.02$).

Figure 4 also shows (lower panel) that acute treatment with LNK or ET had no effect on specific activities of the hepatic enzymes measured. Administration of ET after LNK inhibited both CT (26%, $p<0.05$) and MT (19%, $p<0.025$) L-ALDH from controls. This inhibition was determined as noncompetitive.

Figure 5 illustrates the acute effect of PMZ on voluntary intake of ET by the rat (upper panel) and the effects of equal dose regimens of HAL or PMZ on endogenous female mouse L-ADH and L-ALDH (mid and lower panels). Administration of a single dose of PMZ, 25 mg/kg, i.p., decreased preference to ET drinking from mean base line by approximately 58% ($p<0.01$) for the 24 hrs proceeding drug treatment. This was accompanied by a statistically insignificant small increases in water intake.

Figure 5 (mid panel) also shows that i.p. administration of equal dose regimens of HAL or PMZ, 30 mg/kg/day for 2 days, altered endogenous specific activities of the enzymes measured in the CT but not in the MT liver preparation. This has been shown by inhibition of specific activity of L-ADH by 42% ($p<0.01$) and 51% ($p<0.001$) by HAL or PMZ, respectively. The HAL treatment also inhibited endogenous L-CT-ALDH by 33% ($p<0.01$) from controls. No changes occurred in L-MT-ALDH isoenzymes from mean control as a consequence of either drug treatment.

Table 1 lists the *in vitro* and *in vivo* effect of chlorpromazine (CPZ) on enzymatic activities of L-ADH and L-ALDH reported by various investigators. The *in vitro* use of various CPZ concentrations inhibited L-ADH irrespective of species, strain, gender, enzyme source or the biological preparations used. The L-ADH inhibition by CPZ was determined as noncompetitive in some studies using both rat and mouse liver preparations.

Table 1 also shows that specific activity of L-ADH was not altered from corresponding control by short-term administration of CPZ for as long as 3 consecutive weeks. This is compared to the *in vivo* inhibition of rat brain striatal ADH by CPZ. Mouse L-CT-ALDH but not L-ADH was inhibited by short-term administration of CPZ to animals housed in darkness and under continued exposure to UV illumination.

DISCUSSION

The present results suggest metabolic interrelationships between L-ADH, L-ALDH, BA and major metabolites studied. This may influence certain neurotransmitters function or causes adverse ET-drug interactions. Accordingly, ET consumption in certain clinical conditions in which BA or modifying medications are used should be contraindicated. The short-term treatment with PL, MHPG or VMA exerted inhibitory action on both L-ADH and NADP-linked L-ALDH. Vanillylmandelic acid and MHPG are the respective major acidic and neutral metabolites of the catecholamines norepinephrine and epinephrine. These catecholamines are primarily metabolized to the acidic or neutral metabolites by ALDH or ADH, subsequent oxidative deamination by monoamine oxidase, in the absence or presence of ET, respectively (Davis et al., 1976a-b; Valtaine et al., 1992). Therefore, the inhibition of ET detoxification enzymes by VMA and MHPG may enhance some of the cardiovascular adverse effects of ET due to its action on the sympathetic nervous system and its influence on the adrenal gland catecholamines.

The inhibition of L-ADH and NADP-dependent L-CT-ALDH by PL is of particular interest since the formation of PL, like some of the other O-methylated L-dopa metabolites, is increased subsequent combined administration of L-dopa with carbidopa (Messiha et al., 1972; Sandler et al., 1974), a peripheral aromatic L-amino acid decarboxylase inhibitor, in Parkinson's disease. This suggests that the use of such therapy may sensitize the susceptible parkinsonian patient to ET toxicity particularly since these decarboxylase inhibitors also inhibit L-ALDH (Messiha, 1977; 1978b). Therefore, an inhibition of L-ADH by BA metabolites is more likely to increase blood ET intoxicated levels while inhibition of L-ALDH may produce a disulfiram-like reaction during alcohol consumption. It has been suggested that such metabolic effects might worsen the parkinsonian symptoms (Dougan et al., 1975; Pincus and Tucker, 1978; Messiha, 1983) particularly since ET has been reported to induce parkinsonian-like syndrome in man (Lutz, 1976; Neiman, 1990).

The present results also show that MHPG, PL and VMA inhibited NADP- but not NAD-linked L-ALDH. This suggests a differential isoenzyme(s) sensitivity and cofactor specificity. This may have little metabolic significance since ET-derived acetaldehyde is primarily metabolized by the NAD-dependent ALDH. However, this ALDH inhibition and that produced by PA on NAD-linked ALDH could enhance the formation of BA-derived isoquinolines (McIsaac, 1961; Holtz et al., 1964) which have been hypothesized in the etiology of alcoholism (Davis and Walsh, 1970; Cohen, 1973; Myers, 1989). The isoquinolines formed might then further potentiate the inhibition of L-ALDH since the results show that one of these analogs, THP, inhibited both L-ADH and L-MT-ALDH isoenzymes *in vitro* but not *in vivo*. The difference between the *in vitro* and the *in vivo* effect of THP may be related to the short T½ of THP and/or may be due to its rapid degradation. This may explain the observed lack of THP effect on voluntary intake of ET by the rat which varies according to the route of THP administration, the dose and the regional brain nuclei being affected (Privette and Myers, 1989) which may have contributed to the inconsistent results reported in the ET preference studies (Duncan and Deitrich, 1989).

The determined inhibition of endogenous L-ADH and L-ALDH by two acute dose regimens of the antipsychotic medications HAL or PMZ may explain the PMZ-caused aversion to ET preference. This might have been due to the development of a disulfiram-like adverse behavioral effect as a consequence of L-ALDH inhibition. This antipsychotic medication-ET interaction may have legal implication relevant to aviation and car accidents since ET interferes in various coordinated motor performances at intoxicated blood ET concentrations which might be attained after inhibition of L-ADH by such PA intervention.

The observed *in vitro* inhibition of L-ADH and L-MT-ALDH by THP contrasts with the determined stimulation of L-ADH and subcellular L-ALDH activities by PAP, the O-methylated analog. This implies that PAP may enhance ET elimination and thereby minimizes from ET intoxication and/or may conceivably counteract the inhibitory action of other agents on these enzymes. Therefore, the use of PAP in the management of erectile and psychogenic impotence in man (Virag, 1982; Stief and Wetterauer, 1988; Dhabuwala, 1990) might find clinical application in ET-induced male impotence (Kyrle, 1909; Weichselbaum, 1912; Gerdes, 1978; Van Thiel et al., 1974; Morgan, 1982) caused by ET produced alteration of spermatogenesis due to inhibition of vitamin A catalyzed reaction by testicular ADH (Van Thiel et al., 1974). Moreover, the determined inhibition of L-ADH by HAL, PMZ or CPZ, which is consistent with that produced by CPZ on gonadal ADH (Messiha, 1990b) and its *in vitro* (Khouw et al., 1963; Skursky and Kovar, 1975; Messiha et al., 1983b; Messiha, 1985b; Roig et al., 1991) and *in vivo* (Messiha, 1991b) hepatic effects may contribute to the mechanism underlying antipsychotic medications-produced impotence in man (Shader and DiMascio, 1970, Kogeoros et al., 1986; Abber et al., 1987). These findings may be viewed as indirect evidence to support a role for ADH in reproductive function which can be influenced by ET and other PA. Moreover, the consumption of ET with antipsychotic medications does not only decreases CPZ metabolism, clearance and enhances sedation (Forrest et al., 1971) but also elevates blood ET concentrations (Sutherland et al., 1960; Tipton et al., 1961; Koff and Fitts, 1972), which can be explained by inhibition of L-ADH by CPZ and/or by one of its metabolites. This indicates a metabolic mediated CPZ-ET interaction.

The administration of a large dose of MOPET with a narcotic dose of ET prolonged the duration of ET-mediated narcosis (sleep time) in mice. This is consistent with reports showing similar effects by BA and some major metabolites (Blum et al., 1973; Messiha et al., 1975). This indicates a BA-mediated potentiation of the central depressant action of ET (Holtz et al., 1964; Rosenfield, 1960; Blum et al., 1973). Therefore, the adverse interaction between ET and BA altering drugs or with L-dopa should be considered in alcoholic patient undergoing surgical anesthesia. The ineffectiveness of MOPET on ET narcosis when administered prior to ET may be due to a short T½ of MOPET, a poor CNS penetration rate without ET as the injectable vehicle and/or a faster metabolism by L-ADH which also utilizes MOPET as a substrate (Messiha, 1983).

The results pertaining to the effect of the neuropeptides studied on ET preference show differences between LNK (Tyr-Gly-Gly-Phe-leu) and MNK (Tyr-Gly-Gly-Phe-Met). The LNK treatment increased voluntary consumption of ET compared to lack of action by the MNK treatment. This was unlikely due to any of the individual amino acid constituents, formed subsequent degradation by enkaphalinase and peptidase (Bayon et al., 1983), since the separate or combined administration of leucine and phenylalanine decreased ET consumption which was abolished by glycine (Messiha, 1989). The observed LNK effect on ET preference is more likely mediated by an opiate mechanism since ET consumption alters biosynthesis of various enkephalins, interferes in their binding to opiate receptors and with other neuropeptides activity implicated in ET drinking behavior (Blum et al., 1983; Hynes et al., 1983; Lucchi, et al., 1985; Rapaka et al., 1986; Myers, 1989).

In conclusion, the results demonstrate the variable sensitivity of L-ADH and L-ALDH to the pharmacologic interventions used. This suggests that these enzymes and/or corresponding subcellular isoenzymes possess wide substrate specificity and/or sensitivity to a large number of agents which do not share common chemical structure, or pharmacologic properties. This explains the numerous ET-drug interactions reported (Weber and Preskorn, 1984; Lieber, 1990; Mattila, 1990; Shoaf and Linnoila, 1991). This extends the metabolic importance of these enzymes to cover pharmacological responses triggered secondary to modulation of their action on the various systems involved.

SUMMARY

The results suggest the feasibility of metabolic and behavioral interrelationships between ethanol (ET), certain neurotransmitter substances and major metabolites, some neuropeptides and/or psychoactive agents. This was indicated by the *in vivo* and *in vitro* effects of such authentic compounds on certain ET-elicited behavioral responses and on hepatic ET and acetaldehyde metabolizing enzymes in rats and mice.

ACKNOWLEDGEMENT

All experimental work was supported by and performed at Texas Tech University Health Sciences Center School of Medicine, Lubbock, TX, USA. Travel support was provided in part by the Görlich Foundation, Switzerland. The results derive from previously published materials with permissions from the following: [Pharm. Biochem. Behav. 9, Messiha, F.S., Voluntary drinking of ethanol by the rat: Biogenic amines and possible underlying mechanism], Copyright [1978], Pergamon Press PLC.; [Brain Res. Bull. 11, Messiha, F.S., Extrapyramidal disorders: A possible underlying mechanism], Copyright [1983], Pergamon Press PLC.; [Biogenic Amines, 4, Messiha, F.S., Biogenic amine metabolites and ethanol interaction], Copyright [1987], Pergamon Press PLC.; [Neurosci. Biobehav. Rev., 12, Messiha, F.S., Tourette's medications: Effect on minor oxidative and reduction pathways of biogenic amines], Copyright [1988], Pergamon Press PLC.; [Physiol. Behav. 46, Messiha, F.S., Effect of leucine enkephalin on ethanol produced behavioral depression in the mouse], Copyright [1989], Pergamon Press PLC.; [Physiol. Behav. 46, Messiha, F.S., Enkephalins, their constituents and voluntary drinking of ethanol by the rat], Copyright [1989], Pergamon Press PLC.; [Pharmacology, Messiha, F.S. et al., 13, 340-351, 1975, Ethanol narcosis in mice: Effects of L-dopa, its metabolites and other experimental variables], S. Karger Publishers, Inc.; [Res. Commun. Subst. Abuse, 11, 69-72, 1990, Messiha, F.S., Dopamine O-methylated metabolites and hepatic alcohol dehydrogenase] PTD Publications, Ltd., Westbury, N.Y.; [Vet. Hum. Toxicol. 32, 126-130, 1990, Messiha, F.S., Tetrahydropapverine and ethanol metabolizing enzymes, Comp. Tox. Lab. Manhattan, KA.

REFERENCES

Abber, J.C., Lue, T.F., and Lue, J.A., 1987, Priapism induced by chlorpromazine and trazodone: mechanism of action. *J. Urol.* 137:1039.

Abelin, V.I., Herren, C., and Berli, W., 1958, Ueber die erregende Wirkung des Alkohols auf den Adrenalin und Noradrenalin Haushalt. *Helv. Med. Acta.* 25:591.

Ambroziak, W. and Pietruszko, R., 1991, Humans aldehyde dehydrogenase: Activity with aldehyde metabolites of monoamines, diamines and polyamines. *J. Biol. Chem.* 266:13011.

Axelrod, J., Albers, W., and Clement, C.D., 1959, Distribution of catechol-O-methyltransferase in the nervous system and other tissues. *J. Neurochem.* 5:68.

Bagnall, G., 1991, Alcohol and drug use in a Scotish Cohort: 10 years on. *Br. J. Addict.* 86:859.

Bayon, A., Shoemaker, W.J., McGinty, J.F., and Bloom, F., 1983, Immunodetection of endorphins and enkephalins: a search for reliability. *Jnt. Rev. Neurobiol.* 24:51.

Blair, A. and Bodley, F., 1969, Human liver aldehyde dehydrogenase: Partial purification and properties. *Can. J. Biochem. Physiol.* 47:265.

Blair, A., and Vallee, B., 1966, Some catalytic properties of human liver dehydrogenase. *Biochem.* 5:2026.

Blum, K., Calhoun, W., Meritt, J., and Wallace, J.E., 1973, L-dopa effect on ethanol narcosis and brain biogenic amines in mice. *Nature* 242:407.

Blum, K., Elston, S.F., DeLallo, L., Briggs, A., and Wallace, J., 1983, Ethanol acceptance as function of genotype amounts in the brain of (Met) enkephalin. *Psoc. Nat. Acad. Sci.*80:6510.

Carlsson, A., 1987, Monoamines of the central nervous system: A historical perspective, in: "Psychopharmacology: The Third Generation of Progress," H.Y. Meltzer, ed., Raven Press, New York.
Cohen, G., 1973, A role for tetrahydroisoquinoline alkaloids as a false adrenergic transmitter in alcoholism. *Adv. Exp. Med. Biol.* 35:33.
Davis, V.E., Brown, H., Huff, J.A., and Cashaw, J.L., 1967a, The alteration of serotonin metabolism to 5-hydroxytryptophol by ethanol ingestion in man. *J. Lab. Clin. Med.* 69:132.
Davis, V.E., Brown, H., Huff, J.A., and Cashaw, J.L., 1967b, Ethanol-induced alterations of norepinephrine in man. *J. Lab. Clin. Med.* 69:787.
Davis, V.E. and Walsh, M.J., 1970, Alcohol, amines and alkaloids: a possible biochemical basis for alcohol addiction. *Science* 167:1005.
Dhabuwala, C.B., Kerkar, P., Bhutwala, A., Kumar, A., and Pierce, J.M., 1990, Intracavernous papaverine in the management of psychogenic impotence. *Arch. Androl.* 24:185.
Dougan, D., Wade, D., and Mearrick, P., 1975, Effects of L-dopa metabolites at a dopamine receptor suggest a basis for "on-off" effect in Parkinson's disease. *Nature* 254:70.
Duncan, C. and Deitrich, R.A., 1980, A critical evaluation of tetrahydroisoquinoline-induced ethanol preference in rats. *Pharmacol. Biochem. Behav.* 13:265.
Forrest, F.M., Forrest, I.S., and Finkele, B.S., 1971, Alcohol-chlorpromazine interaction in psychiatric patients. *Aggressologic* 13:67.
Gerdes, H., 1978, Alkohol und Endocrinum, *Internist* 19:89.
Gianoulakis, C., 1989, The effect of ethanol on the biosynthesis and regulation of opioid peptides. *Experientia* 45:428.
Holtz, P., Stock, K., and Westerman, E., 1964, Pharmakoligie des Tetrahydropapaverolins und seine enstehung aus Dopamin. *Arch. Exp. Path. Pharmakol.* 248:387.
Hynes, M.D., Lochner, M.A., Bemis, K.G., and Hymson, D.L., 1983, Chronic ethanol alters the receptor binding characteristics of the delta opioid receptor ligand, D-Ala-2-D-leu-5 enkephalin in mouse brain. *Life Sci.* 33:2331.
Khouw, L.B., Burbridge, T.N., and Sutherland, V.C., 1963, The inhibition of alcohol dehydrogenase: I. Kinetic Studies. *Biochem. Biophys. Acta.* 73:173.
Koff, R.S. and Fitts, J.J., 1972, Chlorpromazine inhibition of ethanol metabolism without prevention of fatty liver. *Biochem. Med.* 6:77.
Kogeorgos, J. and De Alwis, C., 1986, Priapism and psychotropic medicaiton. *Br. J. Psychiat.* 149:241.
Kranzler, H.R. and Orrok, B., 1989, The pharmacotherapy of alcoholism, Rev. Psychiat. 8:231.
Kyrle, J., 1909, Ueber structur nomalien in Menschlichen Hodenparenchym. *Verhand. Deutsch. Pathol. Geselsch.* 13:391.
Lieber, C.S., 1990, Interaction of alcohol with drugs, hepatotoxic agents, carcinogens and vitamins. *Alcohol* 25:157.
Lineweaver, H. and Burk, D., 1934, Determination of enzyme dissociation constants. *J. Am. Chem. Soc.* 56:568.
Lucchi, L., Rivs, R., Govoni, S., and Trabucchi, M., 1985, Chronic ethanol induces changes in opiate receptor function and in met-enkephalin release. *Alcohol* 2:193.
Lutz, E.G., 1976, Neuroleptic-induced akathisia and dystonia triggered by alcohol. *J. Am. Med. Assoc.* 23:2422.
Mattila, M.J., 1990, Alcohol and drug interaction. *Ann. Med.* 22:363.
Mayfield, D., 1968, Psychopharmacology of alcohol: Affective change with intoxication, drinking behavior and affective state. *J. Nerv. Ment. Dis.* 146:314.
Mayfield, D., 1985, Substance abuse in the affective disorders, in: "Psychopathology and Substance Abuse," A. Alterman, ed., Plenum Press, New York.
McIsaac, W.M., 1961, Formation of 1-methyl-6 methoxy-1,2,3,4-tetrahydro-2-carboline under physiological conditions. *Biochim. Biophys. Acta.* 52:607.
Messiha, F.S., Hsu, T.H., and Bianchine, J.R., 1972, Peripheral aromatic L-amino acid decarboxylase inhibitor in parkinsonism. Effect on O-methylated metabolites of L-2-^{14}C-dopa. *J. Clin. Invest.* 51:452.
Messiha, F.S., Morgan, M. and Geller, I., 1975, Ethanol narcosis in mice: Effects of L-dopa, its metabolites and other experimental variables. *Pharmacol.* 13:340.
Messiha, F.S., 1977, Possible mechanism of adverse reaction following levodopa plus benserazide treatment. *Br. J. Pharmacol.* 60:55.

Messiha, F.S., 1978a, Voluntary drinking of ethanol by the rat: Biogenic amines and possible underlying mechanism. *Pharmacol. Biochem. Behav.* 9:379.
Messiha, F.S., 1978b, Modulation of hepatic aldehyde dehydrogenase by carbidopa. *Res. Commun. Clin Path. Pharmacol.*20:601.
Messiha, F.S. and Hughes, M., 1979, Liver alcohol and aldehyde-dehydrogenase: Inhibition and potentiation by histamine agonists and antagonists. *Clin. Exp. Pharmacol. Physiol.* 6:281.
Messiha, F.S., 1983, Extrapyramidal disorders: A possible underlying mechanism. *Brain Res. Bull.* 11:233.
Messiha, F.S., Striegler, R.L., and Sproat, H.F., 1983a, Modification of mouse liver alcohol and aldehyde-dehydrogenase by chlorpromazine. *Drug Chem. Toxicol.* 6:409.
Messiha, F.S., Sproat, H., and Striegler, R.L., 1983b, Chlorpromazine and lithium interaction: A biochemical and histological study. *Br. Res. Bull.*11:249.
Messiha, F.S., 1985a, Effect of centrally acting drugs on ethanol detoxification enzymes in distinct rat brain regions. *Neurobehav. Toxicol. Teratol.* 7:81.
Messiha, F.S., 1985b, Chlorpromazine and ethanol intoxication: An underlying mechanism. *Neurobehav. Toxicol. Teratol.* 7:185.
Messiha, F.S., 1987, Biogenic amine metabolites and ethanol interaction. *Biogenic Amines* 4:265.
Messiha, F.S., 1989, Enkephalins, their constituents and voluntary drinking of ethanol by the rat. *Physiol. Behav.* 46:29.
Messiha, F.S., 1990a, Tetrahydropapaverine and ethanol metabolizing enzymes. *Vet. Hum. Toxicol.* 32:126.
Messiha, F.S., 1990b, Gonadal alcohol and aldehyde dehydrogenase: *In vivo* and *in vitro* effects of psychoactive and endocrine agents. *Pharmacol. Biochem. Behav.* 35:29.
Messiha, F.S., 1991a, Cerebral and peripheral neurotoxicity of chlorpromazine and ethanol interaction: Implication for alcohol and aldehyde dehydrogenase. *Neurotoxicol.* 12:559.
Messiha, F.S., 1991b, Mouse strain-dependent hepatic interaction of psychoactive agents with ethanol and biogenic aldehydes metabolizing enzymes. *Hum. Psychopharmacol.* 6:61.
Morgan, M.Y., 1982, Sex and alcohol. *Br. Med. Bull.* 38:43.
Myers, R. and Holman, R.B., 1966, A procedure for elimination position habit in preference aversion tests for ethanol and other fluids. *Psycho. Sci.* 6:235.
Myers, R.D., 1989, Isoquinolines, beta-carbolines and alcohol drinking: Involvement of opioid and dopaminergic mechasims. *Experientia* 45:436.
Neiman, J., Long, A., Fornzzari L., and Carlen, P., 1990, Movement disorders in alcoholism: A review. *Neurol.* 40:741.
O'Brien, C.P. and Woody, G.E., 1980, The use of Psychotropic medication in the treatment of drug abuse, in: Substance Abuse and Psychiatric Illness," A. Gottheil, A.T. McLellan and K.A. Druley, eds., Pergamon Press, New York.
Palestine, M.L. and Alatorre, E., 1976, Control of acute and alcoholic withdrawal symptoms: A comparative study of haloperidol and chlordiazepoxide. *Curr. Ther. Res.* 20:289.
Pincus, J.H. and Tucker, G.J., 1978, "Behavioral Neurology," Oxford University Press, NY.
Privette, T.H. and Myers, R.D., 1989, Anatomical mapping of tetrahydropapaveroline-reactive sites in brain mediating suppression of alcohol drinking in the rat. *Brain Res. Bull.*22:1039.
Rapaka, R.S., Renugopalakirishnan, Goehl, T., and Collins, B.J., 1986, Ethanol induced conformational changes of the peptide ligands for the opioid receptors and their relevance to receptor interaction. *Life Sci.* 39:837.
Roig, M.G., Bello, F., Burguillo, F.J., Cachaza, J.M., and Kennedy, J.F., 1991, In vitro interaction between psychotropic drugs and alcohol dehydrogenase activity. *J. Pharm. Sci.* 80:267.
Rosenfield, G., 1960, Potentiation of the narcotic action and acute toxicity of alcohol by primary aromatic monoamines. *Q.J. Stud. Alcohol.* 21:584.
Sabelli, H.C., Giardina, W.J., Alivisatos, S.G., Sethi, P., and Unger, F., 1969, Indoleacetaldehydes: serotonin-like effects on the central nervous system. *Nature* 223:73.
Sandler, M., Johnson, R., Ruthven, C., Reid, J., and Calne, D., 1974, Transamination is a major pathway of L-dopa metabolism following peripheral decarboxylase inhibitor. *Nature* 247:364.

Schildkraut, J.J., 1965, The catecholamine hypothesis of affective disorders: A review of supporting evidence. *Am. J. Psychiat.* 122:509.
Shader, R.I. and DiMascio, A., 1970, "Psychotic Drug Side Effects," Williams & Wilkins Co., Baltimore.
Shoaf, S.E. and Linnoila, M., 1991, Interaction of ethanol and smoking on the pharmacodynamics of psychotropic medications. *Psychopharmacol. Bull.* 27:577.
Skursky, L. and Kovar, J., 1975, Interaction of liver alcohol dehydrogenase with neuroleptics chlorprothiexene and chlorpromazine. *FEBS Letts.* 51:297.
Stief, C.G. and Wetterauer, U., 1988, Erectile responses to intravenous papverine and phentolamine: Comparison of single and combined delivery. *J. Urol.* 140:1415.
Summers, J.A., Pullan, P.T., Kril, J.J., and Harper, C.G., 1991. Increased central immunoreactive beta-endorphin content in patients with wernicke-korsakoff syndrome and in alcoholics. *J. Clin. Pathol.* 44:126.
Sutherland, V.C., Burbridge, T., Adams, J., and Simon, A., 1960, Cerebral metabolism in problem drinker under influence of alcohol and chlorpromazine hydrochloride. *J. Appl. Physiol.* 15:189.
Tipton, D.L., Sutherland, V.C., Burbridge, T.N., and Simon, A., 1961, Effect of chlorpromazine on blood level of alcohol in rabbits. *Am. J. Physiol.* 200:1007.
Tottmar, O. and Marchner, H., 1976, Disulfiram as a tool in the studies on the metabolism of acetaldehyde in rats. *Acta. Pharmacol. Toxicol.* 38:366.
Valtaine, A., Beck, O., and Borg, S., 1992, Urinary 5-hydroxytryptophol: A possible marker of recent alcohol consumption. *Alcohol Exp. Clin. Res.* 16:281.
Van Thiel, D.H., Gavaler, J., and Lester, R., 1974, Ethanol inhibition of vitamin A metabolism in the testes: Possible mechanism for sterility in alcoholis. *Science* 186:941.
Virag, R., 1982, Intracavernous injection of papaverine for erectile failure. *Lancet* 2:938.
Weber, R.A. and Preskorn, S.H., 1984, Psychotropic drugs and alcohol: pharmacokinetic and pharmacodynamic interactions. *Psychosomatics* 25:301.
Weichselbaum, A., 1912, Ueber die Veraenderungen des Hodens bei chronischem Alkoholismus, *Verhand. Deutsch. Pathol. Geselsch.* 14:234.

MICROBIAL ALCOHOL, ALDEHYDE AND FORMATE ESTER OXIDOREDUCTASES

Peter W. van Ophem and Johannis A. Duine

Delft University of Technology
Department of Microbiology & Enzymology
Julianalaan 67, 2628 BC Delft, The Netherlands

INTRODUCTION

Formation of alcohols by natural processes takes place in the fermentative breakdown of sugars and the oxidative dissimilation of alkanes. In view of the widespreadness of these processes, it is understandable that many microbial species have the capacity to degrade these compounds. Formaldehyde takes a prominent position among the aldehydes found in Nature. The reason is the frequent occurrance of natural (e.g. methylated amines) as well as man-made C_1-compounds (industrial solvents like DMSO and DMF are used at large scale as well as methylated and methoxylated bulk chemicals, leading to contamination of the environment with these compounds) which are degraded via formaldehyde by a variety of C_1-compounds-utilizing microbes, the so-called methylotrophs. However, also adventitious formaldehyde formation takes place, e.g. in organisms using methylated amines as a nitrogen source or in organisms using pectins, the degradation process liberating methanol from the esterified groups which can be converted to formaldehyde by alcohol oxidizing enzymes. Since formaldehyde is a toxic compound but the ability to assimilate it is confined to methylotrophs, it is obvious that most micro-organisms have developed an oxidative system to get rid of this compound.

Microbial alcohol- and aldehyde-oxidation proceeds via NAD(P)-dependent dehydrogenases, well known from mammals and plants, as well as the so-called dye-

Enzymology and Molecular Biology of Carbonyl Metabolism 4
Edited by H. Weiner, Plenum Press, New York, 1993

linked dehydrogenases (enzymes coupled to the respiratory chain which become detached upon isolation so that they have to be assayed *in vitro* with artificial dyes). Although the latter group is less well known, much progress has been made during the past decade, especially with respect to cofactor identification. Attention will also be paid to the so-called "nicotinoprotein" alcohol/aldehyde oxidoreductases, recently discovered enzymes containing NAD(P) as a cofactor and most of them unable to react with externally added NAD(P).

The classical route for formaldehyde dissimilation consists of the following steps: formaldehyde spontaneously forms an adduct with reduced glutathione (GSH) which becomes oxidized by NAD-, GSH-dependent formaldehyde dehydrogenase (EC 1.2.1.1) after which a specific hydrolase (S-formyl glutathione hydrolase, EC 3.1.2.12) converts the formate ester into GSH and formate; finally it is assumed that formate is oxidized by formate dehydrogenases to CO_2 and H_2O. Recent investigations have revealed that this route is not applicable to Gram-positive bacteria, these organisms having an oxidative route for formate ester conversion. Evidence for this and indications for a wider significance will be discussed.

ALCOHOL OXIDOREDUCTASES

NAD(P)-Dependent Dehydrogenases

Alcohol dehydrogenases using NAD(P) as a coenzyme (EC 1.1.1.1) have been subdivided into 3 groups: the long-chain, zinc-containing group, consisting of 3 classes (I-III); the short-chain, non-zinc-containing group; the Fe-activated group, representatives being found in *Saccharomyces cerevisiae* and *Zymomonas mobilis* (Jörnvall *et al.*, 1987). Although many NAD(P)-dependent alcohol dehydrogenases have been isolated from microbia, the amino acid sequence of most of them is unknown. Comparison of those known with the eukaryotic ones has been carried out (Wales and Fewson, 1990), revealing significant variation in the sequences.

In analogy with that found for NAD/GSH-dependent formaldehyde dehydrogenase (EC 1.2.1.1) in mammals (Koivusalo *et al.*, 1989), NAD/Factor-dependent formaldehyde dehydrogenases from the Gram-positives *Amycolatopsis methanolica* and *Rhodococcus erythropolis* (van Ophem *et al.* 1992b) behave as alcohol dehydrogenase class III, oxidizing and reducing higher aliphatic alcohols and aldehydes, respectively, in the absence of Factor (the unidentified Factor is not GSH, but seems a functional mimick of it, in line with the observation that GSH is absent in Gram-positives (Fahey and Newton, 1983)). They are trimeric enzymes and their dual catalytic abilities might have physiological significance (van Ophem *et al.*, 1992b). Investigations on NAD/GSH-dependent formaldehyde dehydrogenases from *Escherichia coli* (Gutheil *et al.*, 1992), *Thiobacillus versutus*, *Paracoccus denitrificans* and *Candida boidinii* (unpublished results), revealed that these behave as alcohol dehydrogenases class III. A simple explanation for this is to presume that the enzymes act in fact as alcohol dehydrogenases, the ability to oxidize formaldehyde (but not higher aliphatic aldehydes) related to the fact that not formaldehyde but the alcohol-like adduct with GSH or Factor is the genuine substrate. If this also applies to NAD-dependent formaldehyde dehydrogenase (EC 1.2.1.46) from *Pseudomonas putida* C 83 (a structurally different, non-cosubstrate requiring enzyme, also able to oxidize higher aliphatic alcohols but not the aldehydes (Ogushi *et al.*, 1984)), it has to be assumed in this case that either hydrated formaldehyde is the substrate or that an internal complex is formed. The latter possibility seems likely since evidence has been provided that formaldehyde reacts with a cysteinyl residue in the active site (Ogushi *et al.*, 1986). In conclusion, class III alcohol dehydrogenases are widespread in microbia.

Quinoprotein Dehydrogenases

Methanol Dehydrogenase (MDH, EC 1.1.99.8). Details on quinoprotein alcohol dehydrogenase can be found in a number of recent reviews (Anthony, 1992; Duine, 1991). MDH is a so-called quinoprotein as it uses the quinone PQQ (Fig. 1) as cofactor. It catalyzes the reaction:

$$CH_3OH + (\text{cationic})\ \text{dye} \rightarrow CH_2O + \text{dye}H_2$$

Also other primary alcohols (and for some enzymes even secondary alcohols) and formaldehyde are good substrates. The natural electron acceptor appears to be a special cytochrome, named cytochrome c_L. MDH occurs in Gram-negative methanol- and methane-utilizers and is located in the periplasm. It has an $\alpha_2\ \beta_2$-configuration and contains 2 PQQ's and Ca. The genes for several of these enzymes have been cloned and sequenced. Since MDH is situated outside the cytoplasmic membrane, the product, formaldehyde, has (in most cases) to be directed to the dissimilation and to the (cytosol-located) assimilation route. In addition, the enzyme itself is able to oxidize formaldehyde, although a modifier protein seems able to decrease this (Long and

Anthony, 1990). Aspects like the regulation of this delicate balance, the transport mechanism of the toxic formaldehyde, and the assembly of MDH outside the cytoplasmic membrane are completely unknown at this moment.

Ethanol Dehydrogenase (EDH). PQQ-containing EDH has been isolated from ethanol-grown *P. aeruginosa* and *P. putida* strains. The catalytic behaviour, the subunit size and configuration, and the substrate specificity are very similar to those of MDH, except that EDH is unable to convert methanol and uses higher aliphatic aldehydes as

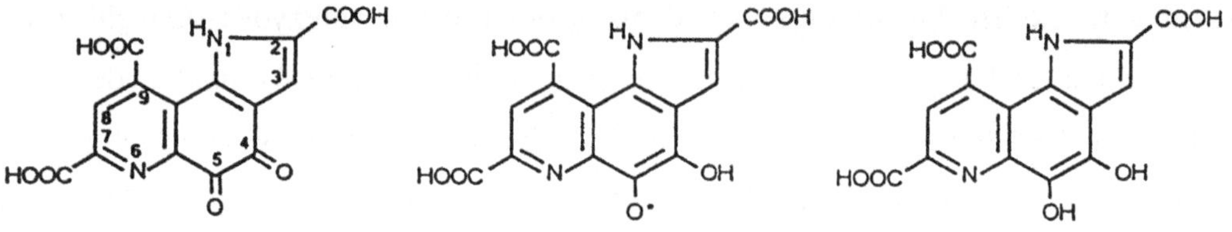

Figure 1. Structure of PQQ and its reduced forms.

substrates. The natural electron acceptor of EDH from *P. aeruginosa* appears to be a novel cytochrome c_{551}, named cytochrome c_{EDH}, and the presence of the small subunit of EDH seems to be essential for electron transfer to take place between them (J.M.J. Schrover and J.A. Duine, unpublished results). The organism contains an NAD-dependent alcohol dehydrogenase as well (Groeneveld *et al.*, 1984), but EDH can be selectively blocked with the inhibitor cyclopropanol.

Quinohaemoprotein Ethanol Dehydrogenase (QH-EDH). QH-EDH has been isolated from *Comamonas testosteroni* (a similar enzyme occurs in *Acetobacter* and *Gluconobacter* species). It contains haem c and PQQ and removal of PQQ yields an inactive enzyme. Evidence has been obtained for an interplay between haem c and PQQ in the catalytic mechanism (A. de Jong, J.A. Jongejan and J.A. Duine, unpublished results). Although the substrate specificity for alcohols and aldehydes is very similar to that of EDH, that for electron acceptors and activators is quite different, pointing to different mechanisms. QH-EDH shows excellent enantioselectivity in the conversion of the C_3-synthons, glycidol and solketal, enabling the production of homochiral products (Geerlof *et al.*, 1992). In this connection it should be mentioned that the organisms in which QH-EDH's occur belong to the fastest (ethanol)-respiring microbes. Therefore, this enzyme seems to be an interesting test case for the application of microbial oxidoreductases as biocatalyst. The amino acid sequences deduced from the genes cloned for QH-EDH from *A. polyoxogenes*, *A. aceti* and *C. testosteroni* show high similarity (Inoue *et al.*, 1990; Tamaki *et al.*, 1991; J. Stoorvogel and J.A. Duine, unpublished results) to each other and to those of MDH (Anthony, 1992), but not at all with NAD(P)-dependent alcohol dehydrogenases. From this it is clear that quinoprotein alcohol dehydrogenases form an independent family of enzymes.

FLAVOPROTEIN ALCOHOL OXIDASES

Flavoprotein alcohol oxidase (EC 1.1.3.13) (see the review by Woodward (1990)) was first discovered in methylotrophic yeasts. The enzyme contains FAD, part of it slightly different from authentic FAD as it has a modified ribitol moiety (Bystrykh *et al.*, 1991). Besides methanol, other primary alcohols and formaldehyde are substrates. Similar enzymes (Table 1) have been isolated from non-methylotrophic yeasts, fungi, and slugs. The enzyme of methylotrophic yeasts is commercially available and has been used in many studies on application, e.g. in analytical determinations and for preparing "biological" aldehydes (Duine and van Dijken, 1991). The long chain alcohol oxidase from alkane-grown *Candida tropicalis* appears to be light-sensitive and to contain an unusual flavin (Dickinson and Wadforth, 1992). From the primary structure determined for the alcohol oxidase from *Hansenula polymorpha* (Ledeboer *et al.*, 1985) it is clear that the enzyme is related to other flavoproteins (glucose dehydrogenase, EC 1.1.99.10) but not to NAD(P)-dependent or quinoprotein alcohol dehydrogenases.

ALDEHYDE OXIDOREDUCTASES

NAD(P)-Dependent Dehydrogenases

These dehydrogenases have been isolated from several bacteria and yeasts and for some of them the primary structures are known (Chalmers *et al.*, 1991; Kok *et al.*, 1989; Heim and Strehler, 1991). Most of the enzymes convert a whole scala of aldehydes,

Table 1. Microbial alcohol oxidoreductases.

ENZYME	COENZYME	COFACTOR	EC NUMBER
NAD(P)-dependent	NAD(P)	(Zn, Fe)	1.1.1.1-x
alcohol dehydrogenases			
(long chain, class I-III			
short-chain			
Fe-activated)			
Nicotinoprotein			
Alcohol dehydrogenase		NAD/NADP(Zn, Mg)	
Quinoprotein			
Methanol dehydrogenase		PQQ(Ca)	1.1.99.8
Ethanol dehydrogenase		PQQ(Ca)	
Ethanol dehydrogenase (quinohaemoprotein)		haem c/PQQ(Ca)	
Polyvinylalcohol dehydrogenase		haem c/PQQ(Ca)	1.1.99.23
Polyethyleneglycol dehydrogenase		quinone	1.1.99.20
Flavoprotein			
Alcohol oxidase		FAD	1.1.3.13
Arylalcohol oxidase			1.1.3.7
Secondary-alcohol oxidase		?(Fe)	1.1.3.18
Long-chain-alcohol oxidase		flavin	1.1.3.20

including formaldehyde, so that the physiological significance for a particular aldehyde is difficult to estimate from this. The NAD-dependent formaldehyde dehydrogenases are discussed in other sections (see Table 4).

Haemoprotein/Quinoprotein Dehydrogenases

Dye-linked, haem-containing aldehyde dehydrogenase has been discovered in a number of methylotrophic bacteria (Patel *et al.*, 1980), and the enzyme originally assigned the EC number 1.1.2.3. Later it was reported that acetic acid bacteria have an aldehyde dehydrogenase which was claimed to contain PQQ (Ameyama and Adachi, 1982) (in this connection, it should be mentioned that no quantification of PQQ in the enzyme has been carried out, that the amino acid sequence of the enzyme does not show homology with those of known PQQ-containing enzymes (Tamaki *et al.*, 1989), and that identification of quinoproteins requires special precautions (Duine, 1991)). Unfortunately, dye-linked aldehyde dehydrogenases are now considered to be quinoproteins to which the EC number 1.1.2.3 has been assigned (Enzyme Nomenclature Commission, supplement), although the number originally belonged to the haemoprotein aldehyde dehydrogenase, and the cofactor identity of these two enzymes is unknown but seems to be different, just as their protein structure. Therefore, it seems justified to regard these aldehyde dehydrogenases as separate enzymes (Table 2).

Methylamine-grown *Hyphomicrobium zavarzinii* contains a formaldehyde dehydrogenase which has a covalently bound o-quinone as cofactor, as established with ESR-spectroscopy, a positive result in an o-quinone assay and the fact that this compound could not be detached by denaturing the enzyme (C.R. Klein, J. Frank, J.A. Duine and A.C. Schwartz, unpublished results). Whether the cofactor is TPQ, TTQ or even a novel o-quinone, remains to be established.

Molybdoprotein Dehydrogenases

The first bacterial molybdoprotein aldehyde dehydrogenase was isolated from ethanol-grown *C. testosteroni* (Poels *et al.*, 1987). It appeared to be quite different from xanthine dehydrogenase, induced in the same organism at different growth conditions. The enzyme is active with a number of artificial dyes, not with O_2, suggesting that it is a genuine dehydrogenase. It contains 1 FAD, 4 Fe, 4 S, 1 Mo, and 1 unidentified pterine per enzyme molecule. A similar enzyme has been recently purified from *A. methanolica* (van Ophem *et al.*, 1992a). In these investigations, it was established that the enzymes have also formate ester dehydrogenase activity, converting the substrate into carbonate ester, the latter decomposing spontaneously into Factor (or alcohol) and CO_2. This seems not to be an artifact of the "hydroxylating" behaviour of a molybdoprotein since bovine milk xanthine oxidase appeared to be inactive in this respect (unpublished results).

Molybdoprotein aldehyde oxidoreductase has also been purified from the anaerobe *Desulfovibrio gigas* (Turner *et al.*, 1987). However, this enzyme lacks FAD so that it is dissimilar. The occurrance of a microbial counterpart of mammalian aldehyde oxidase (EC 1.2.3.1) has so far not been reported.

Tungstoprotein Dehydrogenases

The anaerobe archaebacterium, *Pyrococcus furiosus,* has a dye-linked aldehyde

Table 2. Microbial aldehyde oxidoreductases.

ENZYME	COENZYME	COFACTOR	EC NUMBER
Aldehyde dehydrogenase	NAD(P)		1.2.1.3-x
Formaldehyde dehydrogenase	NAD		1.2.1.46
Aldehyde dehydrogenase	NAD/CoA		1.2.1.10
Formaldehyde dehydrogenase	NAD/GSH		1.2.1.1
Formaldehyde dehydrogenase	NAD/Factor		
Formaldehyde dismutase		NAD	1.2.99.4
Formaldehyde dehydrogenase		quinone	
Aldehyde dehydrogenase		haem c	1.2.99.3
Aldehyde dehydrogenase		quinone	
Aldehyde dehydrogenase (formate ester dehydrogenase)		Fe/S/FAD/ Mo-pterine	
Aldehyde oxidoreductase		Fe/S/ W-pterine	
Carboxylic acid reductase		..S/W..	

oxidoreductase which contains W (Mukund and Adams, 1991). Replacement of Mo by W is well known for other molybdoproteins, but leading to an inactive enzyme form, as was found for the aldehyde dehydrogenases of *A. methanolica* grown on media in which Mo was lacking but W was present (unpublished results). However, since the aldehyde oxidoreductase of *P. furiosis* was active, this suggests that W is a genuine cofactor for this enzyme. W-containing acid reductase has been isolated from the anaerobe, *Clostridium thermoaceticum* (White *et al.*, 1989).

Table 3. Nicotinoprotein alcohol/aldehyde oxidoreductases.

Formaldehyde dismutase (FAlD) (Kato *et al.*, 1986), occurs in *Pseudomonas putida*:

$$2\ CH_2O \rightarrow CH_3OH + HCOOH$$

Methanol dehydrogenase (MeDH) (Arfman *et al.*, 1991), occurs in *Bacillus* spec. C1:

$$CH_3OH + NAD\ (+\ \text{helper protein}) \rightarrow CH_2O + NADH + H^+$$

Methanol/formaldehyde oxidoreductase (MFO) (unpublished), occurs in *Amycolatopsis methanolica, Mycobacterium gastri*:

$$\text{methanol} + NDMA \rightarrow \text{formaldehyde} + NDMAH_2$$

NDMA, p-nitroso-N,N-dimethylaniline.

NDMA-dependent alcohol dehydrogenase (NDMA-ADH) (unpublished), occurs in *Amycolatopsis methanolica*:

$$\text{alcohol} + NDMA \rightarrow \text{aldehyde} + NDMAH_2$$

Enzyme	Cof.	Subun.	ADH[b] act. with		Inh. with		FAlD act.	AlDR[b] act.
			NAD	NDMA	AMP	Pyraz.		
FAlD	NAD	6-7	-	+	-	+	+	-
MeDH	NAD	10	+/-[a]	-			-	+
MFO	NADP	10	-	+		+	+	+
NDMA-ADH	NAD	3	-	+	+	+	-	-

[a] low activity which can be significantly increased, however, by adding helper protein (Arfman *et al.*, 1991). [b]Abbreviations: ADH, alcohol dehydrogenase; AlDR, aldehyde reductase (measured with NADH or NADPH).

NICOTINOPROTEIN ALCOHOL/ALDEHYDE OXIDOREDUCTASES

Recently, a number of alcohol/aldehyde oxidoreductases was discovered containing NAD(P) as cofactor. These so-called nicotinoproteins catalyze a number of activities, the abbreviations of the enzymes and reaction equation indicated in the heading of Table 3.

Structural studies show that the enzymes are dissimilar with respect to primary structure, MeDH showing relatedness to the Fe-activated group of NAD(P)-dependent alcohol dehydrogenases (de Vries *et al.*, 1992) (although the enzyme contains Zn and Mg and is not activated by Fe) and NDMA-ADH to the group of long-chain, zinc-containing NAD(P)-dependent alcohol dehydrogenases, and quaternary structure (Table 3). Thus, tight binding of NAD(P) seems not restricted to a certain type of overall protein structure and these nicotinoproteins do not form a separated family of enzymes in a structural sense. The regeneration of cofactor also appears to be diverse: it is most likely that transfer of reduction equivalents from MeDH occurs to the cytosolic NAD pool with an unknown mechanism in which self exchange may be involved and requiring a helper protein (Arfman *et al.*, 1991)); FAlD does not require removal of or donation of reduction equivalents by an external redox species; it is very likely that MFO and NDMA-ADH deliver their reduction equivalents to the respiratory chain (*vide infra*). Variability also exists with respect to additional activities catalyzed and inhibition exhibited on alcohol dehydrogenase activity (Table 3).

The number of nicotinoprotein alcohol/aldehyde oxidoreductase discovered sofar is low (nicotinoproteins catalyzing other reactions can be found in the literature: Allen, 1982; Caspritz and Radler, 1983; Frey, 1987; Kirkman and Gaetani, 1984; Zachariou and Scopes, 1986). However, it should be realized that: assays with NDMA as electron acceptor and for formaldehyde dismutation are no common practice; enzymes similar to MeDH may have been overlooked as requirement of a helper protein was not realized and not every NAD(P)-dependent alcohol dehydrogenase has been purified to a stage where the bound cofactor should be observable. Thus nicotinoprotein alcohol/aldehyde oxidoreductases may be more frequent than suggested by their present number.

MFO and MeDH have been discovered in several Gram-positive methanol-utilizers, but not in all. *A. methanolica* contains methanol-oxidizing multi-enzyme complexes which can be detected under very uncommon conditions with respect to electron acceptors (Duine *et al.*, 1984; van Ophem *et al.*, 1991). In how far MFO plays a role in these complexes is presently unknown. This organism is the sole Gram-positive methanol-utilizer known which contains PQQ at growth on methanol, however it does not contain the quinoprotein methanol dehydrogenase uniformly used by Gram-negatives. Although it is understandable that the latter enzyme does not occur in the periplasm-lacking Gram-positives, the reason for the variability in structure and catalysis of methanol-oxidizing enzymes in these organism is less clear.

A NOVEL ROUTE FOR FORMALDEHYDE DISSIMILATION

Oxidation of formaldehyde can occur with aspecific and/or with specific dehydrogenases (Table 4), the latter group consisting of some enzymes requiring a helper substrate, resulting in formate ester as a product. Although it is generally assumed that formate ester formation is a variation on the same theme, the ester being hydrolyzed to formate, the absence of a hydrolase and the discovery of formate ester dehydrogenase in *A. methanolica*, are indications that formate ester oxidation may be more widespread. Formate ester dehydrogenase also occurs in *C. testosteroni* (van Ophem *et al.*,1992a) and in *P. denitrificans* (unpublished results), the latter having NAD/GSH-dependent formaldehyde dehydrogenase (Cox and Quayle, 1975). Thus, an oxidative route for formate ester conversion might be an alternative in this case. Finally, methylformate formation has been observed in formaldehyde-dissimilating bacteria (Mason and Sanders, 1989.). This could be due to oxidation of the hemiketal adduct of formaldehyde and methanol by formaldehyde dismutase (and the same enzyme being responsible for methanol production), the methyl formate formed being oxidized by formate ester dehydrogenase to methylcarbonate, decomposing into methanol and CO_2 (van Ophem *et al.*, 1992a).

CONCLUSIONS AND PROSPECTS

An enormous variability exists among the enzymes discussed here not only with respect to protein structure but also to the identity of their coenzyme/cofactor. From this it can be concluded that NAD(P)-dependent alcohol dehydrogenases, quinoprotein alcohol dehydrogenases, and flavoprotein alcohol oxidases are distinct enzyme families. Nicotinoprotein alcohol/aldehyde oxidoreductases are structurally related to the NAD(P)-dependent alcohol dehydrogenases but mechanistically distinct. The present status suggest that the number of tools for alcohol/aldehyde conversion is much larger in microbes than e.g. in mammals, the latter having many isoenzymes not found in

Table 4. Enzymes capable to oxidize formaldehyde.

Conversion into formate by:

Aspecific enzymes

NAD(P)-dependent aldehyde dehydrogenases (EC 1.2.1.x)

Dye-linked aldehyde dehydrogenases (haemoprotein (EC 1.2.99.3), quinoprotein, molybdoprotein)

Specific enzymes

Formaldehyde dehydrogenase (EC 1.2.1.46)

Formaldehyde dismutase (EC 1.2.99.4)

Formaldehyde dehydrogenase (quinoprotein)

Conversion into formyl ester by:

Enzymes requiring a helper substrate

NAD-, and GSH-dependent formaldehyde dehydrogenase (EC 1.2.1.1)

NAD-, and Factor-dependent formaldehyde dehydrogenase (van Ophem *et al.*, 1992b)

microbes. However, since no systematic search has been reported, it cannot be excluded that mammals have a representative of the novel families here discussed.

Although insight into the structure/function relationships scarcely exists, the variability will undoubtedly be connected with catalytic rate and the nature of the successive electron acceptor. The latter has important bio-energetical consequences, as illustrated by summarizing of what is known for alcohol oxidoreductases: the coenzymes NAD(P), used by NAD(P)-dependent dehydrogenases, dissociate from the enzymes so that the reduction equivalents become transferred to the cytosolic redox-pool and ultimately to the respiratory chain at the level of NADH dehydrogenase, providing the organism with the maximal amount of useful energy obtainable from this oxidation step; for the cofactors NAD(P), PQQ and haem c/PQQ, the corresponding enzymes are linked to the respiratory chain at different sites, most probably providing a suboptimal amount of useful energy compared to the first case; for the flavin cofactors, when considering the alcohol oxidases no useful energy is generated (except perhaps a small amount via cytochrome c peroxidase) and the organism has to be equipped with a system to get rid of the dangerous hydrogen peroxide formed. It is expected that ultimately comparative biochemistry will reveal the meaning of this variability in terms of rapidity of substrate conversion and economics of and flexibility in useful energy generation.

Sofar, adequate methanol oxidation has only been observed to occur with enzymes having a bound cofactor (either flavoprotein, quinoprotein or nicotinoprotein). However, this may be fortuitous and is certainly not the sole causitive factor since all of these types of enzymes have counterparts which are able to oxidize alcohols, except methanol. Therefore, the nature of the active site is decisive in this, as is also the case for NAD(P)-dependent alcohol dehydrogenases where insufficient steric interactions and a high pK_a value of methanol seem prohibitive for its adequate oxidation (Pocker and Page, 1990).

Many examples demonstrate now that NAD-dependent formaldehyde dehydrogenase, either GSH- or Factor-requiring, is identical to alcohol dehydrogenase type III (at least with respect to catalytic properties assigned to the latter enzyme). Tentative results on induction of Factor-dependent enzyme in bacteria suggest that the dual catalytic capabilities might have physiological significance.

The presence of the nicotinoprotein formaldehyde dismutase explains why some bacteria are very resistant to formaldehyde as well as the formation of methylformate during this process (although the latter may also be due to formaldehyde dehydrogenases). Some of the nicotinoproteins act as alcohol dehydrogenases, regeneration of the cofactor most probably occurring directly via the respiratory chain. The relationship of these enzymes with the previously detected methanol-oxidizing multi-enzyme complexes of *A. methanolica* and the mechanism of reduction equivalent transfer to external NAD for that occurring in *Bacillus* spec. C1 (assisted with a helper protein), are presently unknown.

Several indications exist to assume that Factor-formate ester conversion in Gram-positive bacteria is not hydrolytic but oxidative. The bio-energetical consequences of this and the physiological relevance as alternative possibility for S-formylglutathione (a compound with a high energy content) conversion in organisms having the well-known hydrolytic route and whose cell-free extracts contain formate ester dehydrogenase activity, remain to be elucidated.

REFERENCES

Allen, S.H.G., 1982, Lactate-oxaloacetate transhydrogenase from *Veillonella alcalescens*, *Methods Enzymol.* 89:367.

Ameyama, M., and Adachi, O., 1982, Alcohol dehydrogenase from acetic acid bacteria, membrane bound, *Methods Enzymol.* 89:450.

Anthony, C., 1992, The structure of bacterial quinoprotein dehydrogenases, *Int. J. Biochem.* 24:29.

Arfman, N., Van Beeumen, J., De Vries, G.E., Harder, W., and Dijkhuizen, L., 1991, Purification and characterization of an activator protein for methanol dehydrogenase from thermotolerant *Bacillus* spp., *J. Biol. Chem.* 266:3955.

Bystrykh, L.V., Dijkhuizen, L., and Harder, W., 1991, Modification of flavin adenine dinucleotide in alcohol oxidase of the yeast *Hansenula polymorpha*, *J. Gen. Microbiol.* 137:2381.

Caspritz, G. and Radler, F., 1983, Malolactic enzyme of *Lactobacillus plantarum*. *J. Biol. Chem.* 258:4907.

Chalmers, R.M. Keen, J.N., and Fewson, C.A., 1991, Comparison of benzylalcohol dehydrogenase and benzaldehyde dehydrogenases from *Acinetobacter calcoaceticus* and from *Pseudomonas putida*, *Biochem. J.* 273:99.

Cox, R.B., and Quayle, J.R., 1975, The autotrophic growth of *Micrococcus denitrificans* on methanol, *Biochem. J.* 150:569.

De Vries, G.E., Arfman, N., Terpstra, P., and Dijkhuizen, L., 1992, Cloning, expression and sequence analysis of the methanol dehydrogenase gene from *Bacillus* sp. strain C_1, *J. Bacteriol*, 174:5346.

Dickinson, F.M., and Wadforth, C., 1992, Purification and some properties of alcohol oxidase from alkane-grown *Candida tropicalis*, *Biochem. J.* 282:325.

Duine, J.A., Frank, J., and Berkhout, M.P.J., 1984, NAD-dependent PQQ-containing methanol dehydrogenase: a bacterial dehydrogenase in a multi-enzyme complex, *FEBS Lett.* 168:217.

Duine, J.A., 1991, Quinoproteins: enzymes containing the quinonoid cofactor pyrroloquinoline quinone (PQQ), topaquinone (TPQ) or tryptophanyl tryptophan quinone (TTQ), *Eur. J. Biochem.* 200:271.

Duine, J.A., Van Dijken, J.P., 1991, Enzymes of industrial potential in methylotrophs, *in*: "Biology of Methylotrophs," J. Goldberg, and J.S. Rokem, eds., Butterworth-Heinemann, Boston, p. 233.

Fahey, R.C., and Newton, G.L., 1983, Occurence of low molecular weight thiols in biological systems, *in*: "Functions of glutathione," A. Larsson *et al.*, ed., Raven Press, New York, p. 251.

Frey, P.A., 1987, Complex pyridine nucleotide-dependent transformations, *in*: "Pyridine nucleotide coenzymes," D. Dolphin *et al.*, ed., John Wiley & Sons, New York.

Geerlof, A., Van Tol, J.B.A., Jongejan, J.A., and Duine J.A., 1992, Microbial alcohol/aldehyde oxidoreductases in enantioselective conversion, *in*: "Microbial Reagents in Organic Synthesis," S. Servi, ed., Kluwer Acad. Publ., Dordrecht, in press.

Groeneveld, A., Dijkstra, M. and Duine, J.A., 1984, Cyclopropanol in the exploration of bacterial alcohol oxidation, *FEMS Microbiol. Lett.* 25:311.

Gutheil, W.G., Holmquist, B., and Vallee, B.L., 1992, Purification, characterization, and partial sequence of the glutathione-dependent formaldehyde dehydrogenase from *Escherichia coli*, *Biochemistry* 31:475.

Heim R., and Strehler, E.E., 1991, Cloning an *Escherichia coli* gene encoding a protein remarkably similar to mammalian aldehyde dehydrogenases, *Gene* 99:15.

Inoue, T., Sunagawa, M., Mori, A., Imai, C., Fukuda, M., Takagi, M., and Yano, K., 1990, Possible functional domains in a quinoprotein alcohol dehydrogenase from *Acetobacter aceti*, *J. Ferm. Bioeng.* 70:58.

Jörnvall, H., Persson, B., and Jeffery, J., 1987, Characteristics of alcohol/polyol dehydrogenases, *Eur. J. Biochem.* 167:195.

Kato, N., Yamagami, T., Shimao, M., and Sakazawa, C., 1986, Formaldehyde dismutase, a novel NAD-binding oxidoreductase from *Pseudomonas putida* F61, *Eur. J. Biochem.* 156:59.

Kirkman, H.N., and Gaetani, G.F., 1984, Catalase: a tetrameric enzyme with four tightly bound molecules of NADPH, *Proc. Natl. Acad. Sci.* 81:4343.

Kok, M., Oldenhuis, R., Van der Linden, M., Meulenberg, C.H.C., Kingma, J., and Witholt, B., 1989, The *Pseudomonas oleovorans* alk BAC operon encodes two structurally related rubredoxins and an aldehyde dehydrogenase, *J. Biol. Chem.*, 264:5442.

Koivusalo, M., Baumann, M., and Uotila, L., 1989, Evidence for the identity of glutathione-dependent formaldehyde dehydrogenase and class III alcohol dehydrogenase, *FEBS Lett.* 257:105.

Ledeboer, A.M., Edens, L., Maat, J., Visser, G., Bos, J.W., Verrips, C.T., Janowicz, Z., Eckart, M. and Hollenberg, C.P., 1985, Molecular cloning and characterization of a gene coding for methanol oxidase in *Hansenula polymorpha*, *Nucleic Acids Res.* 13:3063.

Long, A.R., and Anthony, C., 1990, Modifier protein for methanol dehydrogenase of methylotrophs, *Methods Enzymol.* 188:216.

Mason, R.P., and Sanders, J.K.M., 1989, *In vivo* enzymology: a deuterium NMR study of formaldehyde dismutase in *Pseudomonas putida* F61a and *Staphylococcus aureus*, *Biochemistry* 28:2160.

Mukund, S., and Adams, M.W.W., 1991, The novel tungsten-iron-sulfur protein of the hyperthermophilic archaebacterium *Pyrococcus furiosis*, is an aldehyde ferredoxin oxidoreductase, *J. Biol. Chem.* 266:14208.

Ogushi, S., Ando, M., and Tsuru, D., 1984, Substrate specificity of formaldehyde dehydrogenase from *Pseudomonas putida*, *Agric. Biol. Chem.* 48:597.

Ogushi, S., Ando, M. and Tsuru, D., 1986, Formaldehyde dehydrogenase from *Pseudomonas putida*: the role of a cysteinyl residue in the enzyme activity, *Agric. Biol. Chem.* 50:2503.

Patel, R.N., Hou, C.T., Derelanko, P., and Felix, A., 1980, Purification and properties of a heme-containig aldehyde dehydrogenase from *Methylosinus trichosporium*, *Archiv. Biochem. Biophys.* 203:654.

Pocker, Y, and Page, J.D., 1990, Zinc-activated alcohols in ternary complexes of liver alcohol dehydrogenase, *J. Biol. Chem.*265:22101.

Poels, P.A., Groen, B.W., and Duine, J.A., 1987, NAD(P)-independent aldehyde dehydrogenase from *Pseudomonas testosteroni*, *Eur. J. Biochem.* 166:575.

Tamaki, T., Fukaya, M., Takemura, H., Tayama, K., Okumura, H., Kawamura, Y., Nishiyama, M., Horinouchi, S., and Beppu, T., 1991, Cloning and sequencing of the gene cluster encoding two subunits of membrane-bound alcohol dehydrogenase from *Acetobacter polyoxogenes*, *Biochim. Biophys. Acta* 1088:292.

Tamaki, T., Horinouchi, S., Fukaya, M., Okumura, H., Kawamura, Y., and Beppu, T., 1989, Nucleotide sequence of the membrane-bound aldehyde dehydrogenase gene from *Acetobacter polyoxogenes*, *J. Biochem.* 106:541.

Turner, N., Barata, B., Bray, R.C., Deistung, J., Le Gall, J., and Moura, J.J.G., 1987, The molybdenum iron-sulphur protein from *Desulfovibrio gigas* as a form of aldehyde oxidase, *Biochem. J.* 243:755.

Van Ophem, P.W., Euverink, G.J., Dijkhuizen, L. and Duine, J.A., 1991, A novel dye-linked alcohol dehydrogenase present in some Gram-positive bacteria, *FEMS Microbiol. Lett.* 80:57.

Van Ophem, P.W., Bystrykh, L.V., and Duine, J.A., 1992a, Dye-linked dehydrogenase activities for formate and formate esters in *Amycolatopsis methanolica*, *Eur. J. Biochem.* 206:519.

Van Ophem, P.W., Van Beeumen, J., and Duine, J.A., 1992b, NAD-linked, factor-dependent formaldehyde dehydrogenase or trimeric, zinc-containing, long-chain alcohol dehydrogenase from *Amycolatopsis methanolica*, *Eur. J. Biochem.* 206:511.

Wales, M.R., and Fewson, C.A., 1990, Comparison of the primary studies of NAD(P)-dependent bacterial alcohol dehydrogenases, *in*: "Enzymology and Molecular Biology of Carbonyl Metabolism 3," H. Weiner *et al.*, ed., Plenum Press, New York.

White, H., Strobl, G., Feicht, R., and Simon, H., 1989, Carboxylic acid reductase: a new tungsten enzyme catalyses the reduction of non-activated carboxylic acids to aldehydes, *Eur. J. Biochem.* 184:89.

Woodward, J.R., 1990, *in*: "Autotrophic Microbiology and one-carbon metabolism", Vol I, G.A. Codd, L. Dijkhuizen, and F.R. Tabita, eds., Kluwer Ac. Publ., Dordrecht, p. 193.

Zachariou, M., Scopes, R.K., 1986, Glucose-fructose oxidoreductase, a new enzyme isolated from *Zymomonas mobilis* that is responsible for sorbitol production, *J. Bacteriol.* 167:863.

CARBONYL METABOLISING ENZYMES IN ALKANE-GROWN MICROORGANISMS

Maeve G.A. Fox, Neil Broadway, F. Mark Dickinson and Colin Ratledge

Department of Applied Biology
University of Hull
Hull HU6 7RX, U.K.

INTRODUCTION

The growth of microorganisms on n-alkanes as carbon source requires the presence of enzyme systems capable of oxidising the growth substrate into fatty acids. Some bacteria also produce long-chain dicarboxylic acids, presumably using the same enzyme systems involved in the production of monocarboxylic acids, but with oxidation occurring at both ends of the carbon chain. In the present work we have studied the enzymes found in a yeast (*Candida tropicalis*) and two bacteria (*Acinetobacter calcoaceticus sp. strain HO1-N* and *Corynebacterium sp. strain 7EIC*) which oxidise long-chain alcohols and aldehydes.

METHODS

Alcohol oxidase was prepared from *Candida tropicalis* grown on hexadecane as described by Dickinson and Wadforth (1992). The enzyme was assayed by the method of Kemp *et al.* (1988) and photoinactivation was performed as described by Dickinson and Wadforth (1992).

Acinetobacter calcoaceticus sp. strain HO1-N and *Corynebacterium sp. strain 7EIC* were grown on a salts medium (Jayasuriya, 1955) containing 0.5% or 0.2% hexadecane. Extracts were prepared from harvested cells by 3 passages through the French pressure cell (35 MPa). DNase was added to degrade DNA and reduce the viscosity of the extracts. Cell debris was removed by centrifugation at 8000 g for 15 min at 4°C.

NAD^+ and $NADP^+$-linked alcohol and aldehyde dehydrogenases of *A. calcoaceticus HO1-N* were assayed fluorimetrically at 25°C in glycine buffer at pH 9.5, with 250 µM $NAD(P^+)$ and 50 µM substrate. The enzyme from *Corynebacterium 7EIC* were assayed at pH 8.0. Wurster's Blue dependent aldehyde dehydrogenase activity was measured spectrophotometrically at 600 nm with approx. 90 µM Wurster's Blue and 300 µM substrate in phosphate buffer, pH 7.5.

Dicarboxylic acids were extracted from acidified cell cultures with diethyl ether. The extracts were evaporated to dryness, dissolved in methanol and methylated with trimethylsulphonium hydroxide (Butte 1983). Analysis was carried out by g.l.c. on a column packed with 10% diethylene glycol succinate on Celite, 100-200 mesh. The column was maintained at 200°C.

Enzymology and Molecular Biology of Carbonyl Metabolism 4
Edited by H. Weiner, Plenum Press, New York, 1993

FATTY ALCOHOL OXIDASE AND ALDEHYDE DEHYDROGENASE IN YEASTS

For *Candida* and other alkane-utilizing yeasts it has been shown that there is a proliferation of peroxisomes after growth on alkanes and the induction of a system of membrane-bound enzymes. Amongst other things these enzymes convert n-alkanes to fatty acids (Scheme 1).

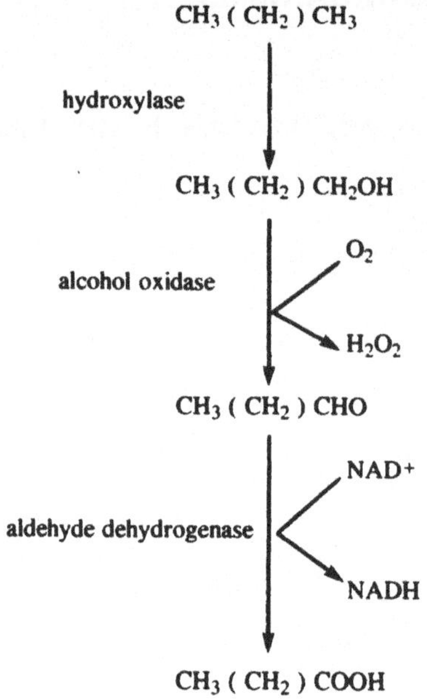

Scheme 1. Alkane oxidation in yeasts

Il'chenko (1984) and Krauzova *et al.* (1985, 1986) showed that in *Torulopsis candida* an alcohol oxidase is involved and not, as had been previously thought, an alcohol dehydrogenase. Alcohol oxidase from alkane-grown *Candida tropicalis* (Kemp *et al.* 1988) has been obtained in highly purified form (Dickinson and Wadforth, 1992) and partially characterised. The substrate specificity of the enzyme from *Candida maltosa* has been described recently (Mauersberger *et al.* 1992). One peculiar property of the alcohol oxidase from *C. tropicalis* is its sensitivity to blue light (Kemp *et al.* 1990, Dickinson and Wadforth, 1992). The enzyme from *Aspergillus flavipes* is also light sensitive (Savitha and Ratledge, 1991). Fig. 1 shows that the *C. tropicalis* enzyme can be protected from light inactivation by inclusion of substrate under anaerobic conditions. This result may explain, at least in part, why the yeast grows perfectly well under bright illumination (Kemp *et al.* 1990) even though the enzyme is thought to be vital for the conversion of alkanes into fatty acids to fuel growth.

Attempts to purify the long-chain aldehyde dehydrogenase from *C. tropicalis* have been uniformly unsuccessful. The enzyme can be solublised readily by Triton X-100 or sodium cholate and is very stable in such extracts in the presence of 100 μM dithiothreitol. However, all attempts at fractionation have been unsuccessful. Binding to column materials (ion-exchange or hydrophobic has always led to complete loss of activity). The enzyme is 95% inhibited by 5 μM of the mixed disulphide diethyl dithiocarbamic acid methanethiol (Kemp 1989) suggesting perhaps the involvement of an active site thiol in the enzyme mechanism.

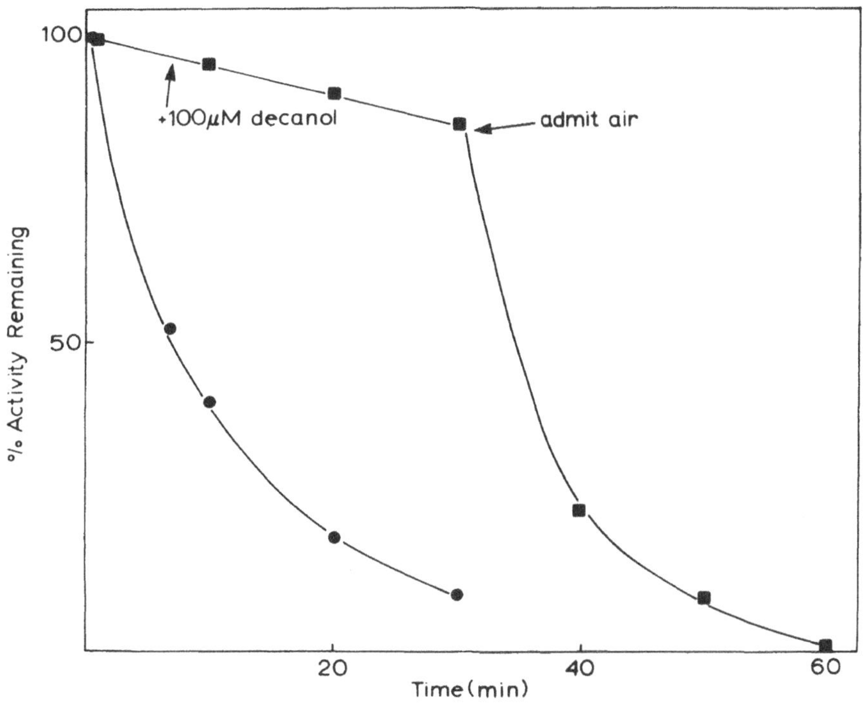

Figure 1. Photoinactivation of alcohol oxidase from alkane-grown *Candida tropicalis* under anaerobic conditions. Incubations were in 50 mM Hepes/NaOH buffer, pH 8.5 at 25°C. The enzyme concentration was 0.1 mg/ml. Anaerobic conditions were achieved by evacuation of the cuvette and flushing with O_2-free N_2. Blue light (λ_{max} = 470 nm) was obtained using a 150 W Xenon lamp and a photographic filter.

ENZYMOLOGY OF DICARBOXYLIC ACID PRODUCTION IN ALKANE-GROWN BACTERIA

Table 1 shows that the growth of strains of *Corynebacteria* and *Acinetobacter calcoaceticus sp. strain HO1-N* on some alkanes leads to the accumulation of dicarboxylic acids.

Table 1. Dicarboxylic acid production by alkane-grown bacteria.

Organism	Growth substrate (20 g/l)		DC produced (mg/l)				
			DC_8	DC_{10}	DC_{12}	DC_{14}	DC_{16}
Acinetobacter calcoaceticus HO1-N	dodecane		13	43	50	n.d.	n.d.
Corynebacterium 21744	dodecane		7	11	590	n.d.	n.d.
Corynebacterium 7EIC	decane	NG	n.d.	n.d.	n.d.	n.d.	n.d.
Corynebacterium 7EIC	dodecane		n.d.	12	300	n.d.	n.d.
Corynebacterium 7EIC	tetradecane		n.d.	10	6	17	n.d.
Corynebacterium 7EIC	hexadecane		n.d.	n.d.	n.d.	n.d.	n.d.

Bacteria were grown for 5 days and the cultures acidified and extracted with diethyl ether. The evaporated residues were methylated and subjected to analysis by g.l.c.

NG - no growth, n.d. - not detected.

It is clear that only a very narrow range of dicarboxylic acids are produced and that the compound with the highest concentration contains the same number of carbon atoms as the growth substrate. In no case does a dicarboxylic acid accumulate with a longer chain length than the alkane substrate. It is likely that dicarboxylic acids arise by diterminal oxidation (a possible mechanism is shown in Scheme 2) with the shorter chain-length dicarboxylic acids arising by subsequent ß-oxidation. It would also seem that the specificities of the enzymes involved in terminal oxidation might explain the range of acids produced. We have therefore studied the alcohol and aldehyde dehydrogenases which are found in these bacteria.

Comparison of alcohol and aldehyde dehydrogenases in alkane-grown *Corynebacterium sp. strain7EIC* and *Acinetobacter calcoaceticus sp. strain HO1-N* shows that the latter contain much higher activities (Table 2). In addition there is a large nucleotide independent aldehyde dehydrogenase activity in *A. calcoaceticus HO1-N* which does not seem to be present in the other organism. The bulk of the enzymological work has been done with *A. calcoaceticus HO1-N.*

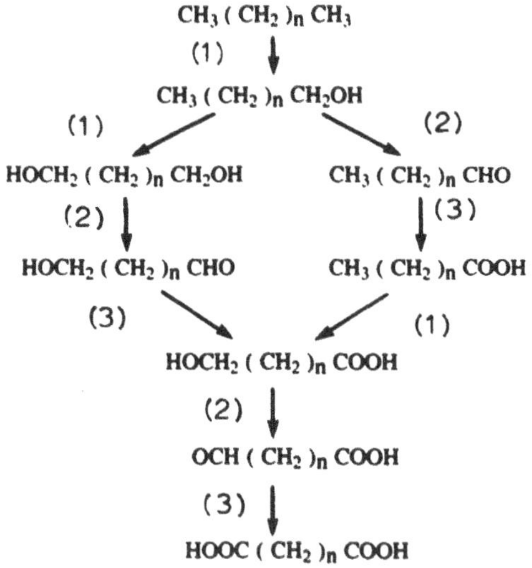

Scheme 2. Pathways for dicarboxylic acid formation.
(1) hydroxylase (2) alcohol dehdrogenase
(3) aldehyde dehydrogenase

Table 2. The specific activities (μmol. min^{-1} mg^{-1}) of alcohol and aldehyde dehydrogenases in cell-free extracts of cells grown on n-hexadecane.

Organism	Cofactors				
	NAD+ substrate		NADP+ substrate		Wurster's Blue substrate
	decanol	decanal	decanol	decanal	decanal
Corynebacterium 7EIC	0.0019	0.002	0.003	0.012	-
Acinetobacter calcoaceticus HO1-N	0.21	0.31	2.0	1.9	15.5

Table 3. Alcohol and aldehyde dehydrogenase activities in extracts of *Acinetobacter calcoaceticus HO1-N* grown on different substrates.

Growth substrate	Cofactors				
	NAD+ substrate		NADP+ substrate		Wurster's Blue substrate
	decanol	decanal	decanol	decanal	decanal
1.0% sodium succinate	1.8	2.7	22	1.0	42
0.5% hexadecane	2.3	3.4	22	20	170

Total activities (nmol min^{-1} per 3 gm wet weight of cells) are recorded.

The activities of alcohol and aldehyde oxidising enzymes found in *Acinetobacter calcoaceticus HO1-N* when grown on two different substrates are shown in Table 3. It is clear that long-chain alcohol dehydrogenases are not induced by growth on hexadecane, but that the $NADP^+$-linked and nucleotide independent aldehyde dehydrogenases are. It is interesting that the activities of the $NAD(P^+)$ linked alcohol and aldehyde dehydrogenases are broadly similar after growth on alkanes. The strong induction of these enzymes suggests that aldehydes are important metabolites in growth on alkanes.

The long-chain alcohol dehydrogenases found in *Corynebacterium 7EIC* show a similar picture to that seen in Table 3. Generally there was no induction by growth on alkanes except for a 2-3-fold increase of a membrane-bound NAD^+-octanol dehydrogenase activity. The latter enzyme appears to have no direct counterpart in *A. calcoaceticus HO1-N*. The *Corynebacterium 7EIC* enzymes are activated strongly by high salt concentrations primarily due to a reduction in Km values under these conditions. Activation by salts is observed with long chain alcohols, diols and ω-hydroxyfatty acids as substrate. Up to 5-fold increases in activity were observed under standard conditions (50 μM substrate) by including 0.8 M potassium phosphate or Na_2SO_4 in assays. However, even when activated the alcohol dehydrogenase activities are still much lower than those found in *A. calcoaceticus HO1-N.*

The properties of the enzymes found in *A. calcoaceticus HO1-N* and listed in Table 3 will now be dealt with in turn.

NAD^+-linked alcohol dehydrogenase

Heat inactivation studies on crude cell extracts at 60°C indicated two different enzyme activities. One was stable (this manifested about 25% of the total activity) and the other was not. The activities are found in the soluble fraction of the cell.

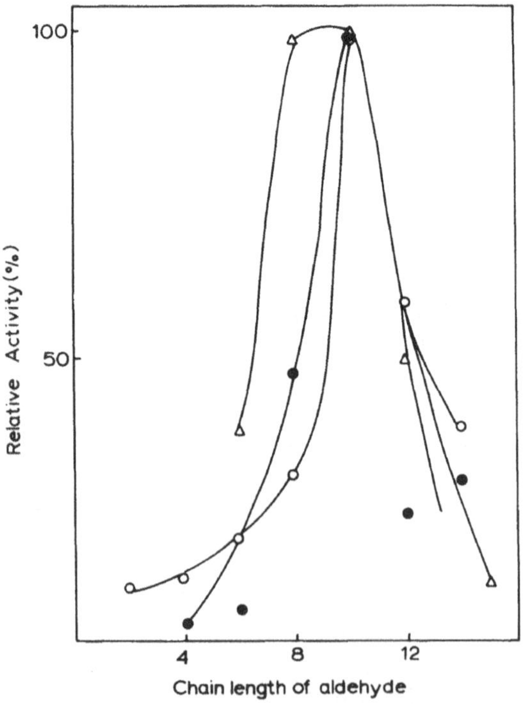

Figure 2. The substrate specificity of enzymes from *A. calcoaceticus*. $NADP^+$-alcohol dehydrogenase (Δ), $NADP^+$-aldehyde dehydrogenase (●), nucleotide independent aldehyde dehydrogenase (O).

PAGE electrophoresis under non-denaturing conditions and activity staining showed only one active band. Because of the relatively low activity of these enzymes no attempt at purification was made.

$NADP^+$-linked alcohol dehydrogenase

This activity is also found in the soluble fraction of crude cell extracts. Heat inactivation studies at 60°C showed that the activity declined in a biphasic manner indicative of two different enzymes. PAGE electrophoresis under non-denaturing conditions and activity staining showed two well separated active species and both have quite different mobilities from the NAD^+ alcohol dehydrogenase described above. The more heat stable $NADP^+$-linked enzyme has been purified 42-fold by gel-filtration and chromatography on Reactive-Red agarose. The enzyme has a molecular weight of 144,000±5000 and a pI = 4.5. The substrate specificity measured under standard conditions is shown in Fig. 2.

NAD^+-linked aldehyde dehydrogenase

This enzyme occurs in the cytosolic fraction of the cells. The lack of inducibility of the enzyme compared to the $NADP^+$-linked aldehyde dehydrogenase (Table 3) shows that the two activities are associated with different proteins. The lack of inducibility and the relatively low activity suggest that this enzyme is not important in alkane metabolism. No attempt at purification has been made.

$NADP^+$-linked aldehyde dehydrogenase

This enzyme appears to be associated with intracellular inclusions of alkane which are found in this organism after growth on alkanes (Scott *et al*, 1976). Application of crude cell extracts to large-pore gel filtration columns leads to emergence of the enzyme in the breakthrough volume $M_R > 10^6$) in milky-white fractions. Extraction of the organic material in these fractions and analysis by g.l.c. indicated about 66% hexadecane (the growth substrate) and 10% hexadecanoic acid. Attempts to solubilise the enzyme by using detergents (1% Triton X-100, sodium cholate etc) were unsuccessful. The treatments resulted in immediate and substantial losses (70-90%) of enzyme activity. Precipitation with 30% v/v ice cold acetone was more successful. This did remove much alkane material and a 6.5-fold purification was achieved. Even so the best preparations were still opalescent indicating that not all the alkane material had been removed. It seems that this enzyme is intimately associated with the intracellular vesicles. It may be one of the proteins found in the membranes surrounding them (Scott *et al*, 1976) and responsible for the hexadecanoic acid found in the enyzme bearing fractions from gel filtration (see above). The substrate specificity of the partially purified enzyme is included in Fig. 2. It should be noted that no $NAD(P^+)$-linked alcohol dehydrogenase, no alcohol oxidase or nucleotide independent alcohol dehydrogenase could be detected in the cell fractions which contain this enzyme.

Nucleotide independent aldehyde dehydrogenase

As is shown in Table 3 this enzyme appears to provide the bulk of aldehyde oxidising activity in *A. calcoaceticus HO1-N*. The enzyme can be assayed using either Wurster's Blue as electron acceptor or a combination of phenazine methosulphate (PMS) and 2:6 dichlorophenol-indophenol (DCPIP). The measured activity with PMS/DCPIP is about half that seen using Wurster's Blue. It is interesting to note that the enzyme can be assayed directly with intact cells and that the activity recorded is about 90% of that seen when the cells are disrupted by sonication or by using the French press. (We make the same observation with the PQQ-containing glucose dehydrogenase of this organism.) The observation indicates that the enzyme is located in the periplasmic space. No alcohol oxidase or nucleotide independent alcohol dehydrogenase was detectable in experiments with whole cells.

When alkane-grown cells were monitored with the oxygen electrode, respiration was stimulated by addition of long-chain aldehyde (decanal). When the experiment was repeated in the presence of PMS the rate of oxygen uptake increased approximately 10-fold. Clearly the aldehyde dehydrogenase reacts much more rapidly in the presence of external artificial electron acceptors. It is possible that the very large relative activities for this enzyme (Table 3) do not at all represent the true physiological activity. In that case the metabolic significance of the enzyme may be much less than it appears to be.

We have attempted to purify the enzyme from extracts of hexadecane-grown cells. About 15% of the enzyme can be released by osmotic shock (Neu and Heppel, 1965). This and the fact that the enzyme can be assayed in whole cells suggests that it occupies a very similar location in the cells to enzymes such as alkaline phosphatase and 5'-nucleotidase in *E. coli* (Neu and Heppel, 1964). On French pressing some 30-40% of the activity is found in the 200,000 g supernatant. The remainder can be eluted from the 200,000 g membrane pellet by extraction with 1% Triton X-100 or 1% sodium cholate. The solubilised enzyme is very stable at 0°C and can be dialysed exhaustively ±10 mM EDTA at pH 7.5 without loss of activity. Further, addition of 100 mM EDTA or 3 mM *o*-phenanthroline or 8-hydroxyquinoline to enzyme solutions and incubation at 25°C for 30 minutes does not affect enzyme activity. There is no requirement for dithiothreitol (or other thiols) to be added to maintain activity, or for exclusion of air.

The solubilised enzyme is readily purified by ion-exchange chromtography on DEAE-cellulose, ammonium sulphate fractionation and gel-filtration chromatography (all in the presence of detergent). The final product is not pure and is some 30-fold purified over the initial solubilised extract. The solution at a concentration of 1 mg/ml is brown-yellow and exhibits a weak green fluorescence. This and its periplasmic location suggests that the prosthetic group may be of the PQQ type, but clearly further work is needed.

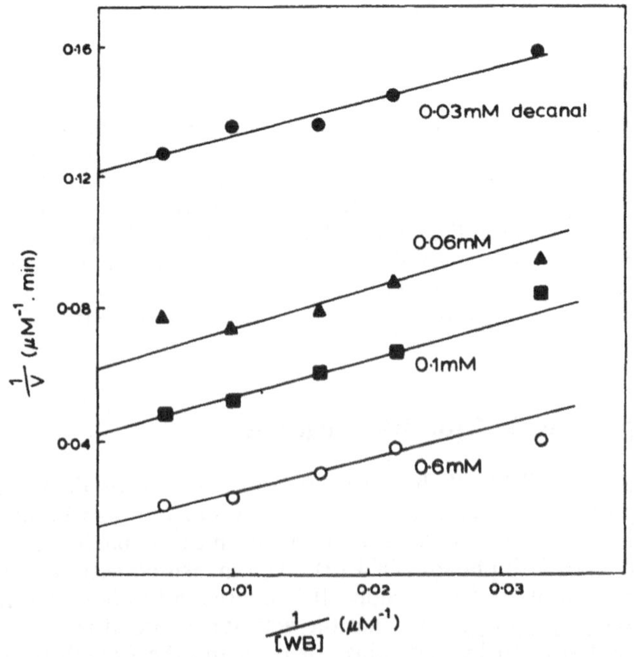

Figure 3. Lineweaver-Burk plot of kinetic data for the nucleotide independent aldehyde dehydrogenase from *A. calcoaceticus HO1-N.*

The substrate specificity of the enzyme is included in the data of Fig. 2 and the results of a kinetic study of the enzyme at 25°C, pH 7.5 are shown in Fig. 3. The enzyme appears to conform to a mechanism of the ping-pong type. The K_m values for decanal obtained from the linear secondary plot of intercepts was 0.45 mM. The K_m for Wurster's Blue was 0.15 mM.

Enzyme specificity and dicarboxylic acid formation

Our studies show that *A. calcoaceticus sp. strain HO1-N* and *Corynebacterium sp. strain 7EIC* contain a range of alcohol and aldehyde dehydrogenase activities which could provide for dicarboxylic acid formation in these organisms (Scheme 2). The specificities of the *A. calcoaceticus HO1-N* enzymes (Fig. 2) clearly favour the C_{12} substrate compared to the C_{16} substrate. Specificity studies with the *Corynebacterium 7EIC* alcohol dehydrogenases show precisely the same kind of behaviour as Fig. 2 using either long-chain alcohols or long-chain diols as substrates. Although assays under standard conditions are not necessarily a reliable guide to intracellular activity it seems at least possible that one reason for the preferential production of C_{12} dicarboxylic acids from dodecane (Table 1) is the specificity of the alcohol and aldehyde dehydrogenases involved. The analysis of course assumes that the hydroxylase is either neutral or also favours the C_{12} substrates. We have no information on this.

Studies on the specificities of the ß-oxidation system of *Corynebacterium 7EIC* have shown that although dicarboxylic acids are attacked they are in general very poor substrates compared to fatty acids of the same chain length (relative rate 1:20). The ß-oxidation system shows quite similar activities with C_{12}, C_{14} and C_{16} fatty acids (C_{14} is slightly preferred), but is less effective against decanoic acid. If fatty acids and dicarboxylic acids are produced together (Scheme 2) it seems that the fatty acids will be oxidised preferentially and dicarboxylic acids could accumulate. If the specificities of the terminal oxidation enzymes favour substantial DC_{12} formation from dodecane, some DC_{14} from tetradecane and very little or no DC_{16} from hexadecane the occurrence of dicarboxylic acids in cultures of alkane-grown organisms (Table 1) can be explained.

Our work provides no explanation of why there is such a multiplicity of alcohol and aldehyde dehydrogenases in these bacteria. Presumably each has its own special role to play, but what they are remains unknown. The two alkane-induced aldehyde dehydrogenases of *A. calcoaceticus HO1-N* are the most interesting of the enzymes identified. They occupy very different locations in the cell (see above) and must have different functions. The $NADP^+$-linked aldehyde dehydrogenase being tightly associated with alkane inclusions which also contain some fatty acid and which are situated in the cytoplasm containing alcohol dehydrogenases could perhaps be principally concerned in the pathways of Scheme 2. The nucleotide-independent aldehyde dehydrogenase is a much more difficult problem. The K_m for aldehyde is quite high (0.46 mM for decanal), there appears to be no alcohol oxidising activity in the same cellular compartment and the activity seen with artificial electron acceptors may grossly overstate its activity under physiological conditions (see above). According to Finnerty *et al.* (1962) *A. calcoaceticus* readily metabolises n-alkyl hydroperoxides and a metabolic route for the terminal oxidation of alkanes involving hydroperoxides and per-acids has been proposed (Finnerty, 1977, 1988). This pathway generates long-chain aldehydes from the per acids and does not involve long-chain alcohols. If operational in the periplasmic space or outer membrane, this pathway could provide a role for the nucleotide independent aldehyde dehydrogenase.

Acknowledgements

This work was supported by the European Community Biotechnology Action Programme (Grant no. 102100).

References

Butte, W., 1983, Rapid method for the determination of fatty acid profiles from fats and oils using trimethylsulphonium hydroxide for transesterification. J. Chromatography 261, 142.

Dickinson, F.M. and Wadforth, C., 1992, Purification and some properties of alcohol oxidase from alkane-grown *Candida tropicalis*, Biochem. J. 282, 325.

Finnerty, W.R., 1977, The biochemistry of microbial alkane oxidation: new insights and perspectives. Trends Biochem. Sci. 2, 73.

Finnerty, W.R., 1988, Lipids of *Acinetobacter*. In Proceedings of the World Conference on Biotechnology for the Fats and Oils Industry pp. 184-188. Edited by T.H. Applewhite. Champaign, Illinois: American Oil Chemists' Society.

Finnerty, W.R., Kallio, R.E., Klimstra, P.D. Wawzonez, S., 1962, Utilization of 1-alkyl hydroperoxides by *Micrococcus cerificans*. Zeit. Allgemeine Mikrobiol. 2, 263.

Il'chenko, A.P., 1984, Oxidase of higher alcohols in the yeast *Torulopsis candida* grown on hexadecane, Mikrobiologiya, 53, 903.

Jayasuriya, G.C.N., 1955, The isolation and characteristics of an oxalate-decomposing organism. J. Gen. Microbiol. 12, 419.

Kemp, G.D., 1989 Ph.D. Thesis, University of Hull, U.K. Fatty alcohol oxidases and fatty aldehyde dehydrogenases in the hydrocarbon utilizing yeasts *Candida tropicalis* and *Yarrowia lipolytica*.

Kemp. G.D., Dickinson, F.M. and Ratledge, C., 1988, Inducible long chain alcohol oxidase from alkane-grown *Candida tropicalis*, Appl. Microbiol. Biotechnol. 29, 370.

Kemp, G.D., Dickinson, F.M. and Ratledge, C., 1990, Light sensitivity of the n-alkane-induced fatty alcohol oxidase from *Candida tropicalis* and *Yarowia lipolytica*, Appl. Microbiol. Biotechnol. 32, 461.

Krauzova, V.I., Il'chenko, A.P., Sharyshev, A.A. and Lozinov, A.B., 1985, Possible pathways of the oxidation of higher alcohols by membrane fractions of yeasts cultured on hexadecane and hexadecanol, Biokhimiya, 50, 726.

Krauzova, V.I., Kuvichkina, T.N., Sharyshev, A.A., Romanova, I.B. and Lozinov, B., 1986, Formation of lauric acid and NADH in the oxidation of dodecanol by membrane fractions of the yeast *Candida maltosa* cultured on hexadecane, Biokhimiya, 51, 23.

Mauersberger, S., Dreschler, H., Oehme, G. and Müller, H-G., 1992, Substrate specificity and stereoselectivity of fatty alcohol oxidase from the yeast *Candida maltosa*, Appl. Microbiol. Biotechnol. 37, 66.

Neu, H.C. and Heppel, L.A., 1964, On the surface localization of enzymes in *E. coli*, Biochem. Biophys. Res. Commun. 17, 215.

Neu, H.C. and Heppel, L.A., 1965, The release of enzymes from *Escherichia coli* by osmotic shock and during the formation of sphaeroplasts, J. Biol. Chem. 240, 3685.

Savitha, J. and Ratledge, C., 1991, Alcohol oxidase of *Aspergillus flavipes* grown on hexadecanol, FEMS Microbiol. Lett. 80, 221.

Scott, C.C.L. and Finnerty, W.R., 1976, Characterization of intracytoplasmic hydrocarbon inclusions from the hydrocarbon-oxidising Acinetobacter species HO1-N, J. Bact. 127, 481.

INDEX

The manufacturer's authorised representative in the EU is Springer Nature Customer Service Centre GmbH, Europaplatz 3, 69115 Heidelberg, Germany. If you have any concerns regarding our products, please contact ProductSafety@springernature.com

Printed and bound by CPI Group (UK) Ltd, Croydon, CR0 4YY
24/04/2026
02096348-0020